History of the European Oil and Gas Industry

Geological Society books refereeing procedures

The Society makes every effort to ensure that the scientific and production quality of its books matches that of its journals. Since 1997, all book proposals have been refereed by specialist reviewers as well as by the Society's Books Editorial Committee. If the referees identify weaknesses in the proposal, these must be addressed before the proposal is accepted.

Once the book is accepted, the Society Book Editors ensure that the volume editors follow strict guidelines on refereeing and quality control. We insist that individual papers can only be accepted after satisfactory review by two independent referees. The questions on the review forms are similar to those for *Journal of the Geological Society*. The referees' forms and comments must be available to the Society's Book Editors on request.

Although many of the books result from meetings, the editors are expected to commission papers that were not presented at the meeting to ensure that the book provides a balanced coverage of the subject. Being accepted for presentation at the meeting does not guarantee inclusion in the book.

More information about submitting a proposal and producing a book for the Society can be found on its website: www.geolsoc.org.uk.

It is recommended that reference to all or part of this book should be made in one of the following ways:

CRAIG, J., GERALI, F., MACAULAY, F. & SORKHABI, R. (eds) 2018. *History of the European Oil and Gas Industry*. Geological Society, London, Special Publications, **465**.

TULUCAN, A.D., SOVEJA-IACOB, L.-E. & KREZSEK, C. 2018. History of the oil and gas industry in Romania. *In*: CRAIG, J., GERALI, F., MACAULAY, F. & SORKHABI, R. (eds) *History of the European Oil and Gas Industry*. Geological Society, London, Special Publications, **465**, 191–200. First published online May 18, 2018, https://doi.org/10.1144/SP465.15

GEOLOGICAL SOCIETY SPECIAL PUBLICATION NO. 465

History of the European Oil and Gas Industry

EDITED BY

J. CRAIG
Eni Upstream & Technical Services, Italy

F. GERALI
University of Oklahoma, USA

F. MACAULAY
Echo Energy plc, UK

and

R. SORKHABI
University of Utah, USA

2018
Published by
The Geological Society
London

THE GEOLOGICAL SOCIETY

The Geological Society of London (GSL) was founded in 1807. It is the oldest national geological society in the world and the largest in Europe. It was incorporated under Royal Charter in 1825 and is Registered Charity 210161.

The Society is the UK national learned and professional society for geology with a worldwide Fellowship (FGS) of over 10 000. The Society has the power to confer Chartered status on suitably qualified Fellows, and about 2000 of the Fellowship carry the title (CGeol). Chartered Geologists may also obtain the equivalent European title, European Geologist (EurGeol). One fifth of the Society's fellowship resides outside the UK. To find out more about the Society, log on to www.geolsoc.org.uk.

The Geological Society Publishing House (Bath, UK) produces the Society's international journals and books, and acts as European distributor for selected publications of the American Association of Petroleum Geologists (AAPG), the Indonesian Petroleum Association (IPA), the Geological Society of America (GSA), the Society for Sedimentary Geology (SEPM) and the Geologists' Association (GA). Joint marketing agreements ensure that GSL Fellows may purchase these societies' publications at a discount. The Society's online bookshop (accessible from www.geolsoc.org.uk) offers secure book purchasing with your credit or debit card.

To find out about joining the Society and benefiting from substantial discounts on publications of GSL and other societies worldwide, consult www.geolsoc.org.uk, or contact the Fellowship Department at: The Geological Society, Burlington House, Piccadilly, London W1J 0BG: Tel. +44 (0)20 7434 9944; Fax +44 (0)20 7439 8975; E-mail: enquiries@geolsoc.org.uk.

For information about the Society's meetings, consult *Events* on www.geolsoc.org.uk. To find out more about the Society's Corporate Affiliates Scheme, write to enquiries@geolsoc.org.uk.

Published by The Geological Society from:
The Geological Society Publishing House, Unit 7, Brassmill Enterprise Centre, Brassmill Lane, Bath BA1 3JN, UK

The Lyell Collection: www.lyellcollection.org
Online bookshop: www.geolsoc.org.uk/bookshop
Orders: Tel. +44 (0)1225 445046, Fax +44 (0)1225 442836

British Library Cataloguing in Publication Data

A catalogue record for this book is available from the British Library.
ISBN 978-1-78620-363-2
ISSN 0305-8719

Distributors
For details of international agents and distributors see:
www.geolsoc.org.uk/agentsdistributors

Typeset by Nova Techset Private Limited, Bengaluru & Chennai, India
Printed and bound by CPI Group (UK) Ltd, Croydon CR0 4YY

Contents

History of the petroleum industry and petroleum geologists

Acknowledgements

A great many people and organizations assisted the editors with the development and production of this volume. In particular, we would like to thank The University of Oklahoma School of Library and Information Studies and History of Science Collections and College of Law; The Library of the American Philosophical Society; Prof. Barry Cooper, President of the International Commission on the History of the Geological Sciences (INHIGEO); Prof. Slawomir Lotysz, President of the International Committee for the History of Technology (ICOHTEC); Prof. William Brice, Petroleum History Institute; Prof. Elena Helerea, University of Brasov; Prof. Stephen Weldon, The University of Oklahoma and Chief Editor of the ISIS Current Bibliography on the History of Science. We also thank Ben Dewhirst and David McCann for assistance in reviewing the drafts of the papers.

The following individuals generously devoted their time and their expertise to reviewing the papers and suggesting improvements:

Uwe Altenberger, Ian Andrews, Mustafa As-Saruri, Balázs Badics, Alain Beltran, Roberto Bencini, Bernard Besly, William B. Brice, Tim Bushell, Roberto Cantoni, Ferdinando Cazzini, Anna Cazzola, Robin Chesters, Jonathan Craig, Paul Culpin, Arnoud De Fejiter, Anthony Doré, Simone Fari, Francesco Gerali, Jon Gluyas, Graham Goffey, John Harper, Barbra Harvie, Alan Heward, Per Högselius, Hans Krause, Piotr Krzywiec, Richard Law, Cliff Lovelock, Fiona MacAulay, Agata Madericova, Richard J. Milton-Worssell, Marco Pantaloni, Emil Platon, Iain Scotchman, Matthew Silverman, Ray Sorenson, Jeff A. Spencer, Randell Stephenson, John Summers, Daniel Truempy, Filippos Tsikalas, Mario Wannier, Lucy Williams, Andrew Zolnai

The editors particularly wish to acknowledge the contribution of Angharad Hills, Tamzin Anderson, Jo Armstrong, Sarah Gibbs, Hannah Turner and the other staff of the Geological Society Publishing House for their guidance, patience and all their hard work during the production of this volume.

The Geological Society thanks the following companies for their generous sponsorship of the Petroleum Group and the conference that led to this volume.

Petroleum Group corporate sponsors at the time of the conference:

Sponsors of European Oil and Gas Industry History conference:

The history of the European oil and gas industry (1600s–2000s)

JONATHAN CRAIG[1]*, FRANCESCO GERALI[2,3], FIONA MacAULAY[4] & RASOUL SORKHABI[5]

[1]*Eni Upstream & Technical Services, Via Emilia 1, 20097 San Donato Milanese, Milan, Italy*

[2]*The University of Oklahoma School of Law, Oil & Gas, Natural Resources, Energy Center, 300 Timberdell Road, Norman, OK 73019-5081, USA*

[3]*The University of Western Australia, FABLE, 35 Stirling Highway, Crawley, Perth, WA 6009, Australia*

[4]*Echo Energy plc, 40 George Street, London W1U 7DW, UK*

[5]*Energy & Geoscience Institute, University of Utah, 423 Wakara Way, Salt Lake City, UT 84108, USA*

**Correspondence: jonathan.craig@eni.it*

Abstract: The history of the European oil and gas industry reflects local and global political events, economic constraints, and the personal endeavours of individual petroleum geoscientists, as much as it does the development of technologies and the underlying geology of the region. Europe and Europeans played a disproportionately large role in the development of the modern global oil and gas industry. From at least the Iron Age until the 1850s, the use of oil in Europe was limited, and the oil was obtained almost exclusively from surface seeps and mine workings. The use of oil increased in the 1860s with the introduction of new technologies in both production and refining. Shale oil was distilled on a commercial scale in various parts of Europe in the late eighteenth century and throughout the nineteenth century but, in the second half of the nineteenth century, the mineral oils and gas produced primarily from shale and coal could no longer satisfy demand, and oil produced directly from conventional oil fields began to dominate the European market. The first commercial oil wells in Europe were manually dug in Poland in 1853, Romania in 1857, Germany in 1859 and Italy in 1860, before the gradual introduction of mechanical cable drilling rigs started in the early 1860s. In the late nineteenth century, the northern part of the Carpathian Mountains in what is now Poland and Ukraine was one of the most prolific hydrocarbon provinces in the world. The Bóbrka Field in the Carpathian foothills of Poland, discovered in 1853, is still producing and is now the oldest industrial oil field in the world.

The 1914–18 and 1939–45 world wars were both major drivers in exploration for and exploitation of Europe's oil resources and in the development of technologies to produce synthetic fuels from the liquefaction of bituminous coal and the combination of carbon monoxide and hydrogen as the Allied and Axis governments struggled to maintain adequate supplies of fuel for their war efforts. In Britain, the first 'accidental' discovery of gas was made in 1875 in the Weald Basin, but it was not until 1919 that Britain's first oil field was discovered at Hardstoft, in Derbyshire, as a result of a government-funded exploration drilling campaign, triggered by the need to find indigenous supplies of oil during World War I. The period of reconstruction after World War II was also critical for the European oil and gas industry with further successful exploration for oil and gas in the East Midlands of England resulting in Britain's first 'oil boom', and the discovery and development of deep gas fields in the Po Valley in northern Italy fuelling the Italian economy for the next 50 years.

Drilling technologies developed during Britain's first oil boom, together with the extrapolation of the onshore geology of the East Midlands oil fields and of the Dutch gas fields, led to the discovery of the huge oil and gas resources beneath the North Sea in the 1960s and 1970s, which enabled Britain, Norway, Denmark and The Netherlands to be largely self-sufficient in oil and gas from the late 1970s until production began to decline rapidly in the early 2000s. Today, oil and gas production in most European countries is at an historical low. Exploration for new sources of oil and gas in Europe continues, although increasingly hampered by the maturity of many of the conventional oil and gas plays, but European companies and European citizens continue to play a major role in the global oil and gas industry.

The establishment of petroleum as a tradable commodity, as well as a natural resource, triggered a paradigm shift in the history of civilization, introducing a new set of values, opportunities and forces, shaping the course of social endeavours as diverse as science, technology, energy production and consumption, and

From: Craig, J., Gerali, F., MacAulay, F. & Sorkhabi, R. (eds) 2018. *History of the European Oil and Gas Industry*. Geological Society, London, Special Publications, **465**, 1–24.
First published online June 21, 2018, https://doi.org/10.1144/SP465.23

Published by The Geological Society of London. Publishing disclaimer: www.geolsoc.org.uk/pub_ethics

leading to associated improvements in living standards throughout the world. Our 160 year-old hydrocarbon-based civilization is founded on this paradigm and, despite many criticisms and shortcomings, hydrocarbons will continue to be a key component of our lives for decades to come (Sertorio 2002; Smil 2017).

Europe is highly dependent on oil and gas as its primary sources of energy and as a basic feedstock for the chemical and manufacturing industries. Natural gas is widely used to generate electricity and for heating, while petroleum products derived from oil (petrol, diesel and kerosene) are essential fuels for transportation. Oil and gas are also used for domestic heating, and as the raw material for producing millions of items ranging from fertilizers, plastics, detergents, preservatives, life-saving medicines, clothing and cosmetics, through to appliances including computers, insulation, inks and toners, paints, glues, solvents and antiseptics to golf balls, CDs, bin bags, nail polish, and chewing gum.

The beginning of the mass consumption of oil was the result of a transnational and cumulative process of trial and error that lasted for centuries. The inception of the modern oil industry gained momentum in the first half of the nineteenth century as a result of the definition of some 'hydro-carbon' compounds obtained by heating oil given by the British scientist Michael Faraday in 1825, the introduction of fractional distillation to refine oil by the American Chemist Benjamin Silliman Jr and the systematic application of mechanized drilling by means of the cable tool system operated by steam power in NW Pennsylvania. After centuries of stagnation, discontinuous production and low consumption, the use of oil increased in the 1860s with the introduction of new technologies in both production and refining. In the second half of the nineteenth century, the mineral oils and gas produced primarily from shale and coal could no longer satisfy demand, and the oil produced directly from conventional oil fields in Eastern Europe, Russia, the USA and, eventually, the Middle East fairly rapidly began to dominate the global fuel market (Jevons 1865; Gerali 2012; Gerali & Gregory 2017*a*). Oil has been the world's most important source of energy since the mid-1950s, largely because of its high energy content, transportability and relative abundance.

Several European countries, among them France, Germany and Italy, have an important oil heritage, rich in invention and technology, despite the fact that they never achieved globally significant levels of conventional oil production. This lack of conventional oil resources influenced both their attempts to develop alternative sources of energy – France worked on oil shales, Germany on coal hydrogenation, and Italy had diversified strategies involving lignite and hydroelectric – and, in the second half of the twentieth century, the success of their citizens as 'oil finders' in the oil fields of the world (Yanarella & Green 1987; Bardini 1998; Pozzi 2009; Beltran & Carré 2016).

Britain, Norway, Denmark and The Netherlands were largely self-sufficient in oil and gas from the late 1970s and early 1980s, following the discovery of oil and gas under the North Sea (Fig. 1) in the 1960s and the subsequent rapid build-up in offshore production, until offshore production started to decline rapidly in the early 2000s. There is, however, a much longer and much less well-known history of oil and gas exploration and production in Europe. In fact, natural seepages of oil and outcrops of oil-bearing shale across Europe have been exploited since the Iron Age, and shale oil was distilled on a commercial scale in various parts of Europe in the late eighteenth century and throughout the nineteenth century (Forbes-Leslie 1917; Strachan 1920; Gallois 2012).

According to the *BP Statistical Review of Energy 2017*, the world's proved oil reserves (excluding self-sourced shale plays) are 1.71 Tbbl (trillion barrels), of which Europe (except Russia) contains a 'mere' 11.6 Bbbl (billion barrels). The five European countries with appreciable oil reserves today are, in decreasing order of reserves at end-2015, Norway (8.0 Bbbl), the UK (2.5 Bbbl), Romania (0.6 Bbbl), Italy (0.6 Bbbl) and Denmark (0.5 Bbbl). Out of the world's total oil production of 92 MMBOPD (million barrels of oil per day) in 2016, these five European countries produced just 3.3 MMBOPD or 3.5% of global production.

The Western and Central European countries with the largest total proved gas reserves at the end of 2016 were Norway (1.8 Tcf (trillion cubic ft)), The Netherlands (0.7 Tcf), Ukraine (0.6 Tcf) and the UK (0.2 Tcf), while Turkmenistan (17.5 Tcf), Azerbaijan (1.7 Tcf) and Kazakhstan (1.0 Tcf) hosted the largest reserves of the countries belonging to the Former Soviet Union (FSU), outside the current Russian Federation. The largest European and FSU gas producers in 2016 were Norway (116.6 Bcm (billion m^3)), Turkmenistan (66.8 Bcm), Uzbekistan (62.8 Bcm), the UK (41.0 Bcm) and The Netherlands (40.2 Bcm), but Europe and the FSU (excluding Russia) as a whole today accounts for only 11% of global gas production (British Petroleum 2017). Despite these figures, there is a long history of exploration, production and consumption of oil and gas in many parts of Europe, as well as of oil- and gas-related science and technology, much of which had a profound influence on the oil and gas industry worldwide. In the history of the oil and gas industry in many European countries, there are important episodes and players that influenced the course of events; some of these are already well documented, while others have gone unnoticed or even forgotten. Nevertheless, each episode is important because, beyond the well-known stories of successes and failures, there are various elements that enable us to gain

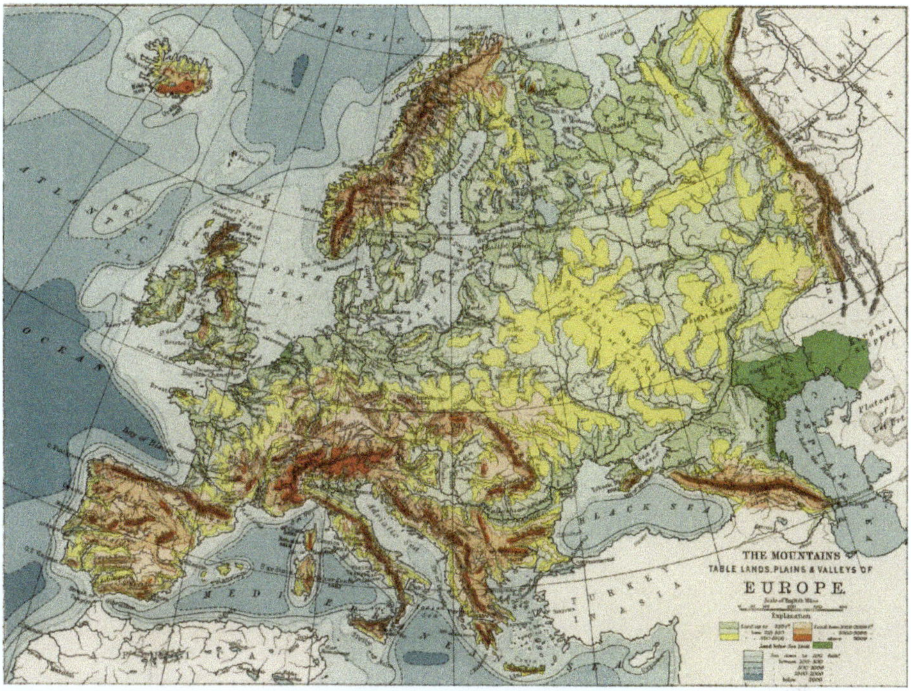

Fig. 1. Physical features of Europe. Published in 1906 by W. & A. K. Johnston Limited, Edinburgh and London. Available from the David Rumsey Historical Map Collection.

new insights into the process of the development of the oil and gas industry in Europe.

This Special Publication brings together a series of important papers dealing with different aspects of the history of the oil and gas industry in the UK, Norway, France, Spain, Italy, the Czech Republic, Hungary, Slovakia, Poland and Romania (Fig. 2), and examines the role of some of the key technologies and of the petroleum geoscientists who influenced the development of the European oil and gas industry between the sixteenth century and the present day.

A bibliography of selected published papers and books, primary and secondary sources, dealing with the history of the European oil and gas industry is provided at the end of this volume. It is not comprehensive, and is intended only as a source of further reading. There is also a comprehensive gazetteer of the main oil and gas industry museums in Europe at the end of the volume. These can be an invaluable source of additional information for those who wish to develop a deeper knowledge of the development of the oil and gas industry in specific parts of Europe.

Oil in the sixteenth, seventeenth and eighteenth centuries: an introduction

The manual gathering of petroleum from surface seeps, rock crevices or shallow shafts was recorded in the chronicles of Europe during the late Medieval Age, but had been occurring since the classical age with oil obtained from such surface seeps being used by Greeks and Romans for several purposes. Although petroleum was used in Europe for many centuries, it was only during the sixteenth century that alchemists, natural philosophers and, later, chemists gradually developed scientific knowledge about the nature and potential uses of oil (Huguenet & Frojo 1856; Novelli & Sella 2009). In 1556, the German mining technologist Georges Agricola published *De Re Metallica*, in which he documented the use of heated petroleum. Forty years later, in 1601, Andreas Libavius, a German alchemist, published in the third volume of his *Singolarium*, the most detailed account on every known occurrence of light and heavy oil in Europe at that time. In 1625, Johan Volck published a 30 page pamphlet, *Description of the Hanau earth balsam, petroleum and soft agate stone*, in which he detailed the advantages of refined oils. Many other contemporary scholars in Europe documented methods of refining petroleum, but these scientific advances had little impact on the socio-economic development of Europe at the time.

The methods used to produce petroleum in Europe remained almost unchanged between the sixteenth and the first half of the nineteenth century.

Fig. 2. General map of Europe showing the areas covered by the papers in this volume.

Basically, it was a craft activity, carried out without specific mining tools that yielded limited quantities of petroleum. Unlike coal, copper and iron, petroleum was not produced by a skilled labour force. Until the mid-nineteenth century, oil exploration was a premature concept; oil was not sought out, but simply found in the ground. Before oil products came into general commercial use during the last half of the nineteenth century, the exploitation of petroleum resources in Britain was restricted to a few surface oil seeps, such as those found in Shropshire, near Formby in Lancashire, in the coal mines of Nottinghamshire and along the south coast of England. For example, oil was distilled from oil-impregnated Carboniferous sandstone quarried at Row Brook in Shropshire during the seventeenth century (Plot 1684; Ele 1697). The use of 'seepage oil' in Britain increased considerably when it was found that oil-soaked peat dug at Formby in Lancashire (**Bunce 2018**) made good fire-lighters, and the heavy tars from the seeps were excellent for caulking boats and for medicinal purposes on livestock. The waxy green oil that seeped steadily into some of the Nottinghamshire coal mines was even used to grease the axles of colliery wagons, but the production from these 'surface' sources was always, necessarily, extremely limited. Seeps of natural gas, both at the surface and in mines, have also been recorded in Great Britain for centuries. One particularly early example, near Wigan in Lancashire (Selley 2012), was recorded by Camden in his book '*Britannia or a Chorographical Description of Great Britain and Ireland*' published in 1586 and is often referred to colloquially as 'Camden's cooker' (Camden 1586).

One of the earliest attempts to produce indigenous oil in Britain occurred on the banks of the River

Severn in Shropshire. In January 1694, British Crown Patent No. 405 was granted to Martin Eele, Thomas Hancock and William Portlock of the British Pitch Works in Bentham, Shropshire who had found a 'way to extract and make great quantityes of pitch, tarr and oyle out of a sort of stone' (Torrens 1994; Moody 2007). In 1716, another patent was taken out in Britain by one Talbot Edwards who claimed that, since the expiry of the 1694 patent, he had found a new and better way to extract these materials 'by fluxing with fire only … the roch or roofe stone' extracted from the local Shropshire coal mines. The market for the products was rather limited at the time and the oil was soon promoted as a medicine for 'the Cure of Rheumatic and Scorbutic Affections'. A third patent was issued in 1742, this one to Michael Betton, glazier of Wellington and his brother Thomas, shoemaker from Shrewsbury, again citing the medicinal use of the oil (Torrens 1994). The manufacture of this 'British Oil' later continued under the famous Darby family, the descendants of Abraham Darby who first introduced coke smelting at Coalbrookedale on the River Severn in 1709, thereby triggering the Industrial Revolution.

Elsewhere in Europe, a Russian-born Swiss doctor and Greek teacher called Eyrini d'Eyrinis (also spelt as Eirini d'Eirinis) discovered asphaltum at Val-de-Travers near Neuchàtel in Switzerland in 1710. He was the author of a landmark volume on the bitumen of the area, in which he presented all the possible uses of natural and refined bitumen. He established a bitumen mine called 'de la Presta' that operated until 1986. In France, the light and valuable oil called naphtha was collected in the region of Alsace, near the village of Pechelbronn (which means Pitch Spring). The locals had scooped up the oil from the water spring since the fifteenth century and exploitation of the local tar sands started in 1627. Then, in 1745, the mining of oil sands began in Merkwiller-Pechelbronn under the direction of Louis Pierre Ancillon de la Sablonnière, by special appointment of King Louis XV. In the 1810s, some wells were drilled to guide the location of the mine galleries. One of these wells penetrated a layer of sand containing oil at a depth of 77 m. However, the first important oil discovery was only achieved 135 years after the beginning of the oil mining in the area, when some wells dug mechanically in 1880 yielded oil in commercial quantities: 2860 wells had been drilled by 1916 and these produced 720 000 t (tons) of oil (Heritier 1994). The oil fields of Pechelbronn were active until 1970 and were the birthplace of present-day companies such as Antar and Schlumberger.

In 1786, a canal tunnel was driven into the side of the Severn Gorge, near Ironbridge in Shropshire, England, by a Quaker ironmaster, William Reynolds, to bring out coal from the local pits (Fig. 3). The tunnel struck a spring of natural bitumen in Carboniferous sandstones about 300 yards from the entrance and seeps along the length of the tunnel yielded as much as 450 gallons of bitumen a day (Trinder 1981). Some was sold as medicine for the treatment of rheumatic and skin complaints, and some was processed for burning in lamps or converted into varnish. The rest was boiled in large cauldrons at the mouth of the tunnel to produce pitch that was used as a preservative for timber, for treating ropes and for caulking ships decks (Morton 2014).

Numerous oil and gas seeps were recorded in the Weald Basin of southern England during the nineteenth century, and gas was encountered during the sinking of a water well at Hawkhurst in Sussex in 1836. The gas was ignited by a lantern, causing an explosion that killed the two labourers who were sinking the well (Evans 2009; Wigley 2015).

In the territories of present-day Germany, petroleum production in the eighteenth century was mainly concentrated in the vicinity of Hannover, Brunswick, Wietz and Celle (Rinehart 1930). In France, thick and sticky asphalt was mined in the vicinity of Gabian, near the city of Clermont-Ferrand, in the region of Auvergne. The town of Gabian was also renowned for La Font de l'Oly (the oil spring), which produced light naphtha. Oil was recovered from La Font from 1608 onwards and the average production was 36 quintals per year. There was another seep, associated with a mineral water spring and a volcanic intrusion, at Le Puy de La Poix (the pitch mount) in the Limagne Basin. Gummy bitumen was collected in the region of Guyana (Schmitz 1938; Heritier 1994; Smith 2006). In Italy, the largest oil provinces were located in the territories of the duchies of Parma and Piacenza, and Modena and Reggio Emilia – stretching from NW to SE along the Northern Apennines (Fig. 1; Fairman 1868; Bonariva 1869; **Cazzini 2018**; **Gerali *et al.* 2018**).

Further scientific studies at this time paved the way to a better understanding of oil as a mineral and as a source of energy. In France, in the *Dictionnaire de Chymie* (1769), Pierre Joseph Macquer (1718–84) provided new evidence for the organic origin of oil. George Balthazar Sage (1740–1824) set out a classification of the different types of bitumen with reference to their density in *Eléments de Minéralogie Docimastique* (1772) while, in 1784, the Irish mineralogist Richard Kirwan (1733–1812) proposed a category of substances called 'inflammables' in his treatise *Elements of Mineralogy*. In 1796, Kirwan also published a study on the composition of coal and bitumen in the essay *Of the Composition and Proportion of Carbon in Bitumen and Mineral Coal*, in which he supported the idea of the major commercial utilization of petroleum.

Fig. 3. The Coalport Tar Tunnel, Ironbridge, Shropshire. (**a**) The Tar Tunnel with a drainage pipe on the left and wells cut into the tunnel walls. (**b**) Bitumen oozing out of the brick-lined walls. (**c**) A pool of bitumen in one of the wells (http://geotopoi.wordpress.com/2012/09/13/coalport-tar-tunnel/).

In the Austro-Hungarian dominions, petroleum was produced along the northern slope of the Carpathian chain (Fig. 1), in the vicinity of the village of Boryslaw, in the region of Galicia – today in Ukraine. In Romania, the core production of oil was concentrated near the city of Ploieşti, in the foreland basin of the Southern Carpathian chain, in the principality of Wallachia. The earliest written accounts of natural occurrences of hydrocarbons in the Carpathians date back to the sixteenth century, but in the eighteenth century and the early nineteenth century accounts of the practical uses of oil, based on its occurrence in the Carpathians, were published by Rzączyński, Kluk, Hacquest and Staszic. Stanislaw Staszic, a Polish priest, philosopher, statesman, geologist, scholar, poet and writer, produced a geological map of the region, dated 1806, showing the numerous oil seeps and different rock types containing hydrocarbons (**Krzywiec 2018**) – probably the first map to specifically document surface hydrocarbon occurrences in Europe.

In the SE part of what is today Moldova, ozokerite, or earth wax (a kind of natural paraffin), had been mined for centuries (Pearton 1971), and 'mineral resin' was collected on the banks of some rivers. In Tsarist Russia, oil production was a burgeoning activity in the Trans-Caucasian area straddling Europe and Asia corresponding to today's Azerbaijan. The economy of the port city of Baku, on the Caspian Sea, was largely based on oil production and trading (Henry 1905).

The nineteenth century: the beginning of the modern age of oil

In 1816–17, the Swiss naturalist Nicholas Theodore de Saussure made a comparative analysis of the composition and properties of petroleum samples collected in several European countries (Saussure 1816, 1817). Other scientists had previously carried out similar work, but Saussure's work has the merit of delivering a comprehensive study that went further than the mere annotation of physical characteristics of the different petroleum samples. Instead, his work highlighted the chemical properties and commercial applications of petroleum depending on its quality.

During the late eighteenth century and the nineteenth century, Scottish chemists were foremost in the development of British coal-gas, coal-tar and shale-oil industries. William Murdoch (1754–1839) and Archibold Cochrane (1749–1831), ninth Earl of Dundonald, carried out pioneering work on the destructive distillation of coal and shale oil. Archibald Cochrane developed, and in 1871 patented, a method 'for extracting tar, pitch, essential oils and cinders from pitcoal, whereby the coals are made to ignite without flaming' (Conacher 1938*b*) and in 1782 he lit Culross Abbey with gas from produced from tar. Cochrane was a brilliant chemist, but a poor businessman. He foolishly failed to repay massive loans made to his wife and died in poverty in Paris in 1831 (Tomory 2012).

Dr James 'Paraffin' Young, the founder of the Scottish oil industry, was much more astute. He rose from a humble background to become a noted and very successful chemist and entrepreneur. Young was born in Glasgow and trained as a chemist (Conacher 1938*a*). He was the manager of an alkali works in Manchester when, late in 1847, he was approached by James Oakes, the owner of the 'New Deeps' coal mine at Riddings in Derbyshire, to help him find a use for the thick, black crude oil that was seeping into his colliery workings from a small 'petroleum spring' (Fowler 1940; **Lea 2018**). In the autumn of 1848, Oakes and Young signed a partnership agreement and set up the first plant to produce paraffin oil for lamps. Oakes provided the crude oil from the mine and the plant to refine it, while Young was responsible for the manufacture and sale of the products (Beaton 1959). The crude oil production from the mine was only about 300 gallons a day and by 1850 the supply of oil had run out. Young returned to Scotland in 1851 having patented methods to extract paraffin from bituminous coal and oil shale, treating the extract and purifying the solid paraffin wax. He built a large plant to process the Carboniferous oil shales of West Calder, establishing an industry that was to dominate the West Lothian region of Scotland for much of the next century, and his patented technology was eventually adopted by many early commercial oil refineries worldwide (Carruthers *et al.* 1927; Conacher 1928; **Dean 2018**). Over the next 100 years, an estimated 75 MMbbl (million barrels) of oil and 500 Bcf (billion cubic ft) of gas were produced from Scottish shales.

In the early nineteenth century, small artisanal refineries with low, and variable, yields were scattered across Europe. Romania in the 1840s was the only European country with documented and consistent commercial-scale refining activity. Romania has a long oil industry history (the use of bitumen in the country dates back to the fifth century AD), and it has the best-preserved official records of oil production, refining and trade in Europe. These confirm the importance of the oil business both for the country and for European oil industry heritage (**Tulucan *et al.* 2018**). In 1840, a refinery with the capacity of 300 t per year was active in Lucăceşti, Bacău (at that time still included in the principality of Moldavia) to produce lamp fuel and greases; in 1856, a 2710 t per year plant was operating in Rafov, in the area of Prahova, and between 1857 and 1862 four more refineries opened in the area of Bacău, with a capacity ranging from 873 to 2446 t per year (Pizanty 1947; Forbes 1958).

During the second half of the nineteenth century, oil and gas produced primarily from shale and coal could not satisfy the increasing demand in the European fuel market, and new technologies for oil production (and, later, gas production) developed rapidly (Jevons 1865; Gerali 2012). In 1853, the Polish chemist Ignacy Lukasiewicz discovered how to make kerosene from crude oil. On the night of 31 July 1853, Lukasiewicz was called to his local hospital in the then Polish town of Lvov (now in Ukraine) to provide light from one of his new kerosene lamps for an emergency surgical operation. The hospital authorities were clearly impressed with his invention because they ordered several of his lamps and 500 kg of kerosene. Soon after, Lukasiewicz travelled to Vienna, the capital of the Austro-Hungarian Empire, to register his distillation process in the Vienna patent office with his co-worker, J. Zeh, on the 31 December 1853. After Lukasiewicz adapted Gesner's distillation method in 1852 to refine kerosene from more readily available 'rock oil' seeps, Lukasiewicz and his business partners, Titus Trzecieski and Mikolaj Klobassa, established the first rock oil mine in Bobrka, near Krosno in Central European Galicia (Poland/Ukraine) in 1853. This led to the rapid growth of the oil industry in the Carpathians, with many shallow wells (30–50 m) dug manually on oil shows associated with the major anticlines in the region. Then, in 1856, Lukasiewicz set up the world's first crude oil refinery at Ulaszowice in Poland.

The world's first commercial oil wells were dug in Bibi-Heybat (Bibi-Eibat) near Baku in Azerbaijan in 1846, in Poland in 1853 and in Romania in 1857 (still 2 years before the drilling of the seminal 'Drake well' in NW Pennsylvania). In the late nineteenth century, the northern part of the Carpathian Mountains between Limanowa (Poland) and Kosów (Ukraine) was one of the most prolific hydrocarbon provinces in the world (**Krzywiec 2018**) (Fig. 4). The Bóbrka Field in the Carpathian foothills of Poland, discovered in 1853, is still producing and is now the oldest industrial oil field in the world (Koszarski & Bally 1994).

Fig. 4. Postcard showing an early oil field in the Northern Carpathians (from **Krzywiec 2018**).

Romania was the first country in the world to have its crude oil output officially recorded in international statistics with an amount of 275 tonnes recorded for 1857. Oil production in Romania had started in the middle of the nineteenth century, mainly from primitive hand-dug and wood-lined wells. An oil refinery was already in operation in Ploieşti in Romania in 1856 and one year later its product illuminated the capital city, Bucharest, with 1000 oil lamps placed along the main streets (Schleder *et al.* 2016) – the first use of refined oil for street lighting anywhere in the world. The first oil discovery in NW Europe was made when the Wietze oil field near Hannover in Germany was discovered in 1859 (Fig. 5), followed soon after by a discovery near Ozzano in the Parma region of northern Italy in 1860 (**Cazzini 2018**).

From the 1860s onwards, the Majella oil district of Abruzzo in Italy played a key role in the development of the Italian oil industry through the first detailed analysis of the associated petroleum systems and the introduction of mechanized drilling (**Gerali & Lipparini 2018**). In 1863, Merzoeff built the first modern refinery in the town of Sourachany, close to the mature oil fields of Baku in Azerbaijan, at a time when they supplied about 90% of the oil used in Europe. Only a year later, the area of Baku had 23 fully equipped refineries (Forbes 1959). At about the same time, in Spain there was early exploitation of oil from bituminous shales, asphalt impregnations or tar sands by subsurface mining and distilling, but with limited economic success (**Navarro Comet 2018**). Also in 1861, the sailing ship *Elizabeth Watts* carried a cargo of 1329 bbl of oil for lamps from Philadelphia to London – the first recorded shipping of Pennsylvanian oil to be sold in Europe.

Exploration for oil started again in the Gabian area in France in the early 1860, and numerous shallow wells were drilled around the old oil spring. In 1884, a well drilled 300 m from the old oil spring hit an oil reservoir at a depth of 68 m but, with the exception of two other dry wells, no further exploration work was undertaken in the area until 1923 (Heritier 1994).

Four of the first 10 giant oil fields to be discovered in the world were found in the late 1880s in what was then the Russian Empire, three of them (Balakhany, Surakhany and Bibi-Heybat) within 25 km of Baku in Azerbaijan (Halbouty 2003; Sorenson 2014) (Fig. 6). The Balakhany Field on the Apsheron Peninsula in Azerbaijan was the largest producing oil field in the world in the 1870s and had 375 producing wells in 1883. It was succeeded as the world's largest producing field by the Bibi-Heybat Field on the south side of the Apsheron Peninsula in the mid-1880s, although this had certainly been in production since 1803 (Sorenson 2014).

The first deep well to encounter gas in Britain was drilled on the crest of the Battle Anticline at Netherfield in West Sussex by the Subwealden Exploration Company in 1875. The first well was abandoned at

Fig. 5. The first commercially productive oil well drilled in the Wietze Field, Hannover, Germany. Drilled by the Hannovarian government in 1859, the well yielded a small amount of oil at a depth of about 100 m. Until the 1930s, it was occasionally pumped by hand. From Rinehart (1930).

a depth of 1018 ft for mechanical reasons, but a second well was drilled nearby and reached a total depth of 1905 ft. While measuring the bottom-hole temperature of the well, a 'naked light was injudiciously lowered down the hole' (Selley 2011) which resulted in an explosion (Willet 1875). There was another accidental discovery of natural gas in 1895 when a 228 ft deep 'artesian bore-tube' was sunk in the stable yard of the New Heathfield Hotel at Heathfield in East Sussex (Dawson 1898). The first commercial discovery of gas in Britain occurred in August 1896, when gas was again encountered, accidentally, close to Heathfield Railway Station by the London, Brighton and South Coast Railway while searching for an underground supply of water for its locomotives (Dawson 1898). The Heathfield discovery set off a widespread search for gas in commercial quantities during the years to 1910. Six more wells were drilled at Heathfield, and by 1903 some 70–80 houses and the railway station in Heathfield were lit by gas. The station continued to be lit by gas until 1930.

Fig. 6. A Baku 'spouter'. The derrick is almost buried in sand ejected from the well (from Beeby-Thompson 1904).

In the late nineteenth and early twentieth centuries, most domestic gas was produced by the destructive distillation of bituminous coal (so-called 'town gas'). The gas was driven off by heating the coal in air-tight chambers, storing it in 'gasometers', and distributing it through a network of pipes to houses and other buildings where it was used for lighting and, much later, for cooking and heating. Gas manufacturing also produced a range of by-products which later became important feedstock for the chemical industry. Manufactured gas allowed the development of what was the first integrated utility network, and its success in Britain led to its rapid spread across Europe (**Thomas 2018**).

The early twentieth century and World War I

In February 1899, Paul Dvorkovitz (1857–1929), a Russian petrochemist in the service of Britain's Petroleum Storage Tanks and Transportation Co. Ltd, published the first issue of *The Petroleum Industrial Technical Review* in London, with the support of British oil entrepreneurs. This was the first European technical journal devoted entirely to the oil industry and soon had a worldwide distribution. Similar editorial initiatives were also active in the USA, but were focused mostly on business and the national industry. Dvorkovitz's journal introduced a new way to communicate the latest oil insights to industry practitioners. It led the market in oil information until the early 1910s and affirmed itself as the 'voice' of the European oil industry. Dvorkovitz and his firms were also the major promoters of the first four International Petroleum Congresses held in Paris (1900), Liège (1905), Bucharest (1907) and Lvov (1910). Prior to this, oil was discussed as a 'branch' discipline in general geology, engineering and chemistry conferences. Dvorkovitz's Congress was the first international forum for global oil industry representatives organized in Europe.

By 1906, there were more than 7000 people working in the oil industry in the Prahova District of Romania (Fig. 7), the first area in the country to power oil rigs with electric engines fed by the local hydroelectric plant. Oil production exceeded 1 Mt (million tons) per year (Schleder *et al.* 2016).

Oil production in Austrian Galicia (an area stretching from present-day Poland to present-day Ukraine) peaked in 1910 at 2 Mt of crude oil (**Krzywiec 2018**), making this the third most prolific oil province in the world after Russia and the USA (Krzywiec 2016). Exploration drilling in Spain started in 1900, but the wells were usually simply located in the vicinity of surface oil seeps and impregnations with little or no geological input (**Navarro Comet & Puche Riart 2018**).

Among the many individuals who contributed to the development of the oil and gas industry in the late nineteenth and early twentieth centuries, Sir Thomas Boverton Redwood (1846–1919) stands out as, perhaps, the greatest petroleum engineer and scientist of his generation (**Sorkhabi 2018**). Redwood, together with his many collaborators and co-authors, contributed, either directly or indirectly, to petroleum exploration and production in many parts of the world where Britain had colonial and commercial interests, and played a leading role in the development of petroleum technology and its application to Britain's oil operations around the world from the 1880s to the 1910s. He established one of the first and most successful petroleum consulting companies, trained generations of petroleum geologists and engineers, served as a technical advisor on many government committees and to many oil companies, and played a key role in securing Britain's oil supply during World War I.

Fig. 7. 'Gusher' in the Campina Field, Prahova District, Romania (from Suciu & Luzzati 1923).

At the beginning of the twentieth century, Britain was importing nearly all of its oil from America and Iran. The Scottish oil-shale industry was making a modest, but important, contribution to meeting the domestic oil demand but, by 1918, Britain was importing over 100 000 BOPD (Hallett *et al.* 1985). Oil imports had increased by over 250% since 1913, in large part due to the increase in consumption required by the Allied Forces during World War I (Giffard 1923), but only about one-seventh of the oil imported in 1918 came from countries within the British Empire.

Access to oil was a major strategic factor in World War I (1914–18). The German invasion of Russia during World War I had as one of its objectives the capture of the Baku oil fields so that these could provide much-needed oil supplies for the German military. By the start of World War I, approximately 1335 km of pipeline had been laid in Romania to transport oil, and Germany's invasion of Romania in 1916 was driven by the desire to have access to the country's rich oil fields. The Romanians, following a request from Britain, destroyed more than 1500 wells to prevent them falling into German hands and, as a result, Romanian production dropped from 1 783 947 t in 1914 to 518 460 t in 1917 during the German occupation (**Tulucan *et al.* 2018**). However, after the conclusion of the war, the damaged fields were quickly recovered and in 1918 the production had already recovered to 1 222 582 t. From 1857 to 1918, Romania produced 21 376 295 t of oil (Suciu & Luzzati 1923). The Gbely oil field in Galicia also played an important role in supplying oil to the German military during World War I (Benadová 2016), which was much more accessible because of the alliance between Germany and the Austro-Hungarian Empire.

Britain's first oil discovery at Hardstoft in Derbyshire in 1919 (Giffard 1923; Brentnall 1995) was made during a government-funded exploration programme to find indigenous oil supplies as a direct result of difficulties encountered in obtaining supplies from overseas during World War I (Craig *et al.* 2013; **Corfield 2018**). The exploration programme actually began shortly after World War I, but it was prompted by the concerns of the British government about the German submarine threat to imported supplies of essential petroleum products.

Fig. 8. The Hardstoft No. 1 well, Tibshelf, Derbyshire (© Devonshire Mss., Chatsworth, Tibshelf/Hardstoft Oilwell papers).

Fig. 9. View of a hilly oil field in Sarata, Romania, 1899 (source: *Petroleum Industrial and Technical Review*, vol. 1, 30 March 1899).

Between 1918 and 1922, 11 wells were drilled (nine in England and two in Scotland), two of which proved commercial volumes of oil (Lees & Cox 1937; Brentnall 1995). The first well was drilled on a prominent faulted surface anticline at Hardstoft, near Bolsover in Derbyshire (Fig. 8), and produced some 29 000 bbl of oil before it was finally abandoned in 1945 (Craig *et al.* 2013). The second discovery, at D'Arcy in Lothian, 10 miles SE of Edinburgh, produced only 50 bbl of waxy oil before it was abandoned. Two wells were drilled in Staffordshire – one on the western side and the other on the eastern side of the Staffordshire Coalfield – and the other seven English wells were drilled in a roughly north–south line along the east side of the Pennines in Derbyshire, but all these except the Hardstoft well were dry (**Corfield 2018**).

Oil was struck at Hardstoft on 27 May 1919 and production commenced in June 1919. Powers to enter onto land for the purpose of drilling for oil (but not ownership of any oil found) were provided under the Defence of the Realm Act of January 1918 and the government used these powers to take the land necessary for the nine well sites. These powers lapsed on the cessation of hostilities at the end of World War I (Hallett *et al.* 1985), with rather embarrassing consequences for the government.

The inter-war years

Although modern rotary drilling techniques were first introduced in Texas in the early 1890s, most wells in Europe were still drilled using the older, slower and potentially more dangerous 'cable tool' method until the 1930s. Cable tool drilling did not allow the pressure in the well to be controlled, and the first indication that a high-pressure productive reservoir had been penetrated was usually the appearance of oil and gas jetting through the rig floor, which often then ignited causing dramatic oil field fires. One such spectacular 'blowout' occurred on 27 May 1929 at Moreni in Romania (Fig. 9) when the Romano American well No. 160 blew-out after encountering high-pressure gas (Spencer & Furcata 2016). The entire 662 m drill string was ejected from the well and sparks produced by the friction ignited the gas escaping from the well. The derrick was thrown 300 m by the explosion. Numerous attempts to extinguish the fire and cap the well, including drilling a tunnel 60 m below the casing head, were ultimately unsuccessful. The light from the burning well was visible from Bucharest, 70 km away and the blowout became internationally famous as the 'Torch of Moreni' (**Spencer & Furcuta 2018**). The fire finally went out naturally in November 1932, after burning for 2 years and 4 months, by which time the blowout and the subsequent attempts to extinguish the fire, including by the pioneer of oil-well firefighting, Myron Kinley (**Spencer & Furcuta 2018**), had resulted in the death of 14 workers and the injury of more than 100 others.

After World War I, interest in exploring for oil in Britain declined when an Act of Parliament (the Petroleum Production Bill) in 1918 conferred on landowners the rights of ownership of any minerals found on their land. This discouraged oil companies from exploration. Cheap supplies of foreign crude oil also reduced the incentives to search for oil in Britain. Oil was produced at Hardstoft at an average rate of 6 BOPD until December 1927 (Craig *et al.* 2013). The well was prone to waxing and silting, and never flowed freely. Production increased to 14.6 BOPD during a pumping test carried out by the government in 1921, but the pumping was not continued because the government did not own the oil it produced. Several of the other wells drilled during this campaign on other structures encountered gas (Veatch 1923) but, ultimately, not in commercial quantities (**Corfield 2018**). On 23 March 1923, the Hardstoft well was taken over by the Duke of Devonshire, who owned the mineral rights and the land on which it was drilled. Eventually, the Petroleum Prospecting Act of 1934 gave the Crown the rights of ownership of any future discoveries of naturally occurring oil or gas and made it possible for companies to acquire exploration licences in the UK for the first time. With the new legislation, interest in exploration was renewed. The Petroleum Prospecting Act was not retrospective; the Duke of Devonshire retained ownership of the oil discovered at Hardstoft and he remains the only individual in the UK today who owns the rights to the oil under his land.

Initially, applications for licences to carry out exploration were concentrated mainly over parts of southern England, particularly Sussex and Hampshire. The first deep oil exploration well to be drilled in Britain, Portsdown No. 1, was spudded by D'Arcy Exploration Co. Ltd (the exploration subsidiary of the Anglo-Iranian Oil Company Ltd) in January 1936 on a site on a hill overlooking Portsmouth Harbour on the south coast (Fig. 10), but it found only a small quantity of oil. Several other wells were drilled in Hampshire, Berkshire, Dorset and Sussex in the following years (including wells at Henfield in Sussex, which encountered a small amount of gas, and at Kingsclere near Newbury in Berkshire), and a little oil was found, but all the wells were abandoned due to the poor quality of the reservoir (Lees & Cox 1937). In 1937, the D'Arcy Exploration Co. Ltd (later to become BP) found gas at Eskdale in North Yorkshire (Anglo-Iranian Oil Company Ltd 1947; **Haarhoff *et al.* 2018**).

In 1937, the Anglo-American Oil Company (a subsidiary of Standard Oil) made an unsuccessful attempt to find oil at Hellingley in Sussex (Anon 1937). The search for oil then moved to the Midlands and the north of England, and particularly to southern Scotland, where, during 1937, D'Arcy made significant gas discoveries (with minor amounts of oil) in the northern culmination of the D'Arcy–Cousland Anticline at Dalkeith, SE of Edinburgh and at Cousland (Longhurst 1959). The gas was piped to the local gasworks during World War II and BP developed the field in the late 1950s. Also in 1937, Anglo-American re-drilled the southern culmination of the D'Arcy–Cousland Anticline, close to the original 1922 D'Arcy discovery well. This led to the later development of the Midlothian oil field (Underhill *et al.* 2008). Steel Brothers & Co. Ltd, whose association with the oil industry was primarily as the managing agents in London to the Attock Oil Company operating in India, completed an exploration well at Edale in Derbyshire in 1938 and, when this was unsuccessful, they carried out a general investigation of areas in Yorkshire, Derbyshire, Cheshire and Lancashire. Further wells drilled in the East Midlands, at Gun Hill, north Staffordshire and at Alport in the Peak District also gave disappointing results (Anglo-Iranian Oil Company Ltd 1947), but a small oil accumulation was discovered at Formby in Lancashire, at a depth of only 18 ft, in 1939 (**Bunce 2018**). Shortly after the Formby discovery, the D'Arcy Exploration Co. made much more significant discoveries in the Carboniferous succession at Eakring (in 1939), Kelham Hills (in 1941), Dukes Wood (in 1941) and Caunton (in 1943) in Nottinghamshire, and at Nocton (1943) in Lincolnshire (Lees & Taitt 1946) (Fig. 7). In 1940–41, Gulf Exploration (Great Britain) drilled a deep exploration well at Penhurst in Kent and then a second well in the Cleveland Hills of Yorkshire, with unsuccessful results in both cases.

World War II

During World War II, oil facilities were a major strategic asset and were extensively bombed. German U-boats again threatened British oil supplies from Persia and this led to the rapid, and top secret, development of domestic oil fields at Eakring and

Fig. 10. Portsdown No. 1 Well. *Source*: BP.

Duke's Wood in the East Midlands of England. Given the critical need for an indigenous supply of oil during World War II, the D'Arcy Exploration Company decided that, in the national interest, it would suspend further speculative exploration drilling in order to focus on developing the two fields in Nottinghamshire. The development of the fields under wartime conditions was hampered by the lack of available labour because the majority of skilled and able-bodied men had been 'called-up' and by the lack of steel, which was required for munitions and shipbuilding. Delays were also incurred because of the restricted lighting conditions permitted during black-out hours and, in the early stages, by periodic air-raid warnings during which drilling had to be suspended. The development of the Duke's Wood Field was further hampered by the shortage of labour and of road-building equipment need to construct access roads in the hilly and wooded area. The various drilling rigs operated by D'Arcy Exploration were moved to Eakring to begin development work, but these heavy-duty rigs proved unsuitable (Southwell 1945). Although improvements were made, by August 1942 the situation was critical – Britain had just 2 months' stock of oil remaining and supplies from overseas were being repeatedly disrupted by German U-boats. Britain turned to its American ally, importing, in great secrecy, American rigs and American drilling crews to drill up the Eakring and Duke's Wood fields. The new American Unitized rigs with jacknife portable derricks were designed to give maximum mobility (Day 1946). What 44 American roughnecks were doing in the middle of Sherwood Forest was kept a secret – apparently, by the simple expedient of spreading a rumour that they were making a Western film starring John Wayne! The new rigs and their American drilling crews greatly reduced the time taken to drill the wells and to move the rigs between drilling locations. The average time taken to rig-up, drill and complete a production well was reduced to 7 days (Southwell 1945) and, with experience and changes in drilling practices, it became possible to drill 800 ft with a single drill bit, thereby reducing the average number of drill bits (which were in short supply) required to drill a production well from 13 to six. By the end of the war, the crews had drilled 380 wells of varying depths (amounting to 735 000 ft of drilling) resulting in 250 successful production wells, which together had produced more than 2.25 MMbbl of oil. When refined, this oil was particularly suited to power the Rolls Royce Merlin engine used in most of the Royal Air Force's high-performance fighters and bombers like the Supermarine Spitfire and the Avro 683 Lancaster. This provided Britain with a distinct advantage when compared with the low octane content of the synthetic aviation fuels produced in Germany by the synthesis of coal through the Fischer–Tropsch process. Before World War II, the German oil industry consisted of a weak upstream sector, mostly run by small- to medium-sized companies with limited financial resources, and a strong and competitive downstream sector. Because of the relatively small number of wells and the outdated technology (wooden rigs and manual gathering of the oil), the Wietze fields, in the Hannover district which is the richest oil area of Germany, produced only 11 773 567 t of crude between 1873 and 1929; while, in contrast, in the 1930s Germany had 27 efficient refineries that refined and re-sold foreign oil. It was this lack of domestic oil resources and the pressing war needs that motivated Germany to turn to the production of very expensive synthetic fuels to sustain their war effort (Rinehart 1930; Dewey 1946).

As the war progressed, supplies of oil to the 'Axis' troops also became increasingly vulnerable as the Allies mounted numerous attacks on German, Italian and Japanese oil resources. In 1943, some 35% of all 'Axis' oil supplies came from the Ploieşti oil fields in Romania, seized by the Germans in late 1940. Attacks on the oil refineries at Ploieşti culminated in the infamous 'Operation Tidal Wave', launched on 1 August 1943, during which 54 out of the 117 B-24 Liberator bombers that flew the low-altitude mission from Benghazi in Libya were shot down with the loss of 532 men (Al-Ani 2015). Several more raids on the Ploieşti oil fields in 1944 were more successful, resulting in substantial damage (Fig. 11) and making the refineries inoperable by August 1944, thereby substantially weakening the German war effort. At the end of the war, Romania was occupied by Russia, and the Russian influence on the Romanian oil industry remained for decades, even lasting until the fall of the Soviet Union in 1991 (**Tulucan *et al.* 2018**).

Near the end of the war in 1944, during the 'Italian Civil War', the Caviaga gas field was discovered in the province of Piacenza, near Milan in the Po Valley area of northern Italy (Fig. 1), following a seismic survey that started in 1940 with American personnel and machinery, but was stopped temporarily in 1941 when the USA entered the war. Italian technicians replaced the Americans on the seismic crew and the survey was completed (**Cazzini 2018**). Caviaga was one of the largest gas fields in Europe at the time, second only to the Deleni Field in Romania, and it played a key role in the rejuvenation of the Italian industrial economy after the war.

Fig. 11. 5 April 1944, B-24 Liberators in formation, bombing the Ploieşti oil installations in Romania.

Pioneering petroleum geologists from Europe

Although the majority of European countries are not major producers of oil and gas, many have produced world-class petroleum geologists, explorers and engineers. Cesare Porro (1865–1940) was the first modern oil geologist in Italy. In contrast to his senior colleagues Giovanni Capellini (1833–1922), Antonio Stoppani (1824–91) and Dante Pantanelli (1844–1913), who focused on oil exploration for just a small part of their careers, Porro worked consistently in the oil sector from 1898 to 1931. He started his career with Royal Dutch Shell in Sumatra and eastern Europe. A few years later, he was hired by Deutsche Bank to work in what was then called Mesopotamia. In 1905, he documented the oil potential of the Mosul region for the first time. Hired by a Dutch private consulting firm in 1906, he was sent to explore areas of Wyoming and Oklahoma. In the early 1920s, he returned to Italy to work on oil exploration in the peninsula and then became a member of the Technical Committee of the newly-formed AGIP. He left in 1931, but was asked again in 1937 to explore the territories of Libya. Porro refused and the task was then offered to Ardito Desio (1897–2001), who between 1938 and 1939 discovered oil for the very first time in the oil formations in the province of Fezzan (Desio 1941; Candela 2014; **Gerali & Lipparini 2018**).

Three German geologists, Alfred Benz (1897–1964), August Moos (1893–1944) and Karl Krejci-Graf (1898–1986), made particularly important contributions to petroleum geology and exploration in the 1930s and 1940s (**Kölbl-Ebert 2018**). Bentz and Krejci-Graf offered their expertise to Hitler's regime, whereas Moos, because of his Jewish background, was murdered. Their biographies shed new light on how German petroleum geologists lived and worked during the span of two world wars in the first half of the twentieth century.

Switzerland has also produced many excellent petroleum geologists, including Carl Schmidt (1862–1923), Josef Erb (1874–1934), Hans Hirschi (1876–1965) and Arnold Heim (1882–1965) (Gisler 2014, **2018**; Gisler & Trümpy 2016). Schmidt and Erb worked for Royal Dutch Shell, mainly in SE Asia, while Hirschi and Heim were both students of Albert Heim (1849–1937), Arnold Heim's father, at Zurich. For a while, Hirschi also mentored Arnold Heim. Arnold Heim, who could not succeed to his father's position at Zurich because of a conflict of interest, led an almost 'nomadic' life, working in various parts of the world. His only permanent position came in 1949 (at the age of 70) when he put together a group of Swiss geologists to work in Iran. Among these were Jovan Stöcklin (1921–2008) and Heinrich Huber (1917–92) who later led the mapping work at the Geological Survey of Iran and the National Iranian Oil Company, respectively, until the 1970s.

Another well-known Swiss geologist during the pre- and post-war period was Augustus Gansser (1910–2012). He worked for Royal Dutch Shell in South America during World War II, was another student of Arnold Heim at Zurich and later was a colleague of Albert Heim while working in the Himalaya in the 1930s and Iran in the 1950s (Sorkhabi 2012). The Swiss petroleum geologists working for Shell in South America during World War II, including Gansser and the two brothers Daniel Trümpy (1893–1971) and Eduard Trümpy (1903–66), remained in the region because of the war and undertook extensive geological mapping. Josef Theodor Erb (1874–1934) was the Chief geologist for Royal Dutch Shell, where he introduced rules and suggestions to standardize the writing of petroleum geological reports (**Wannier 2018**). Erb's 'Uniformity in Geological Reports' was written as an internal Shell report in 1917 and it eventually formed the basis of the 'Shell Standard Legend'.

Another pioneering European petroleum geologist during and after World War II was Simon Papp (1886–1970), who is regarded as the Father of petroleum exploration in his country, Hungary. Papp was an internationally renowned petroleum geologist, who contributed enormously to oil and gas exploration in Hungary. In 1948, soon after the end of World War II, the communist regime in Hungary imprisoned Papp; nevertheless, needing his expertise, the regime arranged for him to continue his geological work from his prison cell (**Tari & Bérczi 2018**).

Developments in the post-war period

The Eakring Field in the East Midlands was the first of 25 onshore discoveries made in Britain over the next 30 years (Falcon & Kent 1960) (Figs 12 & 13), a period that defined what was, effectively, Britain's first 'oil boom'. After World War II, both D'Arcy Exploration Co. Ltd and Anglo-American Oil Co. Ltd resumed their oil exploration efforts in Britain, but production declined to 5500 bbl per year by 1948. The search for other oil fields continued and, in 1953, BP found oil at Plungar in the Vale of Belvoir in Leicestershire – the first post-war oil field discovered in Britain. Following this, new fields were found in Nottinghamshire at Egmanton in 1955, Bothamsall in 1958 and South Leverton in 1960 (Fig. 11). The search in Lincolnshire also proved fruitful, and four discoveries were made between 1958 and 1962.

In 1952, the Gas Council (GC) joined forces with BP in the search for natural gas and for structures suitable for the storage of peak-load gas. In 1958, gas from a discovery at Calow in Derbyshire was

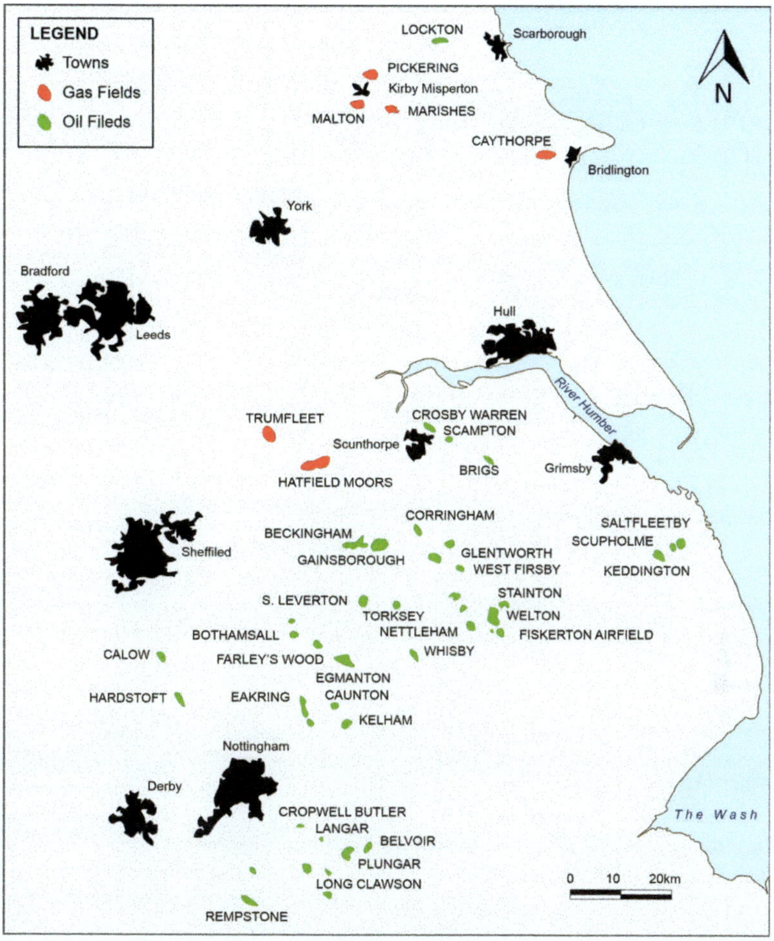

Fig. 12. Oil and gas fields in northern and eastern England (modified from Gluyas & Bowman 1997; Craig *et al.* 2013).

fed to the Chesterfield gas works, and a similar link was made from the Eskdale gas field to the Whitby gas works in 1960.

Elsewhere in Europe, the widespread use of seismic data from the 1950s onwards triggered an intense period of exploration. The first significant oil discoveries in Italy were the Ragusa heavy oil field, discovered in 1954 and the Gela Field, discovered in 1956, both in Sicily (**Cazzini 2018**). By the mid-1950s, the first offshore seismic survey had started in the Adriatic Sea (Fig. 1) and, in 1959, the drilling of the Gela 21 well, the first offshore well in Europe, boosted further exploration possibilities in Italy. In Spain, despite systematic exploration in 1940, it took more than 20 years to find a commercial hydrocarbon discovery. The Castillo gas discovery in the onshore Basque–Cantabrian Basin in Spain was later complemented in 1964 by the oil discovery at Ayoluengo (Burgos) in the southern part of the same basin, which today, more than 50 years later and after intense drilling activity during the 1960s, still remains as the only commercial oil field discovered onshore Spain, with production varying between 100 and 150 BOPD and a total cumulative production of 17 MMbbl of oil (**Navarro Comet 2018**).

Some exploration activity continued onshore in Britain during the late 1950s and the early 1960s (Fig. 13). BP made significant commercial oil discoveries at Kimmeridge in 1959 and subsequently at Wareham in 1964. However, in 1964, the British government ended the preferential tax treatment for producers of indigenous light oils in order to meet an obligation entered into under the European Free Trade Agreement. This had a devastating effect on the economic viability of Britain's onshore oil fields. BP's oil production from the East Midlands fell by a third and development work on several significant

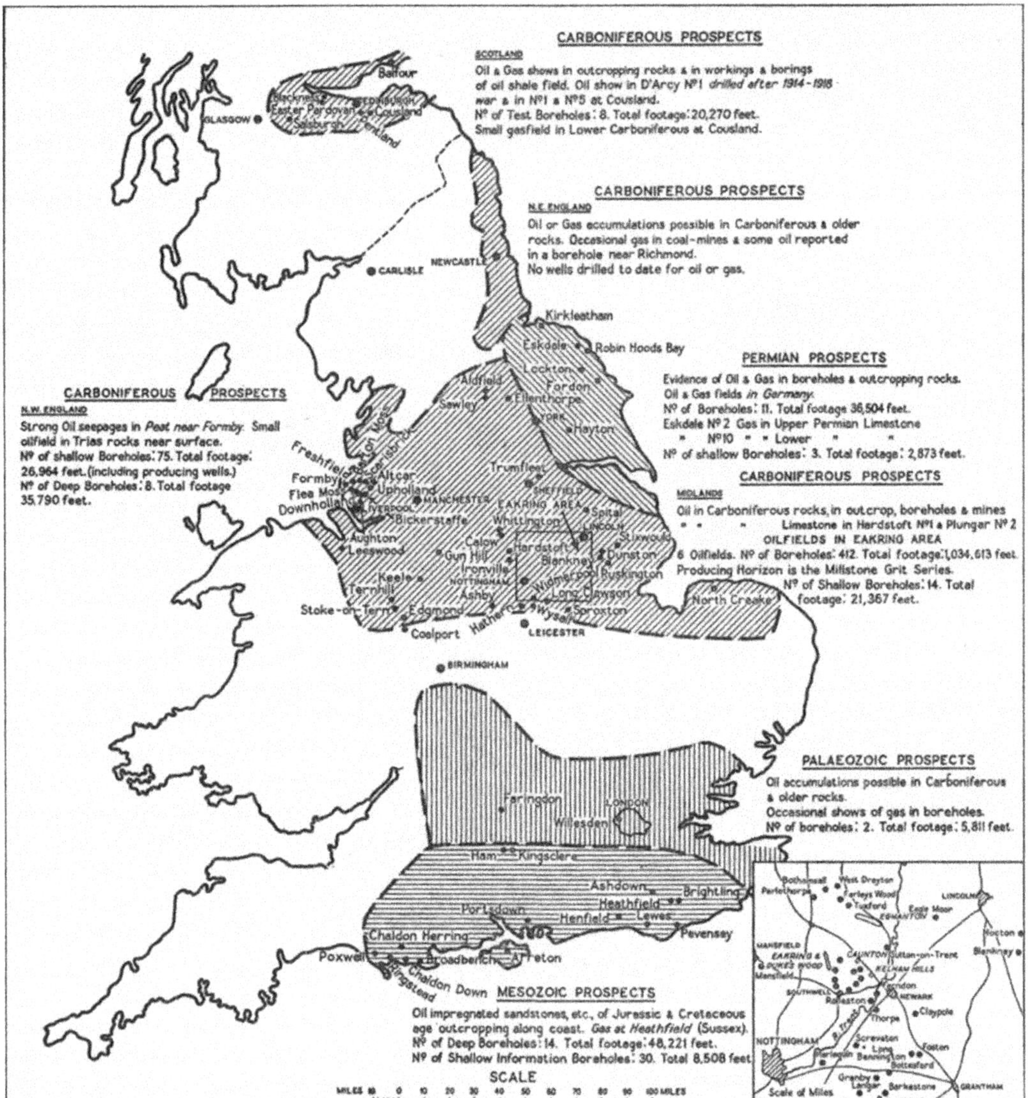

Fig. 13. General map of areas of Britain explored to the end of 1957 (from Falcon & Kent 1960).

discoveries ceased. In 1965, Home Oil drilled Lockton 2a in the North Yorkshire Moors which tested gas at very impressive rates. Subsequent wells, Lockton Nos 3, 4 and 5, all proved dry, but Lockton was still the most important onshore gas discovery of the decade in Britain. It came on stream commercially in 1971. Gas was later also discovered in the same horizon at Ralph Cross, west of Eskdale, and at Malton to the SW of Lockton (**Haarhoff *et al.* 2018**).

In the late 1950s and the early 1960s, gas was discovered onshore at Trumfleet in Yorkshire and at Ironville in Nottinghamshire. The former was found to contain some oil during subsequent appraisal drilling in 1965. Esso made a further gas discovery at Bletchingley in Surrey in 1966, but the field was not developed commercially due to its low calorific value. All of these discoveries were small, and interest in exploring further for oil and gas onshore waned. There followed a period of inactivity while attention switched to the North Sea.

In 1958, a United Nations treaty divided the North Sea into economic zones by country and from this point onwards exploration gradually moved from onshore to offshore. The first discovery in the North Sea, the Ekofisk Field, was made in the Norwegian sector in 1969 (**Jakobsson 2018**), while the first discovery in the UK sector, the Forties Field, was made the following year. Oil was first produced from the UK North Sea from the Argyll Field in June 1975 (**Gluyas *et al.* 2018**).

With the increased focus on the North Sea, interest in onshore exploration declined rapidly during the early 1970s and might have faded entirely but for the events of 1973–74, when the oil supply crisis led to a sharp rise in oil prices, thereby improving the economics of exploration and development onshore in Britain. It was against this political and economic background that, in 1973, the Gas Council discovered Wytch Farm; Britain's largest, and eventually Europe's largest, onshore field, with estimated initial reserves of 500 MMbbl of oil (Morton 2014). In 1998, when it was at peak production, the Wytch Farm Field produced 110 000 BOPD.

However, it was the steep rise in prices in 1979–80 following the Iranian revolution when the price of oil again rose that really rejuvenated exploration for oil and gas in Europe, and elsewhere in the world. In 1980, 10 exploration/appraisal wells were drilled onshore in Britain, compared to five during the previous 3 years. In 1981, a gas discovery was made at Hatfield Moor in Yorkshire, and Carless Capel made an oil discovery in the Weald Basin at Humbly Grove near Basingstoke. Hatfield Moor came on stream in December 1985 with the gas transported to the nearby Belton Brickworks, and Carless Capel made two further oil discoveries at Herriard and Horndean during 1982–83. The success at Hatfield Moor was followed by another discovery at Godley Bridge in Surrey during 1983. Also in the early 1980s, another oil field was discovered near the village of Scothern in Lincolnshire and this eventually led to the development of Britain's second largest onshore oil field, at Welton. In addition to the discoveries at Horndean and Herriard, other oil discoveries were made at Farleys Wood in the East Midlands, and at Hemswell and Nettleham No. 2 in Lincolnshire. The onshore exploration success continued in 1984, with further discoveries at Cropwell Butler, Broughton and Stainton. In addition, an oil discovery was made at Larkwhistle Farm in Hampshire, a find that confirmed the attractions of the Wessex Basin as an oil-bearing area, and adding to existing discoveries in the area at Kimmeridge, Arne, Wareham and Wytch Farm, and also at Stockbridge, Hampshire and Palmers Wood in Surrey. Taylor Woodrow made another gas discovery at Kirby Misperton in North Yorkshire in 1985 (**Haarhoff *et al.* 2018**).

Further onshore exploration success came in 1985, with discoveries at Whisby in the East Midlands, and of heavy oil in the Carboniferous at Milton of Balgonie in the Midland Valley of Scotland. There was also an encouraging start to 1986 with finds at Rempstone, Kirklington and Kinoulton in the East Midlands. BP's Eakring Field in Nottinghamshire, discovered in 1939, produced more than 7 MMbbl, before it was abandoned in 1986 (Fig. 14).

In October 1987, a group headed by RTZ was awarded the very first new-style Development Licence to develop the Crosby Warren Field on the north side of the Humber estuary.

With the discovery and subsequent development of North Sea oil, Britain became self-sufficient in oil from the early 1980s and a net exporter of oil in 1981. Exports peaked in 1985 and production peaked in 1999. With declining North Sea production, the UK became a net importer of oil in 2004. As a result of the discovery and subsequent development of vast gas reserves offshore, mainly in the North Sea and in Liverpool Bay (**Bunce 2018**), the UK became a net exporter of gas in 1997, but, as these reserves declined, the UK became a net importer of gas in 2004. Overall, domestic gas production in the UK (onshore and offshore combined) began to decline in 2000 and by the early years of the second decade of the twenty-first century it was declining at a rate of about 6% per year, resulting in the UK importing 45% of the gas needed to meet domestic demand via pipeline and liquefied natural gas (LNG). In contrast, exploration in Norway continued to spread from the early discoveries in the Norwegian North Sea to the Norwegian Sea (**Jakobsson 2018**), where exploration started in 1980, and the Norwegian Barents Sea (**Jakobsson 2018**), where the first exploration licences were also awarded in 1980, with continuing success.

The development of the retail petrol industry in the UK followed a similar pattern of expansion followed by progressive contraction as oil exploration and production (**Ritson *et al.* 2018**). In 1939, there were approximately 37 000 service stations in the UK, a figure which grew to just over 41 000 in mid-1965, before the industry began a downwards trend, with about 26 500 stations in 1980, under 17 000 in 1995 and 8600 to the end of 2014. Sales of petrol in the UK peaked in 2007 and diesel sales peaked in 2011 (Energy Institute 2015).

This decline of UK gas production and the consequent implications for Britain's national energy security focused attention on the possibility of developing domestic unconventional gas resources, primarily shale gas and coal-bed methane (**Dean 2018**), despite the significant impediments to developing such resources in a highly populated country. The decline in UK oil production throughout the early years of the twenty-first century as the fields in the North Sea became increasingly mature also led to renewed interest in the possibility of redeveloping some of the early fields, such as the Argyll

Fig. 14. The Eakring No. 1 well, Nottinghamshire in June 1964, when it was still producing.

Field (then re-named the Ardmore Field), in an attempt to produce additional previously untapped reserves (**Gluyas *et al.* 2018**).

The successful exploitation of tight oil and tight gas resources in the USA, coupled with concerns about future energy security, also led to an increased focus on the potential for such 'unconventional' oil and gas resources in many European countries in the first decade of the twenty-first century. Poland, in particular, took a leading role in the evaluation of its potential shale-gas resources and, from 2007 onwards, the Polish government started to assign shale-gas exploration licences to both national and foreign companies in the hope of reducing Poland's dependence on Russian gas. Unfortunately, due to a combination of less than favourable geological, legislative and macroeconomic conditions, this early phase of shale-gas exploration in Poland was less successful than initially hoped (**Cantoni 2018**).

Exploration for new sources of oil and gas in Europe continues, but is increasingly hampered by the maturity of many of the conventional oil and gas plays in onshore areas such as in Britain, France and Italy that already have a long history of exploitation since the mid-1800s, the increasing maturity of some offshore areas, notably the UK sector of the North Sea, that have been explored extensively since the early 1960s, and increasing public opposition to the perceived environmental impact of the oil and gas industry in general, and of the technologies required to produce new tight oil and tight gas resources in particular.

Conclusions

The quest for oil and gas has generated geological insights and driven technological innovation since the beginning of the modern oil industry in the mid-nineteenth century. Oil and gas are natural resources, but their exploration and production are only partly controlled by the geology. Exploration is ignited by the demand for these resources, and is regulated by careful evaluations of the costs and likely revenues that will be generated. The rigorous application of petroleum geoscience reduces the risk inherent in oil and gas exploration and production, and so provides investors with improved financial returns.

Europe and Europeans played a disproportionately large role in the development of the modern global oil and gas industry. From at least the Iron Age until the 1850s, the use of oil in Europe was limited, and the oil was obtained almost exclusively from surface seeps and mine workings. The use of oil increased in the 1860s with the introduction of new technologies in both production and refining, but, in the second half of the nineteenth century, the mineral oils and gas then produced primarily from shale and coal could no longer satisfy demand, and oil produced directly from conventional oil fields began to dominate the European market.

The papers in this volume clearly show how the history of the European oil and gas industry has been controlled as much by political and economic conditions as it has by geology, and the historical insights they provide have important implications for current petroleum resource evaluations. In 1902, the State Geologist of Western Australia, Andrew Gibb Maitland, received a letter from a man interested in entering the oil business who wrote: 'Sir, may I ask if you would kindly tell me something of the geology of petroleum and the best means of prospecting for it?'. Maitland answered that the knowledge on the topic was so vast that a library would not have enough space to contain it (Gerali & Gregory 2017*b*). Today, the same question would illicit an answer, not in terms of library shelves and stacks, but in terms of thousands of terabytes of petroleum geoscience data, and in the knowledge and expertise of thousands of petroleum geoscientists.

The discovery of the huge oil and gas resources beneath the North Sea in the 1960s and 1970s enabled Britain, Norway, Denmark and The Netherlands to be largely self-sufficient in oil and gas from the late 1970s until production began to decline rapidly in the early 2000s. Oil and gas production in most European countries is now at an historical low, but exploration for new sources of both conventional and unconventional oil and gas in Europe continues, although increasingly hampered by the maturity of many of the conventional oil and gas plays. Despite the decline in domestic oil and gas production in Europe – or, perhaps, because of it – European companies and European citizens continue to play a major role in the oil and gas industry worldwide.

Acknowledgements Jonathan Craig thanks the management of Eni Upstream & Technical Services in Milan for permission to publish this paper and Massimo Pedini of the Eni Graphics Department for drafting some of the figures.

Funding This research received no specific grant from any funding agency in the public, commercial or not-for-profit sectors.

References

Al-Ani, S. 2015. The Allies find Achilles' heel of the axis. *The Petro-Philatelist*, **36**, 5–6.

Anglo-Iranian Oil Company Ltd 1947. *Petroleum in England*. Anglo-Iranian Oil Company Ltd, London.

Anon 1937. Searching for oil in Great Britain – how a rotary drill works. *The Meccano Magazine*, **22** (November), 647.

Bardini, C. 1998. *Senza carbone dell'età del vapore: gli inizi dell'industrializzazione italiana*. Bruno Mondadori, Milano, Italy.

Beaton, K. 1959. The high cost of whale oil – and what it led to. *World Petroleum reprinted in Oilfield Journal*, **4**, 10–21.

Beeby-Thompson, A. 1904. *Oil Fields of Russia*. Crosby, Lookwood & Son, London.

Beltran, A. & Carré, P. 2016. *La vie électrique. Histoire et imaginaire XVIII-XXIè siècle*. Editions Belin, Paris.

Benadová, P. 2016. History of oil industry in Czech-Slovak region. Abstracts presented at the Geological Society Conference on European Oil & Gas Industry History, 3–4 March 2016, Burlington House, London, 36–39.

Bonariva, A. 1869. *Memorie per la Coltivazione dei Petrolej nelle Province Parmensi. Seconda Edizione Riveduta e Corretta Aggiuntavi con una Indicazione di Varie Località delle Provincie Stesse nelle Quali si Riscontrano Proprie Specie Minerali*. Tipografia G. Ferrari e figli, Parma, Italy.

Brentnall, I. 1995. Britain's first mainland oil well at Tibshelf/Hardstoft. *Ashfield Historian*, **3**, August.

British Petroleum 2017. *BP Statistical Review of World Energy 2017*. 66th edn. British Petroleum (BP), London, https://www.bp.com/content/dam/bp/en/corporate/pdf/energy-economics/statistical-review-2017/bp-statistical-review-of-world-energy-2017-full-report.pdf

Bunce, J. 2018. The history of exploration and development of the Liverpool Bay fields and the East Irish Sea Basin. *In*: Craig, J., Gerali, F., MacAulay, F. & Sorkhabi, R. (eds) *History of the European Oil and Gas Industry*. Geological Society, London, Special Publications, **465**. First published online April 11, 2018, https://doi.org/10.1144/SP465.6

Camden, W. 1586. *Britannia or a Chorographical Description of Great Britain and Ireland*. 1st edn. Printed by Mary Matthews, London.

Candela, A. 2014. 'Dante Pantanelli'. *Dizionario Biografico degli Italiani*, **81**, Roma, Treccani.

Cantoni, R. 2018. Second Galicia? Poland's shale gas rush through historical lenses. *In*: Craig, J., Gerali, F., MacAulay, F. & Sorkhabi, R. (eds) *History of the European Oil and Gas Industry*. Geological Society, London, Special Publications, **465**. First published online May 14, 2018, https://doi.org/10.1144/SP465.16

Carruthers, R.G., Caldwell, W., Bailey, E.M. & Conacher, J.R.H. 1927. *The History of the Scottish Oil-Shale Industry. The Oil-Shales of the Lothians*. 3rd edn. Memoirs of the Geological Survey, Scotland. HMSO, Edinburgh.

Cazzini, F.F. 2018. The history of the upstream oil and gas industry in Italy. *In*: Craig, J., Gerali, F., MacAulay, F. & Sorkhabi, R. (eds) *History of the European Oil and Gas Industry*. Geological Society, London, Special Publications, **465**. First published online April 11, 2018, https://doi.org/10.1144/SP465.2

Conacher, H.R.J. 1928. *The Oil Shales of the Lothians*. Memoirs of the Geological Survey, Scotland. HMSO, London.

Conacher, H.R.J. 1938*a*. Dr. James young – a brief biography. *In*: Crookshank, H.F.C. (ed.) *Oil Shale and Cannel Coal: Proceedings of a Conference Held in Scotland, June 1938*. Institute of Petroleum, London, 1–5.

Conacher, H.R.J. 1938*b*. The mineral-oil industry in Scotland, its raw materials and methods. *In*: Crookshank, H.F.C. (ed.) *Oil Shale and Cannel Coal: Proceedings of a Conference Held in Scotland, June 1938*. Institute of Petroleum, London, 300–313.

Corfield, S.M. 2018. The first oil exploration campaign in the UK, 1918–22. *In*: Craig, J., Gerali, F., MacAulay, F. & Sorkhabi, R. (eds) *History of the European Oil and Gas Industry*. Geological Society, London, Special Publications, **465**. First published online May 14, 2018, https://doi.org/10.1144/SP465.11

Craig, J., Gluyas, J., Laing, C. & Schofield, P. 2013. Hardstoft – Britain's first oil field. *Oil-Industry History*, **14**, 97–116.

Dawson, C. 1898. On the discovery of natural gas in East Sussex. *Proceedings of the Geological Society of London*, **54**, 564–574, https://doi.org/10.1144/GSL.JGS.1898.054.01-04.38

Day, A.H. 1946. Continuing search for oil in Britain. *World Petroleum*, **XVII**, (September), 47–48.

Dean, G.A. 2018. The Scottish oil-shale industry from the viewpoint of the modern-day shale-gas industry. *In*: Craig, J., Gerali, F., MacAulay, F. & Sorkhabi, R. (eds) *History of the European Oil and Gas Industry*. Geological Society, London, Special Publications, **465**. First published online May 8, 2018, https://doi.org/10.1144/SP465.13

Desio, A. 1941. 'Cesare Porro'. *Bollettino della Società Geologica Italiana*, **40**, 46–56.

Dewey, D.R. 1946. *Wartime Research on Synthetic Fuels: Kaiser Wilhelm Institute fur Kohlnforschung*. Department of Commerce, Washington, DC.

Ele, M. 1697. An account of the making pitch, tar, and oil out of a blackish stone in Shropshire. *Philosophical Transactions*, **19**, 544.

Energy Institute 2015. *Retail Marketing Survey – Supplement to Petroleum Review*. Energy Institute, London.

Evans, D.J. 2009. A review of underground fuel storage events and putting risk into perspective with other areas of the energy supply chain. *In*: Evans, D.J. & Chadwick, R.A. (eds) *Underground Gas Storage: Worldwide Experiences and Future Development in the UK and Europe*. Geological Society, London, Special Publications, **313**, 173–216, https://doi.org/10.1144/SP313.12

Fairman, J. 1868. *A Treatise on the Petroleum Zones of Italy*. E. & F.N. Spon, London.

Falcon, N.L. & Kent, P.E. 1960. *Geological Results of Petroleum Exploration in Britain 1945–57*. Geological Society, London, Memoirs, **2**, https://doi.org/10.1144/GSL.MEM.1960.002.01.01

Forbes, J. 1959. *More Studies in Early Petroleum History*. E.J. Brill, Leiden, The Netherlands.

Forbes, R.J. 1958. *Studies in Early Petroleum History*. E.J. Brill, Leiden, The Netherlands.

Forbes-Leslie, W. 1917. The Norfolk oil shales. *Journal of the Institute of Petroleum Technology*, **3**, 3–35.

Fowler, T.B. 1940. The paraffin and petroleum industries pioneer work of Mr. James Young. *Journal of the Society of Chemical Industry*, **59**, 70–72.

Gallois, R.W. 2012. The Norfolk oil-shale rush, 1916–21. *Proceedings of the Geologists' Association*, **123**, 64–73.

Gerali, F. 2012. Scientific maturation and production modernization. Notes on the Italian oil industry in the 20th century. *Oil-Industry History*, **12**, 89–109.

Gerali, F. & Gregory, J. 2017*a*. Understanding and finding oil over the centuries: the case of the Wallachian Petroleum Company in Romania. *Earth Science History*, **36**, 41–62, https://doi.org/10.17704/1944-6178-36.1.41

Gerali, F. & Gregory, J. 2017*b*. Harsh oil: finding petroleum in early twentieth century Western Australia. *In*: Mayer, W., Clary, R., Azuela, L.F., Mota, T.S. & Wołkowicz, S. (eds) *History of Geoscience: Celebrating 50 Years of INHIGEO*. Geological Society, London, Special Publications, **442**, 425–436, https://doi.org/10.1144/SP442.33

Gerali, F. & Lipparini, L. 2018. Maiella, an oil massif in the Central Apennines ridge of Italy: exploration, production and innovation in the oil fields of Abruzzo across the nineteenth and twentieth centuries. *In*: Craig, J., Gerali, F., MacAulay, F. & Sorkhabi, R. (eds) *History of the European Oil and Gas Industry*. Geological Society, London, Special Publications, **465**. First published online May 25, 2018, https://doi.org/10.1144/SP465.20

Gerali, F., Macini, P. & Mesini, E. 2018. Historical study of geosciences and engineering in the oilfields of the Emilia-Romagna region in the socio-economic context of post-Unitarian Italy (1861–1914). *In*: Craig, J., Gerali, F., MacAulay, F. & Sorkhabi, R. (eds) *History of the European Oil and Gas Industry*. Geological Society, London, Special Publications, **465**. First published online May 30, 2018, https://doi.org/10.1144/SP465.21

Giffard, H.P.W. 1923. The recent search for oil in Great Britain. *Transactions of the Institute of Mining Engineers*, **LXV**, 222–250.

Gisler, M. 2014. *Die 'Swiss Gang'. Schweizer Pioniere der Erdölforschung*. Verein für Wirtschaftshistorische Studien, Zurich, Switzerland.

Gisler, M. 2018. Entangled between worlds: Swiss petroleum geologists, *c*. 1900–50. *In*: Craig, J., Gerali, F., MacAulay, F. & Sorkhabi, R. (eds) *History of the European Oil and Gas Industry*. Geological Society, London, Special Publications, **465**. First published online May 14, 2018, https://doi.org/10.1144/SP465.17

Gisler, M. & Trümpy, D. 2016. Historical highlights: the rise of the Swiss Mafia. *AAPG Explorer* (April), 24–27, http://www.aapg.org/publications/news/explorer/column/articleid/28474/the-rise-of-the-swiss-mafia

Gluyas, J. & Bowman, M. 1997. Edale No. 1 oilwell, Derbyshire, UK, 1938. *Marine & Petroleum Geology*, **14**, 191–199.

Gluyas, J., Longxun, T. & Jones, S. 2018. Argyll Field; the first oil field to be developed on the UK Continental Shelf. *In*: Craig, J., Gerali, F., MacAulay, F. & Sorkhabi, R. (eds) *History of the European Oil and Gas Industry*. Geological Society, London, Special Publications, **465**. First published online April 11, 2018, https://doi.org/10.1144/SP465.3

Haarhoff, M.Q., Hughes, F. *et al.* 2018. The history of hydrocarbon exploration and development in North Yorkshire. *In*: Craig, J., Gerali, F., MacAulay, F. & Sorkhabi, R. (eds) *History of the European Oil and

Gas Industry. Geological Society, London, Special Publications, **465**. First published online May 8, 2018, https://doi.org/10.1144/SP465.12

HALBOUTY, M.T. (ed.) 2003. *Giant Oil and Gas Fields of the Decade 1990–99*. AAPG Memoirs, **78**.

HALLETT, D., DURANT, G.P. & FARROW, G.E. 1985. Oil exploration and production in Scotland. *Scottish Journal of Geology*, **21**, 547–570, https://doi.org/10.1144/sjg21040547

HENRY, J.D. 1905. *Baku: An Eventful History*. A. Constable & Co., Ltd., London.

HERITIER, F. 1994. A history of petroleum exploration in France. *In*: MASCLE, A. (ed.) *Hydrocarbon and Petroleum Geology of France*. European Association of Petroleum Geoscientists (EAGE), Special Publications, **4**, 29–45.

HUGUENET, I. & FROJO, G. 1856. *Asfalto. Considerazioni Generali sull'Origine e Formazione degli Asfalti e del Loro Uso come Cemento Naturale Applicati ai Lavori di Utilità Pubblica e Privata Arricchita di Note ed Accresciute di un Appendice sugli Asfalti del Regno di Napoli*. Stabilimeto tipografico di Vincenzo Priggiobba, Napoli, Italy.

JAKOBSSON, K.H. 2018. A history of exploration offshore Norway: the Barents Sea. *In*: CRAIG, J., GERALI, F., MACAULAY, F. & SORKHABI, R. (eds) *History of the European Oil and Gas Industry*. Geological Society, London, Special Publications, **465**. First published online June 6, 2018, https://doi.org/10.1144/SP465.18

JEVONS, W.S. 1865. *The Coal Questions: An Inquiry Concerning Progress of the Nation, and the Probable Exhaustion of Our Coal-Mines*. Macmillan, Basingstoke, UK.

KÖLBL-EBERT, M. 2018. German petroleum geologists and World War II. *In*: CRAIG, J., GERALI, F., MACAULAY, F. & SORKHABI, R. (eds) *History of the European Oil and Gas Industry*. Geological Society, London, Special Publications, **465**. First published online April 11, 2018, https://doi.org/10.1144/SP465.1

KOSZARSKI, A. & BALLY, A. 1994. The world's oldest industrial oil field. Poland relights the fire at Bóbrka. *AAPG Explorer*, **12**, 26–27.

KRZYWIEC, P. 2016. Birth of the oil industry in the northern Carpathians. Abstracts presented at the Geological Society Conference on European Oil & Gas Industry History, 3–4 March 2016, Burlington House, London, 32–33.

KRZYWIEC, P. 2018. The birth and development of the oil and gas industry in the Northern Carpathians (up until 1939). *In*: CRAIG, J., GERALI, F., MACAULAY, F. & SORKHABI, R. (eds) *History of the European Oil and Gas Industry*. Geological Society, London, Special Publications, **465**. First published online May 18, 2018, https://doi.org/10.1144/SP465.24

LEA, C. 2018. Derbyshire's oil and refining history: the James 'Paraffin' Young connection. *In*: CRAIG, J., GERALI, F., MACAULAY, F. & SORKHABI, R. (eds) *History of the European Oil and Gas Industry*. Geological Society, London, Special Publications, **465**. First published online April 11, 2018, https://doi.org/10.1144/SP465.4

LEES, G.M. & COX, P.T. 1937. The geological basis for the search for oil in Great Britain by the D'Arcy Exploration Company, Limited. *Quarterly Journal of the Geological Society*, **93**, 156–190, https://doi.org/10.1144/GSL.JGS.1937.093.01-04.09

LEES, G.M. & TAITT, A.H. 1946. The geological results of the search for oilfields in Great Britain. *Quarterly Journal of the Geological Society*, **101**, 255–317, https://doi.org/10.1144/GSL.JGS.1945.101.01-04.12

LONGHURST, H. 1959. *Adventure in Oil: The Story of British Petroleum*. Sidgwick and Jackson, London.

MOODY, R. (compiler) 2007. *The History of On-Shore Hydrocarbon Use in the UK (From Oil Shales and Seeps to 'Shaleopolis' and Roughnecks!). Abstracts. Petroleum Exploration Society of Great Britain Newsletter*.

MORTON, M.Q. 2014. 'What oilfields?' – onshore oil in the UK. *GEOExPro*, **11**, 70–74.

NAVARRO COMET, J. 2018. Ayoluengo's history – Spain's only onshore oil field. *In*: CRAIG, J., GERALI, F., MACAULAY, F. & SORKHABI, R. (eds) *History of the European Oil and Gas Industry*. Geological Society, London, Special Publications, **465**. First published online May 8, 2018, https://doi.org/10.1144/SP465.9

NAVARRO COMET, J. & PUCHE RIART, O. 2018. A century of hydrocarbon exploration and production in Spain (1860–1960). *In*: CRAIG, J., GERALI, F., MACAULAY, F. & SORKHABI, R. (eds) *History of the European Oil and Gas Industry*. Geological Society, London, Special Publications, **465**. First published online May 8, 2018, https://doi.org/10.1144/SP465.10

NOVELLI, L. & SELLA, M. 2009. *Il Petrolio: Una Storia Antica*. Silvana Editoriale, Cinisello Balsamo, Italy.

PEARTON, M. 1971. *Oil and the Romanian State*. Clarendon Press, Oxford.

PIZANTY, M. 1947. *Consideratiuni Preliminare Asupra Sondajelor de Explorare Efectuate in România in Perioada 1900–47*. Monitorul Petrolului Român, Bucarest.

PLOT, R. 1684. A discourse concerning the sepulchral lamps of the ancients, shewing the possibility of their made divers waies. *Philosophical Transactions*, **14**, 806–811.

POZZI, D. 2009. *Dai gatti selvaggi al cane a sei zampe: tecnologia, conoscenza e organizzazione nell'Agip e nell'Eni di Enrico Mattei*. Marsilio, Venezia.

RINEHART, I. 1930. *Report on the Oil Fields of Northwest Germany*. Lord Baltimore Press, Baltimore, MD, USA.

RITSON, N.H., BYRNE, I. & COHEN, D.A. 2018. UK petrol retailing – competitive rivalry and the decline of the oil majors in the twentieth century. *In*: CRAIG, J., GERALI, F., MACAULAY, F. & SORKHABI, R. (eds) *History of the European Oil and Gas Industry*. Geological Society, London, Special Publications, **465**. First published online April 11, 2018, https://doi.org/10.1144/SP465.5

SAUSSURE, T. DE 1816. Recherches sur la composition et les propriétés du naphte d'Amiano dans les etates de Parme. *Annales de Chimie et de Physique*, **3–4**, 314–321.

SAUSSURE, T. DE 1817. Experiment on the composition and properties of the Naphtha of Amiano. Read to the society of natural philosophy and natural history of Geneva. *Annals of Philosophy*, **10**, 118–127.

SCHLEDER, Z., MAN, S. & TAMAS, D. 2016. History of salt tectonics in the Carpathian bend zone, Romania. Guide for the Pre-Conference Field Trip for the

AAPG European Regional Conference & Exhibition, 18 May 2016, Bucharest, Romania.

Schmitz, P.-M.E. 1938. Les mines d'Asphalte, origine du mot Pechelbronn. *Revue Pétrolifère*, **806**, 1297–1299.

Selley, R.C. 2011. Shale gas: blessing or curse? *Geoscientist*, **21**, 14–19.

Selley, R.C. 2012. UK shale gas: the story so far. *In*: Horsfield, B. & Schulz, H.-M. (eds) *Insights into Shale Gas Exploration and Exploitation*. Marine and Petroleum Geology, **31**, 100–109.

Sertorio, L. 2002. *Storia dell'abbondanza*. Bollati Boringhieri, Torino, Italy.

Smil, V. 2017. *Energy and Civilization: A History*. MIT Press, Cambridge, MA, USA.

Smith, M.S. 2006. *The Emergence of the Modern Business Enterprise in France, 1800–1930*. Harvard University Press, London.

Sorenson, R.P. 2014. 19th century Russian petroleum from Baku: the American perspective. *Oil-Industry History*, **15**, 91–112.

Sorkhabi, R. 2012. Memorial to Augusto Gansser (1910–2012). *Geological Society of America Memorials*, **41** (December), 15–21.

Sorkhabi, R. 2018. Sir Thomas Boverton Redwood (1846–1919): a watershed in the British oil industry. *In*: Craig, J., Gerali, F., MacAulay, F. & Sorkhabi, R. (eds) *History of the European Oil and Gas Industry*. Geological Society, London, Special Publications, **465**. First published online May 25, 2018, https://doi.org/10.1144/SP465.22

Southwell, C.A.P. 1945. Petroleum in England. *Journal of the Institute of Petroleum*, **31** (February), 27–39.

Spencer, J. & Furcata, M. 2016. Myron Kinley and the 'Torch of Moreni'. Abstracts presented at the Geological Society Conference on European Oil and Gas Industry History, 3–4 March 2016, Burlington House, London, 56–57.

Spencer, J. & Furcuta, M. 2018. Myron Kinley and the 'Torch of Moreni'. *In*: Craig, J., Gerali, F., MacAulay, F. & Sorkhabi, R. (eds) *History of the European Oil and Gas Industry*. Geological Society, London, Special Publications, **465**. First published online May 8, 2018, https://doi.org/10.1144/SP465.8

Strachan, A. 1920. *Mineral Oil, Kimmeridge Oil-Shale, Lignites, Jets, Cannel Coals etc*. 2nd edn. Memoirs of the Geological Survey Special Reports on the Mineral Resources of Great Britain, **VII**.

Suciu, P.H. & Luzzati, L. 1923. *L'Italia e il Petrolio Romeno*. Unione Arti Grafiche, Città di Castello, Italy.

Tari, G. & Bérczi, I. 2018. Simon Papp, a prominent Hungarian petroleum geologist: how to run exploration projects from a prison cell. *In*: Craig, J., Gerali, F., MacAulay, F. & Sorkhabi, R. (eds) *History of the European Oil and Gas Industry*. Geological Society, London, Special Publications, **465**. First published online May 18, 2018, https://doi.org/10.1144/SP465.19

Thomas, R. 2018. The development of the manufactured gas industry in Europe. *In*: Craig, J., Gerali, F., MacAulay, F. & Sorkhabi, R. (eds) *History of the European Oil and Gas Industry*. Geological Society, London, Special Publications, **465**. First published online May 14, 2018, https://doi.org/10.1144/SP465.14

Tomory, L. 2012. *Progressive Enlightenment: The Origins of the Gaslight Industry, 1780–1820*. MIT Press, Cambridge, MA, USA.

Torrens, H. 1994. 300 years of oil: mirrored by developments in the West Midlands. *In*: *The British Association Lectures 1993*. Geological Society, London, 4–8.

Trinder, B. 1981. *The Industrial Revolution in Shropshire*. 2nd edn. Phillimore, Chichester, UK.

Tulucan, A.D., Soveja-Iacob, L.-E. & Krezsek, C. 2018. History of the oil and gas industry in Romania. *In*: Craig, J., Gerali, F., MacAulay, F. & Sorkhabi, R. (eds) *History of the European Oil and Gas Industry*. Geological Society, London, Special Publications, **465**. First published online May 18, 2018, https://doi.org/10.1144/SP465.15

Underhill, J.R., Monaghan, A.A. & Browne, M.A.E. 2008. Controls on structural styles, basin development and petroleum prospectivity in the Midland Valley of Scotland. *Marine and Petroleum Geology*, **25**, 1000–1022.

Veatch, A.C. 1923. Petroleum resources in Great Britain. *American Institute of Mining and Metallurgical Engineers*, **LXV**, 3–7.

Wannier, M.M.A. 2018. 'Uniformity in Geological Reports' (1917) by Josef Theodor Erb, petroleum geologist and manager (1874–1934). *In*: Craig, J., Gerali, F., MacAulay, F. & Sorkhabi, R. (eds) *History of the European Oil and Gas Industry*. Geological Society, London, Special Publications, **465**. First published online May 8, 2018, https://doi.org/10.1144/SP465.7

Wigley, P. 2015. Déjà vu: weald exploration a familiar tale. *AAPG Explorer*, **36** (August), 38–39.

Willet, H. 1875. *Quarterley Reports of the Subwealden Exploration Company Ltd*. Smith, Brighton, UK.

Yanarella, E.J. & Green, W.C. 1987. *The Unfulfilled Promise of Synthetic Fuels: Technological Failure, Policy Immobilism, or Commercial Illusion. Contributions in Political Science, No. 179*. Greenwood Press, Westport, CT.

UK petrol retailing: competitive rivalry and the decline of the oil majors in the twentieth century

NEIL H. RITSON[1]*, IAN BYRNE[2] & DAVID A. COHEN[3]

[1]*Department of Global Value Chains and Trade, Faculty of Agribusiness and Commerce, Lincoln University, Christchurch 7647, New Zealand*

[2]*National Energy Foundation, The National Energy Centre, Davy Avenue Knowlhill, Milton Keynes MK5 8NG, UK*

[3]*Department of Agribusiness and Markets, Faculty of Agribusiness and Commerce, Lincoln University, Christchurch 7647, New Zealand*

N.H.R., 0000-0001-7650-7498; I.B., 0000-0001-9782-7343; D.A.C., 0000-0001-9323-6984

*Correspondence: *neil.ritson@lincoln.ac.nz*

Abstract: The development of the retail petroleum industry in the UK has received only sporadic attention by academics. This is curious, as the industry has vital importance to modern commerce. The industry is usually thought of as an oligarchy, with a small number of companies controlling the production and movement of products through vertical integration. However, there is evidence of change in the industry: there has been a significant number of exchanges of retailing sites between the major entities themselves, and a rise in the number and influence of competitors. These constitute both multiple retailers such as supermarket chains and, at the 'competitive fringe', a rise in the number of smaller independents. This paper identifies the historical changes and analyses the factors underlying them.

Studies of the history of industries and the operations of businesses within them have not received due attention. Over 15 years ago, Alexander & Akehurst (1998, p. 1) declaimed the paucity of historical studies on retail marketing, identifying three justifications for increased research attention:

> First, it is recognised that retailing has received limited attention within the discipline of history. Second, where retailing has become an area considered worthy of academic study, notably within management studies, the historical dimension has been under-valued and under-explored. Third, studies in retail history must draw on both historical and management traditions if valuable research is to be undertaken.

Rosenbloom & Dimitrova (2011) suggest there is a 'Dixonian' systems perspective – a paradigm that presents marketing as a phenomenon that reaches far beyond the micro/managerial marketing mix paradigm. Thus, history is important in understanding where we are now, as technological, social, political and economic factors (Kotler 1998) interact over time to create the market and retailing structures of today.

Furthermore, not all developments in industry are based on internal efficiency or an external customer focus: many decisions are based on managers' bounded rationality (Simon 1957). Risk-avoidance creates long-term path-dependency and results in industrial inertia – as we document below, this is a key aspect of the oil companies' behaviour. When we look to discover the factors in the historical development of retailing of petrol in the UK, there are only a few studies upon which to draw. This is surprising, considering the central position of fuel in the everyday lives of UK citizens, whether for personal or commercial needs. Given this, an examination of the industry is warranted. As a former Editor of *Business History* noted, a study focused on the petrol marketplace in the twentieth century 'fills a yawning gap in the literature' (Wilson, pers. comm. 2012). Thus, the role and activities of petrol retailers and related merchants over time will provide useful insight into the structural development of the sector. Sparks (pers. comm. 2011) raised the question of 'how and why we got from there to here', which the present industry analysis addresses.

We begin with the observation that there has been a significant shift in the petroleum industry's primary business model in the UK. This is reflected in the drastic decrease in the number of service stations open to the public. In 1939, there were approximately 37 000 stations (Ferrier 1986), a figure which grew to just over 41 000 in mid-1965. The industry then began to trend downwards, with about 26 500 stations in 1980, under 17 000 in 1995 and 8600 at 31 December

From: Craig, J., Gerali, F., MacAulay, F. & Sorkhabi, R. (eds) 2018. *History of the European Oil and Gas Industry*. Geological Society, London, Special Publications, **465**, 25–38.
First published online April 11, 2018, https://doi.org/10.1144/SP465.5

2014. At the same time, overall fuel sales declined only slightly; petrol sales peaked in 2007 and diesel in 2011 (*Petroleum Review* 1993, 2001; Energy Institute 2015). The fact that volumes have remained relatively stable in the face of a substantial decline in the number of stations documents a dramatic increase in the efficiency of the supply chain. Boyle (1999) reported on these changes, noting that the majors sold off smaller stations, consolidated ownership (e.g. BP moved from owning a third to half of its outlets) and managed franchises more strictly over the first part of this period. In a major volte-face, they sold larger numbers of stations to a new breed of independent operators such as the Co-operative Group, the Motor Fuel Group, MRH and Rontec.

From this starting point, we suggest a set of relevant factors that might be underlying the decline. Note that for this study, we consider only the role of petrol. Other fuels – diesel and liquid petroleum gas (LPG) – have been omitted as petrol is the most publically visible of these and is likely to be seen as the most socially important.

There are recent studies on the extension of forecourt retailing by companies such as RoadChef (Denning & Freathy 1996) and on the franchising of petrol-retailing stations – in particular Shell Select (Boyle 1999, 2002*a*). These were attempts, in the face of increased competition by grocery multiples (supermarkets), at differentiation by creating convenience stores to supplement fuel sales in their petrol stations. However, these analyses had no theoretical underpinning – they remain descriptive. Additionally, the analysis in these studies is at the lowest – functional/operational – level of strategy. At the higher business level of strategy we can analyse how the companies overall are managing the retail side of the oil industry. We note that the problems now besetting the retail petrol industry in the UK have a long history and are bound to earlier conceptions of the market, in particular the role of service or filling stations in the distribution of the products.

One critical goal is to discover and explain why, when petrol is a commodity, there exists such specialist retail outlets. Dixon (1965) claimed that they were always unpopular. He asserted that discussions of merchandising practices in the oil industry have often proceeded from the premise that the existing distributive channel structure for gasoline is a millstone around the industry's neck. One writer has proposed a drastic marketing innovation: the industry must eventually get rid of the service station because the public appears to dislike it intensely. Others have suggested that the industry reject its specialized distributive agencies and recognize the modern trend of retailing to one-stop shopping. Such a view is remarkably prescient for 1965.

At the corporate level, a few studies in the literature have considered more recent changes, such as the move to the upstream end of the oil industry, and the decline of refining (UKPIA 2011). At the industry level, Cohen's (1998) and Cohen & Bradfield's (2001) studies concluded that a single attribute (service at the forecourt) was utilized to develop sales. This was in contrast to Ritson's (1999) argument that the supply chain and the infrastructure of the major oil companies precluded their competitive advantage against the supermarkets. This literature, while useful, lacks specifics.

In this paper, our method is to include earlier academic literature and supplement that with data from sources such as government reports, particularly those of the Monopolies and Mergers Commission (MMC). We have also included data from documents in the archives of the former Institute of Petroleum (now the Energy Institute), whose publication *Petroleum Review* includes an annual retail marketing survey. Beyond these, we were also able to access data from the UK Petroleum Industry Association (UKPIA), the Petroleum Information Bureau and several individual company archives. Data were also drawn from annual surveys conducted by the Energy Institute and via figures reported by Experian Catalyst, which maintains a private database of petrol stations. The data available were not without limitations, however; even today, there is no comprehensive central register of outlets licensed to sell petrol to the public. This paper aims to provide empirical data and offer explanations as to the causes of the changes to the industry we see over time.

After this introduction, the paper is divided into two parts: Part 1 considers the evidence provided by the rich history of the industry. Part 2 analyses a variety of the factors that appear to have led to change in the industry. These include such influences as politics, culture, capacity imbalances and companies' bounded rationality.

Part 1: The historical evidence

Petrol

We begin by defining the area of discussion – the marketing and retailing of petrol. Yergin's (1990) history of the oil industry argued that the initial development of kerosene in the 1850s was as a lighting fuel, and there was slow development of motor vehicles running on petrol in the 1880s. Kerosene was initially for lighting and heating, and is now mainly for jet aircraft engines; diesel was initially designated for heavy commercial vehicles and is now used for a variety of other motor vehicles. Petrol, however, appears to have been the over-riding profit contributor in comparison to other fuels over the years. The filling station came to dominate retailing in both urban and rural areas. Pre-war, the predominant fuel was called 'motor spirit', as Carless,

Capel and Leonard had trademarked the term petrol, although they did not actually enforce their right to the name against competitors (Dixon 1997). We use the UK version of the name for the product, petrol. We chose petrol retailing in the UK as our core interest as this has not been subject to extensive research and yet has been subject to significant changes over time.

Initially, petrol was offered as one rather poor grade (rated by its thermal efficiency or 'octane' rating), with only a few brands such as National Benzole offering a premium grade. The introduction on 1 February 1953 of premium grades of petrol amplified competition in the marketplace. Esso's expansion of premium sales was dramatic – National Benzole had been the most prominent premium brand before the war, when all brands were replaced by a single 'pool' product, but then declined. Shell-Mex and BP's premium grades grew more slowly, so losing some of its market share to Esso and Regent (Dixon 1962). Petrol grades proliferated in the post-war years in an octane battle, triggering the introduction of 100 octane products like 'BP Super Plus' and 'Esso Golden Extra' among others (Bamberg 2000 p. 230). Eventually, standardization was required. This politicized the grading of petrol, as in 1967 four individually rated 'star' grades (numbered two to five–one star was never sold on the UK market) were introduced (BSI 1967). These reflected the octane/power ratings for fuel under British Standards system BS 4040 and were largely adopted by 1971. They became mandatory in 1976 when the government introduced a statutory instrument (1976 SI number 1866). By the late 1980s, political influences in Europe demanded the availability of unleaded petrol – thus removing the 'anti-knock' agent, lead tetra-ethyl, which smoothed the ignition of the fuel within the engine. This meant that the majors had to spend capital on new 'isomerization' units (Ritson & Chamberlin 2012). The number of grades then reduced to at most two (91 and 95 octane), while additives were used to promote some degree of differentiation between brands.

Distribution

Having discussed the product, petrol, we now outline the method of its distribution. Dixon (1997) stated that in the UK, petrol was delivered in bulk, mainly to commercial users. Due to the highly inflammable nature of the fuel, this gave way to delivery via 2 gallon cans following an Act of Parliament in 1901 (Boyle 2002*a*), followed by hand-operated (later electrical) pumps. In the 1920s, BP was first to establish its own extensive and expensive distribution system by providing pumps to independently owned filling or service stations (Ferrier 1986). The UK's retail scene was transformed by the new petrol filling stations which had arisen in the USA. This was probably as early as 1905, although sources vary on the date, and added to the ability of organizations to differentiate what is essentially a commodity. The ability to charge high prices reflected the dominance of certain companies. Responding to the uncontrolled growth in numbers of filling stations (especially on main roads) and following pressure from the Council for the Protection of Rural England, the Roadside Petrol Pumps Act 1927 and the Petroleum Consolidation Act 1928 (Morris 1985; Byrne & Ritson 2012) were passed into law. These allowed local authorities to impose planning regulations on both the numbers and advertising associated with petrol stations.

A change in the market began in the 1950s, when Mobil developed forecourts to a higher standard than traditional dealer-owned sites. That trend continued into the 1960s, when Agip, Amoco, Atlantic, Chevron, Gulf and Total followed suit. Previously, the sale of petrol may have been just one activity alongside car sales, servicing and repairs. Local environmental health or planning officers required petrol pumps with 'swing arms' across the pavement to be removed and few had space to relocate them, along with the stricter planning regulations. This was the death-knell for the old pre-war design of kerbside stations. Older or smaller stations were also unsuited to and unwilling to invest in self-service, which was increasingly expected by motorists. This was first introduced in the mid-1960s, but was initially slow to spread. The proportion of self-service stations rose from 20% in 1980 to 55% in 1990, up to 73% in 2000 and about 98% today. It has been established that sales through self-service stations are proportionately higher (*Petroleum Review* 1980, 1990, 2000; Energy Institute 2015).

Industry structure

The structure of the retail petrol industry was always complex. Both Caffarra (1994) and the Monopolies and Mergers Commission (MMC 1990) have identified the predominant ways in which oil companies in the UK reached the market via retail outlets: company-owned and operated sites; and exclusive supply arrangements with independent dealers, who own the site and sell the wholesaler's brand. Caffarra also noted that four different contractual agreements existed at the time: tenancies, popular until the late 1970s; licenses, aimed at more detailed control over the site as did Jet in 1992; commission agencies which reduced the security of tenure to nil as Fina did in 1991–92; and direct management by the wholesalers' personnel. By 1994, franchising had been added to both company-owned and independents' sites, such as by Texaco in 1992 and BP in 1993 (Caffarra 1994). Detailed franchise agreements required dealers and their staff to promote the brand and stipulated rewards for their efforts. Promotions included signage, display of opening hours, and the presentation and ranges of goods sold. This arrangement had implications, which we discuss later.

Chronology

We now turn to a detailed chronological analysis of the industry. In 1905, the Anglo-American Oil company (AAOC), a subsidiary of Standard Oil, had a virtual monopoly. They claimed a 75% share of the petrol market, which was sold as tins under Pratt's brand. Shell had been unsuccessful in the market, as AAOC had tied 'every agent and distributor … holding up supplies until each had signed a contract undertaking to sell petrol to no other brand but its own' (Henriques 1960, p. 290). However, Shell and AAOC came to a market-sharing agreement and, despite a price war in 1910, by 1912 the 'war' had ended (Dixon 1997). Ferrier (1986) catalogued the history of BP from 1906. It began as the subsidiary of the Europaische Petroleum Union Gesellschaft (EPU). In 1907, BP acquired the agency for the sales of Shell motor spirit up to 1916. With AAOC, BP supplied most of the UK market. There was, to quote BP's Chairman, 'endless scope for the sale of our products' (Ferrier 1986, p. 30). The market was supplied by several refineries: Llandarcy in Wales was authorized in 1916 (but opened in 1922), the Fawley refinery produced from 1921 and the Grangemouth refinery opened in 1924 (Longhurst 1959). Improvements to the Abadan refinery in Persia (Iran) also contributed product to the market. By 1914, BP owned 840 depots, but in 1917 its very substantial assets were acquired by Anglo-Persian Oil Company (APOC) after government intervention – just over half of the shareholders were German and it was therefore classified as an enemy concern (Ferrier 1986).

World War I brought about rationing in 1916 through a Petroleum Pool Board system. A stable market subsequently remained, with the three majors companies – AAOC, Shell and BP: known as 'the Combine' – having a combined 85% share of the market (Fitzgerald 1927). A small shed-like structure built by the Automobile Association in 1919 is generally regarded to be the first UK drive-in filling station (Morris 1985). The 1920s were characterized by a rapid and uncontrolled growth in the number of the new filling stations, although most were simply a row of pumps beside the road. Dedicated filling stations with one or two pumps began to emerge, with prices 2d (2 pennies) per gallon less than the canned version, and were characterized by high profit and low volume. The funds for buying pumps were supplied by the oil companies, effectively tying the station to the brand until the loan was paid off, although stations were free to take pumps (and petrol) from several different brands. From a mere 69 pumps in 1921, APOC grew to 7000 by 1923. AAOC was the pioneer, but BP decided to erect and to own 'depots placed at strategic points as regards motor traffic' (Ferrier 1986, p. 32).

The number of pumps quadrupled between 1925 and 1932. By 1939 there had been a five-fold increase, with 103 000 dealer pumps (Ferrier 1986). Brunner (1930) reported that Esso (AAOC) opened 29 single-brand service stations in the centre of several large cities but the then-powerful Motor Trade Association (MTA), formed in 1910, opposed the move and no company was prepared to take the risk of following AAOC's lead. Independent garages therefore sold most of the fuel and as owner–managers these dealers looked for the best profit from each of the different brands. Thus, travelling sales representatives negotiated supply on an individual basis (Ferrier 1986). The legacy of Petroleum Pool Board rationing – multi-brand sites carrying small stocks – remained until after World War II (Dixon 1997). Dixon asserts that the oil companies did not need vertical integration. As well as the ties to dealers, there was an agreement with the Motor Agents Association (MAA) and the MTA which controlled wholesale and retail prices by way of a 'stop list' of accredited dealers. No one on the stop list could obtain fuel from the Combine. This constraint was entirely legal, as long as it was used 'to protect or promote the legitimate business interest of the association' (Heathcote-Williams *et al.* 1956, pp. 73–74). This was in stark contrast to the anti-trust legislation prohibiting such practices in the USA (US Federal Trade Commission 1952).

A price increase in 1920 led to a London taxi strike and the attention of the government, whose report was critical of excessive prices and the informal market share arrangements between petrol companies. No legal action was taken, but prices fell to very low levels – apparently in an action to discourage new entrants. This did not appear to work as the Russian Oil Products (ROP) and Power Petroleum groups began selling Russian oil. Other new entrants included Cleveland, Dominion and the Texas Oil Company. National Benzole, selling a mixture or petrol and benzole, cooperated with the Combine. In 1924, the Combine introduced regional pricing in an attempt to limit the expansion of the new entrants. This was followed by rebates to dealers and a media effort against Russian oil (Dixon 1997, citing the *Garage and Motor Agent* 13 November 1926 issue) which had little effect, even when prices fell from a high of 4/7d (23p) to 1/ 1½d (6p) a gallon by 1928. Competition was so active that it was unprofitable to sell petrol, yet no company was willing to raise the price in case others failed to follow (Brunner 1930). The now infamous Achnacarry Agreement in 1928 – the 'As Is Agreement' – sought to consolidate the market shares of Standard Oil, AAOC and Shell globally. The agreement restricted production and exports, forming a cartel that limited competition and fixed prices (US Federal Trade Commission 1952). In 1929, agreement was reached with the Russian government to fix the share of ROP in the market, and a 'Junior Combine' of smaller independents was set up which agreed not to open new sites without the approval of the MTA. This included Cities Service, Sealand (selling Power and Dominion), Sinclair Union and United Oil Importers (Albion). The price then rose in 1929. Various changes meant that while some junior combine members went out of business or were taken over by the Combine, others appeared. By 1931, the Combine market share had fallen to 80.9% from 85.7% – ROP's share rose to 5.3% and Standard Oil of Indiana entered with very low prices – but was bought by AAOC. In 1934, independents reached a new agreement with the Combine, extending it in 1938, when the market was seen to be 'well organised' (Dixon 1997, p. 638).

There were different objectives between the companies: AAOC went for gallonage at any price to attain volume; Shell stressed quality and the role of the dealer, and not the station *per se*; whilst BP emphasized the number of stations. The expansion was fuelled (no pun intended) by the increase in registration of motor vehicles (10-fold) and private cars (20-fold) between 1920 and 1939 (Ferrier 1986). However, the crude supply of APOC/BP was inferior on its yield of petrol and its relationship to the quality of other brands – especially as the quality, indicated by the octane rating, needed to increase due to the higher compression ratios of motor cars of the era. In 1933, BP introduced lead tetra ethyl as an additive and advertised it as 'BP plus'; although Shell and as other competitors apparently did not need such an additive, it was advertised by BP as giving the motorist 'something extra'. Even so, Shell's premium petrol sold 12 times as much as the BP offering (Ferrier 1986).

The inability of APOC/BP to compete on price meant that it had to rely on advertising to sell, which was also considered important by the other oil companies. This was in order to encourage consumption, as private motoring was seasonal and many owners only licensed their cars for half a year (Ferrier 1986). Utilizing a similar strategy, BP changed the colours of its pumps from red to green and yellow to emphasize a 'natural harmony for the countryside sites where many pumps were located' (Ferrier 1986, p. 36). Shell-Mex and BP was created over the 1932–39 period by the amalgamation of the distribution arms of Royal Dutch Shell and APOC. This was a cost-cutting strategy and was enacted in many industries due to the Depression and the overcapacity of the industry, which by 1928 was 'greatly exceeding demand' (Ferrier 1986, p. 41). Nevertheless, during the 1920s and 1930s, some 18% of the market had been captured by the low prices of the 'pirate' suppliers such as ROP, with considerable variation in prices over the period from 1912 to 1939 (Ferrier 1986).

In 1939, the petrol companies' competition ended at the outset of World War II. Though there is disagreement about the dates, the Petroleum Board was established; this continued in operation until 30 June 1948 (Dixon 1962), although a single unbranded pool product was sold by all suppliers until 1 February 1953. Shaw (1974) stated that, throughout the 1950s, the market was a concentrated oligopoly dominated both by Shell-Mex and BP ((capturing) 50% of the market) and to a lesser extent Esso (29%). Only Regent/Texaco (13%) and two minor suppliers (Mobil and Petrofina) provided significant competition.

Dixon (1962) labelled the period from 1950 through to 1960 as the strategy of the 'solus', during which the single-brand service station concept developed. He noted that competition had been removed in 1939 by the Petroleum Board in response to the outbreak of World War II, and this was dissolved in 1948. Following this, a 'mutual aid' programme was instituted, along with the rationing of petrol, which ended in May 1950. At this point the market reverted to a conventional model of undifferentiated oligopoly, as there was only the one 'pool grade' of fuel available. That year saw the rise of another 'fringe' market, which included Fina Petroleum Products. Fina was an amalgam of smaller distributing firms totalling about 3% of the market, leaving the bulk of the marketplace to the oligopoly of companies.

The oligopoly was maintained by several factors – the restriction on US dollar purchases required the sourcing of crude from the Middle East, which was in the control of the majors. Anti-trust legislation in the USA precluded market-sharing, there was a severe tanker shortage leading to high rates for spot carriage, and domestic refineries lowered costs for the majors. Esso worked at increasing its market share. In 1951 it expanded its refinery and increased the marketing of its premium grade. As a key part of this strategy, the company challenged the status quo of two-pump sites (constituting about 40% of all sites) which sold mixed brands, but were mainly upplied by Shell-Mex and BP Ltd. Brunner (1930) asserted that the then current 'free-garage system', as opposed to the 'tied-garage principle', favoured Shell and other more popular brands. Esso had experimented with the solus technique in early 1950, which had been successful in other countries. Under this strategy, the company reimbursed the dealers for painting, refurbishment, and co-operation in sales promotion and training of sales attendants (although with a limit of £100). This had significant advantages; the limited ability in the post-war situation to build new sites was substantial and the reduction in advertising during the war years had weakened brands. The combined result of this would lead to reduced costs through the increased volume of a smaller number of sites, which would also have larger storage tanks and so could take larger deliveries at less frequent intervals. These changes in capacities enabled better planning of deliveries.

These developments were opposed by the other companies through a variety of incentive schemes. Regent, for example, lost a great number of its dealers to Esso, despite using cash 'spot concessions' to counter them. Shell-Mex and BP introduced a 'Selected Representation Scheme' for sales promotion. Later, grants were used rather than reimbursement of costs if in arrears, although these could not match the efficiency of the solus system. By April 1951, over 50% of garages had some sort of tie to one company or another, even though many of these were, at first, verbal rather than formal and written. This rapidly increased to around 70% by 1952 and 83% by 1953. On 2 January 1952, competition intensified as Socony-Vacuum, trading as 'Mobilgas', entered the market. Its new refinery was mainly to supply lubricant feedstocks, but one-third of its output was petrol, potentially its most valuable product (Dixon 1997).

Just prior to the war, Petrofina of Belgium had acquired the UK operations of America's Cities Service. As Mobil had constructed a refinery, both companies were keen to develop sales networks and started buying or building their own retail outlets (MMC 1965; Nash & Keegan 1985). Shell-Mex and BP announced the ambitious plan to build 250–1000 new stations in 1954 designed to sell petrol rather than garage services. Regent also planned to grow its outlets. These efforts aligned the new dealers, who were selected by Shell-Mex and BP by virtue of its bargaining power. This allowed Shell-Mex and BP to choose the products to be offered, determine the activities on the forecourt and use its financial resources to develop the sites where appropriate. It also allowed Shell-Mex and BP to buy sites for future development while planning restrictions via the Control of Buildings Operation Order, which was still in force.

By 1956 it is estimated that some 850 stations were in company ownership – 500 by Shell-Mex and BP, 200 by Esso, and the remainder across Regent, Fina and Mobil. Shell-Mex and BP then decided to negotiate terms for 20 year leases, which contained many restrictions and 'roused considerable mistrust'. The Motor Agents Association, Esso and the other companies followed Shell-Mex and BP's lead in 1955, replacing station development grants by loans, a strategy which endangered Regent's financial position. As a consequence, the Oil Company's holdings in Regent were sold to the Texas Company (HM Treasury 1956) in 1956.

Political factors arose again with the Suez Crisis and the re-imposition of petrol rationing in 17 December 1956. This lasted till 14 May 1957, but then renewed competition drove prices down until the end of 1960 (Dixon 1962). This competition resulted in 'astronomical prices' being paid for petrol stations. As Shell-Mex and BP had parted company in 1966, splitting the national chain of filling stations between the brands (Grace's 2015) added to the frenetic state of the market. This was increasingly crowded as Regent, newly financed by Texaco, and Total made higher bids for garages than the others, thus further increasing the competitive atmosphere. However, Dixon (1962) pointed out the lack of real progress in the petrol market in terms of its overall place as a retail operation. This is an important criticism, as we later discover the impact of the supermarkets and multiple retailers in the market.

The solus system of a single-branded service station allowed the dealers to run their businesses efficiently, as they were tied to the brand owners by the substantial rebates. Many observers had been disappointed with the level and quality of the retail service, and the poor results of staff-training programmes. Frequently, the grants had previously not been used to improve the stations but on other activities to which the petrol suppliers had deliberately overlooked. The capital commitments to single-brand stations proved to be a barrier to new entrants: this was to change as political action and supply imbalances altered the nature of the industry.

Thus, although there might have been oligarchy in terms of supply, and as Shaw (1974) suggested there was price leadership, the competition at the retail end meant there had arrived 'a new dimension to oligarchic competition' (Dixon 1962, p. 51). Dealers slowly began to improve their stations and offered price reductions or other incentives such as trading stamps. Lowe (1976) asserted that in 1960 the industry was still an oligarchy, dominated by five major companies. Indeed, two of these, Esso and the Shell-Mex and BP Group, had control of in excess of 75% of the total retail sales in the UK. Refineries were built by both the major corporations (the so-called Group A) and the minors (called Group B), whilst the independents had to buy on the spot market. This latter group had control over 1648 outlets, from a total industry of 32 974 (Lowe 1976; MMC 1979).

In the early 1960s, entry into the petrol industry was achieved through price competition, whereas, later, new firms in Group B exploited distribution and site ownership as a means of gaining market share. The very wide range of price and non-price forms of competition was probably important in breaking down the old oligopolistic structure, although with supply much tighter in the latter part of the period this entry still demanded significant financial resources. The number of promotional offers (e.g. free glasses or competitions such as Shell's 'Match Money', etc.) and trading stamps given with petrol rose sharply as media advertising was falling. Given the variety and the number of such promotions offered, the amount of money spent on them is difficult to estimate. However, most large companies had substantial campaigns at some time during the late 1960s and early 1970s, and these could not have been inexpensive. Esso are quoted as having spent approximately £500 000 on one small campaign in 1970. Interestingly, turnover following this effort increased 15%, but reverted to normal levels after 3 months.

After 1960, however, integrated international oil companies entered the market as Group B, mainly American or European MNCs, and made inroads into the market. These included Conoco's Jet brand, Gulf and Total, and the Group C independents: VIP (which was later sold to US interests), Heron, Curfew and others. Eventually, prices became standardized and competition was often a function of site position, although it also depended to varying degrees on media advertising, price and below-the-line promotions. Further, many petrol outlets were still principally 'garage and service' stations, and petrol sales were both complementary to car repairs and accessory sales, and of secondary importance. This hindered the competition of the established firms for whom this type of organization provided the majority of their outlets, since gasoline sales were not the main interest of the individual entrepreneur (Lowe 1976). Shaw (1974) suggested that by 1969 the small group of five majors were supplemented by 30–40 newcomers, capturing some 20% of the market.

The Monopolies and Mergers Commission Report (MMC 1965) forced the majors to yield absolute control of their stations, allowing sales of third-party oils, and tyres, batteries and accessories (TBAs). Ostensibly, this was to protect smaller suppliers of oil and TBAs but, in effect, it permitted Castrol to retain its position as the *de facto* alternative to lubricants from the main petrol suppliers. The MMC also restricted the length of solus supply arrangements to 5 years in general. Previously, some had been for periods of up to 21 years or more when tied to station development loans. Finally, the MMC limited the number of company-owned stations to a number equivalent to 15% of a company's sales, subject to a minimum threshold level so that smaller company-only chains (such as Heron) could continue to grow. The limit of company stations was dropped in 1968 under pressure from most of the large retailers, including the Group B suppliers (MMC 1990).

In the period immediately following the first MMC report (1964–1983) there was a fall in market share by the majors (Esso, Shell, BP, Regent/Texaco and Mobil), as the Group B and C suppliers rapidly expanded the number of stations, largely through signing up smaller operations in more traditional locations. From the mid-1980s, the rise of the grocery supermarket petrol station (the 'multiple retailers') led to a decline of market share of the oil companies (Shingleton 1998). In 1986 the supermarkets had a combined market share of fuels of the order of 5%; in 1992 alone it rose to nearly 12%. By 1996, this had risen to 25% (Lynn 1996). Supermarket fuel sales continued to grow steadily, reaching a peak of 45.2% in 2013 (*Petroleum Review* 2015). The number of forecourts operated by the supermarkets increased from 294 in 1990 to 870 in 1996 (Shingleton 1998) and up to 1441 by 2014 (*Petroleum Review* 2015). The supermarket fuel era in Europe can be dated to 1963, when the French Carrefour Group opened its first hypermarket with attached filling station. British supermarket operators such as Sainsbury, Tesco, Asda and Woolco noted that French stores with petrol appeared to be more successful than those without. This success was likely to have been based on the public perception that supermarket filling stations were significantly cheaper than traditional service stations. As a consequence of this success, in about 1969 small numbers of English supermarkets also installed pumps. These were initially supplied by major brands such as Mobil who serviced the early Asda sites, as well as the Woolco hypermarkets opened by Woolworths (Byrne & Ritson 2012).

This was subsequently accelerated by geopolitics and supply side issues: the oil crises of 1973 and 1979 altered the situation decidedly, setting in motion a series of petroleum-related world financial traumas. Formed in 1958, OPEC had by the early 1970s wrested dominance from the majors and posted considerable price increases. The Arab–Israeli war of October 1973 and the embargo on exports to the USA forced a supply cut, the first of two oil crises in the 1970s. The price of crude grew by 400% (Sampson 1975). The second oil crisis, precipitated by the Iranian revolution of 1979, yielded a substantial increase of 260% to the price of crude (Boyle 2002*a*). In the 1980s, OPEC members began to exceed their quotas, leading to a glut, which was sold in refined form as petrol on the Rotterdam spot market. Alarmingly for many, prices were lower than the oil companies were paying for crude (Huxley 1986). November crude was \$31.75/bbl (per barrel) in 1985, \$27/bbl in January

1986, \$17/bbl in February, and falling to \$10/bbl in April. Added to this was the rise of the UK's own domestic oil production in the North Sea: encouraged by the government, production only increased the glut (Boyle 2002*b*).

Boyle (1999) showed earlier that after the 1973 crisis, and as early as 1974, the large supermarket chains began selling petrol, led by Tesco. She asserts that these sales were loss leaders, although this is debatable. In the late 1970s the initiative passed to Sainsbury's. In keeping with their focus on its grocery brands, they introduced own-brand fuel sales at selected stores. Although, in 1989, Sainsbury's had only 46 outlets, this was to more than quadruple over the next decade. Their move to own-brand petrol also allowed it greater flexibility in supply. In contrast, Tesco's strategy was initially different; it kept to major brands (mainly Texaco or Gulf, but with smaller numbers of stations selling BP and other big brand name petrol). They also concentrated on adding petrol at more supermarkets. By the time it went the own-brand route in 1991, it had a ready-made chain of 136 service stations that captured a substantial market share (Byrne & Ritson 2012). By 2014, Tesco had become the market leader, selling more fuel from its forecourts than any of the big oil companies through their petrol stations.

Margins were thus squeezed on both sides. The oil companies had little choice but to look for new avenues of revenue – this they found in the emerging 'top-up' society, where major grocery items were bought in a one-stop-shopping trip to supermarkets but the daily topping up of essential supplies was often left to local convenience shopping trips. Here the local petrol station could prosper, as by 1998 convenience shopping presented one-sixth of all shopping sales (Connon 1998). BP was one of the first to operate convenience stores in 1983, and by 1985 it had plans for 100. Total and Tesco were soon to follow; Esso's franchised 'partnership agreement' created Esso Shops. Jet introduced its retail charter and Jiffy Shops, Mobil its Mini-Mart, whilst Shell introduced a fully-managed 'business format' franchise system – Shell Select. Shell's market share had fallen from 20 to 14% over the period of 1985–1991, but Boyle (1999) controversially asserted that Shell Select ultimately failed due to the high set-up costs that dealers incurred.

In 1996, Q8 and Elf established joint ventures with the Budgens and Somerfield supermarkets, and BP with Safeway. Such joint venture sales per square ft were some £22 as opposed to BP's own managed sites of £15 (Boyle 1999). Soon, Esso had become the UK's largest convenience retailer and introduced their 'Fresh ideas' concept of food retailing. Shell became the nation's largest news agency chain, and in many cases non-oil sales were over 50% of profits (Mitchell 1994; Boyle 1999). Other more exotic initiatives followed: Total tried Blockbuster video rental racks (Shingleton 1998). Another response was to launch loyalty card schemes such as Shell Smart: indeed, so many similar concepts were introduced that the market became saturated (Wright & Sparks 1999).

More turbulence was created by the 1990 Gulf war and for a while supermarkets had a price advantage, as Boyle (1999) reports that their prices held steady. However, in 1996, the advent of Esso's Pricewatch campaign eliminated this difference: Esso stations would match the price of any supermarket station within a 3 mile radius. BP's 'Focus on Price' initiative followed. Other major company stations would often cut their prices to match Esso, leaving the net price to the customer lower than smaller firms were able to buy fuel from their wholesaler. Independent service stations, and their small company suppliers, were able to weather the initial supermarket onslaught due to a glut in supply, but Esso's Pricewatch soon proved too hard to beat.

Competitive fringe?

As Byrne & Ritson (2012) reported, these changes should be seen in the context of a retail fuels market that grew rapidly from the end of petrol rationing to 1973, when the first oil price 'shock' (and supply issues from Rotterdam) led to a slight fall in demand. However, by 1976, volumes were again on their post-war growth trajectory. But by the early 1980s, almost a quarter of stations carried the colours of independent British brands, with names such as Thrust, Pace, Anglo, Bulldog or Flare. Added to these were a number of secondary brands such as Butler, Arndale or Paramount, controlled by international suppliers. Petrol was often delivered through independent wholesalers as extensions to longer-standing oil businesses. However, most of these sites were small marginal (and usually rural) locations which proved vulnerable to poor economics and cars' greater fuel efficiency. Absolute numbers peaked around the time of the first oil shock; thereafter, the number of company sites fell, but rather more slowly than dealer locations. When company sites were sold, they were often accompanied by restrictive covenants preventing their continuing use as a retail petrol station. Thus, the growth in the number of independent dealer sites in the 1970s and 1980s came largely from unwanted major company dealer sites; typically, these were the least attractive part of the majors' networks (MMC 1965; *Petroleum Review* 1973, 2012).

At the same time, motorists no longer had to fill up in their home village; they could wait until they next visited a town with more competitively priced petrol, often kept artificially low by a major supplier – Esso predominantly – using selective price support to compete with the supermarkets opening discount forecourts (*Petroleum Review* 1980, 1982, 1991, 2000, 2012; MMC 1990). In 1990 – with little warning – sales stagnated, despite a continuing growth in the number of cars. This was largely due to greater vehicle efficiency, including a long-term switch to diesel cars with inherently better fuel consumption. It may also reflect greater use of public transport, the result of perceived increases in fuel prices and tighter enforcement of speed limits (*Petroleum Review* 1980, 1982, 1991, 2000, 2012). This has led to a continued downward trend: as we saw earlier, by 2014 the number of stations had dropped to 8600, while supermarkets had taken up the balance of demand. In the next section, we consider some of the generic factors affecting the industry.

Part 2: Analysis

Socio-political and industrial factors

We introduce the concepts of oligarchy and the associated lack of competition by one quote and then detail the evidence. Dixon (1962, p. 50) suggests that 'the structure of petrol marketing in the UK has been transformed … by long-term relationships or by vertical integration'. The latter we have already discussed in a previous section, and these relationships can be categorized, as per Kotler (1998), as political and social factors. More complex is the relationship of geopolitical forces to the supply-side economics and subsequent behaviour of companies in the retail sector. We first discuss political and social, following this with consideration of economic and technical factors.

Political factors are present as both direct and indirect forces acting on the industry: indirectly by government reports and geopolitical crises, directly by legislation or the lack of in the case of anti-trust provisions. Cohen (1999*a*) pointed out that petrol, especially as regards its price, 'attracts a lot of media (and Government) attention … and attracts publicity that no oil company would wish to receive' – in this way political and social factors overlap. The government has commissioned various reports – from the MMC, as we have seen, and from other departments such as the Board of Trade (1920, 1931), the Office of Fair Trading 1990 which, according to Harris (1980), rejected the monopolies referral on petrol sales, and the Trade and Industry Select Committee (1988). Underlying the political dimension was the fact that anti-trust law was not a feature – and, as we saw earlier, Dixon (1997) suggested that the ability of the suppliers to fix the UK market between them was a key difference between the UK and the USA industries. At a local level, we noted how local government building legislation affected the way that filling stations were built, even to the extent of the pumps' swing arms, and then planning rules restricted where they were located.

The industry has been affected by the direct interference or indirect pressure from a number of sources such as the Motor Agents Association, the Council for the Protection of Rural England and the Advertising Standards Authority. As we noted earlier, the move to single-brand service stations was opposed by the Motor Trade Association, which restricted these to towns (Ferrier 1986). The Motor Agents Association submission to the Monopolies and Mergers Commission (MMC 1965) concluded that the length of solus ties (sometimes 20 years in duration) meant that new entrants were restricted from entering the industry. Consequently, solus agreements were limited to 5 years, which of itself led to dynamism in the market. Jet reported that this led to switching of supplier – its total losses amounted to 64 sites: 70% went to competitors, the better sites to competitors in Group A and the poorer sites to those in Group C. Group B firms took only a small proportion of the total, which is consistent with them concentrating on new site developments. Jet also acquired a number of sites during this same period, about 42% of which were solus changes, mainly from smaller companies (in Group C) and 58% were virgin site developments. The company emphasized that whilst in the early stages of its development (early 1960s) it gained sites from Group A firms, its major emphasis had now switched to new site development (Lowe 1976). The British Standards Institute (BSI) released accurate octane ratings for all petrols in 1967 (BSI 1967). Shell had been trying to get a clearly non-gullible public to accept that their petrol had a special mileage ingredient – this was as true for Shell petrol as any other and eventually, under pressure from the Independent Broadcasting Authority, the advertisements were withdrawn (Lowe 1976).

Ferrier (1986) reported that Texaco spent large sums of money even in 1972 on pure media advertising, which was a continuation of their policy of getting rid of the old Regent brand name. The brand of petrol was becoming less important to the consumer. Cohen (1999*a*) noted that to them, 'petrol is petrol is petrol' (p. 205). The advertising jingle with the popular crooner Bing Crosby, 'I'm going well I'm going Shell – you can be sure of Shell', may have appealed once, but no longer. Petrol became a commodity despite the effort of Esso with its 'tiger in your tank' slogan and Mobil's flying red horse logo – to make it more desirable through branding (Ritson 1999). Recent market research in *Petroleum Review* (2014) assigned values to factors relevant to petrol sales – price came out at 42%, next was location at 19%, then loyalty cards at 8% and, finally, brand down at only 4%. This contrasts with lay theory: by carrying at least a relatively familiar brand, it was thought that customers gain some assurance that the fuel will come from a reliable and probably consistent source, whereas a wholly 'white' station may take fuel from anywhere and lack consistency of product. Over the years, oil companies have spent many millions of pounds trying to establish and maintain brand loyalty. Techniques utilized have included press and TV advertising, sponsorship of motor racing and other events, and more direct promotional activities, including company-specific trading stamps and electronic loyalty cards. This is an example of bounded rationality and resulting path-dependency – a triumph of gut feel and intuition over facts.

Economic factors usually reflect rates of loan interest, exchange rates and of inflation, but in this market we can identify the problem of price competition as being most important. This is related to social factors: Cohen (1999*b*) made the point that the petrol market possessed specific characteristics and mechanisms which cause it to behave in a distinctive manner:

> The very nature of petrol as a necessary purchase … causes a degree of resentment and a feeling of obligation and restriction of freedom. It is this attitude that causes petrol purchasers to retaliate … and results in them behaving (almost) irrationally … [they] see 'price hunting' as an opportunity to hit back at the oil companies (p. 153).

Purchasers sometimes go considerably out of their way to find the cheapest petrol, forfeiting the savings by the extra distances travelled. Few industries, he asserted, 'exhibit so much obsession among consumers about prices' (Cohen 1999*b*, p. 153).

Furthermore, the economics of petrol trading is a key factor in deciding the behaviour of the companies. Gross margin (GM) – i.e. volume × margin – is vitally important in creating profit, but the classic volume/margin trade-off is not evenly weighted. Leverage comes into effect, whereby any increase in price causes a disproportionate increase in GM; and if the GM is 5% of the selling price, then an increase of 1% of the selling price would have the effect of a decrease in volume of that order. However, this creates a massive increase in GM of 20%. Contrariwise, a 1% decrease requires at least a 25% increase in volume. Despite this, Cohen (1999*a*) asserted that price-cutting 'became almost an obsession as if it were a sign of "virility" rather than an admission of marketing failure' (Cohen 1999*a*, p. 154). Price elasticity of demand should be low in petrol as it is a distress purchase and a commodity. This makes it a non-alternative purchase and so, despite price increases, volume should not decrease radically. Cohen (1999*a*) arrived at an approximate figure of less than 1. Cross-elasticity of demand (CED) refers to the transfer of custom within an industry – in our case, from one service station to another. Major brands, he estimated, had a CED of 10: so a 1% increase in price at one site leads to a 10% change in volume, transferred to other sites. This is extremely high when compared to other industries, reflecting the negative attitudes of customers and the unique situation where prices are normally displayed outside the station. This display enables motorists to compare prices across stations and thus facilitates their purchase decision. Further, Cohen (1999*a*) stated that the CED and speed of customer transfer varied significantly between areas of the UK and even within these, and with the quality of the service station forecourt design.

Geopolitics combines with global economics creating supply-side imbalances. Boyle (2002*b*) quoted Sampson (1975), who stressed the alteration of shortage and glut affected investment decisions in the industry. Company-owned sites were thus considered expensive capital items in a very uncertain market. Between 1960 and 1972, proven reserves of crude oil more than doubled. From the viewpoint of petrol supply, one important development was the discovery of new sources in Libya and Nigeria. Oil from these places was a high-quality type containing a much larger proportion of volatile petroleum products like motor gasoline, kerosene and liquid petroleum gas (LPG). These so-called 'whiter' products could all be obtained from the heavy Middle East crudes. The oil companies, forecasting demand for the lighter products in anticipation of increased motor-car ownership, air transport usage and ethylene consumption (which served as the base product for plastic and other synthetic materials), invested substantial sums of money in new refineries during the late 1950s.

As a consequence, there was substantial excess capacity in 1959–60 and in 1967–69, which was generated by the construction of refineries far outstripping demand for the petroleum products. Of the new refining capacity available in 1960, much was the 'cracking' type, designed to distil out the maximum amount of light volatile products (Lowe 1976). Overall, refining capacity increased by 380% between 1960 and 1972, and so too did consumption of gas and fuel oil. The consumption of motor gasoline over the same period increased by only 250%, however (OECD 1966, 1973). After the Arab–Israeli war, Saudi Arabia stabilized the market with its 'net-back' system until 1990; with the Gulf war increasing the price of Brent Crude to $30, yet a return to peace dropped the price to $11.55 (City Editor 1998). It continued to fall though the 1998 to 1999 period, and by March 1999 it was less than $10/bbl (Tempus 1999). Price continued to be volatile – in June of 1999 it rose to $19/bbl, by year end it stood at $26/bbl, by September 2000 it was over $35/bbl but then fell to $23/bbl by December (Boyle 2002*b*).

The long-term effects of this early period, and especially the shortfall of black products in Europe and the consequent increase in the price of gas, oil, etc., led many American companies for the first time to set up refining capacity in Europe, which was geared much more to the supply of these products. In the UK, Amoco, Burmah, Chevron, Murco and Signal produced a much higher proportion of fuel oil and related fractions than Shell-Mex and BP, Esso, and Mobil. Even so, they must have had an outlet for their petrol by-product, and therefore pursued a policy of either buying up the small independents (as did Conoco and Signal, with Jet and VIP, respectively) or entering the market through site purchase and virgin site development. Without the product differentiation advantage and secure market position of the majors in the UK, the American entrants tried to ensure their position through a policy of direct ownership of sites (Lowe 1976).

Technical issues were more of a gentle and ongoing influence. The basic fuel changed over time: in the 1930s, concern about pinking and anti-knock compression ratios led to more public concern over the cost of motoring (Ferrier 1986). There was also an injection of objective information in the form of the British Standards Institute octane ratings for all petrols in 1967. Not only did this result in a lower product differentiation advantage for existing brands, it also, in the context of quality search theory, represented a tendency for petrol to become less of an experience and more a search product. Together, these three influences were likely to have helped reduce the monopoly power of the existing firms in the industry. Thus, cross-elasticity of demand between brands was relatively high and increasing over the period (Cohen 1999*a*).

Corporate strategies

Given the detail on the external environment above, we now turn to the effect it had on the industry. We emphasized at the outset, the level of strategic action in markets can operate at different levels but these need to be harvested by corporate managers to first decide what business they are or should be in. Initially, of course, the market was carved up by the Combine and its Junior Combine partners. However, the external influence of the 1965 MMC report opened up the market – as did the internal company problem of supply-side imbalances, estimated at approximately 25% more than UK demand (Shingleton 1998). The oil companies, as opposed to the supermarket multiples operating in a retail industry, lacked strategic retail thinking. In part, this is because while the oil companies operate in several vertically integrated but highly specific industries, the multiple supermarket retailers are concentrated in only one sector. Shingleton (1998) claimed that there were several strategic factors which allowed the supermarkets to dominate the petrol-retailing market. First, they had a strategy to incentivize customers to change well-established shopping patterns traditionally centred on the high street by offering petrol. Secondly, they offered lower-priced fuel as a customer service to offset the increased travel costs as customers drove to their out-of-town petrol station locations. Thirdly, they continually sought to establish a brand differentiator and petrol became a key aspect of that effort. Fourthly, they needed a new revenue stream as overall spending in supermarkets had fallen by 1.4% and pressure was applied by investors – despite the comment by Boyle (1999) that there is no evidence that the sale of fuel was a 'loss-leader'.

These strategies were supported by a number of factors: market entry costs were low as fuel sites were lost in the overall capital budgets. Other costs were low in comparison to the oil companies, as they had no upstream exploration and production activities, no refineries or depots nor dedicated distribution systems. Volumes were very high – 9.5 million litres per annum per site on average v. 3 million litres on oil-company managed sites and 1.7 million litres for dealer-managed sites.

The oil companies were poorly placed to compete. Their corporate culture was bureaucratic, inward-looking, focused on upstream supply and product development (such as additives), their image was 'greedy, old-fashioned, out-of-touch, lacking customer focus' (Shingleton 1998, p. 22). It was also speculated that they may genuinely have failed to see the danger or were simply unwilling to accept it and so were slow to respond. They also appeared to be inflexible, even when they saw the danger. For example, they refused to reduce margins in late 1990, whereas the supermarkets cut prices below those offered by the petrol companies at the time of the first oil shock. This contributed to a reduction in brand loyalty, estimated at only 4% of customers. This doubled to 8% when on-site promotions were factored in.

This failure was partly due to the disparate network of sites, and poor and variable service levels, but also to poor forecourt design. Shingleton (1998, p. 22) asserted that the companies' had no ability 'to design service stations which suited high turnover such as wide entry and exit points and inadequate parking facilities by the fuel pumps' (e.g. see Minale 2000). Cohen (1998, 1999*a*, *b*) asserted that sales in attractive upgraded stations were less prone to cross-elasticity than older smaller stations. After its withdrawal from Europe in 1996, Mobil's elegant classic designs were redesigned on its former sites by new owners.

Most of us are likely to be familiar with the common setting of retail outlets: a small building, a line of pumps, usually a canopy, sometimes a shop in addition, all on a bit of tarmac on a road. Although some supermarkets (e.g. ASDA) used kiosks and barriers to prevent theft, this appears to have changed little over the years. However, there seems to be a lack of careful consideration given to the retail presentation of existing sites – even to having lane markings, merchandising and dual tills for separate transaction types. The UK's industry association, the Energy Institute (which incorporated the former Institute of Petroleum) has compiled a so-called 'Blue Book' (covering operations from the design of filling stations to their decommissioning). It also offers a course titled 'Filling Station Design and Engineering' covering 'the technical aspects of handling petroleum products at filling stations and provide[ing] an understanding of how these relate to established good practice and UK legal requirements' (Energy Institute 2016). These include an up-to-date list of standards and regulations affecting all aspects of filling-station equipment, planning, design and construction, and information regarding the issues surrounding storage and handling of vehicle fuels at filling stations (including biofuels). Thus, Boyle (2002*a*, p. 261), in her analysis of Shell Select, summarized the issue as the company's 'lack of understanding of the nature of both forecourt convenience retailing and the forecourt shopping experience'.

Cohen (1999*a*), however, pointed out a more basic reason why the oil companies initially allowed the competition from the supermarket multiple retailers. This relies on two factors: first, the margin on petrol sales was low owing to government duties and taxes, and related price elasticity; and, secondly, the slow uptake by the multiples. With regard to the influence of margin, the concept of the volume ratio means a large loss of volume is not reflected in a large loss of profit. Margins were kept high despite cross-elasticity leading to the transfer of customers to the multiples. These were, however, small in number and, until the mid-1990s, there were few concerns by the 'big five' – Esso, who waited 10 years before its Price Watch response in January 1995, Shell, BP, Texaco and Mobil. Indeed, Mobil supplied ASDA and Woolco directly with petrol. This period of restraint led to 'disastrous long-term consequences' for the oil companies, as by 1995 the multiples had 25% of the market (Cohen 1999*a*, p. 157). Cohen & Bradfield (2001) asserted that the oil companies were conservative, complacent, poor at innovation and marketing, and differentiation of their brands, and since the earliest stages over 100 years ago were inept. They are seen as exploitative, manipulative and focused on profiteering. They behaved as if customers were little more than a nuisance: as a result, refuelling was likely to be a negative experience – an unpleasant, unwelcome disruption to the consumer's time. Yet, it could and should have been an opportunity to add value, reward patronage and build relationships (Barber 2001).

Conclusions

We have charted the development of the petrol-retailing market in the UK, beginning with an oligarchy or cartel permitted by UK law and generating very high profits. This market evolved via a dedicated distribution system to one where price became a key concern of the public and other institutions, including government. We have also shown that lower barriers to entry and increased competition from other majors and independents, and especially supermarket chains, has become a challenge to industry survival. We note the late discovery by the industry that petrol stations are not favoured by the public, and document attempts to create on-site value through sales of low-volume sites and the creation of own-brand convenience stores. These are largely being replaced by alliances with experts – the competitor supermarkets. Our analysis shows the complexity of petrol retailing in terms of both social and political pressures, as well as the internal economics of the petrol market. All of these factors are exacerbated by poor management of petrol retailing by the oil companies, and the excellent strategic analysis and vision of the supermarkets.

Acknowledgements The authors thank the numerous people working at the Geological Society of London for their unfailing courtesy, assistance and forebearance in the preparation of this manuscript for publication.

Funding This work was supported by funding from Lincoln University, New Zealand to Neil H. Ritson.

References

Alexander, N. & Akehurst, G. 1998. Introduction: the emergence of modern retailing, 1750–1950. *Business History*, **40**, 1–15.

Bamberg, J. 2000. *British Petroleum and Global Oil 1950–1975: The Challenge of Nationalism*. Cambridge University Press, Cambridge.

Barber, J. 2001. Petrol retailing within a global context (book reviews). *European Business Journal*, **13**, 202–205.

Board of Trade 1920. *Report of the Committee on Motor Fuel*. Command Paper 596 HMSO, London.

Board of Trade 1931. *Report on the Committee on Restraint of Trade*. HMSO, London.

Boyle, E. 1999. A study of the impact of environmental uncertainty on franchise system: the case of petrol retailing in the UK. *The Journal of Consumer Marketing*, **16**, 181–195.

Boyle, E. 2002*a*. The failure of business format franchising in British forecourt retailing: a case study of the rebranding of Shell Retail's forecourts. *The International Journal of Retail & Distribution Management*, **30**, 251–263.

Boyle, E. 2002*b*. A critical appraisal of the performance of Royal Dutch Shell as a learning organisation in the 1990s. *The Learning Organisation*, **9**, 6–18.

Brunner, C.T. 1930. *The Problem of Oil*. E. Benn Ltd, London.

BSI 1967. *BS 4040-1: 1967, Petrol for Motor Vehicles. Specification and Nomenclature for Grades of Petrol.* British Standards Institution, London.

Byrne, I. & Ritson, N.H. 2012. A history of UK petrol retailing: dynamism v. oligarchy. Presented at the Centre for the History of Retailing and Distribution (CHORD) Conference on Retailing and Distribution History, 5–6 September 2012, University of Wolverhampton, UK.

Caffarra, C. 1994. *Vertical Contracts in Petrol Retailing: Incentives and Risk Sharing*. Oxford Institute of Energy Studies, Oxford, WPM 19, 1–28.

City Editor 1998. Big oil feels the heat. *The Times*, 12 August 1998.

Cohen, M. 1998. The problem of competition in the UK petrol market. *Strategic Change*, **7**, 203–212.

Cohen, M. 1999*a*. Potential pitfalls in the petrol market. *Strategic Change*, **8**, 299–301.

Cohen, M. 1999*b*. Pricing peculiarities of the UK petrol market. *The Journal of Product and Brand Management*, **8**, 153–162.

Cohen, M. & Bradfield, E. 2001. *Petrol Retailing within a Global Context*. Whurr, London.

Connon, H. 1998. Local shop. Sleepy small town. But this is the front line of a retail war. *The Observer*, 30 August 1998.

Denning, J. & Freathy, P. 1996. Retail Strategies for the petrol forecourt: the example of RoadChef Forecourts Ltd. *The International Review of Retail, Distribution and Consumer Research*, **6**, 97–112.

Dixon, D.F. 1962. The development of the solus system of petrol distribution in the United Kingdom, 1950–1960. *Economica, New Series*, **29**, 40–52.

Dixon, D.F. 1965. Innovation in gasoline marketing. *Business Horizons*, **8**, 5–20.

Dixon, D.F. 1997. Inter-war changes in gasoline distribution: a U.S.–U.K. comparison. *Business and Economic History*, **26**, 632–648.

Energy Institute 2015. *Retail Marketing Survey – Supplement to Petroleum Review*. Energy Institute, London.

Energy Institute 2016. *Filling Station Design and Engineering*. Energy Institute, London, https://www.energyinst.org/training/technical-training/filling-stations-training [last accessed 15 June].

Ferrier, R.W. 1986. Petrol advertising in the twenties and thirties: the case of the British Petroleum Company. *European Journal of Marketing*, **20**, 29–51.

Fitzgerald, P. 1927. *Industrial Combination in England.* George Allen & Unwin, London.

Grace's Guide to British Industrial History 2015. *Shell-Mex and B. P.*, https://www.gracesguide.co.uk/Shell-Mex_and_B._P.

Harris, D. 1980. OFT rejects monopolies referral on petrol sales. *The Times*, 3 October 1980.

Heathcote-Williams, H., Roberts, E. & Bernstein, R. 1956. *The Law of Restrictive Practices and Monopolies*. Eyre & Spottiswoode, London.

Henriques, R.D.Q. 1960. *Bearsted: A Biography of Marcus Samuel, First Viscount Bearsted and Founder of the Shell Transport and Trading Company, 1853–1927.* Viking Press, New York.

HM Treasury 1956. *The Trinidad Oil Company; Proposed Purchase by the Texas Company*. Command Paper 9790 HMSO, London.

Huxley, J. 1986. Oil majors strike a nice little earner. *The Sunday Times*, 2 March, 1986.

Kotler, P. 1998. *Marketing Management – Analysis, Planning, Implementation, and Control.* 9th edn. Prentice-Hall, Englewood Cliffs, NJ, USA.

Longhurst, H. 1959. *Adventure in Oil: The Story of British Petroleum.* Sidgwick & Jackson, London.

Lowe, J.F. 1976. Competition in the UK retail petrol market 1960–73. *The Journal of Industrial Economics*, **24**, 203–220.

Lynn, M. 1996. Shoot out at the pumps. *The Sunday Times*, 2 March 1996.

Minale, M. 2000. *How to Design a Successful Petrol Station*. Internos Books, London.

Mitchell, A. 1994. Fill the car with four star food, please. *The Times*, 26 January 1994.

MMC 1965. *A Report on the Supply of Petrol to Retailers in the United Kingdom.* Monopolies and Mergers Commission. HMSO, London.

MMC 1979. *Report The Supply of Petrol in the United Kingdom by Wholesale*. Monopolies and Mergers Commission. HMSO, London.

MMC 1990. *The Supply of Petrol: A Report on the Supply in the United Kingdom of Petrol by Wholesale*. Monopolies and Mergers Commission. HMSO, London.

Morris, L.E. 1985. *The Country Garage*. Osprey Publishing, Aylesbury, UK.

Nash, M. & Keegan, P. (eds) 1985. *Mobil: Into the Second Century*. Mobil Oil/PRS. Jaguar Press, London.

OECD 1966. *Basic Statistics of Energy*. Organization for Economic Co-operation and Development (OECD), Paris.

OECD 1973. *Oil: World Statistics*. Organization for Economic Co-operation and Development (OECD), Paris.

Petroleum Review 1973. *Retail Marketing Survey*. Energy Institute, London.

Petroleum Review 1980. *Retail Marketing Survey*. Energy Institute, London.

Petroleum Review 1982. *Retail Marketing Survey*. Energy Institute, London.

Petroleum Review 1990. *Retail Marketing Survey*. Energy Institute, London.

Petroleum Review 1991. *Retail Marketing Survey*. Energy Institute, London.

Petroleum Review 1993. *Retail Marketing Survey*. Energy Institute, London.

Petroleum Review 2000. *Retail Marketing Survey*. Energy Institute, London.

Petroleum Review 2001. *Retail Marketing Survey*. Energy Institute, London.

PETROLEUM REVIEW 2012. *Retail Marketing Survey*. Energy Institute, London.
PETROLEUM REVIEW 2014. *Retail Marketing Survey*. Energy Institute, London.
PETROLEUM REVIEW 2015. *Retail Marketing Survey*. Energy Institute, London.
RITSON, N.H. 1999. The *real* problem of competition in the UK petrol market: a reply to Cohen. *Strategic Change*, **8**, 235–241.
RITSON, N.H. & CHAMBERLIN, D. 2012. The American challenge in the UK oil refining industry; predation or rescue? *Oil-Industry History*, **13**, 207–224.
ROSENBLOOM, B. & DIMITROVA, B. 2011. The marketing mix paradigm and the Dixonian systems perspective of marketing. *Journal of Historical Marketing*, **3**, 53–66.
SAMPSON, A. 1975. *The Seven Sisters: The Great Oil Companies and the World They Made*. Hodder & Stoughton, London.
SHAW, R.W. 1974. Price leadership and the effect of new entry on the U.K. retail petrol supply market. *The Journal of Industrial Economics*, **23**, 65–79.
SHINGLETON, J. 1998. Development of petrol retailing in the UK. *International Journal of Market Research*, **40**, 13–23.
SIMON, H.A. 1957. *Administrative Behaviour: A Study of Decision-Making Processes in Administrative Organization*. 2nd edn. Macmillan, New York.
TEMPUS 1999. Buys of the black stuff. *The Times*, 15 July 1999.
TRADE AND INDUSTRY SELECT COMMITTEE 1988. *Third Report: Petrol Retailing in the UK*. HMSO, London.
UKPIA 2011. *The Contribution of the Oil Refining Industry in Wales and the UK*. UK Petroleum Industry Association (UKPIA), London.
US FEDERAL TRADE COMMISSION 1952. *The International Petroleum Cartel*. US Government Printing Office, Washington, DC.
WRIGHT, C. & SPARKS, L. 1999. Loyalty saturation in retailing: exploring the end of retail loyalty cards? *International Journal of Retail & Distribution Management*, **27**, 429–440.
YERGIN, D. 1990. *The Prize: The Epic Quest for Oil Money and Power*. Simon and Schuster, New York.

The first oil exploration campaign in the UK, 1918–22

STEPHEN M. CORFIELD

Corfield Geoscience Ltd, 15 Peel Street, Stafford ST16 2DZ, UK

0000-0003-1423-3991

stephen.corfield@corfgeo.co.uk

Abstract: The story of UK onshore exploration goes back to the days of World War I and was prompted by the increasing use of oil for the war effort. The war was drawing to a close as the campaign commenced in 1918. The UK government sponsored the drilling with a budget of £1 000 000 and the work was undertaken by S. Pearson & Sons, a UK engineering company owned by Lord Cowdray (Weetman Pearson). Pearson also had oil interests and he owned the Mexican Eagle Company that had had exploration success in Mexico. Pearson hired a team of American geologists to select suitable drilling locations in the UK. The Carboniferous rocks in the area surrounding the Derbyshire Dome in England and the Midland Valley in Scotland were chosen because of their similarity to the oil-producing areas of Pennsylvania and West Virginia in the USA. Eleven wells were sunk: seven in Derbyshire, two in North Staffordshire and two in Scotland. The first well to be spudded was at Hardstoft in Derbyshire in October 1918 and it was also the first oil discovery. The geological reasoning behind the selection of the drilling sites will be compared with the actual results from 1918–22.

Background to the drilling campaign

The initiative for the first oil exploration campaign in the UK dates back to the days before World War I and is directly related to the decision by Winston Churchill, in his role as First Lord of the Admiralty, to convert the Royal Navy's ships from coal to oil power in 1911. However, the only indigenous oil was produced from the distillation of oil shales in the Midland Valley of Scotland and in 1912, at the peak of production, the output was 213 MMbbl (million barrels) a year (Monaghan 2014). In contrast, 1 865 551 tons (*c.* 13 MMbbl) were imported in 1913, with only 14% being from territories under British control (Giffard 1923). With war threatening, there was a strategic need to gain control of secure supplies and in 1913 John Cadman, Professor of Mining at the University of Birmingham, was appointed to the Royal Commission on Fuel Oil for the Admiralty. He visited the newly discovered oil fields in what was then known as Persia, and the UK government subsequently acquired a controlling interest in the Anglo-Persian Oil Company (later to become BP) in order to secure supplies for the navy (Torrens 1994). Although the oil was under British control, it was vulnerable to supply shortages during wartime due to enemy action and blockades. There was, therefore, a need to explore for an indigenous source of oil. However, it was not until the war was drawing to a close, in 1918, that the decision was made by the government to explore for oil in the British Isles, by which time imports had risen to 5 157 834 tons (*c.* 36.813 MMbbl) a year (Giffard 1923). John Cadman was awarded a knighthood for his contribution to securing the supply of oil to the nation, and he became the government advisor for the UK drilling campaign. S. Pearson & Sons, owned by Lord Cowdray (Weetman Pearson), were the drilling contractors. Weetman Pearson was a British engineer who owned the Mexican Eagle Company, with significant discoveries of oil in Mexico. His team of American geologists had already begun research into the possibility of drilling for oil in the UK when he was approached by Cadman to undertake the drilling programme. He supplied the UK government with his research results, who responded by allocating a sum not exceeding £1 000 000 for the drilling campaign (Giffard 1923).

The selection of the drilling sites

The drilling campaign was not a collection of randomly selected sites and 'wildcat' drilling. Instead, Pearson's team selected the drilling locations using a set of criteria comprising what would now be known as a play (Allen & Allen 2013), and also took into account known shows of oil both at outcrop and the subsurface. The main criteria were:

- 'areas subject to impregnation from the Lower Carboniferous';
- 'of such geological structure as to favour accumulations of oil at commercial drilling depths';
- 'possessing suitable porous beds to hold the oil and impervious beds to prevent its escape' (Giffard 1923).

All of the elements of a petroleum play were considered to be present in the form of trap, source, reservoir and seal in the Carboniferous rocks of North Staffordshire and Derbyshire in central England,

From: CRAIG, J., GERALI, F., MACAULAY, F. & SORKHABI, R. (eds) 2018. *History of the European Oil and Gas Industry*. Geological Society, London, Special Publications, **465**, 39–52.
First published online May 14, 2018, https://doi.org/10.1144/SP465.11

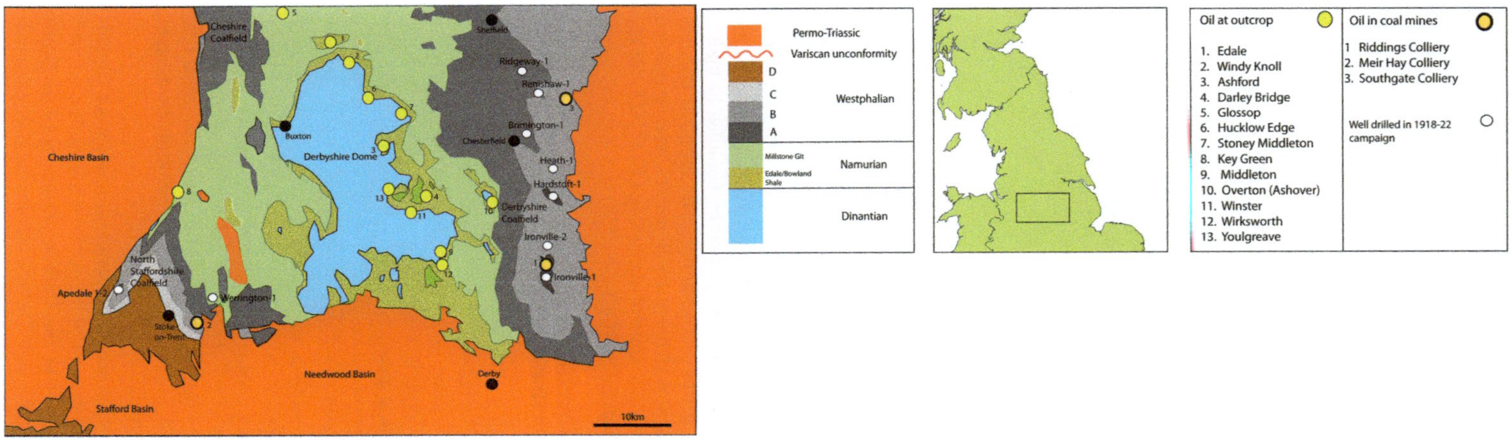

Fig. 1. Geological map of the Derbyshire Dome and surrounding areas showing the location of the wells drilled in Derbyshire and North Staffordshire. Seep data are compiled from Lees & Cox (1937), Kent (1954) and Selley (1992).

Fig. 2. Major folds and faults observed at outcrop in the area of the Derbyshire Dome.

and in the Midland Valley of Scotland (Figs 1 & 2). Furthermore, the Carboniferous of the coalfields of central England was found to be remarkably similar to rocks of the same age in Pennsylvania and West Virginia, where coal mining and oil production had been undertaken in close proximity for many years (Giffard 1923).

Oil shows in central England

The drilling locations chosen by the Pearson team were all in areas of documented oil shows both at outcrop and in the subsurface. Significantly, the influence of James Young and the oil shale industry was apparent in all of the areas. In Derbyshire, a flow of oil into the Riddings Colliery, located on the crest of Alfreton (Ironvillle) Dome, had been sampled by Young in his early distillation experiments in 1847 (Torrens 1994). The flow was estimated at 300 gallons (8.5 bbl) a week but the supply had decreased significantly by 1879 (Kent 1954). Young also sampled oil from the Southgate Colliery in Derbyshire (Fig. 1), where flows averaged 70–100 gallons (2–3 bbl) a day (Kent 1954). In North Staffordshire, a flow of oil into the workings of Meir Hay Colliery in 1874 (Fig. 1) was turned into a commercial operation by a local entrepreneur, William Walker, who refined: 'up to 5 tons of crude oil a week which was seeping from one of the deepest coal seams in these mines' (Torrens 1994). Gas production from Westphalian coals had also started in North Staffordshire in 1858 (Giffard 1938). The introduction of James Young's retorting techniques resulted in a short-lived oil boom and by 1865 eight companies were producing 100 tons of oil a week using 154 retorts (Giffard 1938). Cheap imports of oil from the USA began in 1863 and soon increased to volumes sufficient to force the North Staffordshire oil industry into a steep decline, although oil continued to be produced until about 1880 (Giffard 1938).

Oil at outcrop in central England is largely restricted to heavy oils and bitumens in a NW–SE-trending belt along the NE margin of the Derbyshire Dome (Fig. 1). The oils occur near the top of the Dinantian limestone where it is in unconformable contact with the overlying Edale/Bowland shales (Lees & Cox 1937), suggesting that the Edale Shale formed a regional seal and migration pathway onto the structural high of the Derbyshire Dome. The best-documented example occurs at Windy Knoll near Castleton, where a rubbery bitumen (elaterite) occurs on the crest of a small culmination in the limestone (Selley 1992). There are virtually no shows of oil at outcrop in the Namurian sediments flanking the Derbyshire Dome, apart from the unusual occurrences of liquid oil in goniatite chambers in the Edale and Glossop areas recorded by Lees & Cox (1937). Chemists from Pearson's team had analysed samples of liquid oil and bitumen from Derbyshire, and predicted that the most likely reservoir would be found in the 'Mountain' (Dinantian) Limestone. They also predicted that the source of the oil was in the limestone (Giffard 1923).

Geological setting of the wells in central England

In central England, all of the wells were located on surface anticlines that had been mapped by the Geological Survey, and subsurface information in the form of coal-seam structure maps and boreholes was also available for the structures in the working coalfields. Anticlines flanking the major north–south-trending Derbyshire Dome were formed during the end-Carboniferous Varsican Orogeny, and their orientations reflect the influence of the deeper arcuate trend of a Lower Paleozoic fold and thrust belt (Fraser & Gawthorpe 1990). The folds have a predominant trend of NE–SW in the west and NW–SE in the east (Fig. 2). Bisecting the arcuate trend is the north–south-trending Malvern lineament, which is the dominant structural trend in the North Staffordshire Basin to the west of the Derbyshire Dome (Corfield *et al.* 1996). The Derbyshire wells all reached total depth in Dinantian shelf carbonates (the Mountain Limestone of Giffard 1923) overlain by approximately 100 m of 'Limestone Shales', basinal organic-rich mudstones of early Namurian age (now known as the Bowland Shale; Fig. 3). The overlying Millstone Grit is a coarsening-upwards deltaic sequence with an average thickness of 300 m. The Westphalian A & B Coal Measures complete the drilled Carboniferous sequences and have an average thickness of 700 m. The wells were drilled by American crews using tubular steel derricks powered by steam engines (Craig *et al.* 2013). The drilling was undertaken using a steel-wire rope attached to a percussive chisel-shaped drill bit. All of the wells were cased with six strings starting with 15½ inches and finishing with 6⅝ inches (Giffard 1923).

Derbyshire

Hardstoft-1

The first well dedicated to oil exploration in the UK was also the first discovery well. Hardstoft-1 was drilled on a prominent NW–SE-trending surface anticline on land at Tibshelf belonging to the Duke of Devonshire (Craig *et al.* 2013). The fold is asymmetrical and dips are steeper on the SW limb of the fold due to the Variscan inversion of a

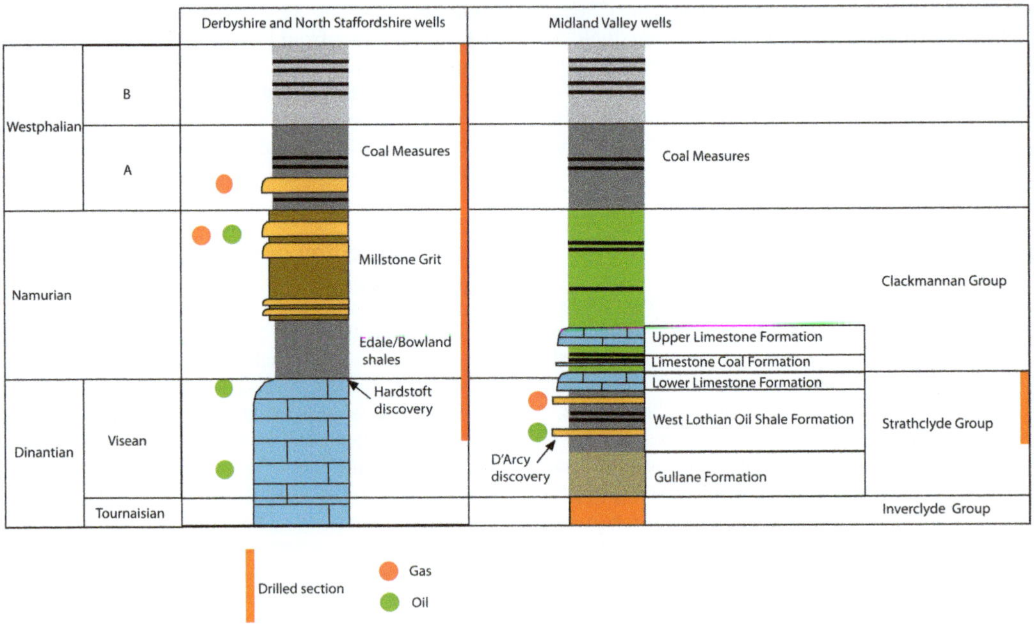

Fig. 3. Stratigraphy of the central England and Midland Valley wells.

major NW–SE-trending fault (Figs 4 & 5). The four-way dip closure at the level of the Westphalian B Silkstone coal is 5 km^2 in area. However, the size of the structure decreases markedly with depth and the trap appears to be a relatively small fault block at the level of the Dinantian limestones (Fig. 5). Drilling commenced with ceremony on the 15 October 1918 with Lord Cowdray, the Marquis of Hartington (the son of the Duke of Devonshire) along with representatives of the government and the oil industry in attendance (Giffard 1923). For the drilling rates at the time, the well progressed rapidly, passing through the Westphalian and Millstone Grit and the Limestone Shale (Bowland Shale) before entering the Dinantian limestone at a depth of 3075 ft (934 m) on 27 May 1919 where oil was encountered (Giffard 1923). The oil rose 90 ft (27 m) within the first hour, accompanied by gas, and was flowing at the surface within a day or two (Giffard 1923). A steady rate of 6–7 bbl/day was obtained between June 1919 and December 1927 (Craig *et al.* 2013).

Heath-1

Heath-1 was located on a north–south- to NW–SE-trending anticline to the north of the Hardstoft structure. It is uncertain if this was a valid trap – the structure map of the Silkstone coal in Giffard (1923) shows a definite four-way dip-closed structure (Fig. 4). However, a later map produced by BP and published by Falcon & Kent (1960) displays a SE-plunging fold with no trap. The well was spudded on 27 February 1919 and immediately ran into problems in the form of an unrecorded mine working which required the placing of concrete pillars to support the weight of the drilling derrick (Giffard 1923). Caving of the wellbore, collapsing casing and the loss of drilling tools added to the problems. Gas was encountered at two levels in the Coal Measures which flowed at a rate of 450 000 cfg/day (cubic feet of gas per day), and this was used to heat the boilers and forge at the drill site (Giffard 1923). A show of oil was encountered in the top of the Dinantian limestone and the decision was made to stimulate the flow of oil using a charge of 110 lb (pounds) of dynamite. This was known as 'shooting the well' and it had been a common practice in the oilfields of the USA since the late 1800s (Schmidt *et al.* 1981). A small amount of oil resulted from the explosion and a second, more ambitious, attempt was made using 10 canisters, 8 ft in length, filled with 1200 lb of nitroglycerine and detonated by a charge of dynamite. The rock was shattered in the area of the blast but, with no further shows of oil, the well was abandoned (Giffard 1923).

Ironville-1 and Ironville-2

The Alfreton Dome (or Ironville Anticline) was the location of the oil seeps from the Riddings Colliery sampled by James Young in 1874. The structure

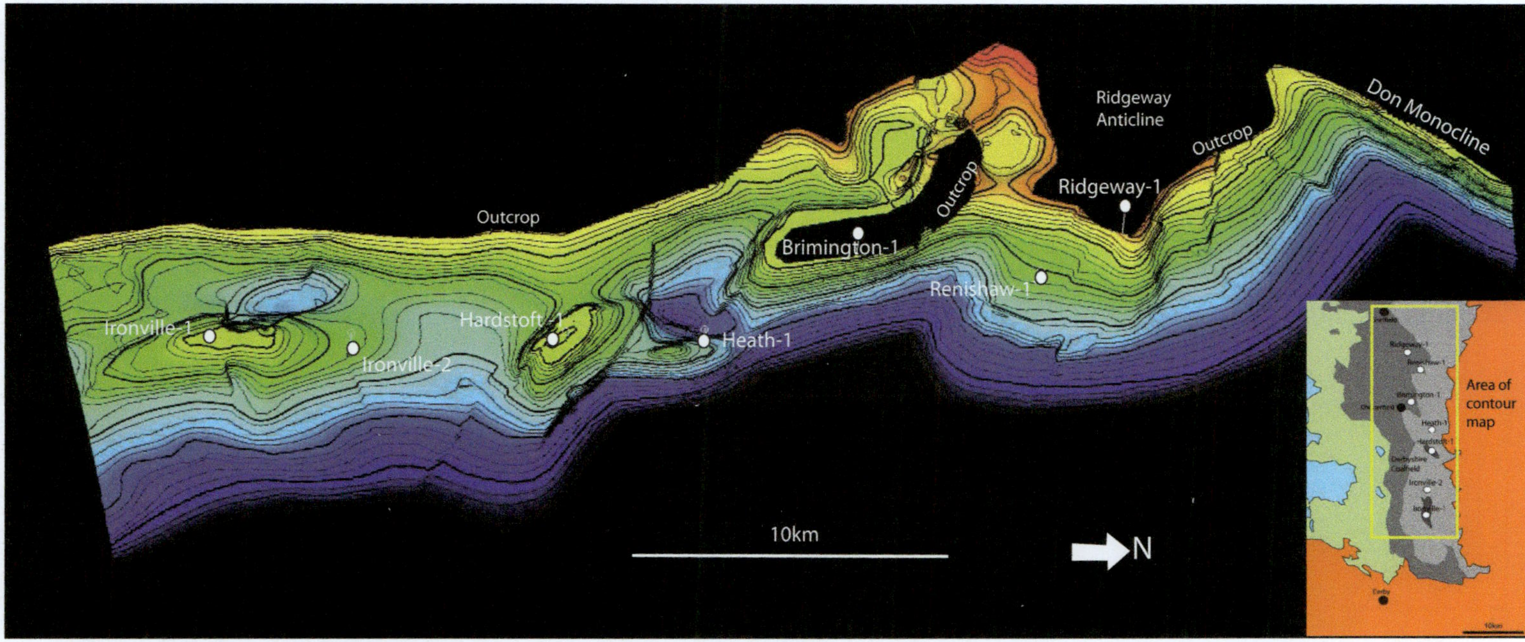

Fig. 4. Contour map of the Top Hard/Silkstone coal in the Derbyshire coalfield showing the location of wells drilled during 1918–22. Compiled from Giffard (1923) and Eden *et al.* (1957).

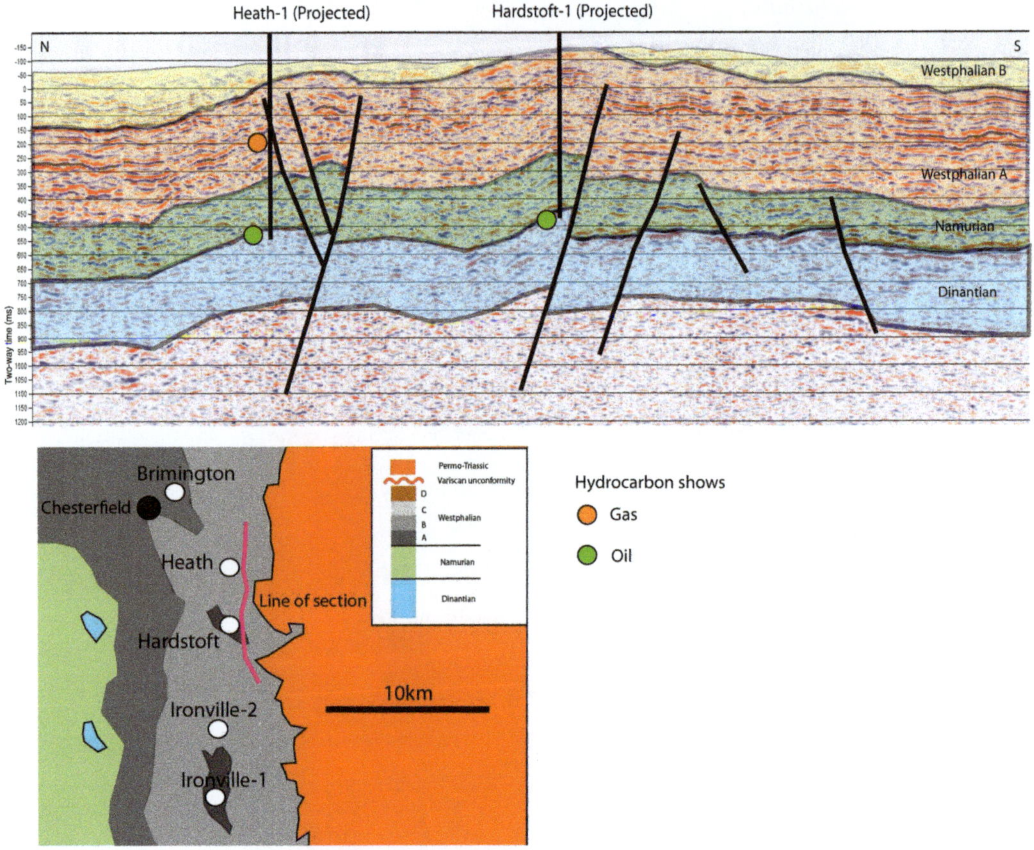

Fig. 5. 1983 seismic line acquired by BP across the Hardstoft Anticline. The wells have been projected onto the line. Note the small size of the fault block at the level of the Hardstoft oil discovery in the Dinantian limestone. From the UK onshore Geophysical Library (UKOGL).

was therefore of great importance to the Pearson team and two wells were drilled (Fig. 4): one on the crest (Ironville-1) and one on the northern limit of the structural closure (Ironville-2). The fold is asymmetrical with a steep west flank and has a structural closure of 19 km^2 at the level of the Silkstone coal (Fig. 4). An east–west-trending normal fault bisects the structure and the Ironville-1 well was spudded to the south of the fault on 3 March 1919 (Giffard 1923). The Millstone Grit and Bowland Shale thin southwards, and the Dinantian limestone was encountered at a shallower level than at Hardstoft. A show of heavy oil was discovered at the top of the limestone at 2031 ft and the drilling continued into the limestone to a depth of 3630 ft, where indications of lighter oil encouraged the drillers to shoot the well with 180 lb of dynamite. This resulted in a small show of oil and the well was subsequently abandoned (Giffard 1923). Shows of oil and gas were encountered in Ironville-2 near the base of the Coal Measures and the top of the Namurian. Oil shows were also seen near the top of the Edale/Bowland shales and in the top of the Dinantian limestone but none were of any importance and the well was abandoned.

Brimington-1

Brimington-1 was located on a well-defined trap at the level of the Silkstone coal which crops out on the flanks of the structure. The fold is arcuate, being north–south-trending in its southern part and NW–SE in the north (Fig. 4). The well was poorly located on the NE flank on the fold, on the limits of the structural closure. Shows of oil and gas were encountered near the top of the Millstone Grit but nothing was found in the rest of the well which continued to a depth of 4040 ft. The well was unusual in that it proved a considerable thickness of lava in the Dinantian limestone together with the first recorded occurrence of anhydrite in the Dinantian of central England (Giffard 1923).

Ridgeway-1 and Renishaw-1

These wells were located on the easterly-plunging nose of the Ridgeway Anticline – a huge structure that extends across the whole exposed coalfield from the Namurian outcrop on the west to where it plunges beneath the cover of Permo-Triassic sediments on the east (Figs 1 & 2). The oil seeps sampled by James Young in the Southgate Colliery were located on a smaller, subsidiary plunging fold on the eastern flank of the anticline. The structure may have been large enough to warrant the drilling of two wells but it lacked a trapping mechanism in the form of a structural closure. There were no oil shows in Ridgeway-1 but the well did encounter a hot brine containing hydrogen sulphide that flowed at a rate of 100 000 gallons/day which resulted in the well being abandoned. Renishaw-1 was located downdip of the Ridgeway-1 well and, despite the lack of a structural trap, it encountered strong flows of gas at two levels in the Coal Measures at a combined rate of 450 000 cfg/day which were cased off (Giffard 1923). No shows of oil were encountered and the well was abandoned.

North Staffordshire

Apedale-2

Apedale-2 was located on the Western Anticline of the North Staffordshire Coalfield. The structure has a NE–SW trend and is an asymmetrical fold with steeply-dipping, locally overturned strata on its NW limb (Figs 6 & 7). There was probably great confidence in this structure, which was well defined by outcrop data and mine plans. There were abundant flows of oil in the coalfield, and the leader of the government team, John Cadman, was familiar with the local geology, having been born in the mining village of Silverdale on the southern flank of the fold. However, the drilling results were a great disappointment. Apedale-1 was the first attempt at drilling the structure and was abandoned due to the loss of the drilling tools at a depth of 1338 ft. Apedale-2 proceeded without difficulty through the Coal Measures. However, the drillers then met a much attenuated Millstone Grit sequence of 118 m overlying a considerable thickness of volcanic tuffs. Volcanic rocks had been encountered in the Dinantian limestone in Ironville-2 and were also present at outcrop on the Derbyshire Dome. The decision was therefore made to continue drilling with the assumption that limestone would underlie the volcanics. Unfortunately, no limestone was encountered and the well was abandoned after 2780 ft (845 m) of volcanic rocks had been penetrated (Giffard 1923).

Werrington-1

This was the only well in the drilling campaign in central England to be located outside a working

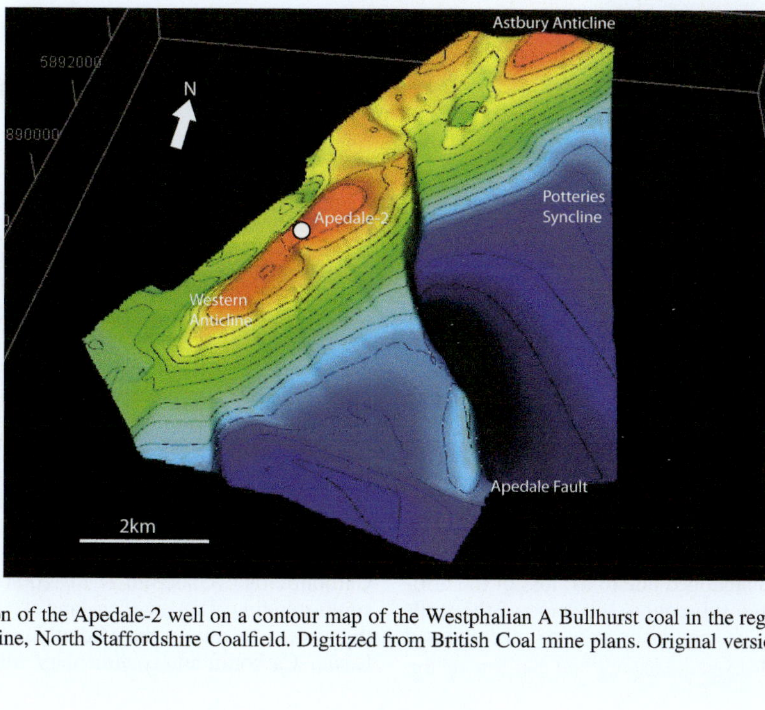

Fig. 6. Location of the Apedale-2 well on a contour map of the Westphalian A Bullhurst coal in the region of the Western Anticline, North Staffordshire Coalfield. Digitized from British Coal mine plans. Original version in Corfield (1991).

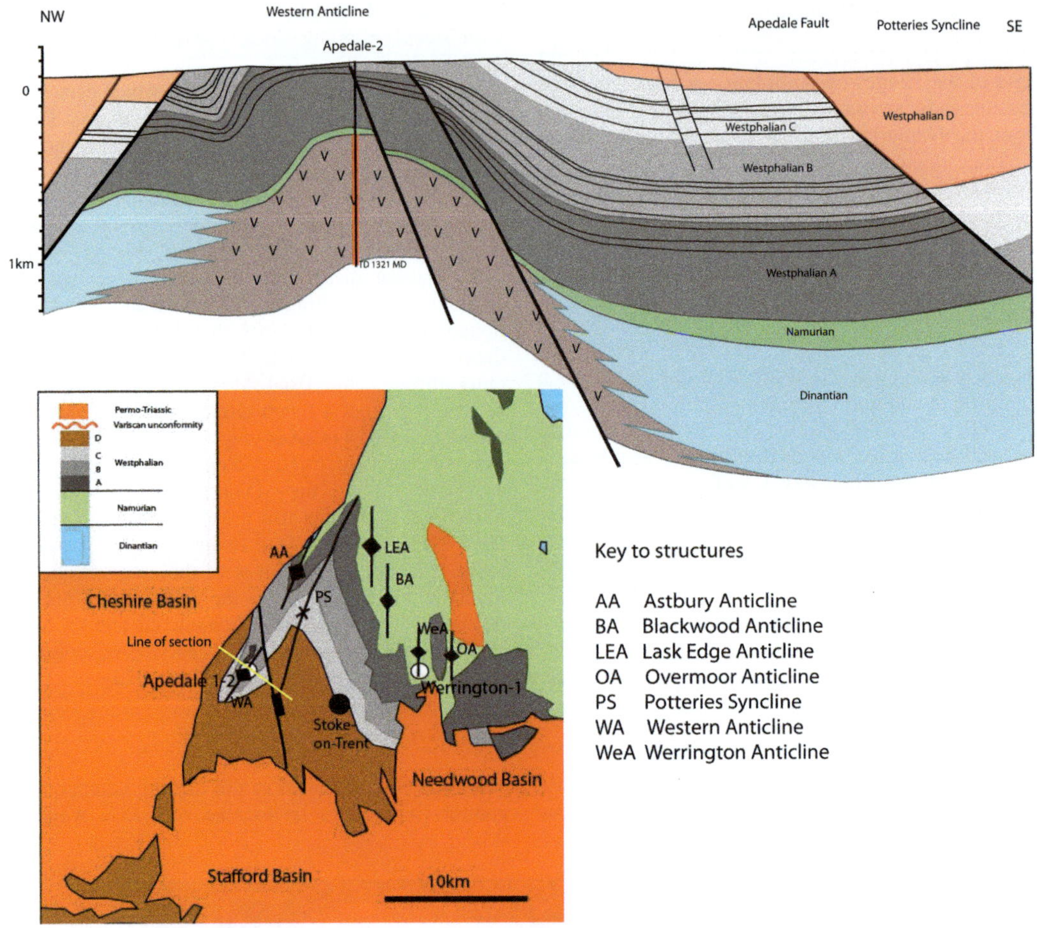

Fig. 7. Cross-section of the Western Anticline at the location of the Apedale-2 well. From Corfield (1991).

coalfield, and, consequently, there was a lack of subsurface data in the form of mine plans and boreholes. The well was located a few kilometres to the north of the seepage of oil in the Meir Hay Colliery on the southerly-plunging nose of the Werrington Anticline (Fig. 8). Namurian basinal sediments are exposed in the cores of four en echelon inversion anticlines on the eastern side of the North Staffordshire coalfield. The Lask Edge Anticline is the only one of these folds that has a trapping mechanism in the form of a four-way dip closure, and all of the other anticlines plunge to the south. Werrington-1 was therefore poorly located on a structure with no closure (Fig. 8). The well was drilled entirely within the Namurian Millstone Grit, and was abandoned due to the loss of the drilling tools and the subsequent blockage of the wellbore. No oil or gas shows were encountered (Giffard 1923).

Geological setting of the wells in the Midland Valley of Scotland

The Midland Valley is bounded by the NE–SW-trending Highland Boundary Fault to the north and the Southern Uplands Fault to the south. Two wells were drilled in the Midlothian region of the Midland Valley where the structure is dominated by the NE–SW-trending intrabasinal high of the Pentland Hills. Silurian–Devonian sediments and volcanics in the Pentland Hills have been uplifted by a major reverse fault, the Pentland Fault, and the footwall of the fault is deformed into a deep fold, the Midlothian Syncline. There is evidence of synsedimentary growth of the syncline during the Carboniferous (Underhill *et al.* 2008). The north–south-trending D'Arcy-Cousland Anticline flanks the eastern margin of the Midlothian Syncline. The Lower Carboniferous sedimentary fill of the basin

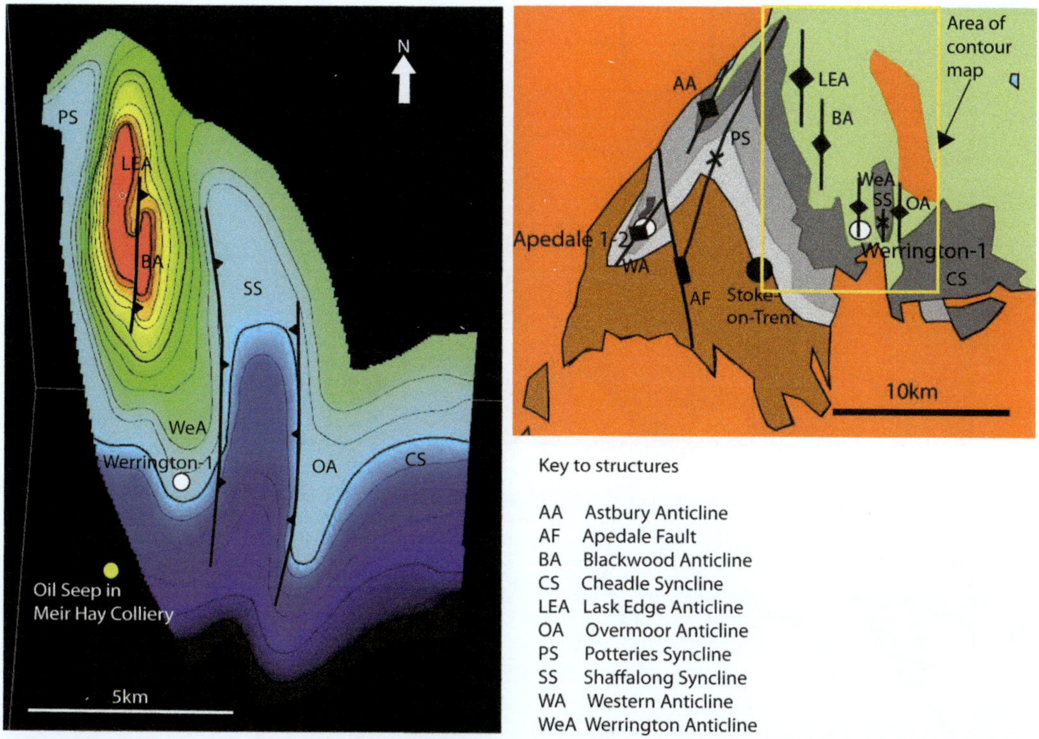

Fig. 8. Werrington-1 well. Contour map of the Lower Namurian Minn Sandstone on the eastern flank of the North Staffordshire Coalfield. Digitized from Corfield (1991).

in the Midlothian region predominantly consists of delta-top fluvial and lacustrine sediments with oil shales and coals. A marine incursion at the top of the Lower Carboniferous is represented by the Lower Limestone Group. The target for both wells in the Midland Valley was the West Lothian Oil Shale Group, a cyclic sequence up to 1120 m thick dominated by sandstones interbedded with siltstones and mudstones. Thin seams of oil shale occur throughout the sequence. Tuffs and ash are also present (Monaghan 2014).

Oil in the Midland Valley

In 1850, James Young moved to the Midland Valley of Scotland where he established an oil shale works at Bathgate, 20 km to the west of Edinburgh (Redwood 1897). By the 1860s there were 120 oil works established in the region, with production of 1 MMbbl/year by 1878. In addition to the oil shale production, there were abundant subsurface seeps of oil in mines, most notably at St Flannan's pit near Kilsyth which had a flow of 56 tons in the year 1905–06 (Giffard 1923). The most celebrated surface seep occurs at St Catherine's well to the south of Edinburgh (Underhill *et al.* 2008), located where the West Lothian Oil Shale Group is overthrust by the Silurian–Devonian rocks of the Pentland Hills fault block (Fig. 9).

West Calder-1

This well was drilled on the West Calder Anticline in the heart of the West Lothian oil shale industry. The 'Five Sisters Bings', a group of spoil tips from the oil shale industry, lay just to the north of the well location and the site of the Bathgate Oil Works, established in 1850 by James Young and his partners, lay 8 km to the NW. There were numerous open-cast and deep-mine workings in the area in addition to boreholes drilled for the oil shale industry. Therefore, the structure could be accurately defined and the well was located on the crest of the fold within the structural closure. The anticline is north–south-trending and is cut by east–west-trending extensional faults (Fig. 10). Uplift of the Silurian–Devonian rocks in the Pentland Hills to the east of the well has resulted in a westerly tilt to the West Calder Anticline (Fig. 9), reducing the area of the structural closure which is relatively

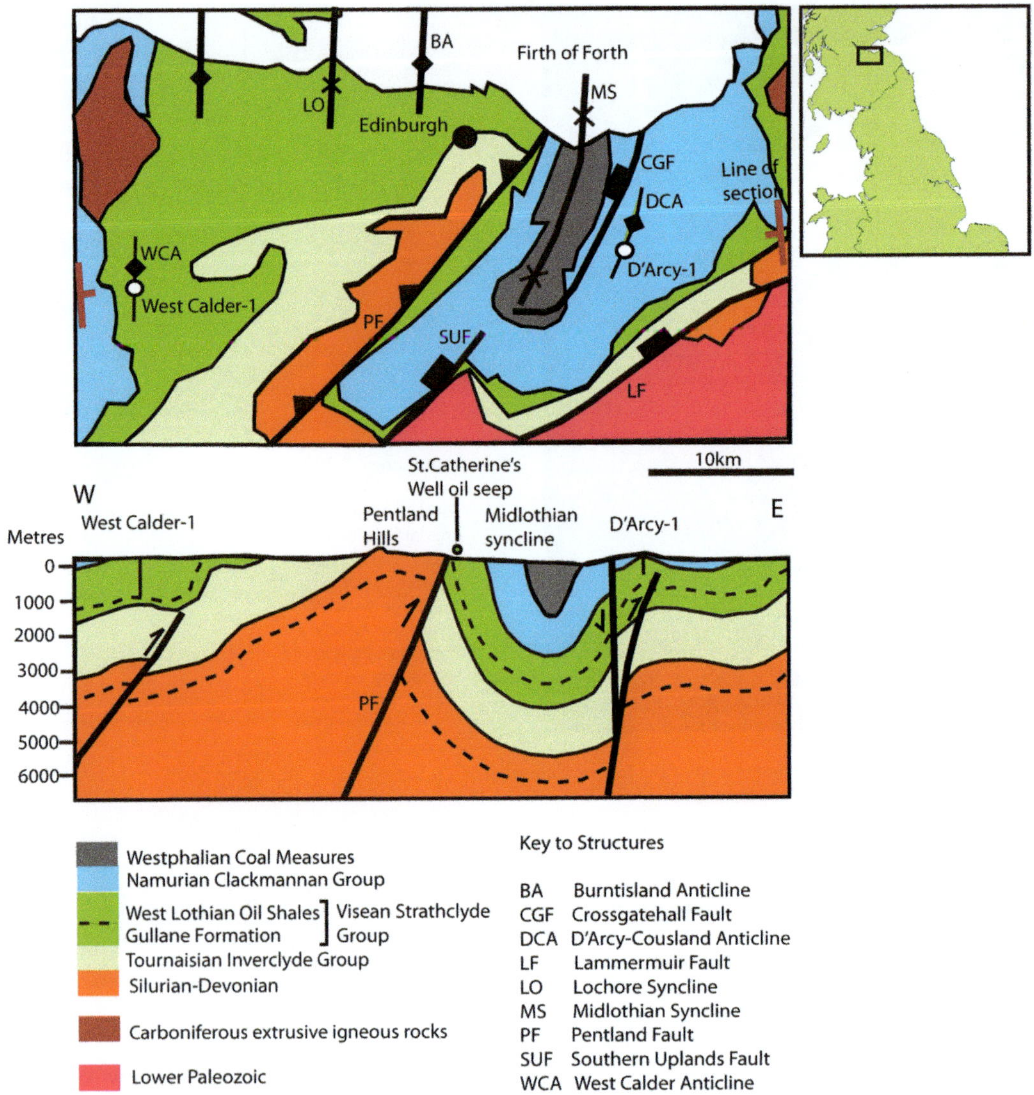

Fig. 9. Geological map and section of the Lothian region in the area around Edinburgh showing the location of the West Calder-1 and D'Arcy-1 wells. Modified from Monaghan (2014).

small (0.7 km^2). The closure is constrained to the east by a spill point across a shallow syncline (Fig. 10). The well was spudded below the level of the Broxburn Shale in the upper part of the West Lothian Oil Shale Group. The well reached total depth at the level of the Pumpherston Shale in the lower part of the Oil Shale Group. The predicted sandstones within the Oil Shale Group were largely absent and there was also a considerable thickness of volcanic ash at two levels in the well (Fig. 10). Officers of the Geological Survey were present at the wellsite and even with their assistance, difficulty was experienced in identifying key oil shale horizons which appeared to be poorly developed at this locality (Giffard 1923). There were shows and oil and gas throughout the section but no significant flows and the well was abandoned.

D'Arcy-1

This well was drilled on the southern culmination of the D'Arcy-Cousland Anticline and was spudded in the Lower Limestone Group (Fig. 11). Gas was encountered near the top of the West Lothian Oil

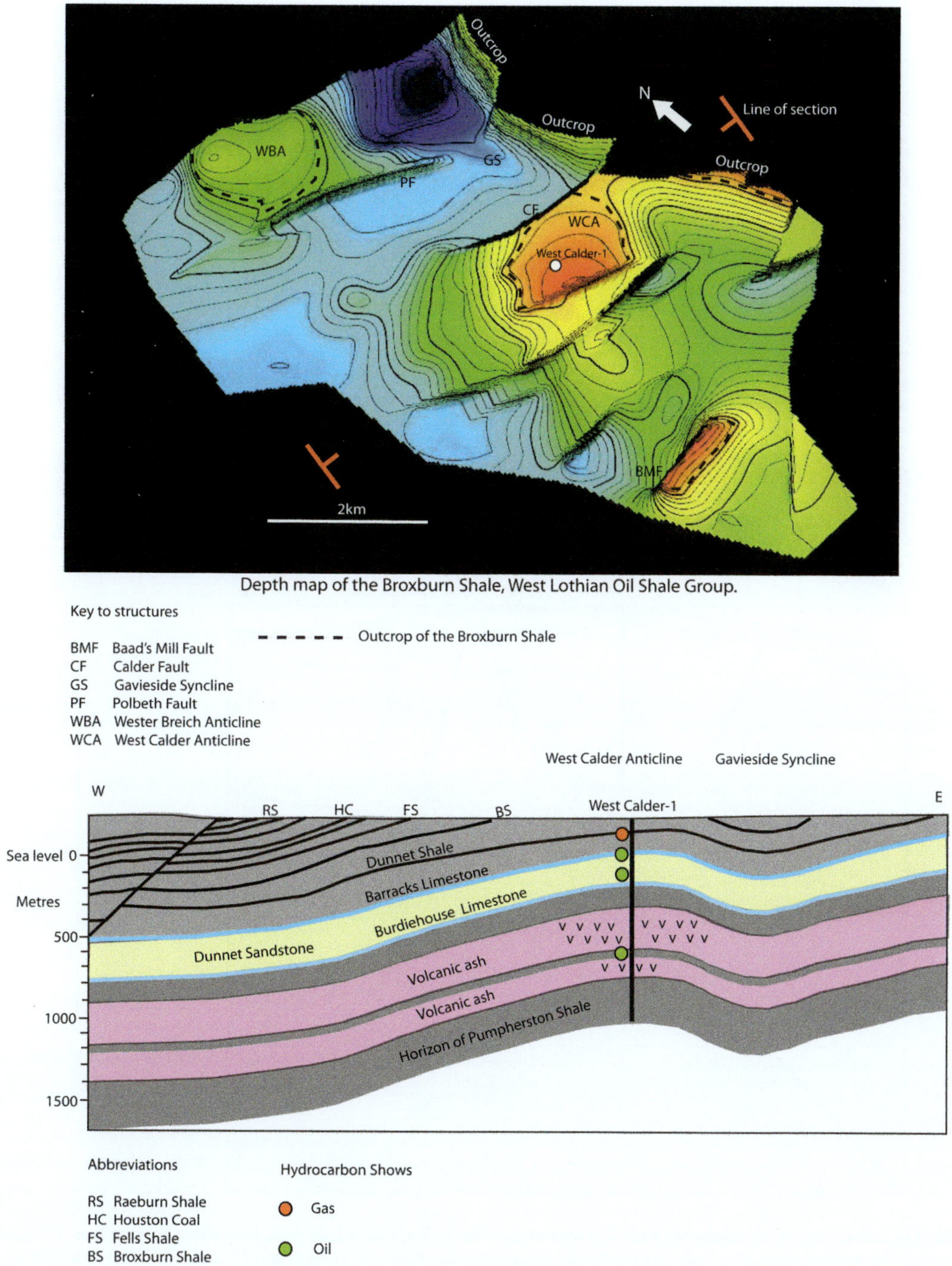

Fig. 10. The West Calder-1 well. Contour map of the top of the West Lothian Oil Shale Group in the region of the West Calder Anticline. Digitized from borehole data and contour maps in Richey (1942).

Shale Group that flowed at a rate of 300 000 cfg/day and was cased off (Giffard 1923). Oil was encountered in what is interpreted to be the Binny Sandstone at 1810 ft and the well was terminated a short distance below the discovery. The oil contained 12.36% paraffin wax. This factor, combined with

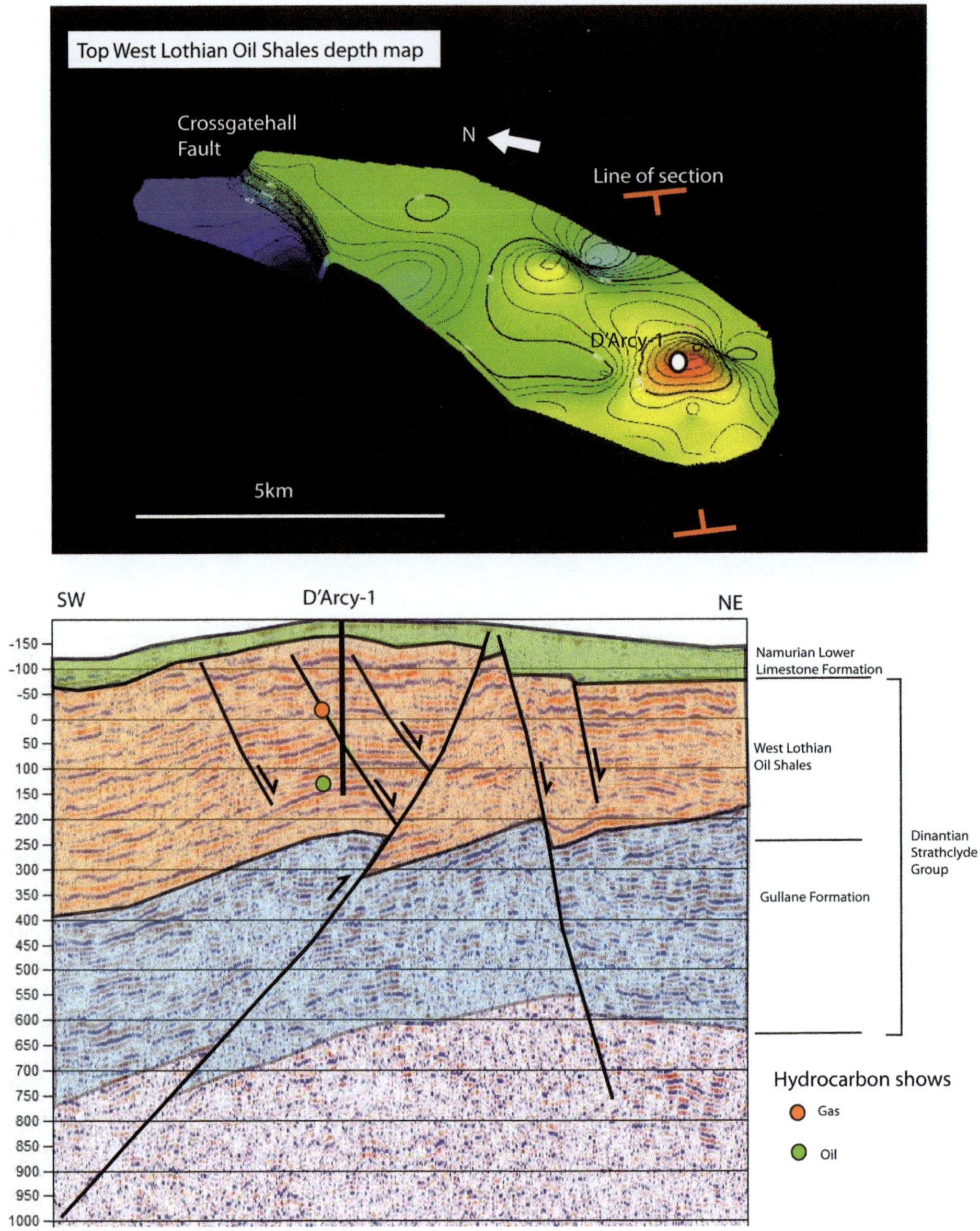

Fig. 11. The D'Arcy-1 oil discovery well. Contour map of the D'Arcy-Cousland Anticline with a seismic line through the well location. The map was contoured using well tops from exploration and production wells (data from the UK onshore Geophysical Library).

the shallow depth and low reservoir temperature, meant that the oil did not flow to the surface. A bailing test over a period of 2 months resulted in the production of 7 tons of oil (*c.* 50 bbl). The well was the last to be drilled in the campaign and was abandoned in September 1922.

End of the drilling campaign

During 1921, the decision was taken that the government would cease operations and any further drilling would be undertaken by private companies. The wells were plugged, with the exception of D'Arcy-1 (which was still being drilled) and Hardstoft-1 (which was producing oil). The method of plugging and abandonment was to seal off water-bearing horizons with oak plugs and cement, which were then weighted down with scrap-iron. The hole was then filled with broken rock and debris (Giffard 1923).

Post-mortem analysis

Reservoir and source rock

The discovery of the Eakring oil field in Nottinghamshire by the D'Arcy Oil Company (the exploration subsidiary of Anglo-Iranian, formerly Anglo-Persian) in 1939 was due to the use of geophysical methods to image the buried structures beneath the Permo-Triassic cover to the east of the exposed coalfield (Lees & Taitt 1945). The Eakring discovery well found oil in the Upper Namurian Rough Rock which flowed at a rate of 12 tons (*c.* 86 bbl) a day (Lees & Taitt 1945). Exploration in the East Midlands in the succeeding years established that the bulk of the oil was to be found in the Upper Namurian and Lower Westphalian delta-top sandstones, and that the source rock was the basinal mudstones of the late Dinantian–early Namurian Bowland Shale (Fraser *et al.* 1990). This should have been apparent to Pearson's geologists from the evidence available from outcrop and the subsurface – the Dinantian limestone at outcrop contained small amounts of heavily biodegraded bitumen and the Westphalian in the subsurface contained liquid oil in quantity. The pre-drilling predictions of the 1918–22 campaign that the Dinantian limestone was both reservoir and source were proved to be false. The oil in the limestone at Hardstoft was serendipitous, even though it was predicted by the team from Pearson. There are a small number of fields that have proved oil in the Dinantian limestone (Eakring, Plungar and Nocton) but in limited amounts. Nocton-2, drilled in Lincolnshire in 1943 by D'Arcy, produced 3 tons (*c.* 21.5 bbl) a day for a short period, which rapidly decreased to 0.5 tons (3.5 bbl) a day with an increasing proportion of water (Lees & Taitt 1945).

Traps

There were 11 wells drilled on nine structures during the 1918–22 campaign (two wells were drilled on each of the Ironville and Ridgeway anticlines). This study has revealed that only six of those structures constituted valid four-way dip-closed traps. The Ridgeway and Werrington anticlines form plunging noses, and it is probable that the Heath structure has the same geometry. Analysis of the size of the structures drilled reveals that the traps were small and therefore the government expectations of a useful indigenous oil supply for the Royal Navy was an impossible target. Based on the structure maps, the largest valid trap drilled was the Ironville Anticline (19 km^2) and the smallest was the West Calder Anticline (0.7 km^2). Subsequent drilling in the East Midlands has revealed that small trap sizes combined with thin reservoirs and low porosities and permeabilities severely limit the field volumes which are in the range of 1–2 MMbbl of recoverable oil (Fraser *et al.* 1990).

Re-drilling the original structures

The view of the government at the end of the drilling campaign was that the results did not justify further expenditure (Giffard 1923), and the general consensus of later authors is that the drilling campaign was not a success with only one producing well producing a few barrels a day (Kent 1985). Many of the wells flowed significant quantities of gas. The gas-producing zones were cased off as gas at that time had little commercial value due to the abundant supplies of gas derived from coal in the UK. However, the relative abundance of shows of oil and gas in many of the wells provided the impetus for companies to re-drill the structures in later years. In the Midland Valley, the northern culmination of the D'Arcy-Cousland Anticline was drilled by the D'Arcy Company in 1937. Three wells were drilled, one of which flowed gas (Lees & Taitt 1945). The Cousland gas field was subsequently developed by BP in the late 1950s and gas was piped to the nearby Musselburgh gas works until closure of the field in 1965 (DECC 2010). The southern culmination of the D'Arcy-Cousland Anticline was re-drilled in 1937 by Anglo-American (a subsidiary of Standard Oil). The well was located a few hundred metres south of the D'Arcy discovery well from 1918 to 1922, and it led to the later development of the Midlothian oil field which operated until the late 1960s (Underhill *et al.* 2008).

In Derbyshire, the Calow Anticline drilled by the Brimington-1 well in 1918–22 was re-drilled by BP in 1957 and was developed as a gas field with three wells. A structure contour map presented in Fraser *et al.* (1990) shows that the structure was underfilled and that Brimington-1 was drilled outside the accumulation. The Ironville Anticline was also re-drilled by BP during this period, with the Ironville-3 well, drilled in 1956, being completed as a gas producer in the Namurian Kinderscout Grit (DECC 2010).

In North Staffordshire, the anticlinorium to the east of the coalfield was revisited by Shell in 1982 and the four-way closure of the Lask Edge Anticline was the location of the Nooks Farm-1A discovery well, which tested gas from late Dinantian sandstones. The well was located approximately 10 km to the NNW of the Werrington-1 well drilled in 1918–22.

Summary

Even though the prevailing view of the first drilling campaign was that it was not a great success, the results of the later wells drilled on the structures selected for the 1918–22 campaign validates the skills of Pearson's team of geologists. The six valid traps drilled subsequently resulted in one more oil field being developed in addition to the Hardstoft discovery and three producing gas fields. The Nooks Farm discovery in North Staffordshire remains to be developed.

Acknowledgements Many thanks to Bernard Besly and an anonymous reviewer for comments that helped to improve the paper.

Funding This research received no specific grant from any funding agency in the public, commercial, or not-for-profit sectors.

References

Allen, P.A. & Allen, J.R. 2013. *Basin Analysis: Principles and Application to Petroleum Play Assessment*. Wiley-Blackwell, Chichester, UK.

Corfield, S.M. 1991. *The Upper Palaeozoic to Mesozoic structural evolution of the North Staffordshire coalfield and adjoining areas*. PhD thesis, University of Keele, Keele, UK.

Corfield, S.M., Gawthorpe, R.L., Gage, M., Fraser, A.J. & Besly, B.M. 1996. Inversion tectonics of the Variscan foreland of the British Isles. *Journal of the Geological Society, London*, **153**, 17–31, https://doi.org/10.1144/gsjgs.153.1.0017

Craig, J., Gluyas, J., Laing, C. & Schofield, P. 2013. Hardstoft – Britain's first oil field. *Oil-Industry History*, **14**, 97–116.

DECC 2010. *The Hydrocarbon Prospectivity of Britain's Onshore Basins*. Department of Energy and Climate Change, London.

Eden, R.A., Stevenson, I.P. & Edwards, W. 1957. *Geology of the Country around Sheffield*. Memoir of the Geological Survey of Great Britain, Sheet 100.

Falcon, N.L. & Kent, P.E. 1960. *Geological Results of Petroleum Exploration in Britain 1945–1957*. Geological Society, London, Memoirs, **2**, https://doi.org/10.1144/GSL.MEM.1960.002.01.01

Fraser, A.J. & Gawthorpe, R.L. 1990. Tectono-stratigraphic development and hydrocarbon habitat of the Carboniferous in northern England. *In*: Hardman, R.F.P. & Brooks, J. (eds) *Tectonic Events Responsible for Britain's Oil and Gas Reserves*. Geological Society, London, Special Publications, **55**, 49–86, https://doi.org/10.1144/GSL.SP.1990.055.01.03

Fraser, A.J., Nash, D.F., Steele, R.P. & Ebdon, C.C. 1990. A regional assessment of the intra-Carboniferous play of Northern England. *In*: Brooks, J. (ed.) *Classic Petroleum Provinces*. Geological Society, London, Special Publications, **50**, 417–440, https://doi.org/10.1144/GSL.SP.1990.050.01.26

Giffard, H.P.W. 1923. The recent search for oil in Great Britain. *Transactions of the Institute of Mining Engineers*, **65**, 222–250.

Giffard, H.P.W. 1938. The former Cannel oil industry in North Wales and Staffordshire. *In*: *Oil Shale & Cannel Coal, Proceedings of a Conference held in Scotland*, June 1938. Institute of Petroleum, London.

Kent, P.E. 1954. Oil occurrences in coal measures of England. *AAPG Bulletin*, **38**, 1699–1713.

Kent, P.E. 1985. UK onshore oil exploration. *Marine and Petroleum Geology*, **2**, 56–64.

Lees, G.M. & Cox, P.T. 1937. The geological basis for the present search for oil in Great Britain by the D'Arcy Exploration Company, Ltd. *Quarterly Journal of the Geological Society*, **93**, 156–194, https://doi.org/10.1144/GSL.JGS.1937.093.01-04.09

Lees, G.M. & Taitt, A.H. 1945. The geological results of the search for oilfields in Great Britain. *Quarterly Journal of the Geological Society*, **101**, 255–317, https://doi.org/10.1144/GSL.JGS.1945.101.01-04.12

Monaghan, A.A. 2014. *The Carboniferous Shales of the Midland Valley of Scotland: Geology and Resource Estimation*. British Geological Survey for the Department of Energy and Climate Change (DECC), London.

Redwood, I.L. 1897. *A Practical Treatise on Mineral Oils and Their By-Products*. E. & F.N. Spon, London.

Richey, J.E. 1942. *The Oil Shales of the Lothians – Structure*. Geological Survey of Great Britain, Wartime Pamphlet, **27**. Geological Survey and Museum, London.

Schmidt, R.A., Boade, R.R. & Bass, R.C. 1981. A new perspective on well shooting-behaviour of deeply buried explosions and deflagrations. *Journal of Petroleum Technology*, **33**, 1305–1311.

Selley, R.C. 1992. Petroleum seepages and impregnations in Great Britain. *Marine and Petroleum Geology*, **9**, 226–244.

Torrens, H. 1994. 300 years of oil: mirrored by developments in the West Midlands. *In*: Curtis, C.D. & Gelling, G. (eds) *The British Association Lectures 1993*. Geological Society, London, 4–8.

Underhill, J.R., Monaghan, A.A. & Browne, M.A.E. 2008. Controls on structural styles, basin development and petroleum prospectivity in the Midland Valley of Scotland. *Marine and Petroleum Geology*, **25**, 1000–1022.

The Scottish oil-shale industry from the viewpoint of the modern-day shale-gas industry

GRAHAM DEAN

ReachCSG, 6 Rubislaw Terrace, Aberdeen AB10 1XE, UK
graham.dean@reachcsg.co.uk

Abstract: Oil production in West Lothian in Scotland started in 1851. It was not the first place to produce oil from shale but it was the largest and most successful. This oil production led to major developments in oil refining and, as a result, a large demand for oil products was created. This new demand for oil products stimulated the search for oil around the world and resulted in the first modern-day oil well being drilled in Titusville Pennsylvania in 1859.

Over the next 100 years, an estimated 75 MMbbl (million barrels) of oil and 500 Bcf (billion cubic ft) of gas were produced from the shales of West Lothian. Initially, the oil was produced from a thin layer of shale at Torbanehill near Bathgate. Later, the much thicker deposits in the Dinantian West Lothian Oil-shale Formation (WLO) were used to produce the oil, although there was also oil produced from other shales mostly in the Lower Coal Measures and Limestone Coal Formation.

This paper looks at the Scottish oil-shale industry from the viewpoint of the modern-day shale-gas industry and highlights the contribution the industry made as the forerunner of the oil industry. The paper attempts to explain why the Scottish oil-shale industry was so successful and influential. Reasons why the oil-shale industry did not develop into modern oil-well production are also put forward.

The Scottish shale-oil industry was successful not only thanks to James Young and his colleagues but also because of the geology of the Central Belt of Scotland and, in particular, the geochemical properties of the WLO.

The shale-oil industry, natural oil seepage, free oil encountered during mining and the historical exploration drilling demonstrate a rich functioning source rock suggesting significant prospectivity for future exploration for both oil fields and shale gas.

Geographical and geological setting

The Midland Valley of Scotland is a fault-bounded, WSW–ENE-trending Late Paleozoic sedimentary basin, bounded by the Caledonide Highland Boundary Fault to the north and the Southern Upland Fault to the south. Carboniferous sedimentary basins and Carboniferous volcanic rocks overlie Lower Paleozoic strata. These rocks form a succession up to locally more than 18 000 ft (5500 m) thick.

The West Lothian Oil-shale Formation (WLO), part of the Strathclyde Group, crops out over much of West Lothian and it was here that much of the oil-shale activity took place. The WLO is described by Cadell (1925), Parnell (1986) and Raymond (1991). Figure 1a shows the geological outcrop map of the Midland Valley of Scotland. Other shales were also used both in West Lothian and elsewhere in the Central Belt of Scotland. The shales mined are highlighted in Figure 2a & b.

For shale gas, prospective Carboniferous shales are buried beneath the Central Belt, the area from Glasgow to Edinburgh, to the Lothians, Falkirk, Clackmannan and Fife. The areas prospective for shale gas as identified in the BGS DECC 2014 report (Monaghan 2014) are shown in Figure 1c. This report estimates the shale oil in-place resource to be 3.2–6.0–11.2 Bbbl (billion barrels) (421–793–1497 Mt (million tonnes)) and the in-place shale-gas resource 49.4–80.3–134.6 Tcf (trillion cubic ft) (1.40–2.27–3.81 Tcm (trillion cubic metres)). The WLO makes the largest contribution to this estimated resource. The Scottish Government has commissioned and published several studies including: a BGS study on faulting and earthquakes (Baptie *et al.* 2016); a study on the economics of shale gas (KPMG 2013); and a report by the Scottish Government Independent Expert Panel on Unconventional Gas (Masters 2014).

The presence of seeps, and oil and gas drilling confirm the presence of mature source rock. The locations of historical oil and gas drilling (Fig. 1a) have mostly targeted sandstones within the WLO on anticlinal structures: Salsburgh, D'Arcy-Cousland, Pumpherston and West Calder. The Bargeddie 1 well targeted a tilted fault block. The boreholes at Airth are drilled into the Bannockburn Coal located in the Limestone Coal Group above the Black Metals (Fig. 2b).

The shales of the Midland Valley of Scotland used in the oil-shale industry

Torbanite or boghead coal

Torbanite is an organic-rich oil shale of lacustrine origin. The organic matter in torbanite is derived

From: Craig, J., Gerali, F., MacAulay, F. & Sorkhabi, R. (eds) 2018. *History of the European Oil and Gas Industry*. Geological Society, London, Special Publications, **465**, 53–69.
First published online May 8, 2018, https://doi.org/10.1144/SP465.13

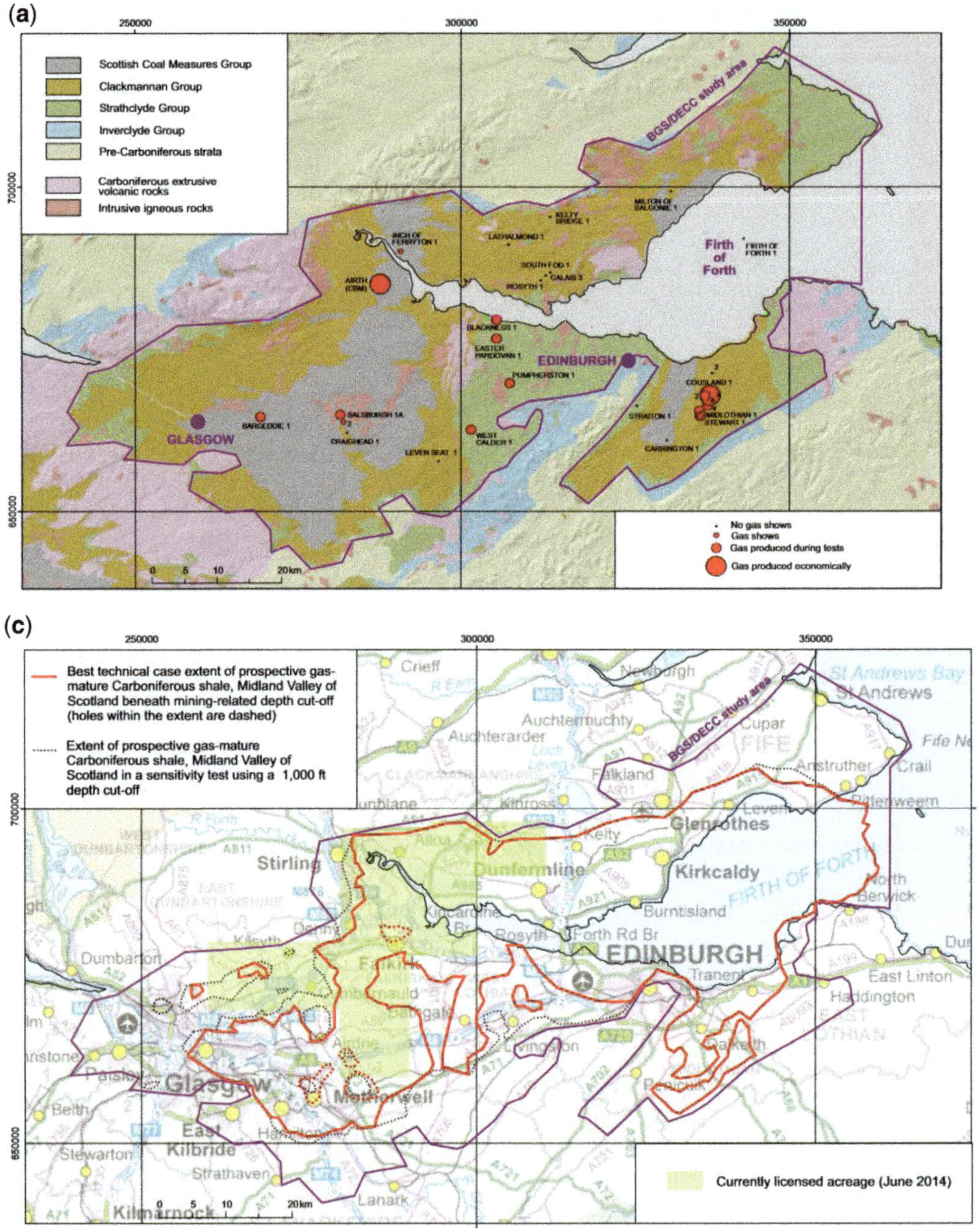

Fig. 1. (**a**) Outcrop geology map showing the location of historical wells with gas shows, tests and production (Monaghan 2014). The historical shale-oil industry was centred in West Lothian (the area west of Edinburgh) where the Strathclyde Group crops out. (**c**) Areas considered prospective for shale gas (Monaghan 2014). The shale-gas and shale-oil potential is mostly in the Strathclyde Group and at depths where it is geochemically mature for shale-gas or shale-oil production. All figure-parts contain public sector information licensed under the Open Government Licence v3.0.

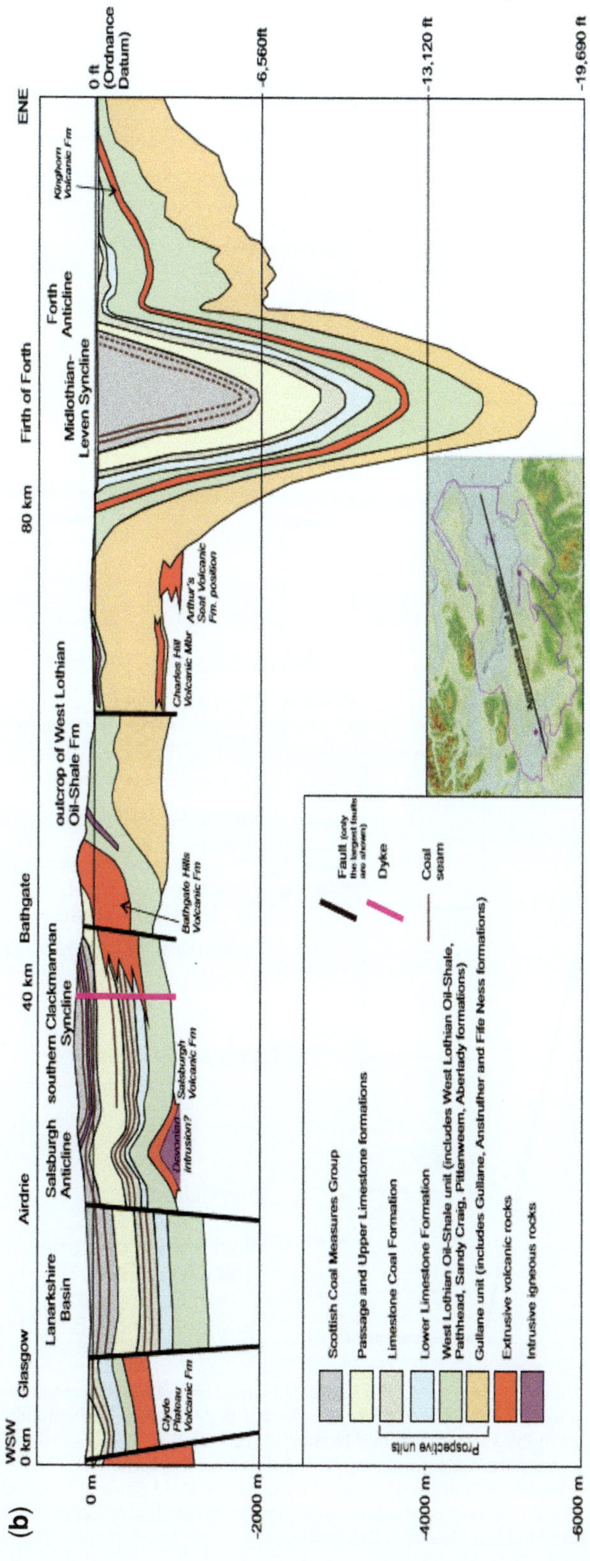

Fig. 1. (*Continued*) **(b)** Schematic cross-section across the Midland Valley of Scotland (Monaghan 2014).

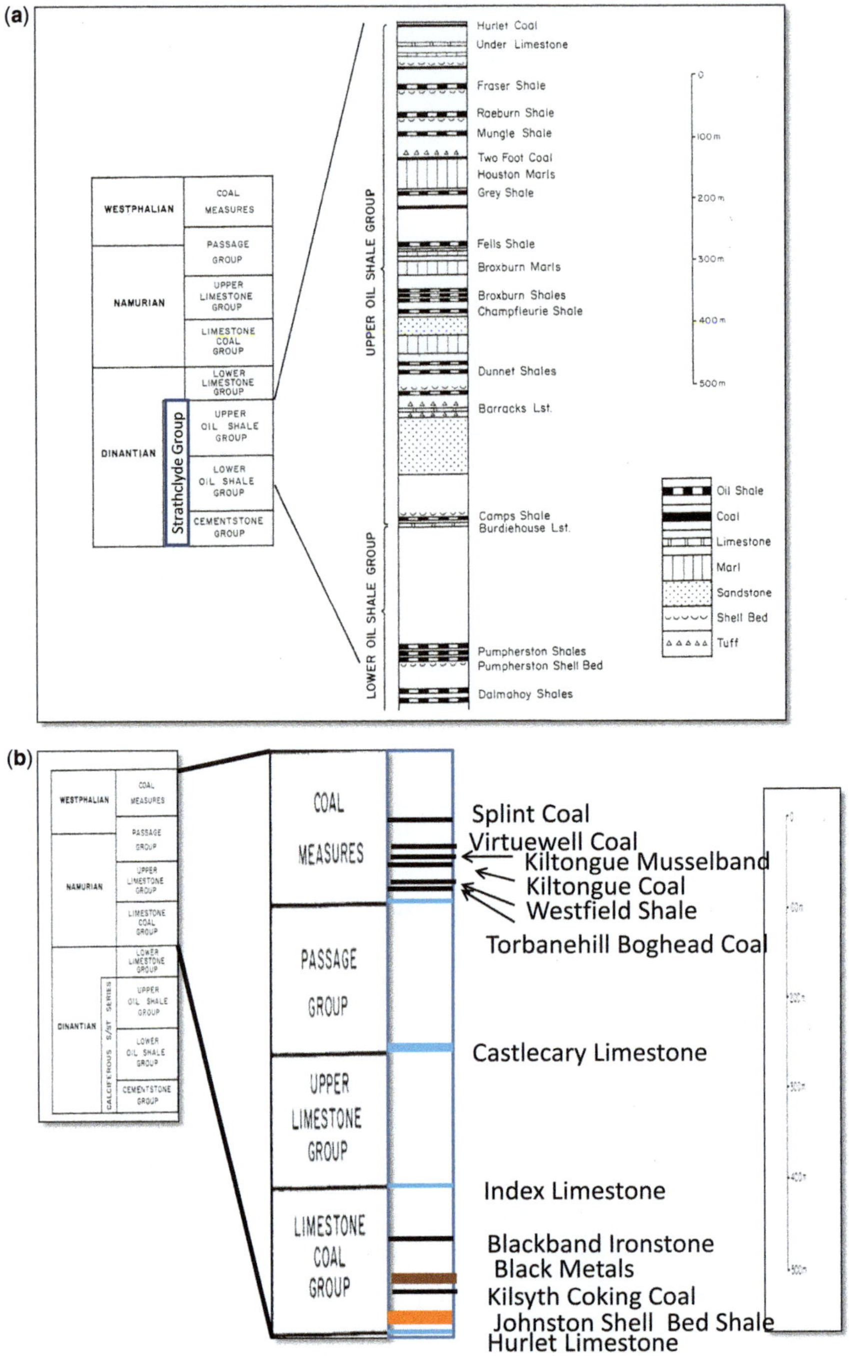

Fig. 2. (**a**) The West Lothian Oil-shale Formation. The intervals marked as 'oil shale' were all mined. The major shales in the nineteenth century were the Broxburn, Fells and Dunnet shales. Later, the Pumpherston shales were extensively exploited. (**b**) Shales in the Westphalian and Namurian were used in oil works (from Redwood 1897, p. 39): the Splint Parrot Coal at Provanhall, Coatbridge; a thin shale above the Virtuewell Coal at Ardon, Airdrie; Kiltongue Musselband shale at Airdrie; Kiltongue gas coal at Airdrie; Westfield Shale at Westfield, Fife; Torbanehill Boghead Coal near Bathgate gave the first oil production; shale above the Castlecary Limestone at Levenseat; Johnstone Shell Bed Shale at Linwood and Houston, Renfrewshire.

from lipid-rich microscopic plant remains similar in appearance to the freshwater colonial green alga *Botryococcus braunii*. Torbanite typically comprises 88% carbon and 11% hydrogen.

Cannel coal or cannel shale

Cannel coal is an organic-rich oil shale similar to torbanite but of terrestrial origin. Like torbanite, cannel coal has a high hydrogen content. It comes from resins, spores, waxes, and cutinaceous and corky materials of terrestrial vascular plants. Cannel coal was accumulated in ponds and shallow lakes in peat-forming swamps and bogs of the Carboniferous age under oxygen-deficient conditions.

The term 'cannel coal' is more widely used than 'cannel shale'. Cannel has the appearance more like coal than shale, and historically and industrially was treated as a coal with a high calorific value. Cannel is classed as a sapropelic coal. Cannel, like oil shale, is a fine-grained sedimentary rock rich in hydrocarbons, whereas in humic coal the organic matter is largely type III and type IV kerogen. In this paper 'cannel shale' is used to emphasize that the exploitation of cannel shale was part of the shale-oil industry. The early patents of James Young referred to coal and so Young's activities were restricted initially to cannel coal. Attempts to circumvent Young's patents led to the first use of other shales that were not called 'coal'.

West Lothian Oil-shale Formation

In the Central Belt, organic-rich mudstones are stacked in sandstone–limestone–shale intervals up to 9800 ft (3000 m) thick. The WLO lies within this sequence and is a Dinantian (Asbian–Brigantian) siliclastic succession up to 1120 m thick (Jones 2005). The base of the formation is the Humble Shell Bed and the top is the Hurlet Limestone. The WLO accumulated in a series of north–south-trending, tectonically and volcanically active basins. There is a complex interaction between freshwater lacustrine, transitional lagoonal and fully marine conditions. At times, these basins were cut-off from the sea creating standing bodies of water named Lake Cadell after the geologist Henry Cadell (1860–1934) noted for his work on the oil shales. Organic-rich mudstones and oil shales were deposited in these ephemeral lakes with anaerobic bottom conditions. Lacustrine facies dominate in the west. In the east, the facies are fluvio-deltaic and lacustrine, with the marine influence varying through time. There is syndepositional volcanism and post-depositional intrusive magmatism which may have contributed to the relatively high maturity-with-depth gradient (Fig. 3). The intrusions also caused localized increases in the maturity of shales and coals.

Geochemistry of the West Lothian Oil-shale Formation

The WLO is geochemically under-mature at the near-surface. So, to produce the oil, the oil-shale industry required mined shale to be heated in retorts in anaerobic conditions to complete the catagenesis of the oil. This process was called 'destructive distillation'. The retorting process distilled the hydrocarbon into its constituent factions. These hydrocarbon products (Fig. 4) were more useful and so more valuable than unrefined oil.

In contrast to the historical oil-shale industry, shale-gas and shale-oil drilling targets mature shales that form the source rocks for the oil and gas in oil and gas fields.

The WLO has moderate–high total organic content (TOC) throughout the sequence (Fig. 5) but only about 3% of the succession has the very high kerogen content required for the shale-oil mining. There were 11 horizons mined for oil shale with thicknesses up to 5 m.

Figure 6 shows the Rock-Eval analysis by Parnell (1988) of outcrop samples of the WLO. The organic content is interpreted to be mostly algae, fungal/bacterial organic matter preserved in a meromictic lake deficient in oxygen at depth. This organic content is often classed as Type I kerogen, a kerogen type more oil-prone than Type II or Type III kerogen. The large shale-gas and shale-oil plays in the USA are usually referred to as Type II shale, as are the Bowland Shale and the Kimmeridge Shale in the UK.

Figure 5 is a Rock-Eval analysis performed on drill cuttings from the Salsburgh-2 well near Airdrie. These measurements were done as part of the studies undertaken for the hydrocarbon licence PEDL162. There is a sharp change in the oil production index (PI) below 3000 ft. This rapid change with depth may be because the kerogen converts over a narrow temperature interval. This is a characteristic of Type I kerogen, which has a low diversity of species forming the organic matter. This relatively precise transformation of the kerogen may have resulted in more efficient retorting of the shale when compared to Type II kerogen shales.

Other studies included vitrinite reflectance (Ro) measurements on drill cuttings from the Salsburgh 2 well (Fig. 3). When the data are extrapolated to surface, the shale has a Ro value of close to 0.6–0.8, indicating approximately 6000 ft of inversion. A vitrinite reflectance of 0.6–0.8 is marginally mature for oil so, when compared with less mature shales, the shale may have required less heat, and so less cost, to be turned into oil. The WLO mines reached considerable depths. The mining at Pumpherston reached depths of over 1000 ft. At this depth, the geochemical maturity of the shales would have been higher than outcropping shale. So

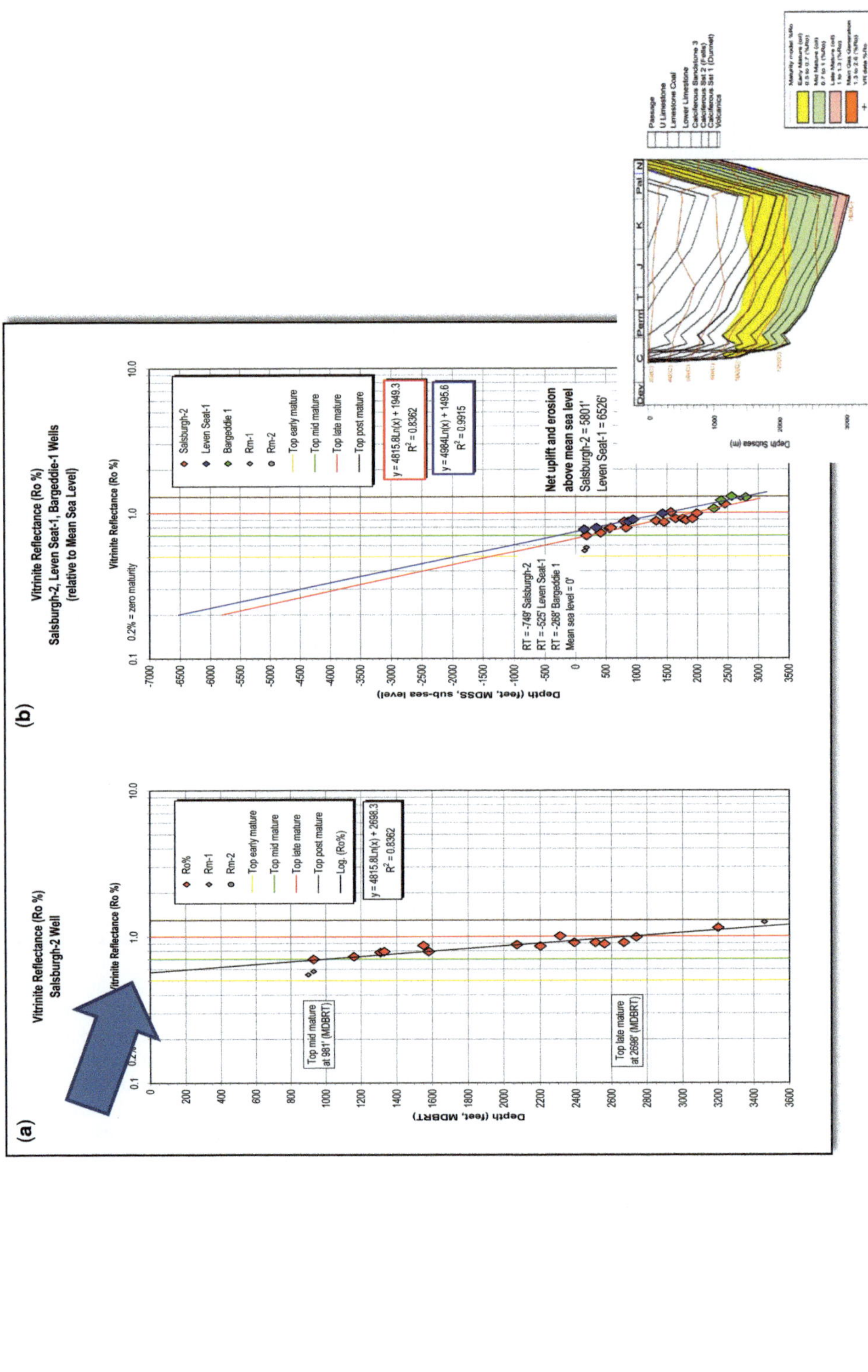

Fig. 3. (**a**) Vitrinite reflectance (Ro) measurements from drill cuttings from Salsburgh 2. (**b**) Vitrinite reflectance measurements from Salsburgh 2, Levenseat 1 and Bargeddie 1. When the data are extrapolated to surface, the Ro value is 0.6–0.8, implying about 6000 ft of inversion (from Reach 2013). MDBRT, measured depth below rotary table.

One ton of shale produced about one barrel of oil which in 1919 was worth 50p/barrel (£25/barrel 2016 money). One barrel of oil was refined into £12 of hydrocarbon products (£600 in 2016 money). The cost of mining the shale was about 30p/ton.

1. Motor spirit of most excellent quality
2. Naphtha for dry cleaning and india rubber manufacture
3. Lamp oils with a seventy-five year record of safety
4. Power oil for the farm tractor and fishing boat
5. Lighthouse and long burning oil for the important duties of lighthouse and railway signal lamps
6. Fuel oil for furnaces and diesel engines
7. Gas oil for the manufacture and enrichment of gas
8. Batching oil for the spinner, the weaver and rope maker
9. Cleaning and lubrication oils for machinery
10. Paraffin wax for candles, tapers and matches, for water proofing and electrical installation
11. Candles of every kind
12. Sulphate of ammonia to manure our fields in peace time and provide explosives in war
13. Paraffin coke, the smokeless fuel

Fig. 4. By 1919 a large number of products were produced.

more frequent occurrences of free oil would be expected and were reported in the deeper mines. The increased maturity with depth is a likely cause of the reported reduction in the oil yield of the shale with depth (Carruthers *et al.* 1927, p. 176).

Seeps in Scotland

Oil has been used and prized since Biblical times when oil was obtained from oil seeps. These oil seeps and gas seeps were often associated with religious sites. Often oil seeps are named after Saint Catherine because, in the Middle Ages, oil was sold by the monks of Saint Catherine's Monastery in Sinai. One such oil seep in Liberton in Edinburgh is called Saint Catherine's well or the Balm well or Saint Margaret's well. The oil from this well was used as a treatment for skin rashes of which there were many in medieval Edinburgh. The oil seeps from the West Lothian Oil-shale deposits close to the Pentland Fault (Fig. 7) (Underhill *et al.* 2008). This fault may be a migration path for the oil originating in deeper, more mature WLO source rocks.

Gas seeps are also common in the Scottish Central Belt, in particular around the Firth of Forth (Appleton *et al.* 1995; Judd *et al.* 1997; Andrews 1988). It has been speculated that St Mungo's sixth century ever-burning tree that forms the Glasgow coat of arms was the result of a gas seep on fire near Culross (Fig. 8). Gilfillan (2016) sampled a gas seep close to Culross and using noble gas fingerprinting showed that the gas was of thermogenic origin.

The oil-shale industry

At the Great Exhibition in London in 1851 (Reports by the Juries 1851) there was great interest in potential sources of supplies of paraffin that could be used for lighting. The French and German sections displayed candles made from paraffin distilled from bituminous shale, and there was also a display of paraffin made from the Kimmeridge shale in Dorset where mining had started in 1849. The jury reported that:

> [E]xamples of paraffin wax are exhibited in the French section which were made from paraffin obtained by distilling schist but far more interesting specimens are sent by James Young, which seem to realise the great problem [of producing clean-burning candles] … the process by which Young obtained paraffin and oils from coal is one of great importance.

James Young was awarded patent No. 13292 in 1850, 'For treating bituminous coal to obtain paraffine and oil containing paraffine therefrom'.

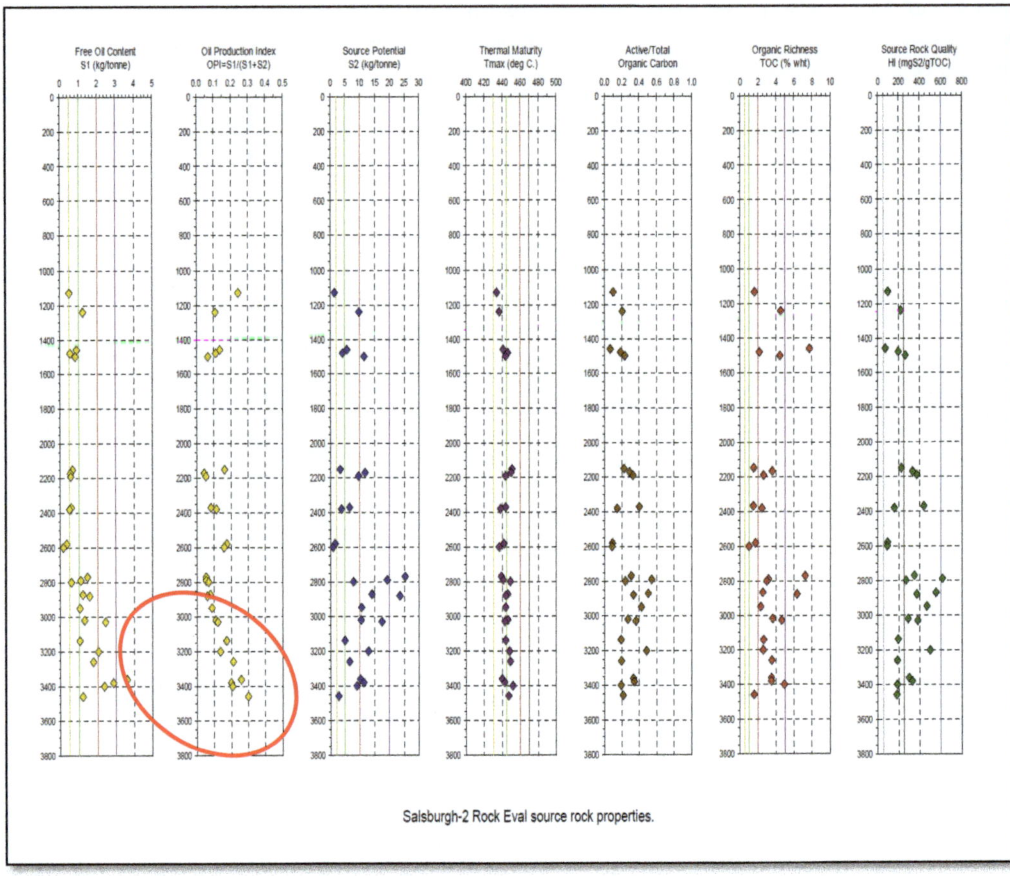

Fig. 5. Rock-Eval analysis of drilling cuttings taken from the Salsburgh 2 well. Note the sharp change in the oil production index (PI) below 3000 ft, circled. Note also the high TOC throughout column 6 (from Reach 2013).

The Scottish shale-oil industry was not the first to produce oil from shale but was the most successful. It was the industry that created much of the demand for oil, the ever-growing hunger for oil that continues to grow today. The first attempt at commercial shale-oil production was in 1837 in Autun, Alsace, France. In the 1840s, attempts were made to commercially produce paraffin wax from peat in Ireland. As early as 1781, Archibald Cochrane, 9th Earl Dundonald, patented a process to retort coals to produce coke, tar, oil and naptha. He is said to have thought of the idea after visiting the pitch lakes in Trinidad when he was a British naval officer. The British Tar Company started production using this process in Culross, Fife and Muirkirk, Ayrshire in Scotland. The coal for the Muirkirk works was supplied by the Kames colliery owned by John Loudon McAdam, who also managed the tar production. McAdam was an engineer and road builder, most famous for his invention of macadam.

The French and German shale industries suffered from military and political insecurity compared to the UK at the time (Thomas 2014). The works in Dorset were located in Wareham but the shale was mined from the cliff-face about 1.5 km east of Kimmeridge Bay at Clavell's Hard (map reference [SY920777]) (Legg 1984; West 2017). From there, the shale was shipped around the Isle of Purbeck to Wareham. The Blackstone Band within the Kimmeridge Shale was worked for many years and Wareham was lit from gas from the works.

Although much modified, the technology used for producing shale oil was taken from the coal-gas industry. The coal-gas industry owes much to William Murdoch who, in 1777, walked from Lugar, near Muirkirk, in Ayrshire to Birmingham to join the Boulton & Watt factory. While he was employed by Boulton & Watt, Murdoch developed the generation of gas from coal and used it to light a factory in Old Cumnock in Ayrshire and also the Boulton & Watt Soho works in 1802.

Stratig.	TOC (%WT)	S1 (mg/g)	S2 (mg/g)	HI	PI
Lower OSG	21.1		44.06	209	
Dalmahoy OS	19.9		117.41	590	
Pumperston OS	9.88	1.17	41.42	419	0.03
Pumperston OS	28.52	1.72	177.71	623	0.01
Camps OS	4.25	1.65	14.64	344	0.1
Dunnet OS	19.59	1.34	124.36	635	0.01
Champfleurie OS	16.3		51.96	319	
Broxburn OS	14.9	0.28	76.43	513	0
Wardie Shales	13.78	0.68	16.69	121	0.04
Under Lst	22.67	2.7	98.36	434	0.03
Under Lst	1.5				
Granton Sst	1.67	0.46	2.56	153	0.15
Granton Sst	3.21	0.39	1.91	59	0.17
Pumperston SB	4.74		3.9	82	
Lower OSG	11.6		30.64	264	
Burdiehouse Lst	1.94	0.6	5.66	292	0.1
Burdiehouse Lst	1.71	0.27	1.81	106	0.13
Burdiehouse Lst	8.1	0.68	1.83	23	0.27

"OS" – Oil-shale

Fig. 6. Rock-Eval measurements performed by Parnell (1988) on outcrop samples. Note the high hydrogen index (HI) and the high TOC. The TOC is high enough on some WLO outcrops that the shale can burn (right). The outcrop sample oil shale is from Linhouse Water, Calder Wood, Mid Calder.

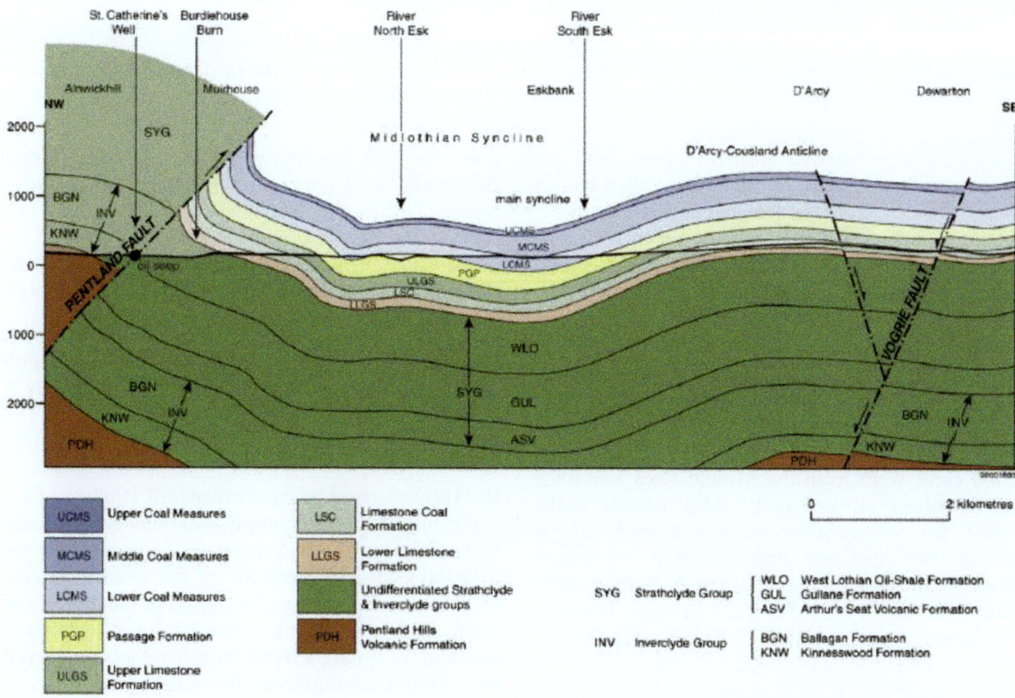

Fig. 7. Cross-section showing the relationship between the St Catherine's oil seep and the D'Arcy-Cousland Anticline (from Underhill *et al.* 2008).

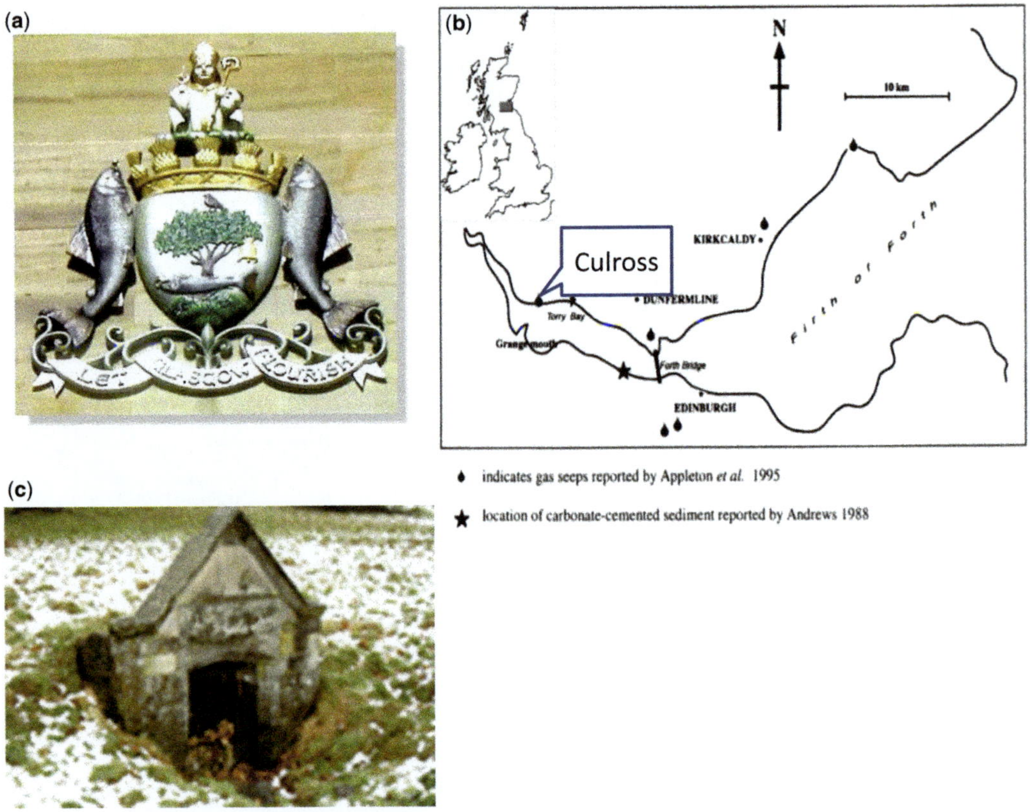

Fig. 8. (**a**) The Glasgow coat of arms showing St Mungo's ever-burning tree at Culross. (**b**) The location of gas seeps near the Firth of Forth (from Judd *et al.* 2002). (**c**) St Catherine's well at Liberton, Edinburgh.

In the early years, the gasworks producing gas from coal relied heavily on oil shales called cannel coals, boghead coals and torbanite. These shales were more extensively mined in the Scottish Central Belt than in England at the time. After the invention of the incandescent mantle, supplies of cannel became less important for gasworks.

It may not have been a coincidence that William Murdoch and Archibald Cochrane both had connections to the Muirkirk area and both were involved in manufacturing gas and tar. Their interest in the subject may have been partly because at Muirkirk the mined coals were from the Middle Coal Measures and contained an unusually thick cannel shale, which: 'was in one place about 6 ft. thick, and is said to have consisted almost entirely of cannel. This seam was at one time treated at the Pathhead oil-works' (Gibson 1922). Muirkirk, Lugar, Patthead and Cumnock are all within 15 km of each other.

The Scottish oil-shale industry

James Young (see Appendix A) started oil production at Bathgate in 1851 having been granted his patent in 1850. The Bathgate works used Torbanehill Boghead Coal, an organic-rich oil shale of lacustrine origin. Torbanehill Boghead Coal was a very thin shale in the Lower Coal Measures (Fig. 2b).

In 1862, Robert Bell started to produce oil by retorting shale at Broxburn. Redwood (1897) reports that Bell said:

> [T]ook a lease of the Broxburn minerals and shortly after, while searching for coal, came across large quantities of shale. When [he] saw the manner in which the shale burned, [he] determined to experiment on it and had experimental retorts erected.

By 1860 the cost of the Torbanehill Boghead Coal had doubled and the supply was rapidly depleting. Young requested advice from the Geological Survey in Edinburgh and chose to use the shale at Addiewell. His Addiewell works started producing in late 1865. The cornerstone of the works at Addiewell was laid by Young's friend from Strathclyde University, David Livingstone, the African explorer. In 1866, Young formed the Young's Paraffin Light and Mineral Oil Company Limited. This company produced about one-third of the Scottish oil production. During the 1860s, there were over 60 oil works in operation across the Scottish Central Belt.

The Scottish Oils Agency was formed in 1918 to market products from the shale-oil companies and in 1919 they were reorganized to form Scottish Oils Ltd, a subsidiary of the largely state-owned Anglo-Persian Oil Company which later became the British Petroleum Company. The managing director of Scottish Oils at the time became the Chairman of BP. He was William Fraser, later Lord Strathalmond, and was yet another alumnus of Strathclyde University (Cook 1971).

Personnel from the Scottish oil-shale industry developed shale oil in far-flung places such as South Africa, New South Wales and South Wales. The refining expertise was put widely to use, enabling commercial oil production in Burma and the creation of the Glasgow company called Burmah Oil.

The Scottish shale-oil industry survived by producing more products (Fig. 4) and staying at the forefront of refining technology: one of the most important by-products was ammonia and ammonia sulphate. The ammonia was used to produce nitrogen-based fertilizer. In 1865, Robert Bell started to produce ammonia sulphate from his shale-oil works in Pumpherston. Robert Bell claimed that this product came about because he was prosecuted by the Rivers Pollution Commissioners for discharging the retort water into the local stream. So, he turned the water onto a field instead. When the water produced rapid growth of grass, he experimented on the waste-water, resulting in the marketing of ammonium sulphate (Redwood 1897, p. 169). Robert Bell was not the first to manufacture ammonia; ammonia had long been produced at the Glasgow gasworks. Manufacture of sal ammoniac in Scotland was started by James Hutton 'the Father of Modern Geology' in 1756, using soot supplied by Edinburgh chimney sweeps.

Shale mining stopped in 1962 largely as a result of the EFTA agreement which prohibited different duty rates for home-produced oil. The chemical works at Pumpherston continued in operation but eventually all refining was moved to nearby Grangemouth.

Oil and gas produced

The estimated oil production from the oil-shale industry in Scotland is 75 MMbbl (million barrels). The amount of gas produced can only be approximately estimated. A reasonable, but very approximate, estimate is 500 Bcf (billion cubic ft) based on the information outlined below.

Gibson reports that, 'during the Great War experiments on cannels as a source for oil were carried out under the Ministry of Munitions at Dundee Gasworks'. 15 357 tons of cannel produced 10 075 bbl (1603 m^3) of oil and 0.24 Bcf (6.7 Bcm) of gas. This is a gas–oil ratio (GOR) of 23.5 Mcf/bbl (where Mcf is thousand cubic ft). If this GOR is applied to the 75 MMbbl of oil, then the gas associated with this amount of oil production would have been 1762 Bcf (50 Bcm).

Redwood (1897, p. 106) gives the gas yields per ton of shale for different retorts. The yields were 3000 cf/ton for a Henderson retort and 7000–10 000 cf/ton for the Young and Beilby retorts. Assuming a ton of shale yielded, on average, 1 bbl of oil, then the gas production associated with 75 MMbbl would have been between 225 and 750 Bcf (6.4–21 Bcm).

The composition of the gas for the Henderson retort was (Redwood 1897, p. 250) 2.19% olefiant gas, 34.2% marsh gas and 64.44% hydrogen gas.

The gas was used mostly as fuel to heat the retorts. In the 1850s, some of the gas from the Bathgate works was used for street lighting for Bathgate. The city of Edinburgh declined an offer to be supplied with gas from the Addiewell works for a sales price of 9p/Mcf; this is approximately 100p/therm (3.4p/kWh) in 2016 money and is approximately the cost today of gas for domestic customers.

The Scottish oil-shale industry changed the world

Young's major achievement was to both refine the oil into marketable products, such as paraffin, and to create a huge demand for these products. Without paraffin and the demand for paraffin, oil drilling in the USA would not have happened when it did.

Prior to the Scottish shale-oil production, oil was used mostly for medicinal purposes. In the USA, oil was marketed as Seneca Oil as a cure for rheumatism, burns and coughs. Often, however, oil was considered useless and an annoyance. Redwood (1896, p. 3) reports that oil: 'was usually found in the brine pits, and was regarded as so great a nuisance that many operators ceased working the salt on account of its presence'.

At first, the main product of the Scottish oil-shale industry was lubricants for the textile industry and a secondary product was naptha used in the paint and rubber industries (McKay 2012). The lighter fractions initially had no market other than as a cheap burning oil. Developments in the mid-1850s of cheap clean-burning lamps made paraffin the most important product. Demand for Young's paraffin grew. It became the premium fuel for lighting displacing whale oil. Whales had been hunted almost to extinction, so the production of paraffin came as a great relief to the whales (Fig. 9). Initially, the burning oil contained lighter fractions that were often a safety hazard. Safety regulations were brought in for quality control but safety issues continued until demand for the gasoline fraction was created by the motor car (Fig. 4).

Fig. 9. *Vanity Fair* cartoon from 1861 'Grand ball given by the whales in honor of the discovery of the oil wells in Pennsylvania'.

It is often stated that the American coal-oil industry developed independently of Young's operations, and that this was the spur to activities at Titusville, Pennsylvania. However, in the USA in the late 1850s there were over 50 refineries producing oil under licence from Young's company. Some of these refineries used cannel shale imported from Scotland. There are no records of large-scale oil refining predating the introduction of Young's process. Abraham Gesner first produced kerosene in 1846 (Gesner 1861) but his patent in the US was awarded in 1854 two years after Young's. By the late 1850s a robust coal-oil industry existed in the USA using locally produced cannel shales.

Young's royalty was 2 cents/gallon. In 1860, Young spent 3 months in the USA to ensure he obtained his royalties of £50 000, which implies that over a quarter of a million barrels of oil were produced under Young's patent. Actual oil production was probably considerably higher because some of the oil works were not paying Young his royalty.

The demand for paraffin – a demand largely created by James Young – and the cost of making oil by retorting mined cannel coal induced a New York lawyer, George H. Bissel, and others to form, in 1858, the Seneca Oil Company. This company took a lease on land at Cherrytree Township near Titusville, Pennsylvania. The leased land was on the banks of the Oil Creek River. The land was chosen because for hundreds of years Native Americans skimmed oil from the surface of the water near the naturally occurring oil seeps.

The company's operations manager was Colonel Edwin L. Drake. Drake tried several ways to access and skim the oil. Eventually, he attempted to dig a deep hole. When a hole collapsed and nearly killed his men, Drake attempted drilling. Drake hired a salt well driller, William A. Smith, in the summer of 1859. After many difficulties, they finally drilled a commercially successful well on 27 August 1859. Drake's well was only 69 ft deep and the oil did not flow to surface. One of the first wells that flowed oil to surface was the nearby Funk Farm Fountain well drilled in June 1861. It was 460 ft deep and flowed 300 bbl/day for 6 months before 'waxing up' and stopping flowing.

By the end of 1862 oil production in Pennsylvania greatly exceeded Scottish oil production. The refineries that had been built in the USA to produce oil from imported cannel coal were reconfigured to refine the oil from the local oil wells. Oil that had sold for \$16/bbl in 1859 (about \$1800/bbl in 2016 money) for the next 80 years had an average price of \$1.6/barrel. The world had changed and would never be the same again.

The US oil industry would have happened whether or not James Young or Colonel Drake had existed; such was the pace of technology and interest in petroleum matters in many parts of the world at that time. But it is widely acknowledged that the industry started when and where it did because of Colonel Drakes's well at Titusville. In contrast, often overlooked is that the motivation for the Titusville drilling was created by the work of James Young and the Scottish shale-oil industry.

Why did the Scottish shale oil not progress to oil production from oil wells?

The early US oil-well production was oil that had migrated from the Marcellus Shale source rock into sandstone reservoirs. The USA switched in 1859 from producing oil from mined shale to producing oil from drilled oil wells drilled into sandstone oil fields. This switch never happened in the Scottish Central Belt despite the WLO being a good source rock, and the presence of sandstones and limestones.

From the early days of the Scottish oil-shale industry, borehole drilling was extensively used to map the shales to aid mining. Boreholes, particularly in the Broxburn and Pumpherston area, often contained gas or oil but these were never commercially exploited. A water borehole drilled around 1884 at Middleton Lodge: 'regularly every month gas was evolved with considerable force and noise, which, on lighting, burned for a day or two as a great column of flame seen from afar' (Carruthers *et al.* 1927, p. 236).

There were many reported cases of oil in the shale mines:

- beneath the Dunnett Shale, in No. 5 or Sandhole Pit, Broxburn (in 1886);
- the dolerite sill below the Dunnet Shale in the Dunnet Mine, Broxburn (in 1911);
- the Whin (dolerite) in No. 1 Mine, Pumpherston, working the Pumpherston Shale (1897);
- the sandstone below the Lower Dunnet Shale in the Breich Pit (1920);
- sandstone beneath the Dunnet Shale in the Westwood Pit (1920).

There were also reports of oil in coal mines in the Central Belt, in particular in mines mining the Kilsyth Coking Coal in North Lanarkshire. The Kilsyth Coking Coal is in the Limestone Coal Formation below the Black Metals Shale and above the Johnstone Shell Bed Shale. In the St Flannan's colliery near Kirkintilloch, oil that trickled from the roof of the workings was collected from 1903 to 1908.

Carruthers *et al.* (1927, p. 230) discussed the theory that the free oil is produced *in situ* by the same method as the oil is produced in the oil-shale industry retorts: that is, from the under-mature shale by heat, with the heat being supplied from intrusions:

> [A] basalt dyke, about 100 ft. broad, runs for a mile or two through the Roman Camp field, where for a distance of about 70 ft. on either side the shale has been rendered useless. It is obvious that the heat from the molten material has distilled the shale. There was no gas in the burned seam, but the cracks and crevices of the igneous rock itself were charged with inflammable gas under considerable pressure. When the dyke was reached in the mine, it was covered by a thin compact layer, and after its removal there was a sudden discharge of brackish water ... When the shale was suddenly distilled by the igneous rock, there would be no escape for the sudden pressure of oil vapours but into the molten material itself, and the temperature would be high enough to convert them into permanent gas. When an intrusive sheet occupies the position of a productive oil-shale, the former often contains pockets or small hollow spaces partly filled with liquid or solid hydrocarbons-evidently the products of distillation. For example, at Albyn Mine, Broxburn, where a basalt sill about 5 ft. thick takes the place of the Broxbum grey seam for a considerable distance, these cavities, partly filled with viscous hydrocarbons, appear in the igneous rock.

Carruthers *et al.* (1927, p. 235) could not apply this theory to all the examples he studied of free oil:

> [R]elating to the occurrences at Westwood and at Breich, all we can say is that the presence of the oil there cannot be accounted for by distillation by intrusive rock in the immediate neighbourhood ... no knowledge is available as to possible fluid migrations therein, which might bring in oil from a distance.

It is likely that some at least of the free oil has migrated from deeper mature source rock. Many of the reported occurrences of free oil are in the Pumpherston/Uphall/Broxburn area. Here, there is faulting that post-dates the intrusions and so the intrusions are steeply dipping. The intrusions may provide a migration path for the hydrocarbons.

A possible reason why Scottish Oils did not attempt to produce oil by drilling was the ownership of the oil. Patents covered the retorting and refining of the shale but not the ownership of any free-flowing oil. In the 1934 Petroleum Act, petroleum 'existing in its natural condition in strata' was nationalized. The Act, however, excluded the oil-shale industry by defining petroleum by saying petroleum:

> [I]ncludes any mineral oil or relative hydrocarbon and natural gas existing in its natural condition in strata, but does not include coal or bituminous shales or other stratified deposits from which oil can be extracted by destructive distillation.

The Petroleum Act has been modified several times since but still includes this definition of petroleum.

Oil exploration drilling

Oil exploration drilling started in June 1919 for the Mineral Oil Production Department of the Ministry of Munitions of War. The work was done by Pearson

Fig. 10. A tanker of oil leaves the Midlothian Oilfield in 1938. The producing well is in the background (from Pathe Newsreel).

& Son Ltd, a family company of Weetman Pearson, Lord Cowdray, Aberdeen's first oil baron. Boreholes were drilled at West Calder and at D'Arcy Farm near Mayfield, Midlothian. Both boreholes took a long time to drill because they discovered gas which was troublesome because they were drilling without

Fig. 11. 28 October 1957, Sir Harold Smith, Chairman of the Gas Council, starts gas production from Cousland (from ITNsource.com).

pressure control. The drill rigs were cable rigs in which the chisel or 'tool' was dropped and then retrieved by a cable recovering about 3 ft of rock with each trip. The West Calder borehole eventually reached a permeable water-bearing sandstone and was abandoned. The D'Arcy well did, however, get some oil flow into the borehole. Figure 7 shows the relationship of the D'Arcy well to the St Catherine's oil seep in Edinburgh (from (Underhill *et al.* 2008).

To date, 103 oil and gas wells have been drilled in the Central Belt but there has been very little production. The first oil production was from wells drilled by Anglo American very close to the D'Arcy well. A well started production in 1938 and some 30 000 bbl of oil are reported to have been produced from the Midlothian oilfield (Fig. 10).

BP (under the name D'Arcy) discovered gas on the same anticlinal structure 1–2 miles further north at Cousland. Studies were made to try to use the gas in manufacturing plants but it was not until 28 October 1957 that the first (shale) gas production in Scotland was started by the Chairman of the Gas Council Sir Harold Smith (Fig. 11). The gas production continued for several years but the total production was less than 1 Bcf of gas. The gas was exported to Musselburgh where it was comingled with the town gas.

In 1962, BP drilled a well near the oil works at Pumpherston, and for a short time there was enough gas to be used in the brickworks associated with the oil plant. In 1964, BP deepened a well that had been drilled in 1945 at Salsburgh, east of Airdrie. The well was fracked and tested gas.

Most of the exploration wells have been drilled on anticline structures identified from the surface outcrop geology. Perhaps the most successful well was drilled in 1989 in Glasgow near the junction of the M8 and M73. Unlike most of the previous exploration wells, this well, Bargeddie 1, was drilled into a large tilted fault block structure. The well was fracked and put on long-term production test with an initial flow rate of 3 MMscf/day (OPC 1989). During the second flow period the well was flowed at a constant rate of 2.05 MMscf/day for 3 days (OPC 1989). The test analysis at the time favoured a small bounded reservoir containing 0.02 Bcf of gas. However, the test report states that other interpretations of the pressure build-up data were possible. These other interpretations would, today, be likened to a shale-gas response. The interval tested was a sandstone below the Hurlet Limestone at the top of the Upper Oil-shale Group.

The efficacy of fracking has improved significantly since this well and the 1964 BP well were drilled. In 1964 and 1989, fracking was able to do little more than increase flow rates by reducing the skin damage caused by drilling. The two major changes to fracking since 1989 are the use of larger volumes of water when fracking, and the combination of horizontal wells and fracking to maximize the surface of the borehole in contact with the productive formation. As a result, fracking is now extensively used for shale gas and shale oil, and widely used on oil fields and gas fields with low-permeability reservoirs. Between 2011 and 2016, a total of 18 fracked horizontal wells have been drilled to develop low-permeability gas fields in the UK Southern North Sea. A paper by Jones *et al.* (2015) gives details of the fracking of one of these wells.

Onshore in Scotland, very few if any of the exploration wells were drilled deep enough to enter fully into the gas-mature window. Future exploration drilling may be more successful if it targets gas fields and shales in deeper fault-block structures, and if completion techniques employ horizontal laterals and modern fracking techniques. Similarly, the history suggests that shale gas and shale oil should be very prospective.

Conclusion

For 100 years, an estimated 75 MMbbl of oil and 500 Bcf of gas were produced from the shales of the Central Belt of Scotland. The Scottish oil-shale industry oil led to major developments in oil refining and, as a result, a large demand for oil products was created. This new demand for oil products stimulated the search for oil around the world. The Scottish shale-oil industry was successful not only thanks to James Young and his colleagues but also because of the wonderful geology of the Central Belt of Scotland and, in particular, the geochemical properties of the West Lothian Oil-shale Formation.

The shale-oil industry, natural oil seepage, free oil encountered during shale mining operations and the historical exploration drilling demonstrate a rich functioning source rock suggesting significant prospectivity for future exploitation of both oil fields and shale gas.

Acknowledgements ReachCSG are thanked for their permission to write this paper. Thanks also to Professor John Underhill for permission to use Figure 7.

Funding This research received no specific grant from any funding agency in the public, commercial or not-for-profit sectors.

Appendix A: Paraffin Young

James Young, Paraffin Young, is the centre figure who established oil production as a major world industry. He was enabled to do this by his connections to Strathclyde University or, more correctly, Anderson's University as it was then called. When James Young attended Anderson's University, the chemistry professor was Thomas Graham, the first chemistry professor at the university, who had been appointed in 1830.

James Young was born in Glasgow in 1811. He was an apprenticed carpenter and cabinet maker when he attended evening lectures at Strathclyde University given by Professor Graham on Chemistry. He then worked for 7 years at Strathclyde University as a demonstrator and assistant to Professor Graham. He became manager of the chemical works of James Muspratt, St Helens, Lancashire in 1838, and moved to Tennants, Clow & Co of Ardwick, Manchester in 1844. James Young was elected Fellow of the Royal Society of Edinburgh in 1861 and of the Royal Society in 1873. From 1868 to 1877 he was President of Anderson's University, and a great benefactor to the university. He founded the Young Chair of Technical Chemistry in 1869. He was a lifelong friend of David Livingstone and funded Livingstone's African expeditions.

While in Manchester, James Young met his friend and fellow alumni of Strathclyde University Lyon Playfair. Lyon Playfair had also worked for Thomas Graham. In 1843, Lyon Playfair was appointed Professor of Chemistry at the Royal Manchester Institution and in 1845 chemist to the Geological Survey.

Playfair's brother-in-law James Oakes owned the Riddings Colliery in Alfreton in Derbyshire. This mine had oil oozing into the mine. James Young recommended to his employer to use this oil but both his employer and Young himself recognized that the stream of oil would soon stop flowing. Young took a lease on the oil production and formed a partnership with a Strathclyde University friend Edward Meldrum. Meldrum ran the operations while Young kept his day-job at the Tennant Works.

Young began experimenting with coal to produce oil and found that he could produce paraffin from cannel coal. In 1850, yet another of Young's colleagues from Strathclyde University, Hugh Bartholomew, sent Young a sample of an oil shale called the Torbanehill Boghead Coal. Hugh Bartholomew was the manager of the Glasgow gasworks. Torbanehill is close to Bathgate near the M8 motorway.

With Meldrum as the operations manager, Young started building works at Bathgate in 1850 and oil production started in 1851. Much of the finance was provided by Edward Binney, who was both a lawyer and a geologist living in Manchester. Binney was particularly interested in oil and in 1843 investigated the oil seeps at Formby.

On 17 October 1850, James Young was granted his patent (No. 13292): 'Improvements in the treatment of certain bituminous mineral substances, and in obtaining products therefrom'. The works initially refined the oil to produce 'burning oil' and lubricating oils, and later paraffin wax was produced by artificially cooling.

James Young himself was active in the industry until 1872 when he registered his last patent. It was the ingenuity and acumen of many others that was responsible for the continued survival and longevity of the industry.

Appendix B: Further information

For general information on the Scottish oil-shale industry, the Museum of the Scottish Shale Oil Industry in Livingston is excellent. The address is: Almond Valley Heritage Trust, Millfield, Livingston, West Lothian EH54 7AR, UK. Their website also is recommended (http://www.scottishshale.co.uk/). For books, *The Scottish Shale Oil Industry & Mineral Railway Lines* by Knox (2013) is entertaining, with a wealth of illustrations. Another recommended book is *Scotland's First Oil Boom: The Scottish Shale-Oil Industry, 1851–1914* by McKay (2012).

References

Andrews, J.E. 1988. Methane-related Mg-calcite cements in recent tidal flat sediments from the Firth of Forth. *Scottish Journal of Geology*, **24**, 233–244, https://doi.org/10.1144/sjg24030233

Appleton, J.D., Hooker, P. & Smith, N.J.P. 1995. *Methane Carbon Dioxide and Oil Seeps from Natural Sources and Mining Area: Characteristics Extent and Relevance to Planning in Great Britain*. British Geological Survey Technical Report **WP/95/1**.

Baptie, B., Segou, M., Ellen, R. & Monaghan, A. 2016. *Unconventional Oil and Gas Development: Understanding and Monitoring Induced Seismic Activity*. British Geological Survey Open Report **OR/16/042**.

Cadell, H.M. 1925. *The Rocks of West Lothian*. Oliver & Boyd, Edinburgh.

Carruthers, R.G., Caldwell, W., Bailey, E.M. & Conacher, H.R.J. 1927. *The Oil-Shales of the Lothians*. Memoirs of the Geological Survey, Scotland Series.

Cook, F.M. 1971. What the oil industry owes to Dr James Young. *Chemistry and Industry*, 585–589.

Gesner, A. 1861. *A Practical Treatise on Coal, Petroleum, and other Distilled Distilled Oils*. Baillière Brothers, New York.

Gibson, W. 1922. *Cannel Coals, Lignite and Mineral Oil in Scotland*. Memoirs of the Geological Survey, Scotland, **24**.

Gilfillan, S. 2016. Resolving the noble gas fingerprint in UK unconventional reservoirs. Presented at PETEX 16, 15–17 November 2016, London.

Jones, N.S. 2005. *The West Lothian Oil-shale Formation: Results of a Sedimentological Study*. British Geological Survey Internal Report **IR/05/046**.

Jones, P., Symonds, R., Talbot, D., Jeffs, P., Kohok, A., Shaoul, J. & Spitzer, W. 2015. Successful hydraulic fracture stimulation of Yoredale Carboniferous sands in the UKCS. Paper SPE 174171 presented at the SPE European Formation Damage Conference and Exhibition, 3–5 June 2015, Budapest, Hungary.

Judd, A., Davies, G., Wilson, J., Holmes, R., Baron, G. & Bryden, I. 1997. Contributions to atmospheric methane by natural seepages on the UK continental shelf. *Marine Geology*, **137**, 165–189.

Judd, A.G., Sim, R., Kingston, P. & McNally, J. 2002. Gas seepage on an intertidal site: Torry Bay, Firth of Forth, Scotland. *Continental Shelf Research*, **22**, 2317–2331.

Knox, H. 2013. *The Scottish Shale Oil Industry & Mineral Railway Lines*. Lightmoor Press, Witney, Oxfordshire, UK.

KPMG 2013. *Economic Impact Assessment and Scenario Development of Unconventional Oil and Gas in Scotland*. A Report for the Scottish Government. KPMG Oil & Gas, London.

Legg, R. 1984. *Guide to Purbeck Coast and Shipwreck*. Dorset Publishing, Isle of Purbeck, Dorset, UK.

Masters, C. 2014. *Independent Expert Scientific Panel – Report on Unconventional Oil and Gas*. Scottish Government, Edinburgh.

McKay, J. 2012. *Scotland's First Oil Boom: The Scottish Shale-Oil Industry, 1851–1914*. John Donald, Edinburgh.

Monaghan, A.A. 2014. *The Carboniferous Shales of the Midland Valley of Scotland: Geology and Resource Estimation*. British Geological Survey for Department of Energy and Climate Change, London, UK.

OPC 1989. *Long Term Test on Bargeddie-1 Post Frac Analysis*.

Parnell, J. 1986. Organic petrography of Lothian oil-shales. Sedimentology and hydrocarbon potential of the Dinantian oil-shales of northern Britain. Presented at the British Sedimentological Research Workshop, 4–5 April 1986, Edinburgh.

Parnell, J. 1988. Lacustrine petroleum source rocks in the Dinantian Oil Shale Group, Scotland: a review. *In*: Fleet, A.J., Kelts, K. & Talbot, M.R. (eds) *Lacustrine Petroleum Source Rocks*. Geological Society, London, Special Publications, **40**, 235–246, https://doi.org/10.1144/GSL.SP.1988.040.01.20

Raymond, A.C. 1991. *Carboniferous Rocks of the Eastern and Central Midland Valley of Scotland: Organic Petrology, Organic Geochemistry and Effects of Igneous Activity*. PhD thesis, University of Newcastle upon Tyne, Newcastle upon Tyne, UK.

Reach 2013. *Geochemical Evaluation Salsburgh-2 Well Midland Valley of Scotland*. Confidential Report for Reach CSG by M.N.D. Kaye, OceanGrove Geoscience Ltd, Project No. GC100119.

Redwood, B. 1896. *Petroleum: A Treatise on the Geographical Distribution and Geological Occurrence of Petroleum*. J.B. Lippincott Company, Philadelphia, PA, USA.

Redwood, I.I. 1897. *A Practical Treatise on Mineral Oils and Their by-Products: Including a Short History of the Scotch Shale Oil Industry, the Geological and Geographical Distribution of the Scotch Shales*. Spon, London.

Reports by the Juries 1851. *Great Exhibition, Volumes 1 and 2*. Spicer Brothers, London, 103 (Vol. 1), 1386 (Vol. 2).

Thomas, R.A.P. 2014. *Gasworks Profile A: The History and Operation of Gasworks*. Contaminated Land: Applications in Real Environments (CL:AIRE), London.

Underhill, J.R., Monaghan, A.A. & Browne, M.A.E. 2008. Controls on structural styles, basin development and petroleum prospectivity in the Midland Valley of Scotland. *Marine and Petroleum Geology*, **25**, 1000–1022.

West, I.M. 2017. *Petroleum Geology of the South of England, Geology of the Wessex Coast*. School of Ocean and Earth Science, Southampton University, Southampton, UK, http://www.southampton.ac.uk/~imw/index.htm

Derbyshire's oil and refining history: the James 'Paraffin' Young connection

CLIFF LEA

North East Derbyshire Industrial Archaeology Society, 15 Kelburn Avenue, Chesterfield, Derbyshire S40 3DG, UK

cliff@nedias.co.uk

Abstract: Following an adventitious oil flow into a coal mine in 1847 in the Riddings area of the county of Derbyshire in the English Midlands, the young Scottish chemist James Young carried out seminal work into the development of oil refining technology. In Derbyshire, by the end of 1848, he set up an early refinery to exploit this oil commercially by distillation, producing both lighting and lubrication fractions which he sold directly to end customers. His findings in Derbyshire led him to move on to investigate production and refining of oil from coal by destructive distillation, technology for which he gained his global reputation; Young's patented technology was adopted by the commercial refineries that were to follow in many countries as the world's oil fields sprang to life.

It may come as a surprise to many that the most fascinating history of the British oil industry occurred in Derbyshire in the English Midlands. There had been many discoveries of mineral oil and bituminous deposits made quite by chance in the county for hundreds, perhaps thousands, of years. One of the earliest known sites, a site close to the Iron Age Mam Tor hill fort, is at Windy Knoll near Castleton. Here, and at the nearby Odin lead mine, Elaterite, which is a naturally occurring brown, soft and sticky semi-solid with a distinct 'oily' smell, composed of a variable mixture of complex hydrocarbons in the mineral oil family, can still be found. The area receives a constant stream of visits from geology students. These are just two instances of many. In the White Peak and Derwent Valley areas of the county there have been countless reports in lead mines and drainage soughs over the centuries; there have been many reports also in the coal measures to the east of the county bordering Nottinghamshire and South Yorkshire.

The earliest known 'written' record of Derbyshire oil was of that found at Eyam, the village more famous for its connections to the plague outbreak of 1665. Dr Thomas Short in 1734, whilst searching the county for mineral water springs, wrote of a lead worker at a mine near Eyam who found bituminous oil deposits in the interior of a boulder which he cracked open; this created great interest and experimentation in the locality. Short (1734, p. 97) wrote:

> Almost opposite these springs is a mine .... In striking the ore, the sudden shaking of the metal .... Gives such a violent motion ... that it makes an explosion like fired gunpowder.

Dr Short went on to describe how a workman took some of the bituminous material that he had found to his home; he put it into a saucepan and melted it, and he described it as a clear balsam, except when thrown into the fire! He wrote:

> The neighbourhood had a merry time for greasing boots and shoes, until they found it quickly shrivelled them up and burnt them to pieces. A great pity such bitumen or Oleum Terrae had thus been wasted
>
> (Short 1734).

It was 100 years later that one of the most interesting early oil finds was stumbled upon quite by chance at Riddings in the east of Derbyshire, and this was to have both national and international significance when very early research led to the establishment of a very early refinery for crude oil producing lighting and lubricating fractions for direct sale.

Riddings

It was in 1847 at one of the ironmaster James Oakes's coal mines, the New Deeps Pit, at Riddings that a remarkable early accidental oil find had occurred (Bailey 1948).

Before 1800, Riddings (Fowkes 1991) was a small hamlet clustered around a market place; the name Riddings has similar derivation to the 'Ridings' of Yorkshire, and refers to a clearing in a forest – in this case Sherwood Forest, which extended westwards to this area at the time. Around 1800, most of the residents were involved either in farming, in textile activity with stocking frames or with local bell-pit mining for coal and ironstone. At Riddings an iron works was already in existence, originally called Alfreton Iron Works, started as a partnership by local men, Saxelby, Oakes, Whylde and Forester. It traded as Thomas Saxelby & Co.

From: Craig, J., Gerali, F., MacAulay, F. & Sorkhabi, R. (eds) 2018. *History of the European Oil and Gas Industry*. Geological Society, London, Special Publications, **465**, 71–76.
First published online April 11, 2018, https://doi.org/10.1144/SP465.4

Ltd, became owned by James Oakes in 1808 and later by his elder son, James Oakes II.

James Oakes was said to have had great interest in geology and, in his quest for raw materials that might expand his business, he speculated that the rich Kilburn coal seam could be reached 250 yards below.

Thomas Haden Oakes, brother of James Oakes II, wrote some 50 years later (Oakes 1884) how, in the 1830s, it had been decided to sink a mine down 240 yards to meet this seam, a level at that time which made the new mine almost the deepest in the county.

In 1847, as the miners were exploiting the coal seam and developing a series of roads underground, they were surprised by a great rush of saltwater brine, severely hindering operations and no doubt causing a great nuisance. They eventually created a sump at the bottom of the New Deeps Mine, collected the brine and transferred it to mining tubs. The tubs were raised up to the surface and the water discharged to the nearby Pinxton Branch of the Cromford Canal, in accordance with agreements with the Cromford Canal Company for disposal of water.

The operation to remove the brine carried on night and day for several months. There was even an attempt at commercial separation of the salt from the brine for onwards sale, but this was uneconomic and the discharge to canal continued.

At the time, the new coal mine was cause for great celebration for the quality of coal produced, and the irritation of liquid ingress was probably regarded as a minor matter of little significance; meanwhile the discharge of water to the canal continued. In time, the vast quantities of water gradually slackened, to be replaced by a more sluggish black-coloured mixture which floated on the water.

It was still brought up and discharged to the canal without thought. Whilst not recognized as such at the time, the dark sluggish material was mineral oil, and at this stage volumes of around 1300 l/day were being brought to the surface and discharged to the canal.

Of course, in the days before invention of paraffin/kerosene lamps and long before the appearance of the internal combustion engine, mineral oil had no special use. The small quantities located around the world, such as in the Tar Tunnel in Shropshire, tended to be used in quack medicines and other doubtful applications.

For how long this discharge carried on is unknown, but Thomas Haden Oakes wrote that the floating oil was carried along the surface of the canal, moved by wind and blown into Pinxton Wharf where it was said to collect. Several incidents (*Heanor and Ripley News* 1982) were reported. A woman threw a pan of red-hot ashes from her fire into the canal, causing the surface to catch fire. Local youths quickly learned that the floating oil on the canal could be ignited and now set it alight for fun. Perhaps more importantly at that time, the local Codnor Park Brewery (Johnson 1970) used water extracted from the canal for its brewing operations, and the contamination was said to have soured the porter. A serious matter!

Oakes discussed this 1847 oil strike and the problems with his son-in-law, Lyon Playfair. He was a highly respected chemist, later to be elected to the Royal Society, knighted as Sir Lyon Playfair and elevated to the peerage as First Baron Playfair. Some years earlier, Playfair had met and stayed with Oakes when he visited the Iron Works with the celebrated German chemist Professor Bunsen (of 'Bunsen Burner' fame), a visit to measure concentration of gases from Oakes's iron furnaces. It was at this time that Playfair (Reid 1899*a*) was introduced to Oakes's sister Margaret and they were to marry in July 1846.

Playfair was intrigued by the Riddings oil strike, recognizing it for the type of liquid hydrocarbon which was already known to come to the surface near Baku in Azerbaijan and elsewhere. On 3 December 1847, Playfair wrote to another chemist colleague from student days, James Young, suggesting that he investigate to see whether it had any commercial use here.

On 3 December 1947, he wrote (Playfair in Reid 1899*b*):

> You know that mineral naphtha is a rare natural product, no spring of it occurring in this country. Lately, a spring … has been discovered on the estate belonging to my brother-in-law. It yields at present about 300 gallons daily.
>
> The naphtha is about the consistency of thin treacle, and with one distillation it gives a brilliant illuminating power. My brother in law intends to set up stills immediately but as they are ironmasters this would be a separate industry, so I have advised them to sell the crude … to chemical manufacturers and thus avoid carrying on an industry foreign to their occupation.
>
> Does this possibly come within the province of your works? If it does I will send you a gallon for examination. Perhaps we could make a capital thing out of this new industry.

At that time oil had little intrinsic value; oil finds in Britain, Europe and America were likely to be used as an ingredient in quack medicines but it was already being considered for addition to coal gas, to improve illuminating power. Oakes had already found, however, that the crude oil itself had excellent lubrication characteristics for his wheeled tubs.

James Young had studied at Anderson's University (now the University of Strathclyde), Glasgow, in the 1830s under the celebrated chemist Thomas Graham, where Playfair was one of his fellow students. Young went on to join the chemical company James

Muspratts chemical works at Newton-le-Willows, Lancashire, in 1838. He later moved to Tennants, Clow & Co. Ltd, Lancashire, in 1844 as Works Manager.

The Tennant Group at the time were headquartered near Glasgow at the St Rollox works, and these works had been described some decades earlier as the largest chemical company in the world. They had grown considerably as Britain's textile industries expanded, and were heavily involved with dyes, bleaching agents, etc.

Initial experimentation at Riddings

Young paid a visit to Riddings in December 1847, and he descended the Deeps Mine on 23 December to investigate and to carry out his first analysis. In his diary, he made the note:

> [O]ut of the crack oozed and dropped this petroleum … the thickness of olive oil, and on a cold day it became nearly solid … and it was a dark coffee colour
>
> (verbatim account by James Young at the Court of Session 1860: Binney & Co. (Young's solicitor) v. The Clydesdale Chemical Co.).

He set up apparatus to distil the oil, probably using the ironworks own laboratory: '[A] little water (distilled) out first, then some Naphtha, not much, then an oil … this looks hopeful' (entry for 24 December 1847 taken from James Young's pocket diary) and he went on to separate the components into its various parts.

On distillation, he collected the pure light fractions; further tests showed that the light fractions had brilliant illuminating power when burned, and could be used in oil lamps. Within a few years the world's paraffin/kerosene lamp industry was to start, and later Young himself would be offering the paraffin and also the lamps for sale from many offices around the globe. Until the advent of paraffin/kerosene, only vegetable oils and whale oil would have been used in oil lamps.

James Oakes had already established that the Riddings crude oil was a valuable lubricant for his wagon axles, replacing animal fats and vegetable oils. After separating out the flammable light fractions from the crude, Young now found great advantages for the refined heavier fraction; he confirmed that it had excellent lubrication qualities.

The new refined lubricant could offer significant advantages for Britain's heavy industries in the 1800s and particularly for textile machinery – the textile industry was Britain's largest foreign currency earner at the time. By the mid-1800s, the ever more advanced machinery being developed (pumps, gears shafts, bearings, etc.) required more subtle lubricants. Thinner oils for finer clearances and faster motion, and thicker oils for heavier, slower moving machinery with greater clearances – mineral oil could now be developed to match the developments in machinery.

After discussion with Playfair, Young recognized the presence of a fine crystalline wax, a commodity already isolated and named 'paraffine' by the German Carl Reichenberg in the 1830s but not yet commercially exploited.

Playfair (in Reid 1899*a*) wrote '[T]his oil, when cold, deposited shining crystals which I recognised to be the paraffin of Reichenberg' (letter of 3 December 1847 to James Young).

Young was later to separate paraffin wax off for making candles, a novel replacement for beeswax or tallow wicks. At a special meeting of the Royal Institution in London, Playfair, by now elevated to the peerage, recounted the technical details of Young's new refining technology as developed at Riddings. It was to be a lecture given for the first time by the light of the new paraffin wax candles. In his memoirs, Playfair, in 1947 (in Reid 1899*a*, pp. 37–38) wrote '[James Young] extracted enough to make two candles at my suggestion, with which I lighted my desk at the Royal Institution during a lecture I gave on petroleum'.

Whilst we take home illumination for granted these days, it was Young's work that led to the establishment and widespread use of paraffin wax candles and paraffin/kerosene lamps. Previously, whilst the rich had expensive beeswax candles, and some people had access to lamps which could burn expensive vegetable or whale oils, Britain's poor had access only to the sputtering light of tallow wicks.

Development of Young's oil refinery at Riddings

It was clear that mineral oil from Oakes coal mine could be exploited and could have considerable value. Young discussed the matter with his then employer, Tennants, Clow & Company Ltd, reviewing whether this was an operation they might wish to develop themselves (Butt 1963).

Tennants' directors declined to take part, but agreed to allow Young to do so whilst still carrying on his position with their subsidiary in Lancashire. The action of Tennants suggests that Young was already well respected and they would not wish to lose him (Butt 1963).

Young now negotiated a contract for the exploitation, refining and sales of products from the Riddings crude oil with James Oakes. He agreed a 7 year lease on the exploitation of the oil spring, and to scale up his earlier laboratory work. He agreed terms in a Deed of Covenant dated 20 September

1848 for the setting-up of a refinery, the main details of which were (Bailey 1948):

(A) James Oakes responsibilities:

- Oakes was to provide an area '*not exceeding 30 perches*' (*c.* 1260 m^2) for storage and buildings to house the works, all with free access.
- Oakes to erect buildings for the refinery.
- Oakes to deliver oil to the refinery from his coal mines, the oil to be contained in casks, tanks or cisterns.
- Oakes to supply at his expense the apparatus, tanks and tools required for the initial set-up.

(B) James Young's responsibilities:

- Young to set up and carry out the refinery operation to produce saleable goods.
- Young to pay for coal, wages, damages and losses.
- Young to pay Oakes 10% per annum on the initial capital expenditure.
- Young to keep commission of 15% on sale of petroleum products.
- Proceeds to be split Five-Eighths to Oakes and Three-Eighths to Young.

In 1848, Young appointed his friend Edward Meldrum as manager of the site. Equipment was installed to deal with approximately 1300 l/day of crude oil, converting it principally into lighting oils, and lubricating fractions on a commercial scale.

There is little information available as to the equipment that Young installed and there are no known plans of the layout of the process or plant equipment. It is, however, reported (Bailey 1948) that there were initially to be two 18 cwt lead-lined tanks. The large tanks would have been used for settlement, to allow water and sediment to separate out by gravity before distillation. There were eight cast-iron stills with glass retorts. The stills might have been of the type commonly used at that time for commercial distillation of alcohol, and readily available from scientific equipment dealers such as Griffin, then in Glasgow – they are still part of one of the world's largest equipment manufacturers in the twenty-first century. A year later, on 22 September 1849, Young recorded in his notebook on 22 September 1849:

> Worms and condensers, cast iron, £50. House £100. Chimney, £15. Pumps, £20. Riddings has three worms.

The reference to 'worms' is to types of coiled condenser. Over the next few decades, the volumes of oil to be refined around the world would increase at a rapid rate, and the scale of refinery operations increased massively. Chemists and chemical engineers of the twenty-first century would regard the small operation at Riddings to be of 'pilot' scale. However, at the time, it was of a scale that treated all the arisings (1300 l/day) from the Deeps Mine source and converting them for onwards commercial sale.

Young's refined products were to be sold commercially. The customers were to be principally in Lancashire, and initially they were cotton spinning and weaving mills where his lubricant found great use in textile machinery. It is likely that Young's sales were within the Lancashire area because these customers would already have been known to him through his earlier business activities in manufacture and supply of dyestuffs and chemical textile treatments through his employer Tennants. Within a year he had appointed a sole sales agency to deal with supply to end users; the agency for both lamp oil and lubricants was J. Hurst of Manchester. The contract was formalized on 3 September 1850, when Hurst was to be responsible for billing, deliveries and debts. Hurst took 10% commission (these details form an entry made in James Young's notebook on 3 September 1850).

The location of the mine where the oil flow had been encountered, the New Deeps, is marked on the 1888 Coal Authority plan of the Alfreton Iron Works site, and this shows a building that it is speculated might have been the location of James Young's laboratory and refinery (Martyn Taylor-Cockayne private comm.). Within a few years, with changes in the types of industry carried out on the site, Alfreton Iron Works went through a number of redevelopments with the introduction of rail lines and sidings, totally altering the topography and eliminating any signs on the surface of the refinery.

Moves to coal distillation

Young might well have continued his works refining mineral oil, but one of his early observations about how the oil was discovered at Riddings took him along another very valuable route. When Young went down the pit on 23 December 1847, he had noted (Martyn Taylor-Cockayne private comm.):

> I found the petroleum dropping from the roof over the coal. The coal was being worked out, and the roof over the coal was sandstone; as the coal was taken away the sandstone broke, and out of the cracked … dropped this petroleum.

He had come to the erroneous conclusion that the oil had been formed underground by thermal cracking of the coal, and he thought the oil produced was then retained within the rocks. He therefore very quickly decided that moves forward should encompass the distillation of coal. Almost immediately he turned his attention to the products that could be obtained from a variety of coals from different sources, and his move to the distillation of coal to produce oil which could be refined using similar techniques is well known. In 1850 Young developed a formal partnership with his attorney Edward William Binney and Edward Meldrum. They developed works at Bathgate, where his tests had shown that the most promising sources of cannel coal for oil production were located. The Bathgate Works were completed and in operation distilling oil from cannel coal by 1851.

So far, Young had ensured commercial confidentiality when he carried out his experimentation and early refining at both Riddings and Bathgate – indeed, no detailed plans have been found of his Riddings site. However, he ensured that his refining technology was protected by a well-reasoned early patent (Patent Office 1850), dated 17 October 1850, and a further 17 patents were to follow. Within a few years, the world's oil industry was to start in earnest, and Young's patents were to have great financial significance; his patented technology – the earliest to cover oil refining – was shortly to be used in the USA and around the world. The *Toronto Globe* estimated in 1862 that Young was making approximately £100 000 per annum from such royalties (*Toronto Globe* 1862) alone.

By 1865, when his early refining patents had expired, the processes carried out in the Bathgate works were described in *The Scotsman* (1865):

> In 1851 the works, which were destined to add threefold to the population and prosperity of Bathgate, were begun. From the hill on which Bathgate is clustered, the distant confusion of roof-tops topped with belching chimneys forms a feature in the surrounding landscape of undulating fields which at once arrests the eye.
>
> The works, mapped out into various sections and departments, are spread over an area of twenty-five acres, and form what seems a huge, bustling, dusky village. Main streets of stills and retorts lead into branch thoroughfares of boilers and tanks.
>
> The strange art of chemistry, which gives to science the power to change and transmute substances, and to reveal their hidden properties, so that they may be made useful to mankind, has seldom perhaps been employed in a more interesting, and in many respects wonderful, manufacture than that which is carried on at Bathgate. Delicately tinted wax candles and oil, almost as transparent as water, would seem to be the last things in the world which one would expect to see derived from coal; and yet this is not only done, but it is achieved in a way that raises our admiration by its perfection and ingenuity.
>
> There are four different articles manufactured by Mr Young – viz., paraffine oil for burning, paraffine oil for lubricating machinery, a light volatile fluid called naphtha, and solid paraffine or wax.

It is commonly reported that by the time Young's Bathgate works opened, refining the oil from boghead coal, the refining operation at Riddings in Derbyshire ceased. Young commented in a lecture on 13 March 1865 (Bailey 1948, p. 358):

> The quantity of petroleum from the mine gradually diminished, and that petroleum work came to an end, but it had served a useful purpose; it had proved that petroleum could be turned to good account.

However, the oil flow did not cease and it continued for some decades. Not all was refined by Young but he was continuing to buy oil from Riddings which he put through his Bathgate refinery until 1859, when barrels of oil were forwarded from Riddings to Bathgate via the port of Hull (details taken from a letter from Alfreton Iron Works, 26 March 1859). Young made a nostalgic return visit to Riddings from Scotland on 1 February 1868 and noted in his pocket diary:

> Wandered about H&R with Mr Oakes looking at old places. The petrol is coming at the rate of 200 galls a week. [T]he old man Sawyers is working it under Mr Horsley senr.

Even later, Bailey (1948) records that in a period of 1873–78, sales of oil from the New Deeps Pit were approximately 30 tons per month but that Young was no longer one of the buyers.

Riddings was still producing oil a decade later. W. Thomson reported (*Derbyshire Times* 1947) that his father had told him that, even as late as the 1880s, oil still flowed into the sump at the bottom of the New Deeps coal mine and was raised to the surface. It was reported then that it was used for lubricating wagons and 'trams at the half dozen collieries of Messrs. J. Oakes & Co. Ltd'.

Meanwhile Playfair had been appointed as scientific adviser for the Great Exhibition at the Crystal Palace of 1851, and he ensured that Young's developments of this new petroleum technology would be on special display there. Whilst Young was involved with the display of material from Tennants, he also had his own oil display (*The Morning Post* 1851). This included lubricating oil from his Derbyshire refinery, as well as light 'paraffin' oil for lamps.

James Young's business had now become global, and his reputation as James 'Paraffin' Young was coined. Within the next few decades, the new refinery technology, which he started in Derbyshire and further exploited at Bathgate, had swept the world and in the early years Young was to receive considerable royalties.

The work to collect the oil from the New Deeps Pit carried on until 1884. The Old Deeps Pit and New Deeps Pit finally closed in 1885, and the site was to be comprehensively developed. The business of James Oakes & Co., Alfreton Iron Works, continued through to the twentieth century (Grace's Guide 2014), expanding into pipes and flanges for power stations, ceramics and clay pipes. Numbers of employees were cited as 800 in the Iron Works and 3000 in Oakes' collieries. Stanton Ironworks Ltd became their major conduit to sales. They were finally purchased by Hepworth Ceramic Ltd in 1972, continuing the manufacture of clay pipes; the Oakes catalogue of 1903–04 contains a drawing of a complex manufacturing site, but now centring on the ceramics side of the business.

When the Americans Kier, Bissel and Evelith started their own refinery operations some years later as the US oil wells sprang to life, they were to pay royalties to Young in order to use his technology.

Conclusion

In conclusion, James Young's Derbyshire oil refinery is all but forgotten, hidden now under the industrial estate which has taken over from the Alfreton Iron Works at Riddings, and the account of how Young had researched and refined mineral oil is little known in the county. It was in 1847 that James Young had carried out laboratory distillation. By 1848 his petroleum oil refinery was in operation at Riddings, refining approximately 1300 l/day, and selling the refined products commercially into industry. Young used the name 'paraffin' to describe other fractions, including lamp oil, and his coal oil refinery was in operation at Bathgate by 1851.

Examination of the chronology shows that Young's refinery at Riddings, Derbyshire, which was set up in 1848, signifies as one of the earliest commercial refineries in the world using crude mineral oil as its feedstock. During his career he was to be awarded 46 patents, of which 18 were to be devoted to oil and refining, and his interests lay in many directions.

Following the great work carried out in Derbyshire before 1850 and later at Bathgate, all eyes were to switch to the USA where the oil rush had started in earnest. The world was to see the rise of major companies, of John D. Rockefeller's Standard Oil which initially eclipsed all others, and in Britain the rise of Shell, Burmah Oil and BP. However, in Derbyshire, the local history of the finding of oil at Riddings and the very early exploitation by Young were to fade from memory as later ventures took over at the Alfreton Iron Works site.

Acknowledgements I would like to thank all those who have assisted me, including the Museum of the Scottish Shale Oil Industry, the Science Museum, the Energy Institute, the University of Strathclyde, Derbyshire County Libraries, the Catalyst Museum and Martyn Taylor-Cockayne of the Jacksdale Area Heritage Group.

Funding This research received no specific grant from any funding agency in the public, commercial or not-for-profit sectors.

References

Bailey, E.M. 1948. The dawn of petroleum refining. *Institute of Petroleum Review*.

Butt, J. 1963. *James Young: Scottish Industrialist and Philanthropist*. PhD thesis, University of Glasgow.

Derbyshire Times 1947. Letter to the editor: Derbyshire oil, Walter Thompson. 21 November 1947.

Fowkes, D. 1991. *History around Alfreton*. Alfreton & District Heritage Trust, Alfreton, Derbyshire, UK.

Grace's Guide 2014. *James Oakes & Co., Alfreton Iron Works, Derbyshire*, https://www.gracesguide.co.uk/James_Oakes_and_Co

Heanor and Ripley News 1982. Canal on fire. The story of the Riddings oil strike. *Heanor and Ripley News*, 15 January 1982.

Johnson, R. 1970. *The History of Alfreton*. Britain & Sons Ltd, Ripley, Derbyshire, UK.

Oakes, T.H. 1884. A short history of Riddings. *Riddings Church Parish Magazine*, 28–31.

Patent Office 1850. *Treating Certain Bituminous Substances and Obtaining Products Therefrom*. Patent number 13,292, 17 October 1850. Patent Office, London.

Reid, T.W. 1899*a*. *Memoirs and Correspondence of Lyon Playfair, First Lord Playfair of St Andrews*. Cassell & Company, London.

Reid, W. 1899*b*. *Memoirs and Correspondence of Lyon Playfair, First Lord Playfair of St Andrews*. Harper & Brothers, New York.

Short, T. 1734. *A History of the Mineral Waters of Derbyshire, Lincolnshire and Yorkshire*. F. Gyles, London.

The Morning Post 1851. The Great Exhibition. 10 May 1851, p. 5.

The Scotsman 1865. Bathgate Paraffin Works. *The Scotsman*, 8 May 1865.

Toronto Globe 1862. 2 May 1862.

Argyll Field: the first oil field to be developed on the UK Continental Shelf

JON GLUYAS*, LONGXUN TANG & STUART JONES

Department of Earth Sciences, Durham University, Elvet Hill, Durham DH1 3LE, UK

J.G., 0000-0002-9386-7206; L.T., 0000-0002-6005-1374

Correspondence: j.g.gluyas@durham.ac.uk

Abstract: In June 1975, oil from the Argyll Field became the first to be produced from the UK North Sea. Seventeen years later, the Argyll Field was abandoned with all production facilities removed.

In 2001, two new companies, Acorn Oil & Gas and Tuscan Energy, had identified Argyll as a potential field redevelopment. The UK's Department of Industry was approached with a request to relicense the Argyll Field out of round in order to redevelop the field. No company previously had sought to obtain a licence for production rather than exploration.

In September 2003, a well was drilled on the renamed Ardmore Field. It flowed at 20 000 barrels of oil per day. However, after 2 months of sustained high rate, the well cut water. With a second well on stream, production peaked at 28 000 barrels of oil for 1 day before the facilities, designed for 50 000 barrels of fluid per day, tripped-out. All was not well; facilities and well issues limited production.

In mid-2005, the field was abandoned again: 5 MMbbl (million barrels) from an expected 25 MMbbl was produced. However, the story was not yet finished. By 2013 EnQuest had acquired the licence and drilled six wells. Production restart began in late 2015.

The Argyll Field in UK Block 30/24 was the first oil field to come on stream in the UK North Sea. It began to export oil via tanker loading in June 1975 and continued in production, albeit with a downgrade to a smaller production vessel, until 1992. During the 17 years it was on stream, 72.6 MMbbl (million barrels) of light 37° API crude was produced but by 1992, with an oil price of under $20 per barrel and daily production dipping below 6000 bopd (barrels of oil per day), the wells were plugged and the field was abandoned.

By the mid-1990s, all data from the field were released into the public domain and had become classic teaching material at Imperial College London. They were used extensively on the Petroleum Geology Masters course. It was while listening to student presentations at Imperial College that one of the authors (JG) became familiar with the field. JG had then just returned from a posting to Venezuela where he worked on the redevelopment of geriatric fields. Might it be possible that Argyll could be redeveloped? Work began. Although abandoned in 1992, it was clear from the outset that Argyll was without doubt a field behind its time. The one near analogue for Argyll, the Auk Field of Block 30/16, had kept pace with technology developments through the 1970s–90s. Auk has Zechstein and Rotliegend reservoirs but no Devonian pay intervals. It was developed with a traditional platform rather than the subsea wells used on Argyll. Auk also had 3D seismic coverage, high-angle wells, water injection and modern well completions. Argyll did not.

Ten years after Argyll was abandoned (2002), two new oil companies, Acorn Oil & Gas and Tuscan Energy, were awarded the licence to redevelop the Argyll Field. Since Argyll had ceased to exist in a legal sense, it was renamed as Ardmore. The first Ardmore well was drilled in the summer of 2003, and oil flowed in again in September 2003. A further 5 MMbbl were produced at high rate before commercial considerations, once again, forced abandonment despite the technical success.

Abandoned twice, Argyll/Ardmore was not yet dead! Ten years after cessation of production from Ardmore, the licence was once again live and EnQuest plc drilled six new high-angle wells. What one might hesitate to call first oil was produced from the field, renamed the Alma Field, in late 2015.

The aim of this paper is to examine the extraordinary development and redevelopment of the Argyll/Ardmore/Alma Field (henceforth referred to Argyll to 1992, Ardmore to 2005 and Alma to date), and determine why it was possible to redevelop the field twice.

Location and discovery

The Argyll Field occurs in a large tilted fault block in a highly elevated position on the western edge of the

From: Craig, J., Gerali, F., MacAulay, F. & Sorkhabi, R. (eds) 2018. *History of the European Oil and Gas Industry*. Geological Society, London, Special Publications, **465**, 77–93.
First published online April 11, 2018, https://doi.org/10.1144/SP465.3

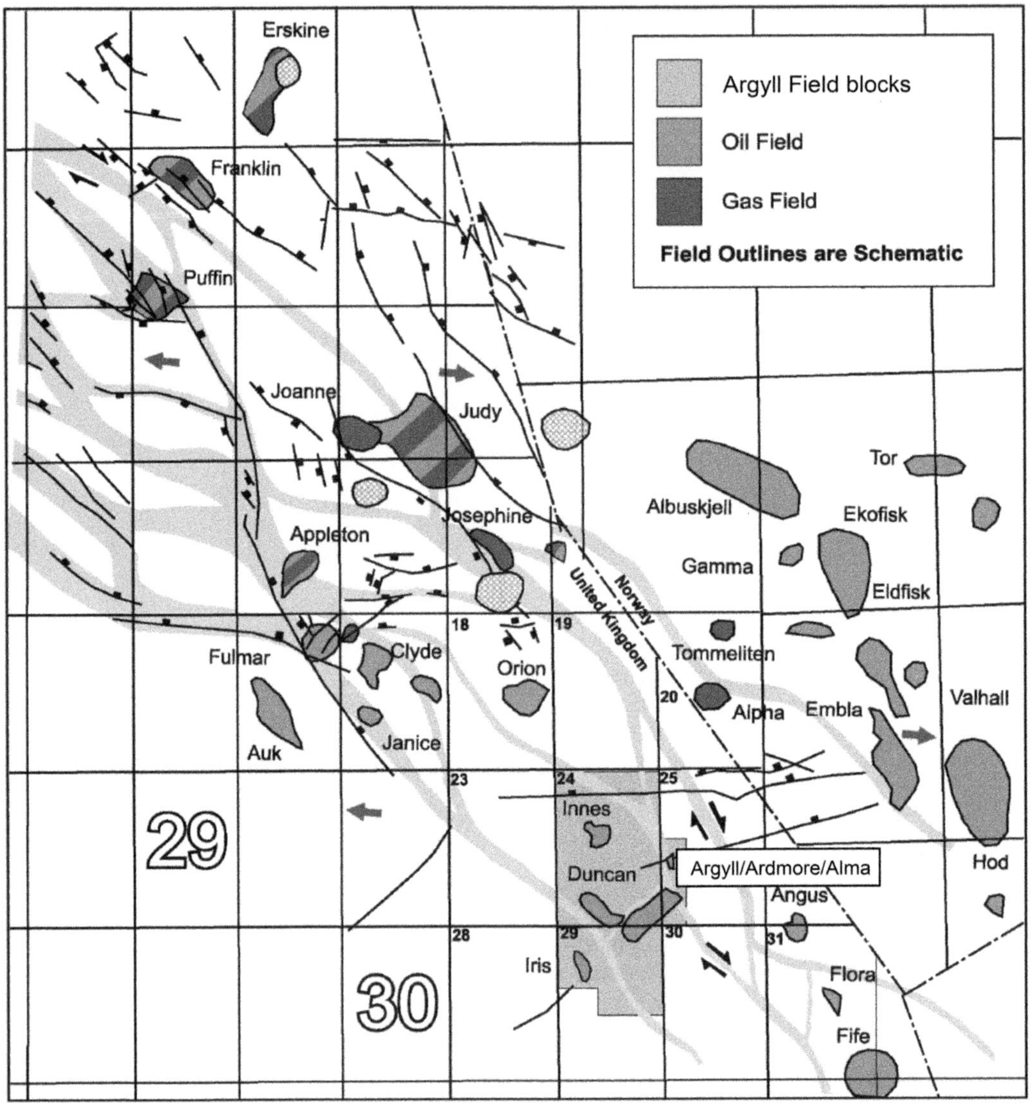

Fig. 1. Regional position of the Ardmore Field showing the Tuscan and Acorn licensed blocks (structure adapted from Gluyas *et al.* 2005).

Central Graben of the North Sea High (Fig. 1). The first well on Block 30/24 was drilled by Hamilton Brothers Oil Company in 1969 and its location was based upon interpretation of sparse 2D seismic data. Indeed, 30/24-1 was plugged and abandoned before the well evaluation was complete and therefore before it was recognized to be a discovery with oil shows in the Zechstein interval. In consequence, a second well 3024 was drilled close to the well 1 location. As with well 1, oil was struck in carbonate rocks at a depth of 8900 ft (true vertical depth subsea (TVDSS)) and flowed oil at 4314 bopd from what was determined to be Permian Zechstein dolomite. The same well also flowed at 1446 bopd from a deeper sandstone reservoir, believed at the time to be the Lower Permian Rotliegend interval but later shown to be Devonian sandstone. Appraisal followed, and in June 1975 the Argyll Field became the first oil field on stream on the UK Continental Shelf (Pennington 1975).

Transworld 58, a converted drilling vessel was used as the production facility. Crude oil was loaded to a tanker offshore and then shipped to market. As the field came on stream, light sweet oil flowed at high rate. Drilling continued to add ever more production wells to the portfolio and, although each new well was drilled deep into the red beds beneath the Zechstein interval, only the Zechstein vuggy and fractured carbonates were ever completed for production. Deeper oil was often proven and flow tested, as well as pressure measurements taken, but wells were effectively plugged beneath the Zechstein. It was about 4 years after production start up, in mid-1979, that a well was completed in the sandstone interval beneath the Zechstein. Hamilton Brothers Oil Company kept very detailed records on the drilling and completion of wells but we have not found any information which allows us to understand why the company only developed one formation per well, and the temporal progression was from completion of the youngest, shallowest Zechstein reservoir early in the production history to deeper and older reservoirs late in field life.

These deeper, oil-bearing sandstones, although described as Rotliegend, were for the most part unlike those seen in the Southern North Sea Gas Basin and onshore England. Eventually, the sandstones were recognized to be a combination of Permian Rotliegend sandstones and Devonian (Old Red) sandstones and siltstones (Heward *et al.* 2003).

Small amounts of oil were found in shallow-marine Upper Jurassic sandstones that only occur on the western flank of the field (30/24-6 and 30/28-8). More surprises were discovered. In the NE of the field, the Zechstein reservoir is completely eroded and the Upper Cretaceous Chalk seals variably fractured but otherwise tight Devonian sandstones that were proven productive by well 30/25a-2 (1982, tested at 2280 bopd) but never completed as a producer. The seal horizon within this well is within the Chalk, the lower part of which is oil bearing. Core from 30/24-20 (1982) revealed a conglomerate facies of Upper Jurassic age containing pebbles of Zechstein dolomite and oil-bearing matrix which had not been described previously. Even the final well drilled by Hamilton on the southern edge of the field (30/24-39, 1989) delivered a further puzzle when volcanic rock (probably lowermost Permian Karl Formation: Stemmerik *et al.* 2000) was encountered at the level where Rotliegend sandstones were expected.

Production peaked in 1976 only 1 year after the field began production (Fig. 2) with a half-year average of about 28 000 bopd. At this time, only the Zechstein interval was on production. Wells completed in the Rotliegend were added in 1979 and the Devonian was brought on stream in 1982. This gave two lesser peaks in production of just over 20 000 bopd in the first halves of 1979 and 1982.

Two smaller discoveries in the block were later tied back to Argyll (Duncan, Jurassic reservoir in 1984; Innes Rotliegend reservoir in 1985). The Innes crude has a higher gas–oil ratio than that of Argyll, and its excess gas production was used for gas-lifting the Argyll wells. Despite the addition of gas-lift, oil production from all three fields continued to fall and by 1992 the daily rate was 6000 bopd. This combined with 14 000 bbl of water production to give a total fluid production of 20 000 bbl/day. A changed production set-up several years previously meant that the total fluid capacity of the vessel was 20 000 bbl/day. Despite there being tested but unproduced wells, the combination of vessels' throughput capacity and insufficient gas for gas-lifting all wells served to limit oil production to 6000 bopd and falling. In the absence of a larger

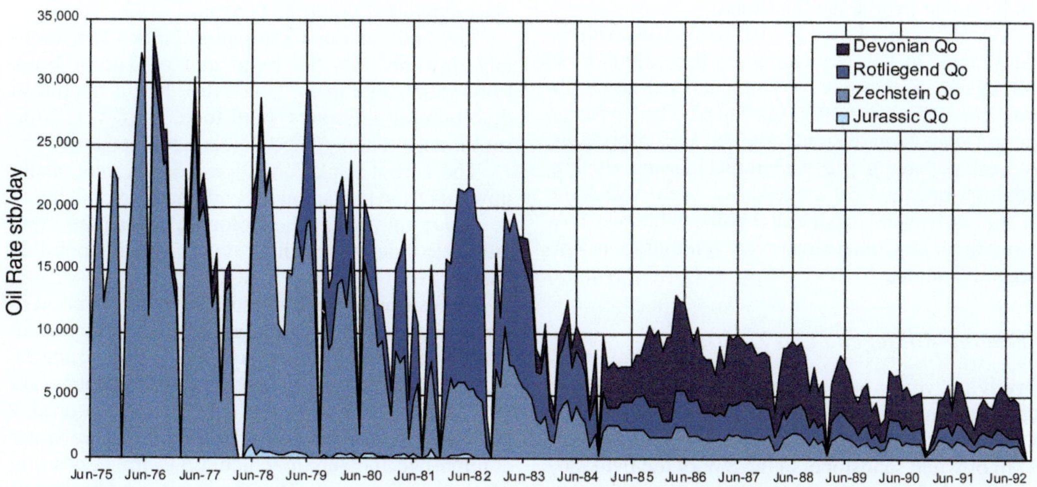

Fig. 2. Production profile for the Zechstein, Rotliegend, Devonian and Jurassic reservoirs of the Argyll Field from start-up in 1975 to abandonment in 1992.

vessel capable of handling more fluids, the field was failing. Hamilton Brothers were in the process of being taken over by BHP and oil was below $20 per barrel. A cessation of production document was lodged with the Department of Trade and Industry, and the Argyll Field along with Duncan and Innes fields were abandoned.

Structure and stratigraphy

The Argyll Field is at the southern end of the Central Graben. It lies largely on the rift shoulder straddling the major bounding fault to the southernmost part of the Central Graben and above a significant transfer zone. Movement within this zone caused repeated inversions of the fault block that contains Argyll, and this led to the complex, residual stratigraphy seen in the area with multiple and composite unconformities (Fig. 3). The most recent inversion appears to have occurred in the early Neogene, giving Argyll its characteristic elongate NE–SW trap geometry (Fig. 4).

The key stratigraphic sections are given in the following subsections.

Devonian

The oldest rocks penetrated in Argyll are Devonian Kyle Group limestones, formed during widespread marine transgression in the Middle Devonian (Ziegler 1990) and informally referred to as the 'Mid Devonian Limestone', a conspicuous, high-amplitude and continuous seismic reflector for recognizing Middle Devonian strata (Milton-Worssell *et al.* 2010). The reservoir interval in Argyll is the Buchan Formation of the Upper Old Red Group, a thick succession of red beds that overlie the limestone.

The Devonian dips at 7°–10° to the SW, with the oldest rock in the NE corner of the Argyll Field (30/25a-2). Progressively younger Devonian subcrops the BPU to the SW (Figs 5 & 6). The Devonian section may contain subtle angular unconformities, as well as having a considerable missing section, although unequivocal evidence of such is lacking. In the early well and field reports, what we now know to be Devonian strata were referred to as Rotliegend Gamma.

Permian–Rotliegend and Zechstein

Overlying what is confidently identified as Devonian sandstone in well 30/24-10 in the north of the field is a rock described as weathered olivine basalt beneath 30 m of white sandstone. In the SW of the field, 30/24-39 includes weathered volcanics or volcaniclastics beneath the Zechstein carbonates. Both intervals may be lowermost Permian Karl Formation (Fig. 7).

Argyll lies within the area known as the Northern Permian Basin, and Early Permian, Rotliegend Auk Formation sandstone overlies the Devonian. This interval was originally referred to as Rotliegend Alpha to distinguish it from what we now know to be Devonian strata and which in the early days was called Rotliegend Gamma. No references have been found to a Rotliegend Beta in the drilling or completion reports or, indeed, the early field appraisals. It is possible that such a term was reserved for a unit that might be seen between the clearly discordant Gamma and Alpha intervals. Rotliegend sandstones *sensu stricto* in Argyll are confined to a low-relief NW–SE syndepositional valley through the centre of the field (Robson 1991). Beyond the immediate area of the Argyll Field, the Rotliegend sandstones are widespread and thicker with a greater range of facies (including fluvial sandstones: Heward *et al.* 2003).

The Kupferschiefer mudstone and Zechstein carbonates overstep the Rotliegend sandstones and rest directly upon the Devonian strata (Fig. 7). The Kupferschiefer mudstone is not everywhere present, and it is possible that it was not deposited over the crest of the partially collapsed Rotliegend sand dunes ahead of the Upper Permian Zechstein marine incursion. Halibut Bank Formation carbonates overlie the Kupferschiefer and are, in turn, overlain by the Sapropelic Dolomite and Turbot Bank Formation carbonates.

The original fabric of the Halibut and Turbot formations is difficult to determine due to pervasive dolomitization, dedolomitization and brecciation. The Halibut Bank interval has a transitional boundary with the Kupferschiefer. It comprises friable and argillaceous dolomite interlaminated on a centimetre scale with non-porous dolomite. This is overlain by an upwards-coarsening, poorly sorted, clast-supported dolomite breccia.

The Halibut Bank Formation passes transitionally upwards into the basal unit of Turbot Bank Formation (Sapropelic Dolomite). It is a laminated dolomitic argillaceous mudstone with very thin (millimetre-scale) graded dolomitized beds.

The Turbot Bank Formation passes transitionally upwards from the laminated carbonate slope facies at the top of the Sapropelic Dolomite into a mesocrystalline to finely crystalline dolomite with possible biomoulds.

Deposition of the carbonate intervals occurred in a shallow-marine environment around what was an intrabasinal high, the embryonic Argyll structure. The carbonates overstep Rotliegend sandstones and over much of the field area rest directly on eroded Devonian sandstone. From time to time, the water column became anoxic which led to the deposition of sapropelic units. The Zechstein succession at Argyll contains no anhydrite and no halite in contrast to elsewhere in the Northern Permian Basin. The

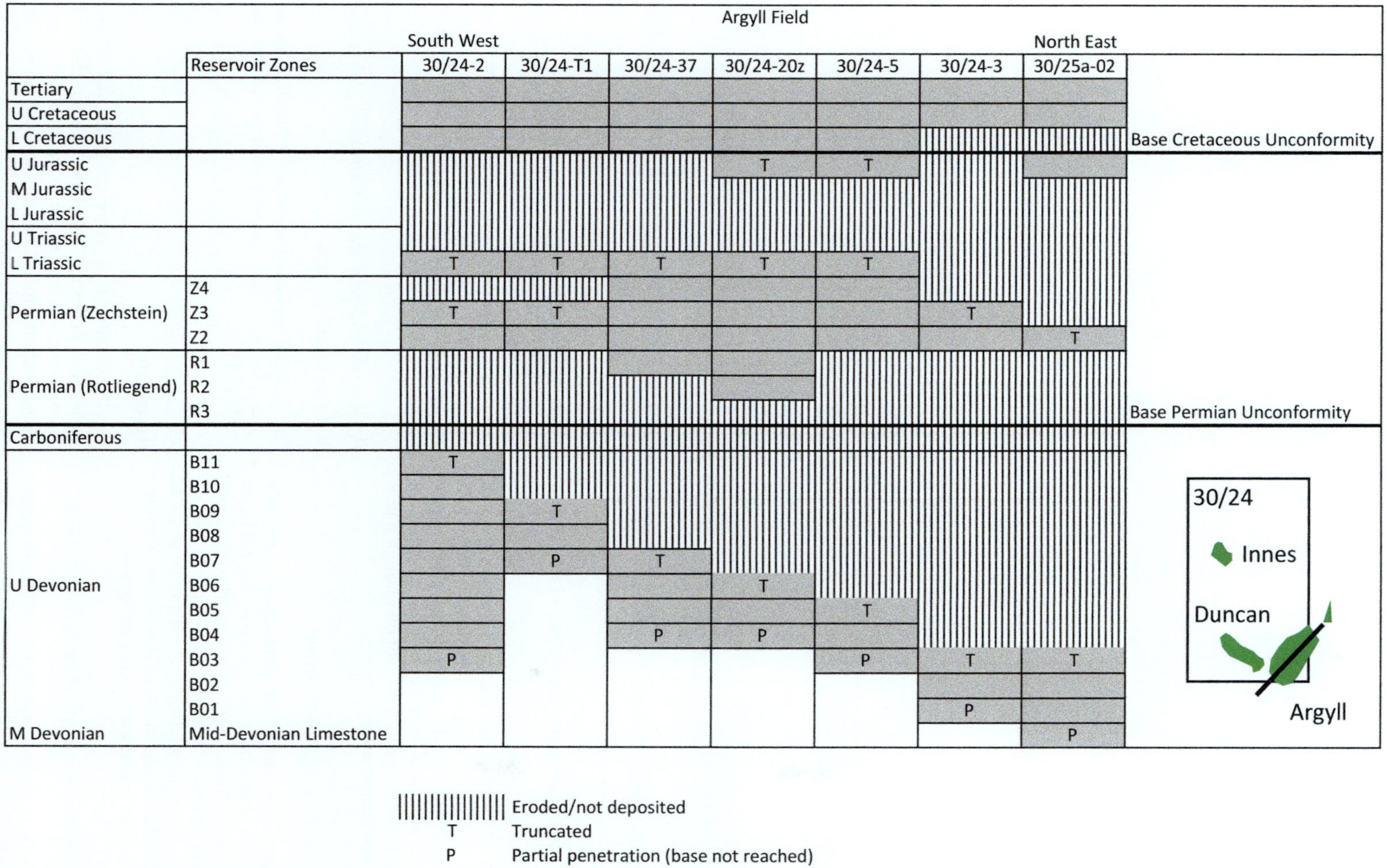

Fig. 3. Simplified chronostratigraphic chart for the Argyll Field. In addition to the unconformities shown, the Upper Cretaceous Chalk onlaps the field crest to provide the top seal in the NE of the field, and it is clear from seismic data that there is also some missing section in the Lower Tertiary.

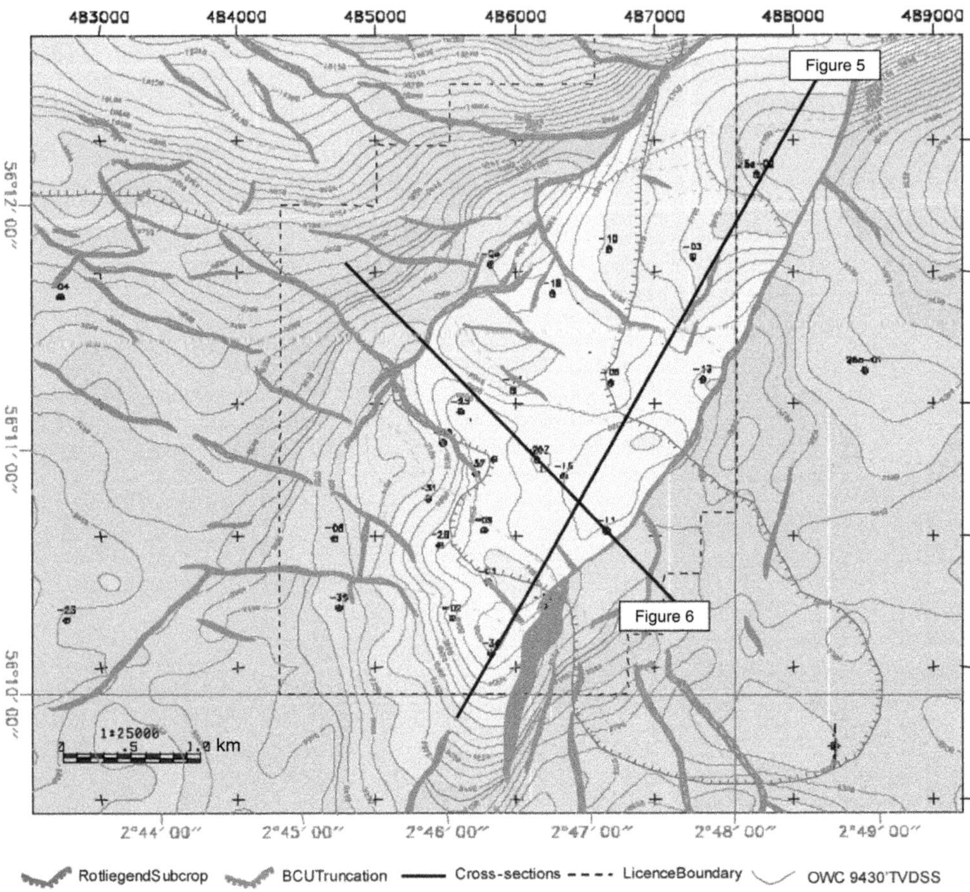

Fig. 4. Top reservoir depth structure map of the Ardmore Field showing the original Argyll Field well locations

presence of collapse breccias at Argyll is taken to indicate that some evaporate minerals may well have been dissolved but it is also likely that the structurally elevated position of the Argyll area in the Late Permian limited evaporite deposition through a lack of accommodation space.

Upper Jurassic Humber Group

The Upper Jurassic Fulmar sandstones occur in a north–south strip across the centre of the field and on the western flank. Here, it only poorly developed as a combination of basal breccia with pebbles of Zechstein dolomite overlain by clean quartzose sandstone. On the western flank of Argyll, the shallow-marine Fulmar sandstone is better developed.

Upper Cretaceous–Paleocene Chalk Group

The Chalk is present across the whole of Quad 30, indicating that by Upper Cretaceous times the area was finally submerged. Oil shows occur within this interval across Argyll, although the same stratum is the ultimate top seal in the NE corner of the field.

Paleocene/Eocene–Recent Stronsay, Westray and Nordland groups

The Tertiary section that overlies Argyll is largely argillaceous and is not considered to be prospective.

Reservoirs

Argyll contains four reservoir units (Devonian, Rotliegend, Zechstein and Late Jurassic: Fig. 7). The three oldest are in pressure and fluid communication, and the fourth, youngest and minor, unit (Jurassic) is independent (Fig. 7).

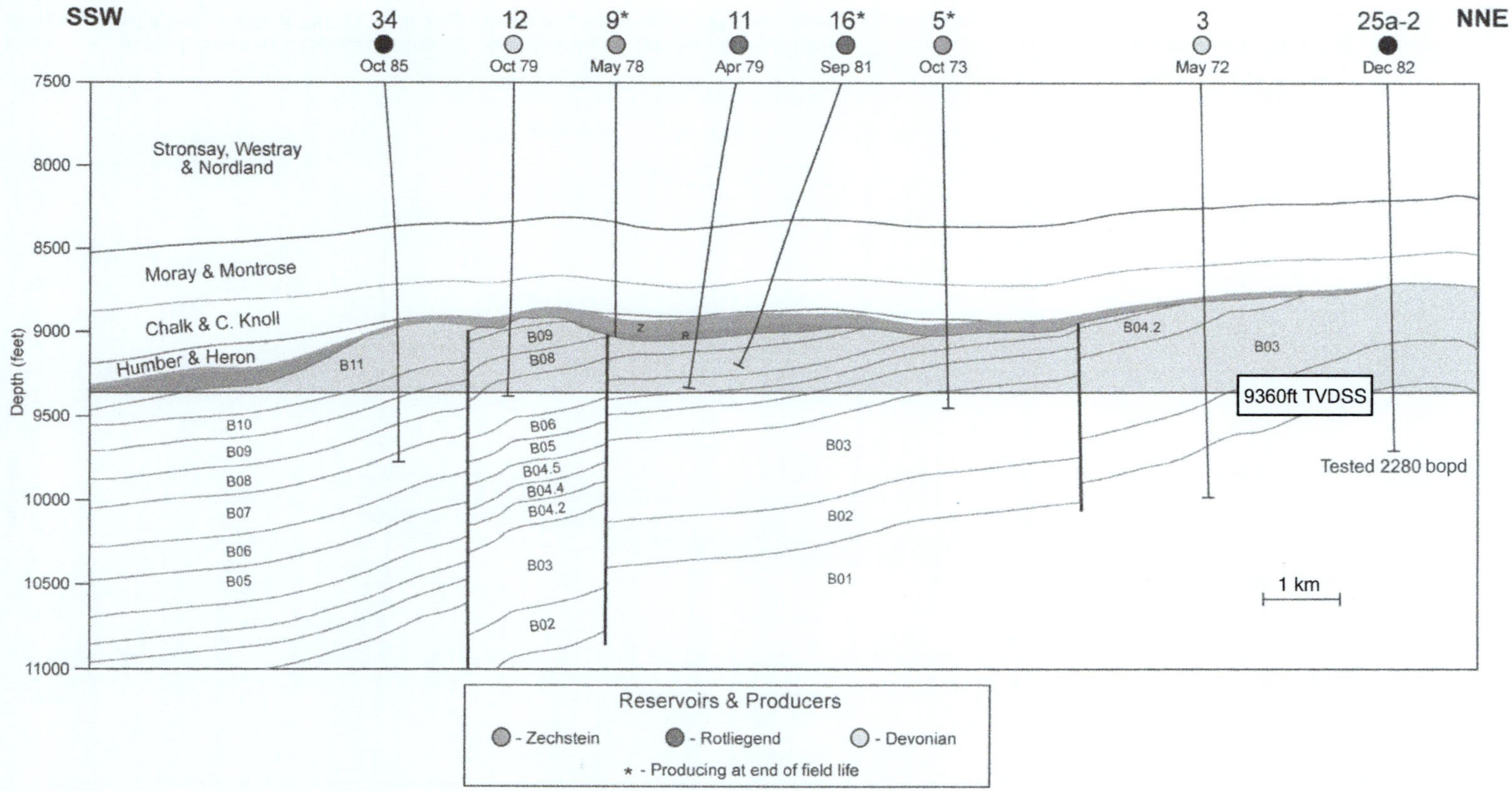

Fig. 5. SW–NE depth cross-section across the Ardmore Field showing the stratigraphic relationships of the productive Jurassic, Zechstein, Rotliegend and Devonian reservoirs above a nominal IOWC of 9360 ft TVDSS.

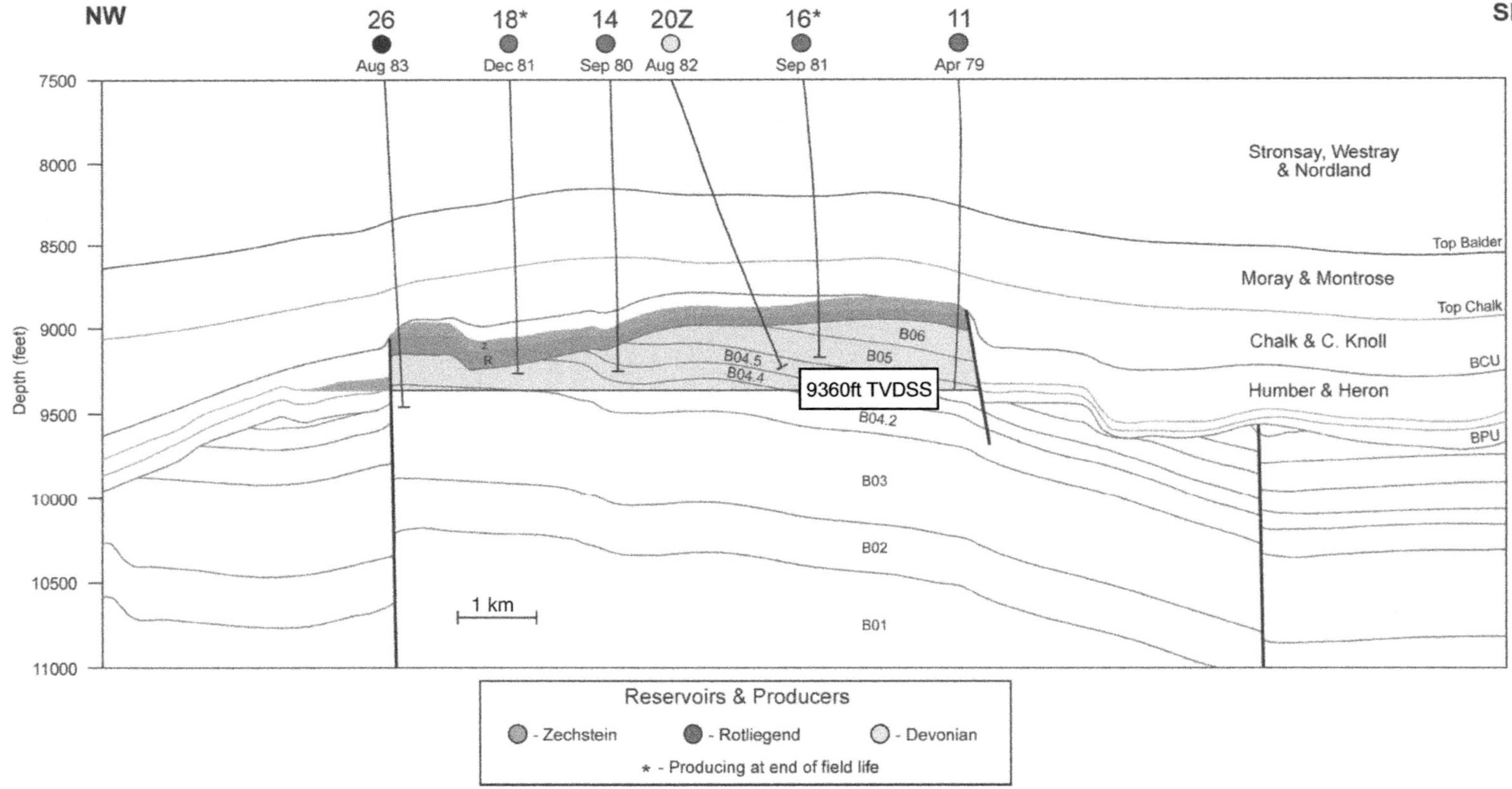

Fig. 6. NW–SE depth cross-section across the Ardmore Field showing the stratigraphic relationships of the productive Jurassic, Zechstein, Rotliegend and Devonian reservoirs above a nominal IOWC of 9630 ft TVDSS. BCU, base Cretaceous Unconformity; BPU, base Permian Unconformity.

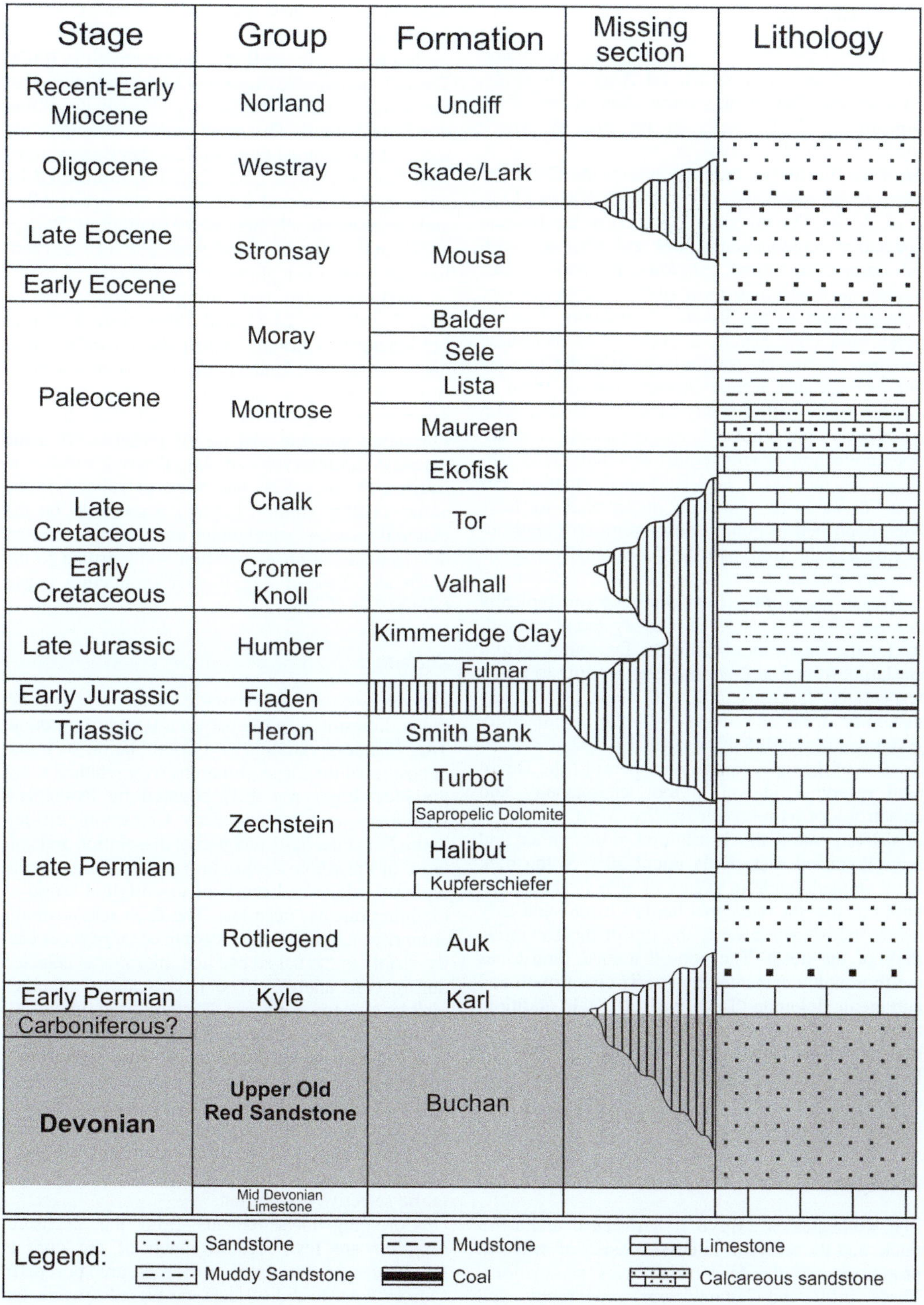

Fig. 7. General stratigraphy of the Ardmore area.

Upper Devonian

The base of the Devonian section is not penetrated by any of the wells drilled on Argyll. However, because the Devonian section dips at about 7° beneath the BPU, wells in the NE of Argyll encountered older Devonian strata than those in the SW. This allows the thickness of the Devonian strata to be calculated to be at least 500 m (Tang *et al.* 2018). The oil-bearing section of the Buchan Formation contains sandstones and siltstones with subordinate mudstones, conglomerates and rare carbonaceous mudstones; these only occur in the very youngest part of the Buchan Formation. The log trends and core exhibit a cyclicity of sandstone and finer sediments for the whole of the interval but there is also a clear overall upwards trend to cleaner better-sorted sandstones interbedded with mudstones from a rather homogenous and monotonous collection of silty sandstones and sandy siltstones at the base. The Buchan Formation has been divided into intervals interpreted to have been deposited in either arid or semi-arid environments, and the wells are then correlated using the wet–dry cyclicity.

Net to gross in the Devonian reservoir typically exceeds 80%. The ranges in porosity and permeability for the sandstones within the Devonian section are large, from 5 to 28% and from <1 mD to >1 D, respectively (porosity arithmetic mean 14.4%; permeability arithmetic mean 340 mD; permeability geometric mean 10.5 mD).

One of the more intriguing aspects of the Devonian reservoir interval is the heterogeneity and anisotropy of its reservoir quality. Most noticeable in slabbed core is the distribution of oil-stained and non-oil-stained rock. Only about 30% of the rock is oil stained; while in excess of 80% is sandstone, with much of the remainder being siltstone and only a few mudstones towards the top of the Devonian (SW of the field). The non-oil-stained sandstones are thoroughly cemented by quartz overgrowth and authigenic dolomite (Tang *et al.* 2018). In addition, there is a progressive diminution in oil saturation with depth/stratigraphy in wells that penetrate the Devonian section. The causes of cementation/non-cementation and their impact on reservoir performance are discussed by Tang *et al.* (2018).

Rotliegend

The Rotliegend in Argyll is between 0 and 34 m thick, and the reservoir interval consists of different proportions of the four main facies associations. These infill end-Carboniferous topography and onlap the Argyll High (Heward *et al.* 2003). The four facies associations include: aeolian slipface sandstones; aeolian ripple-laminated sandstones; waterlain sandstones; and waterlain conglomerates and breccias.

Net to gross in the Rotliegend reservoir is 100%. The most permeable facies within the Rotliegend are the coarse waterlain sandstones (Weissliegend) that dominate Unit 1. The most porous reservoirs are the medium-grained dune slipface sandstones, most abundant in Unit 3. Units 2 and 4, dominated by wind ripple-laminated sands, have low permeability and characteristically high water saturations. Permeability in the waterlain sandstones of Unit 1 and aeolian sandstones is high, typically 1–5 D.

Production logs run during the completion of Argyll wells 30/24-11 and Innes well 30/24-24 (15 km north of Argyll) confirm that the bulk of production was from Unit 1. Seven wells produced oil from the Rotliegend reservoir in Argyll. However, material-balance estimates indicate that the Auk Formation was not just an oil producer. It alone amongst the reservoirs of Argyll has a substantial aquifer. Aquifer size has been calculated to be about 10 Bbbl (billion barrels) based both on the material-balance calculations and on the pressure data gathered from exploration wells drilled to the north and west of Argyll after production began (Heward *et al.* 2003).

Zechstein

The Zechstein interval contains dolomite and subordinate limestone, and is between 0 and 50 m thick. The thickness variation is a combination of initial depositional thickness variations over residual topographical highs not fully covered by Rotliegend sandstones, end Jurassic–early Cretaceous erosion in the NE of the field, and partial dissolution and collapse of the whole Zechstein interval. The dolomite replaces calcite and much of the original fabric of the limestone has been lost. The Zechstein reservoir comprises a dual-porosity system of large pores easily visible to the naked eye and micropores detected during core analysis. The large-scale macropores visible in core comprise fractures and vugs. All porosity in the Zechstein interval is secondary in origin. Some of the fractures are of tectonic origin but most comprise anastomosing fracture systems within collapse breccias. Vugs include irregular 100 μm–5 mm-diameter pores formed by dissolution of allochems during dolomitization and evaporite dissolution. Some vugs are also formed by non-fabric selective dissolution.

Core-plug measurements from the Zechstein reservoir are few because much of the core is fragmented. In consequence, the measured poroperm (porosity–permeability) data are biased towards the poorer-quality reservoir. The highest porosities have been measured in the collapse breccias that have both intercrystalline and vuggy porosity (porosities of up to 20% and permeability up to 1 D). Porosity and permeability are poorly correlated. Indeed, many plugs have only about 5% porosity, while permeability values measured in hundreds of millidarcies have been recorded. The cause of the high permeability at low porosity is fracturing. This appears to be natural and not core-induced. Due to the fracturing, the whole Zechstein reservoir is considered to be a net reservoir. That is to say, net to gross is 100%.

Given the difficulty of quantifying the reservoir quality of the Zechstein from the existing conventional core analysis data, additional information has been taken from the drilling reports. Mud losses were common during drilling of the karstic layers in both the Turbot and Halibut formations, with reported rates ranging from 60 to 200 bbl/min. The effects of this mud invasion are readily seen in both fractures and vugs within the core. Drilling of the Zechstein was also typified by erratic drilling breaks and frequent jamming of the core barrel.

Upper Jurassic

The Upper Jurassic Fulmar Formation reservoir is poorly developed on the west flank of Ardmore where it comprises fine- to medium-grained clean to argillaceous sandstones. Net to gross is typically high (80–100%) and porosity is in the range 16–21%. Permeability varies widely from a few millidarcies in the more argillaceous intervals to a few hundred millidarcies in the cleanest sandstones. Well 30/24-6 produced about a 0.4 MMbbl of oil from this interval. The Jurassic of Argyll was not cored, although that of the immediately adjacent Duncan Field (now called Galia) was cored. Here the Upper Jurassic Fulmar sandstone is interpreted to have been deposited as offshore bar sandstones into which sand-filled tidal channels are cut. Other wells in the 30/24 Block to the north of Argyll often have very poor sandstone development, while a further sand-bar complex forms the oil accumulation informally called Iris in Block 30/29 to the south of Argyll.

Source

The Upper Jurassic Kimmeridge Clay Formation is the source rock for oil within each of the three fields on Block 30/24. The kitchen area was the deep Central Graben to the east and north of Argyll, and the likely sourcing direction was from the north along a broad ramp linking Argyll (and Duncan) with Innes to the north. There may have been several phases of petroleum migration as evidenced by the distribution of a fluorescent dead oil and a non-fluorescent live oil in Innes, and two oil types recorded from the Affleck Field to the north of 30/24. Given that we know the Rotliegend to be the main aquifer in the region, it is reasonable to assume that oil migration occurred through this interval too. The filling of Argyll was likely to have been via the saddle which separates its western flank from the Duncan Field. On entering the Argyll structure, the Devonian section was filled downwards and Zechstein section upwards from the Rotliegend.

STOIIP and reserves

The initial oil in place and reserves figure has been calculated many times for Ardmore and its predecessor Argyll since the field was discovered in 1969. Nine sets of STOIIP (stock tank oil-initially-in-place) calculations from internal company records between 1982 and 2003 have been collated. These range from 185 to 375 MMstb (million stock tank barrels) for the four reservoirs combined (Table 1). A significant uncertainty affecting gross rock volume is the depth of the initial oil–water contact (IOWC). There are few penetrations of a proper contact, with the most common value being around 9360 ft TVDSS as seen in the central and SW parts of the field. Projection of a deeper contact (9430 ft TVDSS) on a field-wide basis certainly accounts for most of the additional oil in the highest STOIIP figure, with most increase occurring in the Devonian reservoir. In the NE part of the field, the quality of the Devonian reservoir diminishes and, in most instances, only 'oil down to' depths that have been measured.

The P_{50} STOIIP estimate is 375 MMstb (summed means). The subdivision by reservoir gives 33 MMstb in the Jurassic, 42 MMstb in the Zechstein, 50 MMstb in the Rotliegend and 250 MMstbl in the Devonian.

It is interesting to note that the STOIIP figure used by Hamilton Brothers Oil Company cessation of production report is the lowest of all STOIIP figures found, while the figure used by Tuscan and Acorn prior to redevelopment of the field as Ardmore is the highest. It is difficult not to think that some influence was brought to bear on the technical evaluation by the circumstances of the two sets of companies; one wishing to abandon the field and the other to redevelop it.

Argyll production

Hamilton Brothers Oil Company kept very detailed well records that cover the drilling and completion phases of the well, together with the production history. However, we have little direct information on the development plan for the Argyll Field. We know that, although Argyll was not the first

Table 1. *Original oil-in-place calculations by different field owners*

Reporting entity	Year	Mean or P50 STOIIP (MMstb)	Notes
Hamilton Brothers Oil Company	January 1982	342	Although the field was on stream by June 1975, these are the earliest STOIIP figures found
Hamilton Brothers Oil Company	February 1984	266	Notes from Technical Committee Meeting
ERC	July 1986	218	Technical consultant report
Hamilton Brothers Oil Company	April 1989	225	Reservoir re-evaluation report
Hamilton Brothers Oil Company	July 1991	198	Cessation of production report
Blackwatch	September 1997	277	Consultant report for Monument Oil & Gas
Atlantis	January 1998	202	Would be redevelopment company
Lasmo	December 2000	185	Consultant report for Lasmo
Tuscan and Acorn	From 2001	375	One of several STOIIP figures generated by Tuscan and Acorn. This one used in the redevelopment plan

commercial oil discovery in the UK North Sea, it was the first field to produce oil. Moreton (1995) reproduced a famous photograph from June 1975, in which the UK's newly appointed Secretary of State for Energy, Tony Benn, appears to be opening a large valve to allow Argyll oil to flow ashore. In the background a hording proclaims, 'UK North Sea Oil, Hamilton Brothers...'. In truth the oil did not flow ashore but was, instead, loaded from the Argyll production facility to a taker that then sailed to port before unloading the oil. Hamilton had won the publicity battle by producing first oil from the North Sea ahead of that from BP's giant Forties Field (Moreton 1995). What is also clear from the old Hamilton company documents we reviewed is that the company geologists and geophysicists were planning, drilling and testing production wells when the production engineers could not then produce them due to a shortage of gas for gas-lift. All wells were drilled as vertical or near vertical; all zones were tested; wells only ever produced from perforations into one reservoir, Zechstein first, Rotliegend second and, eventually, in the eighth year of production, the Devonian. No water injection was ever planned and in many respects the Argyll Field never got beyond the appraisal phase. Despite this, the Argyll Field produced 72.6 MMstb, with 41.1 MMstb flowing from perforations into the Zechstein reservoir, 18.2 MMstb from perforations into the Rotliegend reservoir and 12.8 MMstb from perforations into the Devonian reservoir. The Fulmar sandstone produced about 0.4 MMstb from one well. Comparison of STOIIP and produced reserves appears to indicate that the recovery factor for the Zechstein was almost 100% and about 37% for the Rotliegend, while only 5% of the in-place volume of Devonian oil was recovered. Clearly, the recovery factor for the Zechstein is unreasonable. It is known from the pressure data that the Zechstein, Rotliegend and Devonian reservoirs are in communication, and therefore it is concluded that some of the oil produced from the Zechstein was originally reservoired in either the Rotliegend or Devonian. Analysis of the well performance of those wells completed in the Zechstein reveals a correlation between the volume of oil produced before water breakthrough (measured at a water–oil ratio of 0.5) and the lateral distance to the nearest known aquifer. For early wells, the nearest known aquifer was the field margin, while for later wells it was taken to be any well cutting substantial water (Fig. 8). This simple linear relationship is in contrast to the complex reservoir geology in the Zechstein and it is taken to indicate that the principle transmissive component of the Zechstein reservoir is the pervasive fracture system, and moreover that this trend is observable in the absence of any bottom water to the field.

The remaining reserves for the second phase of development of the field, when it was named Ardmore, were calculated both by extrapolating the water-cut trend for the Ardmore Field and by modelling the reservoir using material balance. Both methods give similar results. When Argyll was abandoned it was producing about 6000 bopd. The total daily fluid rate was 20 000 bfpd, a figure constrained by the production facility then available. Thus, at abandonment, the water–oil ratio was about 2 and extrapolation of the water–oil ratio to 5 (Fig. 9) yields a remaining oil reserve of about 25 MMstb (range 20–40 MMstb).

Ardmore development and production

When in 2001 Tuscan Ltd and Acorn North Sea Ltd applied for a licence to redevelop Argyll, they did so

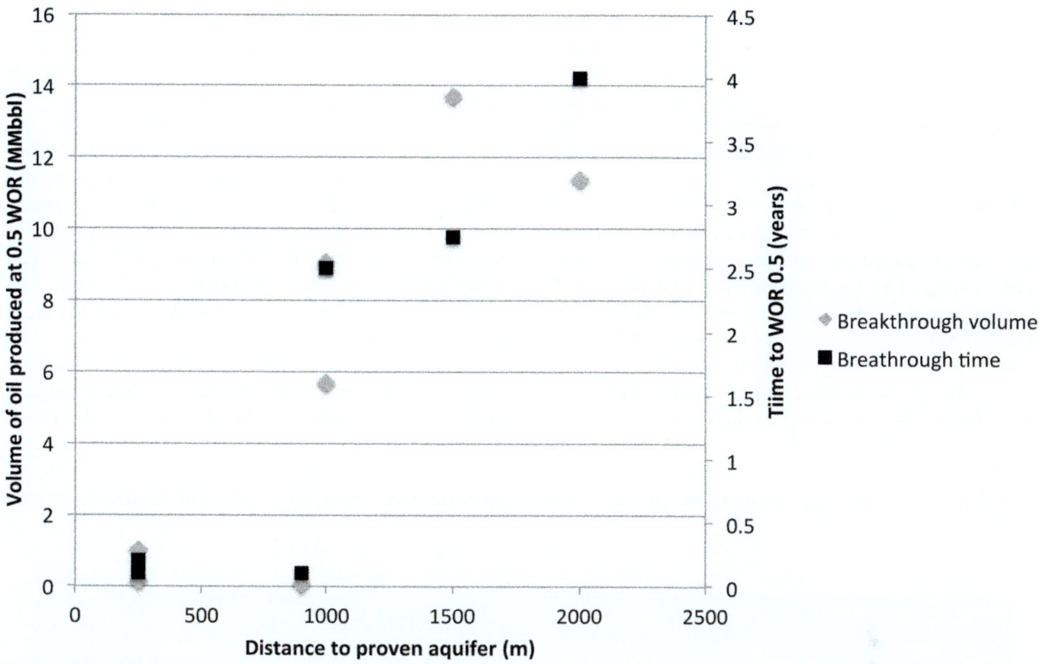

Fig. 8. Relationship between water breakthrough oil volumes and date v. lateral distance to aquifer. WOR, water–oil ratio.

with a very simple development plan constrained by available finances. The approach was to differ little from that used by Hamilton. No water injection was planned on the basis that the Argyll Field had an active lateral aquifer. This together with the use of electro-submersible pumps coupled with moderate-angle production wells (60° for the first well, rising to 80° for subsequent wells) was expected to deliver high-rate sustained production. Thus, for the Ardmore development, three firm wells and

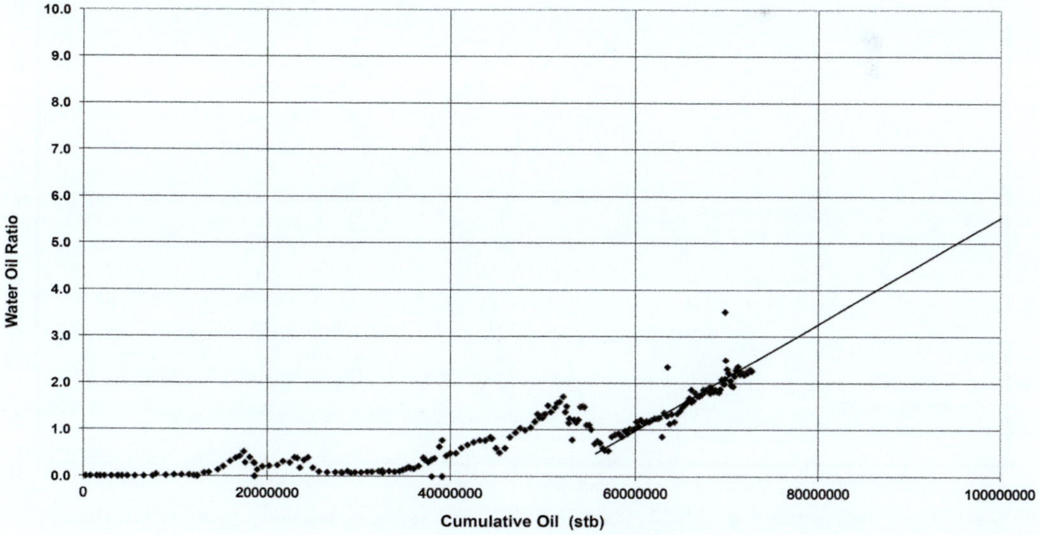

Fig. 9. Water–oil ratio and cumulative oil production plot for the Argyll Field. Projection of the water cut to a water–oil ratio of about 5:1 would deliver an additional 25 MMbbl in the Ardmore redevelopment.

one contingent well were planned for the first phase of field redevelopment to access the remaining 25 MMstb of oil reserves. A further three subsurface locations were identified as possible future well locations dependent upon the outcome of initial drilling and production (Fig. 10). The limited finances available for the Ardmore project led to suboptimal redevelopment of the field. It was recognized by both Tuscan and Acorn that water injection would have been beneficial for pressure maintenance and petroleum sweep. The contracts for the first three wells were turn-key. This then allowed the drilling contractor to define the well geometries and limit the angle through the reservoir section to 60°. The funding model also demanded rapid payback and, hence, wells had to be completed in the highly productive Zechstein interval from day 1, whereas a recovery mechanism optimized to maximizing recovery would have begun with completion of the lower rate and deeper Devonian intervals (see below). By the time the field was abandoned for a second time, three new wells had been drilled and two of these new wells had been sidetracked.

Two types of target were identified for the Ardmore wells. The low-risk targets were Devonian-reservoired oil, proven during the Argyll phase but never produced. The reservoir quality was variable from poor to good, and well rates were expected to be modest relative to the rate that could be achieved from Rotliegend and Zechstein completions. The most likely initial water cut in such wells was calculated to be zero.

The higher risk targets were Zechstein and Rotliegend reservoirs where, because of their excellent properties, unconstrained well rates could exceed 20 000 bopd. The prediction of the initial water cut in such wells and the rate at which water cut would increase was the major risk. Oil production from

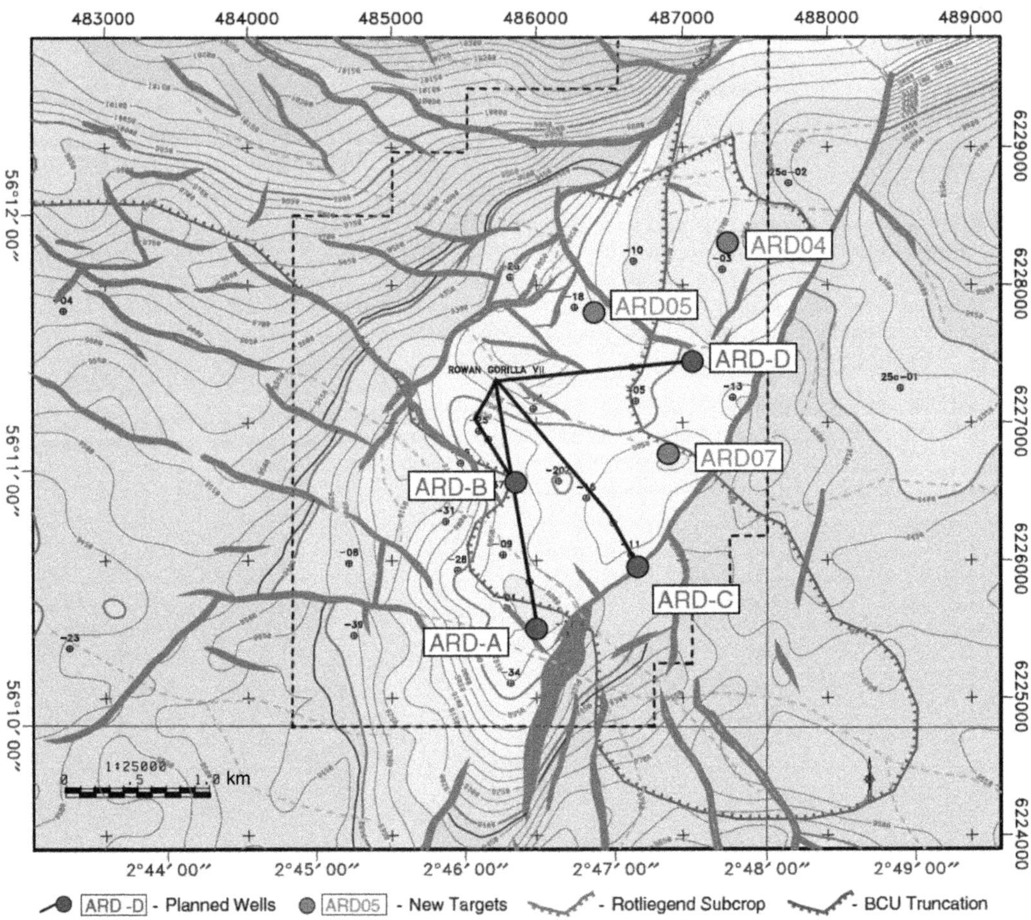

Fig. 10. Possible new well trajectories for the Ardmore Field development. ARD-A became 30/24-T1, ARD-B became 30/24-T2 and ARD-C became 30/24-T3. The other planned wells were not drilled.

Ardmore was not as expected, although it took some time to understand that it was not errors in the reservoir description but, rather, the failure to execute drilling and completions as planned. This was compounded by problems with the production facility that was unable to meet the design criterion and handle 50 000 bwpd (barrels of water per day).

The first well drilled for the Ardmore production phase (30/24-T1, ARD-A) was located updip of one of the best wells from the Argyll production period (30/24-9). Tuscan and Acorn had the benefit of being able to place the well on the basis of interpreted 3D seismic data, while only a sparse 2D dataset was available to Hamilton. T1 was expected to penetrate Zechstein carbonates lying above Devonian sandstones. It was anticipated that flushed horizons could occur anywhere in the section. On the basis of the old wells, it was thought probable that the base of the Zechstein or top of the Devonian were likely to be flushed as this interval was coincident with the top perforations in the old well nearby.

Given the limited funds available to the two companies, it was deemed necessary to produce the well at the highest rate possible: that is, by combined production from the Zechstein and Devonian reservoirs. It was recognized that water ingress to the Zechstein was likely to be rapid, and to this end a different, less simple, completion strategy was employed. A sliding sleeve was used as part of the completion over the Zechstein interval. The idea was that water could be shut off from the Zechstein leaving only the Devonian producing.

Well 30/24-T1 was drilled in August 2003. Petrophysical analysis revealed a full oil column. No water was encountered. The Zechstein was high quality, as was the underlying uppermost Devonian. Pressure data indicated that in 11 years, the field had recovered about two-thirds of the way between the virgin pressure and the field abandonment pressure. All looked good.

In October 2003, production began. The choke was progressively opened and under natural flow the rate increased to over 20 000 bopd – miracle. The rate was about 30% higher than any previous well had ever managed during the Argyll production phase. A production rate in excess of 20 000 bopd was maintained for 2 months without having to resort to switching on the electro-submersible pump. During these 2 months a second well, T2 (ARD B), was drilled in the mid-part of the field. Petrophysical analysis of this well indicated that some intervals in both the Zechstein and the very thin Rotliegend sections encountered were water flushed, and in line with prognosis an initial water–oil ratio of about 1 was anticipated. Preparation was made to bring T2 on stream. However, in the days just before T2 came on, BS&W (basic sediment and water) in T1 began to rise from near zero to a few per cent. Clearly, water breakthrough had begun to occur. T2 came on stream and, despite a 50% water cut, the field production rose to 28 000 bopd: just for 1 day. Then the unexpected happened. Although the production equipment had been designed to handle 50 000 bwpd, it could not, in fact, handle a total fluid of 20 000 bpd– disaster. Day after day the water–oil ratio increased, and day after day the production system tripped. Efforts to reduce the water cut by closing the sliding sleeve failed. Could water be flowing through the Devonian at such a high rate?

Two problems with the production facility were eventually identified and remedial work begun to correct the equipment which had not been built to specification. However, it took about 6 months before the production system could handle the 50 000 bwpd as designed, and an enormous quantity of oil production and, hence, revenue had been lost.

The problem with T1 remained, and as water cut increased so oil production decreased. By mid-2004, the field was producing only 1500 bopd and at that rate it was failing to return sufficient revenue to pay off the loans made to the two companies. Problems with the flow multimeter also meant that it was not possible to be confident about the relative proportions of oil and water production from the two wells. Meanwhile, T3 (ARD C) was being drilled.

Eventually, the decision was made to work over T1 and try to shut off the water. In doing so, the source of the water was revealed. It was coming from the Zechstein but an error in placing the sliding sleeve meant that even when the sleeve was in the closed position it was not covering the perforated interval (Fig. 11). Moreover, it was also discovered that the Devonian section was completely filled with lost circulation material. It had not produced a barrel of oil. The Devonian interval was cleaned and the Zechstein section blanked-off. When the well was brought back into production, it flowed at 11 000 bopd from the Devonian alone; a rate three times better than had ever been seen from the Devonian during the Argyll production period – miracle.

Had well 30/24-T1 been drilled and completed correctly, the well would have produced over 30 000 bopd and history would have been different. However, the damage was done and, despite a further well and two sidetracks being drilled, the field was little more than breaking even. The sidetrack wells were poorly executed as both companies tried to concentrate on delivering new finances but in the end the two companies had insufficient funds to drill their way out of the problem. By mid-2005, Tuscan ceased trading, leaving Acorn to limp on. Acorn did raise sufficient funds to continue but, in an environment of rapidly rising oil price and consequential hike in day rates for contractors,

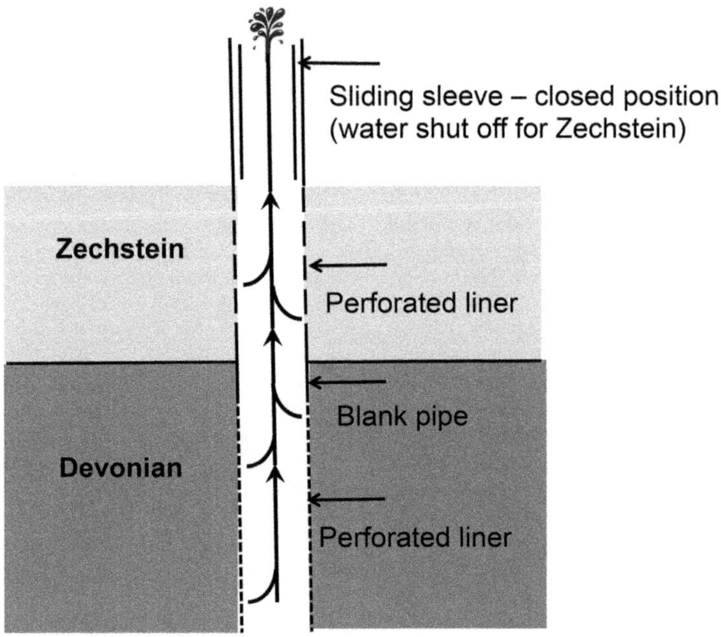

Fig. 11. Relationship between the perforated liner in well 30/24-T1 and the sliding sleeve arrangement. The sliding sleeve is shown in what is supposed to be the 'closed' position in which the perforations to the Zechstein reservoir should be covered by the sleeve. The figure shows the arrangement as executed in which the sliding sleeve was wrongly positioned. In addition, the perforations made to connect the Devonian reservoir to the well bore were found to contain lost circulation material, such that the initial flow from the well contained no contribution from the Devonian reservoir.

contracts were terminated. Acorn oversaw that abandonment of the three wells and closed the second chapter in the Argyll/Ardmore production history. Ardmore produced an additional 5 MMbbl.

Discussion and conclusions

Two small companies with very limited resources brought the abandoned Argyll Field back to life as the Ardmore Field. First oil was achieved in less than 2 years from acquisition of the licence. The limited resources meant that there was no margin for error with the development. However, there were errors and despite those being small relative to many other field start-ups, the two small companies were unable to continue to produce the field. Ardmore was not a commercial success, although the production proved beyond doubt that there are about 20 MMbbl remaining which for a properly capitalized company should make an attractive development.

It is interesting to compare the development schemes of Hamilton for Argyll, and Tuscan and Acorn for Ardmore.

Argyll was developed from a floating facility with very limited options for well intervention. The field development relied on depletion drive and limited natural aquifer inflow. All wells were vertical or nearly so. Gas-lift was used to maintain flow as pressure declined. A change-out of production vessel limited the production capacity as water cut increased and there was insufficient gas to lift all wells with proven and flowed oil. Ultimately, the first life of Argyll was geared towards Hamilton beating BP to first oil from the North Sea and demonstrating high-rate production albeit for a short time.

Ardmore was developed from a fixed platform, with technical options for interventions but low funds to execute any interventions. The field development relied on depletion drive and limited natural aquifer inflow. All wells were drilled as inclined. Initially, wells were allowed to flow naturally and subsequently electro-submersible pumps used to maintain flow. Drilling and production execution errors limited oil production and, hence, income. This, in turn, brought a premature end to production from Ardmore. On reflection, though, the Ardmore phase of development basically used the same strategy as the Argyll phase of development: produce as much oil as possible as fast as possible.

Tuscan and Acorn could not have redeveloped Argyll without access to data from the original development. Finding those data was difficult. The UK government department responsible for oil and gas development and production does not keep such data. All of the companies involved in the original Argyll development had been taken over or ceased to exist by the time Tuscan and Acorn formed. Some data had been acquired by Imperial College London for student projects and this was acquired but most came from individual ex-Hamilton employees who kept copies of data from the Argyll development. They were not asked to keep such data but did nonetheless, and it was just as well for such data formed the basis for the re-evaluation. History was repeated when Tuscan ceased trading. There was no imperative to retain data for the nation. One of us (JG) retained as much data as possible. It then formed the basis for the third development of Argyll as Alma (below).

Postscript

In December 2010, the area containing what had been the Argyll and then the Ardmore Field was relicensed; this time to a far more financially robust company, EnQuest plc. Their plans for a second redevelopment of the Ardmore Field contain, for the first time, water injection and pressure support. Production recommenced in October 2015 and the latest production figures released by EnQuest are for April 2016 with the Alma and Galia (formally Duncan) combined production above 9000 bopd.

Funding L. Tang received funding from EnQuest plc for field studies and analytical work on the Devonian reservoirs in Argyll.

References

Gluyas, J.G., Mair, B., Schofield, P., Arkley, P. & McRae, D. 2005. Ardmore Field: rebirth of the first offshore field UKCS. *In*: Doré, A.G. & Vining, B.A. (eds) *Petroleum Geology: North-West Europe and Global Perspective – Proceedings of the 6th Petroleum Geology Conference*. Geological Society, London, 367–388, https://doi.org/10.1144/0060367

Heward, A.P., Schofield, P. & Gluyas, J.G. 2003. The Rotliegend reservoir in Block 30/24, U.K. Central North Sea: including the Argyll (renamed Ardmore) and Innes fields. *Petroleum Geoscience*, **9**, 295–307, https://doi.org/10.1144/1354-079303-578

Milton-Worssell, R., Smith, K., McGrandle, A., Watson, J. & Cameron, D. 2010. The search for a Carboniferous petroleum system beneath the Central North Sea. *In*: Vining, B.A. & Pickering, S.C. (eds) *Petroleum Geology: From Mature Basins to New Frontiers – Proceedings of the 7th Petroleum Geology Conference*. Geological Society, London, 57–75, https://doi.org/10.1144/0070057

Moreton, R. 1995. *Tales from Early UK Oil Exploration 1960–79: Petroleum Exploration Society of Great Britain 30th Anniversary Book.* Petroleum Exploration Society of Great Britain (PESGB), London.

Pennington, J.J. 1975. The geology of the Argyll field. *In*: Woodland, A.W. (ed.) *Petroleum and the Continental Shelf of North West Europe, Volume 1*. Applied Science, Barking, UK, 285–291.

Robson, D. 1991. The Argyll, Duncan and Innes fields, Blocks 30/24 and 30/25a, UK North Sea. *In*: Abbotts, I.L. (ed.) *United Kingdom Oil and Gas Fields: 25 Years Commemorative Volume*. Geological Society, London, Memoirs, **14**, 219–225, https://doi.org/10.1144/GSL.MEM.1991.014.01.27

Stemmerik, L., Ineson, J.R. & Mitchell, J.G. 2000. Stratigraphy of the Rotliegend Group in the Danish Part of the Northern Permian Basin, North Sea. *Journal of the Geological Society, London*, **157**, 1127–1136, https://doi.org/10.1144/jgs.157.6.1127

Tang, L., Gluyas, J.G. & Jones, S.J. 2018. Diagenetic and geochemical studies of the Buchan Sandstone (Upper Devonian) in the Central North Sea. *Petroleum Science*, in press.

Ziegler, P. 1990. *Geological Atlas of Western and Central Europe*. 2nd edn. Shell Internationale Petroleum, Maatschappij, The Hague.

The history of exploration and development of the Liverpool Bay fields and the East Irish Sea Basin

J. BUNCE

Eni UK Limited, Eni House, Ebury Bridge Road, London SW1W 8PZ, UK
james.bunce@eni.com

Abstract: Hydrocarbon exploration in the East Irish Sea Basin began with the identification of surface oil seeps in peat beds in Lancashire, UK. This precipitated the drilling of the first onshore exploration wells. The discovery of the Formby Field in west Lancashire at the end of the 1930s triggered a wave of further drilling. Wells drilled in west Lancashire had limited success, with only minor hydrocarbon shows, whilst the production from the Formby Field was modest. Nonetheless, the invaluable geological information taken from onshore wells and the ratification of the Continental Shelf Act led to a shift in focus to the offshore and a period of significant interest in the East Irish Sea. Two key periods of oil and gas exploration activity stand out in the history of the offshore basin, the first headed by the Gas Council during the 1970s resulted in the discovery of the gas giants of Morecambe Bay, whilst the second fronted by Hamilton Oil during the 1990s heralded the discovery of oil with the Douglas and Lennox fields in Liverpool Bay. Exploration in the basin has waned during the last decade; however, to date, this mature hydrocarbon province has yielded estimated hydrocarbon reserves of over 1.8 BBOE (billion barrels of oil equivalent).

Hydrocarbon exploration in the East Irish Sea Basin began in the late 1930s. The presence of surface oil seeps in peat beds close to the town of Formby in west Lancashire led to oil company interest in the region and the first exploration well was drilled in 1939. Success quickly followed and oil-bearing sandstones were encountered in the second well, proving the existence of a working petroleum system in the basin and initiating a period of onshore exploration in the region. Reviewing the historical drilling activity in the East Irish Sea Basin highlights how onshore drilling in the West Lancashire Basin dominated throughout the 1940s and 1950s (Fig. 1), prior to the ratification of the Continental Shelf Act in the mid-1960s which opened up the UK Continental Shelf (UKCS) to oil and gas exploration.

Offshore hydrocarbon exploration in the East Irish Sea commenced in the late 1960s, and at the time it was believed that the geology of the offshore basin would be similar to that of the Southern North Sea. Expectations were held for a Carboniferous gas source in the basin, Permian reservoir sandstones, equivalent to the Leman Sandstone Formation of the Rotliegend Group in the Southern Gas Basin, and a Zechstein equivalent cap rock (Colter 1997). However, after the first offshore wells were completed, it was evident that the primary hydrocarbon potential resided in the younger Triassic sandstones; which exhibited better reservoir quality than the Permian sandstones and also showed evidence of hydrocarbons in the form of dead oil staining and fluorescence throughout the interval. The Permian sandstones were water-bearing. Offshore exploration drilling began in the Morecambe Bay area where Gulf Oil spudded the first wildcat well in 1969. The well, which targeted a north–south-trending anticlinal structure at the top Permian, was plugged and abandoned as a dry hole. However, during the 1970s, the Gas Council (a forerunner to British Gas) drilled exploration wells testing structural closures at the top Triassic Sherwood Sandstone Group which resulted in the discoveries of the giant gas fields of South and North Morecambe, and started a period of increased activity in the basin over the next decade. Exploration shifted to the south of the basin during the 1990s, with attention focused in the Liverpool Bay area. The exploration campaign in the area, led by Hamilton Brothers Oil and Gas Limited (hereafter Hamilton Oil), resulted in the discoveries of the Douglas and Lennox oil fields, as well as the Hamilton and Hamilton North gas fields.

Geological setting

The Irish Sea is located off the North Wales coast and the NW coast of England. The East Irish Sea Basin extends west to Anglesey and the Isle of Man, and north to the Solway Firth, bounded by the Lake District Massif, the Anglesey–Isle of Man Uplift and the Solway Firth Basin (Fig. 2). The basin is part of a Permo-Triassic rift system initiated in the early Permian with NE–SW rifting, the onshore West Lancashire Basin forms a sub-basin located on its eastern margin.

The basin was formed during the early Carboniferous due to north–south extension along NE–SW

From: Craig, J., Gerali, F., MacAulay, F. & Sorkhabi, R. (eds) 2018. *History of the European Oil and Gas Industry*. Geological Society, London, Special Publications, **465**, 95–118.
First published online April 11, 2018, https://doi.org/10.1144/SP465.6

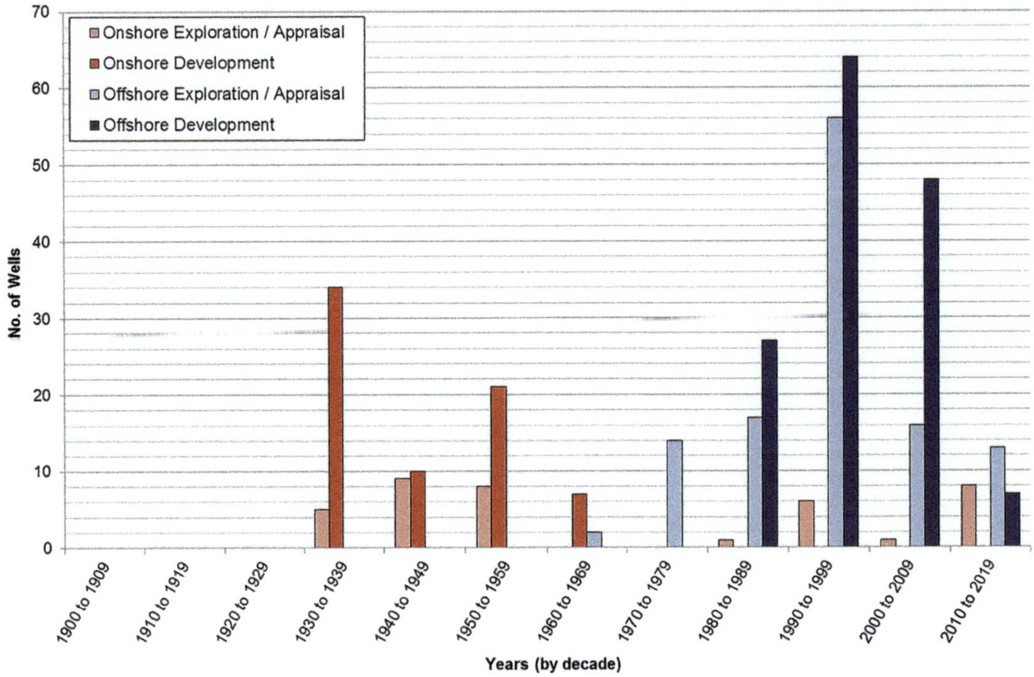

Fig. 1. Drilling activity (by decade) in the East Irish Sea Basin categorized into onshore and offshore wells, and exploration/appraisal wells and development wells. (*source*: IHS Markit).

faults and this Caledonian structural framework has a strong influence on overlying structural trends, with pre-existing faults having been reactivated throughout the evolution of the basin. Three main structural trends exist within the East Irish Sea Basin. A north–south and NE–SW fault trend linked to Permo-Triassic rifting, and a NE–SW trend linked to reactivated Caledonide lineaments (Knipe *et al.* 1993; Johnson & Fisher 1998).

The East Irish Sea Basin is a structurally complex area comprising a number of fault-bound sub-basins and uplifted massifs. The Carboniferous marked a period of basin extension and subsidence which resulted in the deposition of a thick sequence of marine sediments during the Dinantian and Namurian. The Carboniferous ended with the Variscan Orogeny, resulting in the uplift of the basin and erosion of the Carboniferous stratigraphy. Early Permian subsidence due to east–west extension produced north–south-orientated faulting and led to a series of north–south-trending sub-basins. Aeolian desert environments were established over much of Europe throughout the Lower Permian, with uplifted massifs providing a sediment source for subaerial erosion. Continental aeolian–fluvial depositional environments persisted until the middle–late Triassic when fault-related subsidence resulted in marine incursions of the basin. Renewed east–west extension in the Jurassic led to the formation of the hydrocarbon traps in the basin and a sustained period of subsidence and deposition, with sediments buried to a maximum depth in the late Cretaceous (Fig. 3). Tertiary uplift associated with the Alpine Orogeny resulted in a significant amount of erosion removing the entire Jurassic stratigraphy, with estimates of post-Cretaceous uplift and erosion of between approximately 6000 and 8000 ft in the offshore East Irish Sea Basin (Jackson & Mulholland 1993).

In the locality of Liverpool Bay, the East Irish Sea Basin can be divided into two sub-basins: the Gogarth Basin and the East Deemster Basin. The Douglas and Hamilton fields are located within the former, whilst the Lennox Field is located in the latter (Figs 2 & 4). Jackson & Mulholland (1993) defined four structural styles in the Irish Sea and divided the basin accordingly. The Douglas and the Hamilton fields sit within the same structural region, characterized as step-faulted arrays of small tilt blocks, and it is these tilted fault blocks which typify the Douglas, Hamilton and Hamilton North fields. The Lennox Field lies within a region defined by widely-spaced major faults and gentle, long wavelength folds associated with uplift and inversion along pre-existing faults. Jackson & Mulholland (1993) noted that these faults penetrate the full Permo-Trias succession and delineate large

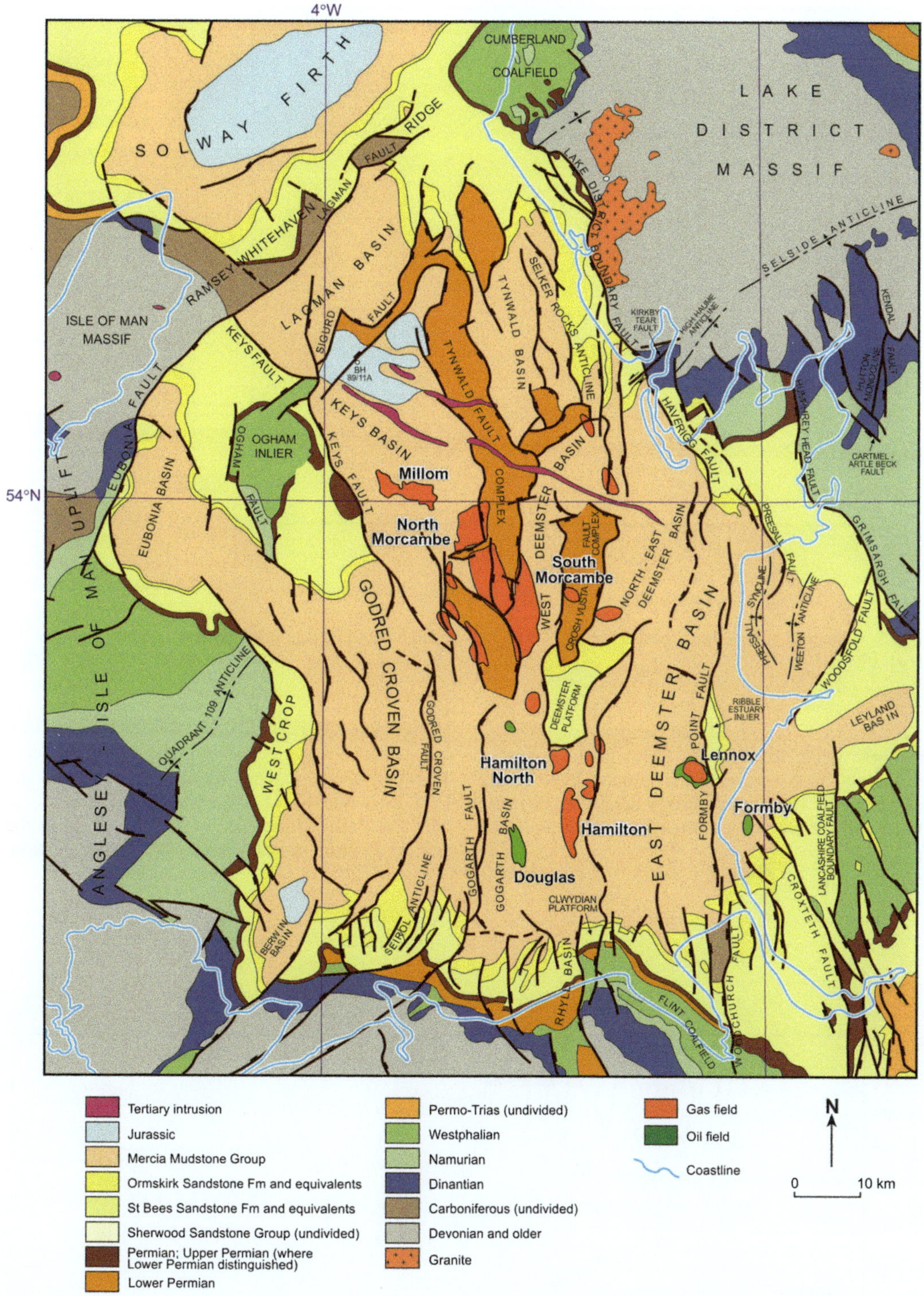

Fig. 2. Major structural trends of the East Irish Sea Basin (Jackson & Mulholland 1993) (data source: C&C Reservoirs, 2016, DAKS™ IQ – Digital Analogs Knowledge System).

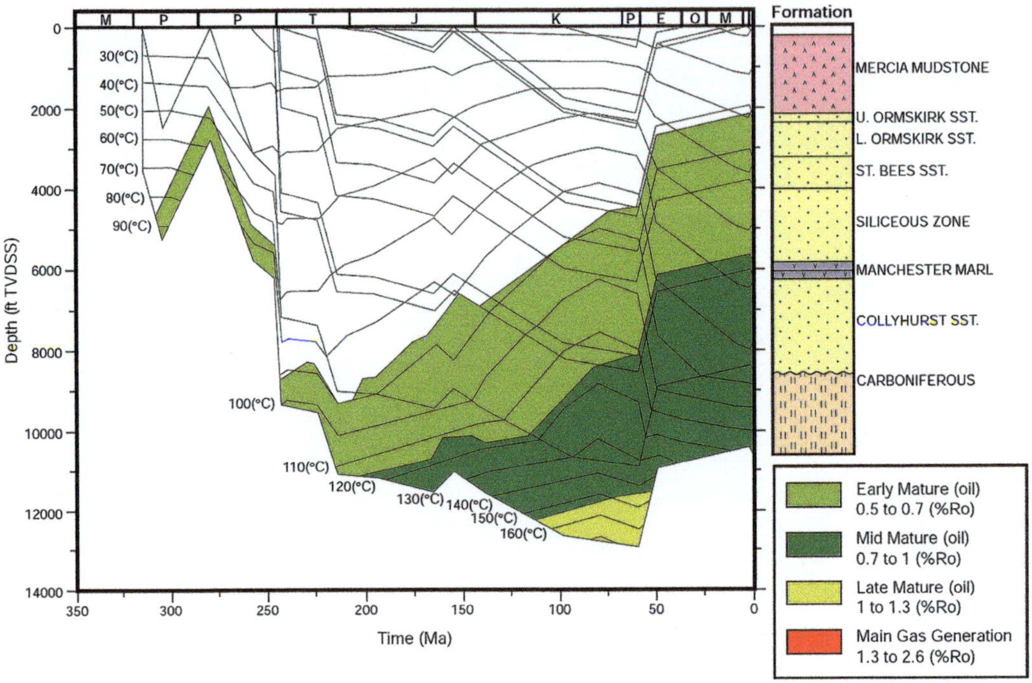

Fig. 3. Burial and thermal history model for the East Irish Sea Basin based on data from well 110/12a-1 (Hardman *et al.* 1993).

fault blocks in the Sherwood Sandstone Group: an example being the Formby Point fault, a synthetic growth fault bounding the Lennox Field to the east (Fig. 4). The Liverpool Bay fields are generally fault- and dip-closed structures formed during Jurassic rifting.

Petroleum system

The petroleum system which developed in the East Irish Sea Basin is Carboniferous–Triassic in age. The Permo-Trias succession (Fig. 5) marks the start of a period of sustained continental deposition.

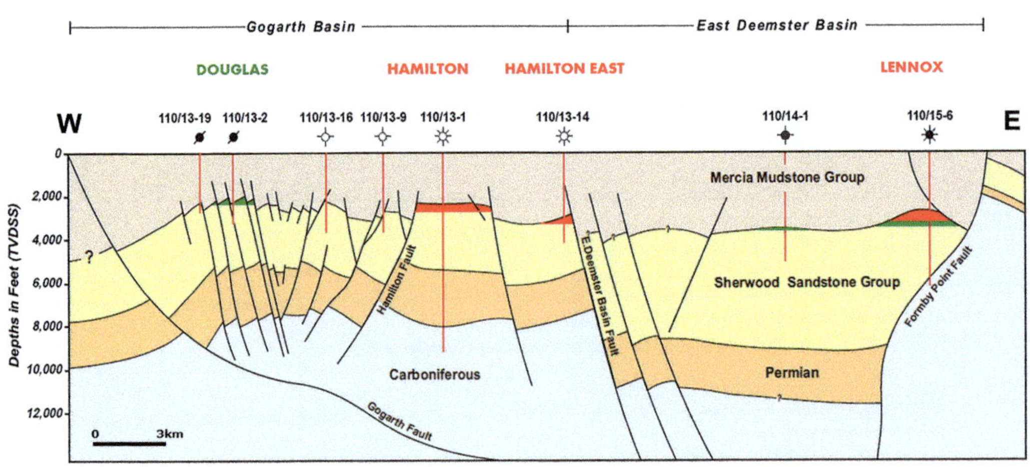

Fig. 4. Structural cross-section (west–east) across the Liverpool Bay area, highlighting the structural configuration of the Douglas, Hamilton and Lennox fields (after Yaliz & McKim 2003).

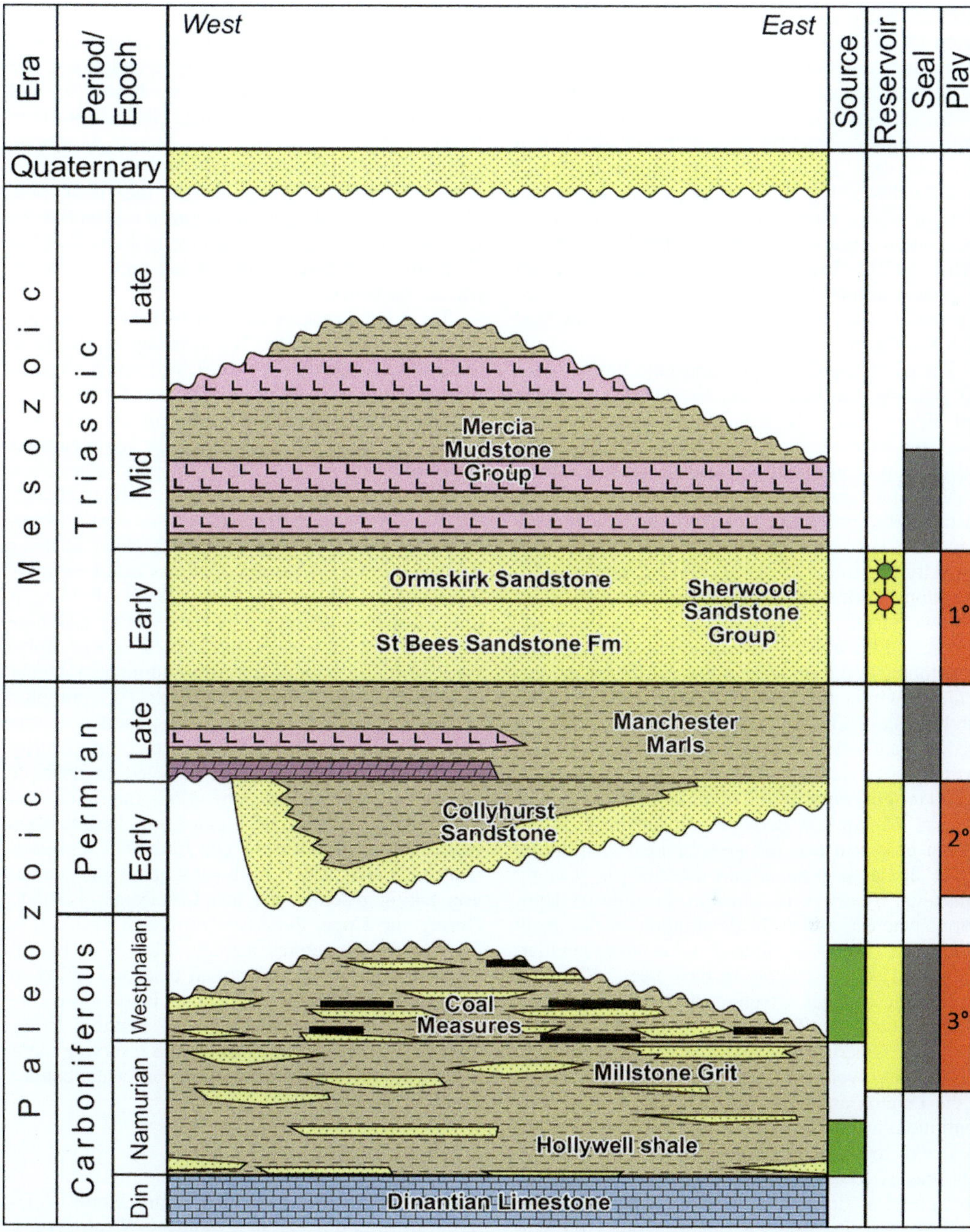

Fig. 5. Generalized stratigraphic column for the East Irish Sea Basin, highlighting the petroleum system and exploration play concepts in the basin (modified from *UK Discovery Digest* 2007).

Following the Variscan Orogeny, the early Permian experiences NE–SW rifting, occurring contemporaneously with the deposition of a thick sequence of continental deposits. The deposition of the Collyhurst Sandstone Formation, equivalent to the Rotliegend Leman Sandstone Formation in the Southern Gas Basin, is a thick and areally extensive aeolian–fluvial sandstone unit. It is overlain by the Manchester Marl Formation, a marine-influenced shale–evaporite sequence, which completes the

Permian succession. The Triassic experiences a return to continental conditions and the deposition of the Sherwood Sandstone Group, which comprises the St Bees Sandstone Formation and the Ormskirk Sandstone Formation. These two formations represent a continued period of continental deposition in a semi-arid fluvio-aeolian environment. The St Bees Sandstone Formation marks a period in which a major north–south fluvial system occupied the basin. This river system, the Budleighensis River (Wills 1970), flowed northwards into the basin from the Cheshire Basin to the south and continued into the Solway Firth Basin to the north of the East Irish Sea Basin.

The overlying Ormskirk Sandstone Formation is the primary reservoir in the East Irish Sea Basin and all commercially producible hydrocarbons to date have been discovered within this interval. The formation was originally subdivided into three stratigraphic members based on the nomenclature used onshore: the Frodsham, Delamere and Thurstaston members. However, an offshore subdivision proposed by Jackson *et al.* (1997) divided the Ormskirk Sandstone Formation into a lower and upper unit: the lower termed OS1 is dominated by aeolian sandstones; and the upper, OS2, is dominated by fluvial sandstones. In the southern region of the basin, particularly within the Liverpool Bay area, a further subdivision of OS2 is proposed: the basal OS2a is predominantly fluvial, whilst the overlying OS2b is more aeolian influenced. This southern subdivision is roughly comparable to the Frodsham, Delamere and Thurstaston division. Meadows & Beach (1993) proposed that the geographical distribution of the fluvial systems within the basin is strongly controlled by active tectonics at the time of deposition, whereas the cyclical variation in dominant facies (i.e. fluvial-dominated v. aeolian/sandflat-dominated) is most likely to have been influenced by climatic changes affecting discharge and height of the water table. The Ormskirk Sandstone varies in thickness across the basin, from approximately 550 ft thick over the Douglas Field to over 800 ft at the Lennox and Morecambe fields, it is laterally continuous and correlatable across the area, although localized tectonic activity and the presence of mini-basins results in local variations. Across the Liverpool Bay fields there is good correlation within the formation. The reservoir quality is good throughout the interval, exhibiting a high net-to-gross, commonly in excess of 80%. Reservoir quality is strongly related to facies, with the best-quality sands associated with the aeolian facies; the typical range in average porosity is from 10 to 20%.

The Mercia Mudstone Group overlies the reservoir sands of the Sherwood Sandstone Group, and consists of a cyclical series of interbedded mudstones, siltstones and evaporites. The Mercia Mudstone Group was deposited in shallow bodies of water which were likely to have been hypersaline and subject to periodic marine inundations; the variations in fluvial discharge, marine influx and evaporation led to the cyclical deposition seen throughout the succession. The Mercia Mudstone Group forms a laterally extensive seal to the Sherwood Sandstone Group across the Irish Sea, although it is likely that the ultimate top seal in the Liverpool Bay area is provided by the Rossall Halite Formation, the first significant salt unit within the Mercia sequence.

The source rock in the East Irish Sea Basin has been identified as the Namurian-aged Holywell Shale. This shale unit is a thick, widespread basinal prodelta mudstone deposited in front of a large prograding delta system active during the Carboniferous. Geochemical analysis of the hydrocarbons from the south of the basin show good correlation with the Holywell Shale samples, which are characterized as an oil-prone Type II kerogen source (Armstrong *et al.* 1997). Recent geochemical analysis on the Holywell Shale of NE Wales again supports the work by Armstrong *et al.* (1997), dividing it into a Lower and Upper unit, representing a change from the strongly marine-influenced Lower Holywell Shale to a more terrestrial-dominated Upper Holywell Shale (Newport 2016). The complex tectonic history of the basin has resulted in the Holywell Shale reaching different stages of maturity in different mini-basins, and also led to multiple phases of hydrocarbon generation, migration and accumulation. The main phase of hydrocarbon generation occurred in the late Jurassic–Cretaceous (Hardman *et al.* 1993), with subsequent gas generation taking place in the late Cretaceous–Tertiary. During the Upper Jurassic Cimmerian uplift, it is thought that structures already filled with hydrocarbons were breached prior to a second influx of hydrocarbons. Evidence for this is seen in core acquired over the Douglas and Lennox fields, which show bitumen staining in the reservoir sands believed to be located at the palaeo-oil–water contact (Yaliz & McKim 2003).

Onshore beginnings

The first phase of conventional hydrocarbon exploration in Britain began in response to World War I. The demands that the Great War put on the UK's oil supplies, particularly because of the adaptation of naval vessels to use oil rather than coal, led the government to look for oil resources on home soil. The first successful exploratory well was drilled at Hardstoft in Derbyshire in May 1919, the well struck oil in Carboniferous limestones and produced oil until 1945. The end of World War I brought with it an end to further oil exploration. It was not until the 1930s that a renewed phase of hydrocarbon exploration began, driven first by the passing of the 1934 Petroleum Act which nationalized the UK's onshore oil and gas resources, and then at the end of the decade by the onset of World War II. Following the passing of the Petroleum Act, Anglo-Persian Oil (later to become BP) launched a major UK onshore oil exploration programme with D'Arcy Exploration Company, named after William Knox D'Arcy the pioneer of petroleum prospecting in Persia (modern-day Iran). Britain's entry into World War II in the autumn of 1939 resulted in the rationing of oil and an increased urgency to find it, particularly on home soil, due to the increased difficulties of importing crude from the Anglo-Persian Oil Company heartland in Iran. This coupling of events led to over 100 onshore exploration wells being drilled throughout the UK in the 1930s alone.

Onshore hydrocarbon exploration started by focusing on occurrences of oil seeps at the surface. In west Lancashire, 1 mile to the east of the town of Formby, active oil seeps had been known about for some time and documented as early as the seventeenth century in William Camden's *Britannia* (1637 edition). Work by F. Wolverson Cope (1939) attempted to map these seeps and derive the source of the oil; two hypothesis existed at the time either favouring a superficial source related to the breakdown of algal material within the peat beds, or a deeper mineral derivation. The distribution of oil at the surface, primarily within peat beds, strongly correlated with mapped faulting within the Triassic; this combined with results of laboratory analysis on the oil samples favoured a deep mineral source for the oil (Cope 1939). The analyses of these active seeps led to hydrocarbon exploration in the West Lancashire Basin and D'Arcy Exploration taking out exploration licence A90 in the Formby area.

Despite the widespread distribution of peat deposits in the Formby area, the subterranean geology was anticipated to consist of Keuper Marl and Keuper sandstones, this was inferred from Triassic outcrops to the south and east of Formby. The hydrocarbon exploration focused on drilling shallow boreholes within areas of known oil seepages, with the aim to discover the immediate, rather than the ultimate, source of the oil and determine the structural configuration of the Triassic stratigraphy (Lees & Taitt 1945). The first well drilled by D'Arcy Exploration in the area was located close to the largest surface oil seep; however, the well reached the top Triassic at 323 ft without encountering significant oil shows. In May 1939, D'Arcy Exploration drilled their second well and discovered oil at the Formby Field, confirming the conclusions drawn by Cope (1939) of a deep mineral petroleum source.

The field is located on a NE–SW-trending faulted anticline structure, dipping to the SW. The lithological sections for the two exploration wells identified that the Formby discovery well was located in the footwall of a normal fault with a displacement of approximately 300 ft; the fault was the likely control on the oil accumulation. The Formby Field is somewhat of a curiosity, with oil accumulated at very shallow depths (*c.* 40–80 ft) within sandstones sealed by Upper Pleistocene boulder clay. The discovery well, Formby-G2 (aka L110/15-2), encountered oil in the Keuper Waterstones Formation, the Ormskirk Sandstone Formation and the Pleistocene Shirdley Sandstone Formation. Correlation of the East Irish Sea Basin stratigraphy between the onshore and offshore is not straightforward when viewing historical well data, particularly because a number of terms (i.e. Keuper) have acquired mixed litho- and chrono-stratigraphic associations, and nomenclature has been revised by different authors. Notwithstanding, the Keuper Waterstones at the Formby location are interpreted as a lateral equivalent to the Triassic Taporley Siltstone Formation seen in the Cheshire Basin (Warrington *et al.* 1980; Jackson *et al.* 1987) and correspond to the uppermost Ormskirk Sandstone Formation in the offshore. The transition between the Sherwood sandstones and the Mercia mudstones in the offshore and onshore is subtly different, with mudstones of the offshore Mercia pinching out onshore where the depositional environment is more proximal.

The main productive interval in the Formby Field is the Keuper Waterstones Formation, with a minor contribution from the shallower Shirdley Sandstone. Oil produced from the Keuper Waterstones sandstone was a waxy, medium gravity oil (38° API) and analysis of the hydrocarbons showed similar crude oil properties to that discovered in Nottinghamshire at the Eakring well (Lees & Taitt 1945; Pinfold 1958). The well flowed at an initial rate of 60 gallons/h (c. 40 bbl/day (barrels per day)) from the Keuper Waterstones, whereas the oil within the Ormskirk Sandstone was heavy and unable to flow, probably due to the effects of water washing in this zone which removed the lighter oil fractions (Roche 2012). The field was rapidly developed during World War II due to its strategic importance and produced hydrocarbons up until 1965, with a total field production of 71 557 bbl.

The field is believed to be the result of a breached trap and vertical migration from a deeper Carboniferous accumulation, and it was this hypothesis that led to the drilling of a number of onshore exploration wells targeting deeper potential Carboniferous traps. D'Arcy Exploration targeted the deeper Carboniferous with five conventional wells drilled between June 1940 and February 1952. The first of these, Formby-1, was drilled directly beneath the

Formby Field and reached a total depth within the Lower Carboniferous shales: the Lower Bowland Shale Formation. At the time, the well was the deepest borehole in Britain, reaching a depth of 7680 ft. Although the well encountered a thick Permo-Triassic sequence, it failed to find hydrocarbons and was plugged and abandoned as a dry hole. Significantly, the well identified the magnitude of the Variscan Unconformity: with the entire Carboniferous Westphalian Coal Measures and a large portion of the Millstone Grit eroded. This further supported the hypothesis of a deeper source of the oil and the migration of hydrocarbons along large faults. The subsequent Formby wells (Formby-3–Formby-6) also failed to encounter any oil shows, indicating that the postulated deeper accumulation from which the Formby oil was sourced was not in the immediate locality of the field. No new hydrocarbon accumulations were discovered in West Lancashire during this deep exploration drilling phase.

Over 90 wells were drilled by D'Arcy Exploration (and later by BP) in the A90 licence between 1939 and 1965; however, apart from minor hydrocarbon shows in the Permian Collyhurst Formation, no further commercial hydrocarbons were encountered. Onshore exploration in the West Lancashire Basin ceased towards the end of the 1950s, and the final development well on the Formby Field was drilled in October 1965 as hydrocarbon exploration began to move offshore.

There was sporadic interest in the onshore East Irish Sea Basin following the cessation of production at the Formby Field, and five further wells were drilled between December 1987 and October 1992 in the West Lancashire Basin, albeit with limited success. British Gas drilled three wells east of Blackpool, the first Thistleton-1 in February 1988, targeted the Triassic sands of the Sherwood Sandstone Group and reached a total depth within Carboniferous Dinantian carbonates. The well encountered gas shows through the Namurian and Dinantian, as well as oil staining within the Namurian sands, although the well was plugged and abandoned as a dry hole. Hesketh-1 was located to the SE of Preston and spudded in March 1990 targeting the Triassic sandstones. The well reached a total depth within the Lower Bowland Shale Formation, encountering minor gas shows but was again plugged and abandoned as a dry hole. The third well, Elswick-1, was located to the east of Thistleton-1 and encountered commercial volumes of hydrocarbons. Spud in April 1990 by British Gas and located close to the nearby village of the same name, the well targeted a surface dome structure and found gas within the Permo-Triassic sands (although the specific age of the reservoir is disputed). The conglomerate sands tested at 68 mscf/day (thousand standard cubic feet per day); however, due to the tight nature of the sands, the well was re-entered in May 1993 in order to perform the first CO_2 fracture job in Europe. This resulted in a significant improvement in well performance with the well flowing at 350 mscf/day. The discovery well was completed as a gas producer and this single well field was brought on production in July 1996, producing over 500 MMscf (million standard cubic feet), which is used for onsite electricity generation. A further exploration well was spud by Fina Petroleum Development Limited in September 1990, located to the south of the Formby Field. The Little Crosby-1 well targeted the Permo-Triassic sequence, reaching a total depth within the Collyhurst sands. The well was unsuccessful and was plugged and abandoned as a dry hole. The last onshore well drilled in the West Lancashire Basin during this period was spudded by Clyde Petroleum in October 1992. This Banks-1 well, located approximately 15 km to the NE of the Formby Field, targeted the Namurian sands but was again plugged and abandoned as a dry hole. The results of exploration wells targeting the Carboniferous in the West Lancashire Basin showed the play was hampered by a number of contributing factors, Tertiary uplift which ended hydrocarbon generation, reactivation and modification of traps, poor reservoir quality, and a lack of efficient seals. This resulted in the Carboniferous of NW England being considered as a low potential hydrocarbon play (Fraser *et al.* 1990).

The move offshore

The impetus for hydrocarbon exploration in UK offshore waters was the discovery of the giant Groningen Field onshore The Netherlands in 1959. The discovery well, Slochteren No.1, encountered gas within the Permian Rotliegend Sandstone with the structure estimated to hold around 100 Tcf (trillion cubic feet). Geologists at the time noted the similarities of the geology at the Groningen discovery with that encountered in the East Midlands; and due to the field's proximity to the North Sea, it was quickly established that the petroleum play could extend offshore into the Southern North Sea. Hydrocarbon exploration in the offshore UKCS was hindered at the time awaiting the division of the continental shelf; however, exploration activity quickly intensified in the mid-late 1960s with the ratification of the Continental Shelf Act in April 1964 and the implementation of the Petroleum Regulations 1964. This led to the UK government opening up virtually the entire North Sea to oil and gas companies during the UKCS 1st Offshore Licence Round in 1964. Offshore exploration in the UK moved from the English Channel to the North Sea and, despite some pessimism over the presence of hydrocarbons in the sedimentary basin, success followed quickly with the discoveries of West Sole and Leman in the Southern Gas Basin, followed by the Arbroath and Forties fields in the Northern North Sea. Contemporaneously in the Irish Sea, Gulf Oil (in partnership with the National Coal Board) had been awarded blocks 110/3, 110/4, 110/8, 110/9 and 110/13 in 1965, after the UKCS 2nd Offshore Licence Round (Fig. 6). The first well in this hitherto frontier basin would spud more than 3 years later in August 1969.

Despite the onshore exploration campaign failing to discover deeper hydrocarbon accumulations in the Permian or Carboniferous strata, the Formby Field had proved the presence of a working petroleum system in West Lancashire. This, combined with the notion that the Irish Sea was analogous to the Southern North Sea, meant there was basis for optimism to explore the offshore East Irish Sea Basin. Well 110/8-1, drilled by Gulf Oil, was the first well drilled in the East Irish Sea and targeted a north–south-trending anticline at the Permian Collyhurst Sandstone Formation. The well was completed in the Carboniferous, with the primary target of the Collyhurst Sandstone water-bearing. Dead oil staining was encountered throughout the Triassic section which exhibited good reservoir quality; however, it was deemed that there was no effective seal overlying the sands and that the section had been an oil migration path. The completion report from the well provides an insight into the role the onshore played in paving the way for offshore exploration success:

> [T]he lithology of the formations penetrated in this well are similar to that found in the onshore wells in Formby which have produced minor amounts of oil from the lower Keuper Sands. The main difference was that our well did not have an effective seal overlying the Keuper Sands.

Two days after plugging and abandoning the 110/8-1 well, Gulf spudded their second well, 110/8-2, on 9 September 1969. It would be this well that would light the blue touchpaper on East Irish Sea hydrocarbon exploration, albeit not for another 5 years.

110/8-2, again, primarily targeted the Permian sands, although the Triassic Lower Keuper (reinterpreted as the Ormskirk Sandstone Formation by Jackson *et al.* 1997), Upper Bunter sandstones (St Bees Sandstone Formation) and Carboniferous Coal Measures were also identified as secondary objectives in the well completion report. The well was located on a large north–south-trending, fault-controlled structure and was completed in the Carboniferous-aged Millstone Grit Formation at a total depth of 10 220 ft. The well encountered sands in both the Permian and Triassic. The Collyhurst sands were poorly developed and the reservoir properties of the interval were not particularly good, with the sandstone grading from 'porous to tight' within only 30 ft of the top of the formation. The Triassic interval was substantially thicker than expected and consisted of 4750 ft of sandstones overlain by approximately 3000 ft of alternating salt, siltstone and shale lithologies comprising the Mercia Mudstone Group. A maximum 'gas show' of 230 000 ppm methane was recorded over the initial 20 ft of reservoir section; however, as later described by Dean (1996), this was, in fact, a gas kick upon entering the gas column. Gulf Oil's interpretation of the log data did not identify any further hydrocarbons throughout the Ormskirk section, with calculations of water saturation varying from an average of 30% in the uppermost 25 ft to 70% below, where the sands were described as 'generally tight'. Despite the presence of gas, the well was plugged and abandoned without testing.

The Gas Council entered the Irish Sea with the award of blocks 110/6, 110/7, 110/12 and 110/14 in the UKCS 3rd Offshore Licence Round in 1970, the blocks were awarded to Hydrocarbons Great Britain Limited (hereafter HGB), a wholly-owned subsidiary of the Gas Council (Fig. 6). In the following round, in 1972, they acquired the adjacent Block 110/2. Around this time there were discussions between Gulf Oil and the Gas Council regarding farm-in opportunities, with Gulf Oil proposing the Gas Council farm-in to their block, 110/8, in which they had drilled two exploration wells. The farm-in deal did not materialize and, as Vic Colter recalls (Moreton 1995), a 'gentleman's agreement' was later informally reached over a lunch between Gulf Oil's manager and the Gas Council's exploration manager, Peter Hinde. The offer was that Gulf Oil would disclose the log data for their exploration wells in the neighbouring licence block (110/8-1 and 110/8-2) in return for access to log data from future Gas Council wells drilled in the basin (Hardman 2003). It was this 'gentleman's agreement' that paved the way for successful hydrocarbon exploration in the basin, and what happened next has been well documented by several authors (Moreton 1995; Dean 1996; Colter 1997; Hardman 2003).

Upon receiving the log data, the Gas Council undertook a reinterpretation of the 110/8-2 well logs. Their log analyst at the time, John Bains (for whom the Bains Field is named in recognition), interpreted a 600 ft gas column in the Triassic sands. The error in the initial interpretation stemmed from the misunderstanding of the formation water salinity and the use of an incorrect value for the formation water resistivity (R_w). Although Colter also suggests a pair of scissors and the inaccurate splicing together of two log runs could have also played a role (Moreton 1995). The water salinity is particularly high in the basin (the nearby Liverpool Bay aquifer salinity

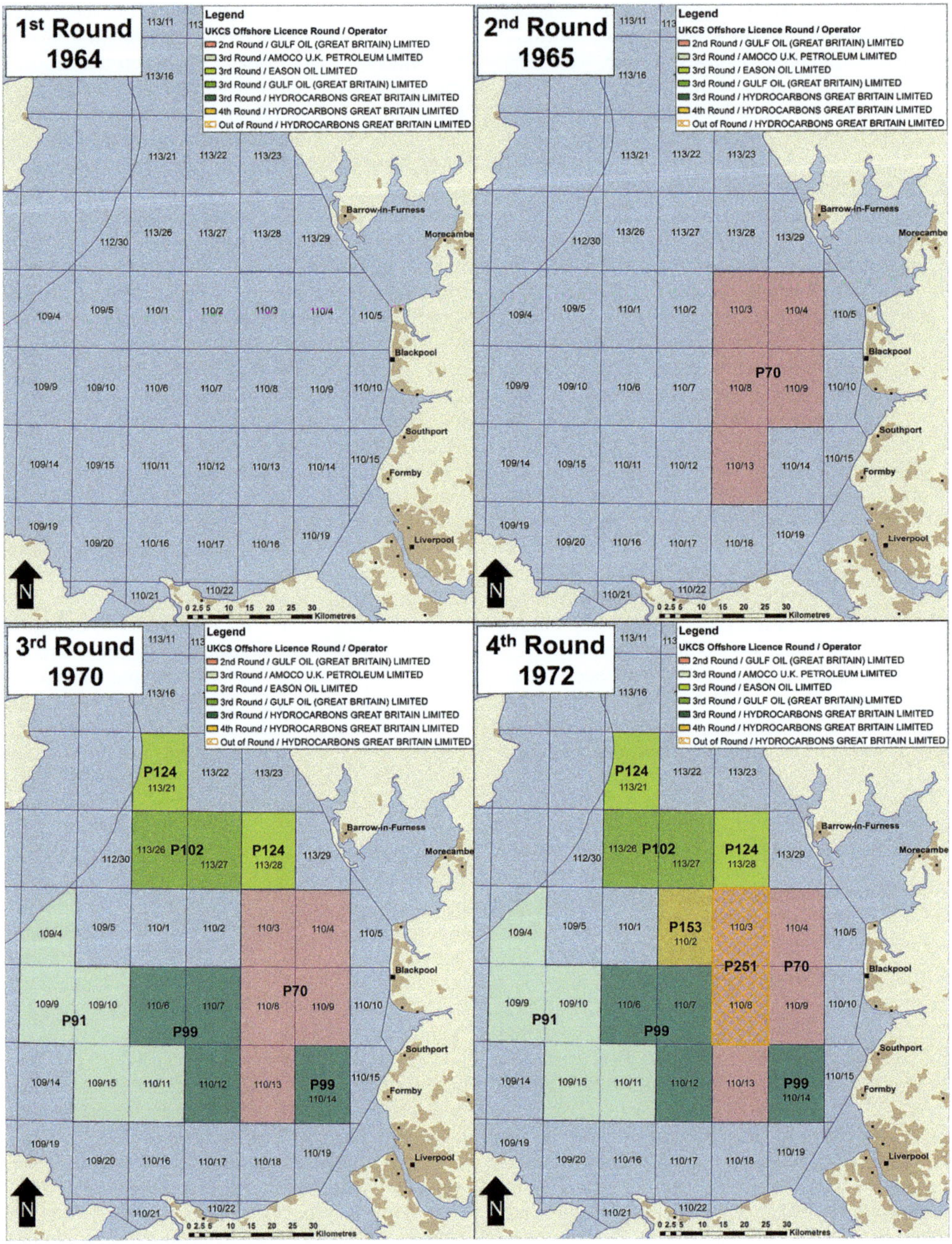

Fig. 6. UKCS Offshore Licence Round awards in the East Irish Sea during the first four offshore licensing rounds (1964–72), including the Out of Round award of the P251 licence area to Hydrocarbons Great Britain Limited in 1976 following the relinquishment by Gulf Oil.

is approximately 270 000 ppm NaCl equivalent) and therefore the use of high R_w values in the initial interpretation resulted in higher calculated water saturations than would be expected close to the top of a significant gas column. The reinterpretation was a major shift from the original interpretation which suggested only a minor 20 ft gas column. The revised water saturation calculation was supported by laterolog resistivity curves, which also showed a gas–water contact at 3750 ft TVDSS (true vertical depth subsea), and by the gas shows and high pressures encountered when entering the reservoir section during drilling.

The Gas Council informed Gulf Oil of their reinterpretation and even enquired about the possibility to farm-in to Block 110/8; however, at the time Gulf Oil had already submitted their intent to relinquish the block to the UK authorities, and they were unable to rescind their notice. In 1976 the blocks were granted to HGB in an out of round award. This now meant that they held key acreage over a significant portion of the East Irish Sea: blocks 110/2, 110/3, 110/6, 110/7, 110/8, 110/12 and 110/14 (Fig. 6).

On the back of Bains' reinterpretation and much persuasion from Peter Hinde (Moreton 1995), HGB drilled 110/2-1 in August 1974, testing a northerly location on the 110/8-2 structure. The well encountered 652 ft of gas-bearing sandstones and tested at a maximum gas rate of 10.25 MMscf/day from the Ormskirk Sandstone Formation (Stuart & Cowan 1991). This confirmed the Gas Council's reinterpretation of 110/8-2, and the discovery of the South Morecambe Field proved the existence of a working and commercial petroleum system within the offshore East Irish Sea Basin, with the Carboniferous providing a gas source in the basin and the Triassic sands sealed by the overlying Mercia Mudstone Group as the reservoir.

Following the successful discovery well, a comprehensive exploration and appraisal well campaign took place in blocks 110/2 and 110/3 between 1975 and 1983 to delineate the field. The first appraisal well on South Morecambe, 110/2-2, was drilled close to an east–west-trending graben which defined the northernmost extent of the structure. The well encountered gas-bearing sands, testing at 23.5 MMscf/day and essentially proving the existence of a single field-wide gas–water contact with the discovery wells to the south (Bastin *et al.* 2003). The next exploration well, 110/2-3, drilled in 1976, targeted a potentially separate structure located north of the graben. The well successfully found gas-bearing Triassic sands to a depth 200 ft deeper than the gas–water contact identified at South Morecambe, proving a separate accumulation existed to the north (Stuart 1993; Cowan & Boycott-Brown 2003). The North Morecambe Field consists of a 975 ft gas column and the discovery well was tested flowing at rates of 14.9 MMscf/day from a 325 ft drill stem test (DST) interval. The South Morecambe Field was the first commercial discovery in the Irish Sea; North Morecambe quickly followed and the two fields would be appraised by a further eight wells. South Morecambe was developed with 34 production wells, whilst North Morecambe was developed with 10 production wells. The fields were brought on stream in January 1985 and September 1994, with field estimates of gas initially-in-place (GIIP) of 5.5 and 1.3 Tcf, respectively.

Following the extraordinary success of the discoveries at Morecambe Bay, exploration throughout the 1980s focused in the neighbouring blocks. HGB held the majority of licence acreage surrounding the Morecambe Bay fields and continued their exploration campaign primarily focusing on structures within the Triassic Sherwood Sandstone. During this period, between 1982 and 1990, four further discoveries were made at Calder, Millom, Bains and Dalton, all of which would go on to be developed (Fig. 7; Table 1) and account for an additional gas volume of over 600 Bcf (billion cubic feet) initially in-place.

Oil discovered in Liverpool Bay

Offshore hydrocarbon exploration in the East Irish Sea began to gain momentum following the discovery of the South Morecambe Field. At the end of the 1980s, exploration focus began to shift away from the Morecambe Bay area and south towards Liverpool Bay. The licence area to the south had been held by Gulf Oil since 1965, after its award in the UKCS 2nd Offshore Licence Round (Fig. 6). It was relinquished between 1971 and 1974 without any well being drilled. In 1989, Block 110/13 was awarded to Hamilton Oil under Licence P710 following the UKCS 11th Offshore Licence Round.

Hamilton Oil, despite being a US company, had been active in UK North Sea hydrocarbon exploration from the outset. The UK division of Hamilton Brothers was headed by Robert Dyk between 1964 and 1980, and recognizing that the most coveted blocks on offer in the UKCS 1st Licence Round would be located in the Southern North Sea (due to the interest generated in the Permian play from the Groningen discovery), he believed that the northern basin offered better opportunities for an unestablished US independent. Hamilton Oil positioned themselves to bid for northerly blocks in search of oil. This maverick move brought with it significant success, discovering the Argyll Field, which became the first field to produce oil in the North Sea using the ingenious method of a floating production facility; beating British Petroleum's Forties Field to first oil in June 1975. The company continued to have success in the North Sea, discovering the Bruce, Crawford, Duncan, Esmond–Forbes–Gordon, Innes and Ravenspurn fields. This early success in UK waters led by Robert Dyk, combined with their bold approach, is likely to have played a role in Hamilton Oil's successful application for the licence block in the Irish Sea.

Hamilton Oil spudded their first well in the East Irish Sea the year after receiving the licence block,

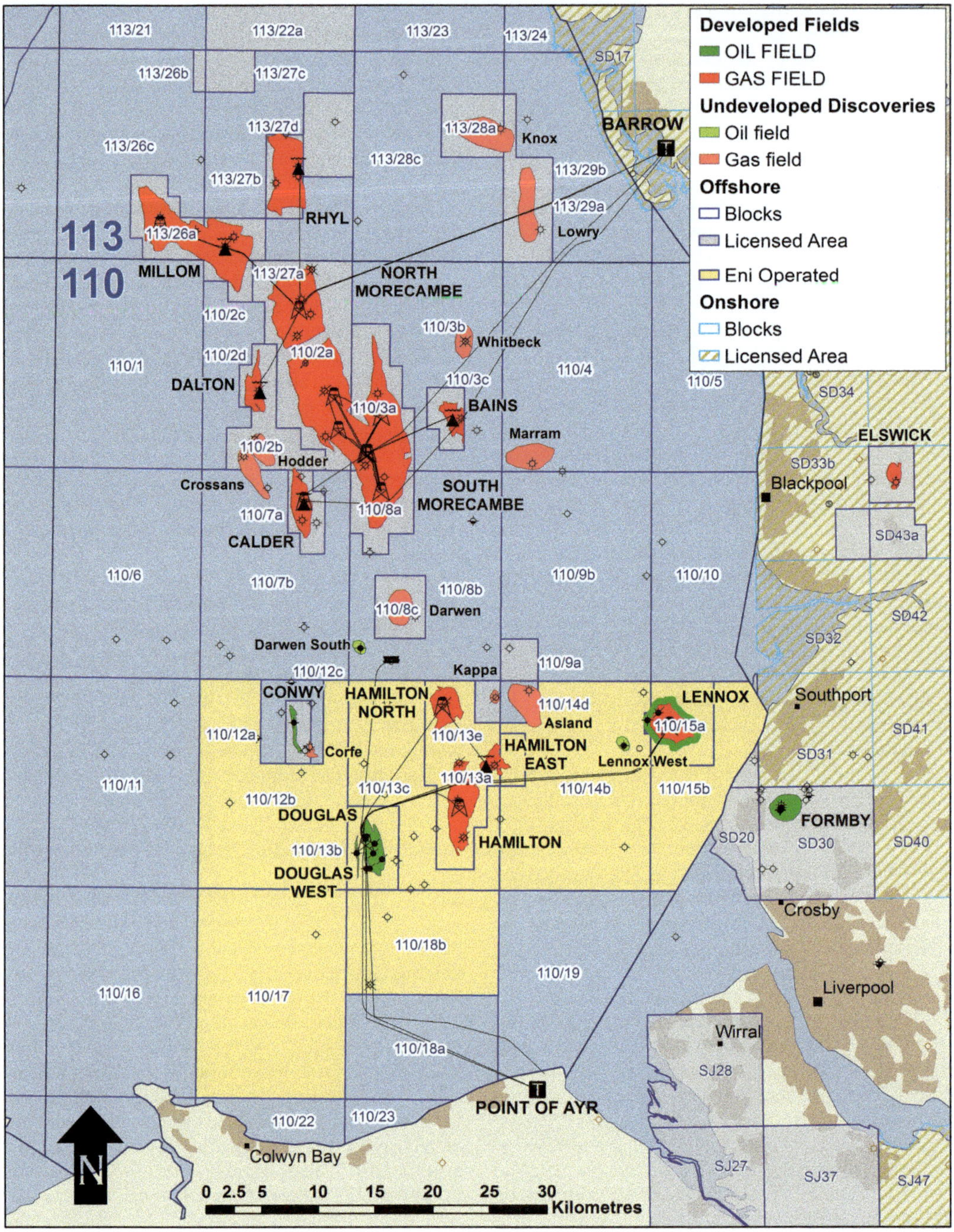

Fig. 7. East Irish Sea Basin exploration and development activity.

on 9 May 1990, with the drilling of 110/13-1. Gulf Oil's woes in the Irish Sea were to continue with the well successfully encountering over 500 ft of gas-bearing sands. The well targeted the Alpha prospect, a north–south-trending Triassic fault block defined on a sparse grid of speculative 2D seismic lines shot in 1986 and 1987 (Yaliz & Taylor 2003). It encountered good reservoir quality sands

Table 1. *Offshore oil and gas discoveries in the East Irish Sea*

Discovery well	Year	Original operator	Hydrocarbon type	Field/discovery name	Field status	Field start-up
110/8-2	1969	Gulf Oil Production Co.	Gas*	South Morecambe	In production	January 1985
110/2-1	1974	Hydrocarbons GB Limited	Gas			
110/2-3	1976	Hydrocarbons GB Limited	Gas	North Morecambe	In production	September 1994
110/7-3	1982	Hydrocarbons Great Britain Limited	Oil and gas	Calder	In production	October 2004
113/26-1	1982	Hydrocarbons Great Britain Limited	Gas	Millom	In production	August 1999
110/12a-1	1988	Hydrocarbons Great Britain Limited	Gas	Corfe	Technical success	
110/3b-4	1988	British Petroleum	Gas	Bains	Ceased production (February 2010)	November 2002
110/2b-9	1990	Gas Council (Exploration) Limited	Gas	Dalton	In production	August 1999
110/13-1	1990	Hamilton Brothers Oil and Gas Limited	Gas	Hamilton	In production	February 1997
110/13-2	1990	Hamilton Brothers Oil and Gas Limited	Oil	Douglas	In production	January 1996
110/14-1	1990	Gas Council (Exploration) Limited	Oil	Lennox West	Technical success	
110/8a-4	1991	British Gas Exploration and Production Limited	Gas	Darwen	Technical success	
110/13-5	1991	Hamilton Brothers Oil and Gas Limited	Gas	Hamilton North	In production	December 1995
110/2b-10	1992	British Gas Exploration and Production Limited	Gas	Crossans	Technical success	
110/15-6	1992	Hamilton Oil Company Limited	Oil and gas	Lennox	In production	February 1996
113/29-2	1992	Amoco (UK) Exploration Co.	Gas	Lowry	Technical success	
110/8a-5	1993	British Gas Exploration and Production Limited	Oil	Darwen South	Technical success	
110/14-2	1993	British Gas Exploration and Production Limited	Gas	Asland	Technical success	
110/13-14	1993	Hamilton Oil Company Limited	Gas	Hamilton East	Ceased production (June 2010)	September 2001
110/2b-11	1993	British Gas Exploration and Production Limited	Gas	Hodder	Technical success	
110/13-17	1994	Hamilton Oil Company Limited	Gas	Kappa	Technical success	
113/28-2	1994	Clyde Expro plc	Gas	Knox	Technical success	
110/13b-19	1997	BHP Petroleum Limited	Oil	Douglas West	In production	April 2003
110/4-1	2009	Ranger Oil (UK) Limited	Gas	Marram	Technical success	
110/12-6	2009	EOG Resources (UK) Limited	Oil	Conwy	In production	March 2016
110/3b-6	2009	Venture Production plc	Gas	Whitbeck	Technical success	
113/27b-6	2009	Hydrocarbon Resources Limited	Gas	Rhyl	In production	March 2013

*Following reinterpretation of log data by the Gas Council in 1974.
Source: Department of Energy and Climate, London.

throughout the Ormskirk Sandstone Formation and was tested, flowing at maximum gas rates of 34.9 MMscf/day during a DST over a 190 ft interval, before being plugged and abandoned. The structure was further appraised with two more wells defining the northern and southern extents of the accumulation: 110/13-3 and 110/13-4, respectively. These three wells delineate the Hamilton Field with initial gas-in-place volumes of 680 Bcf (Fig. 8).

Following on from the successful discovery of the Hamilton Field, Hamilton Oil spudded 110/13-2 later that year in October 1990, again targeting the Triassic Sherwood sandstones. The Gamma prospect was a well-defined, structurally complex series of north–south-trending tilted fault blocks, which were dip-closed to the south and fault-bound in all other directions. 110/13-2 targeted the central of three adjacent fault blocks, encountering the reservoir 103 ft deep to prognosis, primarily due to an issue with the depth conversion, but successfully encountering a 106 ft hydrocarbon column. Unexpectedly, the well identified an oil column rather than the expected fluid of gas, which had been encountered in the Hamilton Field, located

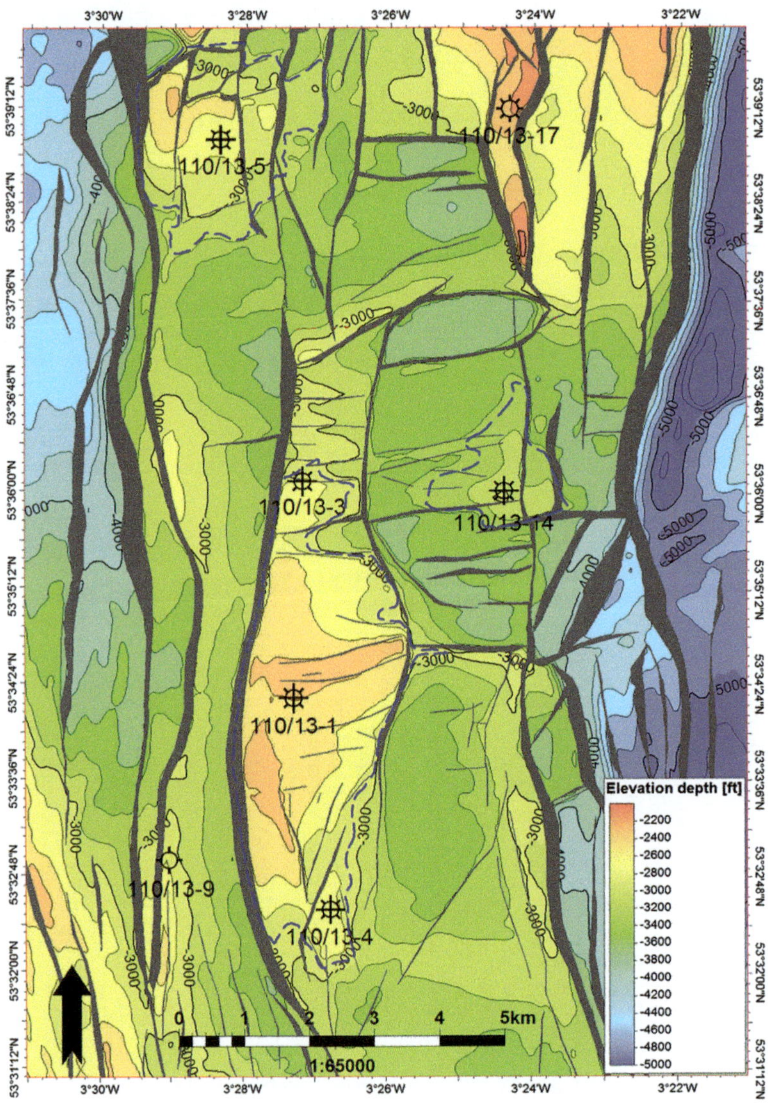

Fig. 8. Top reservoir depth structure map of the Hamilton, Hamilton North and Hamilton East fields, showing the gas–water contact for the individual fields, and the exploration and appraisal well locations.

approximately 10 km to the east. An oil–water contact was interpreted from log data at 2508 ft TVDSS and the well was suspended as an oil discovery. The presence of oil, rather than the anticipated dry gas, was confirmed from RFT (repeat formation tester) pressure data which showed an unambiguous fluid gradient of 0.34 psi/ft, and PVT (pressure–volume–temperature) analysis indicating the oil had an API gravity of 44° and a high H_2S content. The reservoir quality across the oil leg in the discovery well was excellent, with net-to-gross greater than 90% and average porosity values of 22%.

Having had success with their first two exploration wells in the south of the basin, the same licence group from Block 110/13, comprising Hamilton Oil, Ultramar Exploration and Monument Resources, applied for and were awarded Block 110/15 in the UKCS 12th Offshore Licence Round in 1991 under Licence P791. The group, operated by Hamilton Oil, identified a number of exploration prospects (named successively after letters of the Greek alphabet) within blocks 110/13 and 110/15. Hamilton Oil undertook an aggressive, and very successful, exploration drilling campaign throughout the 1990s, drilling 20 exploration and appraisal wells between 1990 and 1995 which bore fruit, with six oil and gas discoveries, four of which were subsequently developed and brought on production before the end of the decade (Table 1). Hamilton Oil's presence and exploration success in the North Sea and East Irish Sea attracted the attention of BHP, who acquired 51% of the company in 1987, before acquiring the remaining 49% in 1991, turning Hamilton Oil into a wholly-owned subsidiary of BHP.

After the drilling of discovery wells at Hamilton (110/13-1) and Douglas (110/13-2), and the subsequent appraisal drilling phase over the Hamilton Field, exploration well 110/13-5 was drilled on the Eta prospect in April 1991. The prospect, to become the Hamilton North Field, was located 5 km north of the Hamilton Field and comprised a north–south-trending horst block, fault-bound to the north and west, and dip-closed to the east and south. The target formation, the Ormskirk Sandstone, was encountered over 360 ft deep to prognosis but, despite this, still found a 475 ft gas column. RFT data from the well confirmed the log-derived gas–water contact and, interestingly, identified that the aquifer was in communication with both the Hamilton and Douglas fields; further pressure data acquisition and analysis has shown that all the Liverpool Bay fields lie on a single, normally pressured aquifer gradient. The well showed similar reservoir properties as seen in the Hamilton Field wells and was tested, flowing a maximum gas rate of 81 MMscf/day from a 130 ft interval within the uppermost Frodsham Member (Zone OS2b in the lithostratigraphic revision by Jackson *et al.* 1997). No further appraisal was deemed necessary on the Hamilton North Field, which is estimated to contain 230 Bcf of gas.

Following the discovery of Hamilton North, the first appraisal well on the Douglas Field was drilled in May 1991; 110/13-6 targeted the fault block immediately to the west of the block tested by the discovery well (Fig. 9). 110/13-6 was located close to the crest of the fault block and encountered the reservoir 81 ft shallow to prognosis and a 318 ft oil column, with a log and pressure-derived oil–water contact at 2536 ft TVDSS. This contact was 28 ft deeper than the contact established in the discovery well; however, the marginal difference in contact depths suggests that the fault blocks are likely to be in communication. The well was tested, achieving an oil rate of 1800 bbl/day; a second flow attempt was made with additional gas injection, and with gas lift the well flowed at a maximum rate of 3400 bbl/day. The field was appraised with a further three wells (110/13-10, 110/13-11 and 110/13-13) prior to field start-up. The first of these, 110/13-10, tested the hitherto undrilled easternmost fault block; it again successfully encountered hydrocarbons and an oil column of 226 ft in good-quality Ormskirk reservoir sands. The contact depth derived from pressure data suggested that the central and eastern fault blocks lay on the same fluid gradients and exhibited a common oil–water contact. Wells 110/13-11 and 110/13-13 further appraised the western and central blocks, respectively.

Between 1991 and 1994 the appraisal drilling on Douglas was interposed by Hamilton Oil's continued exploration campaign; eight of their next 11 wells in Block 110/13 targeted exploration prospects. All their exploration targets had primary objectives within the Triassic Ormskirk Sandstone Formation, and were located in combination fault- and dip-closed structures, characteristic of the extensional structural framework existing across the block; with high-density faulting resulting in the formation of small tilted fault blocks which trend in a general NE–SW direction. Of the eight wells, two successfully encountered hydrocarbon-bearing sands. The first, the Lambda prospect (Hamilton East), was located immediately to the NE of the Hamilton Field and comprised a tilted fault block bound by normal faults to the south and east, and dip-closed in the north and west. The vertical exploration well, 110/13-14, was spudded in November 1993. It reached the Ormskirk Sandstone 27 ft deep to prognosis, encountering a 344 ft gas column in good-quality, high net-to-gross reservoir sands. Log and pressure data identified a clear gas–water contact at 3346 ft TVDSS, which was deeper than the contact depths identified over the Hamilton and Hamilton North fields (Figs 4 & 8). The well was perforated and tested over the uppermost Frodsham Member (Zone OS2b), flowing at a maximum gas rate of 113

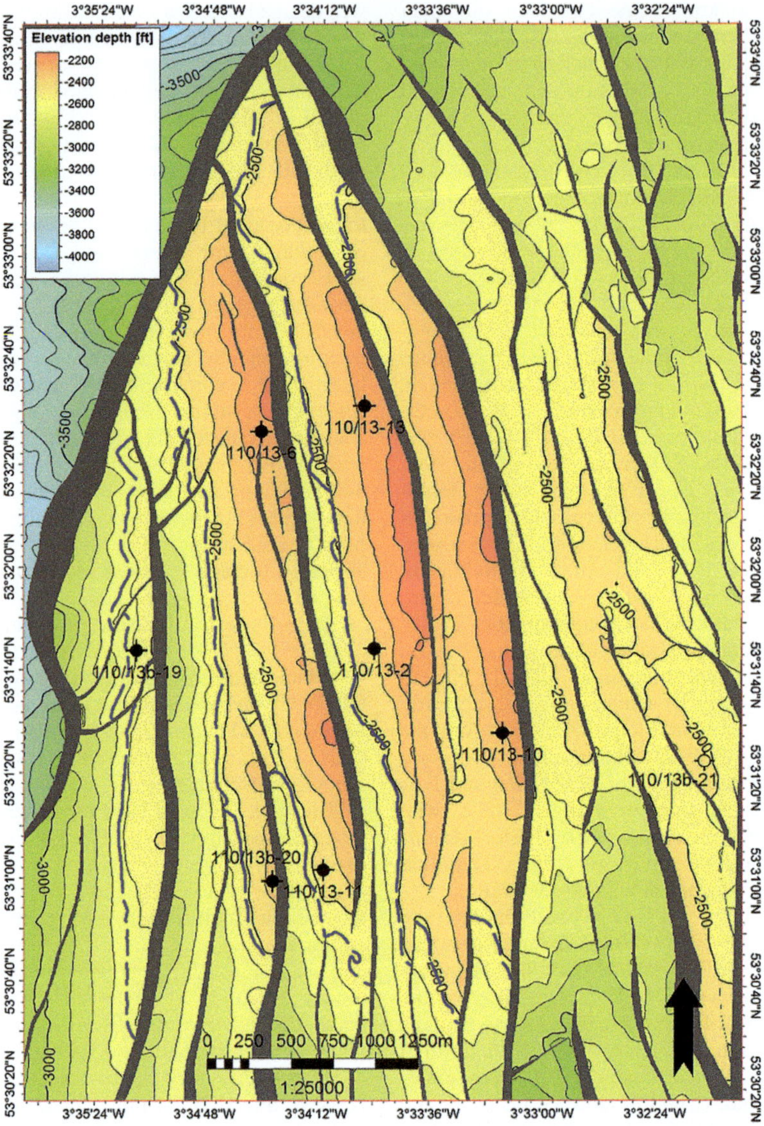

Fig. 9. Top reservoir depth structure map of the Douglas Field, showing the structural configuration, the field oil–water contact, and the exploration and appraisal well locations.

MMscf/day. The Hamilton East Field has GIIP volumes of 63 Bcf and the structure required no further appraisal drilling prior to development. The second successful well targeted the Kappa prospect. 110/13-17 was spudded in April 1994 and targeted another north–south-trending horst block located to the east of Hamilton North. The structure was located close to the block boundary with Block 110/14, which was operated by British Gas at the time and who had drilled the discovery well 110/14-2 on a similar structure 3 km to the east in November 1993. The British Gas well, targeting the Asland prospect, encountered a 133 ft gas column in the Ormskirk Sandstone and flowed at 7.5 MMscf/day. The Kappa well located on the adjacent horst block successfully encountered a 59 ft gas column in good reservoir quality sands of the Ormskirk Sandstone Formation. However, the well was not tested, and plugged back and suspended as a gas discovery. Hydrocarbon charge appears the primary reason for the failure of the remaining six exploration wells; however, the presence of localized bitumen staining and a 'petroliferous odour' identified in wells 110/13-7 and 110/13-9 suggest these structures may have been hydrocarbon-filled at some point, and breached during Cimmerian or Alpine tectonic activity.

Discovery of the Lennox Field

British Gas (through the wholly-owned subsidiary Gas Council (Exploration) Limited) drilled 110/14-1 in October 1990 targeting a faulted rollover anticline structure at the top Sherwood Sandstone level. The well was significant due to its proximity to the neighbouring licence block, 110/15, which at the time of drilling was being offered during the upcoming UKCS 12th Offshore Licence Round. The structural closure targeted by 110/14-1 potentially extended eastwards into the acreage, but the original mapping of the structure by the Gas Council suggested a significant proportion of the closure was located within their 110/14 licence block. The well was drilled to a total depth within the Ormskirk Sandstone, the top reservoir was encountered deep to prognosis but the interval showed good reservoir properties and identified a 26 ft oil column. However, the well was designated a 'tight hole', and was plugged and abandoned without being tested. Hamilton Oil, rather than the Gas Council, would go on to be awarded the neighbouring Block 110/15 in June 1991 following their successful exploration drilling in Block 110/13.

Following the drilling of the unsuccessful Epsilon exploration well (110/13-12) targeting a Triassic faulted horst block, Hamilton Oil moved their attention to exploration drilling in Block 110/15. The Alpha prospect had been identified on good-quality 2D seismic lines shot over 586 km across the block on two surveys in June 1986 and November 1990. The structure was a well-defined four-way dip closure located in the hanging wall of the basin-bounding Formby Point fault to the east (Haig *et al.* 1997), and was the eastern extension of the structure targeted by 110/14-1. 110/15-6 was spudded on 3 June 1992; located at the crest of the structure, it penetrated the top reservoir 113 ft deeper than pre-drill expectations. However, the Triassic sands were hydrocarbon-filled and the well encountered a 888 ft gross hydrocarbon column, with a 744 ft gas column overlying 144 ft of oil-bearing sands. The reservoir quality in 110/15-6 was very good with net-to-gross values of 97 and 98% in the gas and oil legs, respectively, and average porosities of approximately 16% in the gas leg and 20% in the oil leg. The well was tested in the oil leg over a 30 ft perforated interval and flowed 870 bbl/day over 12 h, a second DST was conducted over a 20 ft perforated zone in the gas cap and flowed at rates of 61 MMscf/day.

The Alpha prospect, now assigned the Lennox Field, was immediately appraised via a deviated sidetrack from the 110/15-6 wellbore. 110/15-6z was drilled to test the presence and continuity of hydrocarbons across a significant north–south-trending fault which divides the field into an eastern panel and a downthrown western panel (Fig. 10). The sidetrack into the western panel was approximately 650 m due west of the discovery well (at the top reservoir level), and encountered the Ormskirk Sandstone 320 ft shallower than prognosed. The well confirmed the continuity of the hydrocarbon-bearing sands, with RFT pressure data showing that the two panels were in communication in all three fluids legs, sharing a common contact at the gas–oil (3257 ft TVDSS) and oil–water (3400 ft TVDSS) contacts. The western panel of Lennox was further appraised by well 110/15-7 in March 1993.

Prior to the submission of the Lennox Field Development Plan (FDP) in 1993, British Gas, the licensee of the neighbouring Block 110/14, presented an alternative structural interpretation of the Lennox Field extending the accumulation into Block 110/14. Validation of the interpretation required a further appraisal well and British Gas drilled 110/14-3 adjacent to the block boundary in order to delineate the structural closure. Drilled in December 1993, the well appraised the field extension into Block 110/14 and also aimed to determine whether the minor oil discovery at 110/14-1 was part of the flank of the Lennox Field. The results of 110/14-3 could have had significant repercussions for the field development scheme and a Field Development Plan Addendum was submitted in the event that 110/14-3 was successful in extending the closure to the west. The British Gas interpretation for the western flank of the field presented a low-lying structure above the oil–water contact which would need to be targeted with horizontal production wells drilled from a second drill centre to maximize hydrocarbon recovery. 110/14-3 was drilled to a total depth within the St Bees Sandstone Formation, encountering the top Ormskirk Sandstone reservoir 119 ft deep to prognosis. The reservoir was found to be oil-bearing, with log and MDT (modular dynamic formation tester) pressure data showing a free-water level in agreement with the Lennox oil–water contact, resulting in a thin 30 ft oil column. However, more significantly, the deeper top reservoir showed that the mapped flank extended less than 1 km into Block 110/14 (Fig. 10).

Development of the Liverpool Bay fields

Following their exploration drilling campaign, Hamilton Oil's focus turned to the development of their Liverpool Bay discoveries, and development drilling

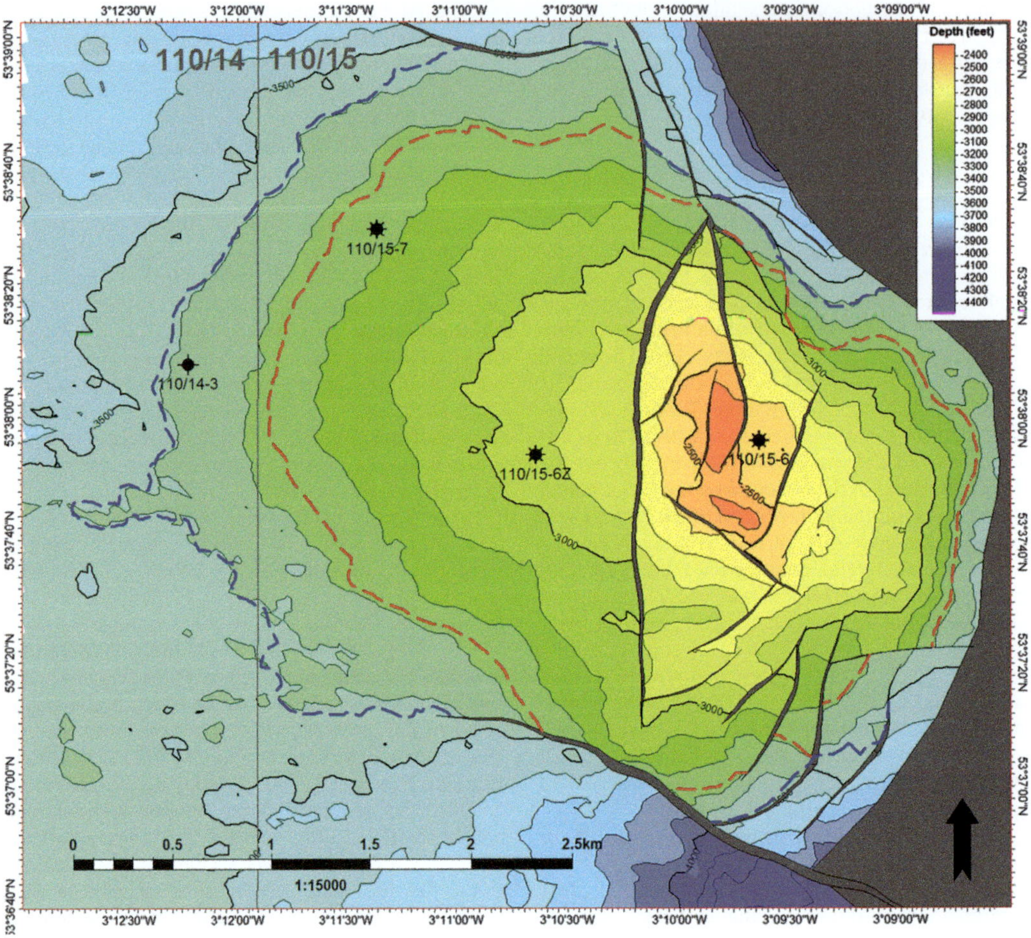

Fig. 10. Top reservoir depth structure map of the Lennox Field, showing the exploration and appraisal well locations, and the field fluid contacts.

on the Douglas Field began in September 1994. The original FDP consisted of 11 production wells and seven water injectors, with injection wells required to provide pressure support to supplement the relatively weak aquifer and improve sweep efficiency. Douglas oil production wells were also equipped with electrical submersible pumps providing artificial lift for the low GOR (gas–oil ratio) fluid. The production wells on the field are located near to the crests of the individual rotated fault blocks, with the injector wells targeting the aquifer in downdip locations. The Douglas Field has an estimated stock tank oil initially in-place (STOIIP) of 250 MMbbl (million barrels). The development scheme consisted of a central 24-slot wellhead platform with oil exported via a 14 inch subsea pipeline to an offshore storage installation (OSI). The Douglas Complex, which comprises a wellhead tower, production platform and bridge-linked accommodation platform, was designed to receive and process hydrocarbons from all the fields in the Liverpool Bay area in an integrated development scheme (Fig. 11): gas from Hamilton and Hamilton North, and oil and gas from Lennox. The gas is received from the Hamilton satellite fields via a 14 inch subsea pipeline and compressed at the Douglas production platform before export onshore to the Point of Ayr Gas Terminal.

The Hamilton North Field was developed with three gas production wells drilled between September 1995 and October 1996 from an unmanned platform located above the field. The platform is tied back to the Douglas facilities and the field came online in December 1995. The Hamilton Field was developed via four production wells, again drilled from a centrally located platform between April

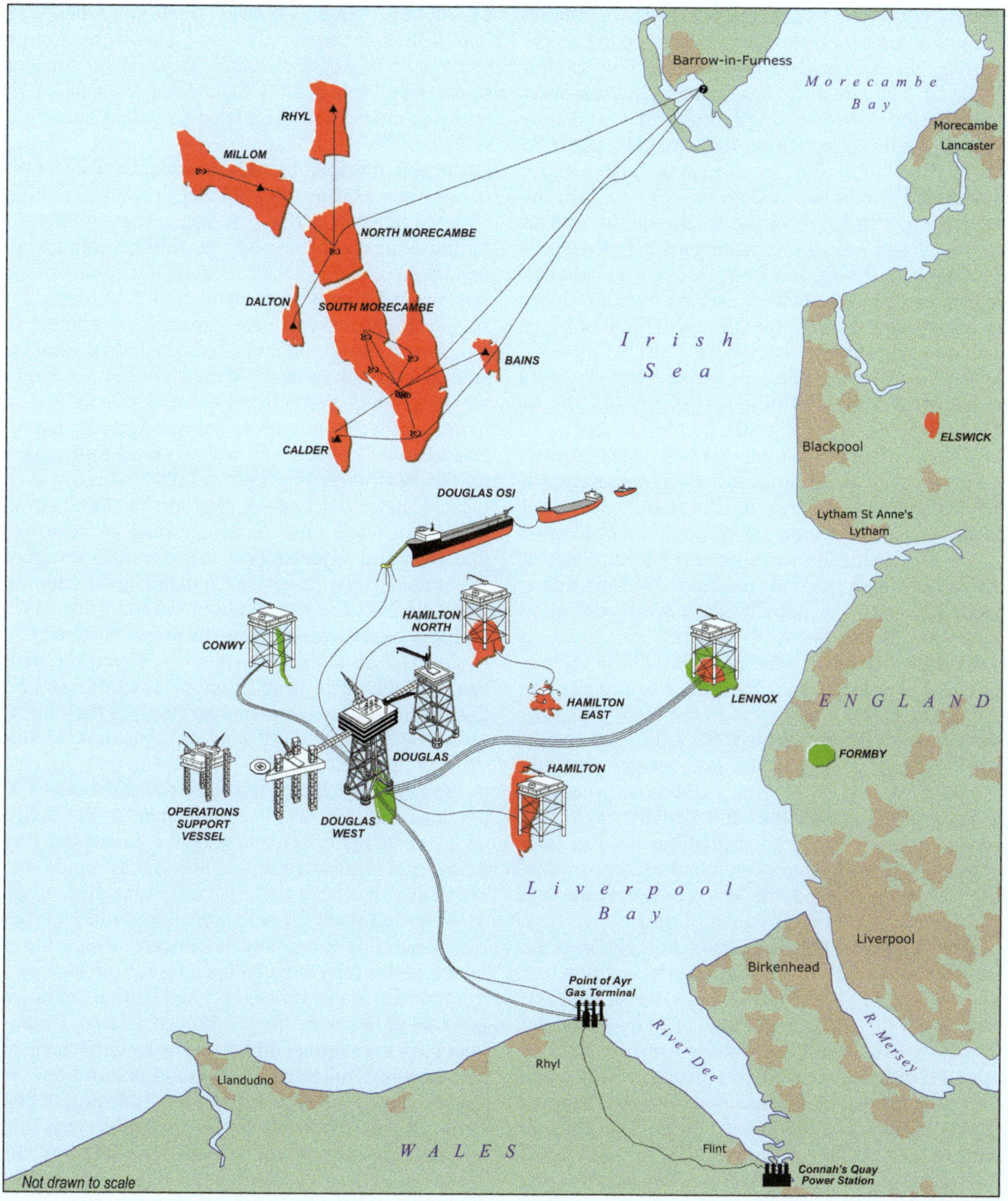

Fig. 11. Facilities schematic for the Liverpool Bay fields, including the surrounding infrastructure of the Morecambe Bay fields.

1996 and August 1998, with first gas from the field in February 1997. Both gas fields produce via natural depletion.

Douglas came on stream in February 1996 with six oil production and five water injection wells (D1–D11); subsequently, a further nine oil producers, one condensate injection and one water injection well have been drilled (D12–D22). Although the field was thought to have been adequately delineated, two further appraisal wells were drilled on Douglas in 1997 and 2000. The first targeted the adjacent fault block to the west of the Douglas Field and was designated the Nu prospect (110/13b-19: Fig. 9). The well, drilled by BHP, successfully found hydrocarbons and an oil column exhibiting a deeper oil–water contact than the neighbouring eastern panel (2641 ft TVDSS v. 2536 ft TVDSS). The structure was termed Douglas West and developed with a single highly deviated production well drilled from the Douglas platform (110/13b-D20y). The second well, 110/13b-20, was drilled within the western panel to confirm the top reservoir structure in the southernmost area of the panel and led to the drilling of a further production well. Production from Douglas has totalled approximately 100 MMbbl, and the field produces at a current rate of approximately 5000 bbl/day with an expected ultimate recovery (EUR) of 46%. At the time of the original FDP, the EUR was estimated at 88 MMbbl with an expected field life of 13 years (v. the 20 year production life to date).

The appraisal of the Lennox Field had confirmed the presence of an oil rim and free gas cap, and field development was planned in two phases: the initial phase focused on the oil rim through the drilling of horizontal production wells radiating from a single, centrally-located, normally unmanned wellhead platform. Lennox was initially developed with nine horizontal production wells and two gas injection wells drilled from the platform. Two-phase separation facilities located on the platform process the hydrocarbons prior to export to the Douglas Complex; liquids are transported via a 14 inch subsea pipeline, and gas via a 16 inch subsea pipeline. During this phase of oil rim production, processed solution gas from Douglas and Lennox crude was exported back to the Lennox platform via a 12 inch subsea pipeline for reinjection into the Lennox reservoir. This maintained the reservoir pressure and maximized production from the oil rim.

The volumes of hydrocarbon initially in-place are estimated from 3D reservoir modelling: Lennox has a STOIIP of 200 MMbbl and GIIP of approximately 400 Bcf (free gas). A total of 14 development wells (L1–L14) have been drilled on the field to date, 11 targeting the oil leg, many of which are dual or multi-lateral producers. Yaliz & Chapman (2003) describe the Lennox Field development prior to gas cap blowdown (GCBD) in more detail than this paper. As gas production began to decline from the Hamilton fields (including Hamilton East, which was brought on stream as a single well development in 2001 to extend the gas production plateau), Lennox gas was required from August 2003 to supplement the total gas volume in order to meet contractual sales gas volumes. Apportioning an increasing volume of produced Lennox gas for sales resulted in a sharper decline in the field reservoir pressure and also increased the movement of both the gas–oil and oil–water levels in the reservoir; this led to an accelerated final phase of oil rim production prior to the initiation of the GCBD. During this period, two infill targets were identified and drilled by wells L13 and L14. The former located in the eastern panel, targeted an area of bypassed oil in a heavily faulted region of the field, and was drilled and completed as a dual lateral oil producer. An additional consequence of the reduced gas injection was an increase in water cut for a number of wells, leading to significant issues when lifting hydrocarbons to surface. A technology review to optimize the completion design identified the implementation of surface-controlled gas-lift completions; this proved highly successful and L13 came on stream producing twice the expected oil volume in the first 5 months (Clarke *et al.* 2006). Gas-lift recompletions were installed on a number of Lennox wells, gas cap perforations were also added to a number of wells in order to naturally gas-lift the oil production. The L14 well was the final well drilled on the field, targeting a potentially undrained area located under the Lennox platform in 2009. However, the initial pilot well identified a shallower than expected producing oil–water contact resulting in a thinner than prognosed remaining oil column. The field has been in a phase of GCBD since 2012, with the conversion of six wells to gas producers.

Eni UK (at the time Agip UK) entered the Irish Sea through their acquisition of LASMO (ULX) Limited in 2001 – Ultramar had changed their name to LASMO in 1992 and went on to acquire Monument Oil and Gas in 1999 – acquiring their interest in the Liverpool Bay fields. In April 2014, Eni UK purchased BHP's 46.1% equity share of the fields, taking over the operatorship of the Liverpool Bay fields and 100% equity. Eni's focus since the acquisition has been on late-life optimization of the existing well stock for gas production. Eni Liverpool Bay Operating Company undertook a significant seven-well intervention campaign in 2015 designed to minimize water production and further optimize gas cap production, the workscope involved plugging back the original oil laterals using a mechanical bridge plug, followed by adding perforations in the gas leg. This resulted in an initial production uplift of approximately 30%, breathing new life into the asset. Extending the field life could be critical in aiding the development of smaller, marginal exploration discoveries which could be tied in to existing infrastructure, as well as deferring the abandonment of the field and decommissioning expenditure.

Recent exploration success

The creaming curve for the East Irish Sea Basin (Fig. 12) exhibits typical behaviour for a mature hydrocarbon province, with the largest discoveries made in the early exploration of the basin (i.e.

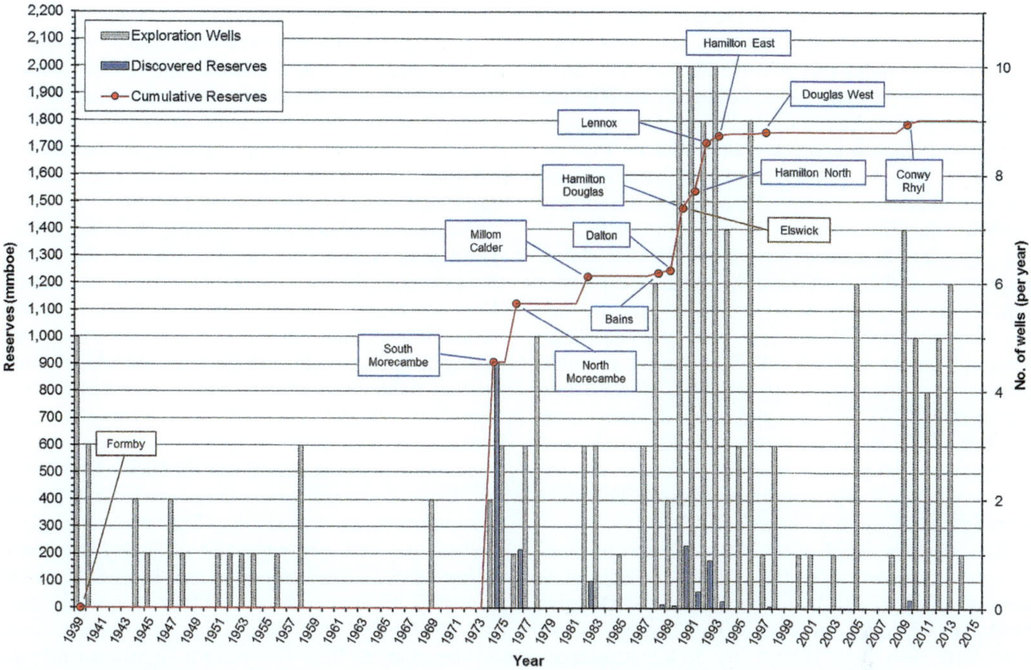

Fig. 12. Creaming curve for the East Irish Sea Basin, highlighting the number of exploration wells drilled per year and the commercial discoveries (i.e. developed fields) including both the onshore and offshore regions: 156 onshore and offshore exploration wells have been drilled between 1939 and 2014, and provide the dataset for the creaming curve (*source*: IHS Markit).

South and North Morecambe, Hamilton, Douglas, and Lennox) before the curve begins to flatten out as exploration success wanes and focuses on smaller accumulations. All the offshore East Irish Sea Basin discoveries are within the Triassic play and the majority were found within about the first 20 years of exploration. The creaming curve for the basin clearly highlights that two periods of exploration and development have dominated the history of the offshore basin, with the mid-1970s and early 1990s dominating exploration activity and success. These periods coincide with the start of exploration in the Morecambe Bay and Liverpool Bay areas, respectively. Somewhat characteristic of exploration on the UKCS in general, the number of exploration wells drilled in the East Irish Sea Basin has begun to decline. Recent exploration success has been dominated by a small number of minor discoveries, which have marginal economics as stand-alone projects and rely on development concepts utilizing existing platform facilities and pipelines through tie-back development schemes.

Of the discoveries made since the Lennox Field in 1992, only four – Hamilton East, Douglas West, and more recently the Rhyl (first gas 31 March 2013) and Conwy (first oil 5 March 2016) – fields have been developed and brought on stream. Twelve other discoveries have yet to be developed (Table 1). Achieving adequate hydrocarbon volumes close to existing infrastructure and being able to develop those volumes with minimal facilities appears crucial.

The Rhyl Field is the most recent commercial discovery in the Irish Sea. The licence acreage was awarded to Hydrocarbon Resources Limited (HRL), a wholly-owned subsidiary of Centrica plc, after the UKCS 24th Offshore Licence Round in 2006. The acreage, Block 113/27b, is located immediately NE of the North Morecambe and Millom fields. The discovery well, 113/27b-6, was spudded on 12 November 2009 targeting a series of fault terraces interpreted on recently reprocessed seismic data, in which a large effort was made to better understand and incorporate the shallow geology into the depth conversion (Ward & Baker 2018). A review by HRL of well 113/27-1, which targeted the ‘Ramsey’ prospect, located immediately to the north of the Rhyl discovery, highlighted the impact of the overburden geology in creating a velocity pull-up effect at the top reservoir. That well, drilled in 1986, targeted a structural closure at top Triassic. It encountered a thick Ormskirk Sandstone interval of good reservoir quality, but the well was deep to prognosis, with low gas saturation, and was abandoned as a dry hole. The post-well review identified that a thick Tertiary dolerite dyke encountered within the Mercia Mudstone Group had created the velocity pull-up and when it was integrated into the velocity modelling (Ward & Baker 2018), the resulting top structure depth map removed the pull-up effect and identified the Rhyl structure to the south of the dry well. 113/27b-6 successfully encountered a 495 ft gas column within the Ormskirk Sandstone Formation. Reservoir quality was good, with porosity values ranging from 8 to 22% and permeabilities ranging from <1 to 1000 mD. The well was tested flowing at a maximum rate of 15.2 MMscf/day. The Rhyl Field is estimated to contain over 100 Bcf of gas reserves (Ward & Baker 2018) and has been developed as a subsea tieback to the unmanned North Morecambe platform. The field started production on 31 March 2013.

The Conwy Field was the most recent field brought on stream in the East Irish Sea Basin, it is located in Block 110/12 to the NW of the Douglas Field, and was discovered in 2009. The exploration well, 110/12-6, was drilled by EOG Resources and targeted Prospect C, a tilted fault block of the Ormskirk Sandstone Formation interpreted from a 3D seismic survey shot in 1995. The well was plugged and abandoned after encountering oil over a 146 ft interval and identifying the oil–water contact. The sands showed typical reservoir properties for the Ormskirk Sandstone, with core analysis showing porosities ranging from 5 to 27% (average of *c.* 19%) with horizontal permeabilities from 0.2 to 4000 mD (average values of 440 mD). The Conwy Field was developed via three oil producers drilled from a central platform, with water and condensate injection used to maintain reservoir pressure and increase oil recovery over the life of the field. The wellhead platform is tied back to the facilities at the Douglas platform and the field came on stream in March 2016, with estimates of oil initially in-place of approximately 20 MMbbl.

Future exploration

Although future exploration in the offshore is likely to continue to focus on the proven Triassic Sherwood Sandstone play, there remains potential in the deeper, largely untested plays associated with the Permian Collyhurst sandstones and the Carboniferous fluvial sandstones. The main risk associated with both these older plays lies in the top seal for the reservoirs, the former requiring a thick and effective Manchester Marl and the latter reliant on intraformational shales forming a competent seal. The deeper reservoir targets are additionally hindered by relatively poor seismic resolution below the Manchester Marl, and a complex overburden consisting of interbedded siltstones, shales and halites which pose significant complications and uncertainties for depth conversion. However, as demonstrated by the Rhyl discovery, taking the time to understand and interpret the overburden geology can make a significant impact to structures at depth.

Near-field exploration utilizing existing infrastructure is likely to play a greater role in the future of exploration in the offshore East Irish Sea Basin. With an aging infrastructure and field decommissioning moving ever closer, the opportunity to reinvigorate the basin and prolong the life of both existing fields and infrastructure is one that is welcomed by oil and gas companies throughout this hydrocarbon province. A number of undeveloped discoveries do exist across the East Irish Sea Basin but, due to their size, conceiving economic development concepts is challenging. Trap size and effectiveness (i.e. seal presence and capacity) is critical for successful exploration in the basin, and development will require existing facilities to quickly and cost-effectively bring discoveries on stream.

The onshore may also have a significant role to play in the future of the region. The West Lancashire Basin has had only modest success since hydrocarbon exploration began there over 70 years ago; however, the influence and learnings from those early exploration wells were instrumental in the later discovery of hydrocarbons in the offshore East Irish Sea Basin. Now, after over 30 years of oil and gas production in the offshore, the onshore basin could return to the forefront of exploration due to the emergence of an unconventional resource play within the Carboniferous of NW England. Organic-rich shales from the Carboniferous-aged Bowland Shale Formation (revised BGS lexicon; equivalent to the Holywell Shale Formation) are of significant interest with the developments in hydraulic fracturing and the successful recovery of both oil and gas from tight shale formations in the USA. In the 2008 UK 13th Onshore Licence Round, a number of blocks were awarded in West Lancashire to Aurora Exploration and Cuadrilla Resources, under licences PEDL 164 and PEDL 165, respectively. Since then, eight onshore wells have been drilled. There has been continued interest in the onshore area, highlighted by the award of several licence blocks in the 2016 UK 14th Onshore Licensing Round, which could see the exploration focus in the East Irish Sea Basin return to the onshore where it began in 1939.

From the humble beginnings of the Formby Field and its 71 557 bbl of oil production, the East Irish Sea Basin went on to see the discoveries of the gas giants of Morecambe Bay and the significant oil fields of Douglas and Lennox. Recent offshore exploration and development successes demonstrate that the East Irish Sea Basin still has opportunities to be exploited. However, it could turn out that hydrocarbon exploration in the basin goes full circle and returns to the onshore. With the rapid growth of the global unconventional industry, hydrocarbon production from the Bowland Shale could become the next significant play in the region.

Acknowledgements This paper is published with permission of Eni UK Limited and Eni Liverpool Bay Operating Company.

Funding This research received no specific grant from any funding agency in the public, commercial or not-for-profit sectors.

References

Armstrong, J.P., Smith, J., D'Elia, A.A. & Trueblood, S.P. 1997. The occurrence and correlation of oils and Namurian source rocks in the Liverpool Bay-North Wales area. *In*: Meadows, N.S., Trueblood, S.P., Hardman, M. & Cowan, G. (eds) *Petroleum Geology of the Irish Sea and Adjacent Areas*. Geological Society, London, Special Publications, **124**, 195–211, https://doi.org/10.1144/GSL.SP.1997.124.01.12

Bastin, J.C., Boycott-Brown, T., Sims, A. & Woodhouse, R. 2003. The South Morecambe Gas Field, Blocks 110/2a, 110/3a, 110/7a and 110/8a, East Irish Sea. *In*: Gluyas, J.G. & Hichens, H.M. (eds) *United Kingdom Oil and Gas Fields, Commemorative Millennium Volume*. Geological Society, London, Memoirs, **20**, 107–118, https://doi.org/10.1144/GSL.MEM.2003.020.01.09

Clarke, A., Ayton, J., Lawton, D., Lean, J. & Burke, K. 2006. Case study: Lennox – the race to produce oil prior to gas cap blowdown. Paper SPE 100126 presented at the SPE Europec/EAGE Annual Conference and Exhibition, 12–15 June 2006, Vienna, Austria.

Colter, V.S. 1997. The East Irish Sea Basin – from caterpillar to butterfly, a thirty-year metamorphosis. *In*: Meadows, N.S., Trueblood, S.P., Hardman, M. & Cowan, G. (eds) *Petroleum Geology of the Irish Sea and Adjacent Areas*. Geological Society, London, Special Publications, **124**, 1–9, https://doi.org/10.1144/GSL.SP.1997.124.01.01

Cope, F.W. 1939. Oil occurrences in south-west Lancashire. *Bulletin of the Geological Survey of Great Britain*, **2**, 18–25.

Cowan, G. & Boycott-Brown, T. 2003. The North Morecambe Field, Block 110/2a, East Irish Sea. *In*: Gluyas, J.G. & Hichens, H.M. (eds) *United Kingdom Oil and Gas Fields, Commemorative Millennium Volume*. Geological Society, London, Memoirs, **20**, 97–105, https://doi.org/10.1144/GSL.MEM.2003.020.01.08

Dean, G. 1996. Undiscovery wells of the UK Continental Shelf. *In*: Glennie, K. & Hurst, A. (eds) *AD 1995: NW Europe's Hydrocarbon Industry*. Geological Society, London, 69–80.

Fraser, A.J., Nash, D.F., Steele, R.P. & Ebdon, C.C. 1990. A regional assessment of the intra-Carboniferous play of northern England. *In*: Brooks, J. (ed.) *Classic Petroleum Provinces*. Geological Society, London, Special Publications, **50**, 417–440, https://doi.org/10.1144/GSL.SP.1990.050.01.26

Haig, D.B., Pickering, S.C. & Probert, R. 1997. The Lennox oil and gas field. *In*: Meadows, N.S., Trueblood, S.P., Hardman, M. & Cowan, G. (eds) *Petroleum Geology of the Irish Sea and Adjacent Areas*. Geological Society, London, Special Publications, **124**, 417–436, https://doi.org/10.1144/GSL.SP.1997.124.01.25

Hardman, M., Buchanan, J., Herrington, P. & Carr, A. 1993. Geochemical modelling of the East Irish Sea Basin: its influence on predicting hydrocarbon type and quality. *In*: Parker, J.R. (ed.) *Petroleum Geology of Northwest Europe: Proceedings of the 4th Conference*. Geological Society, London, 809–821, https://doi.org/10.1144/0040809

Hardman, R.F.P. 2003. Lessons from oil and gas exploration in and around Britain. *In*: Gluyas, J.G. & Hichens, H.M. (eds) *United Kingdom Oil and Gas Fields, Commemorative Millennium Volume*. Geological Society, London, Memoirs, **20**, 5–16, https://doi.org/10.1144/GSL.MEM.2003.020.01.03

Jackson, D.I. & Mulholland, P. 1993. Tectonic and stratigraphic aspects of the East Irish Sea Basin and adjacent areas: contrasts in their post-Carboniferous structural styles. *In*: Parker, J.R. (ed.) *Petroleum Geology of Northwest Europe: Proceedings of the 4th Conference*. Geological Society, London, 791–808, https://doi.org/10.1144/0040791

Jackson, D.I., Mulholland, P., Jones, S.M. & Warrington, G. 1987. The geological framework of the East Irish Sea Basin. *In*: Brooks, J. & Glennie, K.W. (eds) *Petroleum Geology of North-West Europe*. Graham & Trotman, London, 191–203.

Jackson, D.I., Johnson, H. & Smith, N.J.P. 1997. Stratigraphical relationships and a revised lithostratigraphic nomenclature for the Carboniferous, Permian and Triassic rocks of the offshore East Irish Sea Basin. *In*: Meadows, N.S., Trueblood, S.P., Hardman, M. & Cowan, G. (eds) *Petroleum Geology of the Irish Sea and Adjacent Areas*. Geological Society, London, Special Publications, **124**, 11–32, https://doi.org/10.1144/GSL.SP.1997.124.01.02

Johnson, H.D. & Fisher, M.J. 1998. Geological controls of North Sea hydrocarbon plays. *In*: Glennie, K.W. (ed.) *Petroleum Geology of the North Sea: Basic Concepts and Recent Advances*, 4th edn. Blackwell Scientific, Oxford, 463–547.

Knipe, R.J., Cowan, G. & Balendran, V.S. 1993. The tectonic history of the East Irish Sea Basin with reference to the Morecambe Fields. *In*: Parker, J.R. (ed.) *Petroleum Geology of Northwest Europe: Proceedings of the 4th Conference*. Geological Society, London, 857–866, https://doi.org/10.1144/0040857

Lees, G.M. & Taitt, A.H. 1945. The geological results of the search for oilfields in Great Britain. *Quarterly Journal of the Geological Society, London*, **101**, 255–317, https://doi.org/10.1144/GSL.JGS.1945.101.01-04.12

Meadows, N.S. & Beach, A. 1993. Structural and climatic controls on facies distribution in a mixed fluvial and aeolian reservoir: the Triassic Sherwood Sandstone in the

Irish Sea. *In*: North, C.P. & Prosser, D.J. (eds) *Characterisation of Fluvial and Aeolian Reservoirs*. Geological Society, London, Special Publications, **73**, 247–264, https://doi.org/10.1144/GSL.SP.1993.073.01.15

Moreton, R. (ed.) 1995. *Tales from Early UK Oil Exploration 1960–1979*. Petroleum Exploration Society of Great Britain (PESGB), London.

Newport, L.P. 2016. *Petrographic and geochemical analysis of the carboniferous (Namurian) Holywell Shale of northeast Wales*. Durham thesis, Durham University, http://etheses.dur.ac.uk/11467/

Pinfold, E.S. 1958. The search for oil in Lancashire. *Liverpool and Manchester Geological Journal*, **2**, 106–123.

Roche, I. 2012. Lessons from history – unlocking a new UK shale oil play. Paper SPE 150933 presented at the SPE/EAGE European Unconventional Resources Conference & Exhibition – From Potential to Production, 20–22 March 2012, Vienna, Austria.

Stuart, I.A. 1993. The geology of the North Morecambe Gas Field, East Irish Sea Basin. *In*: Parker, J.R. (ed.) *Petroleum Geology of Northwest Europe: Proceedings of the 4th Conference*. Geological Society, London, 883–895, https://doi.org/10.1144/0040883

Stuart, I.A. & Cowan, G. 1991. The South Morecambe Field, Blocks 110/2a, 110/3a, 110/8a, UK East Irish Sea. *In*: Abbotts, I.L. (ed.) *United Kingdom Oil and Gas Fields, 25 Years Commemorative Volume*. Geological Society, London, Memoirs, **14**, 527–541, https://doi.org/10.1144/GSL.MEM.1991.014.01.66

United Kingdom Discovery Digest 2007. Volume 15. Canadian Discovery Limited.

Ward, G. & Baker, D. 2018. Rhyl Field: developing a new structural model by integrating basic geological principles with advanced seismic imaging in the Irish Sea. *In*: Bowman, M. & Levell, B. (eds) *Petroleum Geology of NW Europe: 50 Years of Learning – Proceedings of the 8th Petroluem Geology Conference*. Geological Society, London, 355–371, https://doi.org/10.1144/PGC8.11

Warrington, G., Audley-Charles, M.G. et al. 1980. *A Correlation of Triassic Rocks in the British Isles*. Special Report, Geological Society, London, **17**.

Wills, L.J. 1970. The Triassic succession in the Midlands in its regional setting. *Quarterly Journal of the Geological Society, London*, **126**, 225–285, https://doi.org/10.1144/gsjgs.126.1.0225

Yaliz, A. & Chapman, T. 2003. The Lennox oil and gas field, Block 110/15, East Irish Sea. *In*: Gluyas, J.G. & Hichens, H.M. (eds) *United Kingdom Oil and Gas Fields, Commemorative Millennium Volume*. Geological Society, London, Memoirs, **20**, 87–96, https://doi.org/10.1144/GSL.MEM.2003.020.01.07

Yaliz, A. & McKim, N. 2003. The Douglas oil field, Block 110/13b, East Irish Sea. *In*: Gluyas, J.G. & Hichens, H.M. (eds) *United Kingdom Oil and Gas Fields, Commemorative Millennium Volume*. Geological Society, London, Memoirs, **20**, 61–75, https://doi.org/10.1144/GSL.MEM.2003.020.01.05

Yaliz, A. & Taylor, P. 2003. The Hamilton and Hamilton North Gas Fields, Block 110/13a, East Irish Sea. *In*: Gluyas, J.G. & Hichens, H.M. (eds) *United Kingdom Oil and Gas Fields, Commemorative Millennium Volume*. Geological Society, London, Memoirs, **20**, 77–86, https://doi.org/10.1144/GSL.MEM.2003.020.01.06

The history of hydrocarbon exploration and development in North Yorkshire

M. Q. HAARHOFF[1]*, F. HUGHES[1], M. HEATH-CLARKE[1], D. HARRISON[1], C. TAYLOR[1], D. L. WARE[2], G. G. EMMS[1] & A. MORTIMER[1]

[1]*Third Energy Ltd, 4th Floor, 87–91 Newman Street, London W1T 3EY, UK*

[2]*Independent Consultant*

M.Q.H., 0000-0002-3467-077X; F.H., 0000-0002-9226-7081; M.H-C., 0000-0003-3182-9009; C.T., 0000-0003-4230-2480; G.G.E., 0000-0002-9586-739X

**Correspondence: matthew.haarhoff@third-energy.com*

Abstract: Hydrocarbon exploration in North Yorkshire began in 1937, targeting Triassic and Permian reservoirs below the surface expression of the Cleveland Anticline. D'Arcy drilled the successful well Eskdale-2, marking the first gas discovery in the Zechstein carbonates in the UK. Since then approximately 100 wells have been drilled in the basin with exploration success relatively high. Out of the 25 pure exploration wells in the region, 13 have found hydrocarbon accumulations (flowed gas) and eight of the discoveries have been developed to date. The primary reservoir is the Permian-aged Zechstein carbonate sequence and, more specifically, the Z2, Kirkham Abbey Formation (KAF), which is a tight carbonate reservoir overprinted by a high permeability fracture system. Despite considerable investment and effort over the years, the historical development story of these fields has been very much one of repeated technical and investment failure, with approximately 39 Bcf (billion cubic feet) of the mapped gas initial in-place (GIIP) of *c.* 326 Bcf produced to date, an estimated recovery factor of 12%.

Historical production data show that all the Zechstein reservoirs have experienced early water breakthrough, leading to impaired gas rates and low recoveries. The water influx is due to a highly mobile, but finite aquifer, which under field production conditions preferentially flows through the high permeability fracture system, bypassing the gas stored in the tighter matrix. Third Energy is aiming to resolve the issue of water influx by using artificial lift to encourage the gas to flow. A trial is currently being undertaken at the Pickering gas field and, if this programme is successful, this will provide sufficient confidence for a phased redevelopment programme of surrounding fields.

Whilst North Yorkshire has experienced only limited exploration and production (E&P) activity in the last decade, solving the issue of premature water influx in the KAF fields, combined with the search for unconventional resources in the Bowland section of the Mid and Lower Carboniferous strata will herald a new and exciting phase of E&P activities for this province.

The Cleveland Basin of North Yorkshire (Fig. 1) forms the westernmost extension of the Southern North Sea (SNS) Basin (Smith *et al.* 1985). This province has a long history of hydrocarbon exploration, dating back to the 1930s, and has seen varying degrees of success (Fig. 2).

This paper describes the exploration and production history of North Yorkshire and its immediate surrounding area. The boundary of what is actually called North Yorkshire has changed over the years, and so for this paper is taken to be defined by the current boundary of the county of North Yorkshire. However, this excludes the Kirkleatham gas field to the north and the Caythorpe Field to the south. These fields are, nevertheless, included in the statistics cited below as they lie close by and are in the same geological province.

Approximately 100 wells, only 25 of which are classed as exploration wells, and a multitude of seismic data, including six 3D surveys, have been acquired by various operators since the 1930s in the area. There have been 13 discoveries leading to eight gas field developments, with a further three significant currently undeveloped discoveries (Fordon, Cloughton and Ebberston South). The Fordon gas accumulation, discovered in 1974, has a Sherwood Sandstone reservoir bound within a large, extensional fault block. The Cloughton gas discovery is reservoired in tight sandstones of the Millstone Grit and Bowland sequence within a late Carboniferous anticline. The Ebberston South discovery, initially tested gas by well Wykeham-1 in 1971 and successfully appraised by well EBS-1 in 2009, is a gas-charged Kirkham Abbey Formation (KAF) reservoir in a low-relief west-trending high.

Whilst this activity has obviously led to a better understanding of the subsurface, it has not led to a marked improvement in the reservoir performance.

From: Craig, J., Gerali, F., MacAulay, F. & Sorkhabi, R. (eds) 2018. *History of the European Oil and Gas Industry*. Geological Society, London, Special Publications, **465**, 119–136.
First published online May 8, 2018, https://doi.org/10.1144/SP465.12

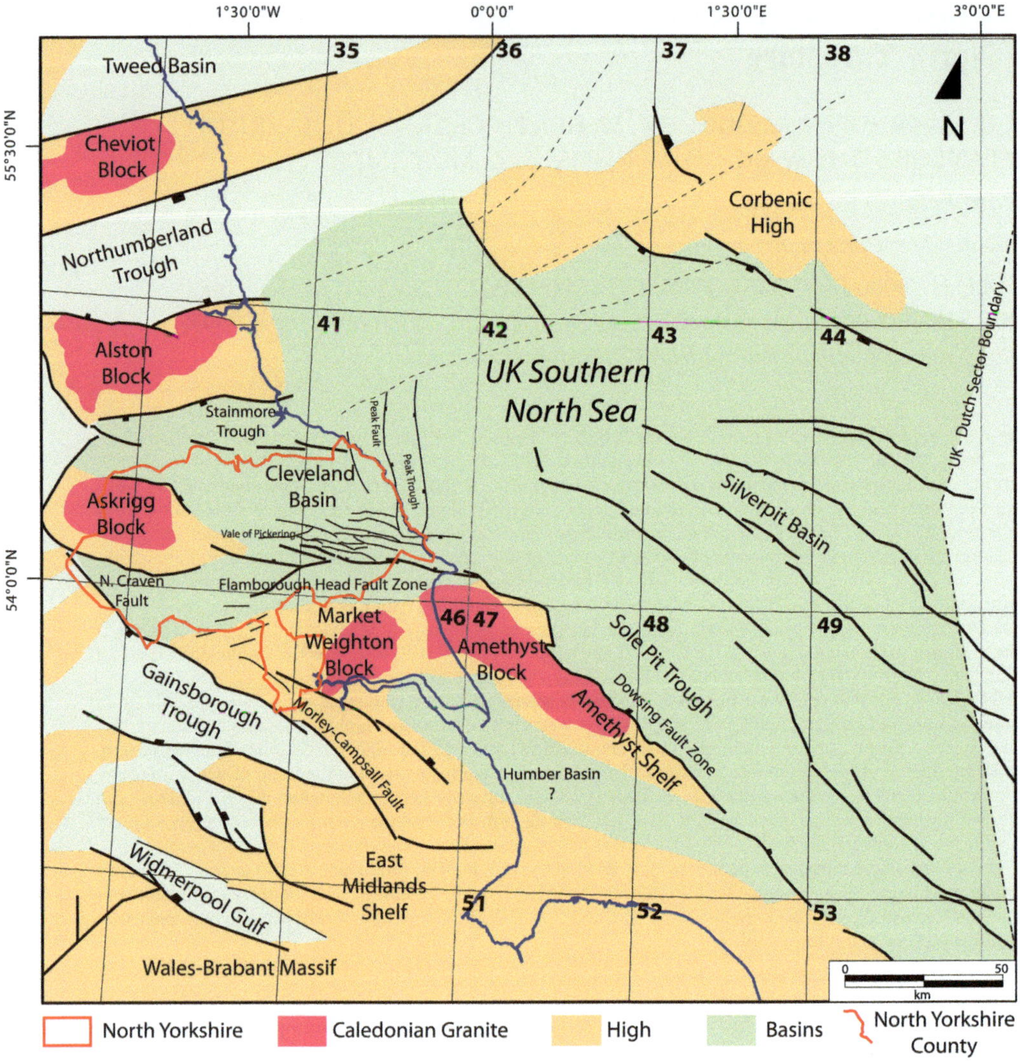

Fig. 1. Structural framework map of northern England and the Southern North Sea. The North Yorkshire county boundary is highlighted in red, illustrating its location with respect to the Cleveland Basin. Adapted from Fraser & Gawthorpe (1990) and Waters *et al.* (2009).

The most productive reservoir in the basin is the Zechstein-aged, shelf-margin carbonates and, in particular, the Z2 or Kirkham Abbey Formation (KAF: once referred to as the Middle Magnesium Limestone). The KAF is a complex reservoir where the primary porosity is overprinted by extensive diagenesis and intense fracturing. Historically, traditional production techniques have been unable to efficiently recover the large volumes of gas in-place (Table 1) due to early water encroachment into the production wells.

The basin is currently the focus of renewed exploration interest, as the boom in production from unconventional resources in the USA has generated considerable interest in the onshore UK: 19 licences in the Cleveland Basin were recently awarded in the 14th Onshore Licensing Round. Adding to the interest is the fact that Third Energy drilled a successful unconventional appraisal well at Kirby Misperton in 2013, with hydraulic stimulation operations planned for early 2017.

Achieving gas production through the existing field facilities would represent the UK's first unconventional resource production. This will be a significant step forward in demonstrating the unconventional resource potential of the region

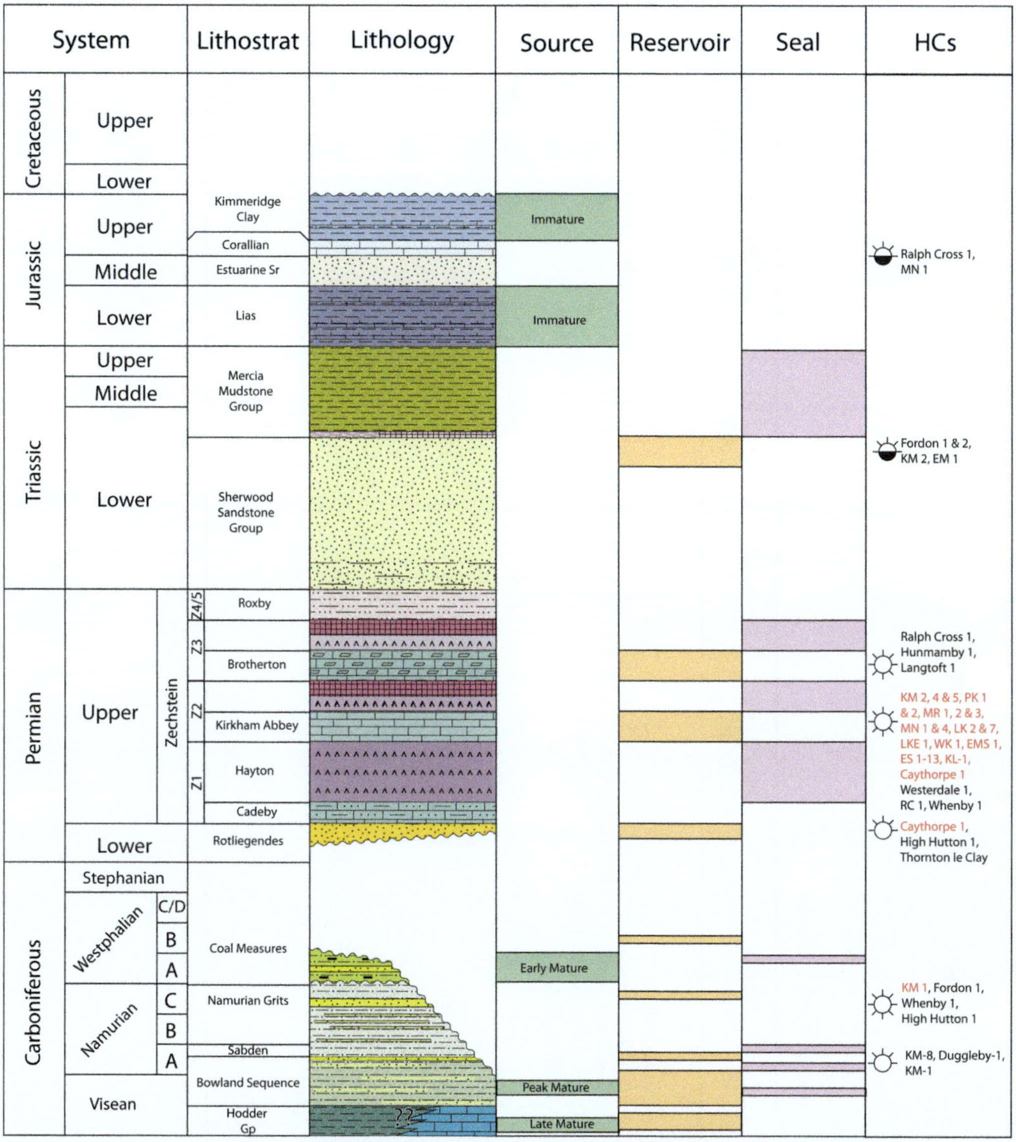

Fig. 2. Map illustrating the distribution of wells drilled within the Cleveland Basin from the late 1930s to present. Modified from Heath-Clarke *et al.* (2016).

and will undoubtedly lead to a major new phase of exploration in North Yorkshire in the coming years.

Structural and stratigraphic basin history

The Cleveland Basin comprises a thick (>3 km), mildly folded and locally heavily eroded Carboniferous succession, overlain by 1.5–2 km of Upper Permian–Upper Jurassic strata. Along the axis of the Vale of Pickering (VoP), these strata are blanketed by a thin (*c.* 10 m) layer of superficial lake and glacial sediments, but across most of the basin the strata crop out at the surface or lie immediately beneath a veneer of top soil.

Currently, the basin is bordered to the north, west and south by the Early Paleozoic Alston, Askrigg and Market Weighton granite blocks, respectively (Smith *et al.* 1985). Offshore to the east, the basin extends to link with the much larger SNS Basin. The faulted margins of the granite blocks, and the

Table 1. *List of Zechstein gas fields and discoveries in North Yorkshire*

Field	Undeveloped discovery	Status	Most likely GIIP	Produced	Current RF (%)	Target RF (%)	Expected remaining rec vol
North Yorkshire NP Fields							
Kirkleatham		Suspended	6.2	0.3	5	75	4
	Ralph Cross	Evaluation ongoing	17.5	0	0	75	13
Eskdale		Decomissioned	40	0.9	2		
Ebberston Moor		Redevelopment approved	88	11.3	13	75	55
	Eberstone South	FDP approved	38	0	0	75	29
Sub total			**189.7**	**12.5**			**101**
Vale of Pickering Fields							
Pickering		Producing	42	4.3	10	75	27
Kirby Misperton		Producing	49.5	6.7	14	75	30
Malton		Producing	19	9.6	51	75	5
Marishes		Producing	12.3	3.6	29	75	6
Sub total			**122.8**	**24.2**			**68**
East Riding Fields							
Caythorpe		Decomissioned	13	2.6	20		
Total			**326**	**39**			**169**

distribution of local depocentres in the intervening Cleveland Basin, are likely to reflect pre-existing basement lineaments.

In the English East Midlands, the Carboniferous basins initially developed as discrete asymmetrical half-graben between carbonate highs, such as the Gainsborough Trough and Widmerpool Gulf (Fraser & Gawthorpe 1990). To the north of this carbonate platform is the Cleveland Basin, a predominately broad open basin. This is supported by seismic mapping constrained by well data (Andrews 2013). By the Late Carboniferous Period, however, the extensional structures were buried beneath a single, rapidly subsiding sag basin (Pennine Basin) that extended over much of northern England, including North Yorkshire, with the centre of maximum thickness centred on the area of the Lancashire coalfield.

The offshore extension of the Cleveland Basin is incorporated into what is now the eastern part of the Mid North Sea High (MNSH), where well data (e.g. 41/1-1, 41/08-2, 41/14-1, 41/15-1, 41/20-1, 41/24-2 & 42/10-2 and 42/16-1) calibrated against seismic data collectively demonstrate over 3000 m of vertical Carboniferous section was present prior to the effects of erosion associated with the Variscan Orogeny. It is likely that a second late Carboniferous depocentre, analogous to that of the Pennine Basin, is developed offshore. This is possibly centred on the southern part of the Quadrant 41 and Quadrant 42 area.

Cessation of rifting during the mid-Visean Stage resulted in a period dominated by thermal sag subsidence (Leeder 1982). This occurred concurrently with transient phases of both localized, but minor, extension and compression: for example, the Craven faults and Ribblesdale folds, respectively, in the Craven Basin. This tectonic activity reactivated the aforementioned pre-existing basement lineaments (Waters *et al.* 2009). Initial sediment influx into the various early Carboniferous sub-basins originated from the north but had changed to the SW by the end of the Westphalian Stage (Hallsworth & Chisholm 2000). During the late Westphalian and Stephanian stages, deformation became increasingly influential across central and northern England and the SNS. This is evident with NW-trending faults reactivated offshore (e.g. the Cavendish–Murdoch Fault Zone) and large, north-striking anticlinal folds of the Baymans Fold Belt developed within the Cleveland Basin and quadrants 41 and 42 of the SNS (Mortimer 1992).

Given the thickness of Carboniferous strata developed immediately offshore of the Cleveland Basin, in what is typically referred to as the MNSH, it can be concluded from the well data calibrated against seismic that at least this part of the regional structural high is not analogous to the onshore Alston and Askrigg Paleozoic granite blocks. Rather, it represents a product of Variscan deformation and potential inversion to a Carboniferous basinal area.

At the end of the Carboniferous Period and into the early Permian Period, much of the SNS Basin became affected by late Variscan post-orogenic tectonism. Regional uplift associated with the northwards-advancing Variscan Front was followed by thermal doming associated with extensive Lower Rotliegendes volcanics centred on the Dutch Sector of the SNS. This caused substantial extensional faulting and widespread erosion (Ziegler 1992) which extended across parts of northern England. It formed a regionally extensive angular unconformity that is commonly referred to as the Base Permian Unconformity (BPU).

Although the Lower Permian Upper Rotliegendes Group is largely absent from the Cleveland Basin, a thin fluvial lag sequence is present south of the west-trending Flamborough Head fault lineament, and forms one of the reservoir units of the Caythorpe gas field. Furthermore, a localized deposit of aeolian and fluvial sandstones, the 'Yellow Sands', is exposed on the Northumberland coast. These deposits probably developed in topographical lows between the uplifted Variscan highs, and the absence of deposition across the bulk of the Cleveland Basin indicates that this area was inverted at this time.

There is little evidence from the well and seismic data to suggest that the Cleveland Basin suffered the same level of extensional tectonism as offshore. The synrift Upper Rotliegendes Formation, developed offshore, is generally absent onshore and many of the underlying Carboniferous faults were not reactivated. As a result, the overlying Late Permian Zechstein succession forms a west-trending embayment that onlaps onto progressively older Carboniferous strata around the basin margins and locally across the crest of the late Carboniferous anticlines.

Renewed basinal subsidence during the Late Permian Period led to the deposition under marine conditions in which the Zechstein succession was deposited. Where fully developed, the succession comprises five cycles of various carbonate facies, anhydrites and halites. Platform carbonates developed principally in the first three cycles of the Zechstein Group (i.e. the Z1, Z2 and Z3 cycles), and are named the Cadby Formation, the KAF and the Brotherton Formation. These platform carbonates developed along the basin margins and prograded basinwards through the development of shoal facies during each cycle. The dolomites of the KAF are the primary conventional reservoir within the Cleveland Basin, where they reach a thickness of up to 220 m, with the best reservoir developed in oolitic shoal facies. This facies extends in a NE–SW trend into Humberside and offshore towards the MNSH where it is also is locally gas charged.

Arid environments dominated the region during the Triassic Period. Initially, this gave rise to a relatively uniform (175 m) sequence of fluvial sandstones, known as the Sherwood Sandstone, which is the lateral equivalent to the Bunter Sandstone developed in the SNS Basin. This is overlain by the playa lake sediments of the Mercia Mudstone Group, referred to as the Haisborough Group offshore.

Erosion has largely removed the Upper Paleozoic and younger sections around the margins of the Cleveland Basin. It therefore remains unclear which, if any, of the margin structures influence the development of this basin. The majority of the intrabasinal faults appear to post-date deposition of the Upper Triassic–Upper Jurassic section. It is therefore likely that this phase of development of the Cleveland Basin formed in response to thermal subsidence taking place offshore which caused the strata overlying the thick Zechstein halites to experience halokinesis driven by the SW basin tilt (Sharp *et al.* 2016). However, regional variations in thickness of the Triassic succession across the Cleveland Basin does suggest that some of the basin-margin faults may have been active and that highs (e.g. the granite blocks to the north and west) were acting as the source for these continental sediments.

During the Jurrassic, the Cleveland Basin was again submerged by cyclic marine conditions. Approximately 1000 m of section is preserved along the axis of the VoP, with an additional post-Kimmeridge section preserved beneath the Base Cretaceous Unconformity as the section dips towards the SNS. The Jurassic section is disrupted by large, west-trending listric faults, parallel to the axis of the VoP, that generally sole out into evaporites of the underlying Zechstein succession. No obvious growth is apparent on these faults, suggesting that these faults post-date the deposition of Upper Jurassic Kimmeridge Clay Formation.

Whilst absent across the bulk of the Cleveland Basin, the presence of a thin Lower Cretaceous succession (Speeton Clay) and thick Upper Cretaceous Chalk succession overstepping the Jurassic section in the SE corner of the basin indicates that the basin was probably covered by marginal coastal plains during the late Early Cretaceous Epoch. It later became fully submerged by shallow-marine 'chalk' seas in the Late Cretaceous Epoch.

Substantial, episodic inversion occurred offshore in the Sole Pit Trough, initially during the Late Cretaceous and then the early Paleogene periods, resulting in basinal uplift removing up to 2500 m of Mesozoic strata, as estimated by apatite fission track data for the Sole Pit (Glennie & Boegner 1981; Green 1986). This zone of inversion extended northwards into quadrants 41 and 42, and to a lesser extent onshore, and resulted in the Cleveland Basin tilting towards the east (Powell 2010), with the late Cretaceous and Paleogene section largely removed from the western VoP during this tectonic episode. Estimates of uplift (and erosion) for the Cleveland Basin vary between 600 and 1500 m (Cope 1986).

The resulting uplift generated the west-trending domal structure onshore, the Cleveland Anticline, which presently underpins the North Yorkshire National Park. The trend of this anticline is in marked contrast to the inversion trend offshore, which is still profoundly aligned north–south: the trend of the Cleveland Anticline almost certainly reflects the trend of the underlying Lower Paleozoic basement and, in turn, Carboniferous structural lineaments. The north–south trend of the Mesozoic inversion in quadrants 41 and 42 offshore is therefore somewhat surprising but, along with the underlying Carboniferous Baymans Fold Belt, could suggest that the basement grain switches abruptly offshore.

Finally, in the Late Paleocene, the Cleveland Dyke was emplaced (*c.* 58.4 ± 1.1 Ma) (Evans *et al.* 1973). This dyke represents the southernmost member of a swarm of east–west- to WNW–ESE-trending tholeiite (quartz dolerite) dykes focused on the Mull Tertiary Central Complex (MacDonald *et al.* 1988). The dyke bisects the Cleveland Anticline and, in en echelon fashion, crosses the UK as far as the Solway Firth.

The area has largely been devoid of active subsidence since the Paleogene Period, with the only sediments laid down being a thin sequence of superficial, glacial till deposits overlain by ancestral VoP lake deposits (Powell *et al.* 2016).

Hydrocarbon system

Exploration activity in the Cleveland Basin has established that a stacked petroleum system exists (Fig. 3) with plays analogous to those seen in the SNS and East Irish Sea basins.

Wells at Ralph Cross and Malton (Fig. 2) encountered oil shows in the Middle Jurassic Ravenscar Group, whilst well Fordon-2 penetrated a 20 ft gas leg in a Triassic Sherwood Sandstone. The Zechstein carbonates, in particular the Z2 cycle locally referred to as the KAF (or Hauptdolomite offshore), is a reservoir in the Eskdale, Ebberston Moor, Kirkleatham, Malton, Kirby Misperton, Pickering, Marishes and Caythorpe gas fields. The Caythorpe gas field has also produced from a thinly developed early Permian-aged Leman Sandstone, although notably this reservoir is absent across most of the Cleveland Basin. If this reservoir had been present, it is likely that this area would have become a more significant gas province.

The coal-prone Westphalian succession, present to the south and across much of the SNS Basin, is largely eroded from the Cleveland Basin. As a result,

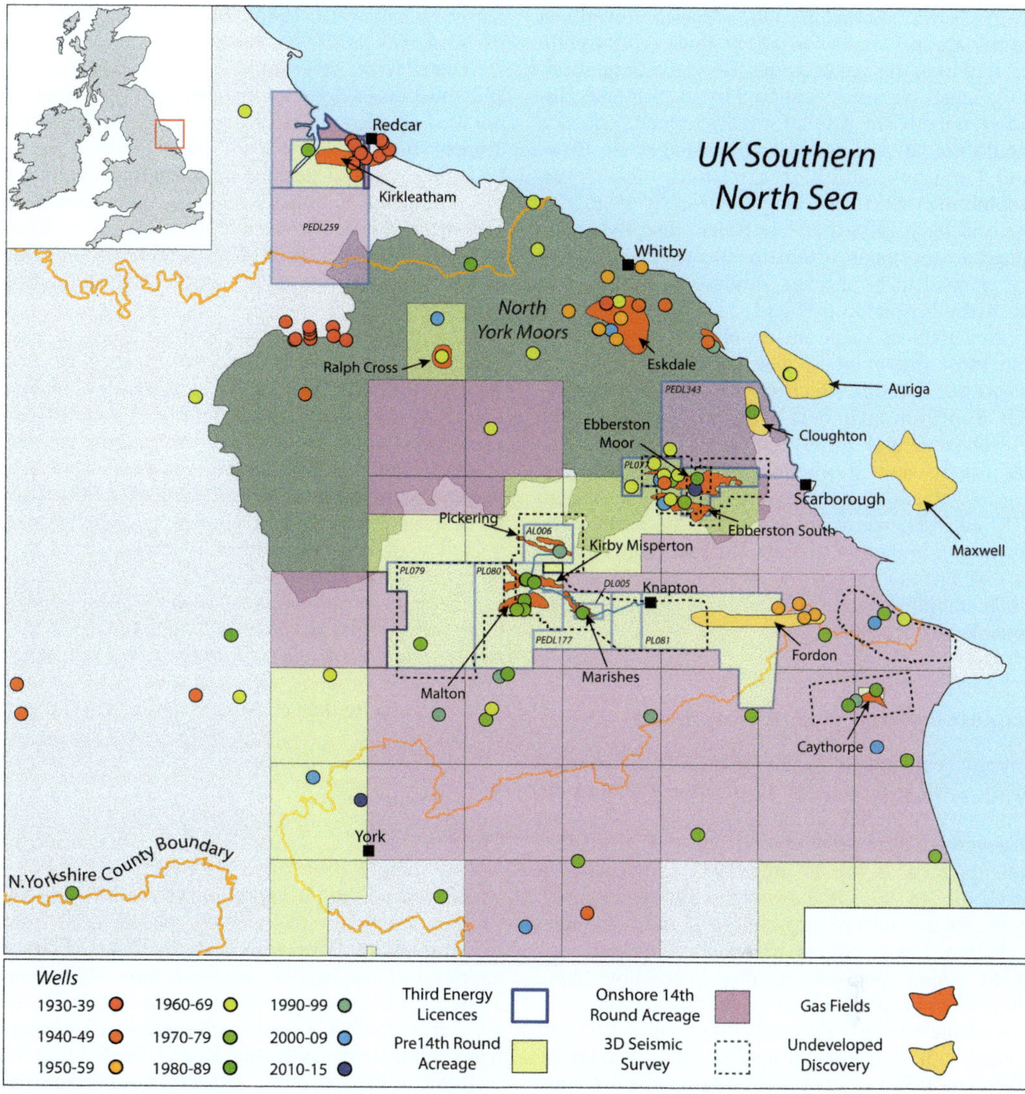

Fig. 3. Hydrocarbon systems chart for the Cleveland Basin. The discovery (black) and production (red) wells, aligned with their producing reservoir interval, are shown on the right.

the Upper Namurian section is truncated by the BPU. However, strong gas shows and modest gas-bearing columns or 'gas legs' are typically encountered where this section is drilled within BPU closure. At Kirby Misperton, 6.8 Bcf (billion cubic feet) of dry gas has been produced from this reservoir after stimulation. This section has also tested gas at the Cloughton-1, Malton-1 and Fordon-1 wells but, in general, represents a poorly appraised reservoir target within the basin.

The thick Visean–early Namurian-aged sequence in the Cleveland Basin comprises relatively organic-rich (2–3% Type III kerogen) mudstones commonly interbedded with sandstones deposited by turbidite and debrite flows (Hughes *et al.* 2018). This kerogen is the most likely source for the gas discoveries in the basin. Further north, in Northumberland, the succession (Yoredale) is dominated by an alternating sequence of deltaic sandstones, shales and occasional coals which are interbedded with thin marine sequences of shale and limestone, typified by the sequence penetrated by the Seal Sands-1 well. The presence of gas at Kirkleatham demonstrates that a viable source rock is present in the Yoredale succession, and in doing so extends hydrocarbon prospectivity into the northern part of the basin.

In North Yorkshire, the Visean–lowermost Namurian succession (>1000 m thick), informally referred to in this paper as the 'Bowland Sequence', is the target for unconventional hydrocarbon exploration in the basin. The informally termed sequence should not be confused with the shale-prone Bowland Formation, which is a relatively thin stratigraphic interval of late Brigantian–early Pendleian age and only 150 m, or less, thick. The tight gas-charged sandstones within the Bowland Sequence are planned to be the next reservoir developed in the Kirby Misperton gas field.

A variety of seals are developed in the basin. The most important to date is the Zechstein Z2 evaporite (Fordon) which overlies the KAF reservoir. However, additional evaporite intervals in the Z1 (Hayton) and Z3 (Brotherton) act as seals to the Cadby and Brotherton carbonate reservoirs respectively. The Z1 also provides a top seal to BPU closures, which are present in the Kirby Misperton and Malton gas fields. The Mercia Mudstone acts as the top seal to the Fordon gas discovery, whilst multiple intra-Carboniferous seals are also demonstrably present.

Exploration and field developments

Initially, exploration in the basin was constrained by the drilling technology of the day. Wells in the 1930s were drilled using a coal-powered steam engine to run the draw works. The well Eskdale-1 was spudded on the 17 July 1937 and reached a depth of 2483 ft (756.8 m) in the Sherwood Sandstone on the surface expression of the Cleveland Anticline. The loss of two tricone bits and numerous failed fishing operations resulted in the hole being plugged and abandoned on 22 February 1938.

Despite this early setback, D'Arcy (now BP) proceeded with the drilling of the Eskdale-2 well on the 16 June 1938, just 39 m east of the Eskdale-1 well. The drilling of the Zechstein salt deposits proved to be a challenge but persistence was rewarded when Eskdale-2 found gas in the KAF. This represents the first gas discovery in Zechstein carbonates in the UK.

After Eskdale, exploratory drilling became more focused on the flanks of the Cleveland Anticline during the late 1960s and early 1970s as drilling technology advanced and the application of 2D seismic data enabled the identification of highs in the subsurface. Exploration success continued with the discovery of the Lockton Field (in 1966), now called Ebberson Moor, and the Malton gas field (in 1970) by Home Oil. Intermittent investment by small exploration and production (E&P) companies and private equity, combined with continued improvements in seismic imaging, led to the subsequent discovery of the Kirby Misperton (in 1986), Marishes (in 1988) and Pickering (in 1992) gas fields.

There were few outright exploration failures and most wells found some level of gas, even if not flowing at commercial gas rates. After the development of the North York Moors and Vale of Pickering fields, it became apparent that they were not producing as anticipated. In order to improve gas production, 3D seismic data were acquired in 2007, and again in 2012 and 2014, but was primarily acquired for field appraisal/development purposes.

The appetite for exploration drilling in this gas-prone basin has diminished in the last few decades. In part, due to the historical development failure, but also due to the highly successful gas exploration offshore. However, overall, the success of exploratory drilling in the area has been high, with an exploration success rate of over 70%. Not all of these discoveries have been developed due to a variety of technical and commercial issues that have ultimately eroded value.

Three major development projects are discussed in the following subsections. Although these developments were undertaken at different times in history, under quite different economic conditions, they have all failed to deliver thc volumes of gas initially envisaged.

Eskdale gas field

The first gas discovery in this province was made by well Eskdale-2, drilled in 1938 by D'Arcy (Fig. 3), which flowed gas at approximately 3 MMscf/day (million standard cubic feet of gas per day) from the Brotherton Formation and the KAF. Further appraisal of the Eskdale structure took place from 1949 to 1954, when wells Eskdale-3–Eskdale-10 were drilled, of which only well Eskdale-10 encountered producible gas in the KAF. The Eskdale-11 well was drilled in 1958, and tested water and gas from the KAF.

Two wells, Eskdale-10 and Eskdale-2, were subsequently re-entered and gas production finally began in mid-1960. Eskdale-10 produced just 0.035 Bcf during production and was abandoned in 1962, whereas Eskdale-2 produced slightly more, around 0.496 Bcf, before it too was killed by water entering the borehole in 1964. The well Eskdale-12 was drilled in 1963 and completed in what is now thought to be the Brotherton Formation in late 1964. This well produced 0.328 Bcf until it could no-longer flow 'unassisted' and was abandoned in 1967.

Although not explicitly stated in the available field or well reports, early water production is likely to have been responsible for the premature abandonment of this field after the production of just 0.859 Bcf. Records of the initial development are difficult to find and much of it is hearsay, but it is believed that the sales gas was piped to Whitby on the coast for use as town gas.

In late 2000, Star Energy drilled well Eskdale-13, 1.2 km east of Eskdale-10, to reappraise the Botherton Formation and the KAF. The KAF was encountered downdip from well Eskdale-10. Despite gas shows observed whilst drilling, petrophysical analysis indicated that the moveable phase was water, suggesting that the mobile gas had been swept during the initial production phase.

Lockton/Ebberston Moor gas field and the Ebberston South discovery

The next significant exploration success was not until 1966, when Home Oil discovered gas in the KAF with well Lockton-2A. The subsequent appraisal programme was largely ineffective because the wells were located using sparse, poor-quality 2D seismic data and sometimes missed the structure entirely (e.g. wells Lockton-5 and Lockton-6) (Fig. 4a). However, the Lockton-7 well, drilled in 1968, successfully appraised the extension of the Lockton structure to the east, testing gas at a rate of 10.6 MMscf/day. Two years later, well Wykeham-1 drilled a structure 3 km to the south and found gas, but this was never brought on production. This well is now recognized to have discovered a separate gas accumulation known as the Ebberston South field. Despite the overall poor level of appraisal success, Lockton was eventually developed by Home Oil, based on the flawed assumptions that the accumulation contained a gas initial in-place (GIIP) of over 300 Bcf and that the recovery factor (RF) would be up to 90%.

The initial contract rate with British Gas was set at 84 MMscf/day, and the capacity at the large, purpose-built Pickering gas-processing plant at Thornton-le-Dale was 100 MMscf/day. The Lockton Field was brought into production in May 1971 with two wells (Lockton-2A and Lockton-7) at an initial, combined peak rate of ±40 MMscf/day (Fig. 5), although clearly there must have been an expectation that additional production wells would be drilled and that the combined production would meet the contracted rate.

Unfortunately, well Lockton-7 had to be abandoned within 1 year due to water loading, after producing just 1.1 Bcf. The Lockton-2A well performed better, flowing gas at over 20 MMscf/day in the first year and ultimately producing 10.2 Bcf (Fig. 5). This superior performance can now, with the aid of 3D data, be attributed to the crestal position of the well on the Lockton structure along with active water management performed by the operator, who successively shut-off deeper perforation zones as water broke through. In contrast, downdip, well Locton-7 intersected a major fault bonding the northern margin of the field. The intense fracture system associated with this fault facilitated the early water influx into this well.

By the start of 1974, the gas rate of well Lockton-2A had declined to less than 1 MMscf/day with a high water cut. The well was plugged and abandoned as the production facilities had not been designed to deal with water. This took place at a time when gas prices were depressed due to the large volumes of gas being found and developed offshore. Drilling new wells and upgrading the processing plant facilities to deal with the produced water was not economic, and as a result the field was prematurely abandoned and the facilities decommissioned. Only 11.3 Bcf was recovered from this field. The story of this development is recounted by Moreton (1995), with personal accounts from those involved.

With the aid of more recent appraisal wells (Lockton-East in 1980, Ebberston Moor-1 in 2007 and Ebberston Moor-2 in 2013) tied to 3D seismic data, the GIIP of this field is now estimated to be approximately 88 Bcf compared to the original 300 Bcf estimate of Home Oil. Nevertheless, this lower value still implies the current recovery factor (RF) is less than 15%, a very low value for a gas field.

As previously mentioned, well Wykeham-1 was drilled by Home Oil in 1971 on a separate accumulation to the south. Two drill stem tests (DSTs) were carried out over the KAF, which initially tested an estimated rate of 1–5 MMscf/day with indications of some formation water. The upper KAF was acidized with 600 gallons of 15% hydrochloric acid and the flow rate improved to 13 MMscf/day. This initial rate, however, gradually decreased to 10 MMscf/day after 3 h with 3–4 bwpd (barrels of water per day).

Since abandonment, the Lockton Field (now renamed Ebberston Moor) and the Ebberston South discovery have been reappraised several times with varied success. In 2007, the Ebberston Moor-1 well was drilled close to the Lockton-2A well (Fig. 4b) and flowed gas at a rate of 4.98 MMscf/day along with 200 bwpd of formation water. In 2009, well Ebberston South-1 was drilled and tested the KAF reservoir in an updip location from well Wykeham-1 (Fig. 4b) and flowed at 9.4 MMscf/day, a rate constrained by a ½ inch choke used because of a planning requirement for the flow to be less than 10 MMscfg/day. Importantly, the well flowed with no associated formation water. In 2013, Ebberston Moor-2 was drilled to appraise an eastern extension of the Ebberston Moor Field (Fig. 4b). This well, when tied to 3D data, proved that the structure extended further to the east than the mapping of

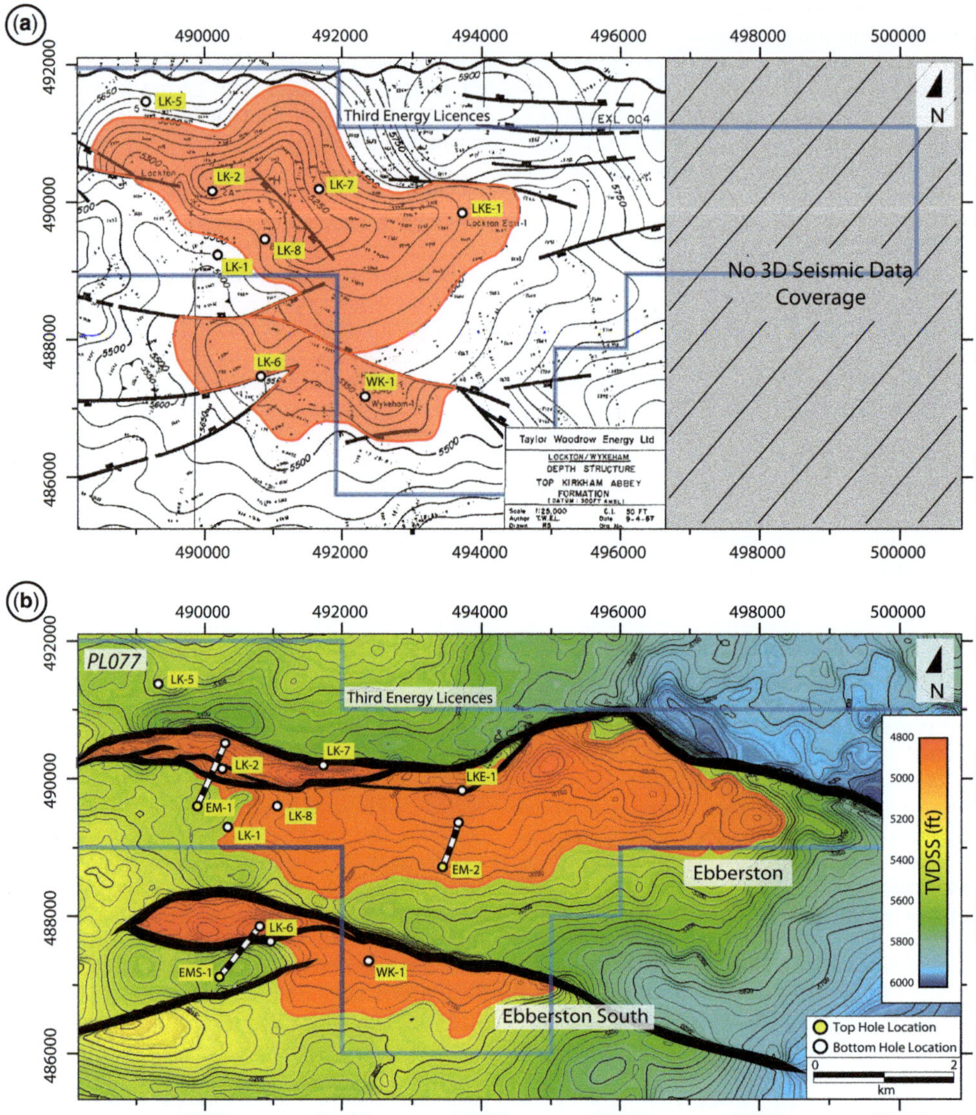

Fig. 4. (**a**) The 1987 depth-structure map of the Ebberston Moor (then Lockton) and Ebberston South fields mapped by Taylor Woodrow Energy Ltd on 2D seismic data. (**b**) Depth-structure map of the Ebberston Moor and Ebberston South fields mapped by Third Energy using 3D seismic data.

2D seismic data by Home Oil had indicated. However, whilst finding movable gas in the reservoir matrix, it also discovered water in the fractures. Any production from this well is likely to be dominated by water without remedial action.

Third Energy has recently received planning approval for a phased development in the North Yorks Moors National Park. The first volume of gas to be developed will be the Ebberson South accumulation, followed by the redevelopment of the Ebberston Moor Field. With the remaining in-place volumes of gas still exceeding 115 Bcf for the two fields, they remain attractive assets, especially considering the nearby infrastructure and low capital expenditure (CapEx) costs relative to offshore. Third Energy aims to achieve a significant improvement to the historical recovery factors through active water management, and a trial to demonstrate the viability is currently underway on the Pickering Field, described in more detail later in this paper.

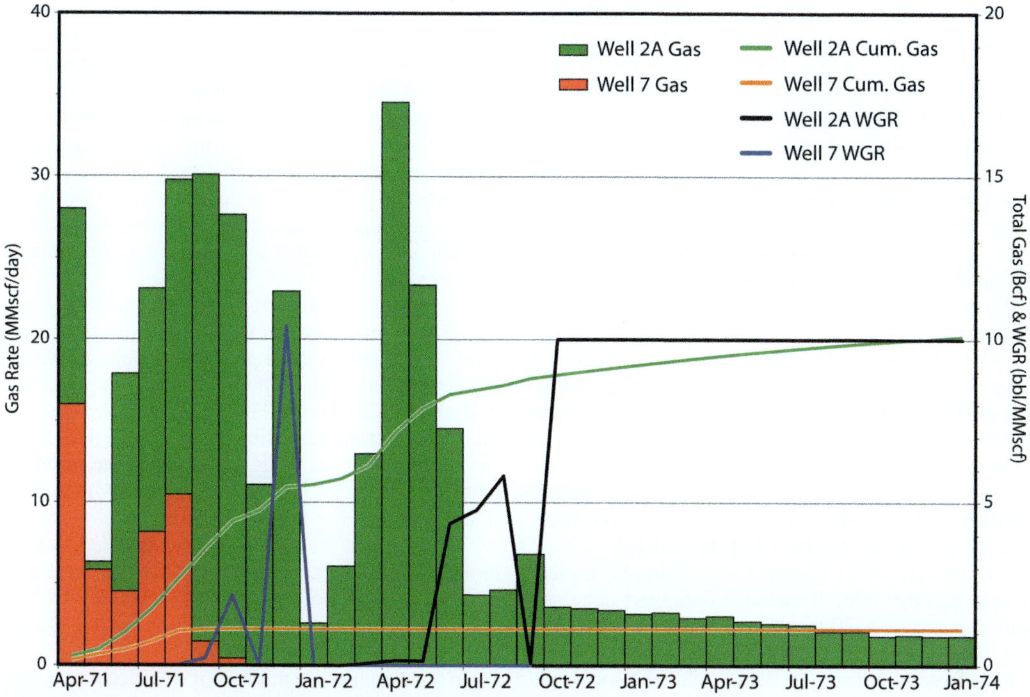

Fig. 5. The production history of the initial development of the Lockton gas field, now known as the Ebberston Moor gas field. TVDSS, true vertical depth subsea.

Vale of Pickering (VoP) gas fields

Four gas fields have been discovered in the VoP. They are Malton (in 1970), Kirby Misperton (in 1986), Marishes (in 1988) and Pickering (in 1992). The Malton, Marishes and Kirby Misperton gas fields were developed by Kelt Energy in 1995, followed by Pickering in 2001 by Tullow Oil. By 2003, the E&P business in the VoP area was owned by Viking UK Gas Ltd.

All four fields comprise the gas-charged KAF trapped in subtle low-relief structures. As with Lockton, the VoP fields were again developed using a sparse grid of poor-resolution 2D seismic data with little regard to the potential of early water influx. Modern 3D seismic data now demonstrate that many of the early production wells are located downdip from the trap crest, close to a gas–water contact (GWC) or near significant faults. The collective GIIPs in the four fields were originally around 120 Bcf. Unfortunately, the production story, with the exception of Malton, has repeated the previous failures of Eskdale and Lockton. Consequently, recovery factors are low (<25%) due mainly to the early water production.

Three of the VoP fields have limited on-site processing capabilities for separating any associated formation water. Critically, the installed facilities rely on reservoir pressures to lift any water reaching the borehole. The water is then transported to facilities at the Kirby Misperton Alpha drill site, where it is reinjected into the KAF using a failed appraisal well (KM-3) for disposal. The Pickering Field, however, currently has no liquid separation equipment and therefore production from this site is sent multiphase to the Kirby Misperton Alpha site for liquid separation and reinjection.

The well KM-3 was originally located on sparse 2D data. It missed the field by drilling downdip with only a thin section of non-effective reservoir above the GWC. New 3D seismic data now demonstrate that this well lies within the overall Kirby Misperton KAF structure and it is likely that the injected water is negatively influencing the gas production. Hence, Third Energy plans to recomplete well KM-3 so that the produced formation water from the fields is injected into the Sherwood Sandstone, instead of the KAF.

The dried gas is co-mingled through a 6 inch pipeline gathering system and transported to a power station at Knapton. The turbine at Knapton has a capacity of around 9.3 MMscf/day, from which it can generate a maximum of 42 MW of electricity for transmission to the National Grid.

The plant, originally owned by Scottish Power, operated at a pressure of around 700 psia, corresponding to the design inlet pressure for its gas turbine for the first 11 years of production. In 2003, the fields were purchased by UK Energy Systems (UKES), a subsidiary of US Energy Systems. UKES consolidated its position in the project by purchasing the power station in 2006 and promptly installed compression. This compression allowed the pipeline operating pressure to be dropped from 700 to 300 psia, with an associated reduction in backpressure on each of the individual production wells.

The wells went into a period of cyclic production after 2009. Production wells were shut-in to allow the well head pressure (WHP) to build up, and then intermittently flowed to flush the gathering system before the WHP drops and water loads into the system again. UKES were by now under severe financial strain as the revenue from the production did not pay for their substantial investment.

In an attempt to improve the technical understanding of the field, 3D seismic data were acquired and processed in 2007 across the VoP fields and the western half of the Ebberston accumulations. Unfortunately, before the data could be processed, the UK subsidiary of US Energy Systems became insolvent and negligible quality control (QC) was applied to the seismic processing. The result was a poor-quality 3D dataset that suffered from severe static problems, and poor imaging of the shallow section and, critically, the KAF objective.

The US Energy Systems business in Yorkshire was put up for sale and well KM-6 was drilled in 2011 in an attempt to improve the sale price by increasing production. Unfortunately, the pre-drill evaluation of the subsurface was incorrect and the resulting well encountered the KAF reservoir downdip of the depleted well KM-4. Poor seismic imaging of the original 3D had led to the incorrect picking of the KAF events across the top of a lens of Fordon evaporites.

After a protracted sales process, Third Energy acquired both the upstream and downstream assets in 2012. Third Energy's expectation was that a detailed evaluation of both the subsurface geology and optimization of the facilities set up to deal with the water influx would enable the recovery factor to be increased.

The journey since 2012 has had its challenges, with the production dynamics of the KAF reservoir harder to understand than expected. Nevertheless, reprocessing of the 3D seismic has taken place to yield a significant improvement in data quality. Also, additional 3D surveys have been acquired across Ebberston Moor and the western VoP fields to fully define the extent of the traps and to match the GWCs to the structural spill-points.

It is now apparent that many of the exploration and production wells are poorly located in downdip positions or on fault zones that partially cut out reservoir and also further exacerbate the early influx of water into the wells. Furthermore, a close match has now been achieved between the geology and geophysics (G&G) calculated GIIPs and the dynamic production-derived connected volumes, providing confidence that the calculated recovery factors are, indeed, low and not a reflection of overestimated volumes.

Expectations that extra gas could be quickly produced from the better-imaged updip areas located away from the historical production were, nevertheless, quickly dashed with the drilling of well KM-7, in the Kirby Misperton Field, and well EM-2 in the undrilled eastern extension of the Ebberston Moor Field. Both of these wells found gas in the reservoir matrix but, crucially, also encountered water-filled fractures above the initial GWC.

These results have necessitated detailed Eclipse models for the reservoir of each field to be developed that not only honours the static data but also the many years of dynamic production data gathered from the VoP and Ebberston fields.

In the meantime, a number of significant planning requirements have been achieved to allow facility optimization. Notably, permission has been secured to reinject the saline KAF formation water, coproduced during gas production at Pickering, into the overlying but, crucially, equally saline Sherwood Sandstone reservoir instead of transporting and injection at KM-3. Injection of formation water into a different aged formation is a planning first for the UK. Furthermore, planning approval has also been achieved for the development for the Ebberston South accumulation and the redevelopment of the Ebberston Moor Field, which, given their location in the North Yorkshire National Park, makes this another notable achievement. Reinjection of the formation water at the Ebberston Field will potentially improve the project economics, as it removes the need for costly surface disposal either via road haulage or pipeline.

Recent activities are now focused on implementing a phased work programme designed to deal with the reservoir formation water and achieve improved long-term reservoir performance. Field trials have already been undertaken at Malton, and now are in place at Pickering, demonstrating that enhanced gas production can be maintained as a minimum if the water is removed from the production well. Importantly, however, it is also expected that by increasing water production, the limited aquifer, including water-filled fractures, will ultimately become depleted.

The prize associated with this enhanced gas recovery project in the VoP alone is substantial.

With only 25 Bcf produced out of the original approximately 120 Bcf GIIP in the VoP, there is scope to recover an extra 65+ Bcf over the next 10 years if recovery factors can be improved to 75%. Similarly, the Ebberston Moor and Ebberston South developments are expected to yield a further approximately 80 Bcf. The cumulative CapEx associated with these projects is modest, and so the economics and materiality compete very favourably with much larger developments offshore.

Future exploration

Exploration and production activities associated with the newly discovered unconventional resources in the area will be transformational. As part of the 14th Onshore Licensing Round, in North Yorkshire, companies have committed to shooting 13 2D and seven 3D seismic surveys, along with the drilling of 10 vertical wells and a further six lateral wells with associated hydraulic fracturing programmes (Gov.uk 2015).

Alongside other operators, Third Energy is also heavily involved in the exploration and appraisal of the unconventional resource within the Cleveland Basin. In 2013, the company drilled production well KM-8 to a depth of 3112 m measured depth (MD). In addition to the conventional gas-charged Namurian section immediately beneath the BPU (primary target), the well encountered over 750 m of gas-charged unconventional Carboniferous ('Bowland') section. Plans are now being put in place to hydraulically fracture this unconventional section to test the potential to flow gas.

The presence of an extensive unconventional resource below the historical reservoirs (Hughes *et al.* 2018) provides a new opportunity for the continued use of the facilities associated with the VoP fields and the Knapton power station.

Field study of the Pickering Field

Early water production

As discussed above, the dynamic behaviour of the KAF reservoir in the producing fields has historically been poorly managed. Gas production has been characterized by high initial rates which decline rapidly with the onset of water production. This characteristic production behaviour is attributed to the dual-permeability system within the KAF, where the matrix has moderate porosity and low permeability, whilst the natural fracture system has low porosity and very high permeability. As a result, wells initially produce at high rates from the natural fractures but, as pressure drops in the reservoir, the aquifer infiltrates the fractures, effectively isolating the bulk of the GIIP in the matrix from the producing wellbore.

To illustrate this point, the production history of the Pickering gas field in the VoP is shown in Figure 6. The other KAF fields have experienced a similar production history, although individual

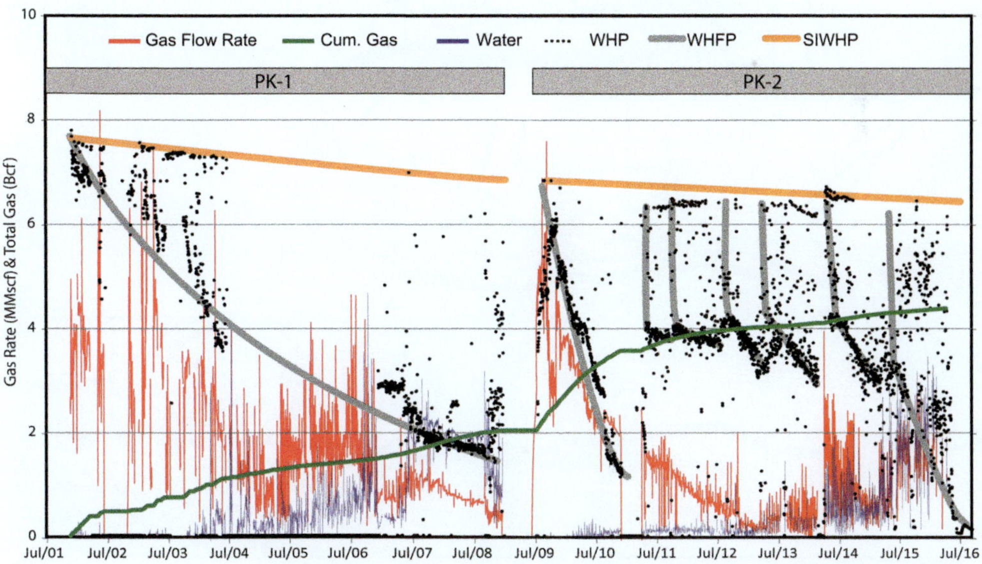

Fig. 6. Pickering gas field production history for the Pickering-1 (PK-1) and Pickering-2 (PK-2) wells: red, gas flow rate; green, cumulative gas produced; blue, produced water rate; black, well head pressure (WHP); grey, trend of the flowing well head pressure (WHFP); orange, trend line showing the shut-in well head pressure (SIWHP).

wells show some variation. This variation is caused by elevation above the field's GWC and local perturbations in reservoir quality.

The Pickering Field has, to date, been produced by only two wells. The Pickering-1 well was drilled by Kelt UK in 1992 using a sparse grid of poor-quality 2D seismic data and, unfortunately, is located downdip of the crestal Pickering horst block (Fig. 7). Pickering-1 was brought on production in November 2001 and cleaned up over a few months to achieve peak rates of 5.5 MMscf/day. The well's initial high gas production rates, however, decreased markedly once water production began in early 2003. Water rates gradually increased to around 200 bwpd before surging in late 2006 to 500 bwpd (Fig. 6). As a result, the water level rose up the well bore, lowering the WHP and reducing the gas rates to a mere 0.5–1.0 MMscf/day. Although the well had increased difficulty in lifting the water, the shut-in top hole pressure (SITHP) remained high at around 1500 psia, indicating substantial residual energy in the reservoir.

By 2008, the operator of the field was Viking UK Gas Ltd. The company acquired and processed the first 3D seismic survey over the VoP fields and the decision was made to drill well Pickering-2 to target a structurally updip location, identified on the new seismic, to access 'attic' gas. The well encountered gas in the KAF, updip of well Pickering-1, and was brought on production in May 2009 with initial gas rates of approximately 5.5 MMscf/day. However, this strategy ultimately proved to be flawed as water quickly reached the new well and the gas rates once again diminished, as illustrated by the production profile of well Pickering-2 in Figure 6.

Water production rates of 50–100 bwpd in early 2010 coincided with a sharp drop in gas production, which has since declined steadily to rates less than 0.3 MMscf/day before being shut-in. Although water production has not exceeded more than

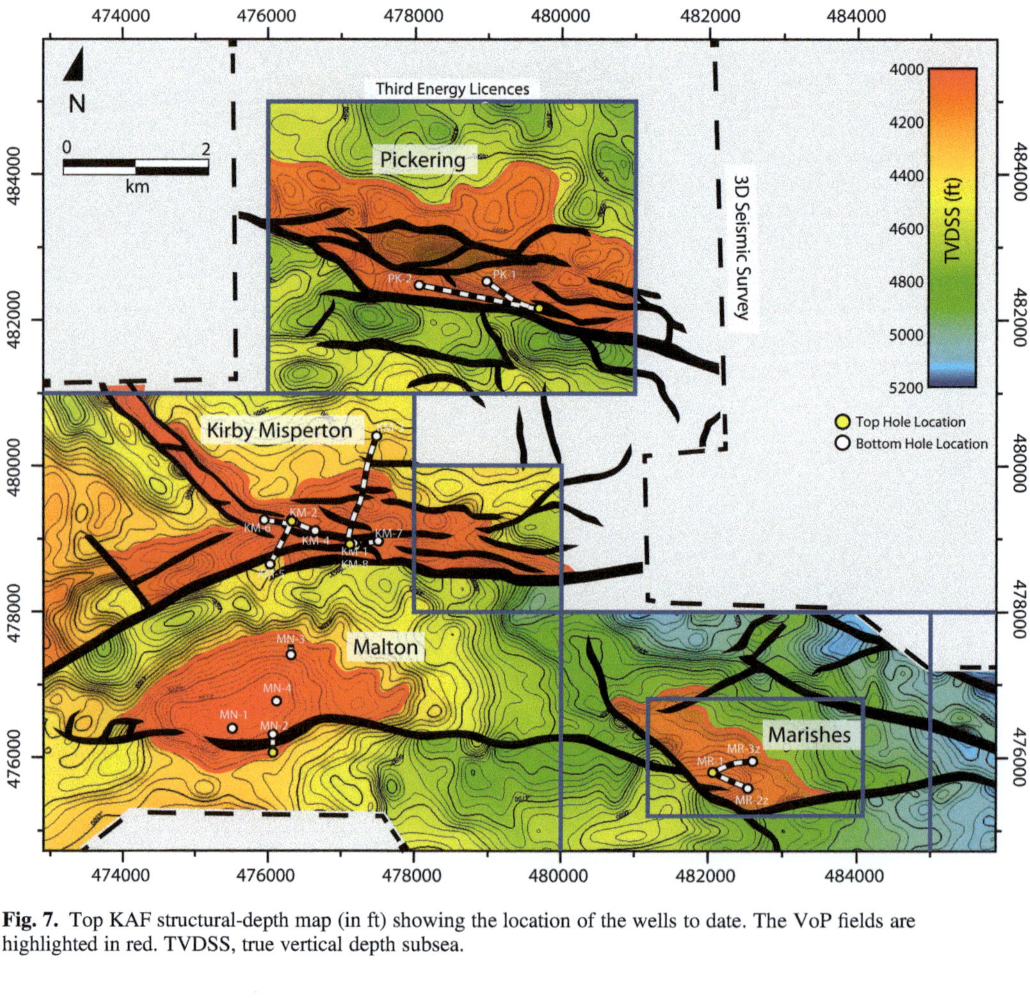

Fig. 7. Top KAF structural-depth map (in ft) showing the location of the wells to date. The VoP fields are highlighted in red. TVDSS, true vertical depth subsea.

500 bwpd, limited water-handling facilities forced the well to be produced intermittently. The mapped GIIP in the Pickering structure is calculated to be between 28 and 62 Bcf, with a P50 GIIP of 42 Bcf. The most likely volume of gas connected to the two wells is in the region of 24 Bcf, which is derived by pressure-decline analysis. The Pickering gas field has only produced around 4 Bcf of gas to date, averaging 2 Bcf per well, which is only 15% of the connected volume. This is similar to the surrounding KAF fields which, with the exception of Malton, have up until now typically recovered 15–20% of the GIIP.

Aquifer support

Modelling the performance of a KAF field, in a scenario where produced water is managed, yields a range of potential recovery factors of between 20 and 80%. The primary uncertainty being the nature (principally size) of the aquifer and the extent to which the influx of water can be managed. Historically, reservoir modelling has assumed large aquifers of up to 3 Bbbl (billion barrels), extensive highly porous fractures (up to 2% of GRV), and a highly permeable fracture system to match the field's production and pressure performance. However, these assumptions are not consistent with the geology. First, the reservoir is bound above and below by Upper and Lower Zechstein evaporites that are sealing with respect to water influx and, therefore, the aquifer is confined to the KAF reservoir. Furthermore, analysis of the log and core data over the KAF indicates that the quality of the rock matrix deteriorates significantly below the GWC, and the fracture porosity is reduced to between 0.01 and 0.2% as a proportion of the gross rock volume (GRV). Therefore, the volume of water trapped in the reservoir below the GWC can only be modest.

The historical performance of wells can be matched with an alternate reservoir model that uses a much smaller aquifer and represents a poor rock storage capacity coupled with the low fracture porosity (0.01–0.2%). This model is analogous to other documented fractured carbonate systems (Weber & Bakker 1981) and is supported by a number of elements confirmed by the historical production. First, the dynamic pressures show that the fields do n't communicate with one another, suggesting that that the large faults are pressure isolating. Secondly, the Malton Field, which had an initial pressure of 1981 psia in 1994, produced around 8.9 Bcf before being shut-in in 2001 with a bottom hole pressure (BHP) of around 1000 psia. An echometer survey acquired at the field in September 2015 showed that the reservoir pressure had equilibrated to 1405 psia after 14 years of being shut-in. This phenomenon is also observed on the Ebberston Moor Field, where Ebberston Moor-1 accessed gas at a lower pressure than the original pressure in the initial Lockton structure, which has not been produced for over 42 years. Both these examples provide evidence that the aquifer had not repressurized the trap.

Collectively, this information proves that the aquifer is finite and has a low transmissibility over large distances. Third Energy therefore believes that it is possible to selectively deplete the aquifer by water production and gradually recover the gas production rates from the wells. Although novel in Europe, this approach has been applied successfully on similar reservoirs in the South-Sichuan gas sector in China since 1969 (Tang 2015).

Historical production mechanism

The KAF exhibits a dual-permeability system that is common in carbonate reservoirs. Prior to field production, the bulk of the gas is within the high-porosity, low-permeability matrix above the GWC, whilst only a minor amount of gas (*c.* 1% GIIP) is stored in the high-permeability, low-porosity fracture system. The aquifer resides in the fracture system and poor-quality matrix below the GWC. With production, reservoir pressure drops and perturbs this state of equilibrium, causing the aquifer to respond.

The aquifer expands and preferentially advances along the fracture system, locally coning around the producing well. Water breaks through into the production well relatively quickly, which begins to load up as water accumulates in the production string. This causes the gas rates to diminish significantly and the production then enters into a long phase characterized by increasing the ratio of water to gas (WGR). With continued production, water expands into almost the whole fracture system above the GWC. Meanwhile, the gas in the matrix remains poorly depleted and unable to effectively drain through the water-saturated fracture system and into the wellbore. Although a water cone may be managed by reducing the drawdown at the well, expansion of the aquifer occurs regionally and isolates the majority of the gas trapped in the matrix. This mechanism is summarized in Figure 8.

It is clear that a 'high' recovery factor can only be reached through reasonably large pressure depletion. In a gas field, this is achieved where you have depletion-drive reservoirs characterized by high average permeabilities (100–1000 mD), not in contact with a large permeable aquifer and relatively homogeneous. Conventional core analysis indicates that the porosity of the KAF matrix ranges from 0 to 21%, averaging 10%, and that the permeabilities of the matrix are 1–100 mD. However, the test-derived permeability is far in excess of this, indicating that the fracture system is the dominant channel for fluid flow. Therefore, to re-establish commercial

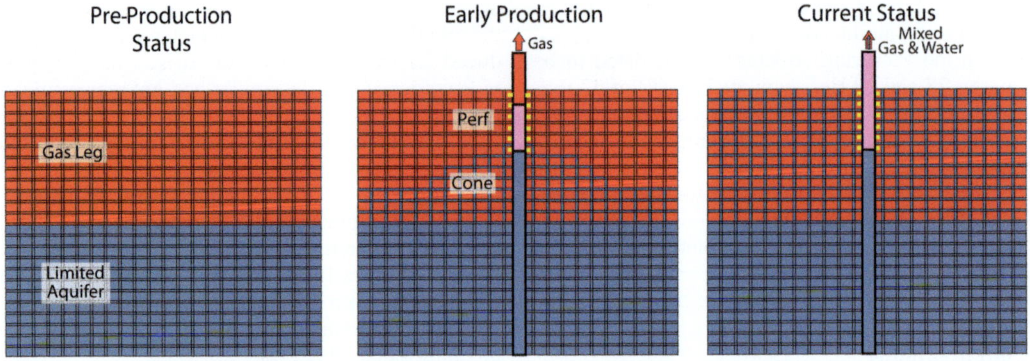

Fig. 8. Sugar cube diagram illustrating the conceptualized production mechanism at work in the VoP gas fields: large squares, reservoir matrix; thin rectangles, fractures; blue, water; red, gas; yellow, perforations.

gas rates and ultimately enhance the recovery, it is imperative to encourage the gas in the matrix to replenish the fracture systems.

In order to do this, the pressure drop between the matrix–fracture interface needs to be maintained to encourage the gas to flow from the matrix blocks into the fracture system. This requires a mechanism by which the pressure of the aquifer in the fracture system is lowered (Fig. 9) and allows the high-pressure gas in the matrix to flow at increasing rates into the production well. Operating under these conditions, the aquifer can be drawndown as it expands into the reservoir. This allows the production wells to effectively drain gas over the field, improving the recovery factors significantly. This mechanism is summarized in Figure 9.

Enhanced gas recovery project (EGRP)

Third Energy implemented a programme of active water production trials in the second quarter (Q2) of 2016. The first was a low-cost N_2 lift operation at the Malton Field and demonstrated that increased gas flow can be achieved through water production. The next, more advanced, trial is taking place at the Pickering gas field and began in Q1 of 2017. The well PK-1 has been reperforated close to the original free-water level in order to access the aquifer and has been fitted with an electric submersible pump (ESP) in the sump of the well to actively remove water in order to drawdown the pressure of the aquifer. The results of reservoir modelling show that this should manage the aquifer's limited expansion, preventing encroachment up the fracture system to above the GWC and allowing overlying gas charged matrix to be drained by the updip well PK-2. The improved gas flow rate is expected to be 5 MMscf/day and will be used for power generation at Knapton. The resulting revenue will be achieved for a very modest CaPex investment.

This production trial is anticipated to show and quantify the reduction in reservoir pressure through

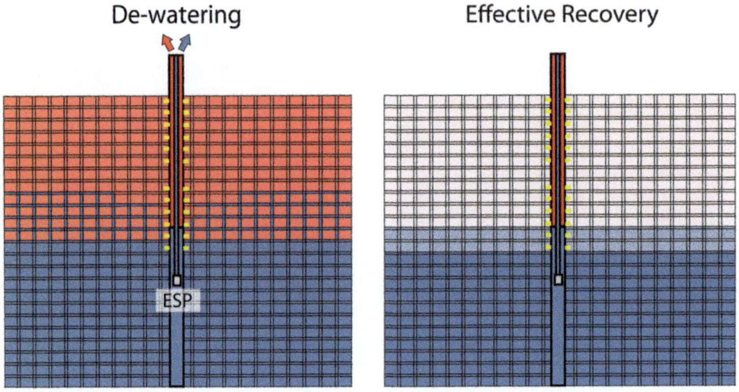

Fig. 9. Sugar cube diagram illustrating the proposed production mechanism: large squares, reservoir matrix; thin rectangles, fractures; blue, water; red, gas; yellow, perforations; ESP, electrical submersible pump.

water production, and will be measured through a downhole gauge in well PK-2. Such a pressure reduction is expected to allow the gas in the bulk of the reservoir to expand and be produced. In addition, water production in well PK-1 should result in water flow *away* from the PK-2 well, thereby beginning the process of reversing the water ingress that has occurred at the Pickering-2 well and restoring the productive capacity.

Importantly, the measurement of pressure response in well PK-2 will allow gas and aquifer volumes to be estimated more accurately. These new pressure measurements are also expected to show the dynamic response of the gas and water system to the pressure reduction. This will enable an assessment of the resultant reservoir pressure response, which can be compared to a longer-term, higher-volume water production project. The associated costs and benefits can then be quantified for the Pickering redevelopment.

Successful implementation of a water management programme will provide a proof of the concept for application to the other fields; in particular, the development of the Ebberston fields. It will also allow the reassignment of the various discovered volumes from prospective and contingent categories to a proved (1P) and proved plus probable (2P) status as the VoP fields are already developed.

Conclusions

Over the last 77 years, the Cleveland Basin has been explored but, relative to many other UK hydrocarbon provinces, is still immature. Drilling activity has demonstrated that a stacked petroleum system is present and significant volumes of gas have been found in the basin.

Despite the exploration success of 12 discoveries, the province has yet to yield a development that can be considered commercially successful, with the Eskdale, Lockton and the VoP fields all failing to produce the forecast reserves that underpinned the original investment.

The reason for failure is that traditional production techniques have been ineffective in dealing with the early ingress of formation water into the reservoir along highly permeable fractures, effectively isolating the bulk of the gas in the less-permeable matrix. To date, only 37 Bcf, or approximately 10%, of the original 370 Bcf field and undeveloped discovered GIIP in this province have been recovered. Consequently, a significant prize still remains, especially as it is onshore and close to both infrastructure and market.

The VoP and Ebberston gas fields have been re-evaluated by Third Energy, and a fully integrated reservoir model has been developed to incorporate the heterogeneity, or dual-porosity system, of these KAF reservoirs.

A low-cost, de-watering programme has been designed, and implementation is in progress. The resulting recovery factors are expected to increase to 70–80% in existing fields and yield a material volume of gas for a relatively low CaPex investment. It is anticipated that through this process of field rehabilitation, the VoP and Ebberston fields could produce approximately 125 Bcf of additional recoverable gas. This approach could also lead to significant interest in a number of other undeveloped KAF discoveries and unappraised structures in North Yorkshire.

Solving the issue of premature water influx in the KAF fields, combined with the search for unconventional resources in the Bowland section of the Mid and Lower Carboniferous strata, will herald a new and exciting phase of E&P activities for this province.

Funding This research received no specific grant from any funding agency in the public, commercial, or not-for-profit sectors.

References

ANDREWS, I.J. 2013. *The Carboniferous Bowland Shale Gas Study: Geology and Resource Estimation.* British Geological Survey for Department of Energy and Climate Change, London, UK.

COPE, M.J. 1986. An interpretation of vitrinite reflectance data from the southern North Sea Basin. *In*: BROOKS, J., GOFF, J.C. & VAN HOORN, B. (eds) *Habitat of Palaeozoic Gas in N.W. Europe*. Geological Society, London, Special Publications, **23**, 85–98, https://doi.org/10.1144/GSL.SP.1986.023.01.06

EVANS, A.L., FITCH, F.J. & MILLER, J.A. 1973. Potassium-argon age determinations on some British Tertiary igneous rocks. *Journal of the Geological Society, London*, **129**, 419–438, https://doi.org/10.1144/gsjgs.129.4.0419

FRASER, A.J. & GAWTHORPE, R.L. 1990. Tectono-stratigraphic development and hydrocarbon habitat of the Carboniferous in northern England. *In*: HARDMAN, R.F.P. & BROOKS, J. (eds) *Tectonic Events Responsible for Britain's Oil and Gas Reserves*. Geological Society, London, Special Publications, **55**, 49–86, https://doi.org/10.1144/GSL.SP.1990.055.01.03

GLENNIE, K.W. & BOEGNER, P.L.E. 1981. Sole Pit inversion tectonics. *In*: ILLING, L.V. & HOBSON, G.D. (eds) *Petroleum Geology of North-West Europe*. Heyden and Sons, London, 110–120.

GOV.UK 2015. [Withdrawn] Oil and gas: licensing rounds – GOV.UK, https://www.gov.uk/guidance/oil-and-gas-licensing-rounds

GREEN, P.F. 1986. On the thermo-tectonic evolution of Northern England: evidence from fission track analysis. *Geological Magazine*, **123**, 493–506.

HALLSWORTH, C.R. & CHISHOLM, J.I. 2000. Stratigraphic evolution of provenance characteristics in Westphalian sandstones of the Yorkshire coalfield. *Proceedings of the Yorkshire Geological Society*, **53**, 43–72, https://doi.org/10.1144/pygs.53.1.43

Heath-Clarke, M., Taylor, K., Harrison, D., Fogg, A., Hughes, F., Haarhoff, M. & Mortimer, A. 2016. The characterization of unconventional reservoirs in the Bowland sequence using onshore 3D seismic data, Cleveland Basin, UK. *First Break*, **34**, 45–52.

Hughes, F., Harrison, D. *et al.* 2018. The unconventional Carboniferous reservoirs of the Greater Kirby Misperton gas field and their potential: North Yorkshire's sleeping giant. *In*: Bowman, M. & Levell, B. (eds) *Petroleum Geology of NW Europe: 50 Years of Learning – Proceedings of the 8th Petroleum Geology Conference*. Geological Society, London, 611–625, https://doi.org/10.1144/PGC8.5

Leeder, M.R. 1982. Upper Palaeozoic basins of the British Isles–Caledonide inheritance versus Hercynian plate margin processes. *Journal of the Geological Society, London*, **139**, 479–491, https://doi.org/10.1144/gsjgs.139.4.0479

MacDonald, R., Wilson, L., Thorpe, R.S. & Martin, A. 1988. Emplacement of the Cleveland Dyke: evidence from geochemistry, mineralogy, and physical modelling. *Journal of Petrology*, **29**, 559–583.

Moreton, R. 1995. Efforts back onshore. *In*: Moreton, R. (ed.) *Tales from Early UK Oil Exploration 1960–1979*. Petroleum Exploration Society of Great Britain (PESGB), London, 95–108.

Mortimer, A. 1992. *Aspects of the Upper Carboniferous structure beneath the UK Sector of the Southern North Sea*. PhD thesis, Institute of Earth Studies, University of Wales.

Powell, J.H. 2010. Jurassic sedimentation in the Cleveland Basin: a review. *Proceedings of the Yorkshire Geological Society*, **58**, 21–72, https://doi.org/10.1144/pygs.58.1.278

Powell, J.H., Ford, J.R. & Riding, J.B. 2016. Diamicton from the Vale of Pickering and Tabular Hills, north-east Yorkshire: evidence for a Middle Pleistocene (MIS 8) glaciation? *Proceedings of the Geologists' Association*, **127**, 575–594.

Sharp, R., Adam, J., Scarselli, N. & Morse, S. 2016. Thick vs thin-skinned deformation of the Sole Pit High (UK Southern North Sea Basin) and its impact on the evolution of supra-salt prospectivity. Paper presented at the PESGB PETEX Conference, 15–17 November 2016, London.

Smith, K., Smith, N.J.P. & Holliday, D.W. 1985. The deep structure of Derbyshire. *Geological Journal*, **20**, 215–225.

Tang, X. 2015. Over four decades' experience with gas dewatering at naturally fractured gas reservoirs in South-Sichuan, China, what have we learned? SPE paper 176394 presented at the SPE/IATMI Asian Pacific Oil & Gas Conference and Exhibition, 20–22 October 2015, Nusa Dua, Bali, Indonesia.

Waters, C.N., Waters, R.A., Barclay, W.J. & Davies, J.R. 2009. *A Lithostratigraphical Framework for the Carboniferous Successions of Southern Great Britain (Onshore)*. British Geological Survey Research Report. British Geological Survey, Keyworth, Nottingham, UK.

Weber, K.J. & Bakker, M. 1981. Fracture and vuggy porosity. SPE paper 10332 presented at the 56th Annual Fall Technical Conference and Exhibition of the Society of Petroleum Engineers of AIME, 4–7 October 1981, San Antonio, Texas, USA.

Ziegler, P.A. 1992. North Sea rift system. *Tectonophysics*, **208**, 55–75.

The development of the manufactured gas industry in Europe

RUSSELL THOMAS

WSP, Kings Orchard, 1 Queen St, Bristol BS2 0HQ, UK

0000-0002-4607-1889

Russ.thomas@wsp.com

Abstract: The manufactured gas industry was one of the great technological innovations of the industrial revolution. Whilst it was developed in Great Britain, this was not in isolation, and required the input of many European engineers, scientists and entrepreneurs for its success. Although the innovation of making inflammable gas for lighting may now seem quite simple, it brought considerable change to society. Initially, gas was used for lighting, replacing candles and oil lamps, greatly improving safety within factories by reducing the occurrence of fires. The concerns over street crime and the potential benefits of brighter gas street lights were key factors in its wider uptake. Lighting was its primary use for the first 70 years. As competition emerged from electricity, the gas industry found new markets in heat and power. Gas manufacturing also produced a range of by-products which were later found to have some very useful properties and became important feedstocks to the chemical industry. Manufactured gas was the first integrated utility energy network and its success in Britain led to a rapid spread across Europe. Since its demise and replacement by natural gas, it has left a shared but unique legacy in each country.

Early uses of gas

'Ex fumo dare lucem' are words of the poet Horace, which roughly translated mean 'to give light from smoke', which was a fair description of how the manufactured gas industry was viewed in the early years and was a motto used by many British gas companies established in the nineteenth century.

There is a general lack of awareness of the source of gas used in Europe prior to the discovery of natural gas beneath the North and Irish seas in the 1960s–70s and elsewhere in Europe. The fact that gas was manufactured from coal and oil for a period of about 150 years is frequently overlooked. A comparison of the chemical composition of coal gas and North Sea natural gas is given in Table 1.

Prior to the discovery of coal gas, there had been an awareness of gas within some societies since ancient times. Most notably, 'Eternal Flames' resulting from gas seeps which had become ignited formed the centrepiece of religious shrines (Hunt 1907). It was not until much later that the great potential of gas was realized, and a practical process for its manufacture developed. Many Europeans had experimented with the distillation of coal, splitting it into its constituent parts of inflammable gas, ammonia-rich water, tar and coke. The Belgian scientist and medic Van Helmont first used the word 'Gas' in the sentence:

> I call this *Spirit*, unknown hitherto, by the new name of *Gas*, which can neither be constrained by vessels, nor reduced into a visible body, unless the seed being first extinguished

in his book *Treatise De Flatibus* published in 1652 (Van Helmont 1652, p. 335).

The French physician Jean Tardin had previously documented in 'Histoire naturelle de la fontaine qui brûle près de Grenoble, avec la recherche de ses causes et principes et ample traitement des feux souterrains' (Tardin 1618) that he had heated crushed coal in a closed vessel, producing flammable smoke. He had assumed that the source of the 'fire well' in Grenoble was smoke escaping from burning coal beds below ground (Tardin 1618). A similar observation was made near Wigan (UK) by Thomas Shirley in 1659 where he identified 'carburetted hydrogen' emanating from a natural spring, which burnt when ignited with a candle. Shirley believed the source of the gas was coal below the ground (Shirley 1867). Shirley's work was continued, years later (1684), by Rev. Dr John Clayton when he excavated the base of the spring to find coal 18 inches below. The gas escaping from the coal measures was inflammable and Clayton correctly assumed that the coal was the source of the gas. Clayton's work continued with the distillation of coal in an open retort, obtaining inflammable gas, tar and water. George Dixon of Durham had built a pilot plant for extracting tar (known to be useful for caulking ship hulls) from coal in 1779 at Cockfield, Co. Durham but, unfortunately, the poor economics of transporting the tar to port and an explosion led to the project being abandoned (Stewart 1958).

From: Craig, J., Gerali, F., MacAulay, F. & Sorkhabi, R. (eds) 2018. *History of the European Oil and Gas Industry*. Geological Society, London, Special Publications, **465**, 137–164.
First published online May 14, 2018, https://doi.org/10.1144/SP465.14

Table 1. *Composition of manufactured coal gas and natural gas (%)*

Source	Non-combustible components			Combustible components			
	Carbon dioxide	Oxygen	Nitrogen	Carbon monoxide	Hydrogen	Methane	Other hydrocarbons (e.g. ethane, propane)
Coal gas	2.0	0.5	5.5	7.5	51.8	27.0	4.7
Natural gas	0.6	–	2.7	–	–	90	6.7

Ward (1965).

The birth of the gas industry

Towards the end of the eighteenth century across Europe, scientist and engineers were active in experimenting with coal distillation and the production of inflammable gas from coal and various other materials. Jan Pieter Minckelers is generally regarded as the Dutch inventor of coal gas. In 1774, he obtained coal gas by distilling powdered coal in the barrel of a shotgun, using the gas to illuminate his lecture room in 1785 (Stewart 1958). Prof. Johan Georg Pickel had lit his chemistry laboratory at the University of Wurzburg, Bavaria in 1786 with gas distilled from bones (Schuster 1939). But, most notably, in France in 1791, Philippe Lebon had developed an apparatus called a thermolampe or 'thermal lamp'. He used this to extract a flammable gas from the distillation of sawdust, the gas being used to light a room. The son of a court official of Louis XV, he was an engineer and scientist of some repute, and devoted great efforts to gas lighting; being awarded a patent in 1799 for the Thermolampe. Lebon's thermolamp was similar in concept to William Murdoch's retort, discussed below, but a major difference was that it used wood instead of coal. Lebon is recognized as the father of the French gas industry, although no practical application of his apparatus continued in France after his death (Hunt 1907).

At the same time in Great Britain, William Murdoch, an engineer from Ayrshire in Scotland, had grown up experimenting with coal distillation. After making his way on foot to the Boulton and Watt factory at Smethwick, Birmingham in 1777, Murdoch found employment as an engineer and later became a partner in the company. His qualities led to him being sent to Cornwall to work with the difficult Cornish mine owners in erecting the large steam engines which powered mine-water-removal pumps built by Boulton and Watt. He was a success and during his time there he experimented further with coal distillation, and lit his house and office with coal gas in 1792. Recalled to Birmingham in 1798, Murdoch continued to experiment with gas lighting. Support was limited until Gregory Watt (James Watt's son) returned from Paris in 1801, having observed the rival work being undertaken by Philippe Lebon. Gregory Watt's intervention persuaded Boulton and Watt to provide Murdoch with the resources he required to further develop the practicality of gas-manufacturing technology (Hunt 1907).

Murdoch was given an assistant in the form of Samuel Clegg and went on to construct a small gas-making plant at the Soho works in 1798. A combination of the death of Lebon in 1804 and the social and political unrest prevented the gas industry establishing in France as early as, perhaps, it could have done otherwise. Lebon's work on the thermolamp cannot be underestimated. It was highly influential in continental Europe, creating a fledgling wood-based gas-manufacturing industry (Hunt 1907). Following Lebon's death, Germany became the main centre for interest in the thermolamp, and a number of books and articles were written on the subject between 1802 and 1812. The most important thermolamps were built by Andreas Winzler, including a large thermolamp in Brno in what is now the Czech Republic. Winzler was a chemist from Unlingen, Germany, who experimented widely with the thermolamp in the first decade of the nineteenth century. Building thermolamps in Brno, Vienna and Blansko, gas lighting was not integral to his work as he had seen wider uses for by-products and intended the gas for heating (Tomory 2012). Winzler wrote the first German treatise on gas manufacture called 'Die Thermolampe in Deutschland' (Winzler 1803). Whilst he met with limited success, he did inspire others in the region, such as Karl Ludwig von Riechenbach, who much improved the airtightness of the design, between the gas generator (retort) and the gas lamp (Tomory 2012).

The development of gas lighting at Boulton and Watt coincided with a time of great change; both Matthew Boulton and James Watt were handing over the reins of the business to their sons, and gas was viewed as an expensive distraction from manufacturing steam engines. The reluctance of the company to enter this new business was challenged by the demands of one of their biggest customers, George Lee of the textile company Philips and Lee, who owned the Salford Twist Mill. The slow progress achieved at Boulton and Watt frustrated Samuel Clegg, who set up as a rival gas engineer in 1804

(Stewart 1958). Boulton and Watt installed gas at the Salford Twist Mill, after first lighting George Lee's house. Clegg installed gas lighting at the mill of Henry Lodge at Sowerby Bridge in Yorkshire (Chandler & Lacey 1949). Both mills were fully lit by gas in 1805, but Clegg is believed to have beaten Murdoch by 2 weeks. Other gas engineers, such as Josiah Pemberton, soon emerged, and hundreds of other mills and factories across Britain were then lit by gas over the next few years, predominantly in the north of England and in the Midlands. But these installations did not supply the public or feed subsequent gas networks (Chandler & Lacey 1949).

From gazomètre to gasholder

The application of the gasholder to the production of manufactured gas is an excellent example of the close ties within the community of natural philosophers, engineers and scientists in the late eighteenth century. At the time there was considerable interest in 'pneumatic' chemistry (the chemistry of gases), and a mixture of friendship, rivalry and co-operation between British scientists, most notably Joseph Priestley, and their French counterparts, especially Antoine-Laurent Lavoisier. Priestley was a member of the famous Lunar Society (a dinner club and informal learned society of prominent figures from the English Midlands), which included amongst its members Matthew Boulton and James Watt (Hunt 1907; Tomory 2012).

French chemists had adopted a more empirical approach to their work, developing accurate tools for the collection, weighing and measurement of gases. Key amongst these was Lavoisier's gazomètre. This simple machine was composed of a counterbalanced cylindrical metal vessel (grand cloche). It was sealed at the top and open at the base, which was placed in a tank of water, with the open side down. The water acted as an airtight seal, which kept the gas in the vessel (Lavoisier 1789). The gazomètre featured in Lavoisier's important book '*Traité Élémentaire de Chimie*' and some of the original equipment survives today in France, thanks to his wife. Lavoisier's fate was not so good, as a 'tax farmer' he was executed during the French Revolution (Tomory 2012).

Lavoisier's gazomètre had been adapted by Boulton in his unsuccessful collaboration with Thomas Beddoes on pneumatic medicine. It was the realization by Boulton and Watt engineers that the gazomètre could be scaled up considerably for the storage of gas on an industrial scale which enabled the gas industry to manage the just-emerging problem of supply and demand (Tomory 2012).

The earliest gasholders, such as those used at Philips and Lee Salford Twist Mill, were cubical containers constructed from heavy iron with a wooden frame, the tanks which held the water were initially built above ground and constructed from wood, sometimes brewer vats were used. Rectangular gasholders continued to be built until around 1815, when they were replaced by the cylindrical design. The gas engineer John Malam improved the cylindrical gasholder design by reducing the weight of the internal framing and using counterbalance weights held by chains. The wooden tanks used were prone to leaking and collapse, and underground brick tanks were introduced in 1818, with stone, iron, steel and concrete tanks following later (Clegg 1841).

The first gasholders using underground tanks were of single-lift construction. The vertical movement of the tank was guided by metal loops on the vessel moving along vertical metal rods on the tank and columns. This was later replaced by wheels running along guided tracks on the supporting columns (Clegg 1841). With the development of the telescopic gasholder by Tait (in 1824), additional storage capacity could be added without increasing the footprint. The telescopic gasholder consisted of vessels situated one inside the other. When the inner vessel was fully extended, it would couple to the outer lift through the engagement of water-sealed cups and dips. Telescopic gasholders could have as many as six lifts (Stewart 1958).

The below-ground water-filled tanks were built from brick, stone or concrete, these tanks were generally made watertight by a layer of puddle clay on the outside face of the tank walls and beneath the base of the tank. Building the tank underground helped to reinforce the tank wall, as it was supported by the compacted ground around it. Due to the difficulty of building below-ground tanks, Clegg devised two other types of gasholder: the rotating (in 1815) and the tent (in 1818). The latter was included in his gasworks design for the Royal Mint in 1817 and the former, although unsuccessful due to its cost and complexity, was the forerunner to the station gas meter (Everard 1949).

The famous gas text 'Kings Treatise' highlighted the importance of the gazomètre in the development of the gas industry, quoting: 'It would be difficult to dispense with the gazomètre since nothing that ingenuity could devise could replace this important device' (Newbigging & Fewtrell 1878).

The rise of the British gas industry

Whilst Britain undeniably took the lead in commercializing manufactured gas, it took a notable European figure to help the industry develop into a position where it could serve the wider public. The business model of William Murdoch and his contemporaries had been to build a gasworks to provide a gas supply to a single factory establishment. This approach only acted to serve the industrialists and provided little benefit to the public, with the exception of cases where a limited public gas supply was provided to light the area just beyond the factory gate (Chandler & Lacey 1949).

The charismatic German entrepreneur Friedrich Albrecht Winzer came to London from Braunschweig (Brunswick) in Germany, where he had connections to the royal court. He had witnessed Lebon's experimental gas lighting with his patented thermolamp at the l'hôtel Seignelay in Paris. Winzer had tried to purchase one of Lebon's thermolamps, but Lebon was not interested. Winzer did, however, assist Lebon in some experiments, which gave him enough of an understanding of the thermolamp that he could have his own one built when he returned to Braunschweig in 1802 (Hunt 1907).

In Braunschweig, Winzer tried to secure the support of the Duke of Brunswick and his royal court to develop the thermolamp technology. His powers of persuasion failed and in 1803 he decided that London was the best place to exploit gas lighting. On moving to England, Winzer anglicized his name to Fredrick Albert Winsor (Stewart 1958). In 1803 and 1804, he promoted his interest in gas lighting by undertaking a popular series of lectures with demonstrations using coal gas at the Lyceum Theatre. On Lebon's death in 1804, Winsor viewed himself as the 'legitimate and successful imitator and proprietor' of Le Bon's ideas (Hunt 1907). In 1804, he took out his first patent (No. 2764) for his gas-making stove or oven (Anon 1860). In 1806, Winsor moved to No. 97 Pall Mall, a very fashionable part of London near to the royal palaces, where he lectured and experimented. From two of his patented stoves, he lit the garden wall between Carlton House and St James Park on the evening of 4 June 1807 to celebrate the King's birthday (Williams 1828). Both of these actions rapidly raised the profile of manufactured gas in Britain, beyond that achieved by Murdoch, Clegg and their contemporaries.

Winsor's vision had become that of centralized gasworks providing gas to multiple establishments through gas mains constructed under the streets. Financially supported by his subscribers, he spent many years campaigning for the creation of a gas company. Firstly, for establishing the National Light and Heat Company with capital of £1 million, but later this was curtailed to a smaller company supplying the cities of London and Westminster and the borough of Southwark. It received its Royal Charter in 1812, and was called the Chartered Gas Light and Coke Company (GLCC), it went on to light the cities of London and Westminster, and the borough of Southwark. After a rocky start with many teething problems due to a lack of engineers and poor management, it eventually became a success.

To resolve the technical problems faced by the GLCC, they recruited the aforementioned Samuel Clegg who was eventually placed in a position where he could gradually rectify many of the technical failings of the company and prepare it for future prosperity. The importance of another notable German chemist, Friedrich Christian Accum, should not be overlooked in the development of the GLCC. Accum had ventured to Britain in 1793, lecturing on chemistry and physics, and started to assist Winsor in 1805; and it is due to him that such early accounts of gas-manufacturing technology exist (Accum 1816). Accum was the GLCC's first chemist and was also an original director of the company, but he left in 1815. Winsor's personal achievement in creating the GLCC was short lived; he was not even selected as an original director of the company, a position he eventually achieved in July 1813 with support of the shareholders (Everard 1949). It became apparent over the following year to the directors and shareholders that Winsor lacked both technical and managerial acumen, and he was forced to flee to France pursued by debtors. In Paris, he established the first commercial gas company; unfortunately, this was short lived and after a few years failed (Everard 1949).

The GLCC was imitated many times, first in London and then further afield, with the first public gas undertaking outside London being established in Preston, Lancashire in 1816, Exeter and Liverpool soon followed. Gas undertakings quickly spread across Britain to light every city, town and many large villages. Gas was used primarily for street lighting and the lighting of public buildings or those wealthy enough to be able to afford it (Stewart 1958). The popularity of gas was due to it being cheaper, brighter and safer than candles or oil lamps. The improved brightness of the light was a key factor, as the light made an effective deterrent against criminals. The adoption of gas was often driven by 'Lighting and Watching Acts of Parliament' whose purpose was to make the principal streets of the town safer (Chandler & Lacey 1949).

Samuel Clegg remained with the GLCC until 1817, pioneering the gas meter, gas governor and new forms of gasholder during his tenure. Clegg fell out with the directors of the GLCC on multiple issues, leaving on 1 April 1817. He went on to act as an engineer for the construction of a gasworks at the Royal Mint, and then left London to assist in the establishment of gas undertakings in Liverpool, Norwich, Bristol, Birmingham and Chester. He returned to London to work for a rival of the GLCC, the Imperial Gas Company, between 1822 and1830. The oldest gasholder in the world, believed to be attributed to Clegg, survives in Fulham as a Grade 2* Listed structure. Clegg later took an appointment with the Portuguese government as an engineer before returning to the UK as a government advisor on new gas companies (Stewart 1958).

Coal gas faced early competition from oil gas. The first patent by Gosling and Taylor was submitted in 1815, and used fish oil in heated retorts to make oil gas. The company Taylor and Martineau produced plants which made gas of 'great illuminating power' from oils and without the sulphurous contamination of coal gas. They supplied the equipment to manufacture gas to many locations in the UK and abroad. Companies were established in Bristol and Edinburgh which produced gas from animal or vegetable oils and distributed it using a network of pipes. Although oil gas was superior to coal gas, the vagaries of obtaining sufficient feedstock oil (e.g. fish oils or vegetable oils) of suitable quality led to variations in production costs which disadvantaged the industry in Britain and it soon died out. Oil gas had greater success abroad where it faced more limited competition from cheaper piped coal gas (Chandler & Lacey 1949).

Another type of oil gas company emerged, which produced the gas and distributed it in a compressed form in cylinders. The London Portable Gas Company was established in 1819 with the aim of exploiting a patent granted to David Gordon and Edward Heard. The London Portable Gas Company supplied oil gas which was 'fixed' and lost little of its illuminating power on compression. One of the four subscribers to the company was John Taylor of Taylor and Martineau, who manufactured oil-gas-making plant. Although the company only lasted until 1838, it did leave a notable legacy as the discovery of benzene was due to the analysis of a liquid residue left within one of the compressed cylinders and was identified by no less than Michael Faraday of the Royal Institution (Stewart 1958).

Gas undertakings had all been privately owned until the Manchester Police Commissioners took an interest; they were the Local Authority responsible for lighting (and watching) the streets with oil lamps. They passed a resolution in 1817 to build a gasworks at Water Street to light the police stations and principal streets, and to sell any surplus to private customers. This was the first example of a municipally owned gas undertaking. Later in the nineteenth century, the municipal ownership of gas undertakings became popular, especially in the north of England, the Midlands and in Scotland. Cities such as Birmingham ploughed the profits from the gas department into major civic projects, such as constructing new sewage and water infrastructure, and Manchester used the profits to finance a new city hall (Hunt 1907).

By 1820, 15 of the principal towns of England and Scotland were lit by gas; by 1830, there were over 200 towns and cities lit by gas in Britain. By 1859, there had been more than a 1000 public gasworks built across Britain. By the 1840s, gas lighting had just started to be adopted in the homes of the wealthy, often only selected rooms and not the entire house. A welcome boost to its popularity came when the New House of Commons adopted gas lighting in 1852, which it used until 1900 (Stewart 1958).

Whilst the demonstration of gas cooking had started as early as the 1820s, cooking by gas did not become popular until the 1870s. A great surge of interest did follow in the late 1840s, mainly due to Alexis Soyers, the celebrity chef of his day, who had used gas cooking extensively in the London Reform Club where he worked (Chandler & Lacey 1949).

The gas industry had largely avoided regulation. However, in 1847, the Gasworks Clauses Act was introduced to regulate the construction of gasworks and the supply of gas to towns. The Act regulated all aspects of the industry from pipe-laying to the waste produced. A number of other important Acts of Parliament were introduced over the next century which would regulate the gas industry. Competition had been the main mechanism for keeping prices low, this was especially true in London where almost all areas had the choice of two potential gas companies to supply them. This system was uneconomic and very inconvenient, as it led to excessive street works. In 1853, London south of the Thames adopted 'districting', with specific companies responsible for supplying certain districts. This concept was later adopted throughout London via the Metropolis Gas Act of 1860 (Stewart 1958).

The longevity of the manufactured gas industry was due to its ability to continually innovate, improving both gas manufacture and its utilization. Between 1850 and 1890, a number of important developments were made by German scientists and engineers. In 1855, the German chemist Robert Bunsen invented the 'aerated burner'. By mixing the air and gas prior to the flame, it could burn hotter and made more efficient use of the gas. This spawned a greater range of uses, most importantly heating applications. This development in time ultimately saved the gas industry, as heating applications became ever more important as the industry evolved (Stewart 1958).

Gas making received a significant improvement when Frederick Siemens adapted the 'gas producer' developed originally by fellow German Bischof, which produced a gas composed primarily of carbon monoxide and carbon dioxide from the incomplete combustion of coal or coke. The gas could be burnt next to the retorts which enabled them to reach a higher temperature. This replaced the previous furnaces which heated the retorts directly by radiant heat from the fuel bed (Thomas 2014*a*).

Siemens combined the producer with a 'regenerative furnace', which could operate at a higher temperature than a standard furnace by recovering the heat from waste gases to preheat incoming air. The combined effects of these technical changes were to increase the temperature of the gas-manufacturing process, which increased the yield of gas made. Continuous regenerators were developed by the Germans Schilling and Klonne, introduced in the 1880s, which simplified the process such that multiple chambers were not required and the hot exit gases could continually heat incoming air (Meade 1921).

A key discovery which aided the utilization of gas was the 'incandescent mantle', developed by the Austrian chemist Carl Auer von Welsbach. He discovered that the oxides of certain rare metals had the ability to emit light when heated to a state of incandescence. Similar discoveries had been made by others before, however; Von Welsbach was the first to have commercial viability, which took many years to perfect. The final article was constructed of fabric soaked in a mixture of 99% thorium dioxide and 1% cerium (IV) oxide, which when placed over an aerated burner produced a brilliant light (Chandler & Lacey 1949).

The incandescent mantle produced much brighter light than previous gas lights and it was timely, with the gas industry facing a new rival: the electric light bulb. The electric arc lamp was demonstrated by Humphrey Davy in 1806, but it was not suitable for widescale use. The more robust light bulb for everyday use came much later and was developed by Joseph Swan (UK) in 1878 and Thomas Edison (USA) in 1879. They later collaborated to form the Edison Swan Electric Company Limited, otherwise known as Ediswan. These bulbs needed further practical refinement to incorporate metal filaments, which was achieved in 1911. Electricity networks took a while to establish first in towns and then, after the Electricity (Supply) Act of 1926, the creation of a National Grid. The gas mantle had enabled gas lighting to compete with electricity for much longer and it persisted in the 1950s. There are still approximately 2000 public gas lights in the Greater London area (Thomas 2014*a*).

Traditionally, gasworks had been constructed within the area they supplied. This, however, became a problem for urban areas which had greatly expanded, such as London, and little or no new space existed for expanding the gasworks. A notable change occurred in November 1869, when the construction of the Beckton gasworks of the Gas Light and Coke Company commenced. Beckton gasworks were built to supply London, but were built outside of the city on the banks of the Thames in Essex. This resulted in moving the polluting gas-making activities outside of the city, improving the quality of the city air, although negating the quality in the vicinity of the Beckton gasworks (Everard 1949).

An oil-rich coal called cannel coal was favoured for use in gasworks, as it was rich in light oils. It could be found in the Midlands, NW and Scottish coal fields. Some gasworks, such as the Western Gas Company based at Kensal Green, London, made gas purely from cannel coal and many small private estate gasworks preferred it because it produced a superior light. In most cases, a small amount of cannel coal was added to traditional gas coals to increase the brightness of the flame, or 'candle-power' (Stewart 1958).

Gas was beyond the means of most of the general public, who could not afford to install gas or purchase the home appliances. This changed with the invention of the prepayment gas meter by T.S. Lacey in 1870. These coin-operated systems could allow the public to pay for gas upfront. It was linked with the renting of gas appliances by gas companies to the public and sparked a rapid increase in gas consumers (Goodhall 1999).

The period from 1870 to 1930 marked a period of considerable development work in gas manufacture. It led to new more efficient and flexible ways of making gas. Whilst the production of 'water gas' had been experimented with since the earliest years of the gas industry, it was difficult to control and had never been a true commercial success. This was until Thaddeus Lowe, an American practical engineer, had in 1873 devised a system which produced gas on a few-minute cyclical basis in a gas machine ('generator set') consisting of three vertical cylinders. First, the system was heated by partially combusting of coke within the 'generator' and, secondly, steam was injected into the hot coke, producing a gas comprised primarily of hydrogen, carbon dioxide and carbon monoxide called 'water gas'. This gas had a low calorific value and poor illuminating power, but could be enriched by the injection of oil: called the 'carburetted water gas' (CWG) process. The oil was sprayed into the hot gas and was fixed into the gas in two chequer brick-filled hot chambers: the carburettor and the superheater. The first major installation of a CWG plant in Britain was at the Beckton gasworks near London. The main benefits of water gas were that its gas-generating heat came from the very same residual coke made at the coal-gas retorts, it provided an alternative method to meet peak demand for gas, the plant could be started in a matter of hours and it minimized the requirement for gas storage to meet peak demand. Two English expatriate American gas engineers, Humphreys and Glasgow, who were employees of the United Gas Improvement Co. (Philadelphia) set up a business of this name to install CWG plants for the European market, and plants were erected across Europe and eventually exported back to the USA (Thomas 2014*b*).

Both the 'inclined' and the 'vertical retorts' were developed during this period to ease the loading of coal and unloading of coke, and to reduce the amount of ground required to manufacture gas, reducing the labour requirements by 75% when employed effectively. The inclined retort was developed based on work undertaken by Monsieur Emile Coze at the Rheims gasworks in France. It was adopted and improved in Britain by engineers such as Frank Morris at Southall and Brentford. The retort was placed at 32° to the horizontal; this is the theoretical angle of repose for coal and required skill to make it work successfully. The inclined retorts gained limited popularity and with the development of the vertical retort they were often superseded (Thomas 2014*a*).

A later development was the 'vertical retort'; as the name suggests, the retort was rotated by 90° so that it was in the vertical plane. The original vertical retort system – the Intermittent Vertical System – was developed by Dr Julius Bueb of the German Continental Gas Company. After trials at their Dessau gasworks, it was patented in England in 1904. It had considerable advantages over the horizontal system; it reduced labour requirements as the coal was delivered to the top of the retort mechanically and then discharged by gravity after carbonization through the retort. The plant extended much higher than horizontal retorts; however, it took up much less ground space. The Dessau vertical retort was used in Britain, with the first installation at the Ayres Quay gasworks in Sunderland (Anon 1909). The original design was much improved in the UK, overtaken by Arthur Duckham's development of the 'continuous vertical retort', in which coal continuously moved slowly downwards, emerging as quenched coke at the base of the retort. The first continuous vertical retorts were built at Bournemouth gasworks by Woodhall–Duckham in 1906, with the rival Glover West Company building their continuous vertical retort plant at St Helens in 1907. Vertical retorts were particularly suited for use with those types of gas coals mined from the English Midlands, Yorkshire and Scotland. Those gasworks which received coal from Durham and Northumberland mines tended to continue modernizing and mechanizing horizontal retorts, as they performed better with these types of coal (Thomas 2014*a*).

There was also another type of plant similar to the 'intermittent vertical retort', called an 'intermittent vertical chamber oven' (IVCO). These IVCOs were also developed by the German Continental Gas Company and their use became popular across the continent, primarily because they were effective with a range of coals including poor-quality slack (a mixture of small fragments of coal, coal dust and non-combustible material). A modest number of such plant were built in Britain, including the White Lund gasworks in Morecambe, the last traditional coal gasworks to be built in England. They were also heated by external gas producers. Another way in which the processes differed was that coke breeze (fine coke) was added to the base of the vertical chamber oven prior to the coal being loaded. This kept the oven's bottom door cool and ensured that the coal was fully carbonized. Towards the end of the carbonization process, the chamber would be steamed. This would allow the IVCO system to produce a portion of 'blue' (uncarburated) water gas. 'Steaming' was widely employed in the same way on other vertical-retort-based processes. The water gas produced had a lower calorific value than the coal gas and would effectively dilute the calorific value of the coal gas down to the 'declared' value. Another similar plant was the 'inclined chamber oven', which was developed in Munich, Germany and worked in a similar manner to that described above but along the principles of an inclined retort (Thomas 2014*a*).

The 'by-product coke oven' was developed in Europe, gaining popularity in Belgium and Germany, and later introduced to Britain. The by-product coke oven produced gas which in many cases was used to supplement or replace the manufacture of gas at gasworks. In the north, the NE and in the East Midlands, coke oven gas eventually replaced about 40% of gasworks output. The largest gasworks in Britain, Beckton and East Greenwich, both installed coke ovens. As the gas industry moved away from gas lighting to heating applications, the Gas Regulation Act was introduced in 1920. It changed the basis of charging for gas from illuminating power to calorific value. It also introduced a national basis for the testing and reporting of gas quality, as opposed to local supervision (Thomas 2014*a*).

World War II had a huge impact on the British gas industry, it led to coal rationing whilst it required a significant increase in gas and by-product (benzene and toluene in particular) output to aid the military effort. The aerial bombing of gasworks and the gas distribution network inflicted huge amounts of damage; skilled staff were lost from the industry to the war effort and funds for new plant were in limited supply but, despite this, gas output was greatly increased. The damage incurred by the gas industry was so severe that it would take major investment to get it back to its pre-war state. In 1944, the Minister of Fuel and Power appointed Geoffrey Heyworth as chairman of a review into how the industry could be rebuilt while reducing the cost of gas for all types of consumers. The Heyworth review highlighted many issues, including the need to improve the gas transmission network by amalgamating to form larger companies. The incoming Labour government decided that nationalization was the best course of action for the gas industry, and the industry was nationalized under the Gas Act of 1948 on 5 January 1949. The 1064 local gas undertakings were vested in 12 autonomous geographical gas boards. The Minister of Fuel and

Power also established the Gas Council to oversee the gas boards (Thomas 2014*a*).

The demise of manufactured gas

During the 1950s the availability of quality gas coals started to diminish, increasing their cost. Some coal fields became exhausted and in others the mines had to reach ever deeper or further under the sea to mine the remaining seams of coal, making it more expensive. The industry had started to look to new technologies which could handle the poor-quality coal available, such as 'complete gasification'. The final chance for coal was when the Gas Council joined forces with the German Lurgi Company to develop new approaches to gasifying lower-grade coal. Two Lurgi plants were built (Westfield and Coleshill) but were expensive to operate, and by this point the future for gas had moved on and the National Coal Board had lost interest (Williams 1981).

One such alternative supply was mine gas, which was rich in methane. The Point of Ayr colliery in North Wales proved to be a valuable source of this gas. The 95% pure methane gas could not be used directly as it had a much higher calorific value than coal gas, but put through a gas-reforming plant converted it to a gas with similar properties to coal gas. Put simply, this process used steam to split the methane into a town gas of hydrogen, carbon monoxide and carbon dioxide (Stewart 1958). Although a useful source, mines gas could only supply a small portion of Britain's requirements and was not a long-term solution.

Early onshore exploration for natural gas in Britain had found small gas fields in Heathfield (Sussex) and Cousland (Scotland), but nothing substantial was discovered, even though major drilling campaigns were mounted during World Wars I and II.

The initial solution to the problem came from the petroleum industry. Britain and the rest of Europe had developed an oil-refining capacity in the twentieth century, which was dependent on imported oil. This provided by-products such as butane and naphtha which could be used in gas-reforming plants providing cheap gas. Since the feedstocks lacked the ammonium and cyanide found in coal gas, it reduced both capital costs and the land take. The processes were highly dependent on the feedstock, and a variety of catalytic and non-catalytic processes were developed in Britain and abroad. Catalytic gas-manufacturing processes were also more widely used in Britain. British-designed plants, such as the SEGAS process, would operate at a low pressure using a hot catalyst to convert the oil to gas. Lime catalysts were used for the range of gas oil to medium fuel oil; nickel catalysts were best suited to light distillate (Thomas 2014*a*).

As technology advanced a range of continuous 'catalytic-reforming' processes were developed. These processes continuously reformed hydrocarbon feedstocks with a low sulphur content at pressures ranging from atmospheric to 40 atmospheres, and at a temperature between 700 and 950°C. The gases produced usually consisted of hydrogen, carbon monoxide, carbon dioxide, some methane and undecomposed steam. The use of reforming plants completely changed the gas industry and its economics at a time when the future had been bleak. This, however, only provided a brief respite and one which became much more expensive during the oil crisis of the early 1970s (Thomas 2014*a*).

The importation of liquefied natural gas (LNG) had first been considered by the North Thames Gas Board in the 1940s. This work continued in the new nationalized gas industry when LNG importation was trialled from the USA (in 1959) and then commercially from Algeria. The expansion in LNG would have continued, was it not for native natural gas reserves being identified and exploited in the North Sea (in 1967). These had followed successful discoveries in The Netherlands at Slochteren. It became apparent at this point that natural gas was the future, and it had the added benefits of being cheaper, safer (less toxic), cleaner-burning and had higher calorific value. During the 10-year period between 1966 and 1976, the gas network and all the appliances on it were converted from town gas to natural gas (Williams 1981).

Most gasworks had closed by the 1970s, with the last gasworks at Millport on the Isle of Crumbrae closing in 1981. The gasworks were then decommissioned in line with standards of the day, which involved removing plant, stripping and disposing of asbestos, demolishing buildings, emptying underground tar tanks, and infilling any resulting voids. Only the operational above-ground plant, such as gasholders, gas-distribution infrastructure and depots were retained; other surplus property was disposed of, where possible (Thomas 2014*a*).

Why Britain?

Britain benefited from a combination of important factors which led to the development of the manufactured gas industry. Being a stable, wealthy, mineral-rich, technologically advanced and politically liberal trading nation, it created a good platform for new industrial technologies to be developed. As the leading nation at the forefront of the Industrial Revolution, the factories and mills acted as test beds for the development of gas-manufacturing technology. At Boulton and Watt, Murdoch had the resources of the leading industrial engineering company of the time, the support of other excellent engineers such as the Creighton Brothers and John Southern, and a network of major clients and industrialists to whom the technology could be sold. The numerous coal fields, development of canals, tramways and the ease of transport using ships via the coastline made it easy to transport coal to the gasworks. This unique combination of factors made Britain the ideal location for the gas industry to flourish.

European expansion

Following the early nineteenth century development of the gas industry in Great Britain, interest grew in gas lighting in the rest of Europe, with accounts of gas lighting taken back by visitors to London and diplomatic staff. The potential for exporting this industry was soon realized by British industrialists and gas engineers alike. This culminated in the formation of the Imperial Continental Gas Association (ICGA), whose emblem is shown in Figure 1. The company was initially formed in 1824 by Major-General Sir William Congreve; he had previously been the Government Inspector of Gasworks. By this time, gas was already starting to be manufactured in some European towns. This was, in part, promoted by success in Britain, but was also due to those continental engineers who followed on from the work of Lebon and Winzler with the thermolamp. The adoption of the thermolamp and oil gas impeded the spread of coal gas in some countries (e.g. Sweden) where coal had to be imported, and wood and wood resin were plentiful and available for making lighting-acceptable wood-gas (Anon 1974).

Fig. 1. The emblem of the Imperial Continental Gas Company.

In 1825, Congreve toured Europe by stagecoach to establish business ventures. Success was mixed, as some places already had established local gas-lighting rivals. His first venture was to purchase a small oil gasworks in Ghent from a local company, Roelandt and Co. who were taken on as agents and the gasworks were later converted to use coal (Anon 1974). A huge business rapidly developed operating gasworks in The Netherlands, Belgium, Germany, Austro-Hungary and France from this small venture. They provided the first commercial-scale public gas lighting in Germany (1825) and Austria (1845). During their history, they lit over 16 towns in France and at one point controlled a significant portion of the gas industry in Belgium. The gas contracts were generally concessions where the ICGA would build and operate the gasworks for a set period of time, after which the contract would be re-let. Due to these types of contracts, political changes and war in Europe, the business was risky and changed considerably from decade to decade. Sometimes contracts were cut short and the ICGA had to hope for some form of compensation, but others were extended many times and the IGCA generally were in a position to turn contracts away (Anon 1974).

The desire for British engineers to export ideas was not unique to gas lighting; there had been a history of British engineers establishing factories in foreign countries. One example was William Wilkinson, who exported British iron-making technology to France and helped to establish an iron foundry in Le Creusot. Aaron Manby, who is mentioned later, also exported British iron-making technology to France, establishing an iron works at La Charenton and later took over the Le Creusot iron works as well. Much European railway construction was undertaken by British engineers, contractors and capitalists.

Independent British gas engineers were important in the transfer of gas-making technology to other European countries: for example, William Richards helped establish many of the gasworks in Catalonia (Moyano-Jiménez & Thomas 2016), and the combined efforts of Malam and Small established a number of gasworks in Norway. Other gas engineering contractors, such as George Bower and William Cartwright Holmes, exported entire gasworks as kits to be reassembled abroad and also prospected for, and invested in, such ventures (Holmes 1874). A new country provided the engineers with an opportunity to establish new markets and potentially make their own fortunes. The British engineers generally sourced their gas-making plant from British companies and also the supply of good-quality British gas coals, which were widely used across Europe. Inevitably, local engineers learned from the British engineers and started to establish themselves, replacing the British and going on to form their own private or municipal gas undertakings.

Other British companies, such as the European Gas Company, the Danish Gas Company, the Malta and Mediterranean Gas Company, the Tuscan Gas Company and the Continental Gas and Water Company, followed the ICGA. It was not just the British who set up these holding companies, as similar companies followed in Belgium (La Compagnie Générale Pour l'éclairage et le Chauffage par le Gaz – Gaz Belge), France (La Compagnie Générale Française et Continentale 'éclairage) and Germany (Deutsche Continental Gas Gesellschaft (Dessau Gas Co.)). All of these companies sought to establish multiple enterprises rather than lighting just a single town or city.

The development of the gas industry in other European countries

This section of the paper provides a brief description of the development of the gas industry in Europe outside of the UK. This is not an exhaustive list, but it covers a diverse range of countries where the gas industry was of some significant importance. It is divided into three groupings, Western and Central Europe; Southern Europe; and the Nordic countries, as some similarities exist between these groupings. Many of the events described during the development of the gas industry in Britain were mirrored in other European countries; however, each country faced its own specific challenges. A key issue was the availability of good-quality gas coals, which influenced the type of gas-making plant employed. British gas coals were exported across Europe, even to the coal-rich areas of the Rhur in Germany. When they became scarce – due to war, strikes or diminishing resources – other gas-manufacturing technologies were developed to make use of those coals or feedstocks which were available.

One of the issues when trying to assess the extent of the gas industry in many parts of Europe are the major geopolitical changes which took place, especially in Germany, Poland, Italy and former countries within the Austro-Hungarian Empire. Those industry registers naming the locations of towns supplied by manufactured gas (primarily documented between 1890 and 1950) can be sporadic and difficult to follow given that the borders and town names regularly changed. The numbers provided in this document reflect listings in the various historical registers and other supporting information.

Western and Central Europe

The Netherlands. The first demonstration of gas for lighting relating to The Netherlands dates back to Jean Pierre Minckelers, a Dutch citizen from Maastricht, who later became a professor working at the University of Louvain in Belgium (Zahn 1911). His interest in gas was derived from his interest in the new sport of air balloons. He was requested to investigate the subject for the Duke of Aremberg, along with his colleagues Thysbaert and van Bouchante. Their aim was to find a cheaper form of gas for air balloons; this led him to investigate distilling gas from coal. He later lit his lecture room by distilling coal within the barrel of a shotgun in 1785. In 1904, to commemorate this event, a statue with a burning gun barrel was erected in his honour at the entrance to Boschstraat in the Market Square in Maastricht (Hijdelaar 1939).

No further development of gas lighting was undertaken in The Netherlands until Bernardus Koning undertook a display of gas lighting in Amsterdam in the Oude Mannen en Vrouwen Huis (Old Men's and Women's House) on 25 May 1816. This demonstration proved the superior quality of gas light over candles and oil lamps, and did much to promote the qualities of gas lighting in The Netherlands. Koning's work was important in promoting the use of gas in The Netherlands (Lintsen 1993).

A committee of the Royal Dutch Institute of Sciences, Letters and Arts was commissioned by the Dutch Department of Education, Arts and Sciences to assess the experiment in 1816; they gave a very positive assessment of gas lighting to the King. The committee determined that his device was different from that described in Frederick Accum's *Practical Treatise on Gas Light*, published in May 1815; Koning's method used a closed retort, but differed in that he fed steam into the hot coal, effectively producing a mixture of coal gas and water gas. Koning promoted his method as providing an increased gas yield due to the additional water gas, as being cleaner and able to produce gas at a higher pressure. Whilst the trial was an undoubted success, the Dutch gas industry was slow to start. This was due to health concerns, the economic policy of the Dutch Government, and the technical uncertainties over the best method and feedstock (e.g. coal or oil) to use (Lintsen 1993).

Koning only constructed two gas plants according to his patented design, built for the factory and home of the Leiden textile manufacturer Koppeschaar and the Dutch parliament in The Hague. The latter plant operated until 1845, when it was replaced by a public supply of gas (Lintsen 1993).

The first public gasworks in The Netherlands, however, were not built until 1826, when gasworks were built in the cities of Amsterdam and afterwards in Rotterdam. Johan Warin established the first small oil gas plant in The Netherlands on Amstelstraat behind the Hoogduitse Theatre in Amsterdam. The plant was supplied by J.P. Taylor and Martineau, and provided gas to light the theatre and neighbouring houses via pipes under the streets. In October 1825, the Amsterdam Pipe-Gaz Company (APGC) was established using this small gasworks, which was expanded slightly in 1826 and used until 1830. The oil gas provided by the APGC was very expensive and attracted only a few wealthy customers. Limitations were placed on which feedstocks could be used in Amsterdam to produce gas, with coal being excluded but oil and resin being acceptable. The APGC started to build a new gasworks in 1830, and for this they paid to use a process developed by the ICGA through their Dutch agent Roelandt & Co, who they had contracted when they purchased their first European gasworks in Ghent. The APCG continued to fail and the shareholders of the company sold out to the IGCA in 1834, who later converted the gasworks to coal-gas manufacture by agreement with the city. The IGCA had a long association with the city, going on to build two large gasworks, the Oestergasfabriek (Eastern gasworks) and Westergasfabriek (Western gasworks): a majority of buildings from the latter still stand as a redeveloped cultural centre. The municipality of Amsterdam took over the IGCA's gas concerns in 1898 (Mac Lean 1977).

In 1824 the Dutch Portable Gas Company (Nederlandsche Vervoerbare Gas Compagnie) was formed as a competitor to the APGC. It produced gas from 'oil and fats from plants and animals'. It initially lit the Hoogduitse Theater, but lost this contract to the APGC in 1826. The oil gas was stored in compressed cylinders and transported to where it was required. The oil gas plants were supplied by the British company J.P. Taylor and Martineau, whose attempted Dutch patents were rejected. Links existed with the London Portable Gas Company who supplied David Gordon's patent lamps. The company was short lived and was liquidated in 1829 (Mac Lean 1977).

Rotterdam's first gasworks was built by the British ICGA and was a conventional coal gasworks. Gasworks construction activity then slowed, with only the construction of Haarlem gasworks by the IGCA in the 1830s. More activity occurred in the 1840s with the construction of gasworks at Gouda, Arnhem, 's-Gravenhage, Kampen, Leeuwarden, Leiden, Maastricht, Utrecht and Zwolle. The gas industry grew gradually with new towns adopting gas lighting until the second decade of the twentieth century. The invention of the gas mantle by Carl Auer von Welsbach gave a significant boost to the industry. This coincided with the availability of simple and reasonably priced gas cookers. Some gas undertaking initially offered special rates for gas cooking, which later became an important part of gas sales, after the use of gas for lighting had diminished.

As in Britain, the prepayment gas meter was introduced in The Netherlands, this brought gas within the reach of the wider public. Other concessions offered included the free installation of gas piping in houses and the possibility to rent gas appliances. The aim of this stretched beyond increasing gas sales, it had a serious public health aspect, with the intention of improving urban living conditions through reducing the production of smoke, soot, dust and ashes. Uses for gas in the home increased, with the provision of water boilers, radiant heaters, tubular radiators and gas refrigerators. By 1916, gas production in The Netherlands had risen to approximately 176 million cubic metres (Hijdelaar 1939).

By 1930, the manufactured gas industry was at its peak, with 195 public gasworks, after which came a gradual decline. The position of the gas industry in The Netherlands prior to World War II was that, with the exception of a few privately-owned gasworks, the entire gas supply of The Netherlands was owned by the municipalities. There were at this time 166 gasworks and 33 gasholder stations for the distribution of gas derived from municipal works. In addition to this, 17 gasholder stations obtained their gas from the coke ovens operated by the government-owned coal mines in the south of the country at Heerlen and four gasholder stations receiving gas from the blast furnaces at Ijmuiden steelworks. At this time, about 71% of the population of The Netherlands (i.e. about 6.1 million people) was able to obtain gas for everyday daily use (Hijdelaar 1939; Peebles 1980).

It was viewed by the Dutch gas industry that it consisted of too many small gasworks, and that through co-operation and improved distribution the number of gasworks could be rationalized into fewer more efficient large gasworks. The Dutch gas industry had entered into a period of gradual change from 1930, which was accelerated due to World War II. The number of gasworks in The Netherlands had reduced to 140 by 1949 (Brandis 1949). It was proposed by the Dutch gas industry that the number of gasworks should be reduced to five or 10 larger works to supply the entire country. Instead, the industry gradually evolved such that the southern provinces were supplied with coke oven gas, while North Holland was supplied with blast furnace gas from the Ijmuiden steelworks. South Holland then started to be supplied with gas made from oil by-products from the refineries in Rotterdam. The 1960s heralded a major change with the discovery of the huge Slochteren gas field in the north of The Netherlands, after which the country gradually converted to natural gas, and exported gas to the rest of Europe (Peebles 1980). In total, approximately 280 gasworks providing a public gas supply had been built in The Netherlands; this included new gasworks built to replace gasworks where the site had become too small or poorly located.

Belgium. Belgium provides an interesting microcosm of the development of the gas industry in Europe, because many of the changes which occurred in Belgium were later reflected elsewhere in Europe. It was one of the earliest continental European countries to adopt gas lighting and, like the UK, it also had extensive coal fields. During its history, about 151 public gas undertakings existed in Belgium; a relatively high number given its size and population.

The early story of gas manufacture in Belgium had close ties with the English Cockerill Brothers who had, in 1802, moved with their father to Liège (at the time, part of The Netherlands). In 1814, the Cockerill's built a machine factory at Seraing near Liège, which included an ironworks. Through the English iron founders James Fenton and Matthew Murray, the Cockerill's imported a gas plant which they used to light their Liège hotel in 1818, and then copied and sold to others. They also obtained drawings of other English gasworks in 1819, the plant from which they also went on to copy and manufacture for others wanting gas plant in the region; they even offered guarantees which English companies would not provide (Mac Lean 1977).

The first gasworks in Belgium (and possibly mainland Europe) was established by an association of people led by Jean Baptiste Meefis who established the 'Société de L'éclairage par le Gaz', which was located in the Saint-Roch district close to the River Senne in Brussels. Members of the association had visited London (a plaque commemorates this site in Brussels today, Fig. 2), Manchester and Birmingham, and had seen the gasworks already

Fig. 2. The plaque in Brussels commemorating the first public gasworks on the continent (*Source*: Pieter de Raedt).

established there. It is interesting to note that one of the members of the association was an Englishman and he had persuaded the company to purchase the gas plant from England rather from a local ironworks run by the Cockerill Brothers. The first building in Brussels lit by gas was the New Theatre, for which a great cost-saving was possible (Mac Lean 1977).

Ghent was the second city lit by gas; this was pioneered by the local Architect Louis Roelandt, who requested powers from the King to light the city with oil gas in 1824. The oil gas being distributed via iron pipes under the streets, objections were received but after consulting the local Professor Johann Karl Friedrich, these were overturned. Establishing the company L. Roelandt and Company, they imported oil gas plant from the London-based Taylor and Martineau Company mentioned earlier. On their first tour of Europe, William Congreve, George Thomas Landsmann and the well-known chemist John Frederic Daniell of the Imperial Continental Gas Association requested permission of the King to establish a gas company in Ghent, looking to borrow and invest 6 million guilders in Dutch countries. Due to a local law forbidding foreign companies to provide insurance against fire and maritime damage, they were prevented from doing so. In June 1825, the ICGA were allowed to purchase the company of L. Roelandt and Company, who then became the agent of the IGCA, assisting their future purchases in Belgium which included in 1844 the purchase of the previously mentioned Brussels gas company (Mac Lean 1977).

The lighting of Liège was a long and protracted affair, and involved L. Roelandt and Company, who first offered to light the city in 1823; however, the city objected. L. Roelandt and Company, which came under the ownership of ICGA in 1825, made another offer to light the city; a site for the gasworks was found but the city objected again, complicated because they already had an 18-year contract to light the city by oil lamps. When another site for a gasworks had been found, the IGCA decided it was too busy and deferred the contract until at least 1830.The city was not lit by gas until 1834 (Brabant *et al.* 1939).

Gasworks were then constructed in Ghent (in 1824), Liège, Charleroi and Fontaine L'Eveque (in 1834), and Tournai and Louvain (in 1835). By 1885, there were 79 gasworks serving 115 boroughs and a population of just over 2 million people. The gas-distribution systems covered a greater number of boroughs than the gasworks serving them. The gas-distribution systems at this time were owned by the following companies:

- Compagnie Continentale du Gaz (part of the IGCA);
- Gaz de Saint-Josse-ten-Noode;
- Sociéte' d'Eclairage du Bassin Houiller de Mons et Extensions;
- Gaz et Electricité du Hainaut;
- Compagnie Générale pour l'Eclairage et le Chauffage par le Gaz.

At this time, there were 12 gasworks for 20 boroughs in the Province of Brabant; 20 gasworks and 44 boroughs in the Province of Hainaut; and, within the other provinces, about as many gasworks as boroughs. In 1905, there was a programme initiated by the Société Generale de Gaz et d'Electricité to interconnect the distribution system in southern Belgium, the result of this was that in 1907 the number of gasworks had fallen to 70, but an increase in both the number of boroughs (190) and population (just over 3 million) served. A similar initiative had been undertaken in Brussels where the 19 boroughs there had been supplied from only two gasworks at Forest and Koekelberg belonging to the Compagnie Continentale du Gaz (Brabant *et al.* 1939).

By 1912, the number of gasworks had increased again to 83 due to the creation of new gas companies supplying 368 boroughs and a population of just over 4.1 million. This was a turning point as the gas network then started to become more fully interconnected, and gas manufacturing was then undertaken by larger more economical gasworks. This progress was curtailed by World War I, in which the gasworks at Ypres, Langemark, Passchendaele, Menin, Nieuport, Roulers and Deynze were completely destroyed. Before reconstruction was started, an attempt was made to find a suitable supply of coke oven gas; successfully achieved, coke oven gas started to predominate in the twentieth century. The mass supply of coke oven gas led to the Belgium gas industry being split into two services: (1) gas production and transmission; and (2) gas distribution (Brabant *et al.* 1939). Three high-pressure gas-distribution systems were developed, these were:

- SAVGAZ: Societe pour L'Achat, la Vente et la Distribution du Gaz – Supplying Liège;
- DISTRIGAZ: Societe de Distribution du Gaz – supplying Brabant, Hainaut, East and West Flanders (part of the ICGA);
- Compagnie Regionale de Transport d'Energie – supplying Zeebrugge to Mouscron, Bruge and Courtrai.

By 1938, only the gasworks in Limbourg, Namur and Luxembourg remained in operation. The gas infrastructure during World War II was heavily damaged. This led to a number of surviving gasworks restarting production, but this was only short lived, as most of the by-product coke ovens escaped the war largely intact and were again able to be used to provide a base-load gas supply. Whilst coke ovens did predominate, water gas plants at some gasworks were retained even after coal-gas production had ceased. Water gas became a popular method to rapidly supplement coke oven gas at times of peak demand. The larger gas companies were also early suppliers of electricity, and obtained the necessary concessions (Brabant *et al.* 1939).

As the availability of coal started to decrease, the gas industry in Belgium started to look for alternatives supplies of gas. It extracted and utilized mine gas (methane), and also experimented with the use of underground coal gasification, although this was not commercialized. With the discovery of natural gas in The Netherlands, Belgium switched to natural gas, taking a gas supply from The Netherlands. The pre-existing gas transmission system, which made the conversion from town gas to natural gas an easier task, was completed in the early 1970s.

The Imperial Continental Gas Association played a long and important role in the gas industry in Belgium, from its purchase of the gasworks in Ghent until its eventual demerger into the Calor Group and Contibel in 1974. Distrigas, which was founded in 1929 by the ICGA, became the main gas distributor in Belgium; the gas transmission business exists today in a modernized form under the name of Fluxys.

France. The development of the gas industry in France cannot be discussed without mentioning Philippe Lebon, who is recognized as the father of the French gas industry and whose exploits are mentioned in a previous section of this paper. Friedrich Winzer had witnessed Lebon's early work in Paris, which inspired his activities to promote gas in Britain. After leaving Britain, Winsor established a gas company in Paris which lit the Passage de Panoramas and Luxembourg Palace (Stoss 1939).

British companies played an important role in introducing manufactured gas to France. Aaron Manby, a Staffordshire engineer who was a partner in the Horseley ironworks, played an important early role, taking out a patent for gas lighting in France on 1821. In 1819, Manby set up an engineering works with Daniel Wilson at Charenton, near Paris, which made France largely able to build its own steam engines, much to the annoyance of the British. In 1823, Manby and Wilson were among the first to obtain concessions to provide gas lighting for several streets in Paris. His son Charles Manby undertook the construction of the gasworks located at Les Ternes in 1825 and the operation of laying the gas mains in the streets. Their company, which was the first successful gas company after the failure of Winsor's initial attempt to establish a gas company in Paris, was known as the Compagnie Anglaise (English Company), a name which survived until 1847 when all the Paris gasworks were amalgamated (Anon 1885). The first provincial gasworks were established in 1825; by 1837, a total 14 gasworks had been established; and by 1845 this had grown to 100 (Stoss 1939).

As the French gas industry developed, French gas engineers and companies in turn exported their technology and experience abroad. For example, Charles Lebon was important in the development of the original gasworks in Barcelona (Moyano-Jiménez & Thomas 2016). French gas engineers were also responsible for the gasworks constructed in Greece, including the well-preserved gasworks in Athens. The French gas industry grew rapidly and by about 1880 around 1000 gasworks had been built across the country, at all scales from large cities to small towns.

Similarly to Belgium, those gasworks that were in the World War I battlefields of northern France were heavily damaged and required reconstruction or replacement where possible with coke oven gas. The IGCA had been very badly affected by World War I as many of their gasworks were in this area. They had, at some point, provided a gas supply to 16 towns and cities during their history, the first being Bordeaux, and in the 1920s a decision was made to amalgamate their French gasworks holdings into the Compagnie Continentale du Gaz (Anon 1974).

Localized gas distribution networks were established in the north of France, around Paris, and in the Massif Central where the coke ovens had been established to support the steel industry. By 1938, prior to World War II, the number of gasworks were 700, many of the smaller gasworks closing on being taken over and receiving a cheaper gas supply from a larger neighbouring gas companies.

The French gas industry was again damaged during World War II and was in a poor state at the end of the war. Similarly to Britain, a decision was made to nationalize the French gas industry, which occurred earlier in 1946. Most of the French gasworks were absorbed into the new organization 'Gaz de France', although small gas companies with less than 100 customers and some large regional gas companies (e.g. Bordeaux) were allowed to remain independent. On nationalization, in 1946, there were 546 gasworks, which had belonged to about 300 companies, operated by Gaz de France; this had reduced to 210 gasworks by 1959 (Le Guellec 1959). An increasing diversity of gas supplies was introduced, with coke oven gas piped in from the Saar region in eastern France and petroleum-based gases from refineries being used, such that, in 1959, 38% of gas was being purchased from third parties (Le Guellec 1959).

Given the limited coal resources available in France, the French gas industry started to look for alternative sources of gas. French gas engineers made important developments in the catalytic gasification of oil. A collaboration between the French 'Office National Industriel de l'Azote' with the 'Gaz à l'Eau et Gaz Industriels' developed the ONIA-GEGI process, and Gaz de France and Messrs Stein and Roubaix developed the Microsimplex process. They were both low-pressure processes which gasified oil using a hot nickel catalyst and found use in Britain (Ward 1965).

The discovery of the Lacq natural gas field in 1951, which was located between Pau and Orthez, heralded a major change in the French gas industry; it was put into operation in 1957. France had also found large gas reserves in Algeria, and the LNG tanker *Jules Verne* was commissioned to import LNG between Algeria and Le Havre. By 1961, France was using gas from a mixture of sources, including gas from coal at gasworks, coke ovens and coal mines, gas from oil at gasworks, gas produced from refinery by-products, butane and propane, and natural gas. The arrival of natural gas led to the construction of major gas transmission systems across France connecting major population centres and, as a result, natural gas replaced other forms of gas (Blears 1966). Therefore, gasworks started to close, with the last one closing in Belfort (Franche Comté) in 1971.

Germany. The first example of German interest in coal distillation can be traced back to 19 August 1681, when J. Becher and H. Serle applied for a joint British patent (No. 214), describing the evolution of combustible gas during distillation of coal. It was, however, a long time until the first example of gas lighting was attempted in Germany, by the Bavarian Pharmacist J.G. Pickel who, in 1786, lit his laboratory in Wurzburg with gas derived from the dry distillation of bones (Schuster 1939).

Andreas Zacharias Winzler, mentioned earlier in this paper, heralded from Unlingen in southern Germany (at the time part of Further Austria). Winzler wrote the first German treatise on gas manufacture called *Die Thermolampe in Deutschland* (Winzler 1803) and inspired much of the later development of the thermolampe in continental Europe. Two other names mentioned earlier in this text were also key to the introduction of gas lighting in Britain: Friedrich Christian Accum and Friedrich Albrecht Winzer, both of whom came from Germany.

Wilhelm August Lampadius is recognized as the being the first true exponent of gas lighting in Germany. In 1812, he built in his apartment on Fischergasse, in Freiberg, a prototype gasworks which supplied gas light to a lantern outside his house. In 1815, he went on to build the first factory gasworks in Germany at the Saxon Amalgamier-Werken at the Halsbrucke in Frieberg. Whilst Lampadius was unsuccessful in persuading Frieburg to adopt street lighting, he went on to build small gas plants across Germany, and notably in Dresden Castle and Opera House (Schuster 1939).

The establishment of the first public gasworks in Germany fell to the IGCA, who built gasworks at Berlin and Hannover between 1825 and 1826. Possibly inspired by Lampadius's work in Dresden, Rudolf Sigismund Blochmann founded the first German public gasworks which was independent of foreign control in 1828, this was built in Dresden. Gas was initially used to light popular areas of the city centre, but later spread much wider. Blochmann used the Lauchhammer factory to build his gas plant and imported any additional items from Britain (Wengenroth 2013).

In 1828, Johann Knoblauch and Johann Schiele built an oil gasworks to light the city of Frankfort-am-Main, the gasworks only lasted 6 months before it closed. It acquired a licence from the IGCA and switched to coal gas, reopening in 1829 under the name Frankfurter Gasbereitungsgesellschaft. In 1844, the IGCA obtained a 25-year concession for the sole production of gas from coal from the city of Frankfurt, sparking a period of competition until the two firms merged in 1909 (Wengenroth 2013).

Blochmann built other gasworks in Aue (in 1834), in Zwickau (in 1835), and in the nearby Saxon city of Leipzig between 1837 and 1838. The IGCA also built a gasworks in Aachen (in 1839) and in Cologne (in 1840). Blochmann then was commissioned to build a municipal gasworks for Berlin to rival the British-owned gasworks constructed by the IGCA. Blochmann and his son Georg Moritz Sigismund Blochmann went on to build further gasworks in Germany, and abroad in Prague and Wroclaw. Other notable German cities lit by gas were Stuttgart (in 1845), Hamburg (in 1846), Nuremberg (in 1847) and Munich (in 1850) (Schuster 1939). By 1850 there were still only 23 public gasworks built in Germany compared to 700 gas undertakings which were in operation in Britain by this time (Schilling & Knapp 1878).

Around 1850, Germany attempted to replace coal with wood; this move was particularly popular in areas that lacked coal resources. Gasworks were built which used wood in Bayreuth and Darmstadt in 1854, and by 1870 up to 60 such wood-utilizing gasworks had been constructed, with Riechenhall in Bavaria the last to be constructed. Their demise was largely brought about by the growing railway network, which could rapidly transport coal into areas where it would otherwise have been too difficult or too expensive. Wood did make a return to the German gas industry during the world wars as a matter of necessity, as it did in other countries during the war.

One of the key German gas companies was the Deutsche Continental-Gas-Gesellschaft A.G. otherwise known as Dessau Gas Co. or Contigas; it was established in Dessau on 12 March 1855 by Duke Leopold IV Friedrich of Anhalt-Dessau. The company initially established a gasworks in Dessau in 1856, lighting the city street in that year. This was followed by gasworks in several dozen cities in Germany and abroad: for example, in Mönchengladbach, Magdeburg, Frankfurt (Oder), Mülheim an der Ruhr, Potsdam, Warsaw and Lviv (Körting 1963). By 1870 there were over 340 public gasworks in Germany making gas from coal, wood, peat and other carbonaceous materials (Peebles 1980).

Germany played a very important role in the development of gas technology. The German Bischof undertook the early pioneering work on the development of the gas producer. Bischof, from Magdeburg in the Saxony-Anhalt region of Germany, constructed the first gas producer in 1839. The gas producer was a simple system whereby coal would be incompletely combusted, producing a gas primarily composed of carbon monoxide and carbon dioxide. This could then be mixed with secondary air and burnt in the position required. In the case of gas manufacture, this was around the retorts: it allowed higher temperatures of carbonization to be reached. In 1857, Frederick Siemens developed a combined gas producer and the regenerative furnace. This was a furnace which could operate at a high temperature by using the regenerative preheating of air. In a traditional furnace, a large part of the heat derived from combustion was lost up the chimney. The regenerative furnace used the hot gases to heat a brick-lined chamber (a regenerator) which absorbed the heat. The hot chamber was then used to heat the incoming air for combustion. Other regenerator designs were developed in Germany, including those developed by Schilling and Klonne which used the hot exit gases to continually heat incoming air (Thomas 2014*a*).

In 1855, Robert Wilhelm von Bunsen, a famous German chemist working in Heidelberg, invented the atmospheric gas burner known to many as the 'Bunsen burner'. This revolutionary design premixed the air and gas prior to entering the burner; this allowed the flame to burn with an intense heat never previously achieved. It made more efficient use of the gas and allowed a greater range of uses, especially in commercial and domestic heating applications. It was also important in the later development of the gas mantle which used the aerated burner to heat the incandescent mantle to produce a much brighter light than previously obtained (Thomas 2014*a*).

The German gas industry grew differently to that of Britain, driven primarily by the availability of coal; although parts of Germany had plentiful supplies of coal, it was short of good gas coal. This had been imported from Britain in ships along the coast and the great rivers, with Ruhr bunker coal exported in return. As this arrangement started to wane, German gas engineers developed technologies to better carbonize German coals. A range of chamber oven designs were developed, these included the intermittent vertical chamber and inclined chamber oven. The intermittent vertical chamber was developed by the Dassau Gas Company and improved by the Stettiner Chamotte-Fabrik, it had the appearance of a very large vertical retort (it was equivalent to three vertical retorts). The inclined chamber oven was original developed in Munich (known as the Munich oven) and also improved by Stettiner Chamotte Fabrik, this had the appearance of a very large inclined retort. An example of this type of gas manufacturing plant at Wismar, Germany is shown in Figure 3. Both were able to handle a wider range of coal, including some deemed poor for gas making and both took up less space than horizontal retorts or coke ovens (Anon 1928). This explains in part why the vertical retort developed in Germany was little employed there, with only three gasworks using vertical retorts in Germany in 1949 (Korting 1949).

There are various records of the number of gasworks in Germany. This is difficult as the boundaries of the country have changed over time. At the end of the nineteenth century, one record suggests there were 1200 gasworks in Germany (Anon 1958). From a review of *Gas World* year books and associated publications from 1898 to 1938, there were 992 cities and towns lit by gas. Most of these numbers would have been small towns. Many of the large towns and cities, however, are known to have had at least two gasworks, and a more accurate figure would be likely to be closer to 1700 public gasworks. The peak in the number of gasworks occurred in the first decade of the twentieth century, before the gas companies started to amalgamate their gas distribution networks and close down the smaller gasworks.

With the predominance of locally available coking coals and the growth of the German steel industry and its associated by-product coke ovens, this provided a surplus of coke oven gas. By 1926 this had stimulated the need for a gas transmission system for the long-distance transport of coke oven gas at high pressure in Germany. This allowed it to be shipped and transported to where it was needed (Korting 1949). By 1957 this gas transmission system was 4700 km long and the number of gasworks in western Germany had fallen to 288, mainly due to the interlinking of the gas network, making some of the smaller gasworks obsolete. As with the rest of Europe, Germany was also looking to oil and natural gas to meet its future gas demand (Anon 1860).

It should be noted that whilst gas manufacturing continued following World War II, the partitioning of the country led to two different approaches being

Fig. 3. The Stettiner Chamotte-Fabrik inclined chamber oven at Wismar, Germany.

taken. West Germany built a gas network around the coking industry of the Ruhr and Saar, and East Germany became heavily dependent on brown coals (lignite) and processes such as those developed by Lurgi to produce gas under higher pressures from these low-grade coals. The last gasworks to close down was located in the city of Leipzig, which had been formerly in East Germany. Like the rest of Europe, Germany is now using natural gas.

Poland. At the time the gas industry was developing in the UK and Western Europe, Poland was partitioned between three countries – Russia, Prussia and Austria – and the history of the gas industry in Poland is therefore fragmented. The major limitation on the construction of gasworks was the financial cost of construction, but Prussian-controlled areas benefited from an already advanced German gas industry (Budziński *et al.* 2002). Within the current boundary of Poland there were believed to have been about 230 municipal gasworks. Many of the larger cities had more than one gasworks, the old gasworks being replaced when they could no longer be expanded. The country was split between large advanced gasworks in the cities and smaller simple country gasworks as may have been found in Britain, France or Germany. German gas companies were very active in Poland: in particular, the German Continental Gas Association of Dessau (Barszcz 2004). The growth of the Polish gas industry benefited from the country's plentiful supplies of coal, which were first exploited in 1767 from a mine at Szczakowa, and the long history of Poland's metal industry.

One of the pioneers of manufactured gas in Poland was Professor Charles Mohr who worked at the Institute of Technology in Krakow. In 1830, he built a small experimental gas plant for teaching his students. This gas plant was used to light a few gas lanterns on Gołębia Street, in the old town of Krakow. Other demonstrations of gas lighting were undertaken across Poland, this included the lighting of Castle Square and the University Yard in Warsaw. These were lit using gas manufactured at the Solec Machine Factory and transported in 'reservoirs' to the gas lanterns (Budziński *et al.* 2002).

The first public gasworks in Poland was built in the port town of Szczecin and Wroclaw between 1846 and 1848 (Palacz 2006). Initially, the town of Szczecin had approached the ICGA to provide a gas supply, but talks failed. The German engineer Georg Moritz Sigmund Blochmann of Dresden was then approached and an agreement was reached to light the town with gas. The gas supply started on 23 April 1848, when the first gas lanterns were lit in the city. The gasworks were gradually expanded to meet increasing gas demand and were damaged during World War II by Allied bombing. The gasworks finally closed in 1987 when the gas supply was switched to natural gas. Gas manufacture was first demonstrated in Wroclaw in 1843, when gas lighting was supplied to the 'Golden Goose' restaurant by Heinrich Mainecke. Shortly afterwards, in 1847, a gasworks was built on Rainbow Street and the city was lit with 857 gas lights. It was the first of four gasworks to be built in the city. The last and largest was built in 1906 on Gas Tarnogaju Street, which closed in 1990 (Zelecki 1995).

The expansion of the coke-making industry in Poland led to a supply of surplus coke oven gas becoming available. Gas transmission systems were constructed to supply gas to customers and this invariably led to some gasworks closing or reducing production. During World War II, many of the gasworks in Poland were destroyed or damaged, the programme of rebuilding lasted until at least the end of the 1950s. Another notable challenge was the effect of Soviet control of Poland and the resulting nationalization of the gas industry in 1947; private enterprise was abolished and new management structures put in place, all new plant being sourced from Soviet-controlled sources. Poland followed the same trends as the other countries mentioned: that is, the gas was used initially for street lighting, and over time began to be widely used for space heating, water heating, cooking and other heating-based applications.

Around 1949, the massive increase in the iron industry in Silesia created a vast coke oven gas transmission network which operated for about 50 years supplying gas to the cities of Krakow, Lodz and Warsaw, during which time it had transported about 65 billion cubic metres of gas. The Polish manufactured gas industry was at its peak in 1969, after which it went into gradual decline. The gasworks at Bystrzyca Kłodzka was the last to close in 1998, thus ending the 150-year history of manufactured gas in Poland. Poland remains one of the best locations to find former gasworks structures in Europe, with three museums and numerous other gasworks buildings, both derelict and redeveloped, still in existence. The following provides a brief history of manufactured gas in a few Polish towns and cities.

The city of Krakow signed a 25-year gas lighting contract with the Dessau Continental Gas Association in 1855. A gasworks was built in Kazimierz and the streets were lit by gas in December 1857. Following ongoing conflict between the city and Dessau, their contract was terminated in 1886. The gasworks then transferred to municipal ownership. A new gasworks was planned to be built in 1910, but the advent of World War I scrapped these plans and the existing gasworks were rebuilt, as they were again following damage caused during World War II. Coke oven gas started to be supplied in 1955 and further coke oven gas was supplied in 1968, leading to the closure of the Kazimiers gasworks. Gas was provided from coke ovens until conversion to natural gas in 1982.

In 1856, the Warsaw City Hall awarded a long-term licence to the Dessau Continental Gas Association for the construction and operation of the Warsaw gasworks. They built a gasworks on Ludna Street which was operated by 'Gas Company Powiśle'. The agreement, which was signed on 1 May 1856, gave the licence holder an exclusive right to light the streets of Warsaw for a period of 25 years. In December 1856, Warsaw's 92 gas lanterns lit the streets for the very first time, the number doubling a year later. By 1858, gas lighting was installed at the Polish Bank, railway workshops, the Warsaw arsenal, hospitals, army barracks and private businesses. With an increasing demand for gas, a second gasworks was opened in 1888, located on Dworska Street in the Wola district (Fig. 4). The gas network and consumption continued to grow and, in 1892, a chemical works was opened to utilize the by-products produced. In 1925, the ownership of the gasworks switched from the Dessau Continental Gas Association to the Municipality of Warsaw; the area around the gasworks became known as the 'Gasworks City'. The Ludna Street gasworks closed in 1930 (Barszcz 2004).

Escaping damage during World War I, the municipal gasworks were critically damaged in World War II and survived an attempt by the retreating German army to destroy the gasworks. Gas

Fig. 4. The former gasworks in Warsaw, showing the laboratory and gasholder house in 1936.

production was restored in June 1945. In 1961, Warsaw started to receive a supply of coke gas from Upper Silesia, leading to the gradual redundancy of the gas-manufacturing plant. Not long after, in 1966, natural gas started to be supplied and a mixed gas was supplied until the conversion to natural gas was completed in 1978. Many of the original gasworks buildings still remain, some of which form the impressive Warsaw Gas Museum.

Gdansk was first lit in 1853, with the gasworks built on the left bank of the Motlawa River. The increase in demand for gas and the limited space available for gas production led to a new gasworks being built on Walowa Street in 1904. These gasworks operated until 1991, when they closed and Gdansk switched to natural gas.

The gasworks in the city of Poznan were commissioned by Mayor Eugene Naumann and designed by the British architect John Moore. Gas production started on 14 November 1856. Coal was initially imported by ship from England to Szczecin, where it was transferred on barges. This was later replaced by Silesian coal transported by rail. By the end of 1856, Poznan had 1500 gas lamps. The end of the nineteenth and early twentieth century was a difficult time for the gas company, facing competition from the rival electricity company. This required the gas plant to be improved to ensure that the quality of the gas was maintained. In 1966, a second gasworks was built in Poznan – a modern, automated plant located in Dwugazu Street, Gdynia. In 1973, after 116 years of operation, the original gasworks closed. In the 1980s, Poznan received a supply of natural gas and the second gasworks closed (Palacz 2006).

Switzerland. The first gasworks built in Switzerland was built in Berne in 1841, followed by the construction of gasworks at Geneva (in 1844), Lausanne (in 1848), Basle (in 1852), Zurich (in 1852), St Gall (in 1857) and Vevey-Montreux (in 1861). Gasworks were built across the whole of the country but they predominated within the low-lying and more densely populated industrial areas of Switzerland (Fig. 4). The gas industry preceded the construction of railways in Switzerland, with the first railway not opening until 1847 (Tobler 1939).

The first of the Swiss gasworks produced gas from wood and peat, as coal was both difficult and expensive to obtain. This was due to the lack of rail infrastructure. As the rail infrastructure developed, however, so did the number of gasworks, as shown in Table 2. Within the *Gas World* year books and the Gas Industry Register listings between 1898 and 1937, 103 towns were listed as

Table 2. *The number of public gasworks constructed in Switzerland up to 1938*

Date period	Number of gasworks constructed
Up to 1860	17
1861–70	19
1871–80	17
1881–90	3
1891–1900	4
1901–10	26
1911–20	15
1920–38	–
Total	101

receiving a gas supply, the discrepancy due to some towns receiving a gas supply from gasworks outside the town. In the 1920s, the Swiss started to invest in expanding their distribution pipelines (up to 30 miles), this allowed the gradual rationalization of smaller gasworks, replaced with cheaper gas supplies from larger more efficient gasworks. Switzerland did not build long-distance gas transmission networks during the coal-gas era. One issue encountered in Switzerland which was seldom encountered to the same degree elsewhere was the relative height difference of the gas network. One of the most extreme examples was the gasworks at Sion (450 m above sea level), which distributed gas to Montana, 1000 m higher up the valley. Given that coal gas is lighter than air, this is problematic and the gas network was designed so the flow of gas was carefully regulated (Tobler 1939).

The Swiss gas industry consisted mainly of small gas undertakings, although a few big undertakings were based in cities such as Geneva, Bern and Basel. Originally, most of the Swiss gas undertakings were private ventures, which took the risk of financial failure. As they grew, most of them transferred to ownership by the municipalities. In 1947, 54 of the gas undertaking were municipally owned and 20 were privately owned, but the private gasworks were generally the smaller undertakings (Zollikofer 1949).

Despite the restrictions of both world wars and often having tight finances, the Swiss gasworks managed to maintain efficient gas manufacture with up-to-date plant. Switzerland also became self-sufficient in manufacturing its own gas-making plant. As well as supplying gas, the Swiss gas industry had a key role in the provision of metallurgical coke to the metal industry. Two of the larger gasworks were equipped with coke ovens. Due to a combination of variable coal supplies and the importance of coke manufacture, many of the Swiss gasworks were equipped with chamber ovens which were better able to produce high-quality coke. The larger gasworks employed vertical chamber ovens and the medium-sized gasworks used horizontal chamber ovens. Small gasworks used horizontal retorts and work of all scales employed a high-degree of mechanization. High-pressure gasholders (bullets and spheres) had also become popular within Switzerland, replacing traditional gasholders as early as 1947 (Zollikofer 1949).

Carburetted water gas (CWG) plants were introduced and widely used after 1900, and the recovery of benzole was widely practised. The Swiss gas industry had tough competition from the emerging hydroelectric power industry. Both industries faced differing challenges: hydroelectric was very expensive to construct and consumed very large areas of land, and, although water was generally abundant and locally available, transferring the power from the mountainous areas to the areas where it was required was a major feat; gasworks took up less space, and could produce and store gas near to where it was required, but Switzerland had no useable gas coals and so became dependent on the importation of coals from other European coal-producing nations, and so dependent on price, quality and availability. Britain became a key supplier of coal; this, however, became particularly problematic during periods of war in Europe. In later years, for reliability reasons, coal was imported from the USA (Zollikofer 1949).

By 1958 there were still 71 gasworks producing gas. The 1960s witnessed the demise of the production of gas from coal and, as with other countries described, Switzerland moved towards the production of reformed gas from petroleum products. This was later replaced by conversion to natural gas which was imported into Switzerland via pipelines, as was the trend across much of Europe (Anon 1958).

Southern Europe

Italy. Italy adopted gas lighting relatively late on, this was probably due to the limited local coal supplies and the time taken for the technology to transfer gradually eastwards from Great Britain. The first example of gas lighting is regarded as the lighting of a small theatre in Milan by Chevalier Aldani using a Lebon thermolamp in 1818. In the same year, demonstrations of street lighting in Milan and Florence were undertaken by Carlotti and Ridolfi, respectively. Limited further development occurred until 1838 when the first factory was lit by gas in Turin, which led to the formation of the Italian Gas Lighting Company. In 1839, the Municipality of Venice agreed a contract with the French company Compagnie Lyonnaise du Gaz for the lighting of the city. Some remains of the Venice gasworks still stand today, including its gasholders. The cities of Milan, Verona, Genoa, Florence, Livorno and Palermo were all lit by gas in 1845. The city of Turin had been fully lit by this time but Rome was not lit by gas until 1853 (Rampone 1939).

Gas was initially used for the lighting of streets and public buildings, with slow domestic uptake, but gradually it became popular for lighting and cooking in the home, especially in northern Italy. By the end of the nineteenth century there were 160 gasworks in Italy. The gasworks were built and operated mainly by foreign companies, primarily English (e.g. Tuscan Gas Company) and French, although some were owned by municipalities or Italian companies. Between 1900 and 1930, there was a rapid transfer of the gas undertakings from foreign companies to municipalities or private Italian companies (Rampone 1939).

By the end of World War I, the Italian gas industry was in a poor condition and great investment was needed to replace obsolete plant, with installations having to be completely replaced in Florence. The industry grew rapidly during the interwar period. By 1938 there were 190 companies producing and distributing gas; of these, 42 were operated by the municipality and 148 were operated by private companies under a negotiated concession. These gasworks were distributing about 600 million cubic metres of gas between them, primarily in the industrial cities of northern Italy. The largest gasworks were located in Milan, Rome, Turin, Trieste, Florence and Venice (Rampone 1939).

The gasworks evolved from using retort-based systems initially to using the chamber-oven-based systems developed and used elsewhere in continental Europe. By 1938 the large Italian gasworks were equipped with up-to-date vertical and horizontal chamber oven processes, with some adopting continuous and intermittent systems, which were highly mechanized and efficient. Italy had limited indigenous coal that was sourced from Sardinia (Pozzi 2010) and it was mixed with coal imported from Germany, Poland or Britain. Use was made of the by-products with coke grading, tar distillation, sulphate of ammonia production and benzole production widely practised. Italy also had a good market for coke, especially during the cold winters experienced in the north of the country (Rampone 1939).

Gas was being distributed at high pressure over long distances by 1938, with over 40 towns being supplied with gas produced in another town. Use had also been made of surplus coke oven gas in Venice, Trieste, Vado Ligure and Genoa to supplement or replace gasworks supplies. Italy was one of the first European countries to identify significant amounts of natural gas within its borders in the 1930s and 1940s, the largest discoveries being made by AGIP in the Po Valley in 1946. By 1958 Italy was using gas from the following sources in this order of preference: natural gas, coke oven gas, liquefied petroleum gas, blast furnace gas, gasworks and oil refinery gas (Pozzi 2010). Natural gas soon replaced coal gas as the gas transmission infrastructure was built. The 1960s and 1970s marked the switch to natural gas which was either produced in Italy or, as local resources were depleted, was imported as LNG from Libya or The Netherlands and Russia via pipeline.

Spain. Gas was introduced into Spain by British and French gas engineers. Barcelona was the first Spanish town or city to be lit by gas, but the uptake of gas was slow across the rest of the country. Early experiments with gas lighting were undertaken in the Llotja del Mar, which increased the interest in gas lighting. As a result, the city council decided to open a public tender to provide gas lighting of the city in 1841. This contract was won by Frenchman Charles Lebon with financial backing from the Barcelona-based bankers Pedro Gil and his three sons. Charles Lebon constructed the factory in the Barceloneta district of Barcelona, on land owned by the Gil family. The first gas public lights were installed in 1842, becoming the first public utility service in Barcelona, predating water distribution by 25 years. La Sociedad Catalana para el Alumbrado por Gas, which is translated as ‘the Catalan Society for gas light’ also known as ‘La Catalana’, was incorporated on 23 January 1843. It was one of the first companies listed on the Spanish stock exchange. The society constructed a network of gas mains throughout the city supplying gas for street lighting and domestic customers (Moyano-Jiménez & Thomas 2016). Lebon continued on to Valencia where he established the gas company Sociedad Valenciana para el Alumbrado de Valencia in 1843 (Garcia de la Fuente 1984). Cadiz was the next town to be lit by gas with a contract initially awarded to Federico Gregory in 1845 (Fernández-Paradas 2015). Despite attempts which stretched back to 1831, gas street lighting was eventually introduced to Madrid in 1846. It was an English company (Manby & Partington), who acquired the concession and created the ‘La Sociedad Madrileña para el Alumbrado de Gas’ with José de Salamanca and other partners (Simón-Palmer 1989). This was the same Manby family that had been involved in the creation of the first gasworks in Paris, mentioned earlier. A gas company in Bilbao was established via the French Compagnie Lyonnaise du Gaz which obtained a concession through Louis Laty in 1845. The Société pour l’Éclairage du Gaz de la Ville of Bilbao was established in 1846 and started to supply gas in 1847.

The British engineer William Richards played a key role in the development of the gas industry in Spain. In 1848, he was employed as the manager of the gasworks in Barcelona by La Catalana, a job he held until 1861 when he returned home to Britain. During this time Richard was responsible for the construction of the gasworks and the laying of gas pipes at Sabadell (in 1851–52), Villanova (in 1852), Reus (in 1854), Tarragona (in 1858) and Manresa (in 1858) (Moyano-Jiménez & Thomas 2016).

By 1861 there were 25 public gasworks in Spain, with 11 of these located in Catalonia, equivalent to 44%. This was part of an uneven expansion as Catalonia always had the majority of the gas undertakings in Spain. By 1937 there were 54 public gasworks in Spain and 25 of these were located in Catalonia, eight were located in the provinces of the northern Cantabrian coast, four in Valencia, eight in Andalucía and the south, and four on the Balearic and Canary Island. This left only five gasworks within the interior of Spain, located in Madrid, Zaragoza,

Valladolid, Burgos and Logrono. Part of the reason for this poor uptake in land was the lack of infrastructure (e.g. railways or canals) to transport the coal to the gasworks; it was also hampered by a lack of funding and infrastructure to transport.

The majority of these gasworks were established in the second half of the nineteenth century, but the industry came up against considerable opposition from electricity, principally hydroelectricity. The competition posed by electricity in the late nineteenth century led to a feeling that gas had no future, and by 1914 the competition had meant that 14 gasworks had closed. Many of the gas companies started to produce electricity and their names changed to reflect this, La Sociedad Catalana para el Alumbrado por Gas changing its name to Catalana de Gas y Electricidad in 1912. Unlike many other parts of Europe, the Spanish gas industry remained mostly privately owned throughout its history.

During World War I, the British coals that the Spanish gas industry had relied upon became both hard to obtain and expensive. Although Spain had its own natural coal resources, it was expensive and unsuited to gas manufacture. The gas industry was particularly badly affected during the Spanish Civil War of the 1930s, especially in Catalonia where some of the gasworks were damaged during conflict. The gas industry struggled following the war, with the Franco government adopting a policy of national self-sufficiency, with closed borders for raw materials, technology and finance. This policy made it hard to maintain old equipment and investment in new, and the available Spanish coals were not best suited to gas manufacture.

In 1963, gas production in Spain started to switch from coal-gas manufacture to the catalytic cracking of light naphtha available from refineries. This era of producing gas from oil was relatively short lived, and in 1969 the era of natural gas started in Spain. Catalana de Gas y Electricidad, in collaboration with Exxon, began importing LNG from Libya and Algeria. The first Spanish methane tanker, *Laietà*, began importing LNG to a regasification plant built in Barcelona. Throughout its history there had been in total 91 public gasworks which had operated in Spain.

The Nordic countries

Finland. Only the cities of Helsinki, Vyborg and Turku in Finland had public gas supplies, these were originally used for gas lighting. A number of private gasworks were constructed to light factories prior to the establishment of the public gasworks. The first factory to be lit by gas was a cotton factory owned by Rauch and Nottbeck in the city of Tampere in southern Finland. Gas lighting was introduced by the factory's English technical director John Barker, after the first factory had burnt down in 1841. The factory gas lighting was supplied from an oil gas plant. In partnership with August Wilhelm Wahren, Barker opened his own factory in 1850 in the city of Turku. Together they visited England, viewing the water gas plant developed by Stephen White; this plant, which produced water gas that could be enriched with resin, grease, tar or oil, was chosen for their gasworks in preference to more conventional coal-based systems. This was because they believed it could produce a brighter light at a lower cost and was also able to use wood, which was plentiful in the area. The gasworks operated until Turku received its own gasworks in 1862.White's gas plant was also installed in another factory belonging to Wahren in the town of Forssa (1851) and also at the Wicksberg's weaving mill in the southern Finnish town of Tammela (Kaukovalta 1934).

The most notable figure in the creation of the Finnish gas industry was Gustaf Magnus Schwartz, the son of a German merchant, who was born in Helsinki when Finland was still part of Sweden. He studied mining in Turku, later obtaining a job at the Mining Board of Control, becoming its Director General. In 1809, he was also appointed Professor of Physics and Technology at the Royal Academy of Sciences. A prolific inventor, he devised a system of producing gas from the oil distilled from wood tar, which he used to light buildings in Stockholm in 1821 (Herranen 1985).

The first five public gas lights in Helsinki were illuminated on 14 November 1860; for the next 40 years the gasworks were privately owned. The Helsinki gasworks manufactured gas from wood until 1882, because it was cheap and readily available (Herranen 1985). The first buildings to be lit by gas were the Staudinger building and the new theatre (Wuolle 1936). The original gasworks were located in the area of the main post office building, in the city centre. When the city took over the gasworks on 1 October 1900, they decided to move to a site called Suvilahti. The gasworks in Suvilahti started to manufacture gas in August 1910 and were in operation until 1994, having undergoing a number of improvements (Wuolle 1936). By 1943 gas lighting had been almost totally replaced by electric lighting. Manufactured gas was now mainly used for cookers and as a source of heat for buildings. In the 1940s, prepayment meters were introduced to over 50 000 properties in and around Helsinki, bringing gas to a wider population who could previously not afford to install gas. In 1991, the Helsinki gas network switched to natural gas. The main uses for natural gas now in Helsinki are in combined heat and power production (CHP) and industry.

The gasworks in Vyborg and Turku manufactured gas from coal. The gas company in Turku was originally a private company, which first lit the town on 21 November 1862. Gas was used for lighting until the end of the 1800s, when gas started to be used for heating as well. The gas company transferred to municipal ownership after around 1910. Gas lighting started to be in use in private homes, and by the end of the 1800s gas was used for heating and shortly afterwards for cooking. To cope with increasing demand, the construction of a new gasworks started in 1912, with the construction of a new housed gasholder which still stands today (Wuolle 1936). Gas manufacturing ceased in Turku in 1973. In Vyborg, the first gas lights were lit on 15 November 1860. The gasworks were gradually enlarged and old plant updated, with the gasworks undergoing a major rebuild in the first half of the 1930s. The last gasometer was demolished when natural gas reached the town in 1974.

Sweden. Experiments with Lebon's well-known 'thermolamp' had been started in Stockholm as early as 1802, but did not produce any practical results. In 1819, the aforementioned Gustaf Schwartz designed a gasworks which used oil distilled from wood tar, although coal could also be used. Several of the gasworks built by Schwartz worked for a long time, one operating as late as 1875, long after the municipal gasworks had been established. The relatively late development of the gas Industry in Sweden was at least partly due to the establishment of several small Swedish-built gasworks in the capital, which produced their gas from wood or oil and supplied gas for illumination purposes.

The first public gasworks established in Sweden were built in Gothenburg in 1846 (Hultman 1939), and in a period of 20 years following this date many other gasworks were constructed. This included those built at Norrkoping (in 1851), Stockholm (in 1853), Malmo (in 1854), Landskrona (in 1856) and Uddevalla (in 1858). The Swedish gas industry suffered during the two world wars when obtaining coal became very difficult due to war restrictions as Sweden had no native coal suitable for gas production. This meant that substitutes like wood and peat had to be used to replace the coal. Where wood was used, the gas produced was very rich in carbon dioxide. It was therefore passed through a layer of incandescent charcoal, which reduced the carbon dioxide to combustible but toxic monoxide. The charcoal, however, also caused a partial breakdown of other products of carbonization, which formed acetic acid. Coal was used to supplement the wood in the retort in order to neutralize the remainder of the acetic acid, and so protect the furnace fittings and pipelines. Whilst wood and peat were carbonized in both horizontal and vertical retorts, it was not used in chamber ovens (Sjolund 1949).

The effect of World War I was that at the end of 1918 about 44% of the materials carbonized by Swedish gasworks consisted of wood, peat, coke and shale (with the greater part of this being wood). The remaining 56% used was coal. The Swedish gasworks had until 1918 almost exclusively used horizontal, vertical or inclined retort settings. In that year, however, the Stockholm gasworks started to use the first horizontal chamber oven plant, proceeded in 1925 by the first vertical chamber oven. By 1938 eight of the larger gasworks and four of the smaller gasworks had been equipped with horizontal chamber ovens. Vertical chamber oven plants were installed on 16 of the large and medium-sized gasworks. Only the smallest gasworks retained horizontal retorts (Sjolund 1949).

The most important driver for the Swedish gas industry was the demand by the municipal central heating plants for the maximum production of a high strength coke, which had to be of uniform quality to ensure efficient operation. As a result, by 1934 as much as 95% of all the gasworks coke produced in Sweden was produced in chamber ovens. A small chamber oven plant from the city of Lund is shown in Figure 5.

By 1937 there were 37 Swedish gasworks, of which 30 were owned by the municipalities and the remaining were split between two private companies with concessions; this remained the same in 1949. They supplied gas to a population of about 1 743 000 people, distributing 212.4 million cubic metres of gas. The gasworks varied considerably in size. In 1949, the largest of them, the Stockholm gasworks, distributed 120 million cubic metres of gas, whilst the smallest gasworks at Mariestad distributed only 0.25 million cubic metres of gas (Sjolund 1949).

Coke sales for domestic and local heating schemes provided security for the Swedish gas industry but due to the cost of importing of coal, the high price of gas had made it less competitive. During the 1930s Sweden started to invest heavily in electricity production and especially in hydroelectric power. This marked a gradual move towards electricity in Sweden.

By 1958, there were still 24 gasworks operating in central and southern Sweden. Some of the smaller gasworks had closed down because they had become uneconomic, but two were producing oil gas. At this time, 375 million cubic metres of gas was manufactured (half of which was made in Stockholm alone). Both propane and butane had started to be used to enrich producer gas or mixed to produce butane air gas.

Stockholm's first gasworks were the Klara gasworks, built at Klara sjö. These were originally privately owned but taken over by the city in 1884. These soon became too small and a new and much larger gasworks, the Värtagasverket, was built in Värtan, opening in 1893. This gasworks used coal

Fig. 5. Woodhall–Duckham intermittent chamber ovens at Lund gasworks, Sweden.

until conversion to oil reforming in 1971, and continued to manufacture gas until its closure in 2011, being one of the last gas-manufacturing plants in Europe to close. A similar situation occurred at the other larger Swedish gasworks.

Norway. The development of the gas industry in Norway was facilitated by British gas engineers. The first gasworks built in Norway were built in Oslo (then known as Christiania) during the years 1846–48 by the British gas engineer James Malam. Assisted by James Small, Malam had obtained a concession of 30 years to build and operate the gasworks. Small promoted gas through the publication of a pamphlet in 1848, describing the advantages of gas lighting. The Christiania gasworks were taken over by the city corporation at the end of the concession in 1878 (Irminger 1949).

The work of Malam and Small was not limited to Christiania; they also looked at the possibility of exploiting gas lighting in other Norwegian towns. Next was Trondheim, where they obtained a 25-year concession, with the gasworks being erected under the supervision of Small during the years 1851–55.

In 1851, a company obtained a 30-year concession for a gasworks in Fredrikshald. Malam commenced construction in 1850, but died in November of the same year and another engineer completed the construction. Small had also visited Bergen in December 1849, where he installed a miniature gasworks in a chemist's shop in order to be able to demonstrate gas lighting to the worthy city fathers. After some years of negotiation, a company was formed with a 30-year concession and the gasworks was erected between 1854 and 1856 by A.L. Pihl, then Manager of the Christiania gasworks.

Other towns followed in adopting gas lighting, including Drammen, Moss and Kristiansand South, which completed their works in 1857, Skien in 1865, and Stavanger and Arendal in 1868. The gasworks were generally built by private companies who obtained concessions for a fixed number of years, the only exception being Moss, which was built by the municipal corporation. A lull in the construction of gasworks in Norway then occurred until the beginning of the twentieth century, after the dissolution of Norway's union with Sweden. The gasworks in Alesund was built in 1905, Kristiansand North in 1908, Tonsberg in 1911, Sandefjord and Tromso in 1912, and Larvik in 1913. All of these gasworks were owned by private companies but gradually transferred to municipal ownership. The gasworks of Tromso had closed by 1923 (Irminger 1949).

The development of the gas industry in Norway was hard hit during World War I, when coal became both very expensive and at times difficult to obtain. This was because Norway had depended on Great Britain for its coal supply. The result was that both gas and coke prices had to increase and many gasworks fell into financial difficulties. Unfortunately, for the Norwegian gas industry, these difficulties occurred at a time when generating electricity from the country's abundant supply of water power became apparent, and the government, as well as most municipalities, invested in hydroelectric power stations, sufficient to satisfy the entire lighting load and much of the heating and power load too. Many of the calculations regarding these schemes were based on competition with high coal prices and when the war ended the price of coal dropped to its pre-war levels again. The hydroelectric companies had to lower prices to an entirely uneconomic level to compete with gas. As a result, municipalities defaulted and the debt had to then be written off. The effect of writing off the debt enabled some hydroelectric plants to deliver electricity at a very cheap rate.

As coal became more expensive, gas gradually lost out to electricity. Some of the larger gasworks converted to oil-based gas production. The last gasworks to close in Norway was that of Bergen, which closed in 1984, although the majority of the gasworks had closed in the 1950s and 1960s. When the supply was given up, energy demands were fulfilled by electricity.

Conclusion

The European gas industry has a long and complex history. The industry started in Britain, but involved a number of key European figure to catalyse the creation of the first public gas company. The industry gradually spread east and south, firstly to Belgium, France and Germany, spread primarily by British engineers. The adoption of coal gas in some countries (e.g. Sweden) was sometimes slowed by the formation of local gas companies which had developed wood- and oil-based gas-making plants. The Imperial Continental Gas Association was key to the development of the gas industries in The Netherlands, Germany, Austria and Belgium. The British gas engineers were followed by those from Belgium, France and Germany, who gradually took control of gasworks in their home markets and provided competition to the British engineers and gas companies in the new emerging markets abroad. European gasworks originally used imported British gas-making technologies (horizontal retorts) and British coal. However, as the industry developed and British coal became harder to obtain, technologies were developed which were better suited to continental coals. As gas-making technologies advanced, the retort-based systems (vertical, inclined and horizontal) that were preferred for British coals diverged from those chamber- and oven-based processes which were more suited to the continental coals. The carburetted water gas process was adapted in most European countries to meet the demand for gas at peak times. In those countries where coke was manufactured using by-product coke ovens, the surplus gas produced provided a base-load gas supply and through this long-distance gas transmission systems were developed. War had a significant effect on the gas industry across Europe, particularly the two world wars. As coal supplies started to diminish, the gas industry moved away from coal gas and started to produce gas from refinery by-products, through oil gas and gas-reforming processes. The discovery of significant European natural gas fields and the ability to import gas via pipelines and as LNG brought an end to the manufactured gas era in Europe. The historical remains of the manufactured gas industry are fragmented across Europe, from the oldest surviving gasholder, gasholder No. 2 at Fulham, to the Assisi panometers in the former gasholder houses of Dresden and the gas museums in Poland to the Gasometer City in Vienna; they all serve as a reminder of this once important and long-lived industry.

The remaining gas industry heritage in Europe

The manufactured gas era has passed, the last gas-manufacturing plants having closed as recently as 2011. In many countries, the low-pressure telescopic gasholders and buildings associated with gas manufacturing have long since been demolished. Many of the best-preserved structures associated with gas manufacturing can be found in Germany, the Nordic countries and the countries of Eastern Europe, Poland in particular. Whilst the UK is where the industry started, only fragments of this industry remain; most of the buildings associated with gas manufacture were demolished shortly after they were taken out of service, and the low-pressure gasholders, the last remaining part of the legacies are now being removed. Relatively little remains of the manufactured gas industry in both Great Britain and Western Europe, and those sites that do remain have been converted to museums, repurposed as offices or housing projects or, in many cases, just left derelict. Primarily, those former gasworks that have survived have tended to be those sites which operated for longer (e.g. the Nordic countries and Eastern Europe), where there has been little impetus to demolish the former gasworks and where the inherent heritage value of the site has been realized: for example, the Fakenham Gas Museum, the Suvilahti gasworks in Helsinki and the Danish Gas Museum in Hobro.

Accounts of the manufactured gas industry are fragmented and the subject matter has been excluded from many of the main displays of national science museums. There are a number of important museums, sites and archives associated with the manufactured gas industry which still remain and some of these are listed in Table 3.

Table 3. *Some museums, sites and archives associated with the manufactured gas industry*

Site	Address	Type of site	Website
Fakenham Museum of Gas and Local History	Fakenham, Hempton Road, Fakenham, Norfolk NR21 7LA, UK	Museum	http://www.fakenhamgasmuseum.com
Biggar Gasworks Museum	Gasworks Road, ML12 6BZ, Biggar, Lanarkshire, UK	Museum	https://www.historicenvironment.scot/visit-a-place/places/biggar-gasworks-museum/
Flame – Gasworks Museum of Ireland	Carrickfergus, County Antrim, UK	Museum	http://www.flamegasworks.co.uk
National Gas Museum, Leicester, UK	Aylestone Road, Leicester, UK	Museum	http://www.nationalgasmuseum.org.uk
National Grid Gas Archive	Unit 1, Europa Court Europa Boulevard Warrington WA5 7TN, UK	Gas archive	https://extranet.nationalgrid.com/GasArchive/Research.htm
Danish Gas Museum	Gasværksvej 2, 9500 Hobro, Denmark	Energy museum	http://www.gasmuseet.dk/
Paczkow Gas Museum (Muzeum Gazownictwa w Paczkowie)	ul. Mailbox 6 48-370 Paczków, province. Opole, Poland	Museum	http://www.muzeumgazownictwa.pl/
Warsaw Gas Museum (Muzeum Gazownictwa w Warszawie)	ul. Kasprzaka 25, 01-224 Warszawa, Poland	Museum	http://muzeum.gazownictwa.pl
Górowo Iławeckie Gas Museum (Muzeum Gazownictwa w Górowie Iławieckim)	ul. Wyszyńskiego 20, 11-220 Górowo Iławeckie, Poland	Museum	http://sitpnig.net.pl/warto-zwiedzic/muzeum-gazownictwa-w-gorowie-ilawieckim/
Dresden Energie Museum	Drewag Kraftwerk, Dresdener Energie-Museum, Dresden, Germany	Museum	http://www.kraftwerk-museum.de/gas.html
Berlin Gas Light Museum	Straße des 17 Juni/Ecke Klopstockstraße, 10557 Berlin, Germany	A collection of 90 gas lights	http://www.museumsportal-berlin.de/de/museen/gaslaternen-freilichtmuseum-berlin/
Gasometer Oberhausen	Arenastraße 11, 46047, Oberhausen, NRW, Germany	Cultural centre	http://www.gasometer.de/de
Gaswerk Technischer Denkmal	Havelberger Straße 25, 16845 Neustadt (Dosse), Germany	Museum	N/A
Gasworks Augsberg	August-Wessels-Straße 30, 86156 Augsburg, Germany	Museum and preserved structures	http://www.gaswerk-augsburg.de/
Industrial Gas Museum	Gazi Pireos 100, 11854 Athens, Greece	Gas museum and cultural centre	http://www.technopolis-athens.com/web/guest/home
Museo Del Gas	Sabadell, Catalunya, Spain	Energy museum	http://www.fundaciongasnaturalfenosa.org/museo/
Gázmúzeum – Budapest	1081, Budapest Köztársaság tér 20	Museum	http://www.museum.hu/muzeum/740/Gazmuzeum
Westergasfactory/Westergasfabriek	Haarlemmerweg 8–10, 1014 BE Amsterdam-Westerpark, Amsterdam, Noord-Holland, The Netherlands	Cultural centre	http://www.westergasfabriek.nl/
Musée de l'Éclairage au Gaz et d'Histoire du Luminaire Ancien	Rue Henri Maus, 92 – 4000 Liège, Belgium	Museum	http://www.liege.be/culture/musees/les-musees-de-liege/musee-de-leclairage-au-gaz-et-dhistoire-du-luminaire-ancien

Acknowledgements I would like to acknowledge the support and assistance of Tom Speight, Allen Hathaway, Katriina Etholen, the National Grid Gas Archive and the Institute of Gas Engineers and Managers History Panel in the preparation of this paper.

Funding This research received no specific grant from any funding agency in the public, commercial or not-for-profit sectors.

References

Accum, F.C. 1816. *Practical Treatise on Gas Light*. R. Ackerman, London.

Anon 1860. *Abridgements of the Specification relating to the Production and Applications of Gas (excepting gas engines)*. Eyre and Spottiswoode for Commissioners of Patents, London.

Anon 1885. Obituary, Charles Manby FRS. *Minutes of the Proceedings of the Institution of Civil Engineers*, **81**, 327–334.

Anon 1909. The first installation of Dassau vertical retorts in England. *The Journal of Gas Lighting, Water Supply & Sanitary Improvement*, **107**, 6 July.

Anon 1928. *The Book of the Woodhall–Duckham Companies*. The Woodhall–Duckham Vertical Retort & Oven Construction Co. (1920) Ltd, London.

Anon 1958. *Gas in Europe – Production, Availabilities, Consumption*. Organization for European Economic Cooperation, Paris.

Anon 1974. *Imperial Continental Gas Association*. Printed privately by the Imperial Continental Gas Association, Westerham Press, London.

Barszcz, M. 2004. Laternie Gazowe Warszawy. *Ochrona Zabytków*, **1**, 122–131.

Blears, D.K. 1966. *Continental Large-Scale Storage Installation*. Fourth Dempster Travelling Fellowship Report. Institute of Gas Engineers Communication, **699**.

Brabant, M., De Croes, J., Dorzee, P. & Beliens, A. 1939. The gas industry overseas, II. Belgium. *Transactions of the Institution of Gas Engineers*, **88**, 793–811.

Brandis, B.Á. 1949. *Overseas News and Views – Netherlands, Gas Journal Centenary Edition 1849–1949*. Walter King, London.

Budziński, Z., Wójtowicz, A. & Nawrocka-Fuchs, B. 2002. *Historia gazownictwa polskiego: od połowy XIX wieku porok 2000*. Polskie Zrzeszenie Inz ynieró w i Technikó w Sanitarnych. Zarząd Główny, Warszawa, Poland.

Chandler, D. & Lacey, A.D. 1949. *The Rise of the Gas Industry in Britain*. British Gas Council , London.

Clegg, S., Jr. 1841. *A Treatise on Gas Works and the Practice of Manufacturing and Distributing Coal Gas*. John Weale, London.

Everard, S. 1949. *The History of the Gas Light and Coke Company 1812–1949*. Earnest Benn Ltd, London.

Fernández-Paradas, M. 2015. *La industria del gas en Cádiz (1845–2012)*. Fundación Gas Natural Fenosa/ LID Editorial, Madrid.

Garcia de la Fuente, D. 1984. *La Compañía Española de Gas, S.A. Más de cien años de historia*. CEGAS, Valencia, Spain.

Goodhall, F. 1999. *Burning to Serve, Selling Gas in Competitive Markets*. Landmark Publishing, Ashbourne, UK.

Herranen, T. 1985. *Kaasulaitostoiminataa Helsingissä 1860–1985*. Energy Utility of Helsinki, Helsinki.

Hijdelaar, P. 1939. The gas industry overseas. VI. Holland. *Transactions of the Institution of Gas Engineers*, **88**, 835–838.

Holmes, W.C. 1874. *Instructions for the Management of Gas Work*. E. & F.N. Spon, London.

Hultman, G.H. 1939. The gas industry overseas. IX. Sweden. *Transactions of the Institution of Gas Engineers*, **88**, 851–856.

Hunt, C. 1907. *A History of the Introduction of Gas Lighting*. Walter King, London.

Irminger, J. 1949. *Norway, Gas Journal Centenary Edition 1849–1949*. Walter King, London.

Kaukovalta, K.V. 1934. *Forssan puuvillatehtaan historia 1847–1934*. Osakeyhtiö Forssa, Forssa, Finland.

Korting, J. 1949. *Germany, Gas Journal Centenary Edition 1849–1949*. Walter King, London.

Körting, J. 1963. *Geschichte der Deutschen Gasindustrie: Mit Vorgeschichte und bestimmenden Einflüssen des Auslandes*. Vulkan-Dr W. Classen, Essen, Germany.

Lavoisier, A.L. 1789. *Traité élémentaire de chimie*. Chez Cuchet, Paris.

Le Guellec, J. 1959. Gas industry in France. Speech to the American Gas Association, American Gas Association Meeting, May 1959, Dallas, Texas, USA.

Lintsen, H.W. 1993. *Geschiedenis van de techniek in Nederland. De wording van een moderne samenleving 1800–90. Deel III (1993) – Textiel. Gas, licht en elektriciteit. Bouw*. Walburg, Zutphen, The Netherlands.

Mac Lean, J. 1977. *Geschiedenis der gasverlichting in Nederland 1809–1850*. Walburg, Zutphen, The Netherlands.

Meade, A. 1921. *Modern Gas Works Practice*. 2nd edn. Benn Brothers, London.

Moyano-Jiménez, F. & Thomas, R.A.P. 2016. William Richards and the unrecorded success abroad: a British engineer in the Spanish Gas Industry in the Nineteenth Century. *The International Journal for the History of Engineering & Technology*, **86**, 124–146.

Newbigging, T. & Fewtrell, W. 1878. *King's Treatise on the Science & Practice of the Manufacture & Distribution of Gas, Volume 2*. Walter King, London.

Palacz, T. 2006. Wielkopolskie Gazownie. *Ochrona Zabytkow*, **3**, 97–110.

Peebles, M. 1980. *Evolution of the Gas Industry*. Macmillan Press, London.

Pozzi, D. 2010. The natural gas industry in Italy since autarky until Eni's hegemony, 1935–65. *In*: Beltran, A. (ed.) *A Comparative History of National Oil Companies*. P.I.E. Peter Lang, Brussels, Belgium, 233–263.

Rampone, C. 1939. The gas industry overseas. VII. Italy. *Transactions of the Institution of Gas Engineers*, **88**, 839–843.

Schilling, N.H. & Knapp, F.L. 1878. *Handbuch fur steinkohlengas-Beleuchtung*. Nabu Press, München, Germany.

Schuster, F. 1939. The gas industry overseas. V. Germany. *Transactions of the Institution of Gas Engineers*, **88**, 822–834.

SHIRLEY, T. 1867. The description of a well and earth in Lancashire, taking fire by a candle approached to it. *Philosophical Transactions*, **2**, 482–484.

SIMÓN-PALMER, M.C. 1989. *El gas y los madrileños*. Gas Madrid/Espasa Calpe, Madrid.

SJOLUND, T. 1949. *Sweden, Gas Journal Centenary Edition 1849–1949*. Walter King, London.

STEWART, E.G. 1958. *Town Gas – Its Manufacture and Distribution*. Science Museum. HMSO, London.

STOSS, M.L. 1939. The gas industry overseas. V. France. *Transactions of the Institution of Gas Engineers*, **88**, 816.

TARDIN, J. 1618. *Histoire naturelle de la fontaine qui brûle près de Grenoble, avec la recherche de ses causes et principes et ample traitement des feux souterrains*. Docteur en Medecine, Tournon, France.

THOMAS, R.A.P. 2014*a*. *Gasworks Profile A: The History and Operation of Gasworks*. Contaminated Land: Applications in Real Environments (CL:AIRE), London.

THOMAS, R.A.P. 2014*b*. *Gasworks Profile C: Water Gas Plants*. Contaminated Land: Applications in Real Environments (CL:AIRE), London.

TOBLER, W. 1939. The gas industry overseas. VII. Italy. *Transactions of the Institution of Gas Engineers*, **88**, 857–869.

TOMORY, L. 2012. *Progressive Enlightenment: The Origins of the Gaslight Industry, 1780–1820*. MIT Press, Cambridge, Massachusetts, USA.

VAN HELMONT, J.B. 1652. *Treatise De Flatibus: Ortus*. Louis Elzevir, Amsterdam.

WARD, E.R. 1965. *Gasmaking*. 2nd edn. The British Petroleum Company Ltd, London.

WENGENROTH, U. 2013. *Technik und Wirtschaft, Vol 8. Wirtschaft*. Springer, Berlin.

WILLIAMS, J. 1828. *An Historical Account of Sub-Ways in the British Metropolis*. S. McDowell, London.

WILLIAMS, T. 1981. *A History of the British Gas Industry*. Oxford University Press, Oxford.

WINZLER, Z.A. 1803. *Die Thermolampe in Deutschland, Pappband*. Karl Franz Siedler, Buchdrucker, Buch- und Kunsthändler, Brünn (Brno).

WUOLLE, B. 1936. *Helsingin kaupungin kaasulaitos 1860–1935*. Helsingin kaupunki, Helsinki, 108–109.

ZAHN, G.P., JR. 1911. *De Geschiedenis Der Verlicting van Amsterdam*. Schetema & Hokema's Boekhandel, K.Groesbeek & Paul Nijhoff, Amsterdam.

ZELECKI, J. 1995. Rozwéj i perspektywy gazyfikacji Wroclawia. *Ochrona Srodowiska*, **2**, 19–95.

ZOLLIKOFER, H. 1949. *Switzerland, Gas Journal Centenary Edition 1849–1949*. Walter King, London.

The birth and development of the oil and gas industry in the Northern Carpathians (up until 1939)

PIOTR KRZYWIEC

Institute of Geological Sciences, Polish Academy of Sciences, 51/55 Twarda Street, 00-818 Warsaw, Poland

piotr.krzywiec@twarda.pan.pl

Abstract: The northern segments of the Carpathians, stretching between Limanowa (Poland) and Kosów (Ukraine), belonged to the most prolific hydrocarbon province in the world in the late nineteenth and early twentieth centuries. The earliest written accounts of natural occurrences of hydrocarbons in the Carpathians date back to the sixteenth century. In the eighteenth and early nineteenth centuries, Rzączyński, Kluk, Hacquest and Staszic provided accounts on methods of practical use related to oil. Staszic's geological map shows numerous oil seeps and different rock types containing hydrocarbons. The development of the oil industry was triggered by Łukasiewicz's discovery of an oil-distillation process and the construction of a kerosene lamp. Following this, the oil industry flourished in the Northern Carpathians. Oil production peaked at 2 Mt (million tons) of crude oil in 1910. In subsequent years, the level of oil production steadily decreased due to a turbulent economy. Exploration for oil, gas and ozokerite resulted in the development of modern micropalaeontology and geological mapping, with a prime example being the regional coverage of almost the entire Northern Carpathians provided by the *Atlas Geologiczny Galicyi* (*Geological Atlas of Galicia*), which consisted of 99 high-quality geological maps at a scale of 1:75 000. Geophysical surveying techniques were applied to subsurface mapping, and higher educational institutions were established in order to support exploration efforts.

The northern segments of the Carpathians, presently located in SE Poland and western Ukraine, belonged to the most prolific hydrocarbon province in the world in the late nineteenth and early twentieth centuries (Frank 2005). This part of Europe had a very complex history. Since the Middle Ages it had been part of the Kingdom of Poland (later the Polish–Lithuanian Commonwealth). In the late eighteenth century, after the partition of Poland, it came under the name of the Kingdom of Galicia and Lodomeria, or simply Galicia (not to be mistaken with the region and autonomous community of NW Spain of the same name), part of the Austrian Empire, and later Austria–Hungary. In 1919, after Poland had regained its independence at the end of World War I, it formed its most southeasterly part. During World War II, this region was occupied by German and Russian (Soviet) armies and, in 1945, it became divided between Poland and the Soviet Union. Since 1991, its eastern segment – previously located within the Soviet Union area – has become part of independent Ukraine as a result of the fall of the Soviet Union and the end of Russian domination over Central and Eastern Europe. Oil and gas fields mentioned in this article are located within this area.

Oil and gas fields and commercial accumulations of earth wax (ozokerite) are located within the Outer Carpathian thrust sheets and their foredeep basin, which has also been incorporated partly into the Carpathian fold-and-thrust belt (Fig. 1) (Karnkowski 1993; Oszczypko *et al.* 2006; Ślączka *et al.* 2006). Cretaceous–Paleogene Outer Carpathian thrust sheets (nappes) are predominantly built of the deep-water turbidites called flysch. Fine-grained clastics, rich in organic matter (e.g. Spas, Verovice, variegated and Menilite shales), form source rocks, while coarse-grained sandstones (e.g. Grodziszcze, Lgota, Godula, Istebna, Ciężkowice and Kliwa sandstones) form good-quality reservoirs (Karnkowski 1993; Dziadzio *et al.* 2006). Traps are mostly structural, related to various folds formed during emplacement of the Carpathian fold-and-thrust belt.

Surface hydrocarbon occurrences (oil seeps and gas exhalations) have been known in the Outer Carpathians for centuries. In the mid-nineteenth century, following the discovery of an effective process of kerosene distillation and its various applications, full-scale commercial mining operations began, first focused on natural oil and then also on ozokerite, and eventually on natural gas.

The history of the development of the oil industry in the Northern Carpathians is well known in Poland, and also in Ukraine and other Central European countries, and has been described in numerous publications (Wolwowicz 1994; Nater & Sozański 2002; Mazan & Wolwowicz 2006; Sozański *et al.* 2006; Graniczny *et al.* 2015 among others). Abroad,

From: Craig, J., Gerali, F., MacAulay, F. & Sorkhabi, R. (eds) 2018. *History of the European Oil and Gas Industry*. Geological Society, London, Special Publications, **465**, 165–189.
First published online May 18, 2018, https://doi.org/10.1144/SP465.24

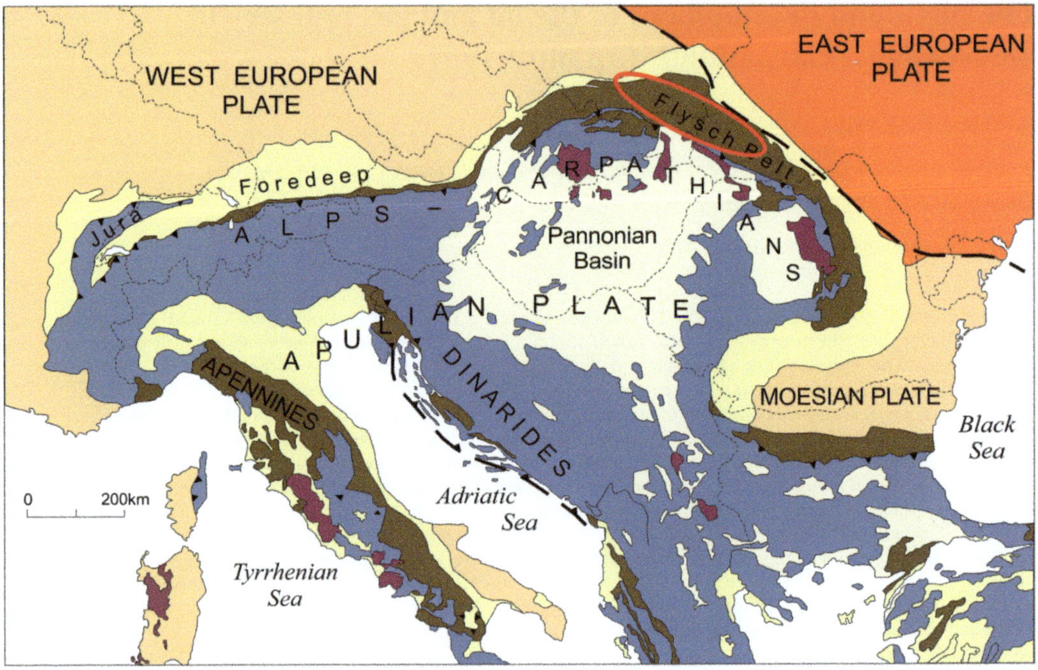

Fig. 1. The Alpine–Carpathian orogenic system (based on Oszczypko *et al.* 2006, modified). The red ellipse is the northern segment of the Outer Carpathians described in this paper.

however, simply due to lack of access to the necessary sources, this knowledge is limited, and usually only short and rather generalized paragraphs in otherwise excellent books on the history of oil industry are devoted to the Northern Carpathian oil province and its role in development of the modern oil industry (e.g. Brice 2009). Over the last few decades, however, foreign readers have become progressively better exposed to the very interesting history of the oil industry in this region (e.g. Koszarski & Bally 1994; Frank 2005; Sozański *et al.* 2006; Mau & Edmundson 2015; Wołkowicz *et al.* 2017).

The main aim of this paper is to summarize and present various aspects of the development of the oil and gas industry in the northern segment of the Carpathian thrust belt – the area that for a certain period of time, brief as it was, constituted the third most prolific oil-producing region in the world (Frank 2005). This is based on various original publications from the nineteenth and twentieth centuries, and synthetic works published more recently, such as the fundamental two-volume magnum opus of the Polish oil industry edited by Wolwowicz (1994). This story starts in the Middle Ages and ends in September 1939; when German and Soviet troops invaded Poland and when World War II began.

The part of Europe described in this paper had a very complex history: for several centuries it was part of the Kingdom of Poland, then for almost 150 years (i.e. during the partitions of Poland) it was under Austrian jurisdiction. After 1919, it again became Polish territory until after World War II when it was divided between the Soviet Union and Poland, with the Soviet Union eventually being replaced by Ukraine. Therefore, cities and villages mentioned in this paper might have Polish, German, Russian and Ukrainian versions of their names; the two latter ones can be written either in the Cyrillic or Latin alphabet. For the sake of simplicity and compatibility, the names of all the towns, etc. mentioned in this paper are given in their Polish versions because these were the ones used in most of the papers, and on most of the maps and cross-sections, published during the time period analysed in this paper. Interested readers can easily find the other versions of the names using the Internet or other resources.

Pre-industrial times or 'oily waters' and 'earthy greasy substances'

The earliest accounts of liquid hydrocarbons (surface oil seeps) in the Carpathians and their practical use have been published in medical books on herbs that have also often contained information about other natural substances such as minerals, etc. (Falimirz 1534; Siennik 1568; Marcin z Urzędowa 1595).

More detailed accounts, including information on the exact locations of particular oil seeps, had been given by Erasm Syxt, a medical doctor from Lwów (Syxt 1617, 1780).

Gabriel Rzączyński, a Polish Jesuit, published two books on the 'natural history' of the Polish territories and their surroundings (Rzączyński 1721, 1742). In both of these books information related to hydrocarbons in the Carpathians could be found; they contained separate chapters in which 'oily waters' (*aquis bituminosis*) are described (Fig. 2) and also included names (given in their Polish versions, although both books were published in Latin) of various places that would be important in the future development of the oil industry, such as Iwonicz (a town that has been known for centuries for its mineral waters, including numerous springs where gaseous hydrocarbons are mixed with water that is constantly 'bubbling', as in the most famous and best known 'Bełkotka' spring), Drohobycz, Jasień and Stebnik (towns known for the surface oil seeps).

Krzysztof Kluk, a Polish naturalist, agronomist and entomologist, in the first volume of his famous book on 'fossil mineral things' (1781), quite extensively dealt with – amongst other things – hydrocarbons (Fig. 3). These were even mentioned in the subtitle of this first volume: 'tłustości ziemne', which could be translated in English as 'earthy greasy substances'. A separate chapter in his book described oil and gas occurrences in various locations, and their practical applications. Kluk, similarly to Rzączyński, mentioned, for example, that the gaseous water springs in Iwonicz could easily burn.

Balthasar Hacquet, a professor of natural history in Lwów (1787–1805) and then, until 1809, in Kraków, published a four-volume book – an account of his extensive travels in the Carpathians (Hacquet 1790–96; cf. Scharr 2004). In this book, he listed sites of natural oil seeps in the vicinity of Drohobycz and described methods for their exploitation. He also mentioned that their origin might be associated with ancient organic matter.

In many aspects, the work and achievements of Stanisław Staszic mark the birth of modern Earth sciences in Poland. Stanisław Staszic (1755–1826) was a Polish priest, philosopher, statesman, geologist, scholar, poet and writer – leader of the Polish Enlightenment. Staszic had graduated from a Jesuit school in Poznań, and then continued his studies in France and Germany, where he was taught by such eminent scholars as Georges-Louis Leclerc Buffon. In the later years of his career, Staszic translated Buffon's seminal work *Les Époques de la Nature* into Polish. Staszic was a very active supporter of industrial development in Poland, had set up the Mining Academy (Akademiczna Szkoła Górnicza) in Kielce (1816–26) and had served as President of the Society of Friends of Sciences (Towarzystwo Przyjaciół Nauk) in Warsaw for almost 20 years, from 1808 to 1826. His first geological publications appeared in 1806 (Staszic 1806*a*). Staszic compiled this and several other essays into his geological magnum opus, a book published in 1815 and entitled *O Ziemiorodztwie Karpatów i innych gór i równin Polski* [*On the Earth-formation of the Carpathians and other Mountains and Lowlands of Poland*]. In this book, Staszic described natural occurrences of oil in the Carpathians and methods to be used for its exploitation, using shovels and hand-dug trenches (Fig. 4). Staszic's book was accompanied by an atlas containing, among other things, a truly impressive map entitled *Carta geologica totius Poloniae, Moldaviae, Transylvaniae, et partis Hungariae, et Valachiae* (cf. Staszic 1806*b*; Grigelis *et al.* 2011; Wołkowicz & Wołkowicz 2014; Krzywiec 2016). This map, dated 1806 and poorly known to geologists and historians of geology outside Poland, was published on four large (66 × 45 cm) sheets and covered approximately 1 million square kilometres – a huge area that stretched between the Baltic Sea and the Black Sea (Fig. 5). Staszic adopted the rock nomenclature of the famous German geologist Abraham Gottlob Werner, and used colours and numbers to indicate approximately 150 various types of rocks and ore deposits. The SW sheet of Staszic's map covers a segment of the Northern Carpathians and is showing numerous oil seeps and different rock types containing hydrocarbons (Fig. 6).

Georg Gottlieb Pusch graduated from the famous Mining Academy in Freiberg under the supervision of Abraham Gottlob Werner. He arrived in Kielce in 1816, following a request from Stanisław Staszic to join his newly established mining academy. Pusch authored a geological field manual dedicated to Staszic and numerous publications on the geology of the Polish territories, including the first modern account of the regional geology of Poland and its surroundings. Its second volume (Pusch 1836*a*) contained a separate chapter on 'springs with oil' (Erdőlquellen), with detailed descriptions of natural oil seeps in the Carpathians, including their location, and also remarks on the practical use of oil and its distillation (Fig. 7). Pusch also published a geological atlas that accompanied his *Geognostische Beschreibung von Polen* – its sheet No. III shows the Northern Carpathians (Fig. 8), including the area of Drohobycz (Fig. 9), which was known for natural oil seeps.

All of these early observations and studies of hydrocarbon occurrences in the Northern Carpathians led to a major breakthrough that took place in the mid-nineteenth century.

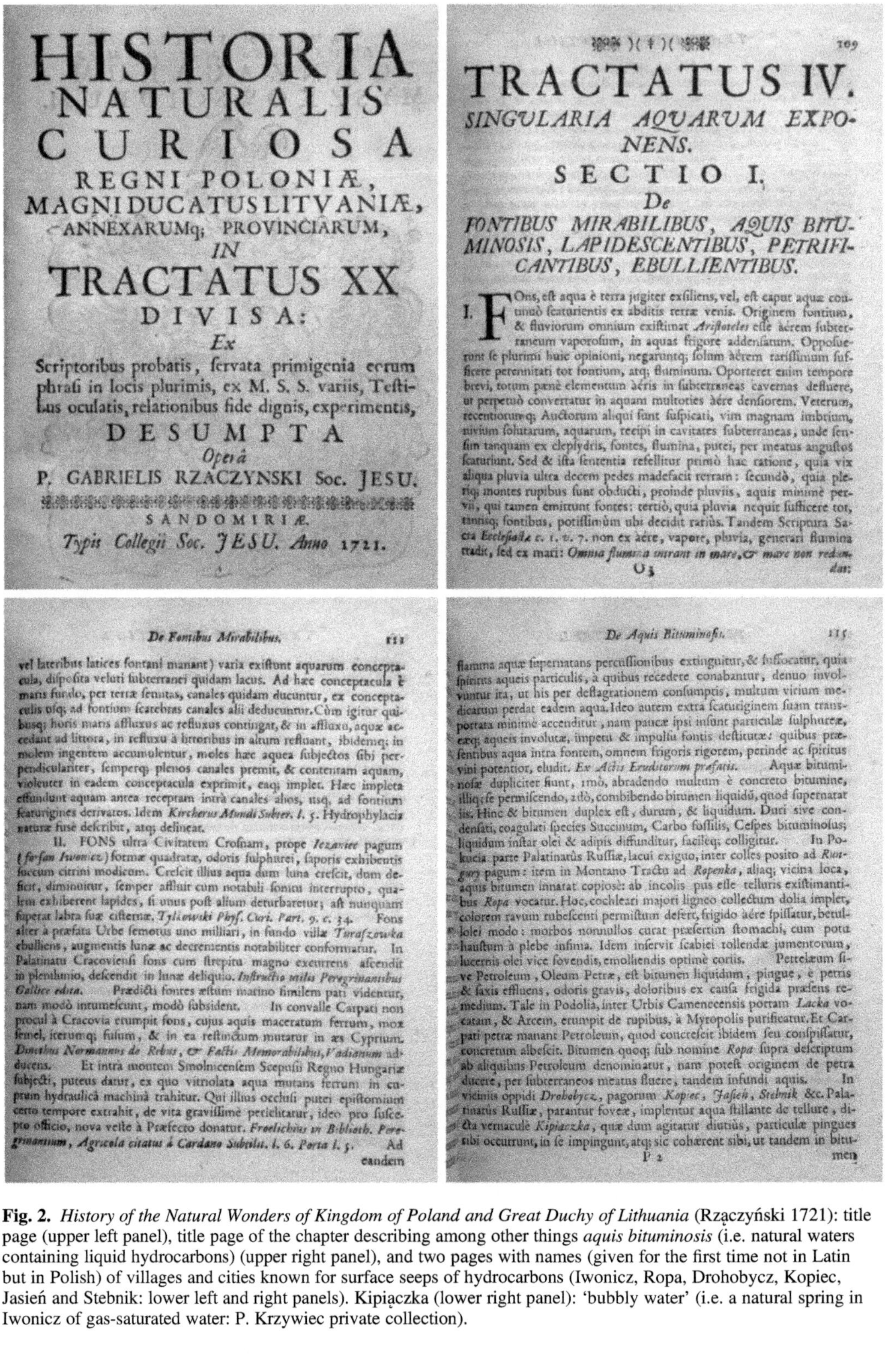
HISTORIA
NATURALIS
CURIOSA
REGNI POLONIÆ,
MAGNI DUCATUS LITUANIÆ,
ANNEXARUMq; PROVINCIARUM,
IN
TRACTATUS XX
DIVISA:
Ex
Scriptoribus probatis, servata primigenia eorum phrasi in locis plurimis, ex M. S. S. variis, Testibus oculatis, relationibus fide dignis, experimentis,
DESUMPTA
Operâ
P. GABRIELIS RZACZYNSKI Soc. JESU.
SANDOMIRIÆ.
Typis Collegii Soc. JESU. Anno 1721.

TRACTATUS IV.
SINGULARIA AQUARUM EXPONENS.
SECTIO I.
De
FONTIBUS MIRABILIBUS, AQUIS BITUMINOSIS, LAPIDESCENTIBUS, PETRIFICANTIBUS, EBULLIENTIBUS.

Fig. 2. *History of the Natural Wonders of Kingdom of Poland and Great Duchy of Lithuania* (Rząnczyński 1721): title page (upper left panel), title page of the chapter describing among other things *aquis bituminosis* (i.e. natural waters containing liquid hydrocarbons) (upper right panel), and two pages with names (given for the first time not in Latin but in Polish) of villages and cities known for surface seeps of hydrocarbons (Iwonicz, Ropa, Drohobycz, Kopiec, Jasień and Stebnik: lower left and right panels). Kipiączka (lower right panel): 'bubbly water' (i.e. a natural spring in Iwonicz of gas-saturated water: P. Krzywiec private collection).

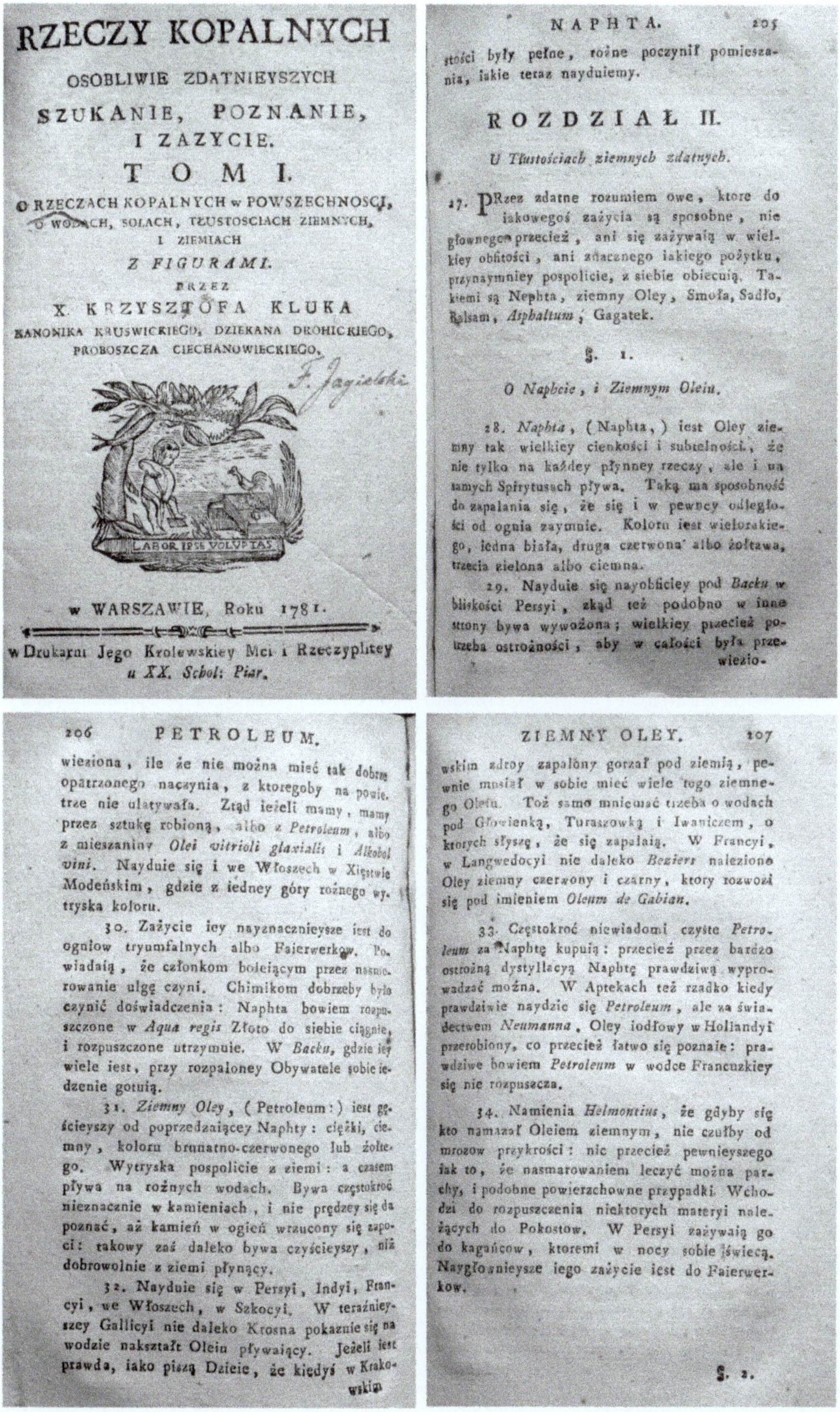

RZECZY KOPALNYCH
OSOBLIWIE ZDATNIEYSZYCH
SZUKANIE, POZNANIE, I ZAZYCIE.
TOM I.
O RZECZACH KOPALNYCH w POWSZECHNOSCI, O WODACH, SOLACH, TŁUSTOSCIACH ZIEMNYCH, I ZIEMIACH
Z FIGURAMI.
PRZEZ
X. KRZYSZTOFA KLUKA
KANONIKA KRUSWICKIEGO, DZIEKANA DROHICKIEGO, PROBOSZCZA CIECHANOWIECKIEGO.

LABOR IPSE VOLUPTAS

w WARSZAWIE, Roku 1781.
w Drukarni Jego Krolewskiey Mci i Rzeczypltey u XX. Schol: Piar.

NAPHTA. 205

stości były pełne, rożne poczynił pomieszania, iakie teraz nayduiemy.

ROZDZIAŁ II.

U Tłustościach ziemnych zdatnych.

27. PRzez zdatne rozumiem owe, ktore do iakowegoś zażycia są sposobne, nie głownego przecież, ani się zażywaią w wielkiey obfitości, ani znacznego iakiego pożytku, przynaymniey pospolicie, z siebie obiecuią. Takiemi są Naphta, ziemny Oley, Smoła, Sadło, Balsam, *Asphaltum*, Gagatek.

§. 1.

O Naphcie, i Ziemnym Oleiu.

28. *Naphta*, (Naphta,) iest Oley ziemny tak wielkiey cienkości i subtelności, że nie tylko na każdey płynney rzeczy, ale i na samych Spirytusach pływa. Taką ma sposobność do zapalania się, że się i w pewney odległości od ognia zaymuie. Koloru iest wielorakiego, iedna biała, druga czerwona albo żołtawa, trzecia zielona albo ciemna.

29. Nayduie się nayobficiey pod *Backu* w bliskości Persyi, zkąd też podobno w inne strony bywa wywożona; wielkiey przecież potrzeba ostrożności, aby w całości była przewiezio-

206 PETROLEUM.

wieziona, ile że nie można mieć tak dobrze opatrzonego naczynia, z ktoregoby na powietrze nie ulatywała. Ztąd ieżeli mamy, mamy przez sztukę robioną, albo z *Petroleum*, albo z mieszaniny *Olei vitrioli glacialis* i *Alkohol vini*. Nayduie się i we Włoszech w Xięstwie Modeńskim, gdzie z iedney góry rożnego wytryska koloru.

30. Zażycie iey nayznacznieysze iest do ogniow tryumfalnych albo Faierwerkow. Powiadaią, że członkom boleiącym przez nasmarowanie ulgę czyni. Chimikom dobrzeby było czynić doświadczenia: Naphta bowiem rozpuszczone w *Aqua regis* Złoto do siebie ciągnie, i rozpuszczone utrzymuie. W *Backu*, gdzie iey wiele iest, przy rozpaloney Obywatele sobie iedzenie gotuią.

31. *Ziemny Oley*, (Petroleum:) iest gęścieyszy od poprzedzaiącey Naphty: cięźki, ciemny, koloru brunatno-czerwonego lub żołtego. Wytryska pospolicie z ziemi: a czasem pływa na rożnych wodach. Bywa częstokroć nieznacznie w kamieniach, i nie prędzey się da poznać, aż kamień w ogień wrzucony się zapoci: takowy zaś daleko bywa czyścieyszy, niż dobrowolnie z ziemi płynący.

32. Nayduie się w Persyi, Indyi, Francyi, we Włoszech, w Szkocyi. W teraźnieyszey Gallicyi nie daleko Krosna pokazuie się na wodzie nakształt Oleiu pływaiący. Jeżeli iest prawda, iako piszą Dzieie, że kiedyś w Krakowskim

ZIEMNY OLEY. 207

wskim zdroy zapalony gorzał pod ziemią, pewnie musiał w sobie mieć wiele tego ziemnego Oleiu. Toż samo mniemać trzeba o wodach pod Głowienką, Turaszowką i Iwaniczem, o ktorych słyszę, że się zapalaią. W Francyi, w Langwedocyi nie daleko *Beziers* naleziono Oley ziemny czerwony i czarny, ktory rozwozi się pod imieniem *Oleum de Gabian*.

33. Częstokroć niewiadomi czyste *Petroleum* za Naphtę kupuią: przecież przez bardzo ostrożną dystyllacyą Naphtę prawdziwą wyprowadzać można. W Aptekach też rzadko kiedy prawdziwie naydzie się *Petroleum*, ale za świadectwem *Neumanna*, Oley iodłowy w Hollandyi przerobiony, co przecież łatwo się poznaie: prawdziwe bowiem *Petroleum* w wodce Francuzkiey się nie rozpuszcza.

34. Namienia *Helmontius*, że gdyby się kto namazał Oleiem ziemnym, nie czułby od mrozow przykrości: nic przecież pewnieyszego iak to, że nasmarowaniem leczyć można parchy, i podobne powierzchowne przypadki. Wchodzi do rozpuszczenia niektorych materyi należących do Pokostow. W Persyi zażywaią go do kagańcow, ktoremi w nocy sobie świecą. Naygłownieysze iego zażycie iest do Faierwerkow.

§. 2.

Fig. 3. K. Kluk's (1781) description of liquid hydrocarbons and their occurrences in, among others, the Northern Carpathians (Krosno, Głowienka, Turoszówka and Iwonicz) (P. Krzywiec private collection).

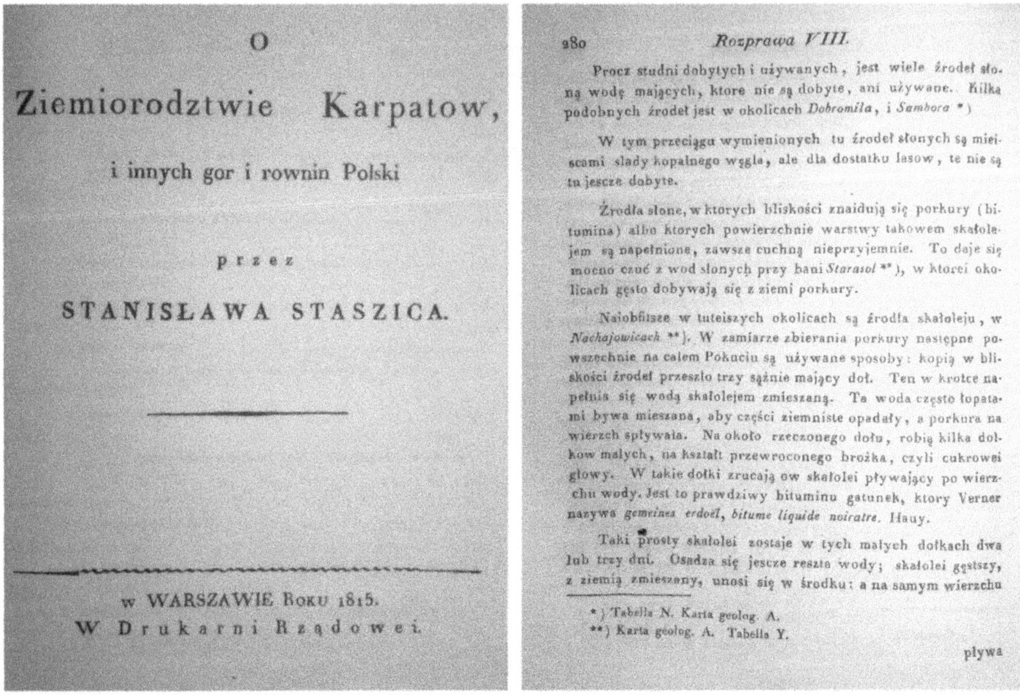

O

Ziemiorodztwie Karpatow,

i innych gor i rownin Polski

przez

STANISŁAWA STASZICA.

w WARSZAWIE Roku 1815.
W Drukarni Rządowei.

280 *Rozprawa VIII.*

Procz studni dobytych i używanych, jest wiele źrodeł słoną wodę maiących, ktore nie są dobyte, ani używane. Kilka podobnych źrodeł jest w okolicach *Dobromila*, i *Sambora* *)

W tym przeciągu wymienionych tu źrodeł słonych są mieiscami slady kopalnego węgla, ale dla dostatku lasow, te nie są tu jescze dobyte.

Źrodła słone, w ktorych bliskości znaidują się porkury (bitumina) albo ktorych powierzchnie warstwy takowem skałolejem są napełnione, zawsze cuchną nieprzyiemnie. To daje się mocno czuć z wod słonych przy bani *Starasol* **), w ktorei okolicach gęsto dobywają się z ziemi porkury.

Naiobfitsze w tuteiszych okolicach są źrodła skałoleju, w *Nachajowicach* **). W zamiarze zbierania porkury następne powszechnie na całem Pokuciu są używane sposoby: kopią w bliskości źrodeł przeszło trzy sążnie maiący doł. Ten w krotce napełnia się wodą skałolejem zmieszaną. Ta woda często łopatami bywa mieszana, aby części ziemniste opadały, a porkura na wierzch spływała. Na około rzeczonego dołu, robią kilka dołkow małych, na kształt przewroconego brożka, czyli cukrowei głowy. W takie dołki zrucają ow skałolei pływający po wierzchu wody. Jest to prawdziwy bituminu gatunek, ktory Verner nazywa *gemeines erdoel, bitume liquide noiratre.* Hauy.

Taki prosty skałolei zostaje w tych małych dołkach dwa lub trzy dni. Osadza się jescze reszta wody; skałolei gęstszy, z ziemią zmieszany, unosi się w środku: a na samym wierzchu

*) Tabella N. Karta geolog. A.
**) Karta geolog. A. Tabella Y.

pływa

Fig. 4. *O Ziemiorodztwie Karpatów i innych gór i równin Polski* [*On the Formation of the Carpathians and Other Mountains and Lowlands of Poland*] by Staszic (1815) contained information on natural oil seeps in the Carpathians, including early methods of oil exploitation, using shovels and hand-dug trenches (P. Krzywiec private collection).

The birth and development of the modern petroleum industry in the Northern Carpathians (1853–1939)

Oil

In the Northern Carpathians, two oil districts could be distinguished, stretching between Limanowa in Poland and Kosów in Ukraine (Fig. 10) (cf. Ślączka *et al.* 2006):

- the eastern district located south of Lwów (Drohobycz, Borysław, Tustanowice, Mraźnica, Daszawa, etc.), currently in Ukraine;
- the western district (Ustrzyki, Sanok, Krosno, Jasło, Gorlice and Limanowa), currently in Poland.

Although surface oil occurrences and shallow oil accumulations have been known for centuries in both districts, the modern oil industry kick-started in the western one. This was due to discoveries made by Ignacy Łukasiewicz and his vision of the future industrial development of Galicia.

Ignacy Łukasiewicz (1822–82) was, without any doubt, one of the founders of the modern petroleum industry. Parallel to Abraham Gesner, who developed a process to refine a liquid fuel (kerosene) from coal, bitumen and oil shale, and Samuel Kier, who established America's first oil refinery in Pittsburgh in 1853, Łukasiewicz, together with Jan Zeh, discovered an effective distillation process to obtain kerosene from crude oil, and then, together with Adam Bratkowski, a local tinsmith, invented an effective and safe kerosene lamp. On 31 July 1853, Łukasiewicz made one of his kerosene lamps available to a local hospital in Lwów to illuminate an emergency surgical operation (cf. Wolwowicz 1994; Sozański *et al.* 2006). In 1854, several years prior to the Titusville discovery by the legendary Colonel Edwin Drake (cf. Koszarski & Bally 1994; Brice 2009), Łukasiewicz, together with his business partners, Tytus Trzecieski and Karol Klobassa, established an oil mine in Bóbrka. This mine first deployed hand-dug oil wells (shafts), then wells drilled using manually operated percussion-type drill bit on rods and a free-fall drilling apparatus, and, finally, cable drilling, eventually (from 1872) powered by a steam engine. These technical achievements place the Bóbrka oil mine amongst the first modern oil mines in the world (Koszarski & Bally 1994). After the establishment of the commercially successful Bóbrka mine, which is very often

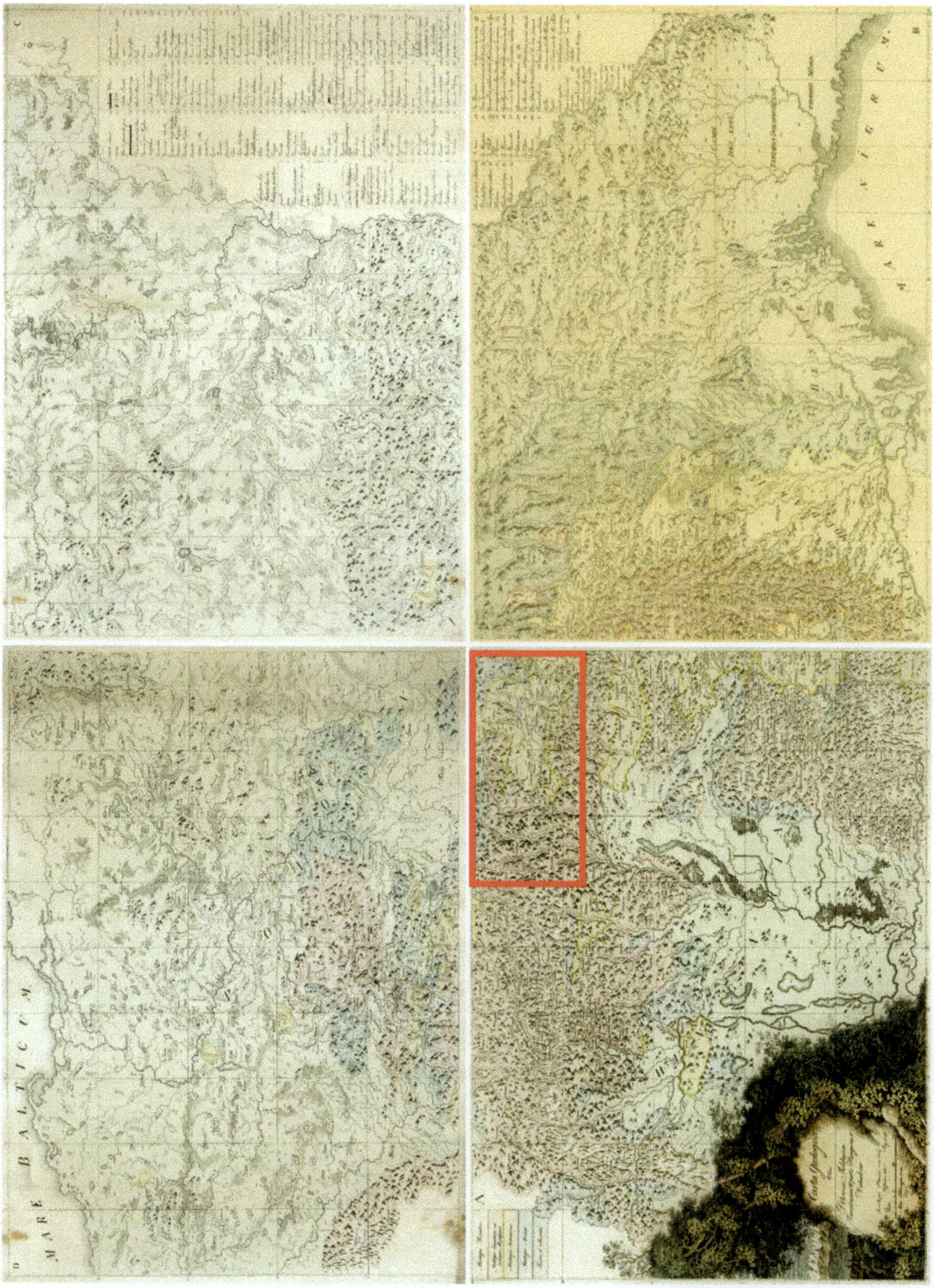

Fig. 5. Staszic's geological map of the whole of Poland, Moldova, Transylvania and parts of Hungary and Wallachia (P. Krzywiec private collection). The red rectangle is the enlarged area shown in Figure 6.

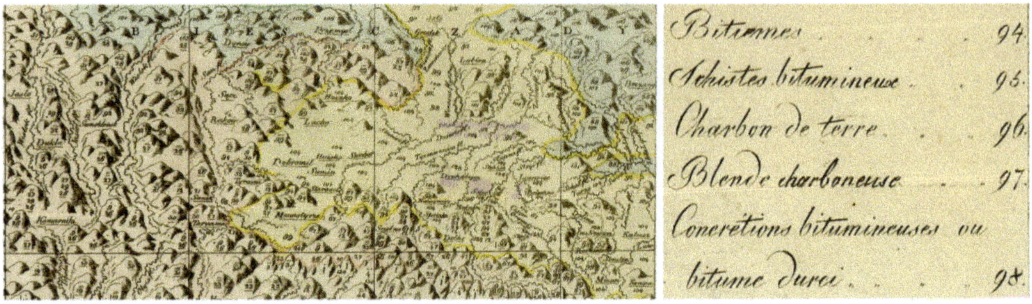

Fig. 6. Numerous oil seeps and different rock types containing hydrocarbons are shown in the Northern Carpathians on Staszic's map (P. Krzywiec private collection).

associated with the beginning of the first phase of development of the modern oil industry in the Northern Carpathians, other producing fields and mines flourished, all located in the Western Carpathian oil district (Siary, Sękowa, Męcina Wielka, Klęczany, etc.). All these technological and industrial achievements that eventually, in early twentieth century, led to the area being established as the third most prolific oil-producing region in the world could be seen as a direct result of the hard work, dedication and persistence of Ignacy Łukasiewicz, one of the 'fathers' of the modern oil industry.

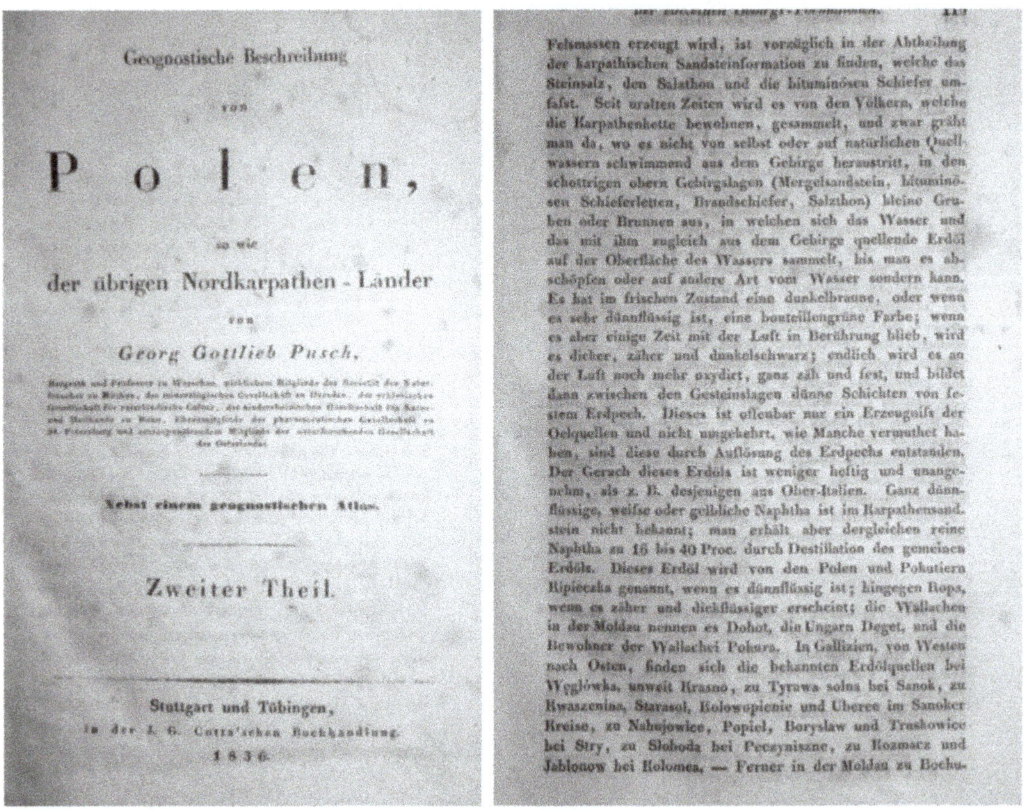

Geognostische Beschreibung

von

Polen,

so wie

der übrigen Nordkarpathen - Länder

von

Georg Gottlieb Pusch,

Nebst einem geognostischen Atlas.

Zweiter Theil.

Stuttgart und Tübingen,

in der J. G. Cotta'schen Buchhandlung.

1836.

Felsmassen erzeugt wird, ist vorzüglich in der Abtheilung der karpathischen Sandsteinformation zu finden, welche das Steinsalz, den Salzthon und die bituminösen Schiefer umfaſst. Seit uralten Zeiten wird es von den Völkern, welche die Karpathenkette bewohnen, gesammelt, und zwar gräbt man da, wo es nicht von selbst oder auf natürlichen Quellwassern schwimmend aus dem Gebirge heraustritt, in den schottrigen obern Gebirgslagen (Mergelsandstein, bituminösen Schieferletten, Brandschiefer, Salzthon) kleine Gruben oder Brunnen aus, in welchen sich das Wasser und das mit ihm zugleich aus dem Gebirge quellende Erdöl auf der Oberfläche des Wassers sammelt, bis man es abschöpfen oder auf andere Art vom Wasser sondern kann. Es hat im frischen Zustand eine dunkelbraune, oder wenn es sehr dünnflüssig ist, eine bouteillengrüne Farbe; wenn es aber einige Zeit mit der Luft in Berührung blieb, wird es dicker, zäher und dunkelschwarz; endlich wird es an der Luft noch mehr oxydirt, ganz zäh und fest, und bildet dann zwischen den Gesteinslagen dünne Schichten von festem Erdpech. Dieses ist offenbar nur ein Erzeugniſs der Oelquellen und nicht umgekehrt, wie Manche vermuthet haben, sind diese durch Auflösung des Erdpechs entstanden. Der Geruch dieses Erdöls ist weniger heftig und unangenehm, als z. B. desjenigen aus Ober-Italien. Ganz dünnflüssige, weiſse oder gelbliche Naphtha ist im Karpathensandstein nicht bekannt; man erhält aber dergleichen reine Naphtha zu 16 bis 40 Proc. durch Destillation des gemeinen Erdöls. Dieses Erdöl wird von den Polen und Pokutiern Ripieczka genannt, wenn es dünnflüssig ist; hingegen Ropa, wenn es zäher und dickflüssiger erscheint; die Wallachen in der Moldau nennen es Dohot, die Ungarn Deget, und die Bewohner der Wallachei Pokura. In Gallizien, von Westen nach Osten, finden sich die bekannten Erdölquellen bei Węglówka, unweit Krasno, zu Tyrawa solna bei Sanok, zu Kwaszenina, Starasol, Kolowopienie und Uherce im Sanoker Kreise, zu Nahujowice, Popiel, Boryslaw und Truskowice bei Stry, zu Sloboda bei Peczyniszne, zu Kozmacz und Jablonow bei Kolomea. — Ferner in der Moldau zu Bochu-

Fig. 7. G.G. Pusch's description of natural oil seeps in the Carpathians (Pusch 1836*a*) (P. Krzywiec private collection).

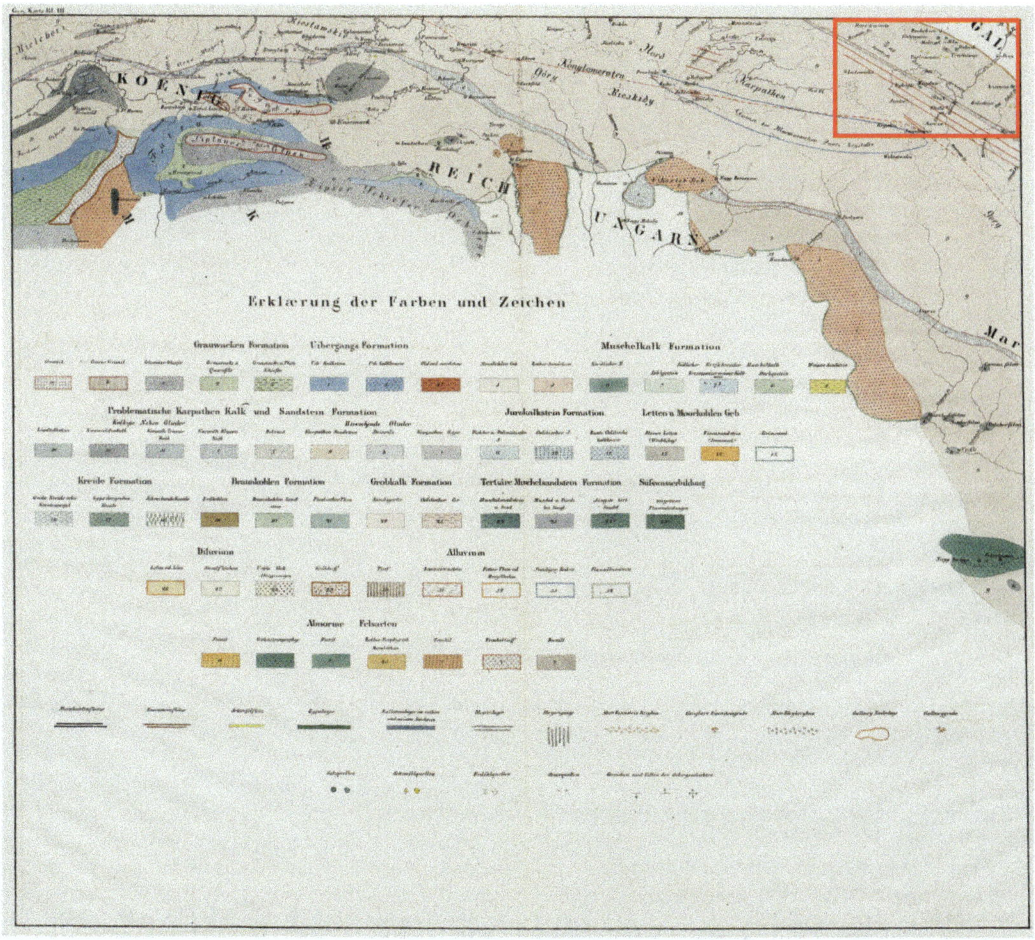

Fig. 8. Geological map of the Northern Carpathians by Pusch (1836*b*). The red rectangle is the enlarged area shown in Figure 9. Erdöhlquellen, oil seeps (Jagiellonian Library collection).

The second phase of development of the modern oil industry in the Northern Carpathians is associated with the completion of a very successful well, 'Na Potoku', by Władysław Długosz. It was drilled in 1896 in Borysław and yielded a daily production of around 40 t (tons) of crude oil. This discovery well was outshone by further wells drilled in this part of the Carpathians, such as the 'Wilno' well, drilled in 1902, that produced up to 400 t of crude oil per day, the 'Litwa' well that produced up to 800 t per day or the famous 'Oil City' well drilled in 1908 with a production that peaked at around 2500 tons per day (Fig. 11). All these discoveries shifted gross production into the eastern oil district.

Following those early discoveries, the last two decades of the nineteenth century and the early twentieth century witnessed a very rapid development in oil exploration and production in the Northern Carpathians. In 1875, there were approximately 900 companies operating at 19 localities. The largest of them was Pierwsza Galicyjska Spółka Oleju Skalnego Gartenberg & Goldhammer, which operated 68 active wells. Another important player was Bergheim & Mac Garvey established in 1884. American oil behemoth, Standard Oil Company, acting through its subsidiary company Vacuum Oil Corporation located in Budapest, started its operations in Galicia in 1908.

In 1919, just after Poland regained its independence at the end of World War I, numerous companies were operating in the Northern Carpathians: Vacuum Oil Corporation (Budapest), Bergheim & Mac Garvey then renamed Galicyjskie Karpackie Towarzystwo Naftowe(Glinik Mariampolski – Gorlice), Aktiongesellschaft für Mineralien Galizische Nafta (Vienna), G. G. 'Galicja' (Drohobycz),

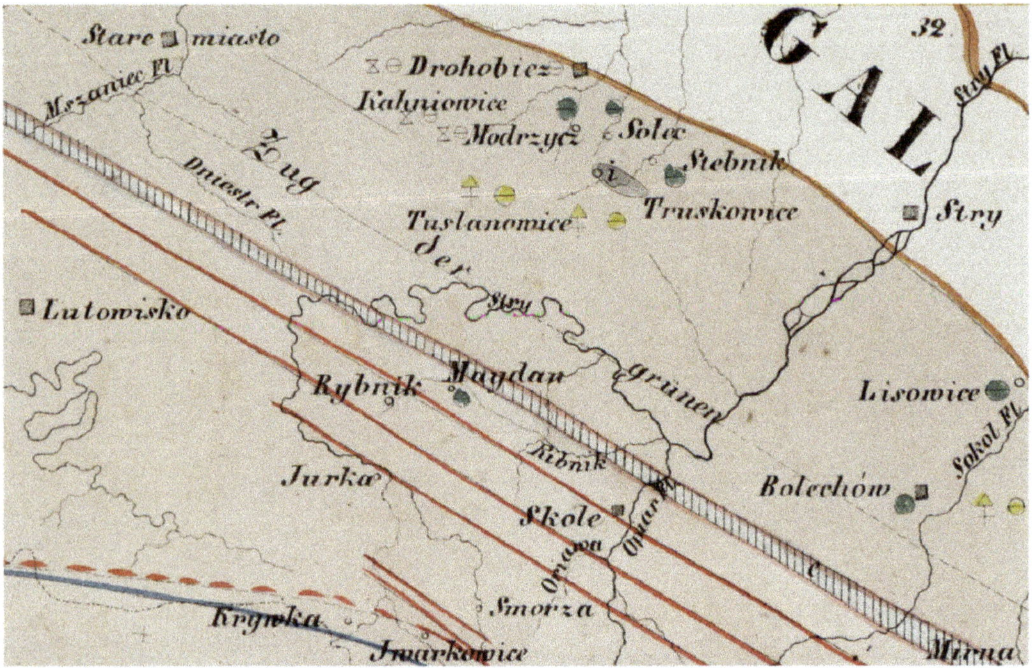

Fig. 9. Oil seeps in the Northern Carpathians, Borysław area (Pusch 1836*b*). See Figure 8 for additional explanations (Jagiellonian Library collection).

Petroleum Licht und Kraftgesellschaft 'Schodnica' A.G. (Vienna), Petroleum Industria (Vienna), Austria Petroleum (Vienna), Mineralölraffinerie A.G. (Budapest), Spółka Akcyjna dla Przemysłu Naftowego 'Trzebinia' (Kraków), Bergheim A.G. (Vienna) and Petroleum Aktiongesellschaft (Vienna). Most of those companies were partly or fully controlled by foreign financial investors. Oil production peaked in 1909 with 2.076 Mt (million tons). Then, a rapid decline occurred in the years 1902–09 due to a significant overproduction of oil that was, to a large degree, caused by a very significant number of companies who were not able to – and often did not want to – co-ordinate their exploration and production efforts. Also, a lack of effective techniques for oil storage forced ineffective sale strategies that led to a rapid decrease in oil prices. From 1908, a positive process of company mergers led to the formation of just four key players: the Galicyjskie Karpackie Towarzystwo Naftowe, Dawid Fanto und Co., Nafta AG and Galicja companies.

Poland, after having been partitioned between Austria, Russia and Prussia for almost 150 years, regained its independence in 1919, as a result of World War I. At that time, there were 28 oil fields located between Limanowa and Kosów (Fig. 10). In 1930, the yearly production reached 662 kt (thousand tons), with 87% of it coming from the Borysława area. The rather unfortunate situation of a dominance of foreign investing capital, known from before the First World War, also continued after it (i.e. after Poland regained its independence). In the years 1923–28, French investors controlled 53% of the oil industry in Poland, Swiss investors controlled 10%, Austrian investors 7%, English investors 4.4%, Dutch investors 4.3% and other investors 3.3%, and so, in consequence, Polish investors controlled only 18% of their industry. Another problem, rooted in Austrian oil law from 1884, was related to the fact that hydrocarbons were treated as a commodity fully controlled by land owners. This led to the establishment of a large number of relatively small companies without the capital and know-how necessary for the modern development of their oil fields.

In 1919, the Polish Geological Institute was established. It became involved in various studies related to the exploration for and production of hydrocarbons through its Department of Salt and Oil (Wydział Naftowo-Solny) located in Borysław. It employed numerous gifted geologists who were responsible for the rapid advancement of the geological recognition of the Northern Outer Carpathians, such as Eugeniusz Jabłoński, Stanisław Weigner, Ludwik Horwitz, Bolesław Bujalski, Zdzisław Opolski, Bohdan Świderski, Jan Nowak and Stanisław

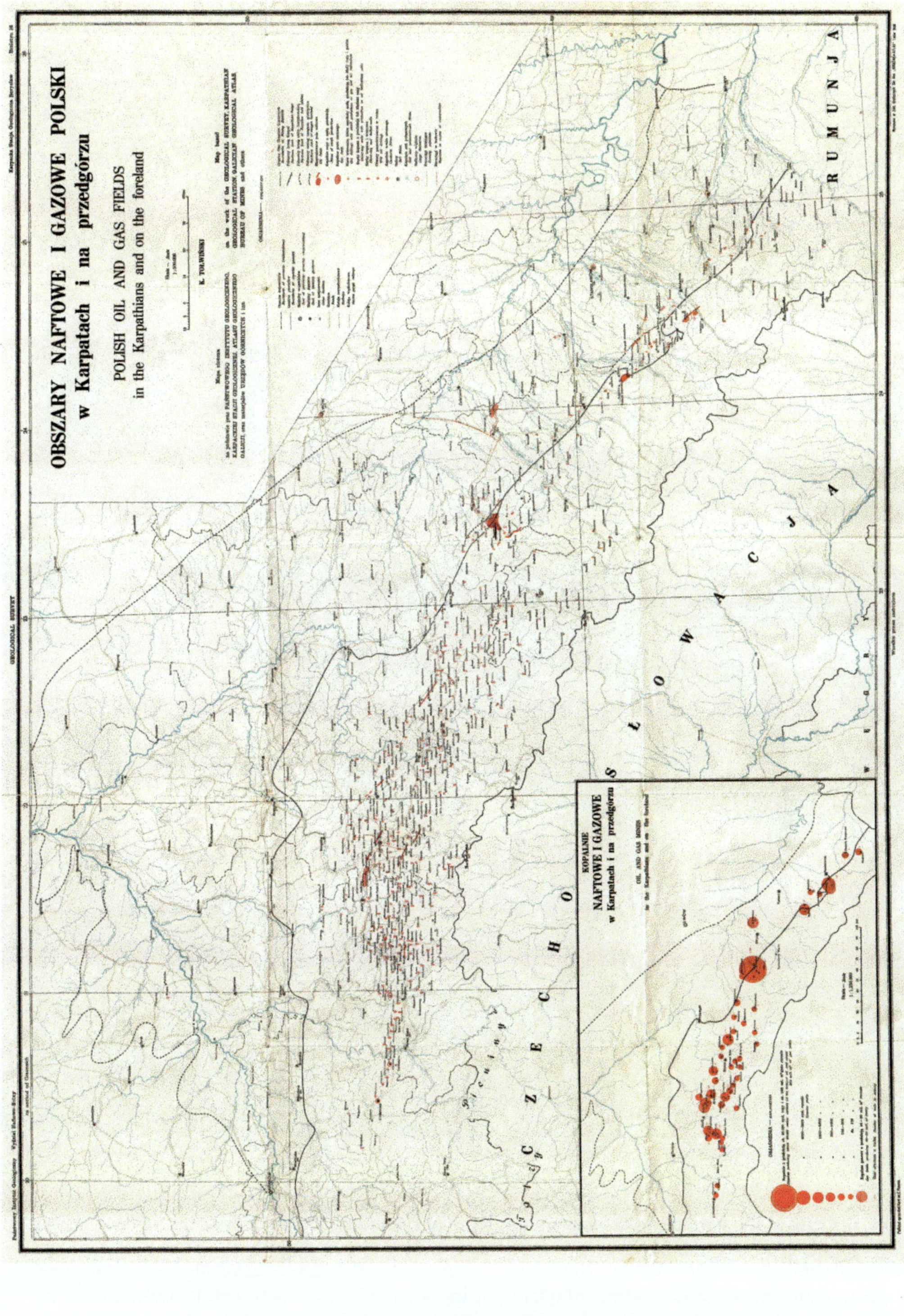

Fig. 10. Map of oil and gas fields in the Northern Carpathians (Tołwiński 1928). Salt and ozokerite (i.e. earth wax) mines are also shown on this map (P. Krzywiec private collection).

Fig. 11. Drilling rigs in the Borysław area.

Krajewski. Because of the unstable economic situation and the depletion of old oil fields, the number of wells was constantly decreasing: in 1926, 20 exploration wells were drilled; in 1928, 17 were drilled; in 1929, 13 wells; and in 1932–34, 18 wells were drilled (including the 2085 m-deep 'Stetland Południe' well in Borysław).

In 1928, the state-controlled exploration company 'Pionier' was established. It started to employ the modern exploration techniques (e.g. geophysical surveying: see later) necessary for new discoveries. The accumulation of an extensive database on oil production and advancements in petroleum geology led to the first regional studies on reserve estimates being published in 1938.

Natural gas

Natural exhalations of natural gas, in many cases associated with 'holy fires', have been known since ancient times from the Middle East, the coasts of the Caspian Sea (Baku, etc.), Burma, Borneo and China, amongst others. Natural gas has also been known from both oil districts in the Northern Carpathians. As was described earlier, numerous natural springs of 'boiling waters' (i.e. water saturated with, often burning, natural gas) were mentioned in ancient texts. Even in the early nineteenth century, they were still treated as some sort of 'miracle' and placed amongst the most interesting natural wonders of the Austrian Empire (Sartori 1810).

During the early days of oil production, natural gas was not used but was released into the atmosphere. In fact, natural gas was treated as a hazard as it was often the reason for dangerous oil eruptions (Fig. 11). Deep drilling for natural gas started in 1896 in the Borysław region, after methods for its commercial usage as an effective energy source had been invented. The most common way to utilize natural gas was to burn it in steam engines used for oil production. In the early twentieth century, gas started to be used in domestic households and the application of this natural commodity proved to be of great commercial value. Following this, increased drilling activities focused on gas accumulations led to the discovery of numerous gas fields: the Białkówka gas field near Krosno (in 1912), the Daszawa gas field near Stryj (in 1921), the Sądkowa gas field near Jasło (in 1922), the Strachocina gas field located between Sanok and Brzozów (in 1928), and the Chodnowice gas field (1938–39) located east of Przemyśl, which after the war was extended towards the west into one of the largest gas fields in Poland (cf. Karnkowski 1993).

Two companies were predominantly involved in the exploration for and production of natural gas: 'Gazolina' located in Tustanowice and 'Polmin' located in Lwów. The 'Gazolina' company was established in 1912, first under the name Zakład Gazu Ziemnego, then as 'Gaz Ziemny' and from 1916 as 'Gazolina'. It built first a short gas pipeline from the 'Klaudiusz' mine in Borysław, then the longer Borysław–Drohobycz gas pipeline. Since 1925 it has operated the large Daszawa gas field, and has built the Daszawa–Stryj–Lwów and Daszawa–Drohobycz gas pipelines. 'Polmin' was a state-controlled company established in 1927. It built a gas pipeline in 1920 from the Roztoki gas field to Gorlice (Glinik Mariampolski). Increased gas production led to the construction of the country's gas pipeline system during the years 1927–39.

Earth wax (ozokerite)

Earth wax or ozokerite was once one of the most important commodities produced in the Northern Carpathians. It is a naturally occurring mineral wax or paraffin. Due to the evaporation and oxidation of petroleum, deposits of ozokerite form in natural open fissures or cracks. Localities in the Northern Carpathians, such as Borysław, Tustanowice, Truskawiec, Nadwórna, Dźwiniacz and Starunia, are famous for very large and relatively shallow accumulations of ozokerite (Windakiewicz 1875; Zuber 1883*a*). It forms veins, sometimes rather thick, that cross-cut the host rock, and can be either very soft or relatively hard, light yellow to dark brown, or sometimes greenish.

Veins of ozokerite known from the Borysław area are located in the proximity of oil accumulations within the deformed Miocene Carpathian foredeep deposits of the Stebnik thrust sheet (nappe). They are often associated with rock salt and could be characterized by significant overpressure which posed a significant threat during mining operations. On several occasions, ozokerite inflow into mining shafts continued for weeks and led to the complete infill of the shaft by a plastic wax.

In 1810, Jan Mitis was granted a production concession by the Austrian mining authorities. Large-scale production in the Borysław region began in 1855, after Robert Doms had initiated his exploration and production activities. In 1865, ozokerite was excluded from the list of restricted minerals controlled by the government, and this led to a rapid increase of its production – in 1886 there were 89 smaller or larger companies producing ozokerite in the Northern Carpathians. This number was reduced to 11 in 1900, and to just two in 1902. In 1899, the Aktien Gesellschaft für Erdwachs und Petroleum Industrie company was established. It controlled a large part of the production fields and dominated exploitation of ozokerite until World War II.

A fast increase in the production of liquid, gaseous and solid hydrocarbons triggered a rapid development of geological and then geophysical exploration techniques.

Search for hydrocarbons – development of petroleum exploration techniques

In the early days, oil production was carried out by means of manual labour, the use of shovels and hand-dug trenches or shallow wells (Fig. 12). Exploration was simple, if not trivial – production

Fig. 12. Early production techniques for oil.

followed the natural surface occurrences of oil, so no specific geological knowledge was required to increase production. With time, however, deep wells started to be used, so a better understanding of the subsurface became a necessity. Experienced on-site geologists were in demand (e.g. Karol Angerman, Edward Widakiewicz and Henryk Walter), consultancy services were also provided by professors of geology working at universities in Lwów and Kraków (e.g. Alojzy Alth, Karol Bohdanowicz, Henryk de Cizancourt, Emil Dunikowski, Józef Grzybowski, Jan Nowak, Wojciech Rogala, Władysław Szajnocha, Leon Syroczyński, Wawrzyniec Teisseyre, Roman Załoziecki and Rudolf Zuber) – some with significant experience from abroad, such as Jan Grzybowski (who had worked in the Middle East, Peru and Italy: e.g. Grzybowski 1899), Henryk de Cizancourt (in South America: e.g. de Cizancourt 1933) or Rudolf Zuber (in South America, Africa and Russia: e.g. Zuber 1887). The involvement of highly qualified and motivated personnel inevitably led to a significant development of various exploration techniques, and this in turn led to new discoveries of deeper and larger oil and gas fields, as well as to a better understanding of the present-day structure of the Carpathian fold-and-thrust belt and its evolution. Of course, this process is far from being completed even today, but its solid foundations were laid down by those who were active during the pioneering days of petroleum and general geology in the Carpathians.

Rapid development of the oil industry in the Carpathians triggered intense geological studies of this thrust belt which resulted in the publication of numerous geological cross-sections, showing the main structural traps and gross lithologies of the reservoir rock (e.g. Bohdanowicz 1932; de Cizancourt 1931; Tołwiński 1934) (cf. Figs 13 & 14). The first geological sketches and maps were usually focused on oil-bearing areas (Figs 15 & 16); however, the necessity for large-scale mapping of larger regions that hosted oil and gas accumulations quickly became obvious (Zuber 1883*b*, 1897). In 1875, the State Mining Council (Krajowa Rada Gornicza), which included experienced and esteemed geologists such as Alojzy Alth, Emil Dunikowski, Feliks Kreutz, Julian Niedźwiedzki and Henryk Walter, was granted the necessary funds to start very ambitious geological cartographic projects – for example, the publication of the *Geological Atlas of Galicia* (*Atlas Geologiczny Galicyi*). Between 1875 and 1912, 25 volumes were published containing 99 maps at a scale of 1:75 000, covering almost the entire Northern Carpathians and their foreland. Published maps were accompanied by explanatory texts, sometimes very detailed and comprehensive. Figure 17 shows the Grybów–Gorlice sheet (Szajnocha 1896*a*), where Łukasiewicz conducted his first exploratory and industrial activities; and Figure 18 shows the Jaslo–Dukla sheet, where the famous Bóbrka oil field is located (cf. Koszarski & Bally 1994).

Geologists working in the late nineteenth and early twentieth centuries with the deep-water flysch deposits of the Outer Carpathians (i.e. turbidites) recognized that microfossils (foraminifera) could be used to correlate particular beds between wells and, as a result, construct more reliable geological cross-sections that were needed to plan deep – and, hence, expensive – exploration wells (e.g. Grzybowski 1895). This led to the rapid development of applied micropalaeontology, with Prof. Józef Grzybowski from Jagiellonian University in Kraków establishing his widely renowned school of stratigraphic micropalaeontology (cf. Kamiński *et al.* 1993), which predated the application of micropalaeontology for oil exploration in the USA (Gries 2017). Apart from Grzybowski, Bolesław Böhm from the Karpacki Instytut Geologiczno-Naftowy (Carpathian Petroleum-Geological Institute) in Borysław and Wilhelm Friedberg from the 'Pionier' company were also heavily involved in early stratigraphic micropalaeontological studies (e.g. Friedberg 1901).

Large amounts of information accumulated during decades of geological studies in the Carpathians allowed more synthetic work to be published on the regional and petroleum geology of the Northern Carpathians (e.g. Walter & Dunikowski 1883*b*; Tołwiński 1922; Nowak 1929). A most comprehensive study including a vast scope of information on the structure of the Outer Carpathians and other petroleum-bearing regions, but also on the origin of hydrocarbons and its relationship with organic matter deposited in sedimentary basins, was published by Rudolf Zuber (1918) (Figs 19 & 20). In 1903, the International Geological Congress was organized in Vienna during which Józef Grzybowski and Władysław Szajnocha led field trips to the oil-bearing areas in the northern Outer Carpathians.

In 1912, the Stacja Geologiczna (the Geological Station) was established in Borysław, which was later renamed as the Karpacka Stacja Geologiczna (the Carpathian Geological Station) and, finally, the Karpacki Instytut Geologiczno-Naftowy (the Carpathian Petroleum-Geological Institute). Its main mission was to conduct various studies related to the regional and petroleum geology of the northern Outer Carpathians. Its personnel were also active in the local school of drilling.

Since 1920, a very important role has been played by geophysical exploration techniques (cf. Kowalczuk 1999). The Department of Geophysics was established in Lwów at the Jan Kazimierz University under the guidance of Prof. Henryk Arctowski. In 1923, the first, very successful, exploration survey was completed by the 'Seismos' company from Hannover in the vicinity of Kraków (the

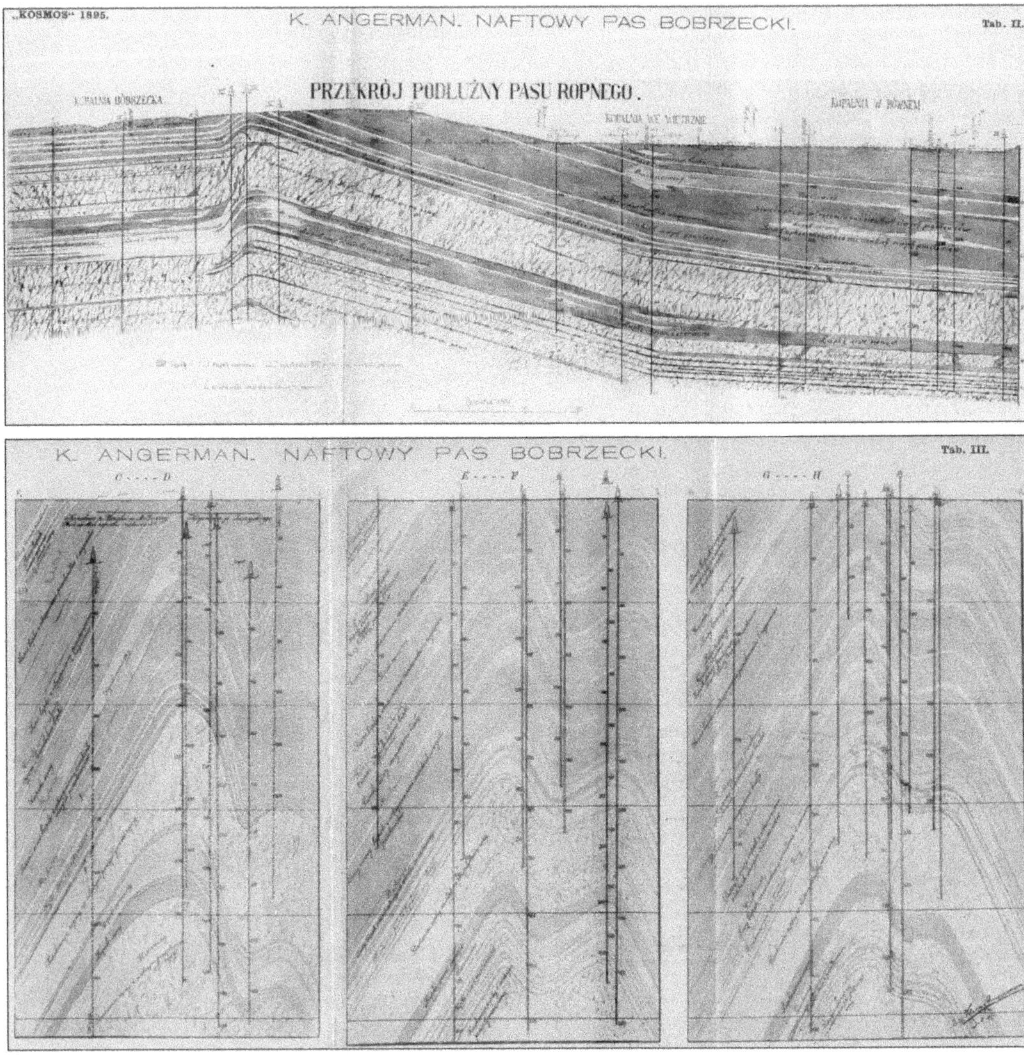

Fig. 13. Geological cross-sections of the Bóbrka oil field (Angerman 1895) (P. Krzywiec private collection).

Barycz–Kosocice area) using a seismic refraction method. Its goal, fully achieved, was to map Miocene salt deposits. Prospecting for potassium salts in the Kałusz area was partly based on a gravity survey completed in 1923. The first geoelectrical surveys for hydrocarbon exploration were commissioned in 1926 by the Vacuum Oil Company; they were acquired in Niebyłów, Synowódzko and Bitków. Prof. Henryk Arctowski and Dr Henryk Orkisz from Jan Kazimierz University in Lwów conducted the first magnetic surveys in the Stryj–Stanisławów–Dolina area in the Outer Carpathians.

In 1928, the Polish government established the ‘Pionier' company, based in Lwów. Fifty-one per cent of shares in this company belonged to the state. Its main goal was the exploration for hydrocarbons in new areas and the development of new, promising exploration techniques, including geophysical surveying. The Advisory Board of this company (Komitet Rzeczoznawców) included very prominent and well-known geologists, such as Prof. Karol Bohdanowicz, Prof. Jan Czarnocki and Prof. Jan Samsonowicz. Within this company, the Department of Geophysics was soon established. It was necessary to employ qualified staff for this department, so Zygmunt Mitera was sent to Germany and the USA, and Afrykan Kisłow and Stanisław Wyrobek went to Strasburg in order to obtain the necessary training and first-hand experience in modern geophysical exploration techniques. The first

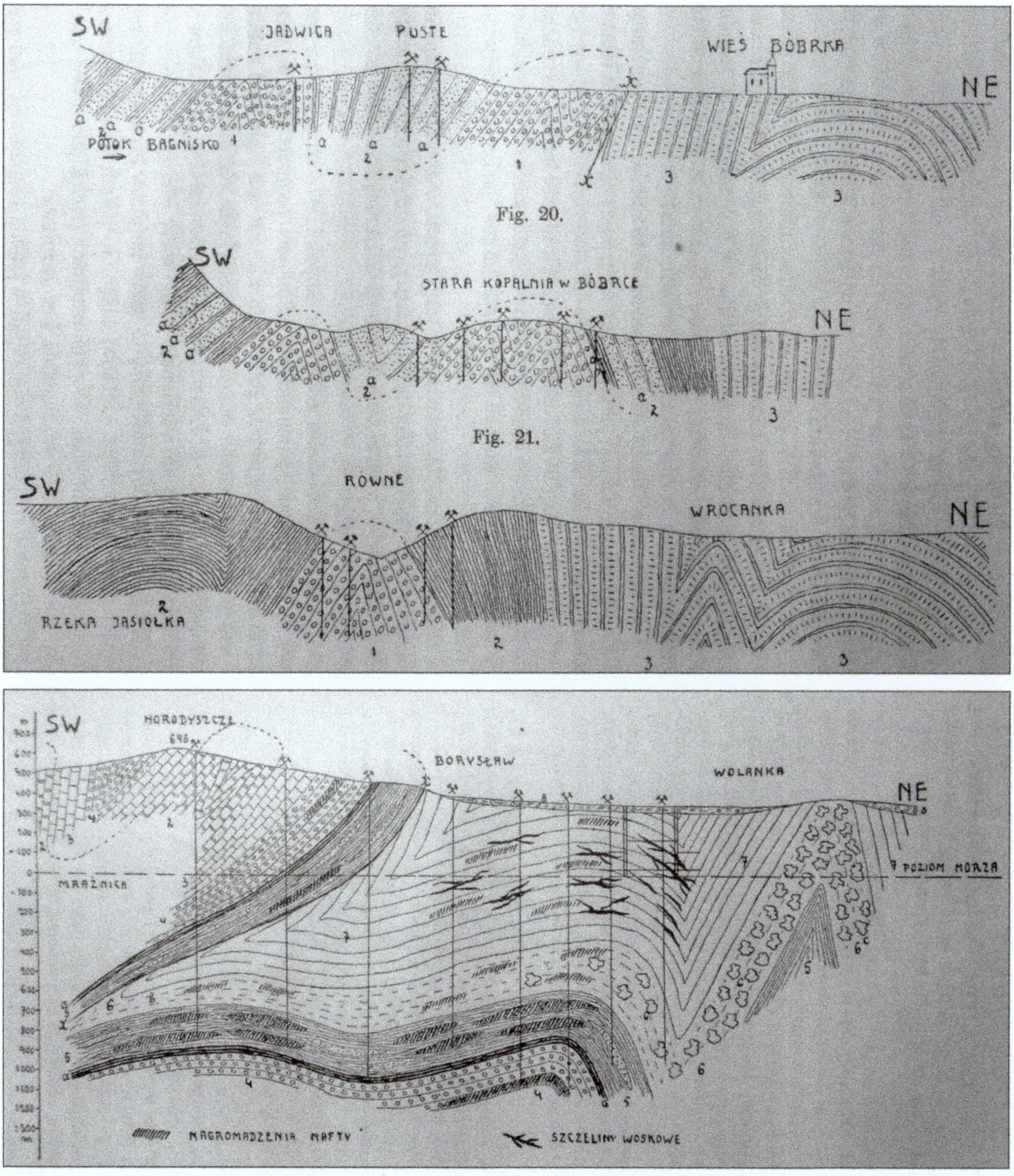

Fig. 14. Zuber's two geological cross-sections of the Bóbrka (upper panel) and Borysław (lower panel) oil fields (Zuber 1918) (P. Krzywiec private collection).

seismic refraction survey was conducted by the 'Pionier' company in 1933 in the vicinity of Truskawiec. In 1933, 'Pionier' purchased a six-channel seismic reflection acquisition system from the Seismograph Service Company (USA). At that time, Poland and Romania were the only countries in Europe where this kind of modern seismic equipment was being used. The first seismic reflection survey was acquired in 1934 in Nahujowice near Borysław, then in the Daszawa, Tustanowice and Stryj vicinities. The main goal of these surveys was to map the Miocene anhydrites that form an excellent seismic marker bed within the frontal Carpathians and their foredeep basin (cf. Krzywiec 2001). In 1936, 'Pionier' purchased another seismic reflection acquisition system, this time from Heiland Research

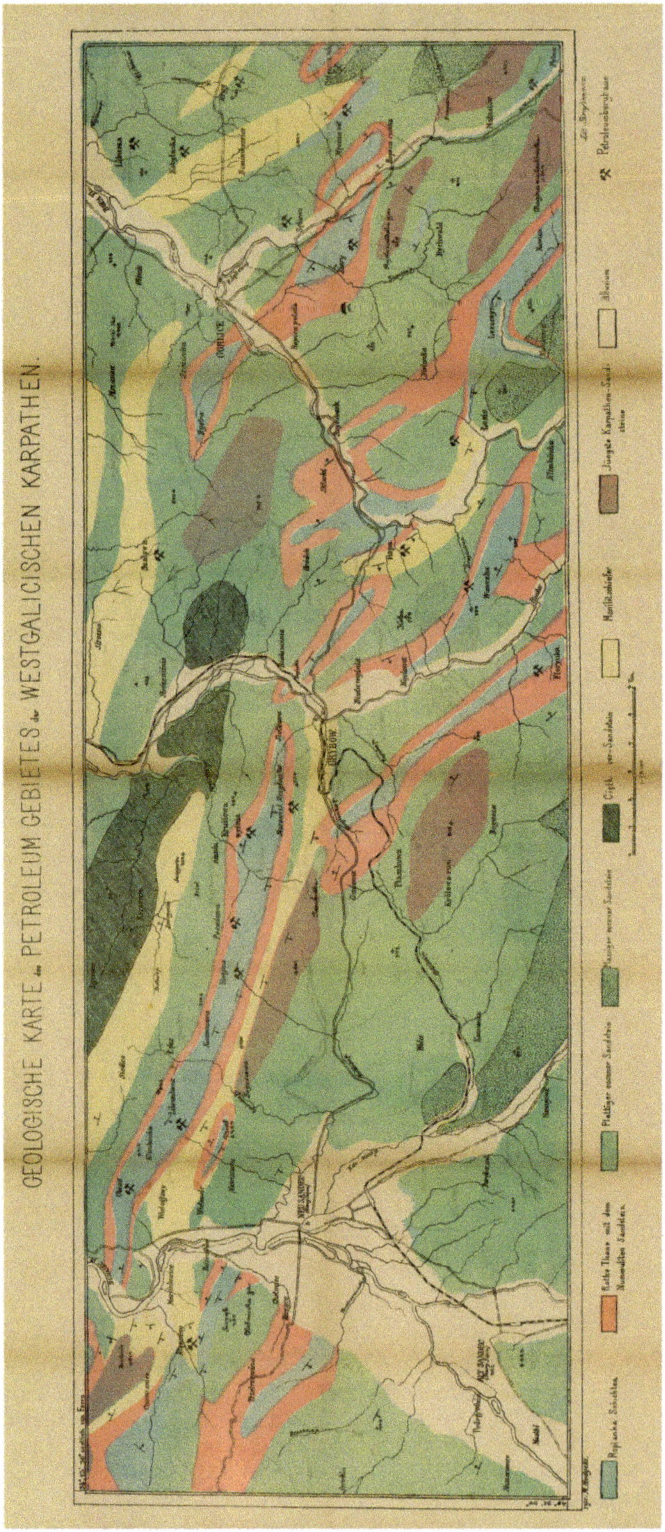

Fig. 15. Geological map of the Carpathian petroleum areas, Western Galicia (Walter & Dunikowski 1883*a*). Petroleumbergbaue, sites (mines) of oil production (P. Krzywiec private collection).

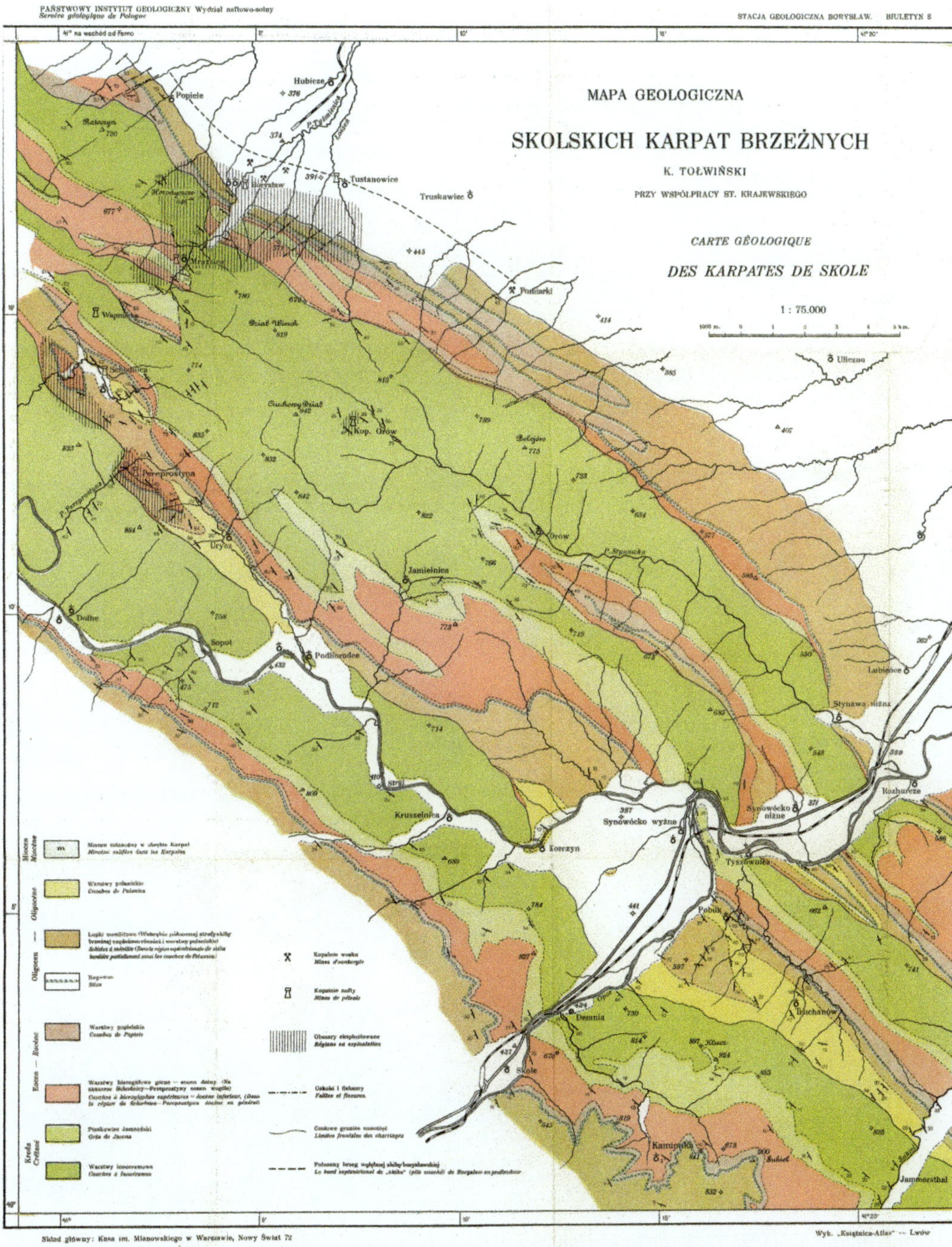

Fig. 16. Geological map of the frontal Skole nappe, Outer Carpathians (Tołwiński & Krajewski 1925). Vertically hatched areas are regions of active hydrocarbon production: Kopalnie wosku (Mines d'ozokeryte), ozokerite (i.e. earth wax) mines; Kopalnie nafty (Mines de pétrole), sites (mines) of oil production (P. Krzywiec private collection).

Corporation; the first data acquisition using this system took place in Tustanowice. In the same year, the production of geophones started in Poland in order to offset the cost of importing expensive geophysical equipment from abroad. Progressively, seismic data acquisition migrated towards the NW: in 1937, new

(a)

(b)

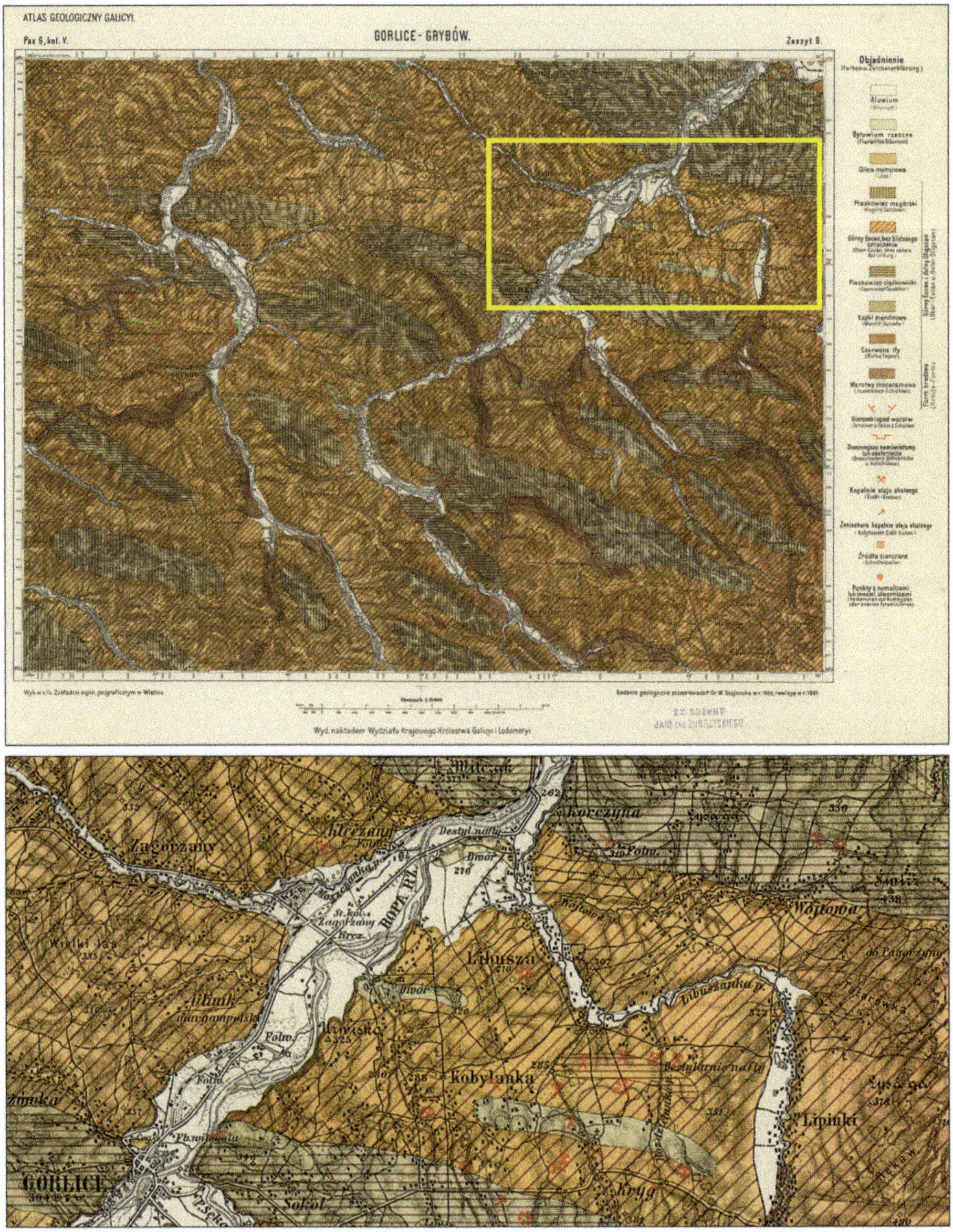

Fig. 17. (**a**) *Atlas Geologiczny Galicyi* [*Geological Atlas of Galicia*], Grybów–Gorlice sheet (Szajnocha 1896*a*). Kopalnie oleju skalnego, oil mines; zaniechane kopalnie oleju skalnego, abandoned oil mines (P. Krzywiec private collection). (**b**) Enlarged area shown by the yellow rectangle in (a) showing the vicinity of Gorlice–Libusza, where Łukasiewicz conducted his first activities.

data were acquired in Medenice–Sambor area, then in the Jarosław–Lubaczów area and, finally, in the Jarosław–Tarnów area. As a result, a regional map of the Miocene anhydrites from Kosów to Tarnów was compiled that was used to identify potential exploration targets within the frontal Carpathians

(a)

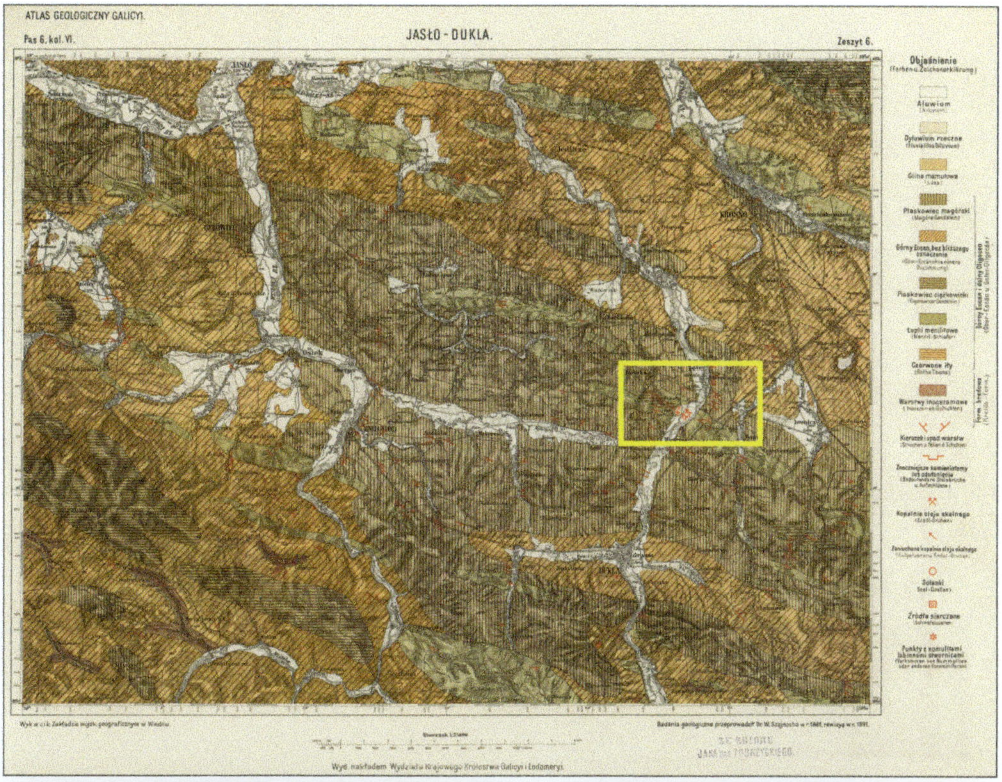

(b)

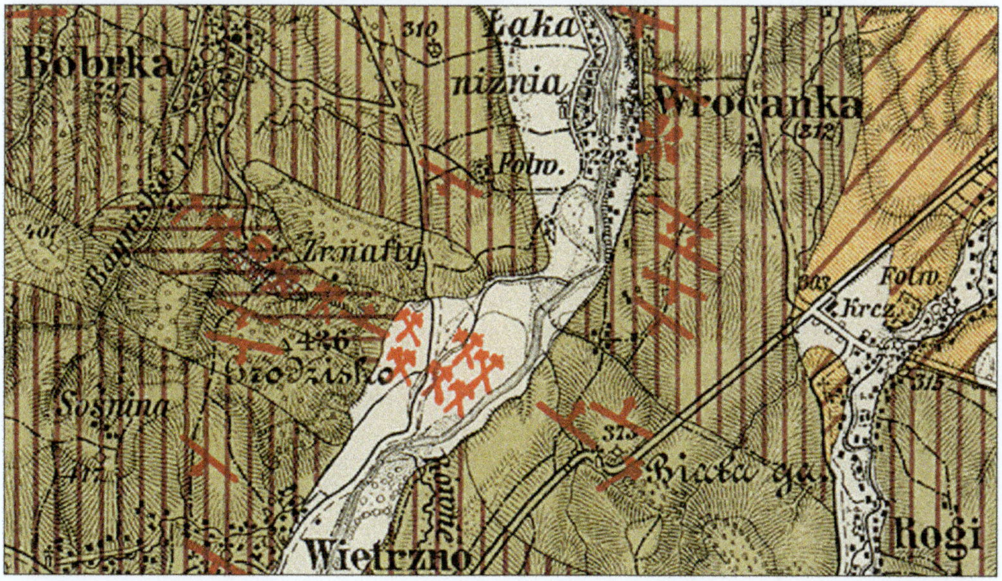

Fig. 18. (**a**) *Atlas Geologiczny Galicyi* [*Geological Atlas of Galicia*], Jasło–Dukla sheet (Szajnocha 1896*b*). Kopalnie oleju skalnego: oil mines, zaniechane kopalnie oleju skalnego: abandoned oil mines (P. Krzywiec private collection). (**b**) Enlarged area shown by the yellow rectangle in (a) showing the location of the Bóbrka oil field.

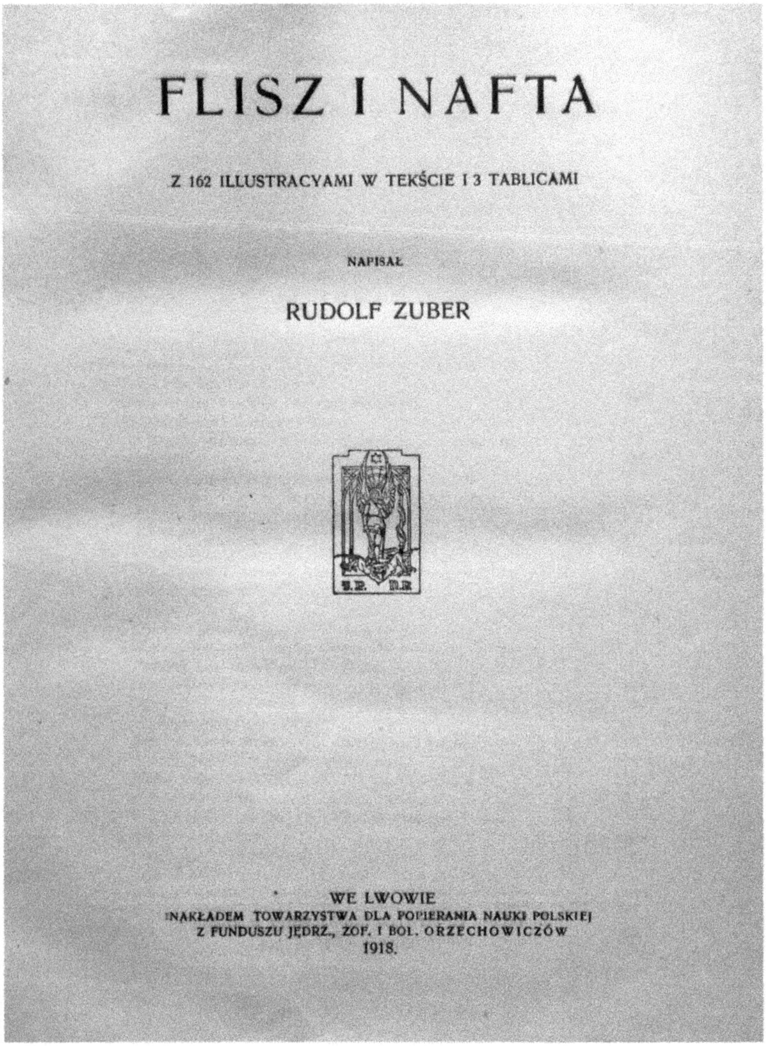

FLISZ I NAFTA

Z 162 ILLUSTRACYAMI W TEKŚCIE I 3 TABLICAMI

NAPISAŁ

RUDOLF ZUBER

WE LWOWIE
NAKŁADEM TOWARZYSTWA DLA POPIERANIA NAUKI POLSKIEJ
Z FUNDUSZU JĘDRZ., ZOF. I BOL. ORZECHOWICZÓW
1918.

Fig. 19. Title page of Rudolf Zuber's *Flisz i nafta* [*Flysch and Petroleum*] (1918) – his magnum opus regarding general and regional petroleum geology (P. Krzywiec private collection).

and their foredeep basin. 'Pionier' continued geophysical surveying in the vicinities of Stryj and Jarosław–Przemyśl.

The first electric well logging in the Northern Carpathians was completed by Schlumberger in 1935 in the 'Wownia 2' well near Stryj.

In 1938, just prior to World War II, Zygmunt Mitera and Stanisław Wyrobek left the 'Pionier' company and established their own private geophysical company 'Geotechnika', which was based in Przemyśl. They purchased a 12-channel seismic reflection acquisition system and started several ambitious exploration projects, not only in the Carpathians but also in central Poland. These promising geophysical activities were abruptly terminated by World War II.

Modern petroleum geology was born in the early nineteenth century in connection with the rapid development of the oil industry in the following decades. This development was accompanied by the establishment of various training centres, schools and university departments that trained highly skilled geologists, drilling engineers, geochemists and geophysicists.

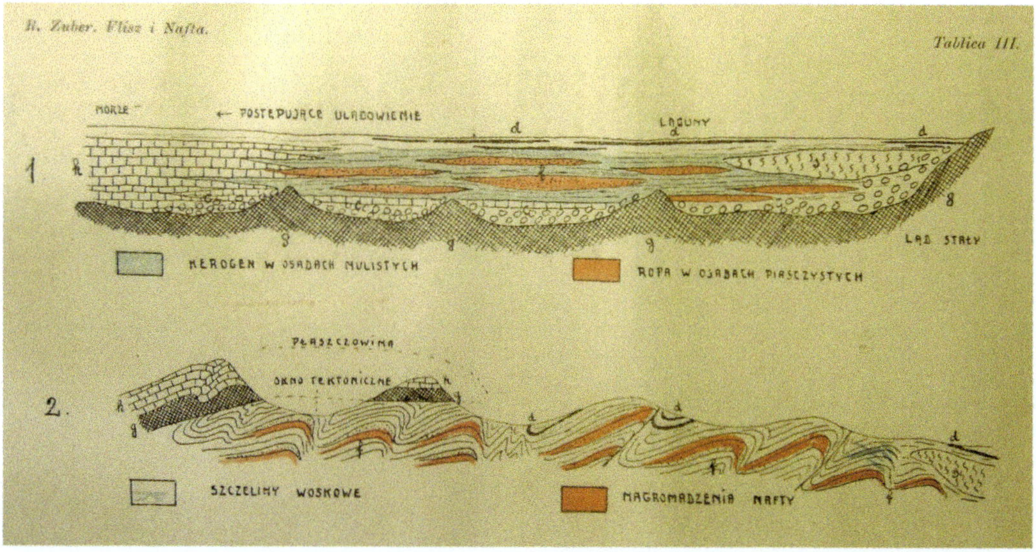

Fig. 20. Zuber's (1918) model of oil formation within the deep-water flysch series (P. Krzywiec private collection).

The petroleum industry requires skilled personnel – education for oil and gas exploration

Modern education in Earth sciences and mining in Poland dates back to 1816–26 when the Mining Academy (Akademiczna Szkoła Górnicza) established by Stanisław Staszic functioned in Kielce. However, this school did not have any lectures even remotely related to the exploration for hydrocarbons.

Starting in 1862, and then in 1871, 1873, 1875 and 1878, petitions to the Sejm Krajowy (Diet of the Kingdom of Galicia and Lodomeria) were submitted requesting the establishment of a mining school at the appropriate level. These were all rejected, however, mainly because the Austrian government did not want to create competition for the well-established mining schools in Loeben (Montanuniversität Leoben) and Przybram (Příbram Mining University). In 1982, lectures on the chemistry of petroleum were initiated and then, in 1891, an independent position for a lecturer in petroleum mining was established at the Szkoła Politechniczna (Polytechnic School) in Lwów. Finally, in 1892, the Austrian Ministry of Religion and Education approved a proposal regarding the establishment of lectures in petroleum mining and the technology of kerosene. The first lecturers, appointed as professors, were Leon Syroczyński and Roman Załoziecki. It took almost 5 years to organize these lectures and formally appoint both professors but, eventually, on 7 December 1897 the inaugural lecture was given by Prof. Leon Syroczyński. He continued his career until his retirement in 1915, and was replaced by Julian Fabiański, who in turn was replaced in 1935 by Stanisław Paraszczak.

Separate attempts were made to establish a mining higher school or university in Kraków. The Austrian ministry responsible for mining activities declared, in 1912, that it intended to establish a 'Mining Academy'. Following this, vigorous preparatory work was initiated; however, everything was halted in 1914 due to the beginning of World War I. The idea of an independent mining university located in Kraków was resurrected after Poland had regained its independence, and on 20 October 1919 the inaugural ceremony of the Akademia Górnicza (Academy of Mining) took place in Kraków. The Department of Drilling and Oil Exploitation was established in 1923 with Prof. Zygmunt Saryusz Bielski as its head; and petroleum geology was taught by renowned and experienced geologists, such as Prof. Karol Bohdanowicz, Prof. Jan Czarnocki and Prof. Wojciech Rogala.

It should also be mentioned that general geology and micropalaeontology were very strong at the oldest Polish university, Jagiellonian University in Kraków. Lectures, very often including various elements of petroleum geology, were being given by Prof. Ludwik Zejszner, Prof. Alojzy Alth, Prof. Jan Nowak, Prof. Józef Grzybowski and Prof. Władysław Szajnocha.

Conclusions

Following the early geological studies of Stanisław Staszic and the discovery of an effective oil-distillation process by Ignacy Łukasiewicz, the petroleum industry flourished in the Northern Carpathians in the second half of the nineteenth century and the beginning of the twentieth century, and for a short period of time gained the status of the third most prolific oil-producing region in the world. Industrial development was associated with rapid advancements in geology, including petroleum geology in its primordial form, pioneering applications of micropalaeontology for oil exploration and the beginning of regional serial mapping projects or early applications of geophysical surveying methods. The importance of the Northern Carpathians as a petroleum province quickly diminished and it has never attained a similar status. Those several decades, however, left important marks on the history of the oil industry. The main aim of this paper was to describe selected keystones in this fascinating history, some technological and scientific advancements, and the key people that were behind them.

Acknowledgements Two anonymous reviewers are thanked for useful suggestions that helped to finally shape this paper, Jonathan Craig (ENI) and Francesco Gerali (University of Western Australia) are thanked for their editorial efforts and advice.

Funding ORLEN Upstream Sp. z o.o. and PGNiG S.A. are cordially thanked for the financial support needed to publish additional colour figures.

References

Angerman, K. 1895. *Naftowy pas bóbrzecki ze stanowiska geologiczno-tektonicznego*. Sanok, Lwów.

Bohdanowicz, K. 1932. Geology and mining of petroleum in Poland. *AAPG Bulletin*, **16**, 1061–1091.

Brice, W. 2009. *Myth, Legend, Reality – Edwin Laurentine Drake and the Early Oil Industry*. Oil Region Alliance, Oil City, PA.

de Cizancourt, H. 1931. Geology of oil fields of Polish Carpathian mountains. *AAPG Bulletin*, **15**, 1–41.

de Cizancourt, H. 1933. Tectonic structure of northern sndes in Colombia and Venezuela. *AAPG Bulletin*, **17**, 211–228.

Dziadzio, P.S., Borys, Z. *et al.* 2006. Hydrocarbon resources of the Polish Outer Carpathians – Reservoir parameters, trap types, and selected hydrocarbon fields: A stratigraphic review. *In*: Golonka, J. & Picha, F.J. (eds) *The Carpathians and their Foreland: Geology and Hydrocarbon Resources*. AAPG Memoirs, **84**, 259–291.

Falimirz, S. 1534. *O ziołach i o mocy ich*. F. Ungler, Kraków.

Frank, A.F. 2005. *Oil Empire. Visions of Prosperity in Austrian Galicia*. Harvard University Press, Cambridge, MA.

Friedberg, W. 1901. Otwornice warstw inoceramowych okolic Rzeszowa i Dębicy (The foraminifera from the Inoceramus beds near Rzeszowa and Dębica). *Rozprawy Akademii Umiejętności, Wydział Matematyczno-Przyrodniczy*, **2**, 601–668.

Graniczny, M., Wołkowicz, S., Urban, H., Wołkowicz, K. & Zdanowski, A. 2015. History of exploration and extraction of crude oil in Poland until 1939. *Przegląd Górniczy*, **12**, 151–156 [in Polish with an English summary].

Gries, R.R. 2017. *Anomalies. Pioneering Women in Petroleum Geology: 1917–2017*. Steuben Press, Longmont, CO, 405.

Grigelis, A., Wójcik, Z., Narębski, W., Gelumbauskaite, L.Z. & Kozák, J. 2011. Stanisław Staszic: an early surveyor of the geology of central and Eastern Europe. *Annals of Science*, **68**, 199–228.

Grzybowski, J. 1895. Dotychczasowe rezultaty badań mikroskopowych namułów wiertniczych galicyjskich kopalń naftowych [Preliminary results of microscopic investigations of drilling muds from the Gallician petroleum mines]. *Kosmos*, **20**, 519–524.

Grzybowski, J. 1899. Die Tertiärablagerungen des nördlichen Peru und ihre Molluskenfauna. *Neues Jahrbuch für Mineralogie, Geologie und Petrographie*, **13**, 610–664.

Hacquet, B. 1790–96. *Hacquet's neueste physikalisch-politische Reise in den Jahren 1788, 89 und 90 durch die Dacischen und Sarmatischen oder Nördlichen Karpathen*. Raspische Buchhandlung, Nürnberg.

Kamiński, M.A., Geroch, S. & Kamiński, D.G. 1993. *The Origins of Applied Micropalentology: The School of Józef Grzybowski*. Alden Press, Oxford.

Karnkowski, P. 1993. *Złoża gazu ziemnego i ropy naftowej w Polsce, Volume 2*. Karpaty i zapadlisko przedkarpackie, Kraków.

Kluk, K. 1781. *Rzeczy kopalnych osobliwie zdatnych szukanie, poznanie i zażycie, Volume I. O rzeczach kopalnych w powszechności, o wodach, solach, tłustościach ziemnych i ziemiach*. W Drukarni Xięży Pijarów, Warszawa.

Koszarski, A. & Bally, A. 1994. The world's oldest industrial oil field. Poland relights the fire at Bóbrka. *AAPG Explorer*, **12**, 26–27.

Kowalczuk, J. 1999. *1895–1995. 100-lecie geofizyki polskiej*. Kalendarium, Kraków.

Krzywiec, P. 2001. Contrasting tectonic and sedimentary history of the central and eastern parts of the Polish Carpathian Foredeep Basin – results of seismic data interpretation. *Marine and Petroleum Geology*, **18**, 13–38.

Krzywiec, P. 2016. The First Geological Map of Eastern Europe. *GeoExpro*, **12**, 68.

Marcin z Urzędowa. 1595. *Herbarz Polski To iest O Przyrodzeniu Ziol y Drzew Rozmaitych, Y Innych Rzeczy Do Lekarzy Należących Księgi Dwoie*. W Drukarni Łazarzowey, Kraków.

Mau, M. & Edmundson, H. 2015. *Groundbreakers: The Story of Oilfield Technology and the People Who Made it Happen*. FastPrint Publishing, Peterborough.

Mazan, L. & Wolwowicz, R. 2006. *A Long Time Ago in the Carpathians. An Account of the Polish Petroleum. A Photo Album 1853–1939*. Wydawca, Kraków.

NATER, R. & SOZAŃSKI, J. 2002. *Nafta. Ludzie i Fakty.* Handel- Usługi- Wydawnictwa Zygmunt Nater, Lesko.

NOWAK, J. 1929. *Die Geologie der Polnischen Ölfelder.* Ferdinand Enke, Stuttgart.

OSZCZYPKO, N., KRZYWIEC, P., POPADYUK, I. & PERYT, T. 2006. Carpathian Foredeep Basin (Poland and Ukraine): Its sedimentary, structural, and geodynamic evolution. *In*: GOLONKA, J. & PICHA, F.J. (eds) *The Carpathians and their Foreland: Geology and Hydrocarbon Resources.* AAPG Memoirs, **84**, 293–350.

PUSCH, G.G. 1836*a*. *Geognostische Beschreibung von Polen so wie der übrigen Nordkarpathen-Länder. Zweiter Theil.* Cotta, Stuttgart-Tübingen.

PUSCH, G.G. 1836*b*. Geognostische General-Karte von dem Königreichen Polen und Galizien mit angrenzenden Theilen von Oberschlesien, Ungarn, Siebenbürgen, Moldau und Podolien, entworfen und gezeichnet von (...). Blatt III. *In*: *Geognostischer Atlas von Polen.* Cotta, Stuttgart-Tübingen.

RZĄCZYŃSKI, G. 1721. *Historia naturalis curiosa Regni Poloniae, Magni Ducatus Lituaniae, annexarumq; provinciarum in tractatus XX divisa: Ex scriptoribus probatis, servata primigenia eorum phrasi in locis plurimis, ex M.S.S. variis, Testibus oculatis, relationibus fide dignis, experimentis, desumpta opera P. Gabrielis Rzączyński Soc. Jesu.* Typis Collegii Soc. Jesu, Sandomirae.

RZĄCZYŃSKI, G. 1742. *Auctuarium Historiae Naturalis Regni Poloniae Magnique Ducatus Lithuniae Annexarumque Provinciarum in Puncta XII.* Gedani.

SARTORI, F. 1810. *Naturwunder des Oesterreichischen Kaiserthumes.* Doll, Wien.

SCHARR, K. 2004. *Die Karpaten. Balthasar Hacquet und das 'vergessene' Gebirge in Europa.* Studien, Innsbruck.

SIENNIK, M. 1568. *Herbarz to iest Ziół tutecznych, postronnych y zamorskich opisanie, co za moc maią, a iako ich używać tak ku przestrzeżeniu zdrowia ludzkiego, iako ku uzdrowieniu rozmaitych chorob.* W Drukarni Mikołaja Szarffenberga, Kraków.

ŚLĄCZKA, A., KRUGŁOV, S., GOLONKA, J., OSZCZYPKO, N. & POPADYUK, I. 2006. Geology and hydrocarbon resources of the Outer Carpathians, Poland, Slovakia, and Ukraine: General geology. *In*: GOLONKA, J. & PICHA, F.J. (eds) *The Carpathians and their Foreland: Geology and Hydrocarbon Resources.* AAPG Memoirs, **84**, 221–258.

SOZAŃSKI, J., KUK, S., JARACZ, C. & DZIADZIO, P.S. 2006. How the modern oil and gas industry was born: Historical remarks. *In*: GOLONKA, J. & PICHA, F.J. (eds) *The Carpathians and their Foreland: Geology and Hydrocarbon Resources.* AAPG Memoirs, **84**, 811–834.

STASZIC, S. 1806*a*. *O Ziemiorodztwie gór dawney Sarmacyi a później Polski.* Drukarnia Xięży Piarów, Warszawa.

STASZIC, S. 1806*b*. *Carta Geologica totus Poloniae, Moldaviae, Transylvaniae, Hungariae et Valachiae.* Drukarnia Rządowa, Warszawa.

STASZIC, S. 1815. *O Ziemiorodztwie Karpatów i innych gór i równin Polski.* Drukarnia Rządowa, Warszawa.

SYXT, E. 1617. *O cieplicach we Skle ksiąg troje.* W Drukarni Akademiey drukował Chrzystoph Wolbramczyk, Zamość.

SYXT, E. 1780. *O cieplicach we Skle ksiąg troje przez Erazma Syxta philozophiey y medycyny Doktora napisana.* Edycya druga, U Jana Augusta Pozera, Warszawa.

SZAJNOCHA, W. 1896*a*. *Atlas Geologiczny Galicyi, zeszyt VI. Grybów–Gorlice.* Wydawnictwo Komisyi Fizjograficznej Akademii Umiejętności, Kraków.

SZAJNOCHA, W. 1896*b*. *Atlas Geologiczny Galicyi, zeszyt VI. Jasło–Dukla.* Wydawnictwo Komisyi Fizjograficznej Akademii Umiejętności, Kraków.

TOŁWIŃSKI, K. 1922. *Złoża ropy i wody podziemne Borysława na tle budowy geologicznej.* Drukarnia Związkowa, Kraków.

TOŁWIŃSKI, K. 1928. *Obszary naftowe i gazowe Polski w Karpatach i na przedgórzu.* Książnica-Atlas, Lwów.

TOŁWIŃSKI, K. 1934. Natural gas in Poland. *AAPG Bulletin*, **18**, 892–907.

TOŁWIŃSKI, K. & KRAJEWSKI, S. 1925. *Mapa geologiczna skolskich Karpat brzeżnych.* Książnica-Atlas, Lwów.

WALTER, H. & DUNIKOWSKI, E. 1883*a*. *Geologische Karte des Petroleum Gebietes der Westgalicischen Karpathen.* Verhandlungen der Kaiserlich Königlichen Geologischen Reichsanstalt, Wien.

WALTER, H. & DUNIKOWSKI, E. 1883*b*. *Das Petroleumgebiet der galizischen Westkarpathen.* Verhandlungen der Kaiserlich Königlichen Geologischen Reichsanstalt, Wien, **8**, 20–25.

WINDAKIEWICZ, E. 1875. *Olej i wosk ziemny w Galicyi.* Nakładem Administracyi Gazety Lowskiej, Lwów.

WOŁKOWICZ, S. & WOŁKOWICZ, K. 2014. Geological cartography in Poland in the 19th century. *Geological Quarterly*, **58**, 623–658, https://doi.org/10.7306/gq.1198

WOŁKOWICZ, S., GRANICZNY, M., WOŁKOWICZ, K. & URBAN, H. 2017. History of the oil industry in Poland until 1939. *In*: MAYER, W., CLARY, R.M., AZUELA, L.F. & WOŁKOWICZ, S. (eds) *History of Geoscience: Celebrating 50 Years of INHIGEO.* Geological Society, London, Special Publications, **442**, 401–411.

WOLWOWICZ, R. (ed.). 1994. *Historia polskiego przemysłu naftowego* (vols 1 & 2). Muzeum Regionalne PTTK im. Adama Fastnachta w Brzozowie, Brzozów-Kraków.

ZUBER, R. 1883*a*. *Nafta i wosk ziemny w Galicyi.* Drukarnia Józefa Bergera, Warszawa.

ZUBER, R. 1883*b*. *Mapa galicyjskiego obszaru naftowego.* Warszawa.

ZUBER, R. 1887. Informe sobre el petróleo de la Laguna de la Brea (provincia de Jujuy, Republica Argentina). *Boletin de la Academia Nacional de Ciencias de Cordoba*, **10**, 442–448.

ZUBER, R. 1897. *Mapa obszarów naftowych w Galicyi.* Nakładem Autora, Skład Główny w Redakcji, Lwów.

ZUBER, R. 1918. *Flisz i nafta.* Nakładem Towarzystwa dla popierania nauki polskiej, Lwów.

History of the oil and gas industry in Romania

ALINA DANA TULUCAN*, LUCIA-ELENA SOVEJA-IACOB & CSABA KREZSEK

OMV Petrom, Coralilor 22, Bucharest, 013329, Romania

A.D.T., 0000-0003-4389-1635; L-E.S-I., 0000-0002-1879-3585; C.K., 0000-0001-8685-0628

**Correspondence: alina.tulucan@petrom.com*

Abstract: Oil and gas activities in Romania cover a long time interval from antiquity until the present day. Most importantly, the year 1857 represents the starting point for industrial production in Romania when three world oil firsts were achieved:

- the first country with a crude oil production formally recorded in domestic and international statistics (275 tons);
- the first petroleum industrial refinery, located in Ploiești, with a processing capacity of 7.5 tons/day;
- Bucharest became the first city in the world to have public illumination, using lamp oils.

After 1857, the evolution of the oil and gas industry in Romania was controlled by important historical events: World Wars I and II, the communist regime (1945–89), and the post-communist period (1990–today).

This chapter presents a brief history of the oil and gas industry in Romania from antiquity to the present time, building on previous works such as, for example, Aldea & Rădulescu (1957), Teodorescu & Ștefănescu (1968), Pătruț *et al.* (1973) and Ivănuș *et al.* (2009), most of which are not in the English language. The following websites provide further background information:

http://furcuta.blogspot.ro/2009/10/romanian-petroleum-history.html

http://peopletales.blogspot.ro/2013/10/oil-industry-in-romania-world-premieres.html

https://ro.wikipedia.org/wiki/Industria_petrolului_%C3%AEn_Rom%C3%A2nia

http://www.ropepca.ro/en/articole/oil-and-gas-history-in-romania/55/

http://www.imra-group.com/employers/local-oil-gas-history

Pre-industrial history (until 1857)

Despite the lack of documents to show oil use in antiquity, archaeological research performed at Târgșor (Prahova County) revealed a large variety of objects dated to the fifth and sixth centuries AD with traces of bitumen. These suggest that black oil was used frequently by that time (Teodorescu & Ștefănescu 1968).

The Middle Ages are better documented. One of the oldest documents found refers to the presence of black oil in an act of donation for the Lucăcești village (Fig. 1), called '*uric*', issued on 4 October 1440 by the Chancellor's Office of Moldavia. In addition, several other donation acts or sale-purchase deeds are known that refer to black oil pits and associated revenue from different areas (e.g. Cricov, Matița and Lucăcești). During this period, the local people exploited the oil using different methods, such as oil pits, mines, washeries and shafts (Ivănuș *et al.* 2009).

Oil pits were first developed in areas where oil reservoirs cropped out – mostly in the East Carpathians and the Carpathians Bend Zone (Fig. 1). The exploitation technique consisted of gathering the oil by natural gravitational flow and then exposing it to solar heat in order to obtain black oil through the evaporation of gas.

Exploitation using mines was developed when the old outcrops started to be depleted, in the second half of the sixteenth century, and continues at a very small scale even today at the Sărata Monteoru oil mine (Fig. 1).

The washeries were initially built as holes dug at the surface and in shaly rocks where the bituminous sands were washed with warm water until the black oil floated to the surface and was then collected using large wooden spoons. Later, these were replaced by wooden washeries (1.5 × 1.2 × 0.5 m) with two unequal compartments; these were still being used during industrial production of crude oil in the case of blowouts and to recover residual water from distilleries.

The shafts were considered to be the best manual technique for oil extraction (Fig. 2), being used from the second half of the eighteenth century and through the entire nineteenth century. Digging was performed using circular sections for hard rocks (Moldavia) and square sections for soft rocks

From: Craig, J., Gerali, F., MacAulay, F. & Sorkhabi, R. (eds) 2018. *History of the European Oil and Gas Industry*. Geological Society, London, Special Publications, **465**, 191–200.
First published online May 18, 2018, https://doi.org/10.1144/SP465.15

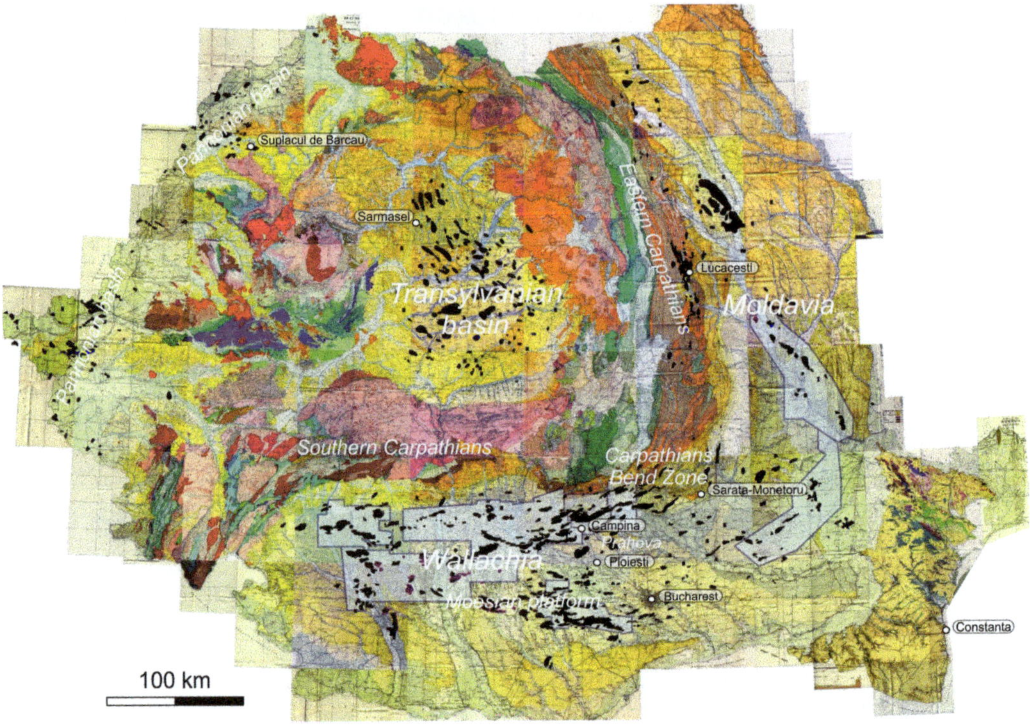

Fig. 1. Geological map of Romania (1:200 000) with oil and gas fields shown (Murgeanu *et al.* 1964–68).

(Wallachia) in primitive conditions of lighting (Fig. 3) and ventilation, which resulted in many accidents. The deepest shaft was 320 m and was dug by the Colombia Company in 2 years (1908–10) at Breaza (near Câmpina: Fig. 1).

Formation of an industry (1857–1916)

The year 1857 marked the inauguration of the formation of Romania's oil industry and the beginning of proper geological research. This later resulted

Fig. 2. Oil extraction from a shaft using a balance (postcard from the National Oil Museum, Ploiești).

Fig. 3. Shaft illumination using a mirror system (Ivănuș *et al.* 2009).

in the first geological map of Romania (1:800 000) being published in 1880 by Drăghiceanu in *Annals of the Geological Institute of Vienna* (Pătruț *et al.* 1973).

The first research in the field of petroleum geology was performed by Pilide in 1877, who studied the productive petroliferous zones from Prahova County (Fig. 1) (Aldea & Rădulescu 1957), and was followed by a rapid development stage through the work of Grigore Cobălcescu (e.g. Cobălcescu 1883) and Gregoriu Ștefănescu. Pilide stated that crude oil was located in the axial zone of anticlines often accompanied by salt, and studied the petroleum formations from the Moldavia and Wallachia provinces (Fig. 1). Then, in 1905, a map of Romania's petroliferous zones was printed (Preda 1957). Starting with 1906, the development of petroleum geology was strongly connected to the activity of Romania's Geological Institute, which was led for some time by Ludovic Mrazec (Mrazec & Teisseyre 1902; Mrazec 1910). The studies issued by the institute until 1916 addressed the origin of petroleum and formation waters, and the location of petroleum outcrops in order to identify new opportunities.

The year 1861 marked the beginning of drilling and oil production through wells in Romania, the first one (150 m depth) being drilled manually using wooden rods and a drill-type bit at Mosoarele (Bacău County). The first mechanized operation took place in Glodeni (Dâmbovița County) in 1890 (Fig. 4). Up until 1912, around 800 wells had been drilled using mainly an improved Canadian mechanical drilling system (92%) and, on a smaller scale, the Pennsylvanian system (3%) and hydraulic percussion system (5%). Rotary drilling was used for the first time in 1907 by the Romanian–American Standard Oil Company to drill well 30 RA Speranța, which unfortunately was dry (Ivănuș *et al.* 2009).

Before 1900, crude oil was exploited by free blowouts (Fig. 5), causing important environmental damage. The free-flowing method was improved first in 1900 by introducing spherical caps, and then, in 1912, by the prototype of a blowout preventer (a hydraulically driven piston valve) experimented in Câmpina by V. Tacit and V. Pușcariu (Ivănuș *et al.* 2009). These first experiments eventually led to the introduction of the first 'Christmas tree'. Until 1916, the oil was exploited mainly by foreign-owned companies (98%). The first foreign investment was recorded in 1895 with the establishment of the great Steaua Română Oil Company.

Natural gas production (113 000 m^3) in Romania was recorded for the first time in 1913 at the Sărmășel site (Fig. 1: Transylvanian Basin). The first gas discovery was in 1909 at the 2 Sărmășel well. Drilling stopped at 302 m due to a strong and permanent gas outburst, which could be sealed only after 27 months of free eruption (Ciupagea *et al.* 1970). In 1915, the first company in Europe (the Hungarian Natural Gas Company – UEG) dedicated exclusively to natural gas exploration, production and distribution was established. UEG drilled about 15 wells up until 1918 in the Transylvanian Basin (Fig. 1), all with gas (Ivănuș *et al.* 2009).

Industrial petroleum processing in Romania started in 1857 when the first refinery in the world was built by the Mehedințeanu Brothers, near Ploiești (Fig. 1). Initially, the crude oil was rudimentary processed, obtaining products such as lamp oil used to illuminate cities, gasoline, oils and paraffin. From 1904 onwards, the quality and number of

Fig. 4. The first mechanized operation at Glodeni (Dâmbovița County) in 1890 (postcard from the National Oil Museum, Ploiești).

refined products increased when the large-capacity refineries were built with modern equipment by companies with foreign capital. In 1908, Lazăr Edeleanu laid the foundation of petrochemistry, obtaining a patent for a selective refinement method of petroleum products with liquid sulphur dioxide (Ivănuș *et al.* 2009).

The crude oil and petroleum products at the beginning were stored in wooden barrels, which were replaced by metal tanks before 1916. The rapid development of railways between 1870 and 1878 from 173 to 1300 km facilitated the transportation of such products by railway tank cars. By 1916, Romania had a well-developed railway network, which was fully exploited by the petroleum industry.

Between 1880 and 1915, approximately 1335 km of pipeline was built to improve the transportation system and in 1909 the Constanța petroleum harbour (Fig. 1) was officially inaugurated for exporting purposes. After 1904, many companies specializing in repairing and manufacturing petroleum equipment were established, starting a new area of development in Romania's petroleum industry.

Romania's petroleum industry during World War I (1916–19)

Romania entered the war on 17 August 1916, joining the Entente and causing serious problems

Fig. 5. Well 1 Româno-Americană in free blowout (postcard from the National Oil Museum, Ploiești).

to the Central Power by depriving them of its petroleum. In the following months, Germany attacked Romania with the purpose of occupying the petroliferous regions and to ensure the sustainability of the war.

Due to the obvious fact that Germany's offensive could not be stopped, in the autumn of 1916 Britain's leaders suggested that Romania's authorities should proceed with the self-destruction of its petroleum industry in order to deprive Germany, at least temporarily, of its petroliferous regions. Compensation being promised, Romania agreed; more than 1500 wells were obstructed, 1000 shafts and wells, and many petroliferous products were set on fire, storage tanks (150 000 m^3) were blown out, and more than 70 refineries were destroyed (Pizanty 1931). On 3 December 1916, the Romanian government left Bucharest and, by 16 December, Germany's military administration was set up in Romania. The self-destruction of the petroleum industry prevented German access only temporarily. By the end of 1917, Germany had managed to establish an oil production similar to that of 1916, and to restore four large and medium-sized refineries (Ivănuș *et al.* 2009).

After the war was lost by the Central Powers and Germany had retreated from Romania, at the end of 1918, the issue of compensation for the self-destruction of its petroleum industry emerged. In August 1919, a commission for the settlement of compensation was set up, which finalized its assessment of the damage in February 1922. After 6 months of extensive debate, the amount of compensation was established at £9 980 527 but the issue of how the payment was to be made remained (Ivănuș *et al.* 2009). The positive outcome of the war for Romania was the unification on, 1 December 1918, of Romania's territories.

Romania's petroleum industry during the interbellum period (1920–39)

After the war, an intensive effort was made to repair the damage created by the self-destruction of its petroleum industry and by Germany's occupation of Romanian territories. During the interbellum period, oil legislation had various effects. The new Mining Law issued in June 1924 created facilities for Romanian capital to become involved in the oil industry, but the nationalization effects were cancelled by the law issued in March 1929 that allowed foreign capital to enter, which had negative consequences for the Romanian petroleum industry. Finally, the law issued in March 1937 granted preferential treatment for companies with Romanian capital and ensured the preservation of resources through a rational exploitation (Ivănuș *et al.* 2009).

The oil- and gas-bearing zones were extended through geological and geophysical research. Petroleum geological research focused on defining the origin of petroleum, the nature of field waters, and on explaining salt diapirism and its role in the formation of oil accumulations.

Ludovic Mrazec was the first Romanian scientist to support the organic theory of petroleum origin, attributing the formation of hydrocarbons to the bituminization process of organic matter. In 1924, I. Popescu-Voitești described the main source rocks in Romania (Ivănuș *et al.* 2009).

Geophysical research consisted of an intensive gravimetric prospecting campaign between 1928 and 1936, using torsional balances. In 1936, using a Thyssen gravimeter, the first Bouger gravimetric anomaly map was obtained, covering a considerably large part of the country. Magnetometric methods were used for salt prospecting in the Florești area (Ștefănescu 1957).

Electric logging (well investigation), which had become a systematic method for underground research since 1931, had been experimented on in Romania between 1928 and 1929.

The rotary drilling system has been widely accepted as starting in 1924. This technique permitted deeper drilling and contributed to the discovery of new petroleum fields. After 1930, the safety of the drilling process was increased by using intermediate casing and installing blowout preventers. In 1934, the second deepest well in the world (1 Chițorani) was drilled in record time (139 days in total), with a total depth of 3382 m (Ivănuș *et al.* 2009).

After the war, oil was produced by naturally controlled flow (bailing and swabbing), replacing the open-flow system and by artificial flow (air-lift and gas-lift injection) methods. The gas-lift method was preferred, being first experimented on in 1921 at the 15 AR Moreni well (near Poliești: Fig. 1). Natural gas production increased by exploiting new gas fields; Romania ranked third in the world by the end of the period in 1938.

Crude oil processing went through an intensive modernization process after the great effort to restore and restart the refineries damaged during the war. The hierarchy of the refineries was continuously modified due to the development of new processing units and the shutdown of others. During this time, crude oil was processed using pipe-still distillation plants and thermal cracking plants. Introduced in 1926 in Romania, the thermal cracking process allowed the fabrication of increased quantities of gasoline with a high octane number for new automobile and aviation engines, representing important progress in crude oil processing.

The transportation of petroleum products was ensured by the railway network and pipeline system, both of them being well developed by the end of the period. In 1935, the pipeline networked included 2600 km of private pipelines (73% for crude oil, 15% for gasoline and 12% for gas) and 750 km of state-owned pipelines (Ivănuș *et al.* 2009).

Regarding petroleum equipment, the existing companies set up numerous workshops for repairing and manufacturing spare parts and components, and for manufacturing drilling production rigs.

Romania's petroleum industry during World War II (1940–44)

In June 1939, Romania signed an economic treaty with Germany in order to gain some time and some economic advantage, avoiding the status of an occupied country. The treaty allowed Germany unrestricted access to Romania's resources (Brestoiu 1986). On 22 June 1941, Romania entered the war on Germany's side; the possession of Romania's oil fields was essential for the Axis countries. Between 1943 and 1944, the Romanian oil industry was severely affected by Germany's failure to provide the necessary equipment and, to a greater extent, by the destruction caused during Anglo-American air bombardments.

On 23 August 1944, Romania turned her arms against her former allies, contributing to shortening the war by 6 months and the Axis capitulation on May 1945. Despite this last action, Romania was treated as an occupied territory by Russia and was forced to sign the Armistice Convention in September 1944. After signing it, Romania's entire economic potential was placed under Soviet control (Ivănuș *et al.* 2009).

Romania's petroleum industry during the communist regime (1945–89)

The year 1944 represented the start of a dark period for Romania's industry, as the country was forced to accept the establishment of the well-known SOVROMS that captured and drained its economy in all sectors. Reconstruction of the petroleum industry that was severely affected during the war was delayed due to lack of indispensable materials. Between 1945 and 1947 significant amounts of petroleum equipment, such as oil drilling systems and refinery machinery, petroleum products and transport facilities (railway engines, tank wagons, river and sea ships), took the road to the USSR as war trophies (Ivănuș *et al.* 2009).

Immediately after the war, the Romanian petroleum industry was controlled by Sovrompetrol (1945–55), established as a Soviet–Romanian joint venture. After 10 years of one-way exploitation of Romanian oil to the benefit of the USSR, the Soviet part in the company was transferred to the Ministry of Industry, Petroleum and Chemistry. In June 1948, local and foreign companies were nationalized and the petroleum industry was reorganized after the Soviet model.

Geological research continued with the same objectives: source rocks, field waters and salt diapirism, with its role in the formation of hydrocarbon accumulations. For a long period, the hypothesis of autochthonous oil was strongly sustained by many specialists.

In a synthesis of petroleum geology in Romania, D. Paraschiv observed the distribution of petroliferous fields in three main geological regions (Fig. 1): pre-Carpathian depressions, with most of the fields belonging to the Tertiary; the Moesian Platform, with the oldest productive geological formations (Devonian); and the Transylvanian Basin with gas fields (Paraschiv 1975).

Regarding the origin of oil-field waters, Gh. Macovei stated that they were not exclusively of marine origin, having a partially organic origin from organisms that generated petroleum (Ivănuş *et al.* 2009). The genetic types of field waters have been defined on hydrogeological maps.

Salt diapirism was intensively studied during the communist regime and two main (Burdigalian–Early Miocene and Badenian–Middle Miocene) salt levels were described. Salt was considered to be a seal rock, contributing to the formation of traps (Pătruţ *et al.* 1973). The richest oil fields in Romania are located on the highest salt diapiric alignments (e.g. Gura Ocniţei–Moreni–Băicoi–Ţintea).

Geophysical research continued with many gravimetric, magnetometic and electrometric surveys covering large areas of the country. During the communist regime, 90% of the petroliferous regions were covered with seismic surveys using initially oscillographic stations, then analogue stations (1969) and, starting in 1970, digital stations for data registration and processing.

After the communist regime consolidated its position by winning the general election in November 1946 and after the Peace Treaty was signed in Paris in February 1947, petroleum companies with western capital suffered obvious discrimination from the Romanian authorities, who were simultaneously offering unconditional support and privileges to Sovrompetrol (Ivănuş *et al.* 2009). These actions had a negative impact on the number of wells drilled after the war.

In 1976, the first exploration well in the Black Sea was drilled to a depth of 5006 m, with water depth conditions of 86 m. The year 1980 marked the drilling of the deepest well (7000 Băicoi) so far in Romania, which was drilled to a total depth of 7025 m. The well had traces of crude oil.

Crude oil production increased under the communist regime (Table 1) by introducing new exploitation methods that gave positive results at greater depths. In 1954, hydraulic cracking was tested for the first time in Romania (deposits of Oligocene age from Buştenari), increasing the well's efficiency. In 1964, steam injection and subsurface combustion were successfully tested on the heavy oil from Pliocene reservoirs at Suplacul de Barcău (Fig. 1).

After the war, natural gas production increased to ensure supplies for the large cities in Romania. Many gas transportation pipelines were built during this period. In 1965, Romania became the first country in Europe to build a station for natural gas production (Ivănuş *et al.* 2009).

Petroleum processing faced many changes between 1945 and 1989. After the reconstruction period, the refineries went through a development and modernization stage in order to increase the processing capacity and product quality.

After nationalization (in 1948), the establishment and organization of the petroleum equipment engineering industry started. Several design-research institutions were founded in all fields of petroleum equipment used for: the manufacture of oil-field equipment; the maintenance and repair of petroleum equipment; and the manufacture of chemical, refining and petrochemical equipment.

The design and research-development activities followed Romanian standards, and the manufacture of petroleum equipment was carried out using Romanian technologies, and obtained exceptional results.

Romania's petroleum industry after December 1989

The petroleum industry during the post-communist period was influenced by the transition to a market economy. Until 1990, approximately 23 000 exploration and production wells have been drilled. After this year, drilling activity significantly declined and crude oil production decreased (Table 1).

The first post-communist bidding round for exploration and production acreage, held by Rompetrol in 1990–91 (Dobrova *et al.* 2004), attracted a number of international companies, but in the years that followed there was little foreign investment. Enterprise Oil Exploration Ltd (Anon 1992*b*) signed an exploration and production sharing agreement with state-owned Rompetrol S.A. for two Black Sea blocks: XIII (3000 km^2) and XV (4000 km^2). Shell Romania Exploration BY (Anon 1992*b*) agreed with Rompetrol S.A. on an exploration and production sharing agreement for deep targets underlying producing gas zones in northern Transylvania (onshore Block 10, with 6150 km^2). An Amoco Production Co. unit signed an agreement with Rompetrol S.A. covering exploration in the Carpathian Mountains – Block 7 (Anon 1992*a*).

After 1990, geological and geophysical research focused on the reserve-enhancement technologies of existing petroleum fields and, to a greater extent, on exploration techniques. The oil systems were intensively studied and 3D seismic data were acquired starting in 1995. Reservoir engineering developed rapidly, with the main target being to improve knowledge of the natural characteristics of hydrocarbon accumulations and reservoir behaviour.

After 1990, petrochemical activity decreased significantly, most of the plants being shut down

Table 1. *Romania's crude oil production 1857–2015*

Year	Production (tons)	Year	Production (tons)	Year	Production (tons)
1857	275	1907	1 129 097	1966	13 100 000
1858	495	1908	1 147 727	1967	13 500 000
1859	605	1909	1 297 257	1968	13 500 000
1860	1188	1910	1 352 407	1969	13 500 000
1861	2403	1911	1 544 847	1970	13 700 000
1862	3226	1912	1 804 761	1971	14 100 000
1863	3886	1913	1 885 619	1972	14 400 000
1864	4591	1914	1 783 947	1973	14 600 000
1865	5426	1915	1 673 145	1974	14 800 000
1866	5915	1916	898 994	1975	15 000 000
1867	7070	1917	724 230	1976	15 100 000
1868	7700	1918	968 611	1977	15 000 000
1869	8140	1919	920 488	1978	14 100 000
1870	11 649	1920	1 034 138	1979	12 700 000
1871	12 520	1921	1 163 315	1980	12 000 000
1872	12 690	1922	1 365 830	1981	12 100 000
1873	14 468	1923	1 515 658	1982	12 200 000
1874	14 350	1924	1 851 303	1983	12 100 000
1875	15 100	1925	2 316 504	1984	11 900 000
1876	15 480	1926	3 241 329	1985	11 300 000
1877	15 100	1927	3 661 358	1986	10 700 000
1878	15 200	1928	4 269 541	1987	10 000 000
1879	15 300	1929	4 827 280	1988	9 800 000
1880	15 900	1930	5 744 000	1989	9 600 000
1881	16 900	1931	6 657 803	1990	8 100 000
1882	19 000	1932	7 350 321	1991	7 000 000
1883	19 400	1933	7 387 000	1992	6 800 000
1884	29 300	1934	8 473 000	1993	6 900 000
1885	26 900	1935	8 384 000	1994	7 000 000
1886	26 000	1936	8 704 000	1995	7 000 000
1887	25 300	1937	7 150 000	1996	6 900 000
1888	30 400	1938	6 610 000	1997	6 800 000
1889	41 400	1939	6 240 000	1998	6 600 000
1890	53 500	1940	5 810 000	1999	6 400 000
1891	67 900	1941	5 577 000	2000	6 300 000
1892	82 500	1942	5 665 000	2001	5 800 000
1893	74 500	1943	5 273 000	2002	5 700 000
1894	70 550	1944	3 525 000	2003	5 600 000
1895	80 000	1945	4 680 000	2004	5 500 000
1896	81 750	1946	4 257 000	2005	5 400 000
1897	110 000	1948	4 149 000	2006	5 000 000
1898	180 000	1950	5 047 000	2007	4 700 000
1899	250 000	1951	6 211 000	2008	4 700 000
1900	250 000	1952	8 002 000	2009	4 500 000
1901	270 000	1953	9 058 000	2010	4 300 000
1902	310 000	1954	9 741 000	2011	4 200 000
1903	384 302	1955	10 555 000	2012	4 000 000
1904	500 561	1960	11 500 000	2013	4 100 000
1905	614 790	1964	12 385 000	2014	4 000 000
1906	887 091	1965	12 800 000	2015	4 000 000

1857–2000, Ivănuș *et al.* (2009); 2001–04, Ministerul Energiei (2016); and 2005–15, BP (2016).

and dismantled. Many refineries were closed, with only the large ones being developed and modernized, such as Petrobrazi (OMV Petrom S.A.), Petromidia Năvodari (initially owned by the Rompetrol Group and at present by Kaz Munay Gaz) and Petrotel Ploiești (Lukoil). The transportation of petroleum products is ensured by CONPET, which operates a vast pipeline network (4500 km), pumping stations, tanks, tank wagons and loading–unloading platforms.

After a significant collapse of the oil-field equipment manufacturing industry, numerous companies survived and gradually recovered and adapted to the new social and economic conditions. The maintenance and repair works of production wells are now performed by specialized departments within the petroleum companies.

The design activity recovered rapidly and was developed to be competitive with that performed world-wide but the research activity did not reach the necessary requirements.

In 1995, the National Agency of Mineral Resources (NAMR) was set up to represent the state's interests (Ivănuș *et al.* 2009) and became the sole licensing authority under Romania's new petroleum law.

In 1997, the national oil company S.N.P. Petrom was founded, taking over the former Autonomous Oil Agency. The newly formed company had to accomplish an internal restructuration in order to optimize assets, stimulate private capital investments, perform onshore and offshore exploration, and modernize the distribution and retail network. On 1 July 2004, the privatization of S.N.P. Petrom took place when Austrian Group OMV acquired 51% of its stock shares. In 1998, the national gas company ROMGAZ S.A. was founded, being reorganized into several smaller companies (in 2000). These national newly formed companies focused their activity on exploration and production of gas deposits, storage, and transportation and distribution of natural gas (Ivănuș *et al.* 2009). Since 2003, ROMGAZ S.A. has had a joint operation agreement with Amromco Energy for 11 exploration and development blocks in the Moesian Platform and Getic Depression.

Over the last few years, several international oil and gas companies have been attracted to invest in Romania. In November 2008, ExxonMobil Exploration and Production Romania Limited acquired a 50% interest in the deep-water sector of Neptun Block in the Black Sea. In 2012, ExxonMobil and OMV Petrom S.A. announced that the Domino-1 exploration well had encountered natural gas in the deep-water sector of the Neptun Block.

Hunt Oil signed a partnership agreement with OMV Petrom S.A. in 2010 for two onshore blocks in the Eastern Carpathians, assuming ownership in 2013. Chevron was interested in shale gas exploration but pulled out in 2015, stating that it did not compete favourably with other investment opportunities in the company's global portfolio.

Progress, prospects and trends

Romania has a long history of oil and gas activities, establishing in 1857 three world oil firsts that marked the beginning of its petroleum industry. The progress of this industry was facilitated by building the first refinery in the world in Ploieşti (in 1857) but was significantly affected by the destructive measures taken during World Wars I and II.

The recovery of the Romanian oil and gas industry after the wars was very slow, after establishment of SOVROMS captured and drained its economy in all sectors. During the communist regime, crude oil production increased by introducing new exploitation methods. After December 1989, drilling activity significantly declined and the progress of the petroleum industry was controlled by a transition to the market economy. Despite this, 3D seismic acquisition helped greatly in unravelling prospectivity in structurally complex areas (e.g. the Carpathians Bend Zone) and elsewhere. This trend is likely to continue in the future.

In addition, recent gas discoveries in the deepwater Black Sea may hold significant gas resources to be developed, which would open up a new development phase for the country. Non-conventional plays are also yet to be developed in Romania. Some studies suggest significant shale gas resources are to be found in Eastern Europe (Klett *et al.* 2015).

All in all, these undiscovered resources (especially gas) may help Romania to decrease its oil and gas imports, and possibly also to export gas to Western Europe.

Acknowledgements Reviews by Jeff Spencer, Emil Platon and Rasoul Sorkhabi greatly improved this paper.

Funding This research received no specific grant from any funding agency in the public, commercial or not-for-profit sectors.

References

Aldea, G. & Rădulescu, G. 1957. Evoluția cercetărilor tehnico-științifice în industria petroliferă din România. *Revista Petrol și Gaze*, **9–10**.

Anon 1992*a*. Amoco plans exploration in Romania. *Oil and Gas Journal*, 10 August 1992, https://www.ogj.com/articles/print/volume-90/issue-32/in-this-issue/exploration/amoco-plans-exploration-in-romania.html

Anon 1992*b*. Enterprise, Shell scheduled to explore Romanian acreage. *Oil and Gas Journal*, 17 August 1992, https://www.ogj.com/articles/print/volume-90/issue-33/in-this-issue/exploration/enterprise-shell-scheduled-to-explore-romanian-acreage.html

BP 2016. *BP Statistical Review of World Energy* (June 2016). BP, London, http://www.bp.com/en/global/corporate/energy-economics/statistical-review-of-world-energy.html

Brestoiu, H. 1986. *Impact la paralela 45, incursiune în culisele bătăliei pentru petrolul românesc.* Editura Junimea, Iaşi, Romania.

Ciupagea, D., Păucă, M. & Ichim, T. 1970. *Geologia Depresiunii Transilvaniei*. Editura Academiei R.S.R., Bucharest.

Cobălcescu, G. 1883. *Studii geologice şi paleontologice asupra unor tărâmuri terţiare din unele părti ale României*. Memoriile Geologice ale Scólei Militare din Iaşi, **1**.

Dobrova, H., Kolly, E. & Sendall, F. 2004. E&P opportunities in ECE – licensing history, foreign companies' experience: success cases. Paper presented at the EAGE 66th Conference & Exhibition, 7–10 June 2004, Paris.

Ivănuş, G., Ştefănescu, I.Ş., Antonescu, N.N., Mocuţa, Ş.T., Stirimin, Ş. & Coloja, M.P. 2009. *The Petroleum and Gas Industry in Romania*. Agir Publishing, Bucharest.

Klett, T.R., Schenk, J.C., Brownfield, M.E., Charpentier, R.R., Mercier, T.J., Leathers-Miller, H.M. & Tennyson, M.E. 2015. *Assessment of Undiscovered Continuous Oil and Gas Resources in the Dnieper-Donets Basin and North Carpathian Basin Provinces, Ukraine, Romania, Moldova, and Poland*. United States Geological Survey Report, https://doi.org/10.3133/fs20163082

Ministerul Energiei 2016. *Analiza stadiului actual. Angajamentele naționale și internaționale ale României în sectorul energetic. February 2016*. Ministerul Energiei, Bucharest, http://energie.gov.ro/strategia-energetica-analiza-stadiului-actual-document-realizat-in-data-de-12-04-2014/

Mrazec, L. 1910. Über die Bildung der rumänischen Petroleumlagerstätten. Third International Petroleum Congress (1907), Bucharest. *Compte Rendu*, **2**, 80–134.

Mrazec, L. & Teisseyre, W. 1902. Privire geologică asupra formaţiunilor salifere şi zăcămintelor de sare din România. *Monitorul Petrolului Român*, **3**, 1–55.

Murgeanu, G., Răileanu, G. *et al.* 1964–68. *Geological Map of Romania (1:200 000)*. Institutul Geologic Român, Bucharest.

Paraschiv, D. 1975. *Geologia zăcămintelor de hidrocarburi din România*. Studii tehnice și economice: Prospecțiuni și explorări geologice, **10**.

Pătruț, I., Paraschiv, D. & Dicea, O. 1973. Considerații asupra modului de formare a structurilor diapire din România. *Revista Petrol și Gaze*, **9**, 533–542.

Pizanty, M. 1931. *Le pétrole en Roumanie*. Ateliérs Graphiques 'Eminesco', Bucharest.

Preda, D.M. 1957. Dezvoltarea geologiei petrolului în România. *Revista Petrol și Gaze*, **9–10**.

Ștefănescu, S. 1957. *Evoluția prospecțiunilor geofizice pe teritoriul R.P.R.* Buletinul Științific al Academiei R.P.R., Secțiunea Geologie și Geografie, **II**.

Teodorescu, V. & Ștefănescu, I. 1968. Contribuții arheologice la istoria petrolului românesc: cele mai vechi dovezi privind cunoașterea și folosirea materiilor petrolifere din țara noastră. *Revista Petrol și Gaze*, **8**, 519–520.

Second Galicia? Poland's shale gas rush through historical lenses

ROBERTO CANTONI

CERI – Sciences Po, UMR 7050, 56 rue Jacob, 75006 Paris, France

0000-0002-2526-2336

roberto.cantoni@sciencespo.fr

Abstract: Since the early 2000s, the exploitation of shale gas has radically modified the US energy scenario. In a number of European countries, the US boom has elicited questions about its repeatability in Europe. Among the staunchest supporters of the development of national shale-gas resources were Polish administrations, which grounded their activism in this domain in the language of energy security, autonomy *vis-à-vis* Russian gas, and in Poland's old oil history. The history of hydrocarbon exploration in the country dates back to the mid-nineteenth century, and is connected to the oil boom that occurred in the region of Galicia. While the boom was over by World War I, promising estimations made in recent years by several agencies about Poland's shale gas reserves have stirred hopes of a 'second Galicia'. From 2007, the Polish government started assigning permits to both national and foreign gas companies. However, factors linked to legislation, geology and macro-economics caused a premature end to hopes of Polish autonomy. After a reconstruction of the history of oil in Galicia and the constitution of the Polish oil and gas sector, this paper narrates the rise and fall of Poland's 'affair' with shale gas.

In April 2011, 1 month after the nuclear incident at the Fukushima Nuclear Power Plant in Japan, the International Energy Agency (IEA) proclaimed in its annual edition of the *World Energy Outlook* an emerging 'golden age' of natural gas to 2035. The IEA's prediction was driven by the following expectations: considerable economic growth in China coupled with significant natural gas consumption; a low share of nuclear energy in electric power generation; an increase in the use of natural gas in the transportation sector; and a boom in unconventional gas and its low prices (IEA 2011*a*). As for the last aspect, the report alluded in particular to the 'shale gas revolution' occurring in the USA, where from the mid-2000s shale gas – that is, gas found trapped within shale geological formations – started to be extracted. Massive findings in the USA led to a radical transformation of the American market: while in 2000 shale gas made up 2% of US natural gas supply, in 2012 this percentage increased to 37% and it is expected to further augment to 65% by 2035 (Yergin 2012, p. 331).

The US shale gas bonanza has quickly modified the whole world gas market, and Europe too is starting to be affected by the repercussions of this phenomenon, following the first deliveries of American shale gas to the continent in mid-2016 (Macalister 2016). Indeed, ever since the inception of the US shale-gas development, a number of EU countries have started to 'think shale': Polish administrations definitely stood out as Europe's firmest shale advocates, together with the British government.

Successive Polish governments were supported in their advocacy of shale gas by prominent national think-tanks. As an example, in May 2011 the neoliberal Kosciuszko Institute, inspired by then-US President Barack Obama's visit to Poland and by his successful 2008 campaign slogan, published a policy brief titled: 'Shale gas? "Yes, we can!"' (Albrycht 2011). In the brief, the Institute's chair, Izabela Albrycht, emphasized prospects for the development of shale gas in Europe, and in Poland in particular: prospects, which appeared rosy in the paper. Albrycht recommended that Poland facilitate the operations of gas companies through a number of liberalization measures and fiscal provisions.

Supported by data published a month earlier by the US Energy Information Administration (EIA), which attributed to Poland Europe's largest shale-gas resources – a fantastic 5.3 Tcm (trillion cubic metres), which the brief's author claimed to be capable of allowing Poland energy autonomy in gas supplies for the next 300 years – Albrycht went on to assert that the development of Polish shale gas would be seen as a 'nightmare for Russia', as it would liberate the country from its energy subjugation to Moscow, and enable Poland to export its shale gas to other EU countries (Albrycht 2011; EIA 2011). The Kosciuszko Institute was not the only Polish institution unreservedly advocating shale gas. Thanks to the establishment of a flourishing shale-gas industry, a Polish chief executive of a Warsaw-based recruiting firm argued in the *Wall Street Journal* in August 2011 that between 50 000

From: Craig, J., Gerali, F., MacAulay, F. & Sorkhabi, R. (eds) 2018. *History of the European Oil and Gas Industry*. Geological Society, London, Special Publications, **465**, 201–217.
First published online May 14, 2018, https://doi.org/10.1144/SP465.16

and 100 000 jobs would be created in Poland in the next 10 years (Kruk 2011).

A mere 3½ years later, the situation looked completely different. On 26 November 2014, the online business magazine *Biztok* published an article entitled: 'Firms are seeking an escape from gas in Poland. We won't be the new Norway' (*Biztok* 2014). The words of the vice-director of the Polish Geological Institute, the academic geologist Grzegorz Pieńkowski, seemed to be the death knell of the whole affair with shale, when in May 2015 he declared: 'There will be no manna from shale' (Hołdys 2015). The high hopes that the American report and Polish opinion-leading media had contributed to raise appeared to crumble as fast as they had emerged. What brought Poland's energy autonomy hopes to such a rapid rise and fall? And why do we need to know anything about Poland's nineteenth-century oil history when seeking to grasp that country's recent developments in the hydrocarbon sector? In the following paragraphs, I will attempt to reply to these questions. I will first resort to published literature to study the '*longue durée*' history of the oil and gas industry in the regions constituting the modern Polish state. I will then analyse the main stages of twentieth-century developments as a preamble to the shale-gas affair. In doing so, I will especially highlight the implicit narrative power of the oil boom that occurred in the second half of the nineteenth century in the East European region of Galicia on current political and industrial discourses. In my analysis of the shale-gas rush, I will focus on uncertainties connected with the production of geological knowledge, the use that policymakers made of such uncertainties, and the importance of the narrative framing energy security and independence. I will conclude with a brief view of Poland's future perspectives in the hydrocarbon sector.

Galician 'El Dorado'

As I will show in the next sections, a boom in the allocation of exploration licences for shale gas in Poland started only in 2007. However, one would be mistaken in thinking that the Central European country is a newbie in the oil world. If there is one thing every Polish person knows about local oil history, it is that it is considerably old. Indeed, one of the reasons we need to look at the history of oil and gas in Poland in the '*longue durée*' is the pride that Polish people take from having been pioneers in this industry (as well as in the industry of coal extraction; which, however, will not be discussed here): pride that informs present discourses about shale gas, and which is not without reason. While claims attributing to Poland the first modern oil well in the world – which I have frequently heard during my interviews with personnel of Polish oil agencies – are exaggerated, as current scholarship assigns such primacy to the area around Baku, Azerbaijan, in 1846 (whereas in terms of oil extraction *tout court*, oil sands had been mined and oil extracted at the Pechelbronn Field in Alsace, France, since the mid-eighteenth century), Poland can certainly be said to have been home to Europe's first modern oil well in 1854 (Akiner & Aldis 2004, p. 5).

Or can it? The answer is less clear than one would think. If we look at the location of Poland's early efforts in the oil industry, we find out that the first oil wells were located at Bóbrka near Jasło, in what is currently the SE area of Poland, the region called Podkarpackie. However, at the time of the drillings, this area was not only part of a different historical region, namely the Kingdom of Galicia and Lodomeria (Fig. 1) – also simply known as Galicia – but even of a different state, namely the Austrian (and not yet Austro-Hungarian: it would only become so in 1867) Empire. Historical intricacy escalates further when one realizes that Galicia itself, whose historical nucleus lay within the modern regions of western Ukraine not Poland, was ethnically divided into western Galicia, mostly populated by ethnic Poles, and eastern Galicia (where Bóbrka is located), mostly populated by ethnic Ukrainians – also called 'Ruthenians' at the time – to the exclusion of the city of Lviv, which had a Polish majority (Wilson 2006, p. 34).

On top of all that, speaking of Poland in 1854 is historically misleading as, at that time, Poland no longer existed as a national entity following the three partitions of the country that occurred in 1772, 1793 and 1795, which had split its original territories between Austria, Prussia and Russia. It is true, however, that for over 400 years before the First Partition of 1772, Galicia had been subject first to the Polish Crown and then, from 1569, to the Polish–Lithuanian Commonwealth. So while the nucleus of Europe's earliest oil industry is, indeed, located in modern Poland, the previous historical sketch highlights the geographical and political complexity of the historical situation at the time the Bóbrka well was discovered.

In addition, it is not without interest to linger on the history of the eighteenth and nineteenth centuries in Poland, and especially on the three partitions as, together with twentieth-century invasions of Poland by the Nazis and the Soviets, those partitions have been fundamental references in Poland's nation building. All these major events leading to territorial dispossession of the Polish population became part of a national narrative that has identified the need for Poland to make decisions for itself and to be autonomous from neighbouring states as the highest national interest. 'The historical narrative is one in which Poland has never been sovereign and able to

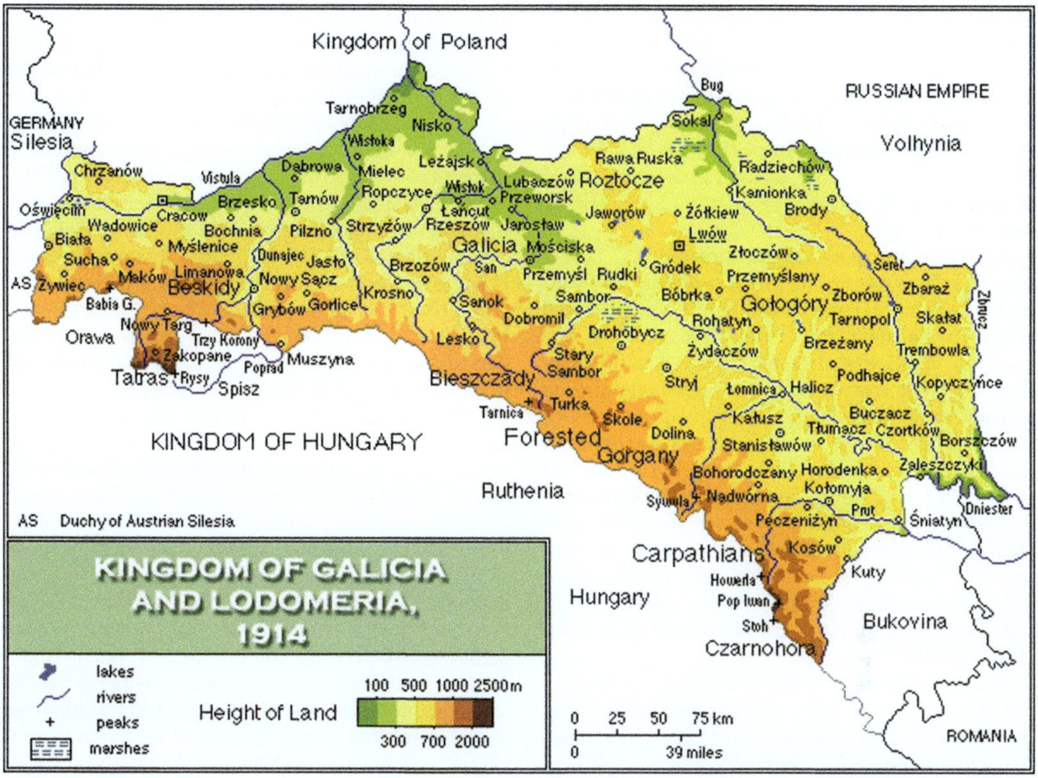

Fig. 1. The Kingdom of Galicia and Lodomeria (1772–1918) (*source*: Wikimedia Commons; map by Mariusz Paździora).

determine its own fate', argued Thomas Risse (2010, p. 77). The argument of national security, in his twenty-first-century framing as energy security, was to be a leitmotif of discourses on shale gas.

Irrespective of issues of national primacy, the Carpathian Mountains have historically abounded in oil seeps, and oil from that area, hand-dipped from pits dug in front of the seeps, was used for burning in street lamps as early as the 1500s to provide light to the Galician town of Krosno. The first mention of 'rock oil' in Poland dates back to even earlier: in the fifteenth century, in the writings of cleric and diplomat Jan Długosz. A second mention is from 1530, when the treasurer of King Sigismund the Old, Seweryn Boner, found oil while searching for gold in the mountain village of Ropa in what is today's southern Poland: it was no accident then that the Polish name for oil – 'ropa' – took origin from that village. The illuminating power of burnt oil may have become known in the next decades, since documents from the sixteenth century certify its use for illuminating the castle of the town of Biecz, and as a warning sign in the event of threats by foreign armies. Records from the seventeenth century are also reported attesting the use of rock oil for medical purposes in the areas of Drohobycz and Krosno (Buhl 2002).

The early extraction of oil on a non-industrial scale created new professional categories, such as prospectors, collectors and crude oil sellers. However, the characteristics of the seep oil – its viscous and sticky consistency – made it unpalatable for alternative uses. The fact that it produced a putrid smell and emitted more smoke and soot than other lamp oils – mainly from animal fat – did not help its commercialization either. In 1835, about 30 surface oil mines operated in Borysław, Galicia, producing less than 20 l of oil a day. While the beginnings of oil extraction may not have allowed the prediction of a promising future for the black matter, by the second half of the nineteenth century things had radically changed, and Galicia had become Europe's main oil region, with all the associated pros and cons.

In the only reference in English available to date for the history of oil in Galicia, Alison Fleig Frank provides numerous descriptive accounts of what came to be known as the 'Galician Hell', and of how life in places like Borysław and Drohobycz was characterized by elements that are typically found in US gold rush accounts, namely widespread drunkenness and the common practice of female prostitution in male-only populated villages, frequent violence, and gambling activities. All of this was made more dangerous by interethnic tensions between Polish, Ukrainian and Jewish worker communities (Frank 2005, p. 7).

On the other hand, Frank stresses that beside these images of misery, there was a widespread feeling in the industrial and cultural élites that oil could be greatly beneficial to Galicia, an area that had remained up to that moment rather economically deprived, prevalently agrarian, politically dominated by landowners and characterized by high rates of outmigration, as well as malnutrition and disease. Local industrialists and techno-scientists saw oil not just as a way out of misery for the region, but also as an opportunity to foster transformation from a traditional agricultural economy to a modern industrial economy (Frank 2005, pp. 12–13). Then 150 years later, the narrative of the salvific power of hydrocarbons for rural and impoverished populations was to be transposed to shale gas.

The breakthrough in the oil history in Galicia was the mentioned 1854 Bóbrka discovery. The invention in 1853 of a new lamp that could burn petroleum odourlessly and safely by Lviv-based pharmacists Jan Zeh and Ignacy Łukasiewicz, caused a boost in interest in oil. Following the successful outcome of the purification process, Zeh and Łukasiewicz sent a large sample of their product to the Lviv General Hospital, which, after successfully testing the new lamps, became the first public building in the world to exclusively use these lamps (Zeh 1889).

The invention of the petroleum lamp and the first oil discoveries in the Bóbrka region triggered the construction of several refineries. Łukasiewicz himself founded several, starting with one in Gorlice – some 50 km west of Bóbrka – in 1854. From 1856, oil extraction developed on a larger scale. Łukasiewicz's successful business attracted competitors, and primitive hand-dug oil wells soon mushroomed in the region. The history of forerunners and pioneers of Galician oil, such as the Polish economist and entrepreneur Stanisław Szczepanowski and the Canadian driller William Henry MacGarvey, has been narrated in Frank's book. Suffice it to say here that they were instrumental in igniting a petroleum fever in Galicia, and in innovating and modernizing drilling techniques, making it possible to considerably increase production (Frank 2005, pp. 79 ff.). Again, a parallel can be drawn with shale gas some 150 years later, and the role of overseas help can be stressed: it has been mainly North American companies flocking to Poland to the shale rush that have contributed to the expansion of the business in that country, as well as to import technological innovations such as hydraulic fracturing and horizontal drilling.

In the second half of the nineteenth century, Galicia became the first European region in terms of oil extraction. Production peaked in the early 1900s. In 1909, Galicia produced over 3 Mt (million tons) of crude oil, accounting for 5% of world production. Thanks to its Galician oil fields, Austria–Hungary became the third-largest oil-producing region in the world (Frank 2005, p. 4). In 1912, the first gas pipeline, used for transporting natural gas from Galician deposits to Europe's first gas plant and crude oil refinery in Borysław, was built. The oil industry in the region employed about 14 000 workers. Initially, activities were performed by a large number of rather small firms, mainly controlled by Jewish entrepreneurs. From 1880, German and Austrian firms started investing in Galician oil, followed by British, French and American concerns.

In the interwar period, however, production declined as a consequence of the exhaustion of oil pools and the limited financial means that most companies could provide. In this period, about 4000 wells dotted Galician territories, together with 34 refineries (Brantly 1971, pp. 186–190). Still, it is useful to remember that the role of oil, up to World War I, was not so much that of fuelling tanks and planes, but rather of illuminating cities. Only with World War I and, to a much greater extent, World War II, the importance of the availability of large oil supplies for the military apparatus and for transport became increasingly manifest. The post-1909 collapse of production was paralleled by the inception of World War I, and the use of oil to fuel the war apparatus led to a massive increase in consumption (Frank 2005, pp. 10, 21).

In 1918, with the end of World War I and the proclamation of the Polish Republic, Poland acquired Galicia but a conflict with Ukrainian nationalists immediately started. It lasted until 1919, and it ultimately sanctioned Poland's sovereignty over Galicia. In that year, an important step in the institutionalization of Polish geological research was made, thanks to the constitution of the Polish Geological Institute (PGI). The PGI was not only assigned the task of conducting comprehensive studies of the geological structure of the country, but was also made responsible for Poland's security in the supply of mineral resources, amongst other things. We should keep in mind the part played by the PGI in the production of geological expertise in Poland. The long experience accumulated by the Institute, and its acknowledged reputation in the field, were to make it a trustworthy source of information in the decades to follow. Its estimations about Poland's shale-gas reserves in 2012, as we will see, were to have resounding repercussions on political expectations on this resource.

While the foundations were therefore laid for the establishment of national geological expertise, the geographical borders of the Polish nation were still rather shaky. The tense situation in eastern Galicia remained one of the thorniest issues in the following decades, even after the international acknowledgement of Polish annexation in 1923. Polish rule lasted until the Molotov–Ribbentrop Pact of 1939 *de facto* authorized the Soviet invasion of eastern Poland. As for Galician oil, starting from the 1920s, it gradually became a thing of the past. By 1925, Galicia's oil ranking had collapsed from its heyday world-third position to world-tenth. Its production now amounted to 0.54% of the total world production. In the interwar period, production continued to be concentrated in the regions of Borysław and Drohobycz. However, political reasons – mainly, French support to the establishment of the Polish Republic – pushed reservoir ownership into French hands. By the end of World War II, however, the Galician oil industry had disappeared (Schwarz 1929, pp. 4–6; source reported in Frank 2005, p. 242).

Post-war developments

World War II completely transformed the geography and political setting of Poland. With the annexation of eastern Galicia to the USSR, Poland retained only poorly productive Carpathian reservoirs. All surviving oil companies and refineries were nationalized. In July 1944, a communist-controlled Polish Committee of National Liberation was established in Lublin, in today's Poland's east. In the same month, a new agency was established whose role was to secure the crude oil infrastructure that had survived the war and German occupation, and commence distribution activities. This was first named the Polish Oil Monopoly ('Polski Monopol Naftowy'), then change its name to Head Office for Petroleum Products ('Centrala Produktów Naftowych'). A decade later the company would undergo further restructuring, and be renamed twice: in 1958 the acronym CPN was added to the original name.

In the second half of the 1940s five refineries were operating in Poland (Niegłowice near Jasło, Jedlicze near Krosno, Glinik Mariampolski near Gorlice, and Trzebinia and Czechowice-Dziedzice): revealingly, all were located in former Galician territory. In the 1950s, the constantly rising demand for fuels and raw materials, a consequence of rapid industrial development and the wider use of vehicles, as well as of insufficient processing capacities at the existing refineries, led the government to the decision to build a further refinery. The construction of this new plant – the Mazovia Refining and Petrochemical Plant ('Mazowieckie Zakłady Rafineryjne i Petrochemiczne') – was authorized in 1959. The facility was located in Płock, in central Poland.

In 1961–65, the first crude oil deposits were found in the Polish Plain. The discoveries initiated exploration and production activities in a region that had not theretofore been involved in the oil history, and in western Poland in particular. The structural consolidation of the Polish oil industry was improved by the constitution of further agencies. The programme for increasing the use of chemicals in the national economy (1968–72) provided for a significant development of the refining and petrochemical industry. The construction of a new refinery in the northern city of Gdańsk, decided in 1971, offered an alternative to the costly imports of petroleum products from Western Europe, in the form of long-term purchases of crude oil from abroad, which were then refined at the Gdańsk plant. A 10 year crude oil supply contract with British Petroleum was signed, providing for annual deliveries of 3 Mt of oil for the new refinery, and the first batch of crude oil was delivered in 1975.

In general, the 25 years following the first oil discoveries constituted a period of rapid expansion of Polish natural gas industrial networks, and of a gas-supply service for cities, industry and households. In 1976, the entire exploration sector was transferred to the state-controlled Oil and Gas Union ('Zjednoczenie Górnictwa Naftowego i Gazowniczego' (ZGNiG)). This would be renamed Polish Oil and Gas Company ('Polskie Górnictwo Naftowe i Gazownictwo' (PGNiG)) in 1982, and this denomination has survived to today. Oil exploration and drilling service companies, such as Petroleum Exploration Companies ('Poszukiwania Naftowe') in Jasło and Piła, were established at that time, as well as the main national geophysical companies 'Geofizyka Kraków' and 'Geofizyka Toruń' (Labuda 2013).

The political transformation that characterized Poland in the 1980s, with the rise of opposition to the communist regime, and the new policies of 'glasnost' ('openness') and 'perestrojka' ('restructuring') launched by Soviet Prime Minister Mikhail Gorbachev, initiated a gradual transition of the country towards a market-orientated economy (Davies 2005; Bafoil 2007). As part of the changes taking place in the Polish economy and especially in the country's public sector, in 1993 the Mazovia refinery was transformed into a joint-stock company and renamed 'Petrochemia Płock'. Five years later the Council of Ministers decided a merger of the Head Office for Petroleum Products CPN and Petrochemia Płock, with a view to establishing a national petroleum company for refining of oil and the manufacture and commercialization of oil products. As a consequence of the merger, in 1999 the Polish Oil Concern

('Polski Koncern Naftowy' (PKN)) was constituted. It was renamed PKN Orlen (from the Polish words for 'eagle' (orzeł) and for 'energy' (energia)) the following year (more information can be found on the PKN Orlen website: http://www.orlen.pl/EN/Company/OurBackground/Pages/PKNOrlenBackground.aspx).

The decade of the 1990s was also a fortunate one for Poland from the point of view of discoveries, as research in the western regions of the country started bearing fruit. Both the Barnówko–Mostno–Buszewo oil field (proven oil reserves: 12 Mt; proven gas reserves: 8 Bcm (billion cubic metres)) and the Lubiatów–Międzychód–Grotów oil field (proven oil reserves: 5 Mt; proven gas reserves: 4.5 Bcm) were discovered in 1993, and production started in 1994. In the early 1990s, an offshore oil and gas field was also discovered by the Petrobaltic Company in the Baltic Sea, 80 km off the Polish coastal town of Rozewie. While small when compared to proven reserves of oil and gas fields found in other regions of the world, these discoveries extended the oil potential of Poland from the historical region of Galicia to the west and north of the country.

In 2004, PGNiG started a policy of divestiture: gas transmission was separated from the holding company, and PGNiG-Przesył (currently OGP-Gaz System) was constituted to manage the transmission network and perform technical operations. Gaz System owns and operates all gas transmission and distribution pipelines in Poland, except for the Yamal–Europe pipeline. The company is also responsible for the construction of the Polskie LNG terminal, which opened in 2015 at Świnoujście in the country's north, close to the German border. Following a 2004 European Union Directive and the Power Energy Law, which imposed a legal obligation of separating gas distribution from commercial activities, six distribution system operators spun off from PGNiG (Ministry of the Treasury 2005; Strzelecki 2015).

In 2003, the Gdańsk refinery plant, which had operated under the name of Gdansk Refinery Inc. until then, took the denomination of Lotos Group ('Grupa Lotos') after acquiring refineries in southern Poland, and incorporating Petrobaltic. Today, the Lotos Group combines several companies operating in manufacturing, trade and service companies, mainly from the exploration, extraction and processing of crude oil and distribution of petroleum products (more information can be found on the Lotos Group website: http://www.lotos.pl/en/869/this_is_lotos/lotos_group/history).

Notwithstanding the efforts made by Polish companies in prospecting for hydrocarbons in their country, these were not paralleled by outcomes. Looking at oil production data for Poland, whereas one does notice an increase in the last few decades from an average of 4.4 Mbbl (thousand barrels) per day in the 1980s to 11.6 Mbbl/day in the 2000s and up to 15.8 Mbbl/day in the first half of the 2010s (Fig. 2), on a world-scale Polish production is rather small and, according to latest available estimates, in 2014 Poland was ranked 74th in world oil production

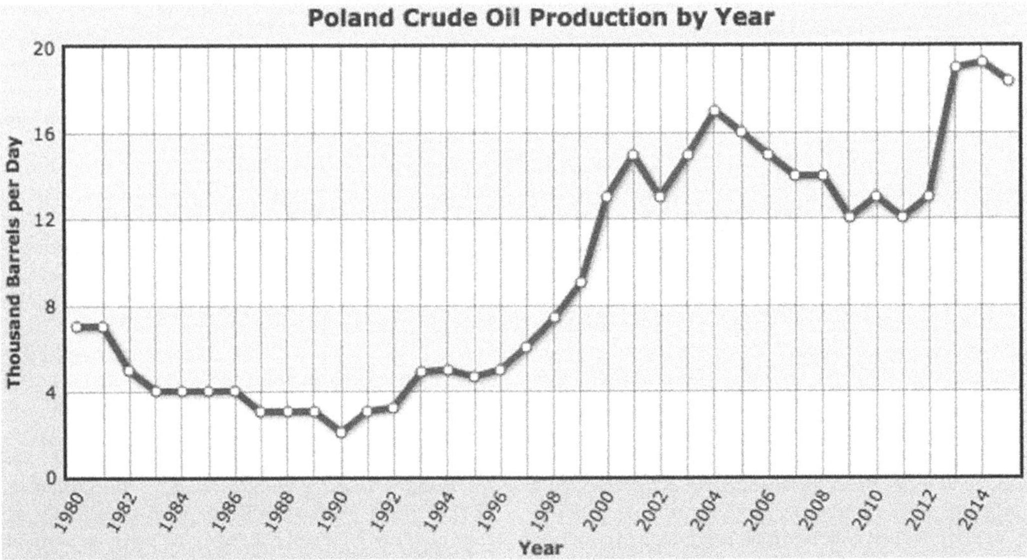

Fig. 2. Poland's daily average oil production (1980–today) (*source*: EIA data and IndexMundi: http://www.indexmundi.com)

with 19.3 Mbbl/day (ranking according to the CIA *The World Factbook*: https://www.cia.gov/library/publications/the-world-factbook/rankorder/2241rank.html; and data from IndexMundi for Poland crude oil production per year: http://www.indexmundi.com/energy/?country=pl&product=oil&graph=production). As a consequence, Poland depends almost totally on imported oil for its economy: a large majority of this oil is imported from Russia. In 2009, Russia provided about 94% of the total crude oil imports; this rate was almost unchanged in 2014 (93%) (data for 2009 are from IEA 2011*b*, p. 115; data for 2014 are from the OEC website: http://atlas.media.mit.edu/en/visualize/tree_map/hs92/import/pol/show/2709/2014/). This is a huge dependence, especially from a country whose relationship with Poland is checkered to say the least.

When we bring together the 'historical' factor of a long tradition in the oil industry on Polish lands, which I have sketched in the previous section, and the somewhat disappointing results obtained in the last decades in Polish exploration, which have made Poland significantly dependent on foreign imports, it does not come as a surprise that the earliest news regarding Poland's potential in shale gas reserves were welcomed with particular enthusiasm in national business circles, and that some commentators went as far as talking about a Second Galicia (Wolniewicz 2012). Before turning to this aspect, however, it is useful to outline briefly Poland's energy dependency problem.

Foreign gas dependence and the problem of shale-gas estimates

When one takes a look at the data on primary energy supply of EU member countries, a striking aspect is the uniqueness of Poland, in that it is the only country that relies heavily on coal. In 2010, coal accounted for 55% of total supply of primary energy in Poland, followed by oil (25%) and natural gas (13%). However, whereas it is true that gas plays a much smaller role than coal in the Polish energy mix, about two-thirds of the gas Poland uses is imported, 90% of it coming from the Russian giant Gazprom. Poland's domestic gas production in 2010 accounted for 6 Bcm and was dominated by PGNiG. But this figure was dwarfed by the amount of gas that Poland imported, which is about 11 Bcm (10 Bcm of which is being imported from Russia and 1 Bcm from Germany) (KPMG 2012). So the situation we have seen regarding Polish oil dependency repeats itself for gas dependency.

Besides the frequent tensions characterizing the Polish–Russian relationship, the fact that these exports must pass through Belarus and Ukraine to be delivered to Poland has caused trouble in the past, since more than once these two countries have failed to respect their status as transit countries. The Ukrainian–Russian gas disputes of 2006, 2009 and 2014 are cases in point, in that following the Russian decision to cut gas supplies to Ukraine, gas supplies to several EU members were also affected (BBC News 2009). In 2009, in particular, Bulgaria and Slovakia suffered a complete cessation of gas deliveries, while a total of 14 EU countries were affected by the cuts to different extents (Godzimirski 2014, p. 2; Stulberg 2015, p. 117). The 2009 episode realized the Polish government's anxiety about too large a dependency on Russian gas. However, Ukrainian–Russian tensions were not the only issue in the energy sector that worried the Polish administration, led at the time by a centre-right coalition formed by Civic Platform ('Platforma Obywatelska' (PO)) and the Polish Peasants' Party ('Polskie Stronnictwo Ludowe' (PSL)).

The perception of big powers negotiating energy affairs to the detriment of Poland was also intensified by the operationalization of the Nord Stream gas pipeline, a 55 Bcm a^{-1} gas pipeline, which became operational in 2011 and transported gas from Russia to Germany via the Baltic Sea, skirting Polish territory as well as those of the Baltic republics. A second pipeline, Yamal, opened in 1997, connects Russia to Germany via Belarus and Poland, and the construction of Nord Stream threatened to decrease the importance of Yamal considerably. The annoyance of the Polish government over the development of the Russo-German gas pipeline was such that at the time of its construction, the Polish Foreign Minister Radosław Sikorski nicknamed it 'the Molotov–Ribbentrop pipeline', referring to the notorious German–Soviet pact of 1939, which gave the go-ahead to the invasion of Poland by the two powers (Kramer 2009; Högselius 2013, pp. 212–216).

As a consequence of these moves that appeared to exclude Poland from any possibility to have its say in gas affairs, successive Polish governments have considered the security of gas supplies as a priority in their agendas, especially because the domestic demand for gas is quickly increasing in Poland – in fact, it already increased by 29% between 2000 and 2010. In addition to this growth in natural gas consumption, the role of natural gas in Poland's energy mix is expected to further increase, particularly due to the growing demand of residential end-users, the expected development of high-efficiency steam and gas co-generation technologies, and the need to provide peak and reserve capacity for wind power plants (IEA 2011*b*, pp. 97–113; KPMG 2012, pp. 38–39, 41) (the 2016 'anti-wind law,' however, threatens the development of wind power in Poland: Harper 2016).

Historical references to the problem of Polish national energy security date back to well before the year 2000. First, as mentioned, in countries such as Poland and Lithuania, the long history of Soviet occupation tends to make security supply one of the first political objectives of all parties. Second, in the post-socialist countries of the EU, energy availability also plays a more important role than in most Western European countries, since the energy bill represents a larger share of monthly household income. In these countries, the liberalization of energy prices has contributed to the problem of energy poverty. Finally, of all European countries supplied by Gazprom, Poland currently pays among the highest prices for its gas (over \$500 m^{-3}, compared with less than \$400 m^{-3} paid by Germany), and this is reflected both in the national balance of trade and in costs that citizens have to pay (Goldthau 2013).

Over the background I have just outlined, from 2009 a number of publications started to appear reporting estimates of Poland's shale-gas reserves. First, the British consultancy agency, Wood Mackenzie, published its figure of 1400 Bcm. In the same year, the USA-based research and consulting company, Advanced Resources Int., released its own estimates, which more than doubled those of Wood Mackenzie, to 3000 Bcm. Less than a year later, the Norwegian oil and gas consulting firm Rystad Energy decreased Advanced Resources Int.'s data by two-thirds, by providing the figure of 1000 Bcm. These estimates were based on a limited amount of public geological data provided by the PGI, which the Institute had collected between 1960 and 1980. While they significantly differed from one another, the figures were still much larger than the volume of officially documented conventional natural gas resources (145.15 Bcm). They therefore led to optimistic conclusions of Poland's potential (Rutkowski 2013).

In 2011, such an optimistic view was further strengthened following the publication of a report by the US EIA, which identified the most promising shale-gas basins with two areas (Fig. 3) – one in the north (the Baltic Basin) and one in the SE (the Lublin Basin) – and attributed to Poland the

Fig. 3. Map of shale gas basins in Poland (*source*: EIA 2011; Wikimedia Commons).

largest technically recoverable resources in Europe (5300 Bcm). This publication marked the consecration of the image of Poland as a 'second Norway', or a 'second Galicia' (*The Economist* 2011; Niemuth & Westphal 2014; Stankiewicz & Suchomska 2014). After the golden times of the nineteenth-century oil boom, Poland appeared to be able to revive its bygone splendours. However, in the following year, a further report was released which seemed to bring hopes and illusions back down to Earth. The PGI's new estimates reduced the shale-gas reserves to 346–768 Bcm: that is to say, one order of magnitude less than the EIA. Coming as they did from the country's most reputable geological institution, these data could not be easily dismissed by the media, and were a hard blow to the aspirations of many an entrepreneur. Yet, even in this scenario, these resources were supposed to be able to provide Poland with enough energy for 35–65 years at the current rate of consumption. Finally, in 2013 a second report by the EIA re-estimated its initial data at 4200 Bcm. The reason for the marked differences in the figures provided by these agencies is to found in the use of markedly different simulation models and methodologies. It is important to notice, however, that none of them was based on test drillings, but on simulations based on US analogues, or on archive sample data (PGI 2012; EIA 2013).

Data on actual reserves were therefore uncertain to say the least, but this uncertainty did not appear in the media. What did appear from 2010 to early 2012 – that is, until the PGI published its estimates – was the narrative of Poland marching toward its 'golden age of gas', to quote the IEA. In particular, the early EIA data were quickly interpreted by most national politicians and the media as *the* solution to the dependence of Poland on foreign gas. Indeed, in terms of party politics, there was a very rapid coming together of a broad consensus on shale gas as the main tool to revolutionize the economy of Poland, make the country a major gas producer and release it from Russian dependence: not only gas dependence, but possibly also oil dependence, by helping Poland shift its fuel mix in the long term to a more gas-based one.

One of the key points in supporting a proactive policy toward shale gas has been the role of such fuel in national plans for energy transition. Within national borders, Poland's coal resources are almost unanimously seen as a guarantor of energy security, as the only indigenous energy the country disposes of in large amounts. However, two sets of issues put coal in the dock: the first one is with regard to the European Union's environmental regulations, which require Poland to gradually phase out coal as a large emitter of greenhouse gases, in favour of lower emitting fuels. In addition, Poland gave an obligation to the EU to decommission old coal-fuelled power blocks with a total capacity of 5000 MW by 2015 and yet others by 2020. The second problem regards the economic viability of Polish coal, as local reserves are fast declining, and shortages are expected soon unless new mines are opened (a plan that meets public opposition). In 2008, Poland became a net importer of hard coal for the first time, and imports from none other than Russia have increased to account for 70% of the total coal imports in 2009 (IEA 2011*b*; Polish Press Agency 2013). Poland's exports of coal have also been dramatically declining over the last 15 years. As a consequence of these factors, in late 2016 the European Commission cleared €1.80 million to support the closing down of Polish now uncompetitive coal mines (European Commission 2016; data on Polish coal exports are from the EIA and are reported by IndexMundi: http://www.indexmundi.com/energy/?country=pl&product=coal&graph=exports).

The consensus on shale gas as a solution to these two sets of problems has therefore been extensive, since shale gas is generally considered a lower emitter of greenhouse gases than coal and oil (Howarth 2014), and would allow a shift to a more competitive source of energy: out of the five parties present in Parliament, four supported the gas industry, including the two largest parties: Civic Platform (representing 41% of the votes in the 2007 elections), and the nationalist right party Law and Justice ('Prawo i Sprawiedliwość' (PiS), representing 32% of the votes). Opposition to shale gas in Parliament only materialized in the following 2011 elections, when the centre-left party Palikot's Movement ('Ruch Palikota', later renamed 'Twój Ruch'), flanked by small environmental groups opposed to shale-gas exploitation, obtained 10% of the votes. However, even the actions implemented by Palikot's Movement were characterized by a certain ambiguity: while they supported the anti-shale protest in the northern region of Pomerania, and helped activists weave a network of contacts with French and British anti-shale movements, they did not do it openly as they were obviously anxious about any criticism of taking an official stance against shale gas, considering the favour with which the majority of the Polish population looked at the promise of the new resource (Friends of the Earth 2012; *Natural Gas World* 2012*a*; Kádár 2014).

That opposition to the shale-gas industry was so minimal in Poland compared to other European countries where shale-gas developments were planned or implemented – France, the UK, The Netherlands, Romania and Bulgaria – finds some rationale in the history of environmentalism in the country. According to the assessment made by Christopher Rootes, while an independent and pluralistic environmental movement emerged in Poland in the 1980s, 'It was not an active participant in the new, post-communist political institutions' (Rootes 1997, p. 335); and while, in 1989, the main movement of anti-communist opposition in Poland – Solidarność – embraced the defence of the environment:

> This inclusion of environmental defence may have owed more to the desire to maintain the broadest possible front of critique of and opposition to the regime than to any profound commitment to environmentalism
>
> (Rootes 1997, p. 339).

See also Waller & Millard (1992) and Glinski (1994).

Solidarność therefore expressed virtually no interest in ecology, so that the environmental movement, already marginalized within the anti-communist opposition, continued to be so even after the transition to democracy, and is still today, albeit with some exceptions. The socio-economic significance and national prestige of the coal sector may have helped to shape the limited concern for environmental policies; also, the stress put by Soviet policies on heavy industrialization as a tool for liberation from capitalist dependence may have left little room for this kind of preoccupation. However, opposition to shale gas did manifest in the northern region of Pomerania and in the SE region of Lublin (in the latter case, the population of the village of Żurawlów succeeded in blocking work by the US company Chevrison for over a year in 2013–14), but it did not ultimately modify nationaisl pro-shale policies (Szydłowski 2014).

Exploration permits started to be assigned as early as 2007. However, it was only in the following years, after the release of the mentioned estimates, that the number of applications skyrocketed: national companies such as PGNiG, Orlen Upstream and Lotos, which were already operating in Poland, were joined by international companies, both small and large, mainly from the USA. Amongst these were also European flagships such as the French Total and the Italian ENI, as well as US majors such as Chevron and Exxon. From 2007 to 2010, the number of assigned exploration permits almost doubled, increasing from 51 to 94: these licences were distributed to more than 20 companies. Most concessions were located in the voivodeships of Pomerania and Lublin. It is not by accident that these were the areas where most protest activities occurred. The number of assigned permits peaked at 115 in 2012 (Fig. 4).

When faced with doubts about whether Polish enterprise would be skilled enough to develop the full potential of its shale-gas reserves, or about whether the administration would be capable of outlining proper legislation for the sector, references to the antiquity of Polish expertise in fossil-fuel extraction dating back to Galician times were common, as were references to local expertise in coal extraction: the implicit argument was that since Poland had been able to secure energy at the time of the Galician boom, and had later been able to develop a coal industry that had become the flagship of Poland's industrial pride, there was no need to worry excessively as far as managing the new kind of fossil fuel was concerned.[1]

Geology, macroeconomics and legislation: the (provisional?) end of the dream

By the end of 2011, however, the very liberal strategy implemented by the PO government, which granted licences on favourable terms to companies, started to be reshaped. The original approach, which had granted foreign companies great freedom of action and which PO had supported in order to attract investment, was gradually replaced by a more nationalist position. This new approach was adopted following the rise in opinion polls of the main opposition party, PiS. PiS representatives harshly criticized the concession policy advocated by the PO government, who, in their view, had been guilty of not imposing sufficient financial constraints on foreign companies in terms of royalties. This, they maintained, had affected the revenues of the Polish state. Adding to the attack led by PiS, in January 2012, a scandal was exposed involving the procedures in awarding concession. Seven people, including three officials from the Ministry of the Environment, an employee of PGI and three representatives from gas companies, were involved in 11 charges of corruption and bribery (Materka 2012*a*, pp. 12–13; Lis & Stankiewicz 2017). This episode, of course, did nothing to improve the credibility of the ruling administration.

Because of the declining popularity of the PO, the Prime Minister Donald Tusk began advocating for partnerships between foreign and domestic firms. In October 2012, the government proposed a project for a new law that would regulate taxation for mining and introduce changes to the current geological and mining regulations. These changes were expected to provide a more adequate level of state control over the exploration companies, and increase the income of communities in relation to the extraction of hydrocarbons. The government also proposed constituting a new agency, the National Mineral

[1]Paweł Poprawa, Consultant at Instytut Studiów Energetycznych, interview with the author, in Cracow, on 8 December 2015; Senior PGI informant, interview with the author, in Warsaw, on 30 October 2015.

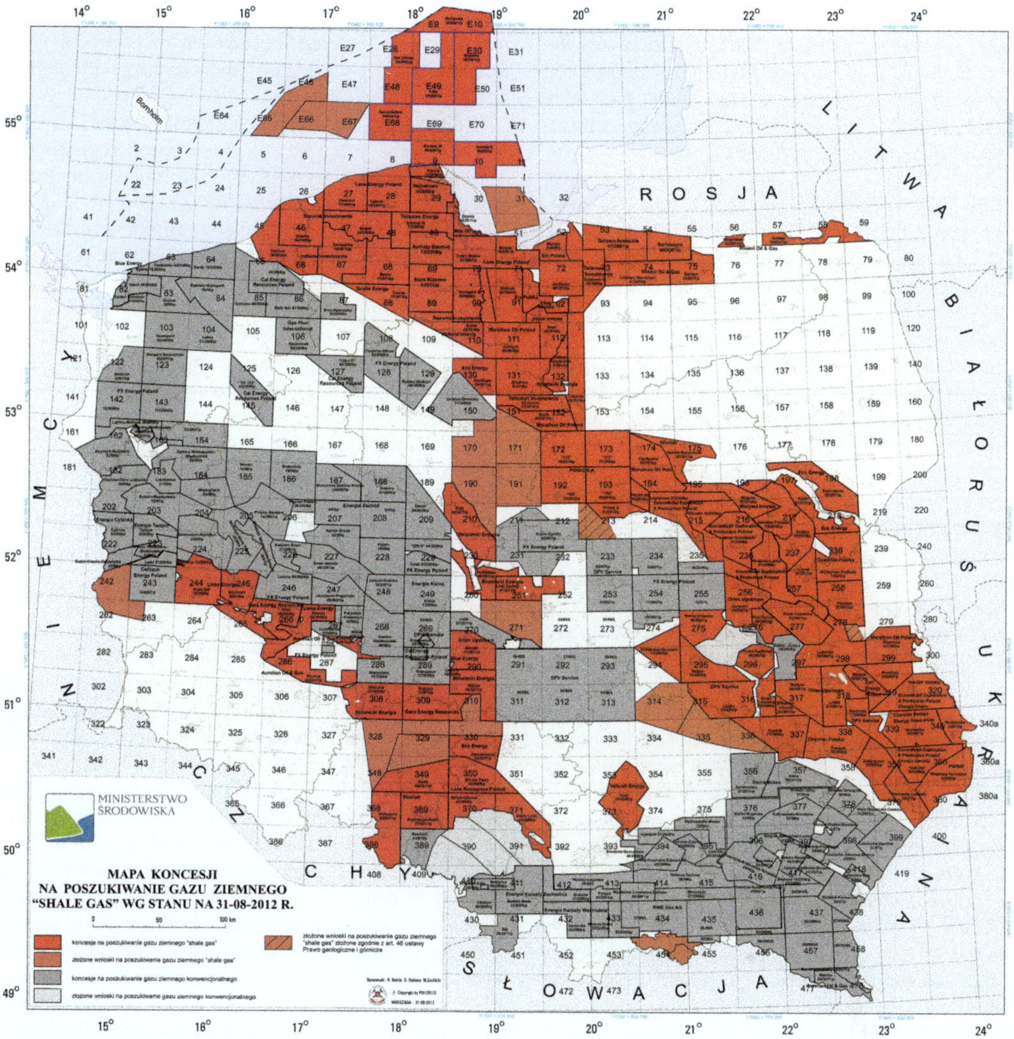

Fig. 4. Map of concessions and applications requests for shale gas by 31 August 2012 (*source*: Polish Ministry of the Environment).

Energy Operator ('Operator Narodowy Kopalin Energetycznych' (NOKE)), which would participate in all activities of exploration companies operating in Poland, within partnerships (Materka 2012*a*, pp. 12–13; Lis & Stankiewicz 2017, p. 8).

The legislative proposal also contained provisions for the establishment of a generational fund, which would invest the profits accruing to NOKE. These revenues would be divided between higher education, science, research and development, and to support the pension system. This solution was based on the Norwegian template; Norway being one of the European countries with the most advanced regulations in the management of the gas sector. However, it was proposed in a situation that differed significantly from that of the Scandinavian countries: in the latter, the gas industry was already well entrenched and operations had long been under way. The situation of contemporary Poland was quite different: the gas industry was in its infancy and the long-past Galician experience would certainly not be of help in this circumstance; and oil exploration on a broad scale was also just beginning (*Natural Gas World* 2012*b*; Pliszczyńska 2013).

In fact, it was not just to NOKE that the gas industry was hostile but, more generally, to an overhaul of regulation in a more protectionist and nationalistic sense, which was in obvious disagreement with the policies that allowed Poland to attract significant foreign investment in many industrial areas in the post-communist era. Foreign companies interpreted the government's turnaround on shale gas as a betrayal of the main tenets of private enterprise, and as an unacceptable intrusion from the state (for information on post-communist transition economies, see Lavigne 1995; De Menil 2003, pp. 271–294; Bafoil 2014, pp. 123–141; see also Dunn 2004). Thus, after the formulation of the idea of NOKE, the Organization of Polish Exploration and Production Industry ('Organizacja Polskiego Przemysłu Poszukiwawczo-Wydobywczego' (OPPPW)) – the employee union that brought together all the main actors involved in the exploration and production of gas – as well as think-tanks such as the Kosciuszko Institute, committed to intense lobbying activity so that the NOKE project would be abandoned. They were ultimately successful: NOKE disappeared from the government agenda.

Still, the system of concessions, complicated by modifications to mining regulations, gradually slowed down applications for, and granting of, new permits; increasing uncertainty surrounding future taxation also contributed to this outcome. Nevertheless, governmental ambiguity on these matters was not the main reason why the dream of Polish energy autonomy started to vanish. The most serious blows to the fledgling Polish shale industry were struck by two other factors: geology and the global hydrocarbon economy (*The Economist* 2013).[2]

As exploration progressed and the first testing wells were drilled, companies became aware that the conditions characterizing Polish subsoil differed significantly from those of the shale-gas plays exploited in the USA. First, the geological layers impregnated with gas were at a greater depth than in the USA. Secondly, Polish geological layers were characterized by a higher clay content than in the USA: this made it harder to apply the known techniques of hydraulic fracturing – the fracturing of rocks on to which the gas is adsorbed through the injection of highly pressurized water and chemicals, in order for gas to be liberated. Indeed, the high clay content was a major issue when trying to keep fractures open. Thirdly, since the number of drilling rigs and other equipment for oil exploration was lower in Poland (and more generally in Europe) than in the USA, the costs involved in drilling a well turned out to be considerably higher in Poland than in the USA. Finally, the organic content of Polish rocks was lower than that of American rocks, which lowered the amount of hydrocarbons that could be extracted by fracking (Materka 2012*b*).[3]

Regarding the macroeconomic factors, from mid-2014 the price of oil experienced a dramatic decline: this not only further contributed to a slowdown of exploration activities in Poland and in other areas of the world where prospecting was in its early stages, but it also caused the bankruptcy of a number of companies operating in fracturing technologies globally (Gandel 2015; for an analysis of the drop in oil price, see *The Economist* 2014). The fall in energy prices convinced international energy companies to exit ambitious projects in which they had been heavily investing in previous years. Ultimately, the ensemble of the three factors that I have outlined, namely the legislative, the geological and the macroeconomic, led many companies to decide that operating conditions would be too complex for an economically viable production of gas to be feasible in Poland.

The Polish parliament's approval in March 2014 of new legislation exempting gas companies from paying royalties until 2020 for the extraction of shale gas, as well as unifying and simplifying procedures for demanding exploration and exploitation rights, which had theretofore been kept separated, was not able to stop the bleeding: companies continued to quit the shale-gas business in Poland (BBC News 2013; *Natural Gas World* 2013; Cienski 2014; *Shale Gas International* 2014; Lowe 2015). In the summer of 2011, gas exploration companies planned to drill 123 test wells in the coming years to reliably assess reserves in Poland. Five years later, the situation looked very different: companies have drilled only 72 test wells since (Pliszczyńska 2013, p. 334). The enthusiasm for the possible new El Dorado of Central Europe had died out, even more rapidly than had the Galician El Dorado in the early twentieth century. In fact, in the case of shale gas, different to the case of Galician oil, there had been no actual boom: the house of cards had folded before it had been built; the promise of new richness had been broken.

According to data from the Polish Ministry of the Environment (2016), in March 2016 only 31 shale-gas exploration permits were still operating in the country (Fig. 5). Compared to the record year of 2012 with 115 active concessions, this figure means that 84 shale-gas exploration concessions

[2]Tomasz Chmal, interview with the author, in Warsaw, on 1 December 2015; Paweł Poprawa, Consultant at Instytut Studiów Energetycznych, interview with the author, in Cracow, on 8 December 2015.

[3]Marcin Zięba, OPPPW General Director, interview with the author, in Warsaw, on 15 October 2015; Senior PGI informant, interview with the author, in Warsaw, on 30 October 2015.

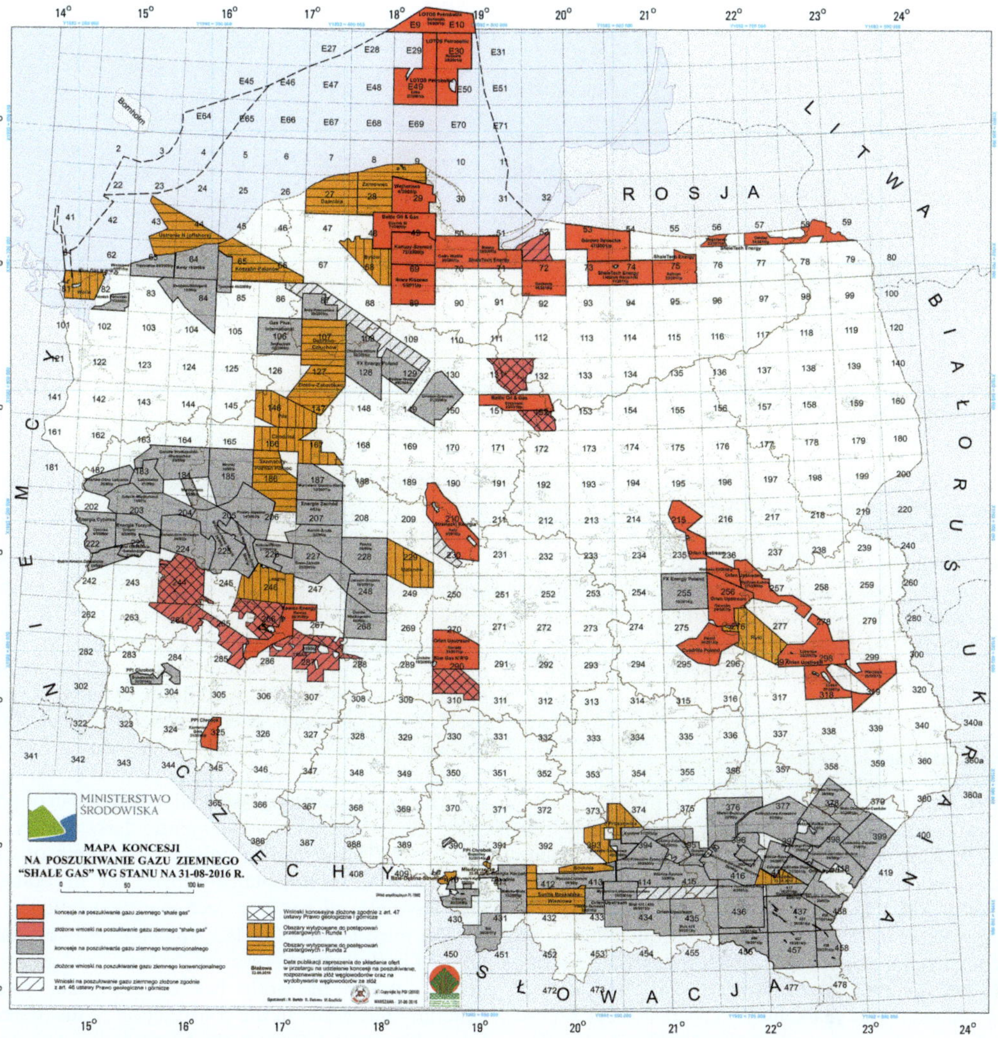

Fig. 5. Map of concessions and application requests for shale gas at 31 August 2016 (*source*: Polish Ministry of the Environment).

have been abandoned in the previous 3 years. The number of companies has dropped from more than 20 to nine, and by late 2016 the only companies operating on Polish territories were the Polish 'big three' (PGNiG, which holds four concessions; Orlen Upstream, with seven concessions; and the Lotos Group, which has seven concessions), and six foreign companies, the largest of which in terms of permits held is ShaleTech Energy, a company that is part of the Luxembourgish Stena Investment group, and which holds seven concessions (Polish Geological Survey 2016). With all probability, these companies will not take the risk of investing in new extraction technologies while macroeconomic conditions stay as they are. The total area of exploration activities has decreased from a maximum of almost 90 000 km^2 (almost 30% of Poland's area) in 2012–13 to the current 24 290 km^2, corresponding to only 7.8% of Poland's area. The yearly number of drilled wells reported in Figure 6 clearly shows the rapid rise and decline of shale-gas activities.

The shale-gas industry is Poland is presently going through an anticlimax. In this regard, it is significant that in 2014 even the Kosciuszko Institute, which since 2011 had been one of the most ardent promoters of shale gas and which had published a series of very hopeful reports on the possibility of an energy boom in Poland, has moved its interests

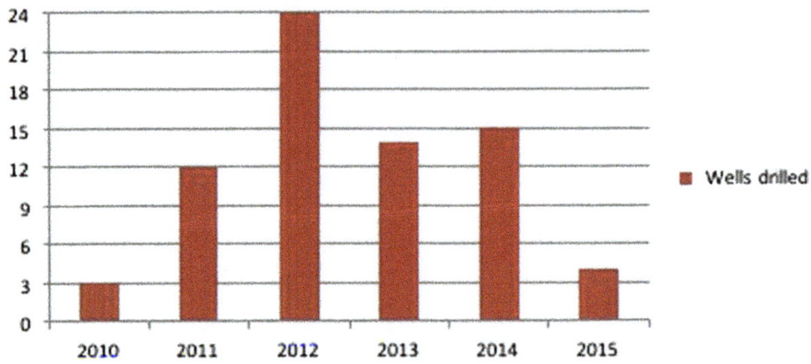

Fig. 6. Shale-gas testing wells drilled in Poland yearly (2010–15).

away from shale gas and into biosecurity.[4] The publication of the *International Shale Gas and Oil Journal*, which the Institute had launched in 2013, was also discontinued in 2014 after the publication of four issues. Today, shale-gas perspectives in Poland look bleak for the industry (and rather good for environmentalist movements): certainly, no one seems to continue to believe the tale of a second Galicia.

Conclusions

When Polish people say that their oil industry is one of the oldest in the world, they are certainly not wrong. But what does the 'antiquity' argument imply for the chances of developing a shale-gas industry today in Poland? Not much, in fact: as it is one thing to have established an oil industry in the nineteenth century and a coal industry even earlier, and a completely different thing to establish a gas industry 150 years later. The historical hiatus in the practices of managing extraction activities from the second half of the twentieth century (when the Galician oil industry had all but disappeared) to the late 2000s (when the Polish government tried to 'import' the shale-gas revolution from the USA) meant that not only had the techno-scientific proficiency that had been achieved during the Galician heyday been lost in the following decades, but also that the managerial proficiency developed at that time had been discontinued and was no longer applicable to a very different historical context. Post-war policies in the oil and gas sector in communist Poland made it possible to establish a few enterprises capable of performing some exploration work, but the sector in which communist-time administrations invested most heavily was refining, not exploration and production. Thus, when shale-gas businesses arrived in Poland, the national industry and the legislative apparatus were unprepared.

The outcome of the Polish shale-gas affair may seem unexpected if one looks exclusively at media representation. Here we find staggering perspectives of hundreds of thousands of jobs, fabulous foreign investments and the image of Poland as a large gas producer, which would, in addition, eventually free itself of the Russian energy yoke. Advocates of shale gas largely won their battle far too easily, by constructing a powerful narrative centred on energy security. This narrative essentially silenced an opposition that was already neither particularly numerous nor vocal. The historical political weakness of the post-1989 environmentalist movement in Poland may have played a role in such marginalization.

The uncertainties that are typical of early estimations of hydrocarbons reserves, at least until an adequate number of wells has been drilled, were generally neglected in the construction of this narrative. Ultimately, however, the boom ended up being a bubble, which burst as soon as companies were faced with the reality of Polish geology and legislation. Once one takes into account the coming together not only of historical but also of geological, geo-economic and legislative factors, which I have analysed in this paper and which have shaped Poland's shale-gas non-revolution, the present outcome appears to make some sense.

The strength of the shale-gas propaganda between 2010 and 2013 was soon repaid by a deception that was equal in magnitude and opposite in direction, and with results that considerably fell short of expectations. From late 2013, shale gas rapidly disappeared from the media and from the hot topics in political debates (the campaign for the

[4]Izabela Albrycht, Kosciuszko Institute's Chair, interview with the author, in Cracow, on 28 October 2015.

2015 parliamentary elections is a case in point). Today, the emphasis on gas remains significant, but the tone of the media is certainly more moderate than in past years. The energy security narrative has shifted from shale gas to, more generally, all sorts of gas. Instead of focusing on the number of test wells drilled, it centres on the construction of pipelines connecting Poland with the Baltic countries, the Czech Republic and Slovakia, and to a liquefied natural gas (LNG) terminal that was inaugurated by the new Prime Minister, Beata Szydło, in late 2015 in Świnoujście, and which received its first commercial deliveries in June 2016 (Strzelecki 2016). These new pathways, it is thought, will improve Poland's energy security by diversifying its supplies, and compensate for the crumbling of the shale-gas dream.

Acknowledgements The author's thanks go to the two anonymous reviewers who provided useful comments and suggestions.

Funding The research for this paper was made possible by a post-doctoral grant assigned to the author by IFRIS, France.

References

Akiner, S. & Aldis, A. 2004. *The Caspian: Politics, Energy and Security*. Routledge, London.

Albrycht, I. 2011. *Shale Gas? 'Yes we can!' Barack Obama in Poland*. The Kosciuszko Institute Policy Brief, May 2011. Kosciuszko Institute, Kraków, Poland.

Bafoil, F. (ed.) 2007. *La Pologne*. Fayard, Paris (chapters VIII (by M. Zaremba) and IX (by J.-Y. Potel)).

Bafoil, F. 2014. *Emerging Capitalism in Central Europe and Southeast Asia. A Comparison of Political Economies*. Palgrave Macmillan, Basingstoke, UK.

BBC News 2009. Ukraine warns EU of gas 'problem', 3 January 2009. BBC, London, http://news.bbc.co.uk/2/hi/europe/7809450.stm

BBC News 2013. North American firms quit shale gas fracking in Poland, 8 May 2013. BBC, London, http://www.bbc.com/news/business-22459629

Biztok 2014. Firmy szukające gazu uciekają z Polski. Nie będziemy nową Norwegią. *Biztok*, 26 November 2014, http://www.biztok.pl/biznes/firmy-szukajace-gazu-uciekaja-z-polski-nie-bedziemy-nowa-norwegia_a18962

Brantly, J.E. 1971. *History of the Oil Well Drilling*. Gulf Publishing Co., Houston, Texas, USA.

Buhl, E. 2002. *Skarb gorlickiej ziemi. Ocalić od zapomnienia*. Urząd Miejski w Gorlicach, Gorlice, Poland.

Cienski, J. 2014. Eni joins shale gas exodus from Poland. *Financial Times*, 15 January 2014, https://www.ft.com/content/8da7841a-7df1-11e3-95dd-00144feabdc0

Davies, N. 2005. *God's Playground. A History of Poland, Volume 2*. Oxford University Press, Oxford.

De Menil, G. 2003. History, policy, and performance in two transition economies: Poland and Romania. *In*: Rodrik, D. (ed.) *Search of Prosperity: Analytic Narratives on Economic Growth*. Princeton University Press, Princeton, NJ, USA, 271–294.

Dunn, E.C. 2004. *Privatizing Poland. Baby Food, Big Business, and the Remaking of Labor*. Cornell University Press, Itacha, NY, USA.

EIA 2011. *World Shale Gas Resources: An Initial Assessment of 14 Regions Outside the United States*. United States Energy Information Administration, Department of Energy, Washington, DC, USA, April 2011.

EIA 2013. *Technically Recoverable Shale Oil and Shale Gas Resources: An Assessment of 137 Shale Formations in 41 Countries Outside the United States*. United States Energy Information Administration, Department of Energy, Washington, DC, USA, June 2013.

European Commission 2016. State aid: Commission clears PLN 7.95 billion Polish support for closing coal mines, press release, 18 November 2016. European Commission, Brussels, http://europa.eu/rapid/press-release_IP-16-3824_en.htm

Frank, A.F. 2005. *Oil Empire: Visions of Prosperity in Austrian Galicia*. Harvard University Press, Cambridge, MA, USA.

Friends of the Earth 2012. *Unconventional and Unwanted: The Case against Shale Gas Extractive Industries: Blessing or Curse?* Friends of the Earth, Amsterdam.

Gandel, S. 2015. Frackers could soon face mass extinction. *Fortune*, 26 September 2015, http://fortune.com/2015/09/26/frackers-could-soon-face-mass-extinction

Glinski, P. 1994. Environmentalism among Polish youth. *Communist and Post-Communist Studies*, **27**, 145–159.

Godzimirski, J.M. 2014. *European Energy Security in the Wake of the Russia-Ukraine Crisis*. Strategic File No. 27 (63), December 2014. Polish Institute of International Affairs (PISM), Warsaw.

Goldthau, A. 2013. *The Geopolitics of Natural Gas. The Politics of Natural Gas Development in the European Union*. Harvard University's Belfer Center, Cambridge, MA, USA & Rice University's Baker Institute Center for Energy Studies, Houston, TX, USA.

Harper, J. 2016. New law threatens wind power in Poland, 10 March 2016. DW (Deutsche Welle), Bonn, Germany, http://www.dw.com/en/new-law-threatens-wind-power-in-poland/a-19106846

Högselius, P. 2013. *Red Gas. Russia and the Origins of Europe's Energy Dependence*. Palgrave Macmillan, New York.

Hołdys, A. 2015. Manny z łupków nie będzie. *Gazeta Wyborcza*, 19 May 2015, http://wyborcza.pl/1,75400,17941067,Manny_z_lupkow_nie_bedzie.html

Howarth, R.W. 2014. A bridge to nowhere: methane emissions and the greenhouse gas footprint of natural gas. *Energy Science & Engineering*, **2**, 47–60.

IEA 2011*a*. *Are We Entering a Golden Age of Gas?* Special Report. World Energy Outlook 2011 OECD/IEA, Paris.

IEA 2011*b*. *Energy Policies of IEA Countries – Poland – 2011 Review*. OECD/IEA, Paris.

Kádár, B.Á. 2014. *The effect of anti-fracking movements on European shale gas policies: Poland and France.*

Master's dissertation. Central European University, Department of International Relations and European Studies.

KPMG 2012. *Central and Eastern European Shale Gas Outlook.* The KPMG Global Energy Institute.

KRAMER, A.E. 2009. Russia gas pipeline heightens East Europe's fears. *The New York Times*, 12 October 2009, http://www.nytimes.com/2009/10/13/world/europe/13pipes.html?_r=0

KRUK, M. 2011. Polish shale sector needs hands. *The Wall Street Journal*, 23 August 2011, http://www.wsj.com/articles/SB10001424053111903327904576524124036057978

LABUDA, W. 2013. Upstream sector in Poland – past, present and future. Gaz i Ropa z formacji łupkowych, news service of the Polish Geological Survey, undated (presumably 2013), http://infolupki.pgi.gov.pl/en/economy/upstream-sector-poland-past-present-and-future

LAVIGNE, M. 1995. *The Economies of Transition: From Socialist Economy to Market Economy.* Macmillan, London.

LIS, A. & STANKIEWICZ, P. 2017. Framing shale gas for policy-making in Poland. *Journal of Environmental Policy & Planning*, **19**, 53–71.

LOWE, C. 2015. Chevron to stop its shale gas exploration in Poland. *Reuters*, 31 January 2015, http://www.reuters.com/article/chevron-poland-shale-idUSL6N0VA08020150131

MACALISTER, T. 2016. First US shale gas sails into Europe as Ineos carrier arrives in Norway. *The Guardian*, 23 March 2016.

MATERKA, E. 2012*a*. End of transition? Expropriation, resource nationalism, fuzzy research, and corruption of environmental institutions in the making of the shale gas revolution in northern Poland. *Debatte: Journal of Contemporary Central and Eastern Europe*, **19**, 1–33.

MATERKA, E. 2012*b*. Poland's quiet revolution: the unfolding of shale gas exploration and its discontents in Pomerania. *CEJISS*, **6**, 180–211.

MINISTRY OF THE TREASURY 2005. *Information of the Ministry of the Treasury Concerning the Polish Oil and Gas Company (PGNiG S.A.).* Ministry of Treasury, Warsaw, 7 September (press release), http://archive.is/H68P

NATURAL GAS WORLD 2012*a*. Poland: Political group quietly supporting anti shale movement. *Natural Gas Europe*, 27 August 2012, http://www.naturalgaseurope.com/poland-political-group-quietlysupporting-anti-shale-movement

NATURAL GAS WORLD 2012*b*. Poland unveils shale gas taxation and regulatory legislation. *Natural Gas Europe*, 16 October 2012, http://www.naturalgaseurope.com/poland-shale-gas-taxation-legislation

NATURAL GAS WORLD 2013. Marathon departure another setback to Polish shale gas industry. *Natural Gas Europe*, 8 May 2013, http://www.naturalgaseurope.com/marathon-departure-another-setback-to-polish-shale-gas-industry

NIEMUTH, S. & WESTPHAL, S. 2014. Blind politics of ambition: shale gas in Poland. *JEMPAS*, **1**, 39–46.

PGI 2012. *Assessment of Shale Gas and Shale Oil Resources of the Lower Paleozoic Baltic-Podlasie-Lublin Basin in Poland – First Report.* PGI, Warsaw.

PLISZCZYŃSKA, K. 2013. Poszukiwanie i rozpoznawanie niekonwencjonalnych złóż węglowodorów – stan prac i działania Ministerstwa Środowiska. *Przegląd Geologiczny*, **61**, 334–337.

POLISH GEOLOGICAL SURVEY 2016. Shale gas exploration results in Poland as of April 2016, 11 April 2016. Gaz i Ropa z formacji łupkowych, news service of the Polish Geological Survey, http://infolupki.pgi.gov.pl/en/exploration-status/news/shale-gas-exploration-status-poland-april-2016

POLISH MINISTRY OF THE ENVIRONMENT 2016. Stan prac w Polsce – Zestawienie koncesji dotyczących gazu z łupków, March 2016. Polish Ministry of the Environment, Warsaw, http://lupki.mos.gov.pl/pliki/201603/03-16-koncesje.pdf

POLISH PRESS AGENCY 2013. Coal extraction in Poland – essential but declining industry, 12 July 2013. Polish Press Agency, Economic Service, Warsaw, https://www.msp.gov.pl/en/polish-economy/economic-news/4447,Coal-extraction-in-Poland-essential-but-declining-industry.html

RISSE, T. 2010. *A Community of Europeans? Transnational Identities and Public Spheres.* Cornell University Press, Ithaca, NY, USA.

ROOTES, C.A. 1997. Environmental movements and green parties in western and eastern Europe. *In*: REDCLIFT, M.R. & WOODGATE, G. (eds) *The International Handbook of Environmental Sociology.* Edward Elgar, Cheltenham, Gloucestershire, UK, 319–348.

RUTKOWSKI, M. 2013. Shale gas in Poland, May 2013. SHIP (Shale Gas Information Platform), GFZ German Research Centre for Geosciences, Potsdam, Germany, http://www.shale-gas-information-platform.org/de/areas/die-debatte/shale-gas-in-poland.html

SCHWARZ, R. 1929. *Petroleum-Vademecum: Tafeln für die Erdölindustrie und den Mineralölhandel*, 6th edn. Verlag für Fachliteratur, Berlin, 4–6 (source reported in Frank 2005, p. 242).

SHALE GAS INTERNATIONAL 2014. Total calls it quits on Polish shale gas. *Shale Gas International*, 14 April 2014, http://www.shalegas.international/2014/04/14/total-calls-it-quits-on-polish-shale-gas/

STANKIEWICZ, P. & SUCHOMSKA, J. 2014. Ćwiczenie z demokracji? Lokalne konflikty na tle gazu łupkowego. *Nowy Obywatel*, **12**, http://nowyobywatel.pl/2014/04/30/cwiczenie-z-demokracji-lokalne-konflikty-na-tle-gazu-lupkowego/

STRZELECKI, M. 2015. Poland opens LNG terminal, pledges to end Russian dependence. *Bloomberg*, 12 October 2015, http://www.bloomberg.com/news/articles/2015-10-12/poland-opens-lng-terminal-pledges-to-end-russian-gas-dependence

STRZELECKI, M. 2016. Commercial LNG Cargo as new supply route open. *Bloomberg*, 17 June 2016, http://www.bloomberg.com/news/articles/2016-06-17/poland-gets-first-commercial-lng-cargo-as-new-supply-route-open

STULBERG, A. 2015. Out of gas? Russia, Ukraine, Europe, and the changing geopolitics of natural gas. *Problems of Post-Communism*, **62**, 112–130.

SZYDŁOWSKI, J. 2014. Żurawlów: Po 400 dniach blokady Chevron wraca do domu. *Dziennik Wschodni*,

7 July 2014, http://www.dziennikwschodni.pl/zamosc/zurawlow-po-400-dniach-blokady-chevron-wraca-do-domu,n,140709712.html

The Economist 2011. Energy in Poland: fracking heaven. *The Economist*, 23 June 2011, http://www.economist.com/node/18867861

The Economist 2013. Mad and messy regulation: shale gas in Poland. *The Economist*, 10 July 2013, http://www.economist.com/blogs/easternapproaches/2013/07/shale-gas-poland

The Economist 2014. The Economist Explains: Why the oil price is falling. *The Economist*, 8 December 2014, http://www.economist.com/blogs/economist-explains/2014/12/economist-explains-4

Waller, M. & Millard, F. 1992. Environmental politics in Eastern Europe. *Environmental Politics*, **1**, 159–185.

Wilson, A. 2006. *Ukraine's Orange Revolution*. Yale University Press, New Haven, CT, USA.

Wolniewicz, P. 2012. Szejkowie z Galicji, czyli naftowa gorączka w Polsce. *Żywa Planeta*, 4 July 2012, https://zywaplaneta.pl/ropa-naftowa-w-polsce/

Yergin, D. 2012. *The Quest: Energy, Security, and the Remaking of the Modern World*. Penguin Boooks, New York.

Zeh, J. 1889. Pierwsze objawy przemysłu naftowego w Galicyi. *Czasopismo Towarzystwa Aptékarskiego*, **18**, 201–205.

A history of exploration offshore Norway: the Barents Sea

K. H. JAKOBSSON

Norwegian Petroleum Directorate, Professor Olav Hanssensvei vei 10, Stavanger 4003, Norway

knut.henrik.jakobsson@npd.no

Abstract: The exploration activities in the Barents Sea started as far back as 1907 with the first geological expedition to the Spitsbergen archipelago. Exploration for oil and gas started with the Kvadhuken well in 1961, followed by several other wells. Fina Exploration drilled two wells on the Island of Hopen in 1973. The NPD acquired the first seismic data in 1969. The universities collected seismic profiles from 1969 as part of the Continental Shelf Project.

The first licences in the Barents Sea were awarded in the 5th concession round in Norway in 1980. The results so far (2016) are two fields in production, Snøhvit and Goliat, and several in the evaluation phase.

The main exploration challenges in the Barents Sea are the Tertiary uplift and erosion of Jurassic and Cretaceous rocks, and the maturation of the source rocks due to the uplift. The discoveries Gotha and Alta in 2013–15 proved the Permian carbonate play. The awards in the 23rd concession round opened the Barents SE area bordering Russia. One well was drilled in 2017 with encouraging results and two more will be drilled in 2018.

The Barents Sea is the offshore area from Novaya Zemlya in the east to the continental slope of the North Atlantic Ocean to the west and north to Spitsbergen–Franz Joseph's Land at 82° N. It is a continental shelf area with sediments from Devonian times to recent. Possible recoverable resources are estimated to be 1000 and 1500 m^3 oil equivalent (OE) of hydrocarbons (6–9Bbbl (billion barrels)), with both oil and gas: The Norwegian part of the Barents Sea consists of two areas:

- The Barents Sea South extends from the Norwegian coast to 74° 30″ N, and in the east to longitude 32° E and in the west to longitude 16° E. The southern Barents Sea has an area of 313 000 km^2.
- The northern Barents Sea is the area to the north up to 80° N and has an area of 460 000 km^2.

Exploration efforts in the Barents Sea started in 1970. The NPD, or the Petroleum Office in the Ministry of Industry, acquired a small regional grid in 1970 outside Tromsø and Hammerfest (Fig. 1). The Norwegian Geological Survey collected gravity and magnetic (grav–mag) data in the same area in 1969.

The geology of the Barents Sea is similar to the geology of Bear Island and on the Spitsbergen archipelago (Horn & Orvin 1928; Orvin 1940). The Norwegian Polar Institute had conducted investigations in these areas since 1907, when the famous geologist Dr Adolf Hoel participated in the first scientific expedition, the Spitsbergen Expedition, financed by Prince Albert of Monaco (see Norsk Polarinstitutts Historie: http://www.npolar.no/no/om-oss/historie.html).

He returned in 1908 to continue geological investigations; this time financed by the University of Oslo and private donors. The Norwegian government financed activities from 1909, and the Norwegian Polar Institute (Norges Svalbard- og Ishavsundersøkelser), established in 1928, continued the investigations of the Arctic and is still active.

The political status of the Spitsbergen archipelago and the surrounding shelf

The Spitsbergen archipelago was considered a no-man's land until 1925 (Tamnes 1991). It was a free-for-all area with no juridical control. The Spitsbergen Treaty of 1920 gave Norway formal rights to the area with restrictions. The signature states had the right to do business in the area based on Norwegian rules and regulations. The area was also to be a non-military zone. Various disputes concerning Norway's rights to the area and whether it was part of Norway or a 'secondary' country took place over the years. Today, however, the Spitsbergen archipelago, including Bear Island and the Jan Mayen, are integrated parts of Norway. The United Nations Convention on the Law of the Sea gave the coastal states the rights to a 200 nautical mile economic zone, and Norway established this around the Spitsbergen Islands in 1977. The coastal nations' rights to their continental shelf became an issue in the 1990s, and Norway submitted its application to the United

From: CRAIG, J., GERALI, F., MACAULAY, F. & SORKHABI, R. (eds) 2018. *History of the European Oil and Gas Industry*. Geological Society, London, Special Publications, **465**, 219–241.
First published online June 6, 2018, https://doi.org/10.1144/SP465.18

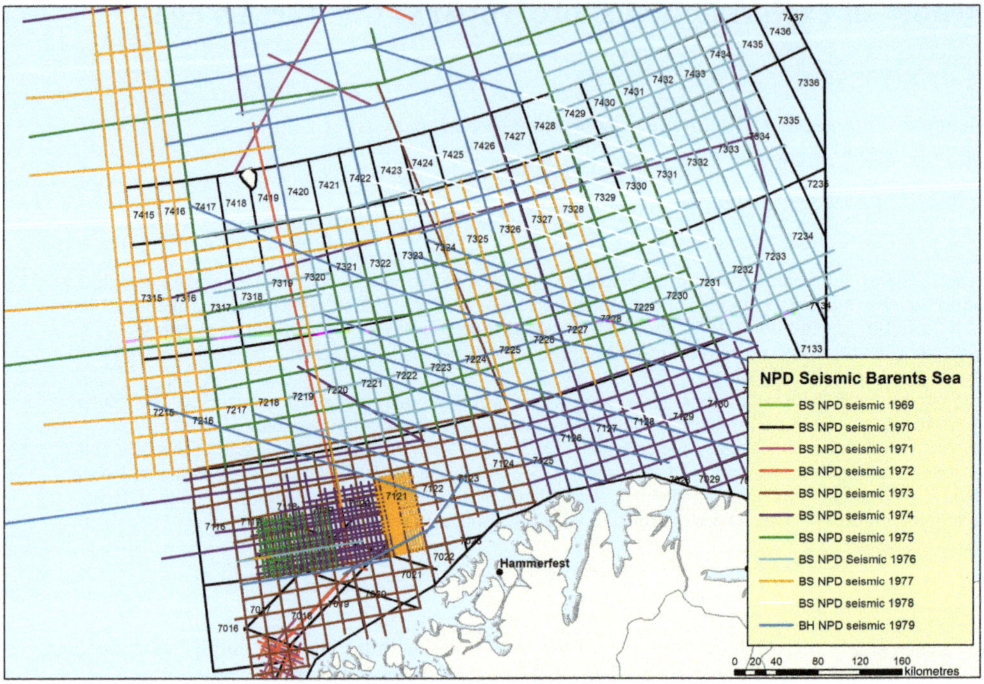

Fig. 1. NPD seismics 1970–79.

Nations in 2006 for the rights to all of the continental shelf area to the agreed borderlines with neighbouring states and to the North Pole (CLCS 2006).

Arctic climate and environmental considerations

The Barents Sea is shallow with an average water depth of 230 m. To the west is the deep Norwegian Sea, which reaches down to 2500 m. The northern Barents Sea north of Bear Island is normally covered by ice for large parts of the year. Climate change, however, has gradually driven the ice edge northwards. The marginal ice zone, the area between permanent ice and ice-free sea, varies over time (Ministry of Climate and Environment 2014–15). A ban on hydrocarbon exploration exists in ice-covered seas; the southern Barents Sea, however, is ice free and exploration drilling is allowed throughout the year. Licences awarded in the 23rd concession rounds in the area to the NE are just south of the marginal ice zone and special environmental considerations are placed on activities there.

Activities prior to the concession awards

The NPD and the University of Bergen acquired the first seismic lines in the Barents Sea in 1969. Seismic surveys were shot each year by the NPD, the universities and the oil companies (Fig. 1).

Seismic acquisition in the early years used dynamite as a source. Norway had an ample supply of German explosives from the war, which were used for this purpose (Fig. 2) (Sellevoll & Sundvor 2001).

The Geological Survey of Norway (NGU) conducted gravimetric and magnetic surveys from 1969 onwards. Data compilations and additional information are available on their homepage (http://www.ngu.no/en/node).

The structural framework of the Barents Sea

The structural framework of the Barents Sea (Fig. 3) is dominated by a large, intra-cratonic platform with several basins and highs formed during late Devonian/Carboniferous–Early Permian tectonic events (Rønnevik & Jacobsen 1984; Gabrielsen *et al.* 1990). The area was further affected by the upper Jurassic tectonic episode, which formed the basins and highs we see today (Figs 4 & 5). The main basins are the Hammerfest Basin and the Nordkapp Basin to the south, with the extensive Bjarmeland Platform to the north. The western area is dominated by the Loppa High and the Western Fault Supercomplex. The Finnmark Fault Complex forms the southern boundary of the basin (Fig. 3).

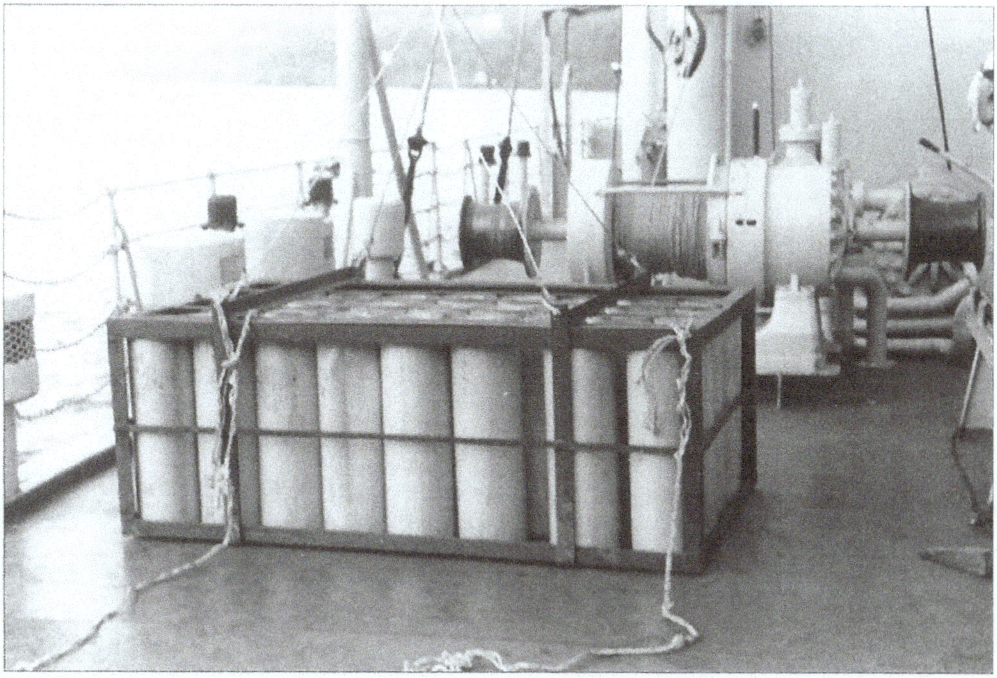

Fig. 2. Dynamite used in the Barents Sea in the early acquisitions (Sellevoll & Sundvor 2001).

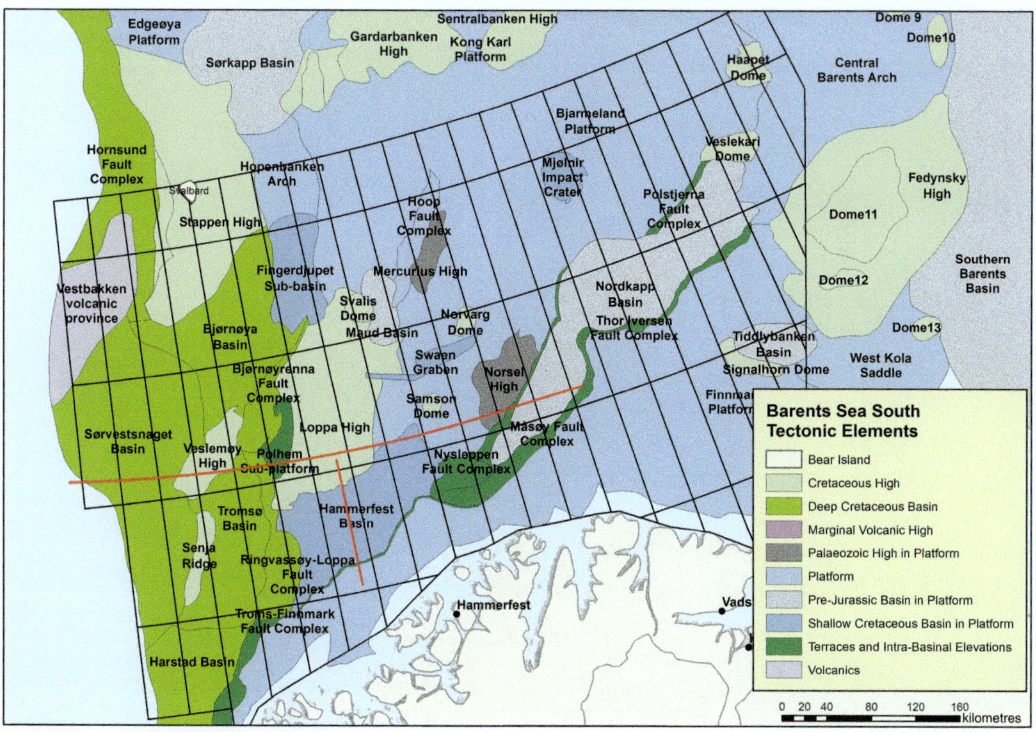

Fig. 3. Structural framework of the southern Barents Sea. The red lines are the locations of the cross-sections.

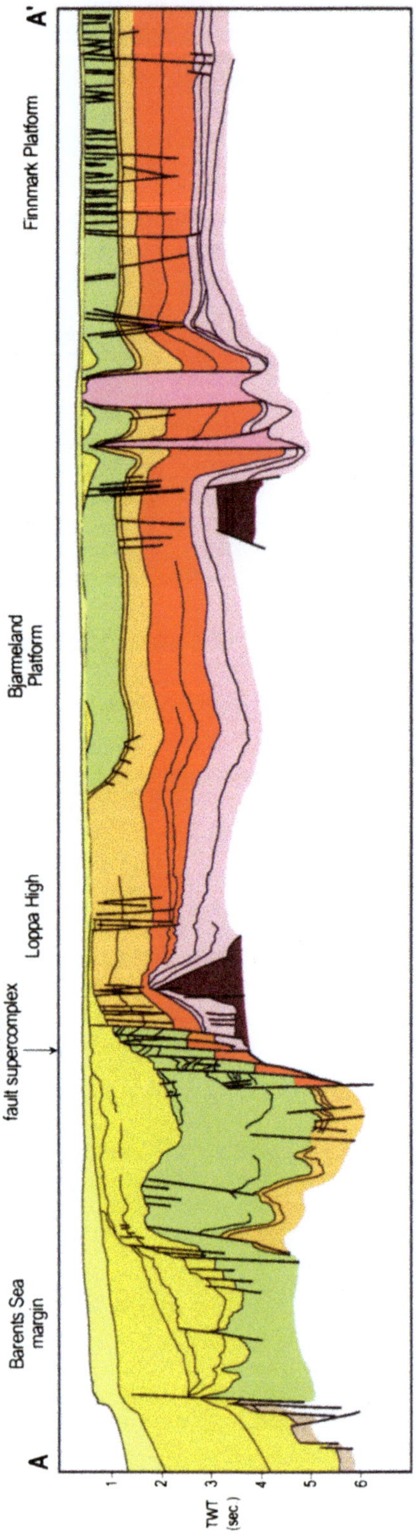

Fig. 4. East–west structural cross-section of the southern Barents Sea (Gabrielsen *et al.* 1990).

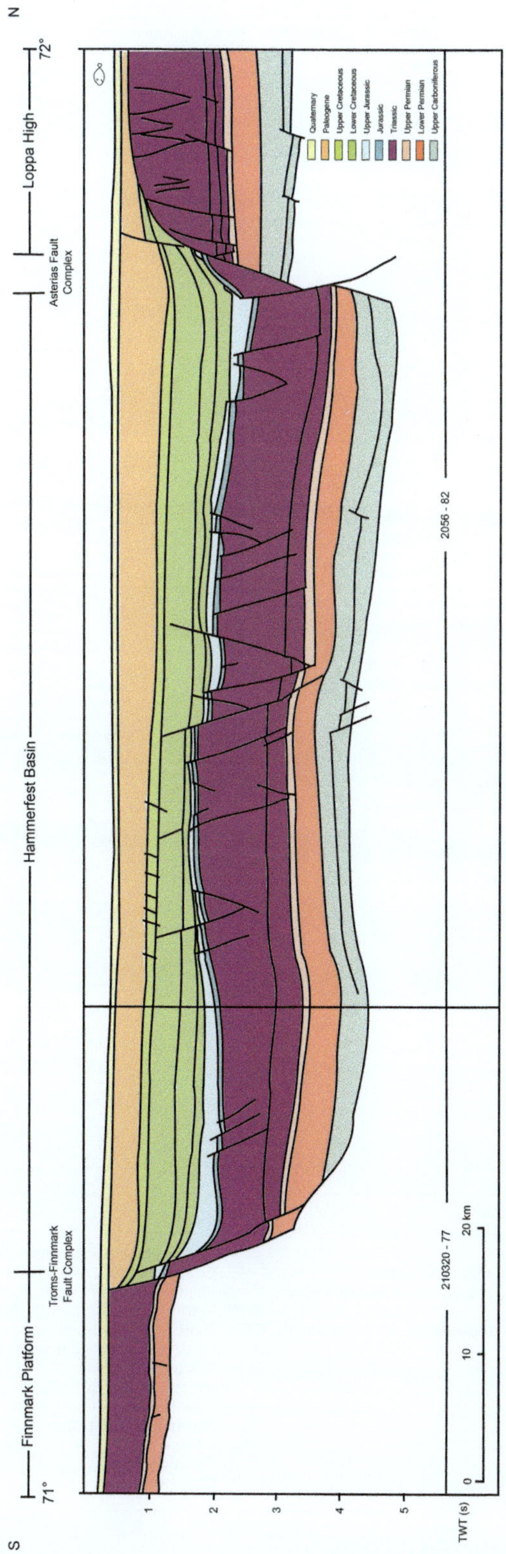

Fig. 5. North–south cross-section of the Hammerfest Basin (Gabrielsen *et al.* 1990).

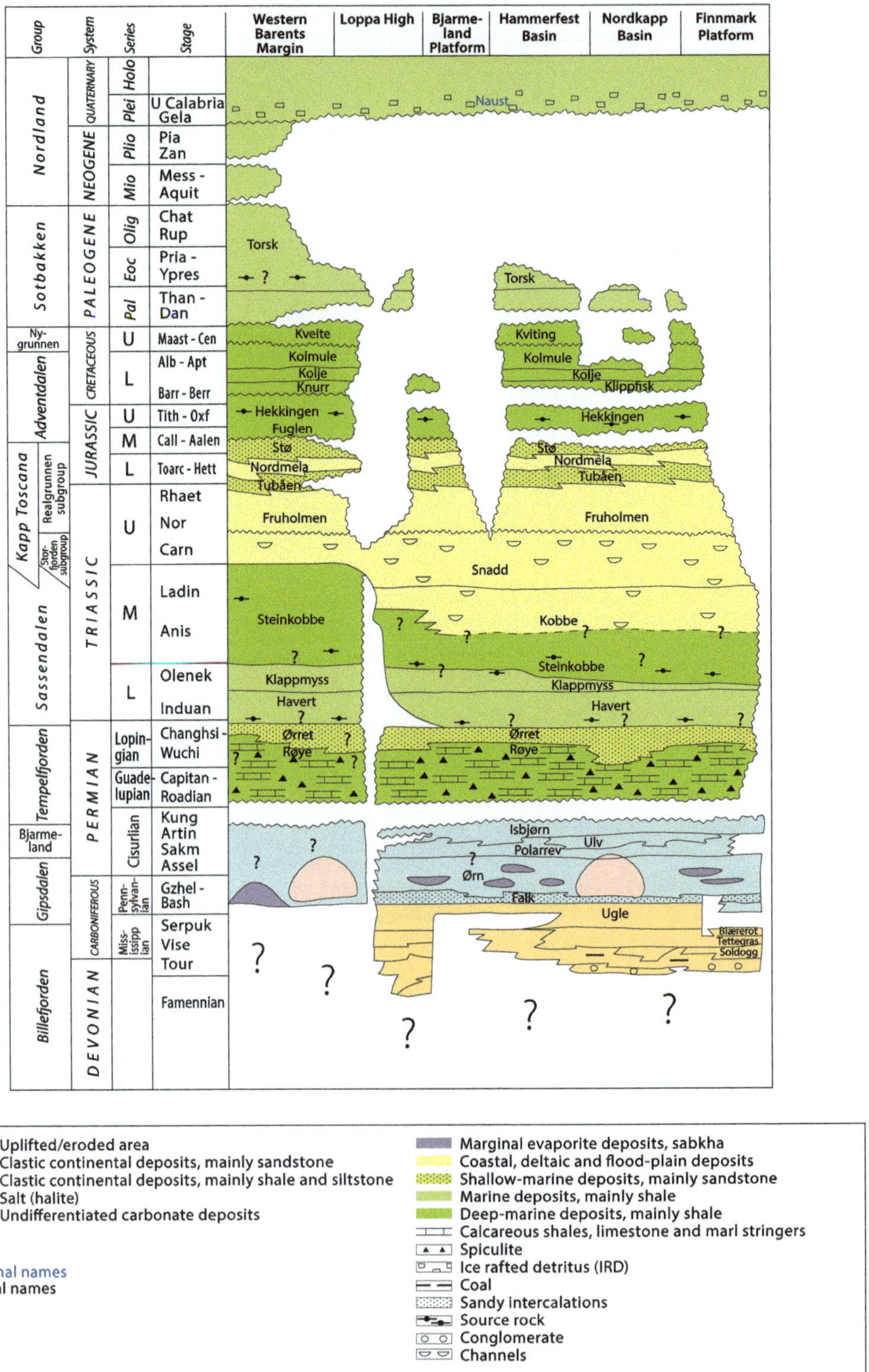

Fig. 6. Lithostratigraphy of the Barents Sea (NPD 2017*a*).

Stratigraphy

The Barents Sea has sediments from at least the Late Devonian to present (Fig. 6). The Devonian comprises a series of sandstones and conglomerates deposited in a terrestrial environment. The eastern part of the Barents Sea, on the Russian Shelf, is dominated by shallow-marine to marsh-deposits, with shelf carbonates and deep-marine clays in the deepest parts.

The Early Carboniferous strata are a series of continental rocks to the west and shallow-marine/lacustrine rocks to the east. The sea transgressed westwards throughout the Carboniferous, and the Norwegian part of the Barents Sea became gradually more marine as the coastline receded.

The Falk Formation of shallow-marine sandstones is a likely reservoir; while source rocks could be the lacustrine black shales and coals from the lower Carboniferous, generating oil and gas. The Permian consists mostly of marine sediments, such as evaporites, salt in the Nordkapp Basin, and calcareous muds and limestones. The Ørret Formation is a deltaic to shallow-marine sandstone deposited from the Finnmark Platform into the Hammerfest Basin, where it is a good reservoir unit. Limestones and chert reservoirs are present in the Tempelfjorden Group Røye Formation both on Spitsbergen and in the Barents Sea in general. The Triassic succession consists of the Havert, Klappmyss and Steinkobbe formations of marine to shallow-marine origin, with the Kobbe, Snadd and Fruholmen formations of coastal to deltaic origin. The sandstones are, in general, deeply buried and with little porosity due to diagenesis and clay minerals. The best reservoirs are along the Finnmark Platform and in the western parts (Larrsen *et al.* 2002; Halland *et al.* 2013).

The Permian Period ended with the collision of the Siberian Plate with the Laurentian continent, and the formation of the Ural Mountains. In the Triassic, the Barents Sea Shelf became a large Laurasian shallow-marine area with black shales of the Botneheia Formation in the NW and a prograding coastline from the east with a delta system moving westwards across the shelf. The result is a good source rock overlain by good reservoir sandstones on the Bjarmeland Platform. The source rocks, however, seem to be insufficiently buried to have reached the maturity level sufficient for enough oil and gas generation.

The area was uplifted in the Early Jurassic and erosion of the upper Triassic rocks occurred over large areas of the Barents Sea, especially on the Loppa High and on the Bjarmeland Platform (Fig. 4). The Early Jurassic became a continental lowland with marine deposits in the west. The Tubåen Formation is an excellent reservoir unit deposited in a

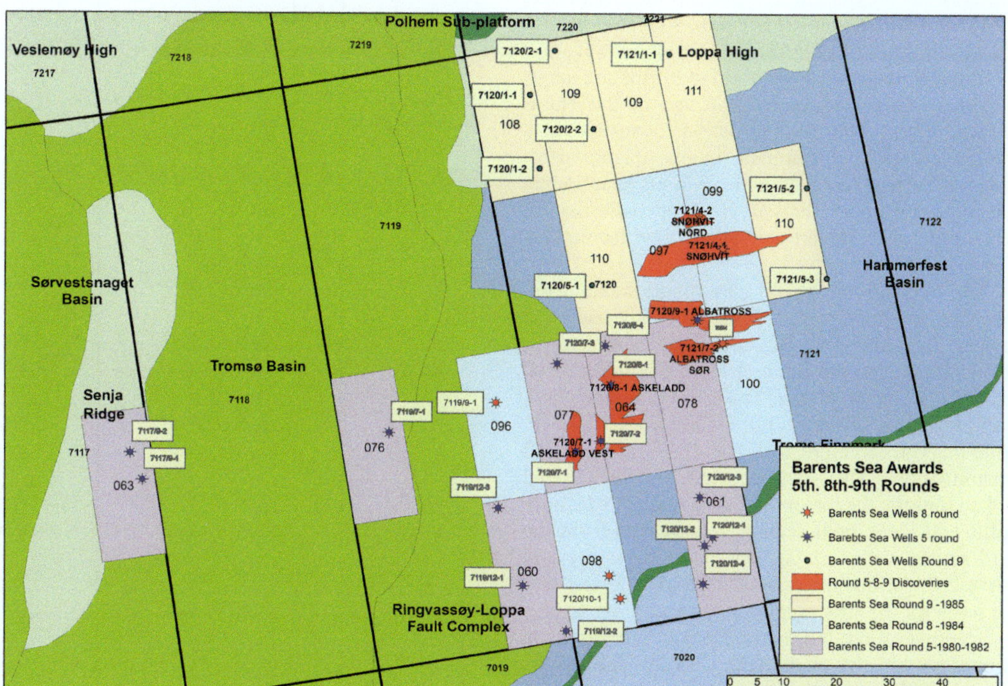

Fig. 7. Barents Sea 5th–9th licensing rounds. Licence awards, wells and discoveries in the Hammerfest Basin.

fluvio-deltaic sandstone environment. In the Middle Jurassic, erosion continued over large areas, and sediments are mostly found in western and eastern areas (Henriksen *et al.* 2011). The Hekkingen Formation of Tithonian–Oxfordian age is an excellent source rock, and similar to the Draupne–Kimmeridge Clay formations of the Haltenbanken–North Sea basins.

The Early–Middle Cretaceous is mostly marine shales in the southern Barents Sea. Gravity-flow deposits are encountered in submarine canyons and could be a model for future discoveries. The Skalle gas discovery in 7120//2-1 is hopefully the first of many in these settings. The area was uplifted in the Late Cretaceous, and either erosion or non-deposition occurred in large parts of the Bjarmeland Platform and the Loppa High. The SW area has marine to deltaic deposits. The Tertiary uplift is a consequence of the Atlantic spreading, and Tertiary sediments are not present in large parts of the Barents Sea. Pleistocene–Holocene sediments cover the area today and act as a seal for the eroded Jurassic reservoirs.

Awards in the 5th Licensing Round 1980

The Norwegian authorities awarded seven new concessions in the Barents Sea in 1980 (Fig. 7). They were in the Tromsø Basin and the Hammerfest Basin. Sixteen exploration wells and three appraisal wells were drilled in these licences, with five discoveries (Tables 1 & 2; Fig. 7).

The discoveries were in the Stø Formation sandstones of early–middle Jurassic age in the Hammerfest Basin. The Stø Formation, deposited in a shallow-marine environment in the Hammerfest Basin, is the best reservoir in the Barents Sea. Source rocks are the upper Jurassic Hekkingen Formation and the Triassic Havert and Steinkobbe formations (Fig. 6).

The five structures with gas, condensate and oil (Fig. 7) comprise parts of the Snøvit Field, and include the Albatross and Askeladden structures, now classified as the Snøhvit Field Complex. Well 7120/12-2, close to the Tromsø Finnmark Fault Complex, had residual oil stains from 1370 to 1690 m. The structure had once been a major oil accumulation. The hydrocarbons, however, had leaked through the major fault system due to Tertiary uplift and weak fault sealing. The results from the exploration wells established the stratigraphy and proved a working petroleum system, and are still the basis for the stratigraphy in the area (Fig. 6).

The first round (5th) in the Barents Sea found the sweet spots in the Hammerfest Basin. The results proved the Tertiary uplift and the challenges with seals and mature source rocks. The industry considered the Barents Sea to be a gas province with

Table 1. *Discoveries in the Barents Sea: 5th Licensing Round*

Block	Licence	Round	Area	Status	Data awarded	Area (km^2)	Status	Field/discovery	Discovery year
7119/12	60	5	Barents Sea	Inactive	18 January 1980	336			
7120/12	61	5	Barents Sea	Inactive	18 January 1980	336			
7117/9	63	5	Barents Sea	Inactive	27 March 1981	332			
7120/8	64	5	Barents Sea	Active	27 March 1981	332	Production	Askeladden/Askeladden Central	1981–83
7119/7	76	5	Barents Sea	Inactive	23 April 1982	332	Production		
7120/7	77	5	Barents Sea	Active	23 April 1982	332	Production	Askeladden West, Askeladden Central	1982
7120/9	78	5	Barents Sea	Active	23 April 1982	332	Production	Albatross, Albatross South	1982–84

Table 2. *Exploration wells and results in the Barents Sea: 5th Licensing Round*

Well name	Licence	Round	Purpose	Content	Year	Field	Depth TVD (m bmsl)*	Age	Formation	First level with HC	First level with HC
7117/9-1	63	5	Wildcat	Dry	1982		3196	Early Cretaceous	Kolmule Formation		
7117/9-2	63	5	Wildcat	Dry	1983		4991	Early Cretaceous	Kolmule Formation		
7119/7-1	76	5	Wildcat	Dry	1983		3134	Permian	Undefined Gp		
7119/12-1	60	5	Wildcat	Oil shows	1980		3087	Early Jurassic	Stø Formation		
7119/12-2	60	5	Wildcat	Shows	1981		1902	Late Triassic	Fruholmen Formation		
7119/12-3	60	5	Wildcat	Gas/ condensates	1983		3308	Early Jurassic	Nordmela Formation	Stø Formation	Middle Jurassic
7120/7-1	77	5	Wildcat	Gas	1982	Snøhvit	2839	Late Triassic	Tubåen Formation	Stø Formation	Middle Jurassic
7120/7-2	77	5	Wildcat	Gas/ condensates	1983	Snøhvit	2522	Late Triassic	Fruholmen Formation	Stø Formation	Middle Jurassic
7120/7-3	77	5	Wildcat	Shows	1984		3061	Early Jurassic	Nordmela Formation		
7120/8-1	64	5	Wildcat	Gas/ condensates	1981	Snøhvit	2610	Late Triassic	Fruholmen Formation	Stø Formation	Jurassic
7120/8-4	64	5	Wildcat	Dry	2007		2696	Late Triassic	Fruholmen Formation		
7120/9-1	78	5	Wildcat	Gas	1982	Snøhvit	2300	Late Triassic	Snadd Formation	Kapp Toscana Gp	Early Jurassic
7120/12-1	61	5	Wildcat	Shows	1980		3516	Middle Triassic	Kobbe Formation		
7120/12-2	61	5	Wildcat	Gas/ condensates	1981		4667	Pre-Devonian	Basement	Stø Formation	Middle Jurassic
7120/12-3	61	5	Wildcat	Gas	1983		2522	Late Triassic	Fruholmen Formation	Stø Formation	Middle Jurassic
7120/12-4	61	5	Wildcat	Dry	1984		2198	Late Carboniferous	Ugle Formation		
7120/8-2	64	5	Appraisal	Gas	1982	Snøhvit	2590	Late Triassic	Fruholmen Formation	Stø Formation	Middle Jurassic
7120/8-3	64	5	Appraisal	Shows	1983	Snøhvit	2325	Early Jurassic	Nordmela Formation		
7120/9-2	78	5	Appraisal	Gas	1984	Snøhvit	5069	Late Permian	Røye Formation	Kapp Toscana Gp	Early Jurassic

*TVD, total vertical depth; m bmsl, metres below mean sea-level.

Table 3. *Barents Sea awards in the 8th and 9th licensing rounds*

Block	Licence	Round	Area	Status	Data awarded	Area (km^2)	Status	Field/discovery
791.00	96	8	Barents Sea	Inactive	9 March 1984	332		
1186.67	97	8	Barents Sea	Active	9 March 1984	327	Production	Snøhvit extension west
712.00	98	8	Barents Sea	Inactive	9 March 1984	336		
1780.25	99	8	Barents Sea	Active	9 March 1984	327	Production	Snøhvit extension east
1017.29	100	8	Barents Sea	Active	9 March 1984	332	Production	Albatross South extension
7120.00	108	9	Barents Sea	Inactive	1 March 1985	323		
7129/2 and 7129/3	109	9	Barents Sea	Inactive	1 March 1985	646		
7120/5 and 7121/5	110	9	Barents Sea	Active	1 March 1985	655	Production	Snøhvit eastern extension
7121.00	111	9	Barents Sea	Inactive	1 March 1985	323		

Table 4. *Wells and results in the Barents Sea: 8th Licensing Round*

Well name	Licence	Round	Purpose	Content	Year	Field	Depth TVD (m bmsl)*	Age	Formation	First level with HC	First level with HC
7119/9-1	96	8	Wildcat	Dry	1984		3248	Late Triassic	Fruholmen Formation		
7120/10-1	98	8	Wildcat	Dry	1984		2000	Late Triassic	Fruholmen Formation		
7120/10-2	98	8	Wildcat	Dry	1990		2497	Late Jurassic	Hekkingen Formation		
7121/4-1	99	8	Wildcat	Oil/gas	1984	Snøhvit	2609	Late Triassic	Fruholmen Formation	Stø Formation	Middle Jurassic
7121/4-2	99	8	Wildcat	Gas/condensate	1985	Snøhvit N.	2799	Late Triassic	Fruholmen Formation	Stø Formation	Middle Jurassic
7121/7-2	100	8	Wildcat	Gas	1986	Albatross S.	2155	Late Triassic	Fruholmen Formation	Stø Formation	Middle Jurassic
7120/6-1	97	8	Appraisal	Oil/gas	1985	Snøhvit	2820	Late Triassic	Tubåen Formation	Stø Formation	Early Jurassic
7121/7-1	100	8	Appraisal	Gas/condensate	1984	Snøhvit	2160	Late Triassic	Fruholmen Formation	Stø Formation	Middle Jurassic
7120/6-2	97	8	Appraisal	Oil/gas	2007	Snøhvit	3035	Late Triassic	Snadd Formation	Stø Formation	Middle Jurassic

*TVD, total vertical depth; m bmsl, metres below mean sea-level.

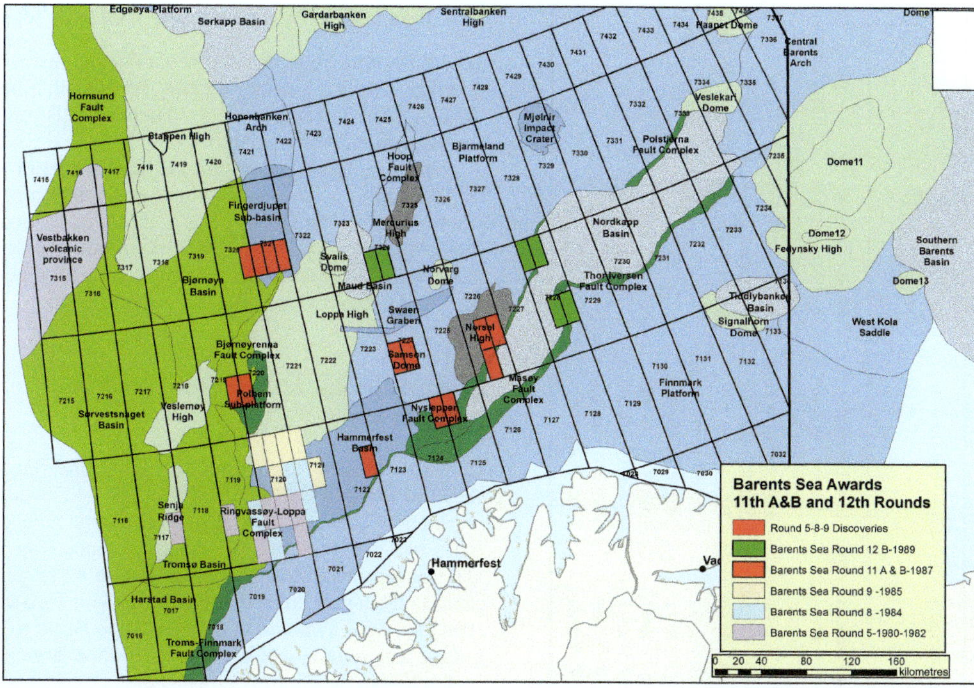

Fig. 8. Barents Sea awards, wells and discoveries in the 11th A & B and 12th licensing rounds.

stranded gas where the oil had leaked due to Tertiary uplift, and the markets were considered too far away. They lost interest in further exploration for some years. The next round of awards was in 1984.

Awards in the 8th Licensing Round 1984

The eighth concession round in 1984 awarded only five blocks in the SW corner of the Hammerfest Basin (Fig. 7). The blocks, awarded to explore for additional resources to the 5th round discoveries, found three additional gas-bearing structures: Snøhvit, Albatross South and Snøhvit North. The Snøhvit discoveries added significant volumes to the previous discoveries and a possible field development became an option. The technology for liquefied natural gas (LNG) transport from this remote location to the market was costly and not very efficient, and foreign oil companies were reluctant to endorse a field development (Jacobsen 2010).

The wells drilled farther east, in PL 086 and 089, were dry. Well 7120/10-1, drilled close to the Troms Finnmark Fault Complex, had stains of dead oil found in 7120/12-2, and again proved the effect of the Tertiary uplift and seal-breach.

Table 5. *Awards in the Barents Sea 11th A & B licensing rounds in 1987 and the 12th Licensing Round in 1989*

Block	Licence	Round	Area	Status	Data awarded	Area (km^2)
7124/3 and 7125/1	135	11-A	Barents Sea	Inactive	10 April 1987	646
7219/9 and 7220/7	136	11-A	Barents Sea	Inactive	10 April 1987	629
7224/7 and 7224/8	137	11-A	Barents Sea	Inactive	10 April 1987	629
7122/6	138	11-B	Barents Sea	Inactive	29 May 1987	327
7226/8-9-11	139	11-B	Barents Sea	Inactive	29 May 1987	948
7320/9 and 7321/7	140	11-B	Barents Sea	Inactive	29 May 1987	594
7321/8 and 7321/9	141	11-B	Barents Sea	Inactive	29 May 1987	594
7228/1 and 7228/2	160	12-B	Barents Sea	Inactive	3 March 1989	612
7228/8 and 7228/9	161	12-B	Barents Sea	Inactive	3 March 1989	629
7324/10 and 7324/11	162	12-B	Barents Sea	Inactive	3 March 1989	603

Table 6. *Wells and results in the Barents Sea: 11th A & B licensing rounds in 1987 and the 12th Licensing Round in 1989*

Well name	Licence	Round	Purpose	Content	Year	Field	Depth TVD (m bmsl)*	Age	Formation	First level with HC	First level with HC
7124/3-1	135	11-A	Wildcat	Oil/gas	1987		4727	Late Carboniferous	Ørn Formation	Tubåen Formation	Early Jurassic
7125/1-1	135	11-A	Wildcat	Oil/gas	1988		2199	Middle Triassic	Kobbe Formation	Stø Formation	Early Jurassic
7219/9-1	136	11-A	Wildcat	Shows	1988		4286	Late Triassic	Snadd Formation		
7224/7-1	137	11-A	Wildcat	Shows	1988		3064	Early Triassic	Havert Formation		
7122/6-1	138	11-B	Wildcat	Gas	1987		2707	Middle Triassic	Snadd Formation	Stø Formation	Early Jurassic
7226/11-1	139	11-B	Wildcat	Gas	1988		5200	Pre-Devonian	Basement	Havert Formation	Early Triassic
7321/7-1	140	11-B	Wildcat	Gas shows	1988		3545	Middle Triassic	Snadd Formation		
7321/8-1	141	11-B	Wildcat	Shows	1987		3470	Late Permian	Røye Formation		
7321/9-1	141	11-B	Wildcat	Shows	1988		1799	Late Triassic	Snadd Formation		
7324/10-1	162	12-B	Wildcat	Shows	1989		2919	Early Triassic	Havert Formation		
7228/2-1	160	12-B	Wildcat	Shows	1989		4000	Early Triassic	Havert Formation		
7228/9-1	161	12-B	Wildcat	Oil/gas shows	1990		4477	Early Permian	Ørn Formation		

*TVD, total vertical depth; m bmsl, metres below mean sea-level.

Awards in the 9th Licensing Round 1985

The 9th round awarded four new concessions in the Hammerfest Basin adjacent to the 5th and 8th rounds blocks (Fig. 7; Tables 3 & 4). The six wells drilled resulted in only one small discovery, Snøhvit Beta.

The disappointing results led to a pause in the award activities from the Norwegian government. Exploration in the Barents Sea had proved to be difficult and new knowledge was required. The NPD analysed all available data, updated the play models and recommended new areas for exploration.

Awards in the 11th A & B Licensing Rounds 1987

The government offered new blocks in the Barents Sea in 1986 in the 11th A & B rounds; and awarded seven new concessions spread over a large area, and with two and three blocks in each licence (Fig. 8). The government strategy was to obtain more information from new areas of the Barents Sea. Two concessions were awarded in the Nordkapp Basin to the NE of the Hammerfest Basin (Fig. 8): one large area to the NW in the Bjørnøya Basin, and one concession to the west on the Bjørnøya Fault Complex/Loppa High area (Table 5).

The 11th round resulted in nine exploration wells. All had traces of hydrocarbons (Table 6). In PL 135 (blocks 7124/3 and 7125/1) (Figs 9 & 10), well 7124/3-1, located east of the Hammerfest Basin on the Nyslepp Fault Complex, tested reservoir rocks of middle Jurassic age and the late Carboniferous Ørn Formation limestones. The verification of the source rock potential of the Triassic was another important goal. The well found 13 m of gas-bearing sandstones in the middle Jurassic Bamse prospect, with 1 m oil lag at 1285 m. The Carboniferous Ørn Formation at 3900 m was dry. The well 7125/1-1 to the east tested the Binne prospect and found oil in the lower Jurassic Stø Formation. The Kobbe Formation (middle Triassic) had gas shows in the sandstone intervals.

PL 139, NE of PL 135 (Fig. 8), located on the Norsel High on the eastern rim of the Bjarmeland Platform (Fig. 3) was awarded to test the lower Jurassic–upper Triassic succession, as well as the Ørn Formation (Carboniferous). As a frontier area, seismic interpretation, structural understanding and palaeontological information were important aspects to better understand the Norsel High.

Well 7226/11-1 found only thin sandstones in the Jurassic–Triassic interval with no shows. The Havert Formation (lower Triassic), however, encountered gas. The Permian carbonates at a depth of 3103 m were dry, but valuable information on the stratigraphy and lithology established the

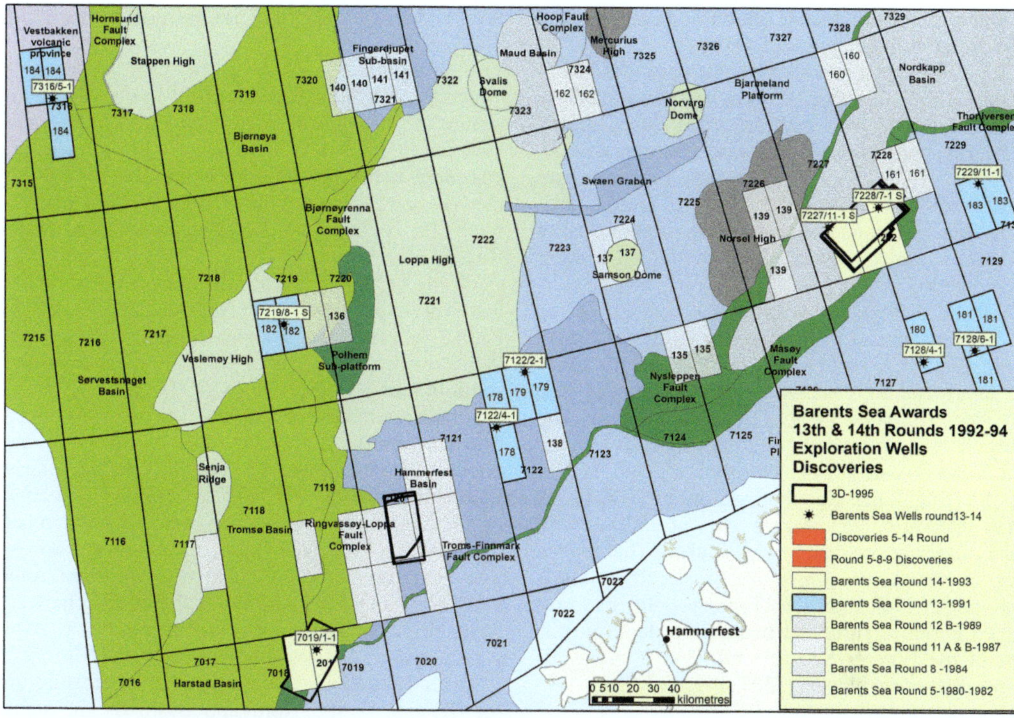

Fig. 9. Barents Sea awards in the 13th licensing rounds 1992–94.

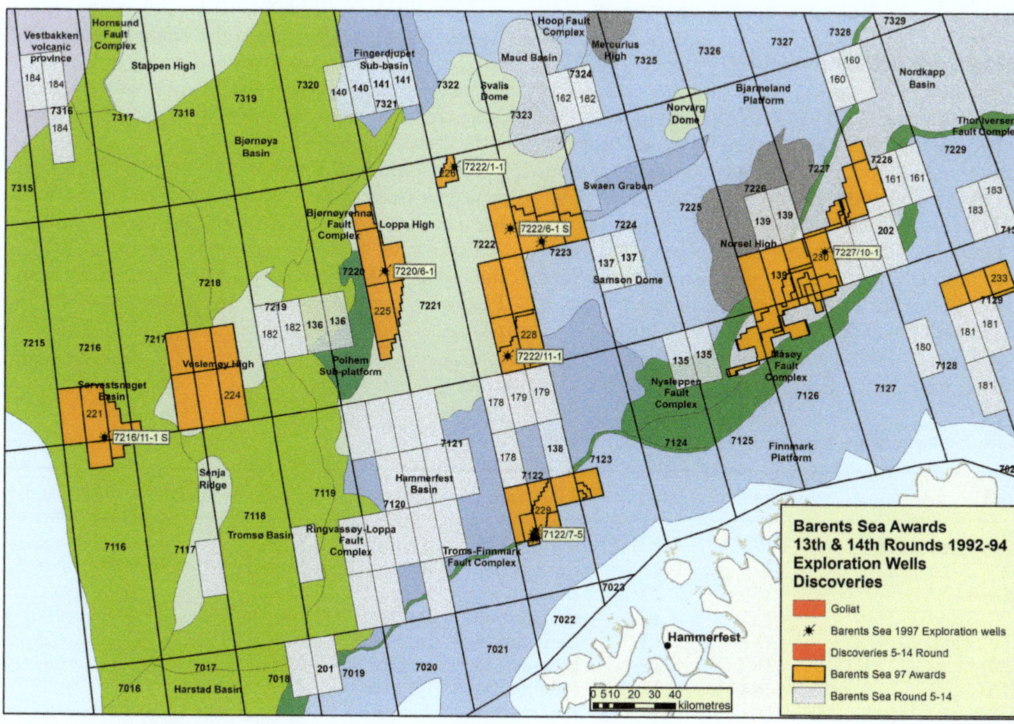

Fig. 10. Barents Sea 1997 awards and previous licensing rounds.

stratigraphy for this area (Larrsen *et al.* 2002). The well drilled to basement at 5200 m.

PL 137 tested the Samson Dome in blocks 7224/7 and 7224/8. The Barents Sea has many of these domes (Figs 3 & 8) which are salt induced, partly with an evaporitic or carbonate core (Gabrielsen *et al.* 1990). Well 7224/7-1 (Fig. 10) tested the structure and found minor gas in the Jurassic–Triassic sections. The sandstone reservoirs were of poor quality and the results, albeit with some gas, were disappointing.

The third frontier area in PL 140 and PL 141 to the NW consists of four blocks in the Fingerdjupet Basin (Fig. 10). Three wells, drilled here to verify the interpretation, established the stratigraphy and provided a seismic tie to future exploration. The prospect was a rotated fault block with Jurassic and Triassic sandstones. The first well, 7321/7-1 (1988), found the Stø Formation at 2000 m. A small interval of gas-bearing sand at 2380 m with good reservoir qualities was found. The source rock intervals were immature or slightly mature, and another indication that Tertiary uplift had prevented burial to sufficient maturation depth in this part of the Barents Sea as well. Two more wells in neighbouring blocks (7321/8-1 and 7321/9-1) had some shows.

The last frontier areas to be investigated were located on the western side of the Loppa Ridge (Fig. 10) in the Bjørnøya Fault Complex. PL 130 consisted of two blocks (7219/1 and 7220/7), and one well was drilled to test the prospect and gain information. The targets were the same Stø and Snadd formations, as well as providing a tie for further seismic interpretations. The usual reservoir sections had good reservoir properties and well-developed sandstones but no hydrocarbons.

The results from the 11th rounds, albeit no discoveries, provided valuable stratigraphic and structural information on the Barents Sea, and provided anchor points for further seismic interpretations, as well as proof of working petroleum systems.

Awards in the 12th B Licensing Round 1989

The disappointing results in the 11th rounds prompted the Norwegian government to continue with awards of strategic licences and three more were awarded in 1989 (Fig. 10; Table 4). PL 160 and PL 161 were in the eastern parts of the Barents Sea on the NW and SE rims of the Nordkapp Basin, respectively. PL 162 explored the Maud Basin to the north (Fig. 8).

Well 7228/2-1 S, on the western margin of the Nordkapp Basin and close to the Nyslepp Fault Complex, was designed to drill a large salt-induced anticline that is bisected by a major north–south-trending fault. The large Triassic throw on the fault defined the targets as two prospects. The downthrown prospect seemed to have the thickest accumulation of Triassic Anisian–Early Ladinian deposits, and was the primary objective of the well. Secondary objectives were other Middle–Upper Triassic and Jurassic sandstones. The well found the expected reservoirs but only small shows and residual oil in the cores. The well confirmed the Hekkingen Formation as a good source rock.

Well 7228/9-1S tested a similar salt feature on the SE side of the Nordkapp Basin on the same play concept. The well was dry for similar reasons as the 7228/2-1S well.

The third well, 7324/10-1, tested the Maud Basin. The play model with the Klappmyss and the Havert formations sandstones as reservoirs, and the Snadd Formation as the source rock, did not work. The well encountered several black shale zones with variable source-rock qualities.

The 11th and 12th rounds provided significant new knowledge to the Barents Sea, and proved viable play models and working petroleum systems. All of the wells had traces of hydrocarbons or gas but none in significant volumes. The Bjarmeland Platform clearly had a hydrocarbon-generating problem.

Table 7. *Barents Sea awards in 13th and 14th rounds*

Block	Licence	Round	Area	Status	Data awarded	Area (km^2)
7122/1 and 7122/4	178	13	Barents Sea	Inactive	1 March 1991	650
7122/2 and 7122/3	179	13	Barents Sea	Inactive	1 March 1991	646
7128/4	180	13	Barents Sea	Inactive	1 March 1991	327
7128/6 and 7128/9	181	13	Barents Sea	Inactive	1 March 1991	986
7219/7 and 7219/8	182	13	Barents Sea	Inactive	1 March 1991	629
7229/11 and 7229/12	183	13	Barents Sea	Inactive	1 March 1991	637
7316/4-5-8	184	13	Barents Sea	Inactive	1 March 1991	883
7018/3 and 7119/1	201	14	Barents Sea	Active	10 September 1993	680
7227/11 and 7228/7	202	14	Barents Sea	Inactive	10 September 1993	1040

Table 8. *Barents Sea wells in the 13th and 14th licensing rounds*

Well name	Licence	Round	Purpose	Content	Year	Field	Depth TVD (m bmsl)*	Age	Formation	First level with HC	First level with HC
7122/2-1	179	13	Wildcat	Dry	1992		2120	Middle Jurassic	Stø Formation		
7122/4-1	178	13	Wildcat	Shows	1992		3015	Late Triassic	Snadd Formation		
7128/4-1	180	13	Wildcat	Oil/gas	1994		2528	Pre-Devonian	Basement	Røye Formation	Late Permian
7128/6-1	181	13	Wildcat	Oil shows	1991		2543	Pre-Devonian	Basement		
7229/11-1	183	13	Wildcat	Dry	1993		4628	Late Carboniferous	Ørn Formation		
7316/5-1	184	13	Wildcat	Gas	1992		4014	Late Cretaceous	Kveite Formation	Torsk Formation	Eocene
7219/8-1	182	13	Wildcat	Dry	1992		4404	Early Jurassic	Stø Formation		
7019/1-1	201	14	Wildcat	Gas	2000		2998	Early Jurassic	Tubåen Formation	Stø Formation	Middle Jurassic
7227/11-1	202	14	Wildcat	Shows	2006		3198	Early Permian	Ørn Formation		
7227/11-1	202	14	Wildcat	Dry	2006		2294	Early Permian	Ørn Formation		
7228/7-1	202	14	Wildcat	Oil/gas	2001		2848	Early Triassic	Klappmyss Formation	Snadd Formation	Late Triassic
7228/7-1	202	14	Appraisal	Oil/gas	2001		2122	Late Triassic	Snadd Formation	Snadd Formation	Late Triassic
7228/7-1	202	14	Wildcat	Dry	2001		1987	Early Permian	Undefined Group		

*TVD, total vertical depth; m bmsl, metres below mean sea-level.

Awards in the 13th Licensing Round 1992–94 and the 14th Licensing Round in 1993

The Norwegian government awarded seven licences in the 13th round in 1991 and two more in the 14th round in 1993 (Fig. 9; Tables 7 & 8). Thirteen wells, seven in the 13th and six in the 14th, were drilled. Again, oil and gas shows and small accumulations of hydrocarbons were found in seven wells, while four were dry. The strategy again was to gain more information, and blocks were awarded in the NW part in the Hornsund Fault Complex–Vestbakken Volcanic Province (Fig. 11). The results were similar to those in previous wells: dead oil, immature source rocks but with good reservoir sandstones in the Jurassic–Triassic sandstones.

Well 7316/5-1, to the NW, tested Paleocene and Cretaceous submarine fan/gravity flow play types. The well encountered 20 m of gas in the Paleocene Sotbakken Group sandstones at around 1300 m depth; a second sandstone at 1442 m was water wet.

PL 182 (blocks 7219/7 and 7219/8), located on the Veslemøy High and in the Bjørnøyrenna Fault Complex, opened a new geological province. Well 7219/8-1 tested a rotated fault block with the middle Jurassic Stø Formation and a secondary target in the upper Jurassic Hekkingen Formation. The well was dry, but valuable information about the source rocks in the area was obtained.

PL 180 and 181 (blocks 7128/4, and 7128/6, 7128/ 9 and 7129/4), which were awarded to test the Finnmark Platform proved the Permian carbonates, the Ørn Formation and the Carboniferous sandstones, and the Falk Formation, on top of the basement. Well 7128/6-1, drilled in 1991, was dry, but the 24 cores (472 m) gave valuable information on the carbonates. The second well (7128/4-1) drilled in 1993, found oil and gas in the Ørn Formation, and proved the Upper Permian carbonate play model and the siliceous spiculite play model.

The exploration efforts in the Barents Sea after 10 years of drilling and 10 rounds of licensing awards resulted in gas discoveries in the Hammerfest Basin (the Snøhvit Field), numerous small oil and gas discoveries and shows, and some dry wells. The major play concepts had been established and the exploration risks well understood. The Tertiary uplift had breached many structures, prevented source rocks from being buried sufficiently to generate enough hydrocarbons and proven variable reservoir sandstone qualities. The Norwegian government decided to postpone further activity. The NPD and the oil companies undertook a comprehensive analysis of the results to prepare for future awards. General interest was low and exploration activities stopped.

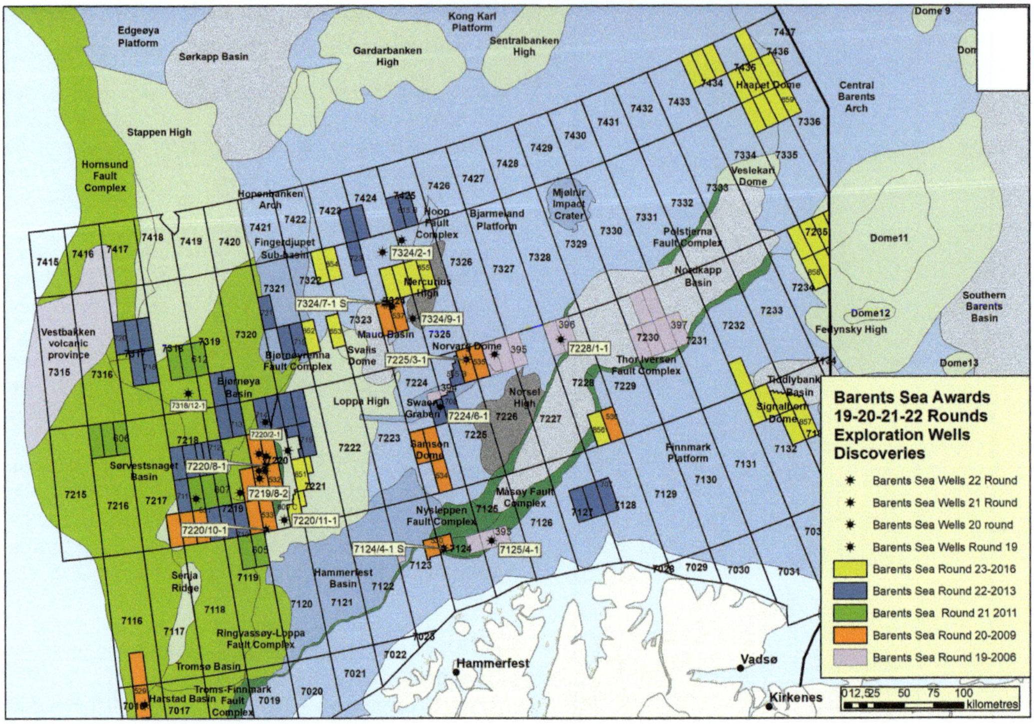

Fig. 11. Barents Sea 19th–22nd licensing rounds awards, wells and discoveries.

Awards in the Barents Sea 1997 Licensing Round

The NPD evaluated the petroleum potential in the Barents Sea in 1995–96 to find the best areas for further exploration. Based on the results, the government invited the industry to nominate blocks for the next concession round. Twelve companies nominated 166 blocks. The Barents Sea Project 1997 awarded 10 new licences in selected areas (Fig. 10; Table 9); some in frontier areas and others adjacent to earlier awards. The activities had operational restrictions to allow for fishing interests and environmental concerns.

PL 221 (Fig. 10) is in the Sørvestnaget Basin. The play model is Paleocene and Eocene gravity-flow sandstones sourced from marine shales of the same age, in addition to Aptian black shales (http://www.npd.no/en/Topics/Geology/Geological-plays/Barents-Sea/Barents-Sea-Paleocene-and-Supra-Paleocene-plays/). The opening of the North Atlantic created rotational fault blocks along the western margin of the Barents Sea due to extension. Well 7216/11-1 S tested these sandstones and found possible traces of hydrocarbons in core No. 1. The turbidite sandstones were of excellent reservoir quality and proved the depositional model.

The second licence, awarded on the Veslemøy High (Fig. 10), was also in a frontier area. Six blocks with a seismic programme were awarded as PL 224. No wells were drilled.

The third area, PL 225, is located on the western margin of the Loppa High. Permian carbonates were the main play model. Exploration risks were mainly reservoir quality and preservation. Tertiary uplift was between 1500 and 2000 m, and reactivation of the faults may have resulted in leaks during the uplift periods. Well 7220/6-12 (2005) confirmed the source/migration concept. Residual hydrocarbons and good oil shows were found in carbonates of the Gipsdalen Group Ørn Formation from 1138 m down to 1430 m.

The third concession, PL 228, located on the Bjarmeland Platform, tested the Snadd Formation and the Kobbe Formation (middle–upper Triassic). Three wells found channel sandstones in the Snadd Formation with good reservoir properties and with small volumes of oil and gas. The wells proved the channel play in the Triassic and gave hope for further exploration of this play concept.

Tertiary uplift resulted in lower pressure and gas expansion in the various structures on the Snøhvit Field. The field itself has a thin oil leg. The oil was believed to have been pushed out of the structures

Table 9. *Wells and results from the BS 1997 Licensing Round*

Well name	Licence	Round	Purpose	Content	Year	Field	Depth TVD (m bmsl)*	Age	Formation	First level with HC	First level with HC
7122/7-1	229	BS-1997	Wildcat	Oil	2000	Goliat	1523	Triassic	Snadd Formation	Tubåen Formation	Early Jurassic
7216/11-1	221	BS-1997	Wildcat	Dry	2000		3733	Paleocene	Torsk Formation		
7122/7-2	229	BS-1997	Appraisal	Oil	2001	Goliat	1417	Triassic	Snadd Formation	Tubåen Formation	Late Triassic
7131/4-1	233	BS-1997	Wildcat	Dry	2005		1295	Middle Triassic	Kobbe Formation		
7220/6-1	225	BS-1997	Wildcat	Oil shows	2005		1540	Pre-Devonian	Basement		
7122/7-3	229	BS-1997	Wildcat	Oil/gas	2006	Goliat	2725	Permian	Tempelfjorden Gp	Tubåen Formation	Late Triassic
7122/7-5	229	BS-1997	Wildcat	Dry	2006		2227	Early Triassic	Klappmyss Formation		
7122/7-4	229	BS-1997	Wildcat	Oil/gas	2006	Goliat	2389	Early Triassic	Havert Formation	Fruholmen Formation	Late Triassic
7122/7-5	229	BS-1997	Appraisal	Oil	2007	Goliat	1973	Middle Triassic	Kobbe Formation	Kobbe Formation	Middle Triassic
7222/11-1	228	BS-1997	Wildcat	Oil/gas	2008		2625	Middle Triassic	Kobbe Formation	Snadd Formation	Middle Triassic
7222/6-1	228	BS-1997	Wildcat	Oil/gas	2008		2848	Early Triassic	Havert Formation	Snadd Formation	Middle Triassic
7223/5-1	228	BS-1997	Wildcat	Gas	2009		2549	Early Triassic	Klappmyss Formation	Snadd Formation	Triassic
7122/7-6	229	BS-1997	Appraisal	Oil/gas	2013	Goliat	2026	Early Triassic	Klappmyss Formation	Fruholmen Formation	Early Triassic
7227/10-1	230	BS-1997	Wildcat	Dry	2014		3135	Middle Triassic			
7222/1-1	226	BS-1997	Wildcat	Dry	2016		2400	Permian			

*TVD, total vertical depth; m bmsl, metres below mean sea-level.

and eastwards into neighbouring structures. PL229, located on the highest point along the Troms–Finnmark Fault Complex was a likely place for the oil to go. Well 7222/7-1, drilled in 2000, found the Goliat Field in oil-bearing upper Triassic Snadd Formation and Kobbe Formation sandstones. The reserves are 20 $MMSm^3$ (million standard cubic metres) of oil with associated gas. The discovery of oil in the Goliath structure established the Barents Sea as an oil province and not just a stranded gas province.

Awards in the Barents 19th Licensing Round 2006

Awards activities stopped for the Barents Sea for almost 10 years. The 19th round, announced in 2006, awarded licences from 2007 to 2012, with an additional award in 2014 in the eastern part of the Barents Sea (Fig. 12; Table 10). In the three licences on the Bjarmeland Platform (PL 394, PL 395 and PL 396), each had one well drilled and gas was found in two prospects, Ververis and Arenaria. PL 397, comprising six blocks, acquired a detailed seismic grid and found no drillable prospects.

Awards in the Barents 20th Licensing Round 2009

The Norwegian government invited interested companies to nominate blocks for the 20th concession round in 2007, and the companies nominated 129 blocks in the Barents Sea. The NPD re-mapped the existing seismic, evaluated the total Barents Sea area and recommended the blocks to be included in the 22th round. The government awarded nine concessions with 34 blocks or part blocks in this round in 2008 (Fig. 11; Table 11).

NDP mapping had identified two concessions, PL 532 and PL 533, on the Loppa High as the most promising, with lower–middle Jurassic sandstones present and a likelihood of sufficient and mature source rocks. Well 7220/8-1 drilled the Skrugard prospect in 2011. The well proved the optimistic evaluation by the NPD and discovered the Johan Castberg Field. Subsequent wells in neighbouring prospects (Table 12) added resources to the field and, the field development plan was submitted to the authorities in December 2017. The development plan was approved by the Norwegian Parliament in 2Q 2018.

PL 537 (blocks 7324/7 and 7324/8) is the northernmost award in the 20th round. Well 7324/8-1 drilled the Wisting prospect in 2013. The Wisting discovery is a middle Jurassic Stø Formation sandstone with good reservoir properties. The prospect is located between the Maud Basin and the Hoop

Fault Complex. The Hansen prospect to the NW also discovered oil. Additional future exploration will probably discover sufficient additional resources for a field development here.

Two prominent dome structures on the Bjarmeland Platform, the Norvarg and Samson domes, were also very attractive exploration targets. Exploration results for the Triassic–lower Jurassic sandstones on the platform had shown a low generation of hydrocarbons (HCs), resulting in small HC columns and poor reservoir properties. The Norvarg Dome, situated farther north, could have different facies and, thus, better prospectivity. The Samson Dome, drilled in 1988, was re-awarded in the 20th round for another try. The area was relinquished without further drilling.

Well 7225/3-1, the first well on the faulted Norvarg Dome, had multiple targets. The primary objectives were to prove hydrocarbon presence, determine the fluid nature and evaluate reservoir characteristics in sandstones of the Late Triassic (Carnian) Lower Snadd Formation and the Early Triassic Kobbe Formation. Secondary objectives were to prove hydrocarbons in the Jurassic Stø Formation, the Early Triassic Havert Formation, and the Permian Tempelfjorden and Bjarmeland groups.

The Jurassic Stø Formation was gas bearing from the top at 726 m to a likely gas–water contact (GWC) at 766.5 m. The Snadd Formation with the top at 804 m was water bearing. An intra-Carnian section, from 1146 to 1521 m, contained gas in two zones with GWCs at 1218 and 1250 m, respectively, and in a third thin sandstone from 1347 to 1357 m. The Kobbe Formation, encountered at 1521 m, was 634 m thick with gas in thin sandstone stringers. The deeper layers were all water wet. The well was a gas discovery and the appraisal well 7225/3-2 confirmed the presence of gas in channel sandstones. The volumes found were insufficient for a development and the licence was relinquished in 2014.

Awards in the Barents 21st Licensing Round 2011

The Norwegian government awarded 12 concessions in the 21st round in 2011.

The 12 concessions consisted of several blocks, and 35 blocks were awarded. Ten exploration wells have resulted in four discoveries: Alta, Iskrystall, Mercury and Atlantis.

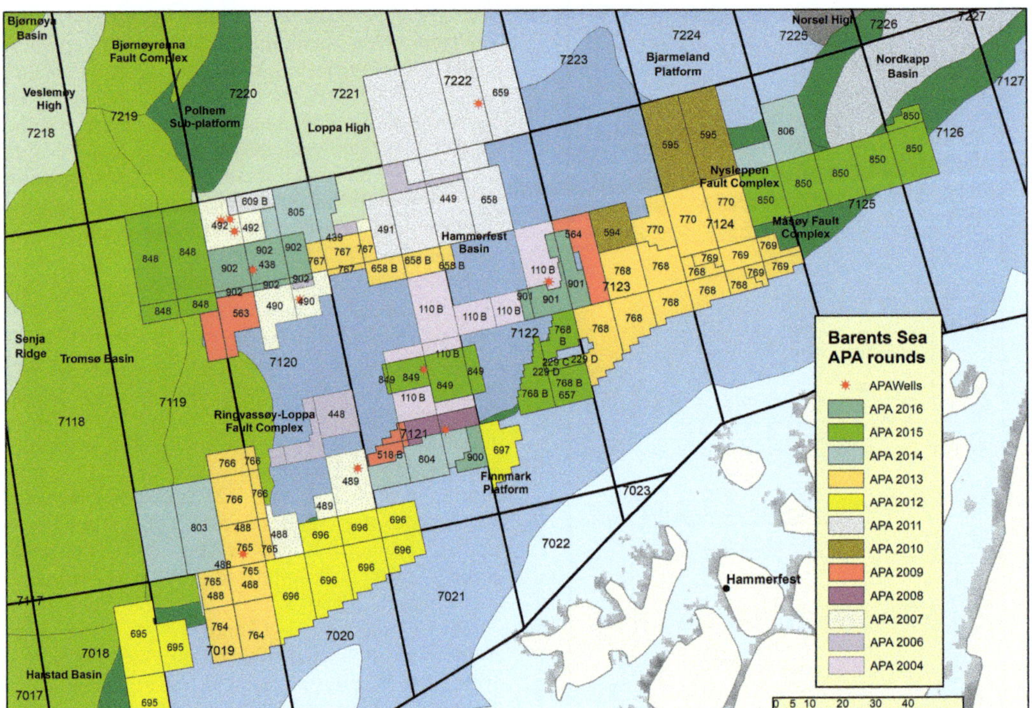

Fig. 12. Barents Sea APA awards 2004–16.

Table 10. *Exploration wells in the Barents Sea: 19th Licensing Round*

Well name	Licence	Round	Purpose	Content	Year	Field	Depth TVD (m bmsl)*	Age	Formation	First level with HC	First level with HC
7125/4-1	393	19	Wildcat	Oil/gas	2007	Nucula	1615	Early Triassic	Klappmyss Formation	Nordmela Formation	Early Jurassic
7125/4-2	393	19	Appraisal	Oil	2008		1750	Early Triassic	Klappmyss Formation	Fruholmen Formation	Late Triassic
7224/6-1	394	19	Wildcat	Gas	2008	Arenaria	2338	Middle Triassic	Kobbe Formation	Kobbe Formation	Middle Triassic
7226/2-1	395	19	Wildcat	Gas	2008	Ververis	2991	Early Triassic	Havert Formation	Kobbe Formation	Middle Triassic
7228/1-1	396	19	Wildcat	Dry	2012		1712	Middle Triassic	Kobbe Formation		
7123/4-1 A	110 C	19	Appraisal	Shows	2008	Tornerose	2626	Late Triassic	Snadd Formation		
7123/4-1 S	110 C	19	Appraisal	Shows	2008	Tornerose	2706	Middle Triassic	Snadd Formation		
7125/4-3	393 B	19+	Wildcat	Dry	2014		1033	Late Jurassic	Hekkingen Formation		

*TVD, total vertical depth; m bmsl, metres below mean sea-level.

The Mercury discovery in PL 615 (blocks 7324/1, 7324/2, 7324/3 and 7325/1) is to the north along the Hoop Fault Complex. Gas sands, encountered in several sand layers in the Snadd, Kobbe and Havert formations, had poor reservoir properties. This discovery is the northernmost in the Barents Sea and opened the northern part of the basin for further exploration.

The Alta discovery in PL 609 (block 7220/11) is located on the southern end of the Loppa High. Oil- and gas-bearing Permian carbonates of the Ørn and Falk formations comprise the reservoir. The Alta discovery is bearing enough to warrant a development together with the neighbouring discoveries.

The 21st round concessions found additional hydrocarbons on the Bjarmeland Platform, and to the north of it. Triassic shallow-marine to coastal/fluvial low-relief sandstones have channel deposits intermixed with coastal/tidal sands in a low-energy environment. Exploration success will depend on a larger channel system and better migration.

The Permian carbonates in the Alta discovery also confirmed this play model, and additional discoveries are expected.

Awards in the 22nd and 23rd Licensing Rounds in 2014 and 2016

Awards in the 22nd round continued the exploration along the western Barents Sea area on the Loppa High. Three wells had been drilled by the end of 2016, all with gas (Table 13; Fig. 11).

One well on the Finnmark Platform in PL 708 (blocks 7130/4 and 7130/7) found oil in Permian carbonates.

The 23rd round awarded blocks in the SE Barents Sea (Fig. 11). The area has unknown potential. The exploration well 7435/12-1 discovered gas in the Korpfjell prospect. Three more wells will be drilled in this area in 2018. Discoveries in the Russian part of the Barents Sea indicate active petroleum systems.

The APA system, introduced in the Barents Sea in 2004 (Fig. 12), offered blocks in the SW area where the geology and potential were known. The eight wells drilled so far have yielded one significant discovery, Gotha. Well 7120/1-3 encountered oil in Permian carbonates and confirmed the validity of this play for further activity. The Alta discovery is another Permian discovery.

Summary and conclusions

Exploration efforts to find economic viable oil and gas fields in the Barents Sea have involved a long and difficult process. From the early start in 1969 until today, 19 concession rounds and several

Table 11. *Wells and discoveries in the Barents Sea 20th Licensing Round*

Well name	Licence	Round	Purpose	Content	Year	Field prospect	Depth TVD (m bmsl)*	Age	Formation	First level with HC	First level with HC
7016/2-1	529	20	Wildcat	Dry	2013		4060	Late Paleocene	Sotbkken Gp		
7218/11-1	531	20	Wildcat	Dry	2013		2540	Early Cretaceous	Kolmule Formation		
7219/12-1	533	20	Wildcat	Oil/gas	2016	Filicudi	2500	Late Triassic	Tubåen Formation	Tubåen Formation	Early Jurassic
7220/4-1	532	20	Wildcat	Gas	2014	Kramsnø	3239	Middle Triassic	Snadd Formation	Stø Formation	Middle Jurassic
7220/5-2	532	20	Wildcat	Gas	2013	Nunatakk	1779	Early Jurassic	Tubåen Formation	Knurr Formation	Early Cretaceous
7220/7-1	532	20	Wildcat	Oil/gas	2012	Havis	2229	Late Triassic	Fruholmen Formation	Stø Formation	Middle Jurassic
7220/8-1	532	20	Wildcat	Oil/gas	2011	Skrugard	2221	Late Triassic	Snadd Formation	Stø Formation	Middle Jurassic
7220/10-1	533	20	Wildcat	Gas	2012	Salina	2405	Late Triassic	Snadd Formation	Kolmule Formation	Early Cretaceous
7225/3-1	535	20	Wildcat	Gas	2011	Norvarg	4147	Permian	Isbjørn Formation	Stø Formation	Middle Jurassic
7324/7-2	537	20	Wildcat	Oil	2014	Hanssen	1719	Middle Triassic	Snadd Formation	Stø Formation	Middle Jurassic
7324/8-1	537	20	Wildcat	Oil	2013	Wisting C.	930	Late Triassic	Snadd Formation	Stø Formation	Middle Jurassic
7324/8-2	537	20	Wildcat	Dry	2015			Late Triassic			
7124/4-1	530	20	Wildcat	Dry	2011		2730	Early Triassic	Havert Formation		
7220/7-2	532	20	Wildcat	Oil/gas	2013	Skavl	1761	Triassic	Snadd Formation	Fruholmen Formation	Late Triassic
7220/7-3	532	20	Wildcat	Oil/gas	2014	Drivis	2059	Late Triassic	Fruholmen Formation	Stø Formation	Middle Jurassic
7324/7-1	537	20	Wildcat	Shows	2013		2477	Middle Triassic	Kobbe Formation		
7220/5-1	532	20	Appraisal	Oil/gas	2012	Skrugard	1740	Late Triassic	Fruholmen Formation	Stø Formation	Middle Jurassic
7225/3-2	535	20	Appraisal	Gas	2013	Norvarg	2209	Triassic	Klappmyss Formation	Kobbe Formation	Triassic
7324/7-3	537	20	Appraisal	Oil	2016	Wisting C.		Middle Jurassic			Middle Jurassic

*TVD, total vertical depth; m bmsl, metres below mean sea-level.

Table 12. *Wells and results in the Barents Sea: 21st and 22nd licensing rounds*

Well name	Licence	Round	Purpose	Content	Year	Field	Depth TVD (m bmsl)*	Age	Formation	First level with HC	First level with HC
7218/8-1	607	21	Wildcat	Dry	2014		3000	Early Cretaceous	Kolmule Formation		
7219/8-2	608	21	Wildcat	Gas	2013	Iskrystall	3413	Late Triassic	Fruholmen Formation	Stø Formation	Middle Jurassic
7220/6-2	609	21	Wildcat	Dry	2015		985				
7220/11-1	609	21	Wildcat	Oil/gas	2014	Alta	2251	Late Carboniferous	Ugle Formation	Undefined GP	Triassic
7224/2-1	611	21	Wildcat	Dry	2016		2944	Early Triassic			
7324/2-1	615	21	Wildcat	Dry	2014		1090	Late Triassic	Snadd Formation		
7324/9-1	614	21	Wildcat	Gas	2014	Mercury	1100	Late Triassic	Snadd Formation	Stø Formation	Middle Jurassic
7325/1-1	615	21	Wildcat	Gas	2014	Atlantis	2865	Early Triassic	Havert Formation	Snadd Formation	Late Triassic
7220/6- 2	609	21	Wildcat		2015		985				
7220/6-2 R	609	21	Wildcat	Oil/gas	2016		1318	Permian	Ørn Formation	Ørn Formation	Permian
7220/11-2	609	21	Appraisal	Oil	2015	Alta	2050	Late Triassic			Triassic
7220/11-2 A	609	21	Appraisal	Oil/gas	2015	Alta	2121	Triassic			
7220/11-3	609	21	Appraisal	Oil/gas	2015	Alta	1956	Permian			Triassic
7220/11-3 AR	609	21	Appraisal	Oil/gas	2016	Alta	2600	Carboniferous			
7130/4-1	708	22	Wildcat	Gas	2016		3184	Carboniferous			Permian
7220/2-1	714	22	Wildcat	Gas	2014		1590	Late Triassic	Snadd Formation	Stø Formation	Middle Jurassic
7319/12-1	713	22	Wildcat	Gas	2014	Pingvin	1540	Late Cretaceous	Kolmule Formation	Torsk Formation	Paleocene

*TVD, total vertical depth; m bmsl, metres below mean sea-level.

Table 13. *Barents Sea APA wells and results*

Well name	Licence	Round	Purpose	Content	Year	Field	Depth TVD (m bmsl)*	Age	Formation	First level with HC	First level with HC
7122/6-2	110 B	APA2004	Appraisal	Gas	2006	Tornerose	3070	Middle Triassic	Kobbe Formation	Snadd Formation	Late Triassic
7120/2-3	438	APA2006	Wildcat	Gas	2011	Skalle	2620	Late Triassic	Snadd Formation	Kolmule Formation	Early
7119/12-4	488	APA2007	Wildcat	Dry	2011		2910	Late Triassic	Fruhclmen Formation		
7120/1-3	492	APA2007	Wildcat	Oil/gas	2013	Gotha	2539	Permian	Røye Formation	Røye Formation	Permian
7120/12-5	489	APA2007	Wildcat	Dry	2011		3628	Triassic	Kobbe Formation		
7120/1-4	492	APA2007	Appraisal	Gas	2014	Gotha	2520	Late Permian	Røye Formation	No formal name	Permian
7120/6-3	490	APA2007	Wildcat	Shows	2012	Juksa	3017	Early Jurassic	Nordmela Formation		
7121/9-1	518	APA2008	Wildcat	Dry	2011		2458	Middle Jurassic	Kapp Toscana Gp		
7222/11-2	659	APA2011	Wildcat	Oil	2014	Langlitinden	2918	Early Triassic	Klappmyss Formation	Kobbe Formation	Middle Triassic

*TVD, total vertical depth; m bmsl, metres below mean sea-level.

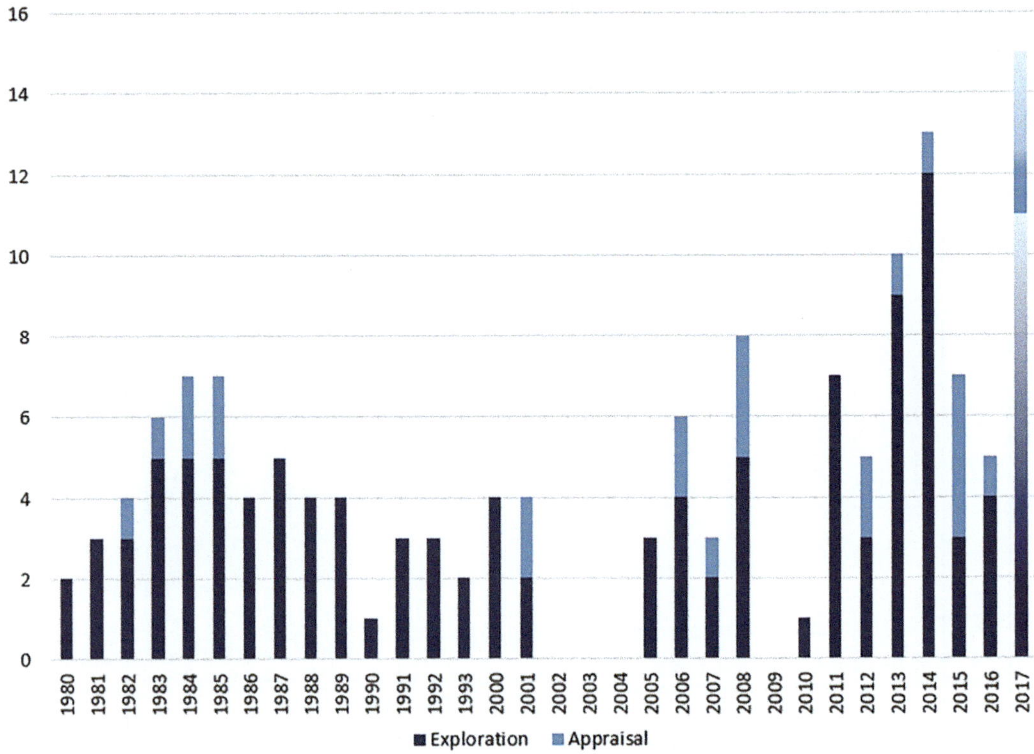

Fig. 13. Exploration and appraisal wells.

APA rounds, two fields are producing – the Snøhvit Field and the Goliat Field – both in the Hammerfest Basin. The Johan Castberg field is approved for development and the Alta and Gotha discoveries are in the planning phase. A development decision is expected in 2017–18. The Wisting and Hanssen discoveries are also substantial, and will be developed when the infrastructure farther south is in place.

Of the 108 exploration wells drilled to 2016, 47 had oil and/or gas or shows in one or several reservoir horizons (Fig. 13).

The exploration wells in the Bjarmeland Platform had shows or small amounts of hydrocarbons. Tertiary uplift had breached the seal or the source rock had been insufficiently buried to generate enough quantities. The reservoir parameters were variable from poor to moderate, low porosity and permeability.

The main play model is that of Jurassic–Triassic shallow-marine sandstones in a rotated fault block sourced from lower–middle Triassic shales.

Permian carbonates in the Hammerfest Basin and on the Finnmark Platform have been proven plays and more exploration will result from this model.

Cretaceous turbidite/gravity-flow fans on the western slope have also been proven, albeit there have been no commercial discoveries as yet.

A better understanding of the channel configurations of the lower Jurassic–Middle Triassic Fruholmen, Snadd and Kobbe formations on the Bjarmeland Platform will eventually lead to commercial discoveries here. New insight into Tertiary uplift and basin modelling will reduce exploration risk in the future.

The NPD's updated resource report for the Barents Sea, updated in 2017, includes the northern Barents Sea (NPD 2017*b*). Substantial yet-to-find (YTF) resources were reported from this area.

Acknowledgements The author would like to thank the Norwegian Petroleum Directorate for their support and my colleagues for reviews and suggestions to improve the paper.

Funding This research received no specific grant from any funding agency in the public, commercial or not-for-profit sectors.

References

CLCS 2006. *Summary of the Recommendations of the Commission on the Limits of the Continental Shelf in Regard to the Submission Made by Norway in Respect of Areas in the Arctic Ocean, the Barents Sea and the Norwegian Sea on 27 November 2006*. Commission of the Limits of the Continental Shelf (CLCS), United Nations, New York, http://www.un.org/Depts/los/clcs_new/submissions_files/nor06/nor_rec_summ.pdf

GABRIELSEN, R.H., FÆRSETH, R.B., JENSEN, L.N., KALHEIM, J.E. & RIIS, F. 1990. *Structural Elements of the Norwegian Continental Shelf. Part 1: The Barents Sea Region*. Norwegian Petroleum Directorate (NPD) Bulletin, **6**, http://www.npd.no/Global/Norsk/3-Publikasjoner/NPD-Bulletin/NPD_BulletinNr6.pdf

HALLAND, E., MUJEZINOVIC, J. & RIIS, F. 2013. *The CO_2 Storage Atlas Barents Sea*. Norwegian Petroleum Directorate (NPD), Stavanger, Norway, http://www.npd.no/en/Publications/Reports/CO2-storage-atlas-Barents-Sea/

HENRIKSEN, E., RYSETH, A.E., LARSSEN, G.B., HEIDE, T., RØNNING, K., SOLLID, K. & STOUPAKOVA, A.V. 2011. Tectonostratigrapy of the greater Barents Sea: implication for petroleum systems. *In*: SPENCER, A.M., EMBRY, A.F., GAUTIER, D.L., STOUPAKOVA, A.V. & SØRENSEN, K. (eds) *Arctic Petroleum Geology*. Geological Society, London, Memoirs, **35**, 163–195, https://doi.org/10.1144/M35.10

HORN, G. & ORVIN, A.K. 1928. *Geology of Bear Island*. Skrifter om Svalbard og Ishavet, **15**.

JACOBSEN, A.R. 2010. *Snøhvit. Historien om olje og gass I Barentshavet*. Statoil, Oslo.

LARRSEN, G.B, ELVEBAKK, G. *ET AL*. 2002. *Upper Palaeozoic Lithostratigraphy of the Southern Norwegian Barents Sea*. Norwegian Petroleum Directorate (NPD) Bulletin, **9**, http://www.npd.no/no/Publikasjoner/NPD-bulletin/259-Bulletin-9/

MINISTRY OF CLIMATE AND ENVIRONMENT 2014–15. *Stotingsmelding No. 20. Update of the Integrated Management Plan for the Barents Sea–Lofoten Area Including an Update of the Delimitation of the Marginal Ice Zone*. White Paper. Norwegian Ministry of Climate and Environment, Oslo.

NPD 2017*a*. *Geological Assessment of The Petroleum Resources in Eastern Parts of Barents Sea North*. Norwegian Petroleum Directorate (NPD), Stavanger, Norway, http://www.npd.no/en/Publications/Reports/Geological-assessment-of-petroleum-resources—Barents-Sea-north-2017/

NPD 2017*b*. *Doubling the Resource Estimate for the Barents Sea*. Norwegian Petroleum Directorate (NPD), Stavanger, Norway, http://www.npd.no/en/news/News/2017/Doubling-the-resource-estimate-for-the-Barents-Sea/

ORVIN, A.K. 1940. *Outline of the Geological History of Spitsbergen*. Skrifter om Svalbard og Ishavet, **78** (2nd edn published 1969).

RØNNEVIK, H. & JACOBSEN, H.-P. 1984. Structural highs and basins in the western Barents Sea. *In*: SPENCER, A.M. (ed.) *Petroleum Geology of the North European Margin*. Graham & Trotman, London, 19–23.

SELLEVOLL, M. & SUNDVOR, E. 2001. *Jordskjelvstasjonen. Institutt for den faste jords fysikk gjennom gjennom ett århundre*. Institutt for den faste jords fysikk, Universitetet i Bergen, Bergen, Norway.

TAMNES, R. 1991. Forsvarsstudier 71/1991. Institutt for Forsvarsstudier.

The history of the upstream oil and gas industry in Italy

FERDINANDO FRANCO CAZZINI

Department of Environment and Earth Sciences, University of Pavia, Via Ferrata 1, 27100 Pavia, Italy

ferdinando.cazzini@gmail.com

Abstract: The aim of this work is to provide a synthetic history of the Italian upstream oil and gas industry, from its early start until today. Among European countries, Italy has one of the richest abundance of evidence of hydrocarbon seepages. The populations that have inhabited the country during the various historical periods took advantage of these phenomena, harvesting oil and bitumen from the surface. A testimony of such activity is an ingot of purified bitumen dated as first century AD found in central Italy. In the nineteenth century, seepages attracted the interest of oil companies that began to explore the Apennines. The drilling of the first modern oil well was realized in 1860 near Ozzano (Parma). The first onshore seismic reflection survey was acquired in 1940. The well Caviaga-1 was drilled in 1944 near Milan, discovering a gas field of 12.5 Bcm. It was one of the major European gas fields and it marked the starting point of the modern Italian petroleum industry. The widespread use of seismic reflection triggered an intense period of exploration. The first material oil discoveries were Ragusa (in 1954) and Gela (in 1956), both in Sicily. By the mid-1950s, the first offshore seismic survey started in the Adriatic Sea. In 1959, the drilling of well Gela 21, the first offshore well in Europe, boosted further exploration possibilities. During the following decades, the level of exploration and production (E&P) activities remained high, while from the beginning of the third millennium there has been a general slowdown. Since 2007, the level of exploration drilling has dropped to under 10 wells per year. In 2014, this negative trend was further confirmed with no exploratory wells drilled. The causes of this decline derive from the exploration maturity of the biogenic gas play and from the heavy bureaucratic processes involved in obtaining authorization. However, the main cause is probably the strong opposition from environmentalist associations to any kind of petroleum activities.

The history of the oil industry in Italy can be divided into four homogenous periods marked by the following milestones (see Fig. 1):

- 1860: the drilling of the Ozzano well marks the division between the Pre-Industrial Period and the beginning of the industrial period. This has the same historical meaning for Italy as the well drilled in 1859 by Colonel Drake in Titusville, Pennsylvania, USA.
- 1944: the discovery of the Caviaga gas field separates the Pioneers' Period from the Growth and Maturity Period.
- 2007: the number of exploratory wells falls below 10 per year, marking the start of the Decline Period that extends until today.

Pre-Industrial Period

The Pre-Industrial Period extends from 1860 back to historical reports of Romans and Greeks. Numerous ancient authors mentioned the various use of oil and bitumen but it is not the aim of this work to analyse all of them. Only a few examples of this long historical period will be highlighted. The use of hydrocarbons developed easily throughout ancient history as it took advantage of the large variety of hydrocarbon seepages occurring in the Peninsula. Martinelli *et al.* (2012) proposed a modern inventory and analysis of natural hydrocarbon occurrences in Italy (Figs 2 & 3).

Gas seepages

The report by Martinelli *et al.* (2012) indicates 169 sites of gas occurrences all over the Italian territory. The highest concentration is in the Northern Apennines (Fig. 2a). Not particularly numerous but of great interest are the so-called 'eternal fires' mentioned since Pliny the Elder. The eternal fire in the area of Barigazzo (near the town of Modena: 29 in Fig. 2a) was described in the eighteenth century by Lazzaro Spallanzani (1729–99) and in the nineteenth century by Antonio Stoppani (1824–91), both professors of Natural Sciences at the University of Pavia, Italy and considered to be the 'fathers' of Italian geology. In the same area of the Apennines, the 'eternal fires' of Monte Busca (near the town of Forlì: 35 in Fig. 2a) have been documented by recent photographs (Fig. 4).

Oil seepages

Petroleum springs are rarer than gas seepages. The inventory reports only 45 sites (Fig. 2b)

From: Craig, J., Gerali, F., MacAulay, F. & Sorkhabi, R. (eds) 2018. *History of the European Oil and Gas Industry*. Geological Society, London, Special Publications, **465**, 243–274.
First published online April 11, 2018, https://doi.org/10.1144/SP465.2

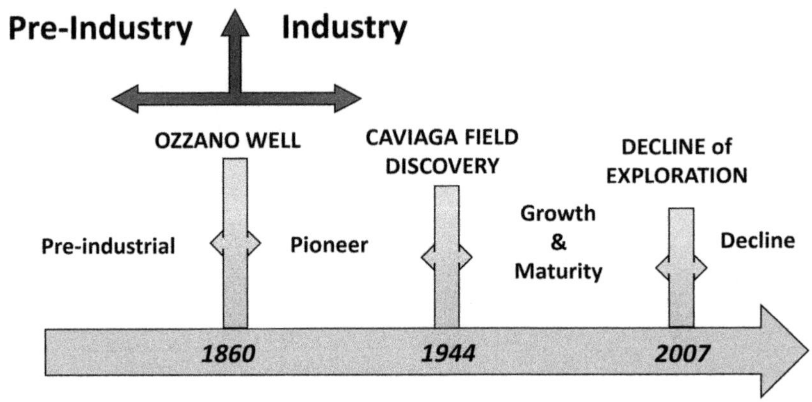

Fig. 1. Milestones in the Italian petroleum industry.

homogeneously distributed from the Alps to Sicily. The Tramutola oil seepages (30 in Fig. 2b) are located 3 km NE of the homonymous village of the Val d'Agri within the Southern Apennines. It is a good example of the exploration strategy applied before the availability of seismic data and is linked, essentially, to drilling wells near surface occurrences. In 1936, the oil seepages (Fig. 5) triggered the discovery of the Tramutola oil field. In 1980, seismic data showed several anticlines at a depth of about 4000 m. The exploration of this new play resulted in the discovery of the Mount Alpi giant oil field in 1988. In 1989, the Tempa Rossa oil field was discovered in the same area. Today, the shallow and small Tramutola oil field is believed to result from the dismigration of the deeper oil pools.

Solid hydrocarbons (bitumen)

This is probably the phase of hydrocarbons more frequently and universally utilized by ancient cultures. Impregnations are present from the Alps to Sicily. The inventory indicates 91 sites where solid hydrocarbons can be found (Fig. 3a). Bitumen mining was a historical ongoing activity that became an industrial practice in several areas during the nineteenth and twentieth centuries. Among the various production zones, two were particularly developed. The first one is located in the Central Apennines in the Pescara River Valley and is known as 'Bitumen of Majella' (Fig. 3a, 40–47), taken from the name of the local dominant mountain. The second one is located in SE Sicily in the Irminio River Valley near the town of Ragusa (Fig. 3a, 86–89) and is known as 'Pietra Pece' (pitch stone) or 'Ragusa Asphalt'. In both sites, the generated oil has migrated upwards, impregnating Tertiary (mainly Miocene) reservoir rocks. The oxidation and biodegradation transformed the original oil in bitumen. The industrial mining activity, carried out by Italian, French, English and German private companies, started during the mid-nineteenth century. In 1913, the total Italian production of bitumen was 117 000 t (tons), of which 59 000 t came from the Majella area. Bitumen mining became uneconomical during the mid-1950s. The richness of the solid hydrocarbons of the Majella is well documented in Figure 6, portraying bitumen pouring down from the walls of the mine. The asphalt mining operations continued over the centuries in several sites, such as Lettomanoppello, Scafa, Roccamorice and San Valentino. Archaeological studies proved that the Romans developed a bitumen mine in the same area. An interesting testimony of such activity is an ingot of purified bitumen dated first century AD (Fig. 7). The words 'ALONI – C (aii) – F (ilii) – ARN (ienses) – SAGITTA', the name of the mine owner, are carved on it. The translation is 'Aloni Sagitta, son of Caius, (of the tribe) Arniense'. The absence of the stamp S. P.Q.R. indicates that the owner of the mine was probably a private citizen and not the Emperor or a member of the Senate. This could prove the existence of concessions for solid hydrocarbon mining, similar to the ones related to metallic substances mined all over the Roman Empire. The bitumen was produced in chunks the size of bricks, which were probably transported to Rome on the back of pack animals. Bitumen was primarily used to caulk the hulls of ships, but it had other uses as well. The presence of solid hydrocarbons has attracted the oil exploration. The results turned out to be modest in the Majella area, with the discoveries of the small Tocco da Casauria (in 1933) and Vallecupa (in 1955) fields. In the Ragusa area, exploration was successful with several material discoveries such as the Ragusa (in 1954) and Gela (in 1956) fields,

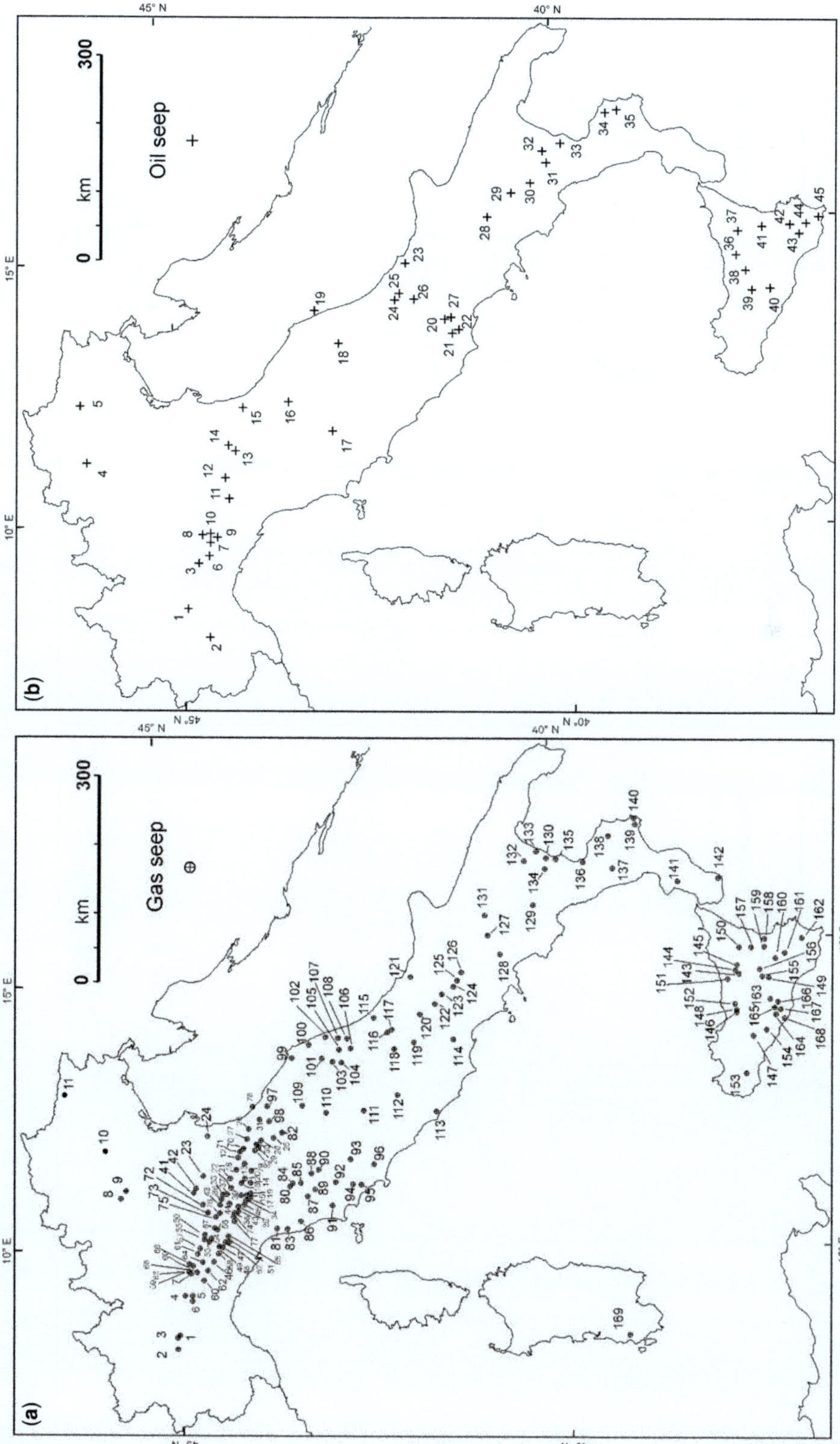

Fig. 2. Natural hydrocarbon seepages in Italy: **(a)** gas seeps; and **(b)** oil seeps (after Martinelli *et al.* 2012).

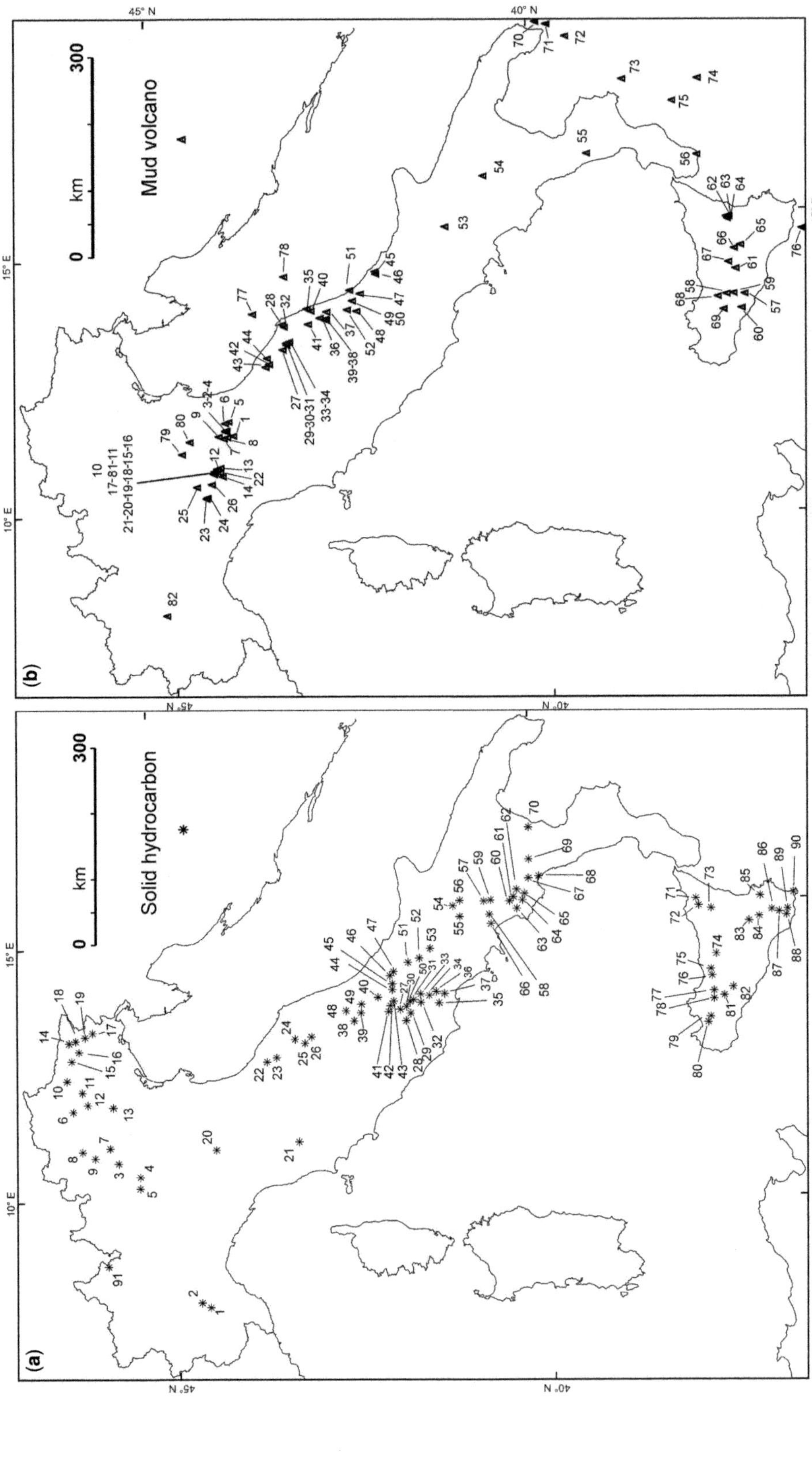

Fig. 3. Natural hydrocarbon seepages in Italy: **(a)** bitumen seeps; and **(b)** mud volcanoes (after Martinelli *et al.* 2012).

Fig. 4. The 'eternal fire' of Monte Busca also known as 'vulcanino' of Monte Busca (Tatinax, https://www.flickr.com/photos/tatinax/2825615864 CC BY-ND 2.0).

and, nowadays, it is the most prolific oil basin in Sicily.

Mud volcanoes

Mud volcanoes are present in the Apennines Thrust Belt, with a concentration found in the Northern and Central parts, as well as in Sicily (Fig. 3b). The reported numbers of sites are 82, of which nine are offshore and recognized by geophysical tools. There are two areas in Italy in which this geological phenomenon is particularly well documented: the first one is in the Northern Apennines and the best example is the 'Salse di Nirano' (11 in Fig. 3b); and the second one is in central Sicily and is the 'Maccalube di Aragona' (57 in Fig. 3b).

The area of the 'Salse di Nirano' is near the town of Modena, a few kilometres from Maranello, home of the Ferrari car factory headquarters. Today, this area is a natural geological park, with a very high concentration and variety of mud volcanoes. The local word 'salse', used to define mud volcanoes, is clearly linked to the word salt. In fact, together with plastic shales, and liquid and gaseous hydrocarbons, saltwater is one of the characteristic emissions of mud volcanoes. Antonio Stoppani gives a detailed description of Nirano's mud volcanoes in his book *Il Bel Paese* (Stoppani 1897), a sort of a geological travel diary of the Italian Peninsula (Fig. 8).

The 'maccalube' is the local word used in Sicily for mud volcanoes and it derives from the Arabic 'maqlub' meaning 'overturning ground'. The 'Maccalube di Aragona', a few kilometres from the town of Agrigento, is, nowadays, a natural conservation area that preserves a specific geological landscape. During the phases of extreme gas and mud emissions, this mud volcano can be quite dangerous. Therefore, for security reasons, it is strongly recommended not to go close to it. It is sad to recall that during the summer of 2015, two children playing near the cone were tragically killed by the mud flux.

Antique petroleum production

In the proximity of the city of Modena, between the already described 'Salse di Nirano' and the village of Sassuolo (the name is from the Latin 'sax oleum' meaning oil from rocks), lies the site of Montegibbio (also known as Monte Zibio) along the tiny valley of the river 'Rio del Petrolio' (meaning oil creek). It is difficult to find other areas with such a concentration of place names that are evocative of petroleum. The oil production was described by Francesco Ariosto

Fig. 5. The oil spring of Tramutola (© Andrea Sabbadini (http://andreasabbadini.photoshelter.com/contact) published on http://www.internazionale.it/reportage/2015/08/15/basilicata-petrolio).

(1430–99) in his book '*de Oleo Montis Zibinii seu Petroleo Agri Mutinensi Libellus*' published in 1690 by O. Jacobeo in Copenhagen. In this particular site, oil was produced for centuries from hand-dug holes. This petroleum was called 'Olio di Santa Caterina' or 'Olio di Montegibbio', and it was traded all over Europe in two different varieties: the white one, which was more valuable and costly, and the red one, which was of less commercial value, both mainly used in medical treatments. An idea of the technology adopted for such production during the sixteenth century is given in Figure 9.

An interesting offshore artisanal oil production (Fig. 10) is described in a print of the Flemish painter Johannes Stradanus (1523–1605). The place portrayed is certainly Sicily, as written on the description. It might be offshore Agrigento near seepage number 76 in Figure 3b. The sailors were collecting oil droplets on the sea surface (on the print, 'Nahptha Bituminis') with natural sponges and then storing it

Fig. 6. Majella, mine of Lettomanoppello – bitumen pouring down from the mine's roof, 1956 (Archivio fotografico Mino Gelsomoro' from 'Progetto VALSIMI' http://www.valsimi.it/ Provincia di Pescara and Regione Abruzzi).

in terracotta jars. The separation of the oil from the water was probably obtained through gravity segregation and occurred in those very jars.

Pioneers' Period

The Pioneers' Period started in 1860 with the drilling of the first modern oil well in Italy and ends in 1944 with the discovery of the Caviaga Field. The drilling of the well by Colonel Drake in Titusville, Pennsylvania, USA, in 1859 is universally recognized as the starting point of the petroleum industry in the world. However, in a number of countries, a similar drilling operation was carried out even before that. In fact, it is historically proved that in 1847 the first petroleum well in Azerbaijan was drilled near Baku; in 1854, the first Galician wells were drilled near Boryslav; also, in 1857, the first Romanian well was drilled near Bucharest. However, when considering the development of the oil industry in the USA after the well of Colonel Drake, it is correct to consider that well as the founding milestone of the modern petroleum industry worldwide.

Fig. 7. Ingot/bricks of bitumen from the Roman mine at Lettomanoppello, dated first century AD; dimensions of 35 × 26 × 10 cm (Museo Naturalistico e Archeologico Paolo Barrasso in Caramanico Terme – Pescara).

The early discoveries

Achille Donzelli, an Italian entrepreneur, drilled the first modern oil well in 1860, in Ozzano near the town of Parma in the Northern Apennines. The well was located in an area (Fig. 11) rich in gas and oil seepages. Two wells were successfully completed at depths of 35 and 45 m with a production of a few tens of kilograms of petroleum per day. Even if there is no historical confirmation of the drilling method used for the Ozzano wells, they are undoubtedly the first industrial attempt at oil production in Italy.

Fig. 8. Nineteenth century print of the Nirano's mud volcano 'le Salse di Nirano' (Stoppani 1897).

Fig. 9. Print of unknown author dated 1540. It describes the Montegibbio (Modena) oil gathering and its storage in special jugs to be ready for transportation by cart or by pack animal (Gorgoni 2003).

Fig. 10. Oil gathering with sponge offshore Sicily – print from the sixteenth century (Jan van der Straet (Joannes Stradanus) 'Using Sponges to Collect Naphtha from the Surface of the Waves', plate 86 from Venationes Ferarum, Avium, Piscium – engraved by Jan Collaert (1566–1628), published by Phillipus Gallaeus of Amsterdam).

Fig. 11. The well drilled near the village of Ozzano (North Apennines) (SPI 1975. *La ricercapetroliferain Italia nel'900*. Internal Report, http://www.pionierieni.it/wp/wp-content/uploads/SNR-62-Societ%C3%A0-Petrolifera-Italiana-1905-1975-SPI-1975.pdf).

During 1863, the exploration moved from the North to the Central Apennines near the aforesaid bitumen occurrences of Majella. The first well was named after Tocco da Casauria, a nearby village. It reached a depth of 60 m and produced 500 kg of oil per day; by the end of 1865, the cumulative production had reached 300 t. In 1868, the Italian government awarded the development concessions of Vallezza (Parma) and Montechino (Piacenza). In the same year, the exploration of the area SE of Rome near the town of Frosinone began. Oil was discovered at Ripi and San Giovanni Incarico. In 1890, the exploration of the valley of the Chero creek, near the city of Piacenza in the North Apennines, resulted in the discovery of the Velleia Field. The number of wells quickly reached five and the production was about 400 l of oil per day. In 1893, the Italian government awarded the production concession of Velleia.

The foundation of the Società Petrolifera Italiana (SPI)

In July 1905, Luigi Scotti (1859–1933) founded the Società Petrolifera Italiana (SPI). The headquarters were established in the small town of Fornovo Taro near Parma (Fig. 12). For decades, SPI has been the largest private and the first integrated oil company of Italy. Oil production was coming from the concessions of Vallezza and Salsomaggiore, with oil being delivered via a pipeline to the refinery at Fornovo Taro, owned by SPI. In 1927, Scotti sold SPI to Standard Oil Company of New Jersey, USA. This major company introduced the advanced tools and technology used in the industry in the USA. In 1932, SPI reached maximum efficiency with the production of 70 000 bbl/year (barrels per year) of oil, which represented 82% of the entire Italian oil production. In 1934, rotary drilling was utilized for the first time in Italy in the Vallezza Field. SPI was then sold to Philips Petroleum in 1966, to Eni in 1987 and, finally, to Gas Plus in 2004. In 2005, SPI celebrated 100 years since its foundation. In 2010, the Petroleum Museum of Vallezza was inaugurated.

The foundation of the Società per Azioni Petroli d'Italia (SPDI)

The Società per Azioni Petroli d'Italia (SPDI) was created in 1906 after the merger of Société Française des Pétroles Zipperlen & Co. and Société Pétroles de Montechino. The assets of the newly formed oil company consisted in the concessions of Montechino–Gratera–Velleia and the refinery at Fiorenzuola d'Arda (Piacenza). At the beginning of the twentieth

Fig. 12. Main fields of the Pioneers' Period.

century, 403 oil wells were operational. In 1958, the yearly production amounted to 300 t of light oil and 500 000 m^3 of gas. The company closed down in the 1960s for bankruptcy, after an accident involving fire destroyed the refinery in 1956.

The foundation of the Società Idrocarburi Nazionali (SIN)

The Società Idrocarburi Nazionali (SIN) was established in 1925, with its headquarters in Florence. In

1938, the Pietramala Field was discovered in the Northern Apennines between the cities of Bologna and Florence. A gas pipeline from the field to Florence was built during World War II to sell methane as a motor fuel.

The foundation of the Azienda Generale Italiana Petroli (AGIP)

The Azienda Generale Italiana Petroli (AGIP) was created as the national Italian oil company on the 3 April 1926 by Mussolini's government, with the aim of promoting the exploration and production of hydrocarbons in Italy to supply the nation with fossil energy. The first period of exploration was not particularly successful, with the discovery of only marginal accumulations: the Podenzano gas field (Piacenza), the Tocco da Casauria deep oil field (Majella), the Tramutola oil field in the Agri Valley and the Ripi deep oil field (Frosinone). Guido Bonarelli, the first exploration manager of the company, suggested moving exploration from the compressional belts to the tectonically undisturbed foredeep/foreland areas. The idea was excellent; however, reliable tools were not yet available to reveal the hidden traps buried by thick alluvium. In 1940–41, AGIP completed the first reflection seismic survey in an area south of Milan. From among the several anticlines resulting from the data interpretation, Caviaga was selected. The site was drilled in 1944 and resulted in the first material gas discovery of Italy. In 1945, Enrico Mattei (1906–62) was charged to liquidate AGIP. As soon as he became aware of the Caviaga discovery, he made a U-turn and decided to go ahead and develop the field. Caviaga changed the future of AGIP forever. In 1953, the Italian government established Ente Nazionale Idrocarburi (Eni) as the new holding in the energy business, with the task of co-ordinating and consolidating all state-owned operating companies. Embedded within this law (number 136, approved on 10 February1953) there was the exploration and production monopoly of Eni over the entire Po Plain, which was considered the most promising area. From 1995, the privatization of Eni took place, with the merging of the various state companies into a single firm, and with the Italian government retaining the 'golden share' of 30%. By 1998, the Po Plain monopoly had ceased. Today, Eni is one of the major oil companies, operating all over the world.

The fields of the Pioneers' Period

The Pioneers' Period was characterized by a dozen minor discoveries concentrated on the Northern, Central and Southern Apennines (Fig. 12). In the early twentieth century, only four oil companies (SPI, SPDI, SIN and AGIP) were producing oil and gas from the various concessions granted by the government.

Velleia–Montechino–Gratera fields. These three fields were, respectively, discovered by SPDI in 1890, 1892 and 1907. Production started rapidly and, by the late 1950s, before being finally abandoned, they had produced about 200 000 t of oil. Velleia reached its production peak in 1895 with 4000 t/year, while Montechino and Gratera had a peak of 3000 t/year in 1911. Approximately 700 productive wells were drilled. The reservoir of the fields consists of fractured limestones of the Cretaceous. The ultimate recoverable reserves (URR) have not yet been published; however, a reasonable estimation is between 1.5 and 2 MMBOE (million barrels of oil equivalent).

Vallezza Field. Several wells were present in this area and had been discontinuously producing oil since 1869. In 1907, SPI took over, and in 1909 the field commerciality was proved and the relative concession granted. In the same year, the production of sweet and light oil (52° API) started. The peak of oil production was reached in 1932 with 8581 t/year. Gas production started in 1930 and condensate production in 1932. The total number of productive wells was 180. Production ended in 1993 with the abandonment of the field. The structure is geologically complex with evident compressional faults and strata repetition. Vallezza was the largest Italian field of the Pioneers' Period with a URR of 2 MMBOE.

Ozzano–Vizzola fields. These small fields were discovered by SPI in 1909 a few kilometres from Vallezza, in the same area of the first modern oil well in Italy. In 1942, a new oil and gas pool was discovered in Miocene sandstones, and it was productive until 1970. The URR are 0.2 MMBOE.

Salsomaggiore Field. This field was discovered by SPI in 1923. The producing area encompasses the three districts of Centopozzi, Salsominore and Rovacchio. Concessions over the first two districts were secured by SPI, while AGIP obtained the concession of the third in 1930. The production peak was reached in 1932 with 12 800 t/year. By the 1950s, oil production ended, while gas production continued in an intermittent way. The URR are 1.2 MMBOE.

Pietramala Field. This small field was discovered in 1938 by SIN. The URR are 0.6 MMBOE.

Fontevivo. This oil and gas field was discovered in 1929 by AGIP near the town of Parma. Production ended in 1953. The URR are 0.2 MMBOE.

Podenzano. This gas field was discovered in 1932 by AGIP near the town of Piacenza. Production ended in 1962. The URR are 0.7 MMBOE.

Tocco da Casauria. The concession was obtained by AGIP in 1934 beside the old bitumen mines. Main production ended during the 1950s, with discontinuous events up until 1963. The URR are 0.1 MMBOE.

Tramutola. In 1937, AGIP discovered a small oil pool in the Tertiary flysch near a natural oil spring. The field was abandoned in 1959. The URR are 0.1 MMBOE.

San Giovanni Incarico. This was discovered by SPDI near an oil seepage in 1915. The field produced small quantities of heavy oil.

Ripi. It was discovered by SPDI in 1924 and production started in 1932. AGIP took over the field in 1939, abandoning it during the 1950s. Today, local oil companies are still obtaining minor oil production. The URR are 0.6 MMBOE.

The unconventional methane production of 'Polesine'

Quaternary biogenic gas is widely present at shallow depths in the eastern part of the Po Plain. It is possible to make a modest discovery wherever a well is drilled. The methane is dissolved in formational water and is naturally bubbling at surface conditions. The best production area occupies a triangular zone of about 2.000 km^2 in the lowlands of the Po Delta known as 'Polesine' (Fig. 12). The first discovery was made in 1934. However, the development of these resources only started in 1939, when the shortage of fuel due to World War II forced Italians to refuel cars, trucks and buses with methane. In 1943, this area was producing more gas than all the other conventional fields put together. An estimated total of 3000 wells (the last reported official number was 2500 in 1951) have been drilled with an average depth of 450 m. The average production life of the wells was about 5 years. The yield per well varied from 600 000 to 800 000 m^3. The operators were small private companies grouped into a consortium named Consorzio Italiano Metano (CIM). Production was 1 m^3 of methane from 1 m^3 of saltwater. The average production per year was calculated to be 350 Mcm (million cubic metres). The environmental impact of this activity was soon recognized. A shallow reservoir production created the conditions for a strong subsidence. The disposal of saltwater within this farmland area was a problem and a dilemma. However, the production continued through the 1940s and the 1950s, with reported local subsidence ranging from 1 to 2 m. It was quite inevitable that a flood would occur sooner or later, and it happened during the winter of 1951. The inhabitants started to strongly oppose the project. The government finally stopped it in 1961. Its negative legacy, today, is the local political opposition to any exploration and production (E&P) initiatives both onshore and offshore. The development of the 'Alto Adriatico Project', which consisted of several conventional offshore gas fields with total reserves of 40 Bcm (billion cubic metres), was stopped by the authorities who feared that possible subsidence would affect the nearby city of Venice. The 'Polesine' subsidence was actually due to the enormous quantity of saltwater produced from a very shallow depth, and the concurrence of methane production certainly had a minor impact.

Conclusion of the period

The Pioneers' Period is characterized by an early start of exploration, guided by oil and gas seepages, in the Northern Apennines and then in the Central Apennines. Several Italian and foreign oil companies were attracted by the promising geology of Italy. The overall results of the exploration were disappointing, as only a dozen small fields were discovered. The thrust belt, which forms the backbone of the Peninsula, preserves poorly the trap's integrity and this is the main reason for the abundance of seepages. Geophysical gravity tools were implemented by AGIP from 1927 using the Eotvos Torsion Balance. Rotary drilling equipment was put forward by SPI in 1934. In 1940, AGIP introduced seismic reflection methods in collaboration with Western Geophysical. Industrial oil production started in the Velleia Field in 1890, quickly followed by others. The Italian consumption of fossil energy was, by far, not satisfied by domestic production, which was only a small percentage of the amount needed. The best production year of the period was 1932, with 27 000 tons/year which corresponds to 540 bbl/day (barrels per day).

The Growth and Maturity Period

This period stretches from 1944 to 2007. It was a 'golden age' for the Italian oil industry, with large investments and the discovery of several important oil and gas fields. It can be divided into two main cycles of exploration (Fig. 13). The first one is known as the 'Biogenic Gas Cycle' and it was the follow-up of the Caviaga discovery. This cycle was characterized by a boom in the exploration in the Po Plain that rapidly expanded all over the peninsula and, finally, into the Adriatic Sea. The activity peak occurred in the 1958–66 decade, with more than

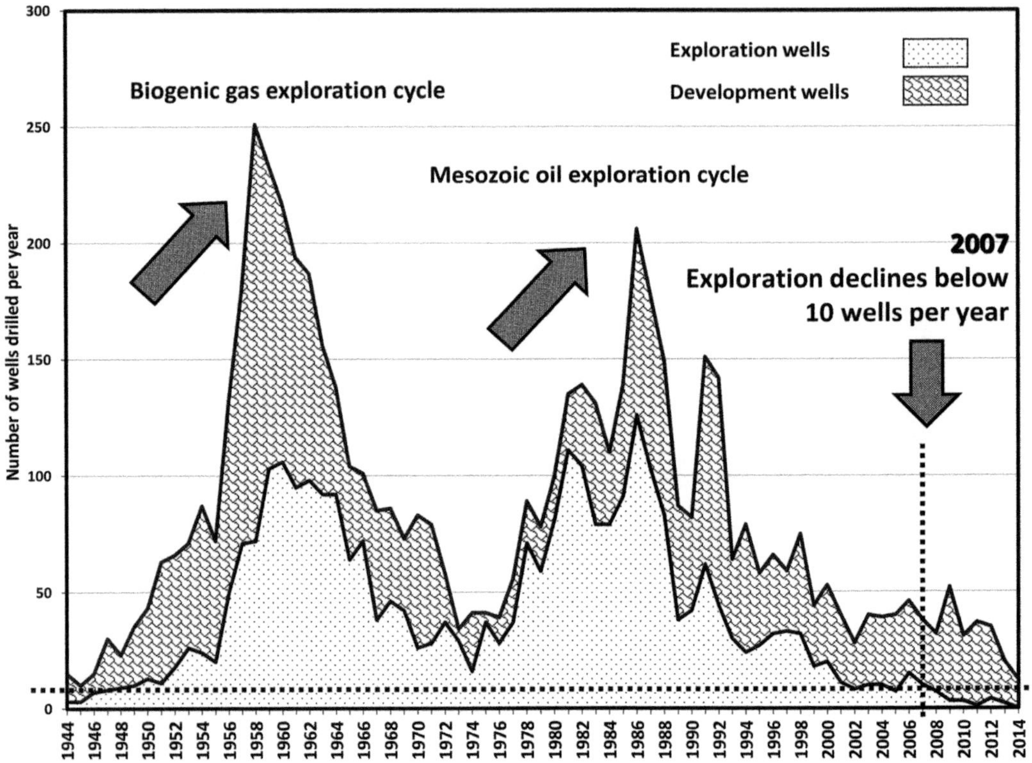

Fig. 13. Exploration and development wells drilled from 1944 to 2014. Note the two cycles of exploration coinciding with more than 100 exploratory wells per year.

250 exploration and development wells drilled. The exploration drilling peak was reached in 1960 with more than 100 wells. The second cycle of exploration is known as 'Mesozoic Oil Cycle'. The drilling peak was reached in 1986, with more than 200 exploratory and development wells, while the exploration activity alone reached two peaks in 1981 and 1986 with more than 100 wells. The larger volumes of gas (Fig. 14a) were discovered in 1968 (in the Agostino-Porto Garibaldi Field) and 1971 (in the Barbara and Luna fields). The main volumes of oil (Fig. 14b) were discovered in 1984 (the Trecate Field), 1988 (the Monte Alpi Field) and 1989 (the Tempa Rossa Field).

The Caviaga Field

At the end of the 1930s, AGIP decided to follow the idea of Guido Bonarelli to move the exploration from the fold belt to the unexplored Po Plain. The target was to find larger traps with a better seal and integrity. In order to do this, a seismic crew of Western Geophysical was recruited. The survey started in 1940 and prematurely ended on 11 December 1941 due to the declaration of war between Italy and the USA. However, in spite of the difficulties, the survey was completed. The interpreted seismic lines displayed several anticline features. Caviaga was the name assigned to one of them, which had a size of 8 × 3 km (Fig. 15). It is located near the town of Lodi, about 30 km SE of Milan. AGIP started drilling the prospect in 1943, reaching total depth (TD) in July 1944 during the dramatic events of the Italian Civil War. The well was tested in Lower Pliocene sandstones (later called Caviaga Sands) from 1384 to 1403 m with a production of 100 000 m^3 of gas per day accompanied by some condensate. The well was completed, and the gas was used to refuel the cars and trucks of the local community. In the meantime, this small production served as a long-term production test to monitor reservoir performance. The well discovered a gas field with reserves of 12.5 Bcm (corresponding to 80 MMBOE), one of the most significant European gas discoveries of the time, second only to the Romanian Deleni gas field. For the first time, the presence of large amounts of hydrocarbons was revealed within the buried anticlines of the Po Plain. In 1945, Enrico Mattei

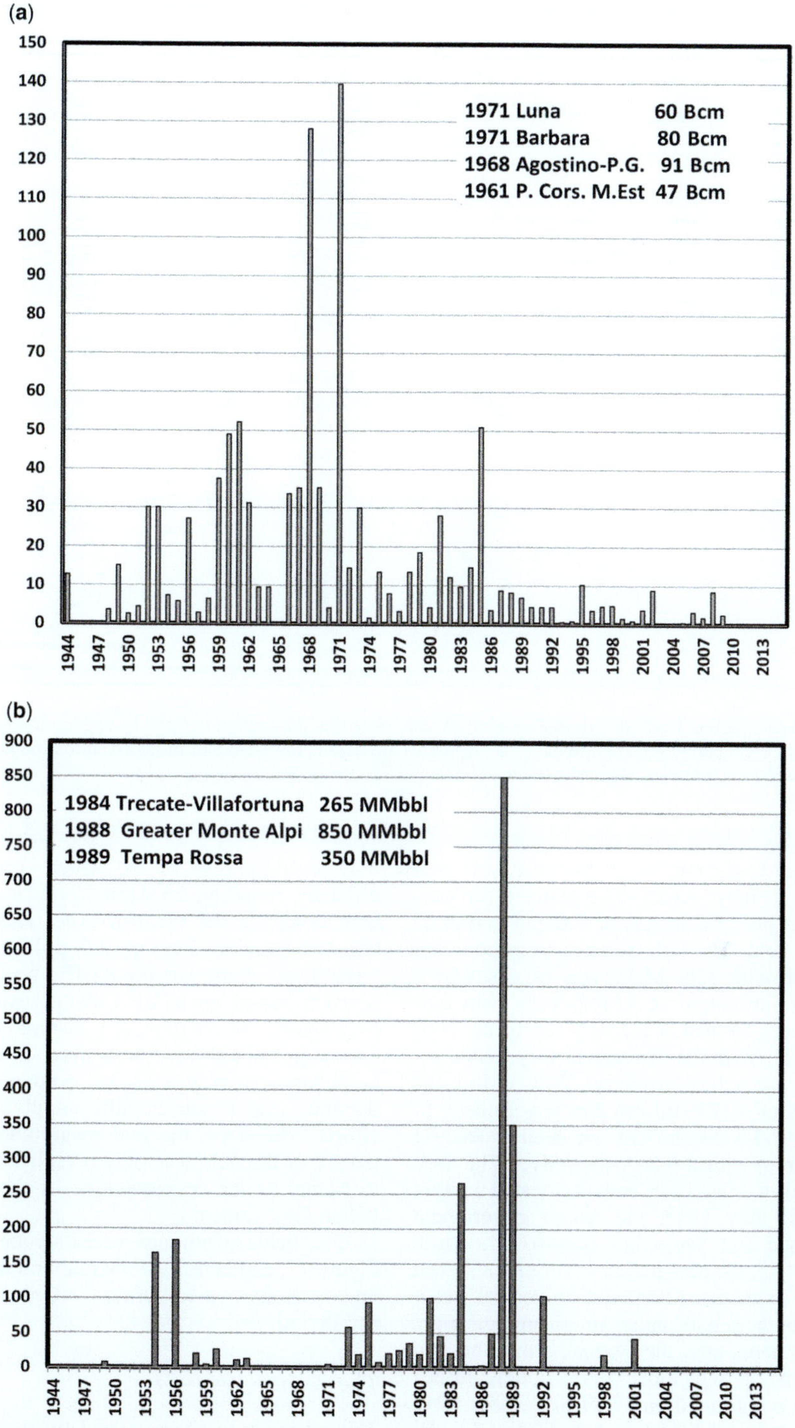

Fig. 14. Discovered reserves per year: (**a**) gas (in Bcm); (**b**) oil (in MMbbl).

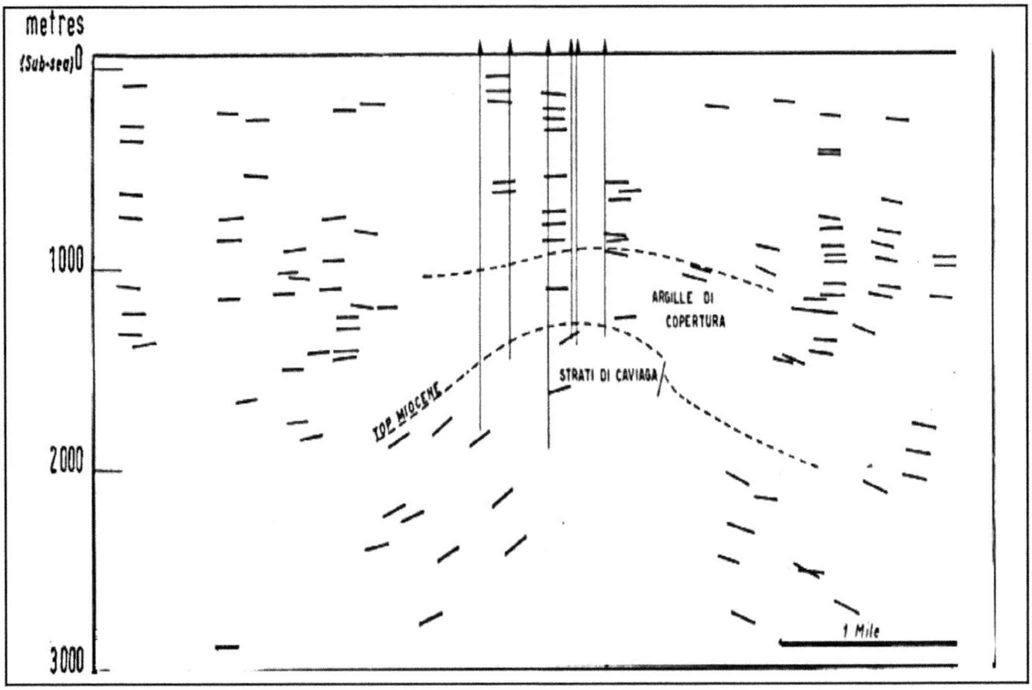

Fig. 15. Caviaga anticline from the original seismic survey. Note that the section is made by reflection segments interpreted from consecutive recorded tracks. The segments are depth converted and migrated following the dip in time of the interpreted horizons/segments (Facca 1951).

was instructed to dismiss AGIP by the Provisional Government of Italy. As soon he realized the economic importance of Caviaga, he decided to instead develop the field. The field reached its production peak in 1956 with 1280 Mcm/year of gas, and it still generates an output of 5 Mcm/year from four producing wells located at the very top of the anticline (Fig. 16). With the Caviaga history, however, there is an unresolved mystery. Why didn't the management of AGIP inform the government of the discovery? To answer that, we need to analyse the matter from a historical perspective. The well was spudded in May 1943 and was tested in July 1944. On 25 July 1943, the fascist government was dismissed and Mussolini imprisoned. On 8 September 1943, the new Italian government signed an armistice with the Allies and moved its headquarters to southern Italy under American protection. On the other hand, after the German army liberated Mussolini on 12 September 1943, a new fascist government was formed on 18 of September 1943 based in the town of Salò (near the lake of Garda) under the protection of the Germans. This was the start of the Italian Civil War. Mussolini and the Germans were holding northern Italy, while the king and his new government were holding southern Italy with the help of the Allies. The Civil War officially ended on 25 April 1945, resulting in the withdrawal of the German army and the execution of Mussolini. During such political and social turmoil, to whom was the AGIP management supposed to report about the Caviaga results? Not an easy answer. In addition, any material discovery of natural gas could have been used by the Germans to enhance their process of synthetic fuels production, vital to successfully supplying their war efforts. Therefore, the real magnitude of the discovery of the field was only revealed in May 1945 to Mattei as the representative of the Provisional Italian Government.

The fields mentioned hereafter are reported in Figure 17, and constitute the main oil and gas accumulations discovered during the Growth and Maturity Period.

The decade 1944–54

This period was characterized by the appraisal of the Caviaga Field through the drilling of well 2 in 1946, followed by the development and production of the full field. During this decade, the first material quantity of oil in the Po Plain was discovered in

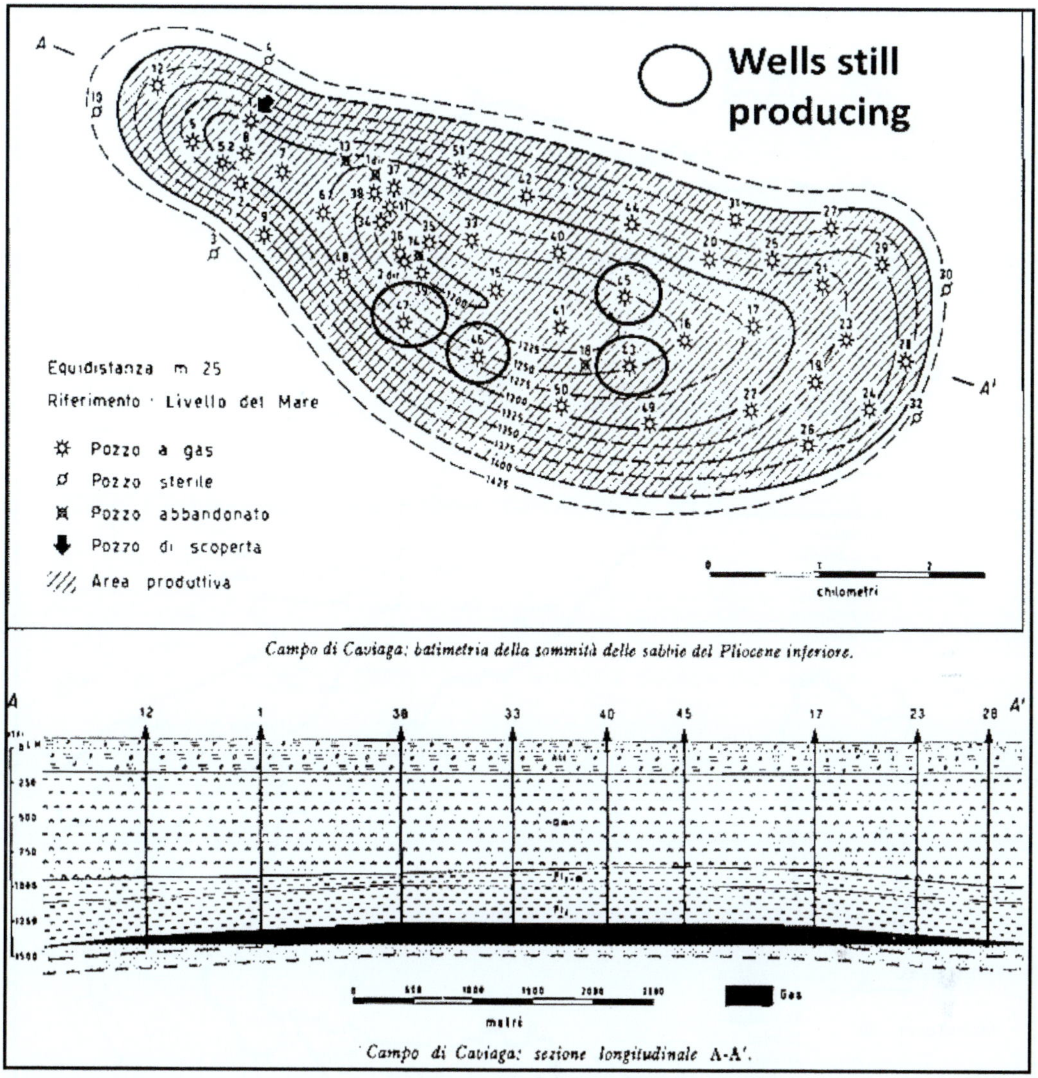

Fig. 16. Top reservoir map and cross-section of Caviaga Field (Accademia Nazionale dei Lincei 1959).

1949 at Cortemaggiore (oil and gas 97 MMBOE). In 1954, the first oil of Sicily was discovered at the Ragusa Field (in 1954, oil 172 MMBOE).The exploration strategy consisted of quickly testing the other structures of the Caviaga trend, such as Ripalta (1948, gas 21 MMBOE), Cornegliano (1950, gas 15 MMBOE), Bordolano (in 1951, gas 27 MMBOE) and Sergnano (1954, gas 18 MMBOE). The following step was moving SE towards the Adriatic Sea, with the discoveries of Correggio (1952, gas 43 MMBOE), Ravenna (1952, gas 135 MMBOE) and Alfonsine (1953, gas 52 MMBOE).

The decade 1955–64

It was characterized by the acquisition of the first offshore seismic survey and by the first offshore drilling (1959, Gela 21). The discovery of the first offshore field was a historic moment, with Ravenna Mare in the North Adriatic Sea (1960, gas 42 MMBOE), followed by the nearby structure of Porto Corsini Mare Est (1961, gas 292 MMBOE). Meanwhile, the inland exploration in the Po Plain was continuing with the discoveries of Spilamberto (1956, gas 68 MMBOE) and Minerbio (1956, gas 80 MMBOE). In Sicily, the exploration was reinforced, leading to the

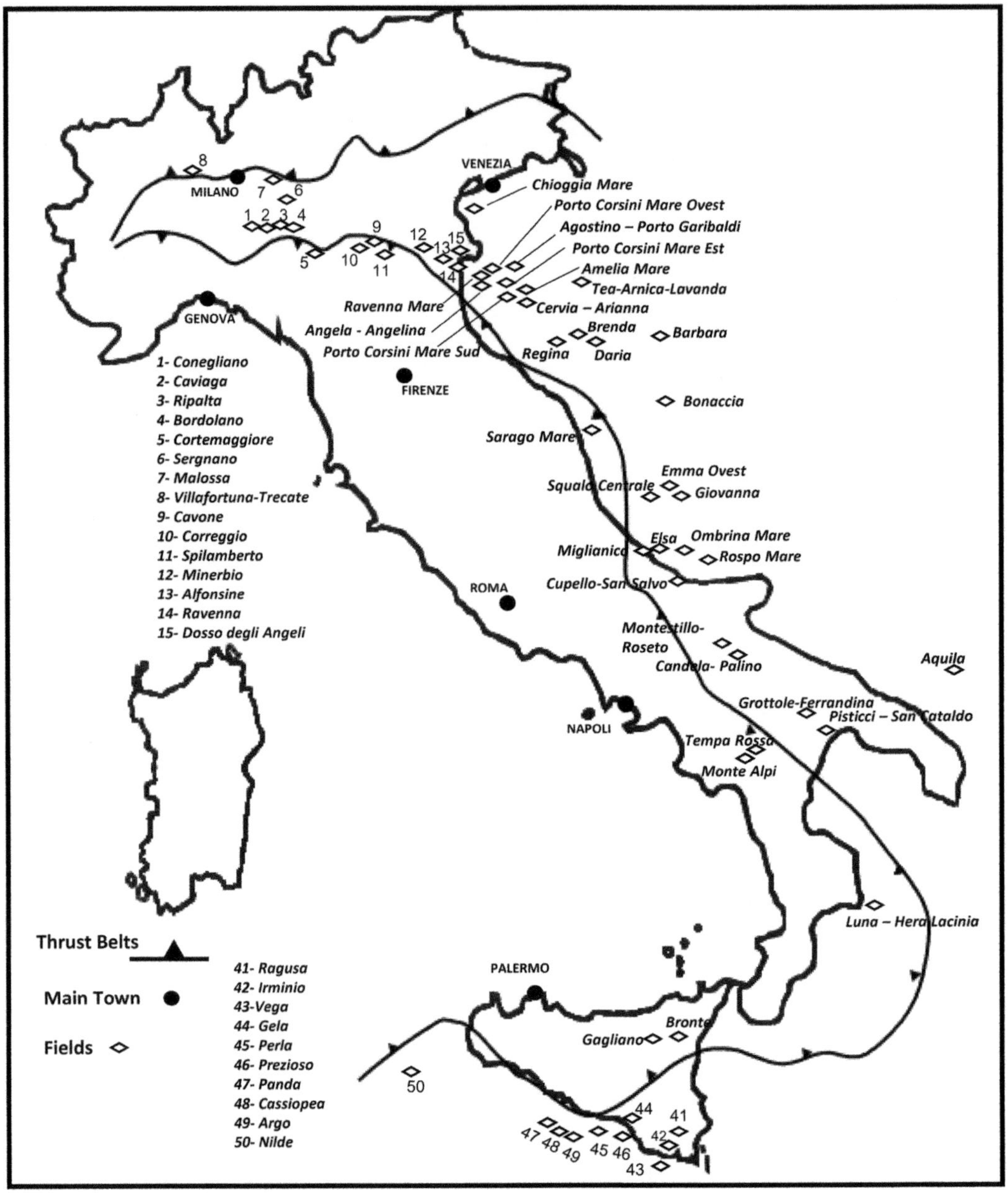

Fig. 17. Main fields of the Growth and Maturity Period.

discoveries of a new play at Bronte (1955, gas/condensates 24 MMBOE) and Gagliano (1960, gas/condensates 79 MMBOE). The Mesozoic oil play was consolidated in Sicily with the discovery of Gela (1956, oil 186 MMBOE). The onshore exploration in the Bradano Basin in southern Italy gave its first significant results with Grottole-Ferrandina (1959, gas 88 MMBOE), Cupello-San Salvo (1959, gas/condensates 115 MMBOE), Pisticci (1960, oil and gas 33 MMBOE), Candela-Palino (1962, gas/condensates 148 MMBOE) and Montestillo-Roseto (1963, gas/condensate 54 MMBOE).

The decade 1965–74

It was fundamental for the offshore exploration of the Adriatic and Ionian seas, and for the deep exploration of the Mesozoic oil play in the Po Plain.

Several offshore fields were discovered, such as Cervia-Arianna (1966, gas 183 MMBOE), Porto Corsini Mare Ovest (1966, gas 183 MMBOE), Amelia (1968, gas 142 MMBOE), Agostino-Porto Garibaldi (1968, gas 600 MMBOE), Barbara (1971, gas 400–500 MMBOE), Luna–Hera Lacinia–Linda (1971, gas 370–400 MMBOE) and Angela-Angelina (1973, gas 100 MMBOE). The onshore biogenic gas play resulted in Dosso degli Angeli (1969, gas 188 MMBOE). The onshore Mesozoic play resulted in Malossa (1973, gas/condensate 57 MMBOE; deepest pool in Europe at more than 5000 m) and Cavone (1973, oil 33 MMBOE).

The decade 1975–84

The most significant result of this period relates to the first deep-water oil discovery of the Aquila Field (1981, 30 MMBOE; water depth 850 m) in the South Adriatic Sea. The booming seismic technologies, known as direct hydrocarbon indicators (DHI), deserve a special mention as they allowed the exploration of subtle gas traps onshore/offshore. The Villafortuna–Trecate Field, the main oil field of the Po Plain (1984, onshore oil and gas 291 MMBOE), revealed the deepest oil pool in Europe: 6300 m. A new and successful exploration campaign took place in Sicily with the discoveries of Nilde, which showed a new oil play within the Sicilian Channel. Vega (1981, offshore oil and gas 72 MMBOE), Irminio (1982, onshore oil and gas 32 MMBOE) and Prezioso (1983, offshore oil and gas 21 MMBOE) were also discovered in Sicily. The main discoveries in the Adriatic Sea were Rospo Mare (1975, oil 95 MMBOE), Squalo Centrale (1975, gas 45 MMBOE), Emma Ovest (1978, gas 59 MMBOE), Sarago Mare (1979, oil 31 MMBOE), Bonaccia (1981, gas 84 MMBOE) and Brenda (1983, gas 22 MMBOE).

The decade 1985–94

It was characterized by oil discoveries in the Southern Apennines where a new oil play was established. Exploration of biogenic gas, based on DHI, continued even if several small discoveries were recorded, thus indicating the maturity of the play. Several discoveries in this period were not put into development/production for technical and political reasons. The Adriatic Sea revealed the Chioggia (1985, gas 148 MMBOE; not in production), Daria (1985, gas 58 MMBOE), Giovanna (1985, gas 75 MMBOE), Ombrina Mare (1987, oil and gas 41 MMBOE; not in production) and Elsa fields (1992, oil 110 MMBOE; not in production). Exploration in the Southern Apennines resulted in the discoveries of Val d'Agri (1988, oil and gas 600–900 MMBOE) and Tempa Rossa (1989, oil and gas 270–400 MMBOE; not in full production).

From 1995 to 2007

The deep-water discovery of Aquila was put in production in 1998 using a floating production storage and offloading (FPSO) unit. The trend of the small gas discoveries in the Adriatic Sea continued, as well as the decline of exploration investments. The few material results in the Adriatic Sea were mainly due to DHI-supported prospects such as Regina (1995, gas 22 MMBOE), Porto Corsini Mare South (1997, gas 17 MMBOE) and Tea–Arnica–Lavanda (1998, gas 17 MMBOE). Miglianico was discovered onshore (2001, oil 52 MMBOE, not in production). The foredeep basin offshore Sicily was finally explored, resulting in the discovery of Panda (2002, gas 48 MMBOE; not yet in production), Argo (2006, gas 14 MMBOE; not yet in production), and Cassiopea (2008, gas 50 MMBOE; not yet in production).

The Italian petroleum systems

The hydrocarbon occurrences in Italy could be grouped into two geological supercycles. The first is Mesozoic in age and is mainly oil-prone. The second is Tertiary in age and is mainly gas-prone:

- The Mesozoic Supercycle is characterized by an extensional tectonic regime with the generation of prevailing heavy and sour oil linked to carbonate source rocks. Sometimes, the phase is light oil and, less frequently, gas and condensate. The supercycle could be subdivided into three families of petroleum systems according to the age of the source rocks.
- The Tertiary Supercycle is characterized by a compressional tectonic regime with the generation of thermogenic gas (poor in condensates) from source rocks of Oligocene–Miocene age. The hydrocarbons are trapped in Miocene sandstone reservoirs within the external part of the Apennines Thrust Belt. However, the main phase of this supercycle is the biogenic gas generated and trapped within the turbidite sequences of the Pliocene–Pleistocene in the foredeep and in the foreland. Therefore, the supercycle could be subdivided into two families of petroleum systems according to the age of the source rocks.

Petroleum system families (PSFs) have been recognized in recent work (Bertello 2008; Bertello *et al.* 2011; Cazzola *et al.* 2011) and have been clustered into five main groups based on the age of the source rocks (Fig. 18).

Mesozoic Supercycle – PSF of the Middle Triassic. This petroleum system family (PSF) is limited to

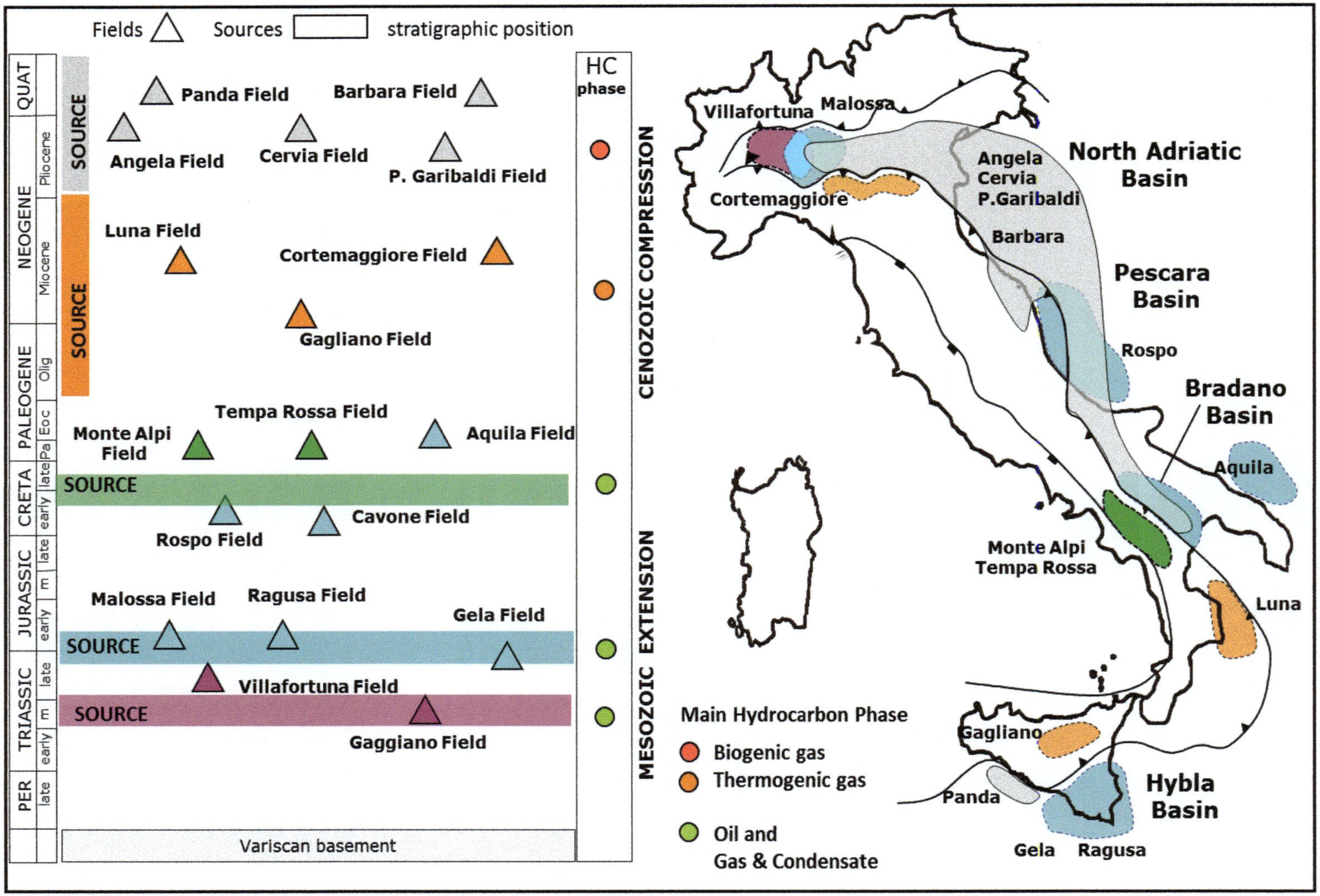

Fig. 18. The Italian petroleum systems families (PSFs). (Left side) Stratigraphic distribution of the five PSF source rocks (coloured strips) and of the main related fields (coloured triangles). (Right side) Areal position of the five PSF source rocks and relevant fields. The colour code is the same for both sides (after Cazzola *et al.* 2011).

the foreland of the western Po Plain that corresponds with the hydrographical basin of the Ticino River. Only the Villafortuna–Trecate and Gaggiano fields have been attributed to this system. The main reservoirs are the limestone rocks of the Main Dolomite Formation, San Giorgio Dolomite Formation and the Conchodon Dolomite Formation. The source rocks are the limestone facies of the Meride Limestone and the Besano Shales formations, both of Middle Triassic age. The kerogen is marine type II with an average total organic carbon (TOC) of 4%. The seal is represented by the tight Medolo Limestone Formation.

Mesozoic Supercycle – PSF of the Late Triassic–Early Jurassic. This PSF is linked to different source rocks that have the same age and same geological depositional setting within separated anoxic basins extending over the entire Italian foreland from the Po Plain to the Middle/South Adriatic and, finally, to Sicily. In the Po Plain, the main source rocks are the Riva di Solto Shale Formation. In the Middle Adriatic (Pescara Basin), the main source rocks are the so-called 'Emma Limestone', an anoxic marls facies found in the Emma-1 well. In the South Adriatic, the source rocks have been localized within the basal facies of the Burano Anhydrite Formation. These sources rocks are also present as anoxic limestone facies in the Sparviero-1 and Grifone-1 wells, in addition to the Delvina-1 well in Albania. In south Sicily (Hybla Basin), the source rocks have been recognized within the Noto Limestone Formation and the Streppenosa Shale Formation. Considering the areal extent of these Late Triassic–Early Jurassic anoxic facies and their fragmentation in different small basins, it is not surprising to find a wide variability in the TOC, which ranges from 1 to 3%. Kerogen is typically of type II from marine marl/limestone, while it is more shale-prone in the Po Plain. The generated phases are also variable, from heavy-sour oil to, less frequently, light oil and gas/condensates. The main field in the Po Plain is Malossa, and the phase is gas and condensates. In the Central Adriatic, the main fields are Rospo Mare, Sarago Mare, Elsa, Miglianico and Ombrina Mare, with a typical heavy-sour oil phase. In Sicily, the main fields are Ragusa, Gela, Irminio, Vega, Perla and Prezioso, again with a typical heavy-sour oil phase. One exception is the South Adriatic Aquila Field, where the oil is light. The bitumen occurrences of Majella and Ragusa described in the Pioneers' Period are related to this petroleum system.

Mesozoic Supercycle – PSF of the Cretaceous. This PSF is limited to the external part of the South Apennine Thrust Belt. The main fields attributed to this system are Val d'Agri (encompassing Monte Alpi, Monte Enoc and Cerro Falcone) and Tempa Rossa. This area yielded in excess of 1 Bbbl (billion barrels) of oil and is ranked among the largest onshore accumulations in Europe. The reservoirs are fractured limestones of the buried Apulia Platform, which extends in time from the Cretaceous to the Miocene. The oil column exceeds 1000 m, sometimes reaching more than 2000 m in Tempa Rossa. The seal is represented by Lower Pliocene shales. The source rocks were not identified immediately after the discoveries. Finally, they were found in a few deep wells of the area. These source rocks are mainly Albian–Cenomanian in age and are marine anoxic carbonates facies containing sulphur. The prevailing kerogen is type II with the TOC varying from 1 to 4%. Cretaceous anoxic strata are quite common in the Apennines, as well as in the Mediterranean region (e.g. the Cenomanian–Turonian Bahloul Formation and the Albian Fahdene Formation of Tunisia). In Italy, two main oceanic anoxic events, known as the 'Selli Event (OAE1a)' of Aptian age and the 'Bonarelli Event (OAE2)' of Cenomanian–Turonian age, have been recognized and described. However, in spite of their high TOC, the reduced thickness of the outcrop did not support the idea of including these levels within the Italian effective petroleum source rocks.

In the Bradano Basin, the geological situation is complex and not yet fully clarified. Within the same field biogenic gas pools in Pleistocene–Pliocene shallow levels, mixed thermogenic–biogenic gas in Middle–Lower Pliocene levels and thermogenic gas or oil within the deeper limestone reservoirs of Miocene–Cretaceous age are present. The oil of the Pisticci Field could belong to the Cretaceous system while, as for Cupello-San Salvo Field, the source attribution is uncertain and could possibly belong to the LateTriassic–Early Jurassic system. To complete the picture of the Cretaceous oil, the Nilde Field is worth mentioning, as well as the Zibibbo, Norma and Naila fields, all of which are located in the Sicily Channel (Fig. 17). The oil of these fields has always been without a confirmed source-rock attribution. It cannot be excluded that they were sourced by the Cretaceous Lower Fahdene Formation of Albian age and finally trapped on Miocene reservoirs, as is the case for the nearby Tunisian fields.

Tertiary Supercycle – Oligocene–Miocene thermogenic gas PSF. This PSF extends along the external portion of the entire Apennines Thrust Belt. The source rocks are the shales of the Oligocene–Miocene turbidite sequences (the 'Marnosa-Arenacea' Formation is a typical example) involved in the thrusts. Kerogen is of type III with a strong terrestrial component, indicating mainly a gas-prone source

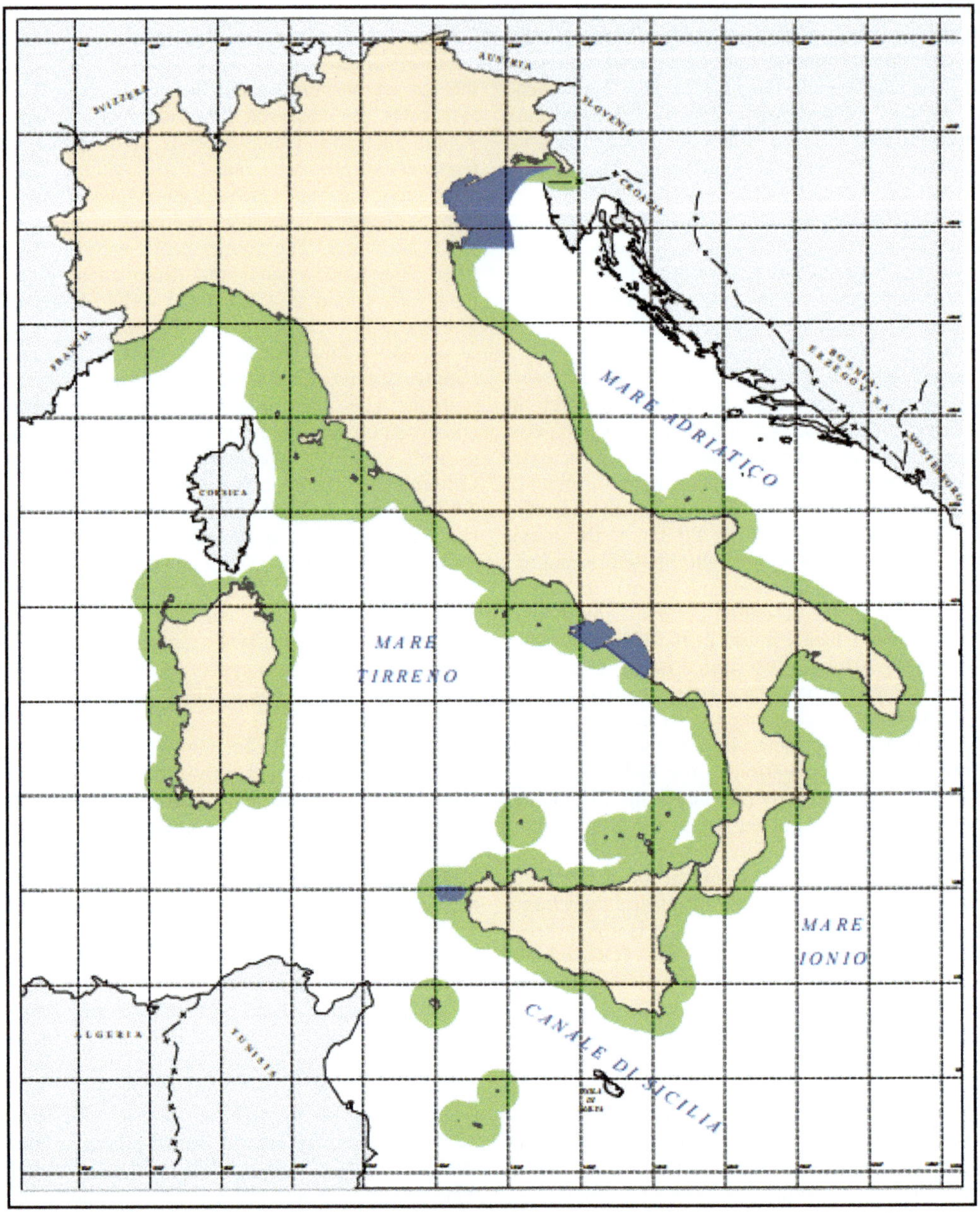

Fig. 19. Offshore Italy: banned areas to mining activities (green/blue colours) (UNMIG 2016: http://unmig.sviluppoeconomico.gov.it/).

with the generation of minor quantities of light oil and condensate. The average TOC is poor and rarely reaches 1%. The shales are acting both as a source and a seal. The reservoirs are the intercalated sandstones of Miocene age. In the Po Plain, the main field is Cortemaggiore. The fields and seepage of the Northern Apennines previously described in the Pioneers' Period belong to this system (i.e. Vallezza, Ozzano, Vizzola, Montechino, Gratera, Velleia, Salsomaggiore, etc.). In offshore Calabria, the gas fields of Luna-Hera Lacinia have also been attributed to this PSF, even if there is no positive laboratory confirmation. In onshore Central Sicily, the gas fields of Bronte and Galliano (as well as

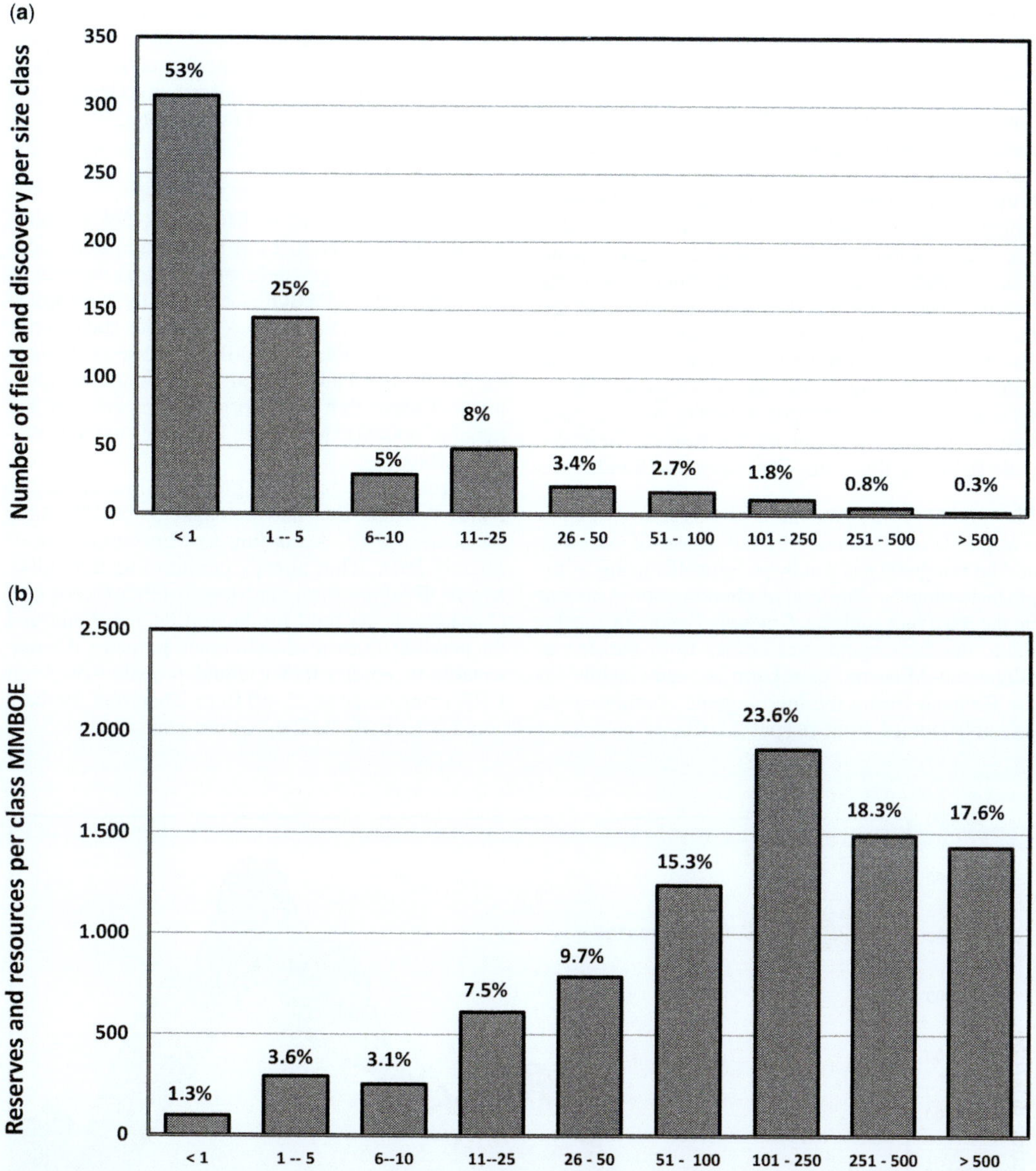

Fig. 20. (**a**) Field and discovery numbers per volumetric size classes. (**b**) Field and discovery volumes (MMBOE) per size classes.

the other nearby discoveries) have also been attributed to this PSF.

Tertiary Supercycle – Pliocene–Pleistocene biogenic gas PSF. This PSF is present along the entire Pliocene–Pleistocene foredeep/foreland of the Alps and Apennines, from the Po Plain through the Adriatic and all the way to Sicily. It is the most widespread petroleum system and the richest in terms of discovered volumes. The shales in the turbidite sequences are both the source rocks (type III with poor TOC) and the seals. The intercalated sandstones in the turbidite are the reservoirs. This system is quite similar to the previous one, with the exception of the temperature being always below the thermogenic threshold even at a depth of more than 3500 m. The formations are only partially involved in the compressional stresses, showing only gentle structures, if any. Consequently, all the geological conditions are present for biogenic dry gas generation. An impressive number of fields belong to this system. In the Po Plain, the largest accumulations are the Alfonsine, Ravenna, Dosso degli Angeli, Sergnano and Settala. In the North Adriatic, the largest ones are the Agostino-Porto Garibaldi, Porto Corsini Ovest and Est, Barbara, Chioggia, Angela-Angelina, and Amelia. In the Middle Adriatic, they are the Giovanna, Emma Ovest, Squalo Centrale and Bonaccia. In the Bradano Basin, the biogenic gas is, by far, the prevalent phase of the basin. Here, the main fields are the Montestillo-Roseto, Pisticci, Grottole-Ferrandina and Candela-Palino. In offshore south Sicily, biogenic gas is present within the Pleistocene–Pliocene foredeep basin developed in front of the thrusts. Here the main fields are the Argo, Cassiopea and Panda.

The mixing of thermogenic and biogenic gas. Several gas fields contain a mixed phase of biogenic and thermogenic gas that is recognizable using isotopic technologies. This kind of phenomenon is known in the Po Plain and the Bradano Basin. In the Po Plain, the thermogenic gas comes from the deeper Oligocene–Miocene petroleum system; while in the Bradano Basin, the thermogenic contamination probably comes from deeper Mesozoic petroleum systems. The mixed-phase fields in the Po Plain are located near the wedge of the main thrusts (e.g. Caviaga, Ripalta, Bordolano, Cornegliano, Minerbio and Spilamberto). In the Bradano Basin, the Grottole and Candela fields can be cited as examples.

Giant fields

Italy has only two recognized giant fields (above 500 MMBOE). The first one is Agostino-Porto Garibaldi, a biogenic gas field with original reserves in the range of 600 MMBOE. Geologically, it is located in the foredeep of the North Adriatic Sea and belongs to the Pliocene–Pleistocene PSF. The second one is the Val d'Agri Field located in the Southern Apennines Thrust Belt with original reserves in the range of 600–900 MMBOE. It belongs to the Cretaceous PSF.

Barbara is a biogenic gas field located in the North Adriatic Sea and belongs to the Pliocene–Pleistocene PSF. According to Ministerial data of January 2016, it has already produced 61 Bcm (400 MMBOE) of gas from sandstones of Pleistocene age. Considering the field production decline trend and the possible ultimate abandonment pressure, it is reasonable to predict that it could probably reach an URR in the range of 75–80 Bcm. Therefore, Barbara could potentially be Italy's third giant field.

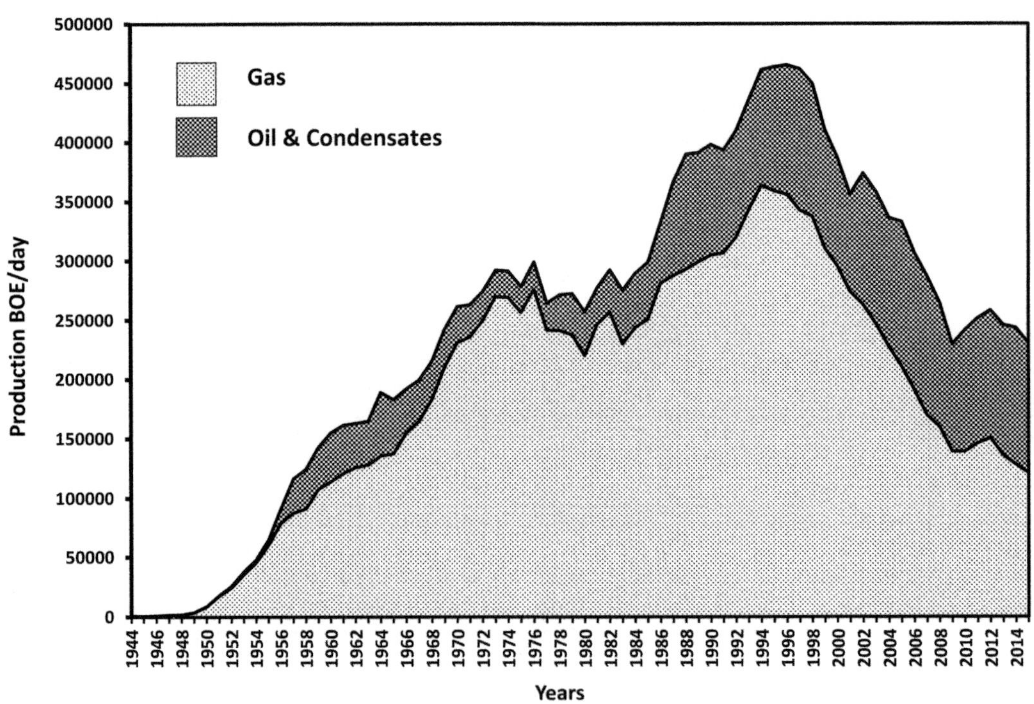

Fig. 21. Italian historical production of oil and gas.

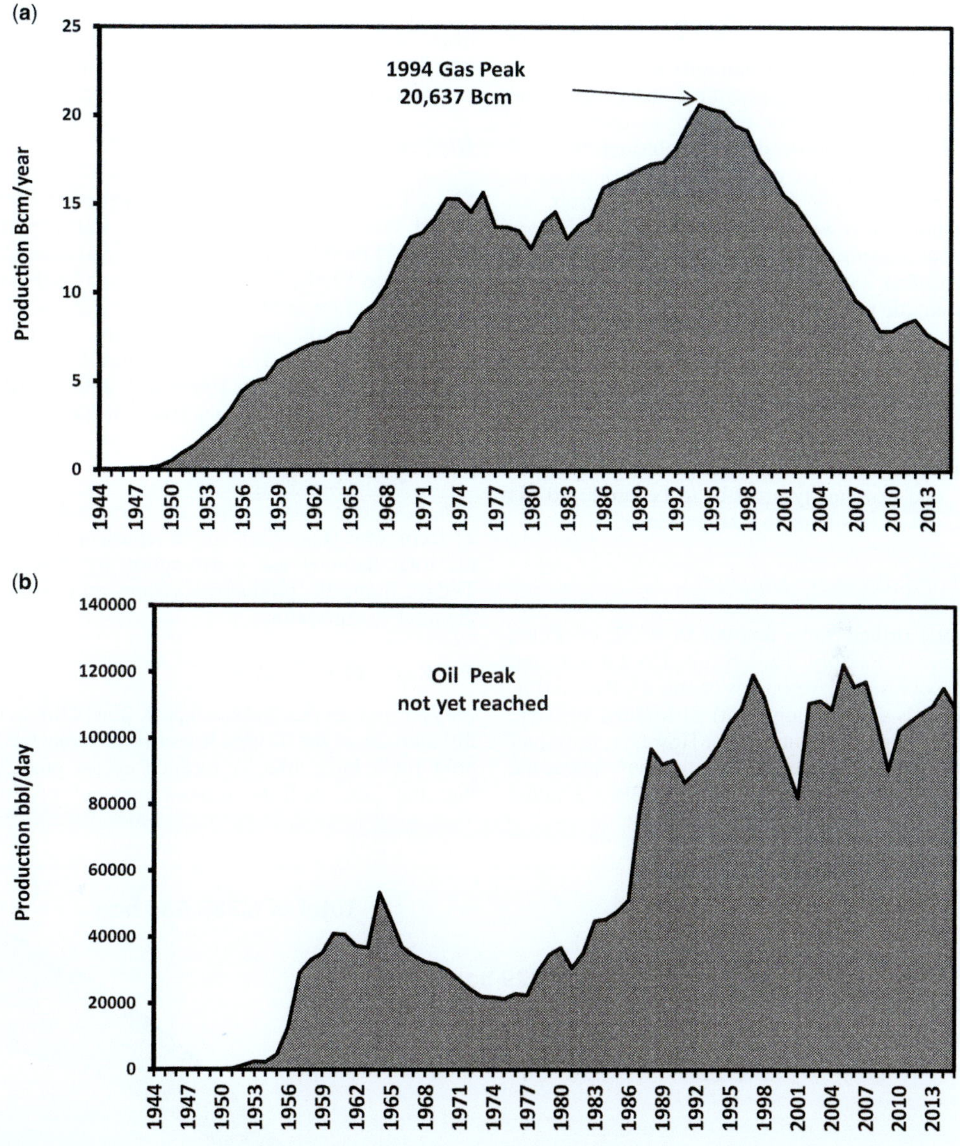

Fig. 22. Historical production rate: (**a**) gas (Bcm per year); (**b**) oil (bbl per day).

Conclusion of the period

This period was the real 'golden age' of the Italian oil industry, during which all of the main objectives in terms of reserves/resources, development and production were achieved. After this, the game was almost over.

The Decline Period

Since 2007, exploration activities have been at their historical minimum. Several discoveries are on the development waiting list, while others are unlikely to be approved for development by the government. This situation is strongly preventing new investment. A law was approved in 2010, after the accident at the Macondo Prospect, to ban drilling within 12 miles of the coast (shown in green in Fig. 19), along with the already established ban in the Special Preservation Areas (shown in blue in Fig. 19). In 2012, the ban on the drilling of new wells within the 12 miles became permanent. The bureaucratic process to obtain government authorization is quite unpredictable as there are different local authorities involved, each having the possibility of veto. Public opinion is largely against any kind of upstream operations both onshore and offshore. It is quite common to hear this hypocritical statement in the media: 'Italy will never become an energy self-sufficient nation. Instead of purchasing 90% of fossil fuel abroad, let us purchase 100% so as to avoid polluting our homeland'. As a result, no exploratory wells were drilled in 2014. The offshore situation is even worse, with the last exploratory wells drilled in 2008 (Cassiopea-1, Argo-2 and Ombrina-2dir).

The results of this long historical journey can be summarized with some statistical data through which the following fundamental questions can be answered: how many fields have been discovered since the beginning of the exploration and how big are they? How many hydrocarbons have been produced and how many are still to be produced? Is Italy a gas or an oil geological province?

Field and discovery statistics

The total drilled wells amount to 7222, of which 3233 are exploratory. They resulted in a discovered volume of reserves/resources of about 9 BBOE (billion barrels of oil equivalent) distributed between 600 oil and gas accumulations. However, only half of these should be considered 'proper' fields, the remaining being immaterial. In fact, 53% are below 1 MMBOE (Fig. 20a). If the situation is examined from the volume perspective, 75% of the total volume belongs to the last four classes starting from 51 MMBOE (Fig. 20b).

Historical production

The oil and gas production was negligible before World War II. For this reason, the statistics start from 1944 until the last release of official data. During the late 1990s, Italy reached a remarkable total production close to 500 MBOE/day (Fig. 21). It is interesting to note that the volume of condensate included with the oil is very limited. Condensate peaked at the end of the 1960s with 2500 bbl/day, while today its contribution is less than 500 bbl/day.

Peak of gas production

The gas peak was reached in 1994 with almost 21 Bcm/year (Fig. 22a), which represented 40% of the total national gas consumption for that year. Today, domestic production covers only 10% of the total consumption.

Peak of oil production

Italy has not yet reached its oil peak (Fig. 22b), as the full start-up of the Tempa Rossa Field remained on hold for a long time. A recent forecast predicted that the peak will be around the end of 2018

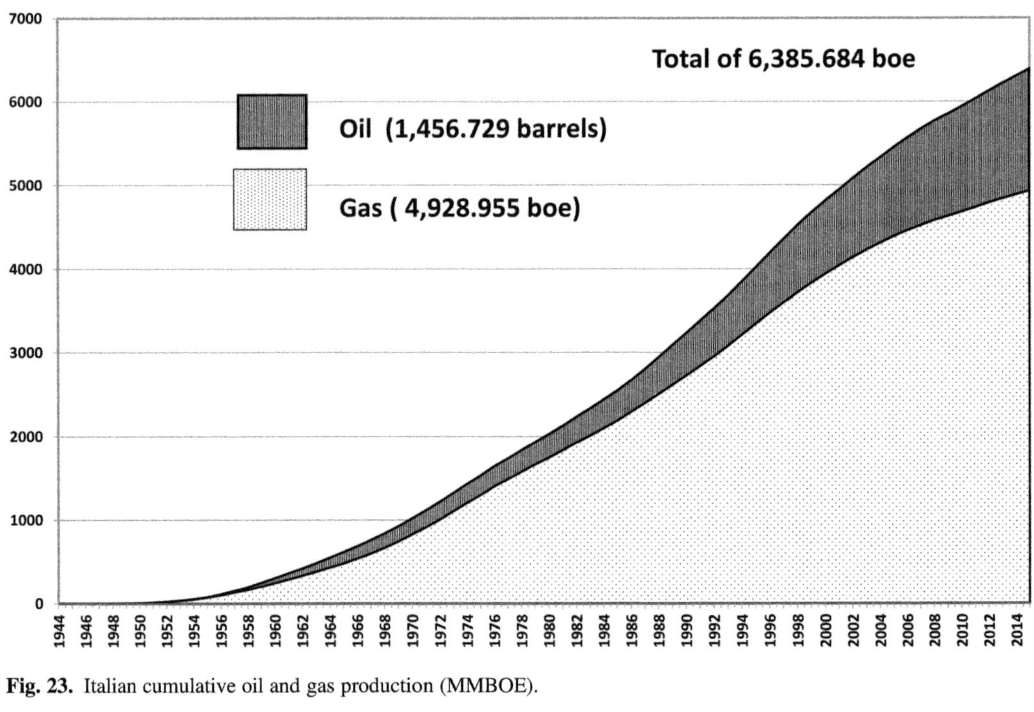

Fig. 23. Italian cumulative oil and gas production (MMBOE).

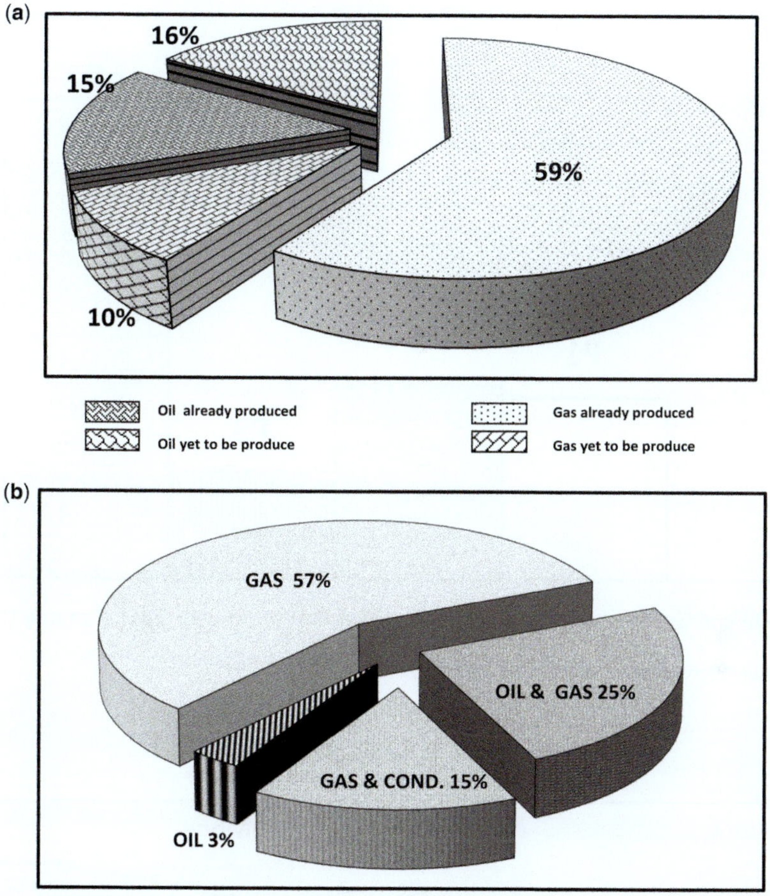

Fig. 24. (**a**) Comparison between the oil and gas production with the remaining 3P reserves. (**b**) Comparison between volumes of the fields/discoveries original hydrocarbon phase.

beginning 2019 with 160,000 bbl/day compared with the current production of 100,000 bbl/day. The domestic oil production covers 10% of the total consumption.

Cumulative production of oil and gas

At the end of 2015, the cumulative oil and gas production corresponded to 6.4 BBOE (Fig. 23). Cumulative gas production amounted to 767 Bcm, corresponding to 4.9 BBOE (77% of the total). Cumulative oil production (including condensates) amounted to 200 Mt (million tons), corresponding to 1.5 Bbbl (billion barrels) (23% of the total). The cumulative condensate production represents only 0.2% of the total.

Prevailing hydrocarbon phase

Italy can be defined as a 'biogenic gas' country, with exploration and production concentrated on Pliocene–Pleistocene foredeep formations. To prove this, we add together and compare the cumulative production and the remaining 3P (Proved plus Probable plus Possible) reserves of oil and gas ('yet to find' volumes are excluded from this analysis). About 70% of the hydrocarbons is gas, while the remaining 30% is oil (Fig. 24a). The total gas phase could be further subdivided: 80% is biogenic dry gas, 10% is thermogenic gas and the remaining 10% is mixed biogenic–thermogenic gas (Mattavelli 1987). Examining the classification of fields and discoveries (oil; gas; oil and gas; and gas and condensates), the data confirm the previous analysis, with 72% as gas + gas and condensates, while oil + oil and gas give the remaining 28% (Fig. 24b).

Remaining reserves

The remaining reserves are officially disclosed yearly by government as P1 (Proved), P2 (Probable) and P3 (Possible). At the end of 2015 the remaining

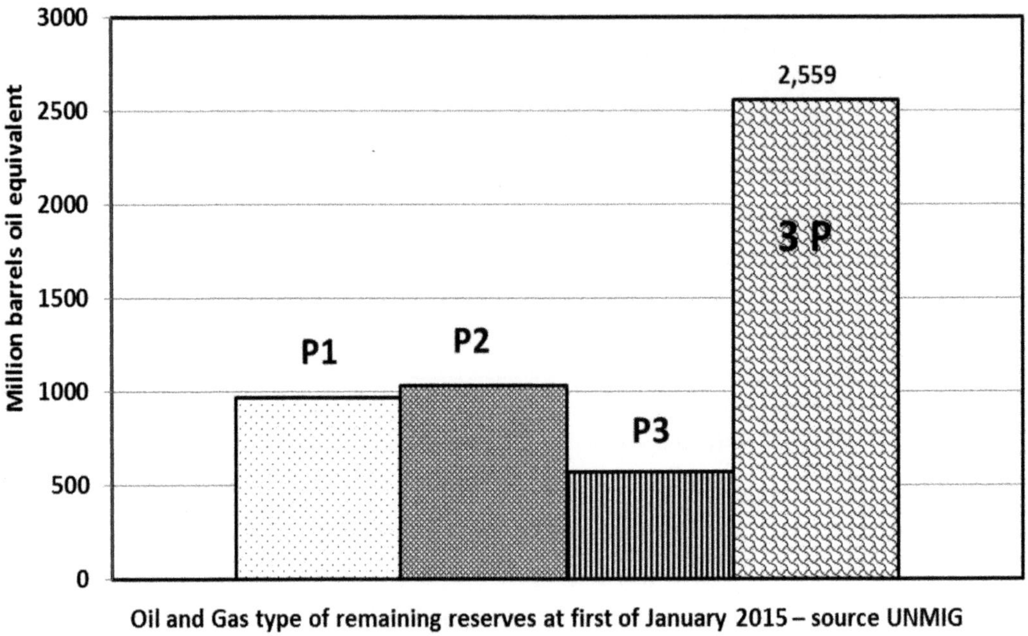

Fig. 25. Remaining 3P reserves as at 1 January 2015.

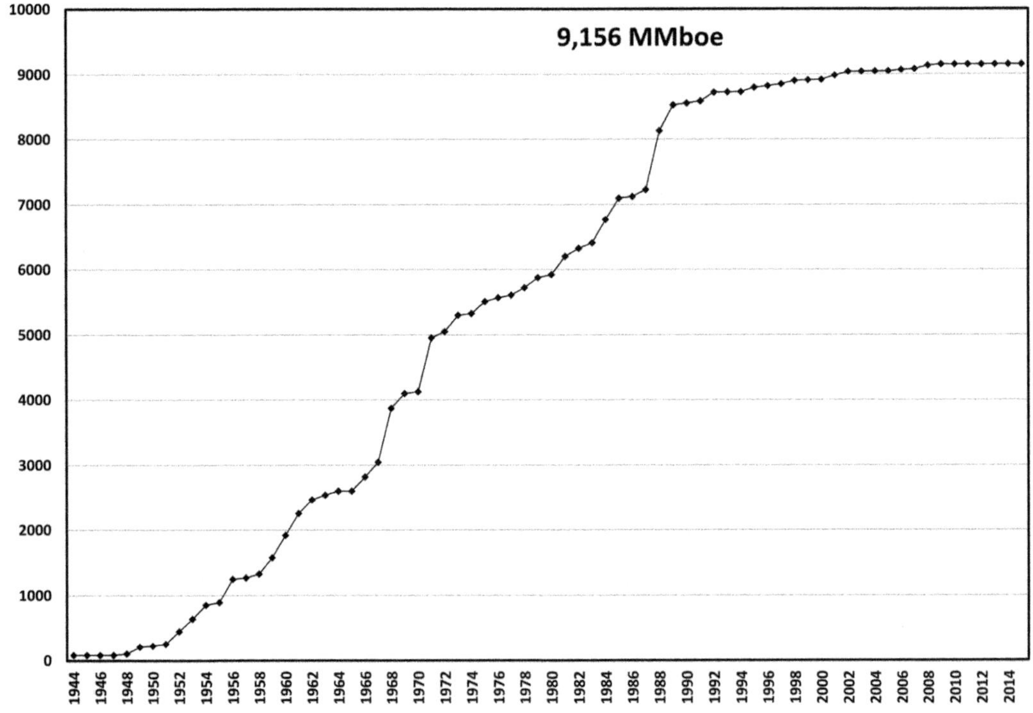

Fig. 26. Creaming curve of discovered reserves and resources of oil and gas (MMBOE).

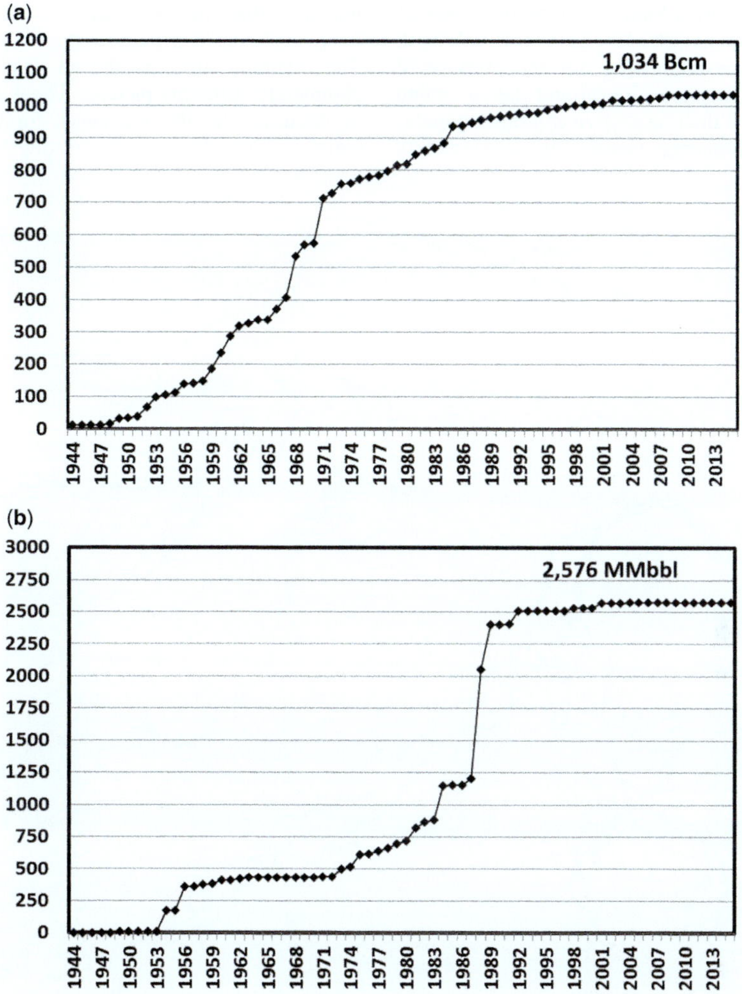

Fig. 27. Creaming curve of discovered reserves and resources of (**a**) gas (Bcm); (**b**) oil (MMbbl).

3P reserves (P1, P2 and P3) for gas and oil were 2.56 BBOE (Fig. 25), of which the remaining oil was 234 Mt (corresponding to 1708 MMbbl (million barrels)) and the remaining gas was 140 Bcm (corresponding to 887 MMBOE).

Total hydrocarbons of Italy, and the oil and gas creaming curve

To evaluate the total hydrocarbons generated and trapped in the complex geology of Italy, we could sum up the historical cumulative production and the 3P reserves. The non-commercial fields and the fields under studies/appraisal not yet approved by the government are excluded from this analysis. The total is 9 BBOE (2559 MMBOE of 3P reserves + 6386 MMBOE already produced). Another way to calculate this total is by using the creaming curve that has the advantage of considering all the fields and discoveries irrespective of their commercial or legal status. This method, however, suffers from a chronic lack of up-to-date data. The total oil and gas creaming curve gives a value of 9.2 BBOE (Fig. 26), which is slightly higher but perfectly comparable to the one obtained with the previous method. Consequently, by subtracting the first from the second, we should obtain the non-commercial resources plus the resources of not-yet-approved development projects. However, the value obtained is 200 MMBOE, which is, by far, too low. In fact, the main projects on hold and not yet approved are: the 'Alto Adriatico' biogenic gas project with 250 MMBOE (it includes the Chioggia Field with 148 MMBOE); the Elsa heavy oil field with 110

MMBOE; the Ombrina Mare heavy oil field with 41 MMBOE; and, finally, the Miglianico oil field with 52 MMBOE. It is interesting, for the purpose of this analysis, to examine the oil and gas creaming curves to evaluate their respective exploration maturity. The gas creaming curve (Fig. 27a) clearly indicates that the exploration of the gas play in Italy has been in a mature stage since the 1990s. The creaming curve for the oil (Fig. 27b) shows a completely different picture. There are no signals of maturity and the flat curve for the year 2000 simply reflects the lack of investment.

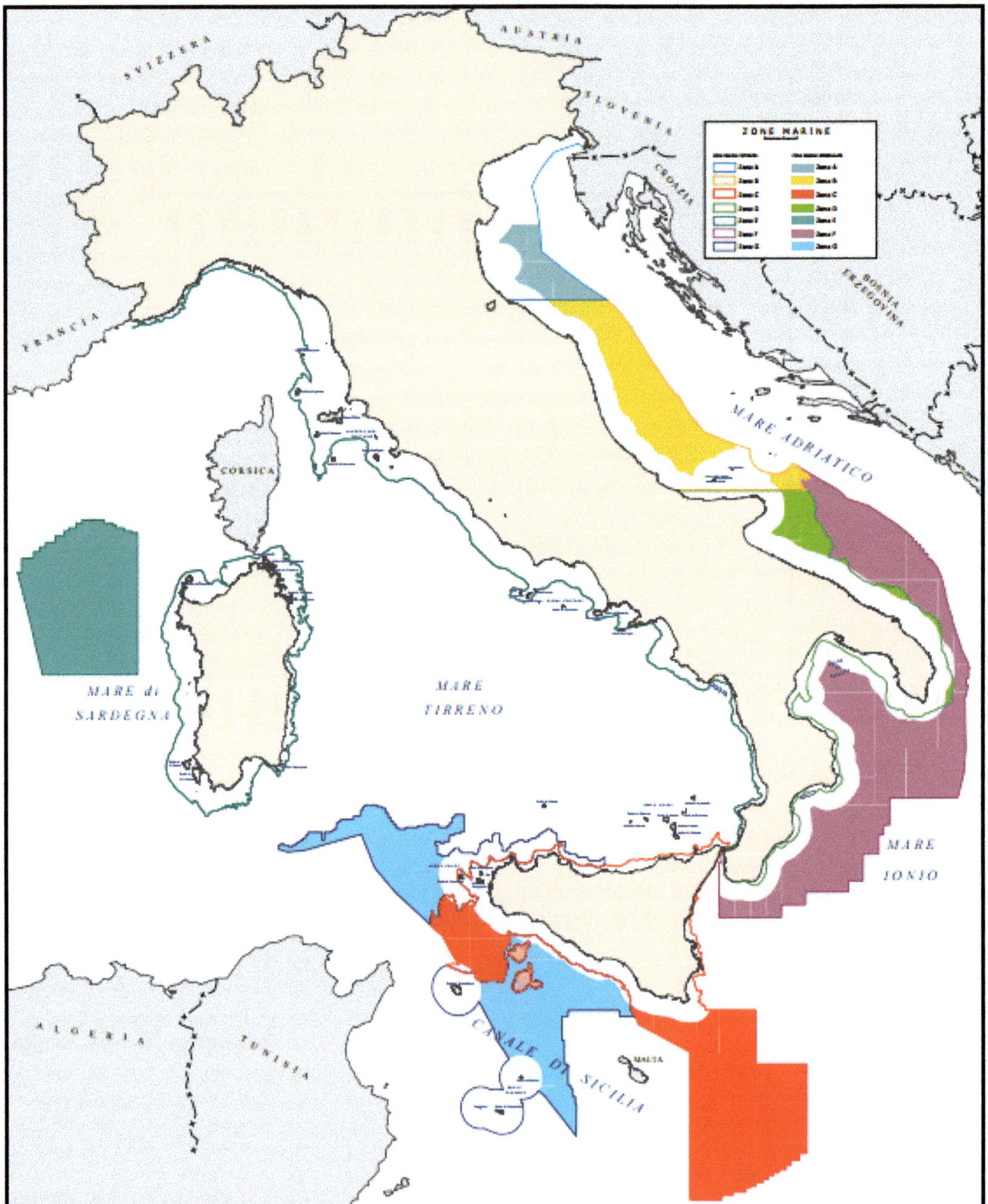

Fig. 28. Italian offshore areas open to exploration (UNMIG 2016: http://unmig.sviluppoeconomico.gov.it/).

Historical recorded accidents in the oil sector

It is interesting to check whether the aversion of the Italian public opinion to oil and gas domestic operations is linked to any major accidents that have occurred in the recent past. Two dramatic accidents took place in the downstream and midstream sectors. In 1991, the oil tanker *Amoco Haven* sank in Genoa harbour which resulted in five fatalities and sea pollution. In 2009, a train transporting liquefied petroleum gas (LPG) exploded in Viareggio Railway Station (31 fatalities). However, limiting our analysis only to the upstream, which is the focus of this work, the main accidents consist of well blow-outs. Most of them are concentrated immediately after World War II, in a period of high activity and poor technology. They include the explosion of Caviaga-13 in 1949, Cortemaggiore-18 in 1950, Bordolano-2 in 1952 and Ragusa-9 in 1955. After this period, as technology improved, the number of accidents significantly decreased: only two have occurred since then. In 1965, during the drilling of Porto Corsini-7, the well blew out causing three fatalities and the loss of the jack-up. In 1994, during the onshore drilling of Trecate-24, the well blew out resulting in no victims but, instead, in the loss of the rig and related pollution of the area. Therefore, there have been no upstream accidents offshore since 1965 and none onshore since 1994. It is thus safe to assume that it is very unlikely that this aversion is linked to the memory of past drilling accidents. Of course, we have to take into consideration the effects of the 'Macondo Syndrome'. The opposition of public opinion has probably been influenced by the genuine belief that a possible sharp change in energy consumption might occur, from fossil to renewable, irrespective of a pragmatic approach to the problem.

Conclusion of the period

There is an energy paradox in Italy: in spite of the proved hydrocarbon potential still existing in the country, the choice is to leave the 'domestic' oil and gas underground. This implies the need to purchase almost all of the necessary fossil energy from abroad with the relevant exposure to supply risks. New investments in the upstream sector could only happen if existing laws were changed. The central government should return to being the only administration entrusted with energy decisions, in order to avoid 'NIMBY' ('Not In My Back Yard') effects. Italy has made a big effort in electrical production and, today, more than 50% comes from renewable sources. However, hydrocarbons remain the main energy source for civil use, as well as for private and public transport. Oil and gas are still essential for Italy, considering the 1987 political decision to abandon nuclear power generation.

The Ministry of Energy has recently offered the international industry some new offshore deep-water acreages. These new blocks are located in the Sardinia Sea (shown in dark blue in Fig. 28) and in the Ionian Sea (shown in violet in Fig. 28). In addition, it has opened the shallow waters between Malta, Italy and Libya. This zone is known as the Medina Bank (shown in red in Fig. 28) and it is historically an area of international disputes. Adopting this strategy, the Ministry expects to attract fresh investments into the financially starving upstream. Only the future will tell whether this is the right key to again unlock oil and gas exploration in Italy.

Funding This research received no specific grant from any funding agency in the public, commercial or not-for-profit sectors.

References

Accademia Nazionale dei Lincei 1959. Attidel Convegnodi Milano sui giacimenti gassiferi dell'Europa Occidentale, Roma.

Ariosto, F. 1690. *de Oleo Montis Zibiniiseu Petroleo Agri Mutinensi Libellus*. Oligero Jacobeo, Copenhagen.

Bertello, F. 2008. Overview of Italy's petroleum systems and related oil and gas occurrences. Presented at the 70th EAGE Conference & Exhibition incorporating SPE EUROPEC 2008, 9–12 June 2008, Rome, Italy.

Bertello, F., Fantoni, R., Franciosi, R., Gatti, V., Ghielmi, M. & Pugliese, A. 2011. From thrust-and-fold belt to foreland: hydrocarbon occurrences in Italy. *In*: Vining, B.A. & Pickering, S.C. (eds) *Petroleum Geology: From Mature Basins to New Frontiers – Proceedings of the 7th Petroleum Geology Conference*. Geological Society, London, 113–126, https://doi.org/10.1144/0070113

Cazzola, A., Fantoni, R., Franciosi, R., Gatti, V., Ghielmi, M. & Pugliese, A. 2011. From thrust and fold belt to foreland basins: hydrocarbon exploration in Italy. Presented at the AAPG International Conference and Exhibition, 23–26 October 2011, Milan, Italy.

Facca, G. 1951. *Synopsis of the 3rd World Petroleum Congress*, 26 May–6 June 1951. The Hague, Netherlands, p. 258.

Gorgoni, C. 2003. *Le salse di Nirano e le altre salse emiliane. I segreti di un fenomeno tra mito e realtà*. Comune di Fiorano Modenese, Tip. ABC, Sesto Fiorentino, Florence, Italy.

Martinelli, G., Cremonini, S. & Samonati, E. 2012. Geological and geochemical setting of natural hydrocarbon emissions in Italy. *In*: Al-Megren, H. (ed.) *Advances in Natural Gas Technology*. InTech, Rijeka, Croatia, http://www.intechopen.com/books/advances-in-natural-gastechnology/geological-and-geochemical-setting-of-natural-hydrocarbon-emissions-in-italy

Mattavelli, L. 1987. Geochemistry and habitat of natural gases in Italy. *In*: *Proceedings of the 13th International Meeting on Organic Geochemistry*, Venice,

Italy, 21–25 September 1987. *Organic Geochemistry*, **13**, 1–13.

SPI 1975. *La ricercapetroliferain Italia nel'900*. Internal Report.

Stoppani, A. 1897. *Il Bel Paese: conversazioni sulle bellezze naturali, la geologia e la geografia fisica d'Italia*. L.F. Cogliati, Milano, Italy.

UNMIG 2016. *'Ufficio Nazionale Minerario per gli Idrocarburi e la Geotermia' and DGRME (Direzione Generale per le Risorse Minerarie ed Energetiche), Annual Reports and Open Files of Geological, Geophysical, Production, Reserves Data*. L'Ufficio nazionale minerario per gli idrocarburi e le georisorse (UNMIG), Rome, http://unmig.sviluppoeconomico.gov.it/

Maiella, an oil massif in the Central Apennines ridge of Italy: exploration, production and innovation in the oil fields of Abruzzo across the nineteenth and twentieth centuries

FRANCESCO GERALI[1,2]* & LORENZO LIPPARINI[3]

[1]*The University of Oklahoma School of Law, Oil & Gas, Natural Resources, and Energy Center, 300 Timberdell Road, Norman, OK 73019-5081, USA*

[2]*The University of Western Australia, FABLE, 35 Stirling Highway, Crawley, Perth, WA 6009, Australia*

[3]*Earth Science Department, 'La Sapienza' University of Rome, Piazzale Aldo Moro 5, 00185 Rome, Italy*

**Correspondence: fgerali@ou.edu*

Abstract: This article aims to present selected episodes of the scientific and entrepreneurial activities realized in the Majella oil district, Abruzzo, central Italy, between the 1830s and the 1940s. Majella had an important role in the early process of modernization of the Italian oil industry. As we will see, the application of science and technology in the wellsite, a pioneering integrated production model, unexpected environmental constraints and the longevity of the business made Abruzzo into a benchmark. Between 1864 and 1865, the pattern of the secular and unchanged local manufacturing of bitumen, based on human labour and the manual harvesting of the mineral from natural outcrops, was outdone by two new conditions: the study of subsoil and the utilization of mechanical drilling. These early records of paid oil consultancies and the utilization of steam power were followed by the impressive, but short and illusory, peak in production never recorded before. In the following decades, the Majella district was very active and drew the attention of several international operators, in spite of the declining production of crude – compensated for by the yield of bitumen and shales – and the unstoppable rising of the Emilian Apennine ridge in the provinces of Pavia, Piacenza and Parma. From the early 1900s, foreign companies gradually reduced investment in new exploration in Maiella, where the core production was now bitumen and asphalts rather than oil and, by the 1920s, most of the industry was controlled by government authorities and local business. The advent of Fascism in the 1930s saw Maiella become a stronghold of the autarchy's policies; later, the improvised and inefficient national fuel planning of Italy at the start of the Second World War saw Abruzzo's oil and bitumen supplies become a strategic resource. The Maiella district has the longest production history in Italy and today geoscientists are surveying and interpreting the geology of the area with a new perspective.

Geological setting and petroleum system of the area

Our research focuses on the central Abruzzo areas of Tocco Casauria (Fig. 1), between the Apennine chain front to the west and the Maiella Mountain to the east, close to the villages of Roccamorice and Lettomanoppello. Geologically, the Maiella Mountain is a NNW–SSE anticline located in the Central Apennines, formed during the Middle–Late Pliocene (Ghisetti & Vezzani 2002): it records a carbonate succession from the Lower Cretaceous to the Miocene, topped by siliciclastic sediments of Messinian–early Pleistocene age (Donzelli 1968; Crescenti *et al.* 1969).

From the petroleum geology perspective, the NW sector of the Maiella structure is particularly interesting due to the numerous oil seepages detectable at surface, mainly associated with the high-porosity carbonate reservoirs of the Bolognano Formation (Oligo-Miocene), but locally also to the Santo Spirito Formation and the Orfento Formation carbonates (Paleocene–Oligocene and Upper Cretaceous in age, respectively) (Brandano *et al.* 2013; Scrocca *et al.* 2013). Evidence of oil occurrences is found over a wide area of more than 20 km^2, mainly between 400 and 1000 m in altitude, with some exceptional oil seeps found above 2000 m (Ferraris 1942).

The nearby Tocco Casauria area is, instead, located about 10 km west of the Maiella Mountain, at the front of the Apenninic chain, within the deformation belt at the frontal thrust of the Morrone Mountain. In this area, the same Bolognano reservoirs known to crop out in the Maiella Mountain have been successfully proven by wells to be oil bearing in the near subsurface. Indeed, a few kilometres NW of the Tocco da Casauria village, a stable natural water spring,

From: Craig, J., Gerali, F., MacAulay, F. & Sorkhabi, R. (eds) 2018. *History of the European Oil and Gas Industry*. Geological Society, London, Special Publications, **465**, 275–303.
First published online May 25, 2018, https://doi.org/10.1144/SP465.20

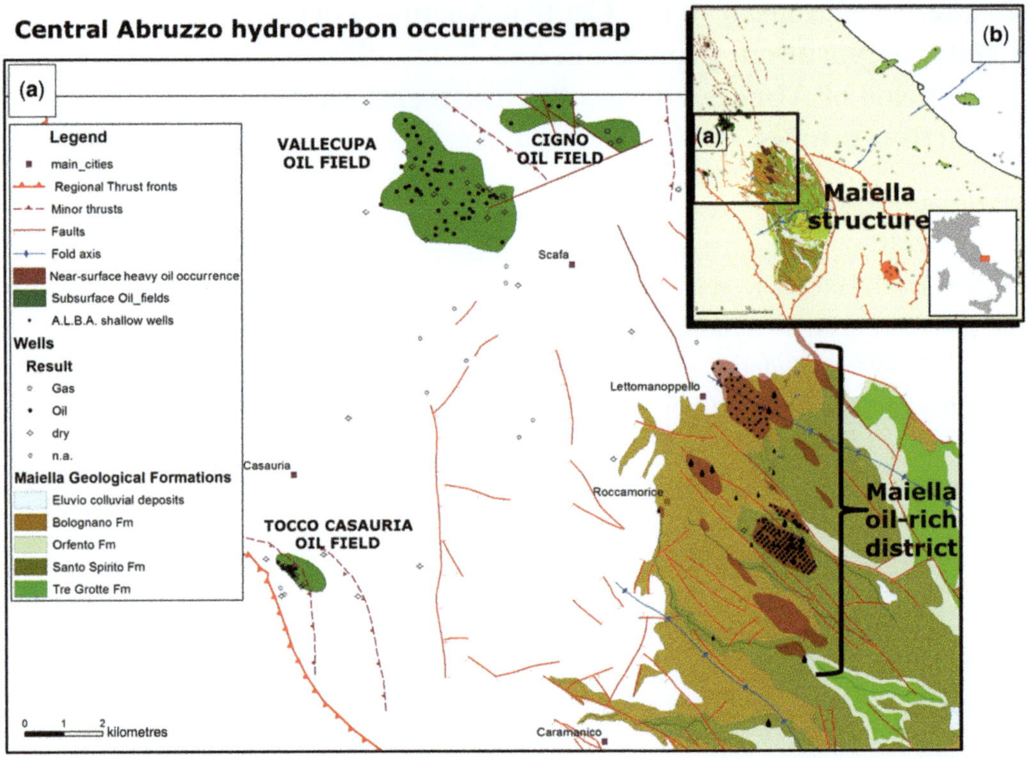

Fig. 1. Map of hydrocarbon occurrences in central Abruzzo, both in the subsurface (in green) and in the near-surface (in brown), including the bitumen/asphalt-rich outcrops of the NW flank of the Maiella Mountain and the Tocco Casauria 'oil district'. Surface geology and main tectonic elements are reported (Vezzani & Ghisetti 1998; VIDEPI Project 2012). Oil occurrences are mostly present within the Oligo-Miocene 'Bolognano Formation' reservoirs, both at surface and in the subsurface, but also in the 'Santo Spirito' Formation and 'Orfento' Formation carbonates (Paleocene–Oligocene and Upper Cretaceous in age, respectively).

containing important quantities of oil, has been known since historical times to emerge and flow into the 'Little Arollo River': this unique 'oil spring' has been interpreted as the spectacular result and evidence of a local hydrogeological drainage pattern of an aquifer crossing the shallow near-surface oil accumulations.

It is important to note that the same reservoirs in the region have also been successfully explored in the subsurface by more recent and modern exploration efforts and deep drilling, especially in the 1980s and 1990s, resulting in a few mid-sized oil discoveries and production fields, such as Tocco Casauria itself, and the Vallecupa and Cigno oil fields, today depleted and abandoned.

These hydrocarbon occurrences have generally been associated with Late Triassic–Early Jurassic (Liassic) carbonate source rocks (Mattavelli & Novelli 1990; Zappaterra 1994; Cazzini *et al.* 2015), while hydrocarbon migration is suggested to have been associated with the main fault systems and fractured corridors within the carbonate succession (Brandano *et al.* 2013; Cazzini *et al.* 2015).

The hydrocarbons found in the study area have generally been classified as heavy oil, in the subsurface, and bitumen/asphalt along the outcrops at surface: from historical photographic evidence (Fig. 2), oil discovered in the shallow subsurface by mining tunnels below the Maiella NW flank did still show a viscosity that can be estimated to be about 4°–8° API by visual comparison with fluid samples. Even today, at surface, when summer temperatures are above about 20°C, bitumen is able to flow from the rock's matrix porosity and fractures.

Notes on the pre-industrial history of the Abruzzo oil provinces up to the 1850s

The historical use of bitumen in Abruzzo was documented for the first time in 1868, close to the village of Lettomanoppello in the locality called Pignatara. Following an archaeological campaign, a workshop for the purification of asphalt dating from Roman times was discovered. Among the remains of several

Fig. 2. Viscous bitumen natural flow in a mine tunnel (courtesy of Lettomanoppello City Hall). Estimated 4°–8° API crude oil by visual comparison with fluid samples.

clay vases containing asphalt, archaeologists found a rectangular loaf of bitumen (35 × 26 × 10 cm (length × width × thickness)) on which was engraved 'ALONI.C.F.ARN.SAGITTAE'. This inscription allowed the artefact to be dated back to about the first century BC (the Ancient Rome Period) (Toppi 2017), and to attribute the ownership of the workshop to a man named Alone Sagitta from the Arniense Tribe (descendant of the old pre-Roman Marrucini), who probably were also the owners of the mine at Pignatara. The finding confirmed the mining of the product, its packaging and its commercialization; and that the Romans were trading bitumen from Maiella as much as the more famous oils from Sicily described by classical authors (Zambra 1869; Massimi *et al.* 2002; Novelli & Sella 2009).

The exploitation and utilization of the bitumen and asphalts of Maiella, near Lettomanoppello and Tocco Casauria, has been reported since the early Middle Ages. In the twelfth century, it was traded with the Republic of Amalfi, where it was used to caulk the ships (Anon 1182; Trapani 2003).

Flavio Biondo da Forlì (1392–63), historian and humanist of the Renaissance period, in his last work *Italiam illustratam* [*Italy Illustrated*], which was published posthumously in 1482, mentions the 'fons olei Petronici perennins scatent' ('the perennial source of oil that flows'), down the hill of the castle in Località Cantalupo (this settlement, now disappeared, was located between the village of Musellaro in the municipality of Bolognano and the municipality of Tocco da Casauria, in the province of Pescara). He refers to the 'Oleum Petronicum' (Medieval Latin, which can be translated as 'rock oil') collected from the spring by the German and Hungarian travellers who appreciated the medical properties of the oil; it was a valuable commodity (Biondo 1482).

Scipione Mazzella's (1586) *Descrizione del Regno di Napoli* [*Description of the Kingdom of*

Naples] (Fig. 3) reports on the 'mine of pitch ... that resembles bitumen, but it finds more applications as pitch' (p. 93).

In 1551, Leandro Alberti in his *Descrittione di tutta l'Italia* [*Description of the Whole of Italy*] talks about a liqueur substance named 'Oglio Petronico' and renowned for its medical properties (Alberti 1551).

Cristoforo Scanello in the *Cronica o ver compendio dell'antiqua regione et provintia del Abruzzo* [*Chronicle* (or *Compendium*) *of the Ancient Region and Province of Abruzzo*], published in 1568, talks about an 'oil given by the Lord' a remedy for every infirmity and called 'Olio Petronico' (Scanello 1568).

The observations from these natural philosophers and geographers, together with many other sources from 1500 to 1800, give us two important pieces of information: first, oils were a mineral resource consistently available over the last two millennia; and, secondly, that heavy oils, asphalts and bitumen were an abundant resource, processed on a large scale, but commercialized in a local/regional range. The more valuable light fractions were scarce, gathered in limited quantities and traded along longer routes (Trapani 2003).

These observations, merged with information available for the other Italian oil regions, show how at the turn of the nineteenth century the 'geography' of the oil resources in Italy was already outlined for the generation of explorers to come. Northern Italy offered lighter and more valuable

DESCRITTIONE
DEL REGNO
DI NAPOLI.
NELLA QVALE S'HA PIENA CONTEZZA, COSI' DEL Sito d'esso, de' nomi delle Prouintie antiche, e moderne, de' costumi de' Popoli, delle qualità de' Paesi, e de gli huomini famosi che l'hanno illustrato; come de' Monti, de' Mari, de' Fiumi, de' Laghi, de' Bagni, delle Minere, e d'altre cose marauigliose, che vi sono.
Con la nota de' Fuochi, dell' Impositioni, de' Donatiui, e dell'entrate che n'hà il Rè.
Et vi si fa mentione de i Rè, con la loro vita, & effigie, che l'han dominato, de' loro titoli, dell'incoronatione, e del lor modo di scriuere à diuersi Prencipi; de' Pontefici e de' Cardinali che vi nacquero, e de' Vicerè statiui, e de gli Arciuescouati, Vescouati, e delle Famiglie Nobili, che vi sono.
Co' Nomi de i Baroni, e loro Arme, e'l diuisamento delle lor Corone; e con vn preambolo de i Rè di Gierusalem, oue si dechiara perche i Rè di Napoli habbiano quel titolo.
Con la Tauola copiosissima, & altre cose notabili, che nella prima Impressione non erano.
DI SCIPIONE MAZZELLA NAPOLITANO.
CON PRIVILEGIO PER ANNI DIECI.
IN NAPOLI, Ad istanza di Gio. Battista Cappello. M.DCI.

Fig. 3. *Descrizione del Regno di Napoli* by Scipione Mazzella (1586).

oils; and central and southern Italy were promising sites for the heavy oil industry – and it is not a coincidence that the historical literature often reports the contiguity of sulphur deposits and seepages of black oils (Tenore 1828; Pilla 1841).

In the early nineteenth century, asphalts were still being dug from shallow mines and bitumen was still collected from the land seepages or skimmed from the waters of creeks. The first was mostly used to pave streets, the latter as insulation in construction; oil was not officially acknowledged in the pharmaceutical industry but the locals were still using it for the same secular medical purposes (Huguenet & Frojo 1856).

Nothing changed in the *modus operandi* of local manufacture until the publication of the observations completed in the 1830s by Baron Giuseppe Nicola Durini (1765–45) on the local sulphur and oils. He was a land owner educated in law, and a science enthusiast affiliated to the Società Reale delle Scienze in Napoli [Royal Society of Sciences in Naples]. With the latter, he published in 1837 the pamphlet *Industria dello zolfo e bitume in Abruzzo Citeriore* [*The Bitumen and Sulphur Industry in the Abruzzo Citeriore*]. Durini's analysis is halfway between the natural sciences and an assessment on the potential of the land resources, which results in the first monographic study ever completed on the bitumen of the area (Verlengia 1956). Durini studied the physical proprieties of the bitumen filtered from the limestones and the clays along the beds of creeks between the Maiella Massif and the Morrone Mountain. He observed the relationship between the rainy season and the increasing flakes of bitumen flowing up into the waters of the Arollo Creek; he hypothesized that the mineral rose from deeper and abundant subterraneous deposits because of the circulation of the waters; he estimated that every year the equivalent of 130 barrels of oil were carried away by the stream. To prevent this waste, Durini proposed building a series of filtering pools into the bed of the smaller creeks to take advantage of the downwards inclination of the river course. He thought that forcing the water into a descending collection point, where wood sticks would form a rudimental filtering grid, would supply enough mineral to be used for heating and illumination. Durini promoted the mineral oil of Maiella as an old resource that belonged to modernity, for which the real potential was still to be unveiled. He completed the pamphlets with general notes on the organic origin of oil, showing his familiarity with the recently established term 'hydrocarbons' (Durini 1837; Anon 1845).

The first local businessman to test Durini's theory was Silvestro Petrini (1812–1912). Between 1840 and 1844, Petrini identified the possible asphalt mines on the western side of Maiella in the districts of Lettomanoppello and San Valentino. He built a small plant to process the mineral, and to produce bitumen and pitch. Heavy oils were mined in the traditional way; no information is available about the existence of possible shafts hand dug to collect oil. The venture was successful and the business took off very well until 1848, when Petrini fell into disgrace for political reasons. The family was isolated and forced to abandon the plant, which shortly after was pillaged (Petrini 1911; Novelli & Sella 2009).

The inception of the modern oil industry in the Maiella district, 1863–70

The illumination system in Italy was based on three fuels: rendered vegetable oils used by the lower classes; gas obtained from lignites, pitch and coal (which also gave a limited yield of dark oils), which was used for homes and public illuminations in urban centres; and, in a more minor way, gas and oils produced by the retorting of shale rocks. Whilst the vegetable, cheap and low-quality oils were abundant, Italy had negligible coal deposits in its boundaries and was dependent on foreign supplies. Coal was quite cheap per volume unit, but had high transport costs: in the harbour of Genoa, in NW Italy, the same quality of coal could cost twice as much as that in Cardiff in the UK. The shale deposits were small and poor in quality, and the Italian shale industry was not well developed (Bardini 1998).

In 1861, the first bulk of refined oils for lamps imported from the USA arrived and were rapidly adopted into the European markets, Italy included. The cheap price and the quality of the product gained the trust of the users, as much as the interest of entrepreneurs who saw in petroleum and its derivatives a lucrative business. Obviously, oil was not a job for everyone. Although the industry in Italy was characterized by a chronic lack of investment capital, it was not so difficult to find investors but much more so to find the operators (Gerali & Gregory 2017).

Between 1864 and 1870, the Maiella oil district – a wide area that includes the villages of Abbateggio, Manoppello, Lettomanopello, Bolognano, San Valentino, Roccamorice, Scafa and, the most important, Tocco Casauria – entered officially into the history of the modern Italian oil industry thanks to two outsider 'wildcatters', but who were not totally unrelated to the business (Fig. 4). The surface evidence of an active petroleum system in the region, represented by the Tocco Casauria oil springs and by the numerous oil seeps along the NW flank of the Maiella Mountain, triggered new interest in the region in the early 1860s. At Tocco Casauria, the oil springs that first attracted the interest of new investors were circumscribed between the two branches of the Arollo Creek, named 'Piccolo' (Little) Arollo and 'Grande' (Big) Arollo which were separated

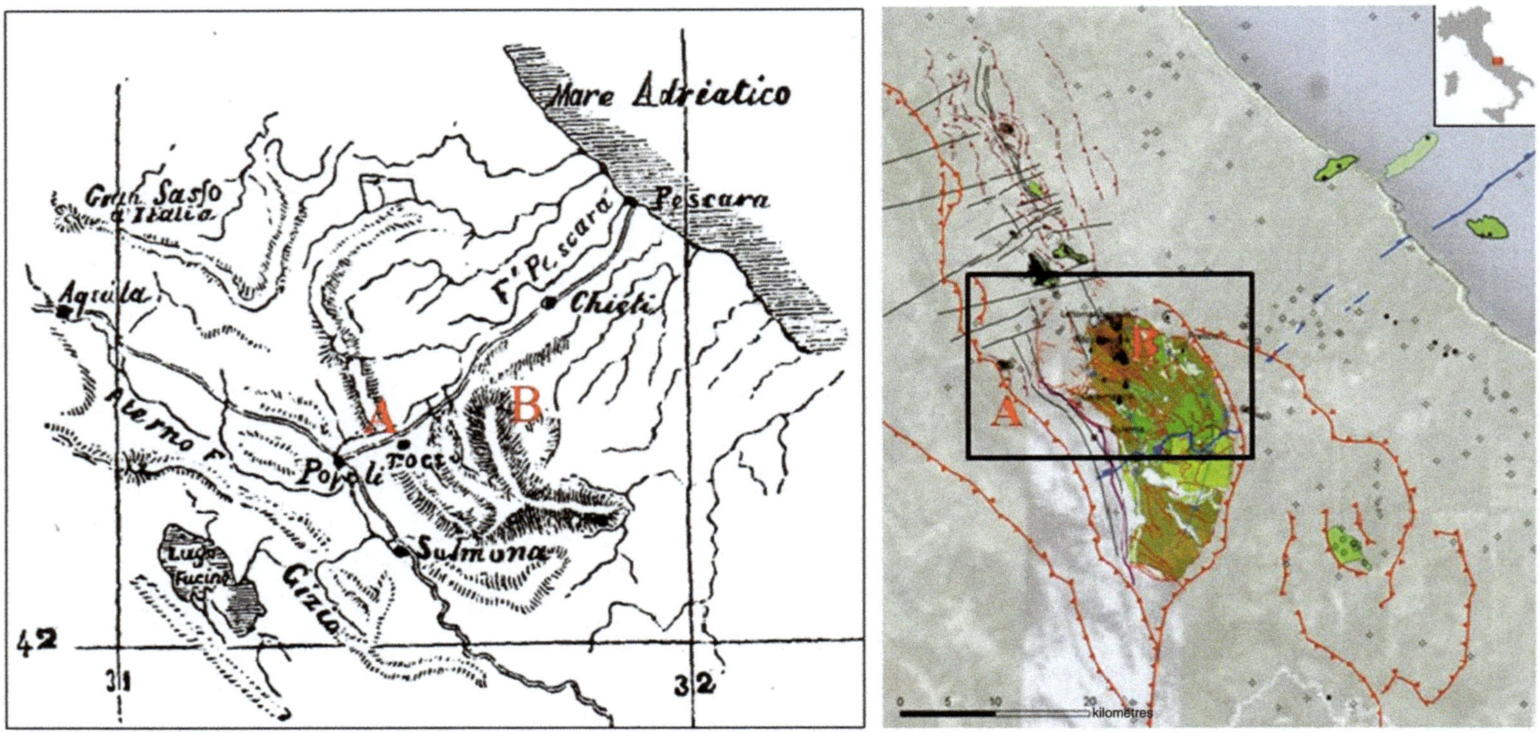

Fig. 4. Yesterday and today: A is the Tocco Casauria 'oil spring' area and B is the Maiella Mountain, where oil seepages occur over extended portions of land.

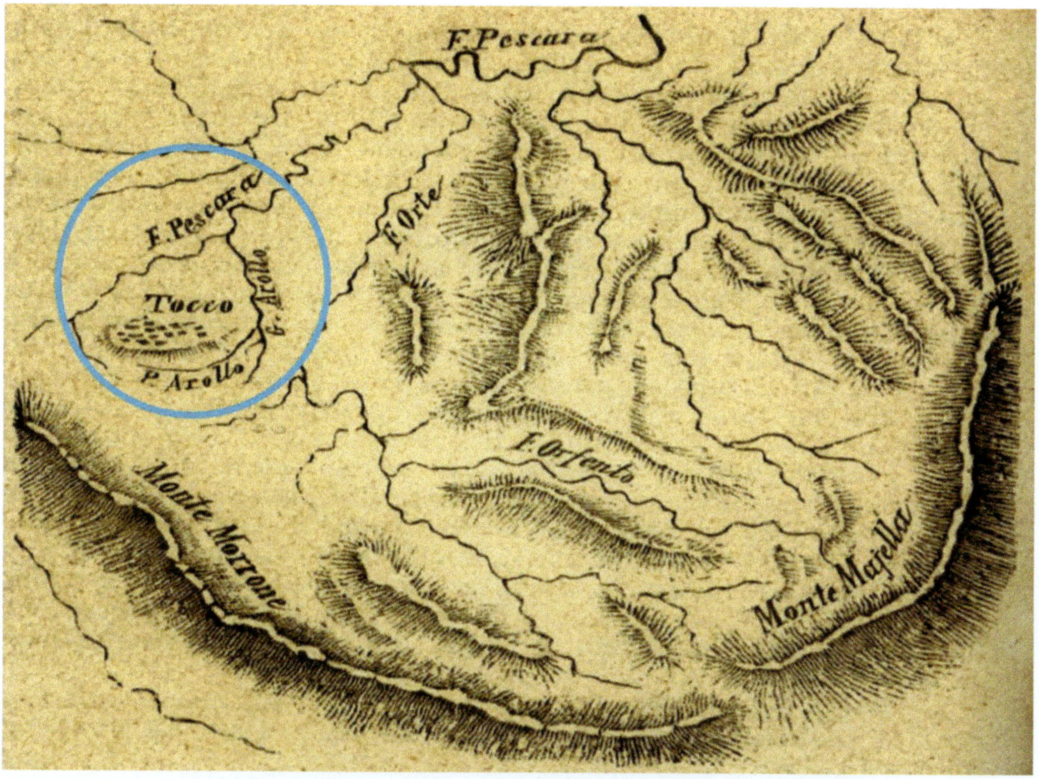

Fig. 5. Tocco Casauria, and the Grande Arollo and the Piccolo Arollo creeks.

by a rocky slope called 'Colle d'oro' ('Golden Hill') (Gerali 2013) (Fig. 5).

Enterpreneurs and geologists in Abruzzo

The inception of mass production in Tocco Casauria commenced in early 1863. The brothers Maurizio and Giuseppe Laschi from Vicenza, NE Italy, together with Siro Trovati and Vittorio Calabi from Milano, founded the Società Laschi e Trovati and acquired the rights of exploitation of lands in the proximity of the Little Arollo – the small outcrop of oil included in the area was known as the 'Spring of the City', because it was owned by the municipal authorities (Fig. 6). The Laschi brothers in Vicenza owned the mining company Società Vicentina Montanistica, as well as a refinery. Since 1839, they had been in the business of gas distillation from lignites and shale rocks, but the poor quality of the mineral produced in some of their mines (e.g. the Monteviale and the Calverina in the proximity of Vicenza) was jeopardizing company interests (Errera 1870; Stoppani 1876). The brothers Laschi travelled almost 600 km to find an alternative fuel to blend and enrich lignites and shales. But, little was done since they arrived in the area of the Little Arollo (Stoppani 1866*a*, *b*).

In September 1864, the Società Vicentina Montanistica hired Abbot Antonio Stoppani (1824–91) to survey the company land. He was a leading scholar in Italy and was also well known in Europe (Fig. 7), and considered by many to be the 'father' of geology in Italy. At this time, he was also working at the University of Pavia and is credited as being the first Italian scholar author of a comprehensive essay on petroleum geology (Stoppani 1864; Zanoni 2014). Stoppani visited the lands of the company for a few weeks and then concluded that the only way to obtain enough petroleum to haul to the refinery in Vicenza and make a profit was to drill using the then 'modern' technologies to a depth of 90 m. He then explained the presence of the bitumen in the creek: he theorized that the waters in the subsoil passed through a 'pocket' of bitumen that was wiped out in different quantities depending on the pressure of the stream. The Società Vicentina Montanistica did not immediately follow the recommendation of the geologist. However, in late 1865, after months of manual digging, at a depth of 32 m, enough oil

Fig. 6. Maurizio Laschi.

emerged to persuade the Società to purchase drilling machinery from the Degousée–Laurent company in France, the leader in Europe in the drilling of artesian waters (Degousée & Laurent 1861, 1868).

The main machine was a drilling-by-percussion system working with hollow rigid rods and a free-fall (sliding) chisel. This machine was similar to the Kind and Fabian systems used to drill artesian wells (the latter was already utilized in precedence in the oil fields of Bobrka and Galicia in the Austro-Hungarian Empire). In principle, and depending on the conditions of operation (e.g. moving the walking beam by means of human labour or steam engine), it involved a greater outlay of capital and staff than manila rope-drilling (better known as the American system, or American standard). In return, it was guaranteed to keep to the vertical of the hole better in soft soils, to maintain the stability of the well walls and to achieve a larger hole, which could range up to 1 m in diameter (Degousée & Laurent 1861, 1868; Neuburger *et al.* 1901). The derrick was set up to drill the first well for the Società by the end of 1865. Unfortunately, the limited set of data available does not allow us to establish with any precision how much oil was produced from those wells (the only definite reference tells of about 14 tons of oil being collected in December 1866) and its influence on the total yield from the Maiella fields (Stoppani 1866*b*).

Fig. 7. Antonio Stoppani (Lecco 1824–Milano 1891).

However, it is known that the production enabled the Società to stay in business until 1870, before having to close (Fig. 8).

In the spring of 1864, the concession for the exploration rights around the Grande Arollo creek was granted to Carlo Ribighini, the owner of the Ancona branch of the Suisse company Blumer & Jenny. Ribighini was an entrepreneur who came from a wealthy family known in Ancona. He was administering a naval trade house and observed the increasing turnover of US-refined oil passing through the docks of Ancona harbour. He was also in touch with the US Consulate in Ancona (where a year later he was also appointed as Vice-Consul), which helped him to realize the magnitude of the oil business. Ribighini himself wrote that the real purpose of his venture in Tocco Casauria was to find crude oil.

The land to which Ribighini had been granted rights had never been of interest to the inhabitants of the village: the bitumen harvested from the stream of the Big Arollo was not dense and, therefore, more

I PETROLII IN ITALIA.

PARTE PRIMA:

IL PETROLIO NEGLI ABRUZZI.

Le industrie native e le industrie importate, 1. — Influenza dell'elemento morale nell'industria, 2. — Speranze per l'Italia, 3. — Un nuovo ramo d'industria mineraria, 4. — Programma dell'autore, 5. — Esistenza de' petrolii in Italia, 6. — Zone petroleifere, 7. — Un viaggio a Tocco, 8. — La valle del Pescara, 9. — Sollevamento dell'Abruzzo citeriore, 10. — Abbondanze de' bitumi, 11. — Sviluppo e singolarità del travertino, 12. — Antichi efflussi di petrolio, 13. — La sorgente petroleifera del commune, 14. — Come se ne conobbe l'esistenza, 15. — Primitivi apparati, 16. — Sgorghi spontanei, 17. — Manovre de' Toccolani, 18. — Deplorabile spreco, 19. — Caratteri del petrolio di Tocco, 20. — Una seconda sorgente petroleifera, 21. — Emanazioni gazose e metamorfismo, 22. — Sorgente degli Anconitani, 23. — Primo pozzo d'assaggio, 24. — Sgorghi invernali, 25. — Una leggenda che conduce ad una scoperta, 26. — Un esperimento fortunato, 27. — Fortuna degli Anconitani, 28. — Spiegazione del fenomeno principale, 29. — Spiegazioni di alcuni accessorii, 30. — Fatto relativo all'ossidazione de' petrolii, 31. — Pronostici, 32. — Migliori apparati intorno alla sorgente del commune, 33. — Il primo traforo, 34. — Comparsa del petrolio, 35. — Traforo artesiano, 36. — Prosperi successi, 37. — Estensione della zona petroleifera, 38. — Indizi nella provincia di Teramo, 39. — Altri nella provincia d'Ascoli, 40. — Avvenire di quelle popolazioni, 41. — Il progresso già installato nel commune di Tocco, 42. —

1. L'agricoltura e l'industria sono le figlie primogenite del suolo, e queste le madri delle grandi società civili. Un popolo stanziato sopra suolo infecondo, sopra una terra nelle cui viscere natura non nascose quelli che gli antichi poeti chiamarono tesori di Pluto, attende da estranie contrade il pane di che nutrirsi e gli strumenti più necessari alla vita: questo popolo vivrà nomade, disperso come le tribù erranti nelle sconfinate lande del Sahara; anzi non avrebbe nemmeno ragione di essere, se esistesse assolutamente una regione arida ed infeconda. È un fatto invece che le ricchezze del suolo crearono, generalmente parlando, i popoli industriali. Gli abitatori dell'Hartz, della Sassonia, della Slesia, divennero popoli metallurgici, perchè ricche di miniere trovarono le loro contrade; e se l'Inghilterra tien lo scettro nell'industria siderurgica, gli è anche vero che non ha che a rimovere le sue zolle per ritrovare annualmente i suoi 10 milioni di tonnelate di minerali di ferro e accanto ad esse i suoi 90 milioni di tonnelate di carbon fossile per ridurli in ghisa (1).

(1) La cifra di 90 milioni di tonnelate di carbon fossile prodotte annualmente dall'Inghilterra, risultò dall'ultima *Esposizione internazionale* del 1862, come si legge nella *Relazione* or ora pubblicata in Torino dal prof. I. Cocchi.

Fig. 8. *Petroli in Italia – Abruzzo* (Stoppani 1866*a*).

difficult to filter and collect from the water with the old method of the descending pools used until then. Ribighini suspected that he was not looking at common bitumen; instead, he believed that he had found the mineral he was looking for to start implementing his ambitious plan. Before proceeding, he decided to invest some money in chemical analyses, and sent some samples to be tested in laboratories in Trieste (NE Italy, at that time it was included in the Austrian Empire) and Marseille in France. The results were positive, and at the end of the summer of 1864 he started to dig manually shallow wells close to the seepages and on the banks of the Grande Arollo creek. Ribighini concluded the year of 1864 by harvesting little more than 3 tons, a pretty miserable quantity of light bitumen filtering the waters of the creek; but the hunt for the 'big' deposit was a failure.

Ribighini heard about the geologist who had been hired on the other side of the Colle d'oro, on Little Arollo. The two companies were pursuing the same business just a few hundred metres from each other, but they were working with 'fair play'. As we have seen, the interests of the companies were different and there was no real conflict of interests; no other large operators were involved in the vicinity and the pressure – or, better, the rush – for oil was low. The only real competitor to contain and overtake was the refined oils (both for home and industrial uses) from the USA. Since 1861, the affordable and good imported oil had gained a greater share of the national fuel market every year, and, subsequently, it was depleting the chances of a successful start-up for the small national firms, such as Blumer & Jenny and the Società Laschi e Trovati.

Ribighini decided in late 1864, on the advice of other engineers, to write to Giovanni Capellini (1833–22), Professor of Geology and Palaeontology at the University of Bologna (Fig. 9), who had previous experience of oil fields. Between 1861 and 1864, Capellini had worked on the palaeontology of the oil regions of Emilia, in northern Italy; he had visited the oil fields of Pennsylvania in the USA; and in 1864 he was the consultant geologist to a British firm in Walacchia, Romania (Gerali & Gregory 2017). Ribighini questioned the nature of the subsoil, and whether it was possible to gauge the size of the deposit and foresee its productivity. Ribighini was

Fig. 9. Giovanni Capellini (1833–1922).

PETROLIO DI TOCCO

E

BITUMI DI LETTO MONOPELLO

RELAZIONE

DEL PROFESSORE

all'Università di Bologna

CAVALIERE G. CAPELLINI

TORINO

TIP. G. FAVALE E COMP.

1866.

Fig. 10. The report of Capellini on Tocco Casauria oil and Lettomanopello asphalts.

looking for exactly the same answer as that sought by today's oil companies: to know how valuable the oil pool could be before investing too much capital. Capellini answered with frank scepticism, maintaining that the samples received were too heavy and were not of competitive oil that could be refined to fuel for lamps, and it was not possible for him to travel to Abruzzo in such a short time (Accademia Lunigianese di Scienze, Personal Archive of Giovanni Capellini, Box XIX, Folder O, papers 1–3; Biblioteca Archiginnasio, Manuscript Section, Fondo 'Giovanni Capellini', Box CXVIII, Folder 4, papers 1–3).

In spite of the discouraging words from the geologist, Ribighini moved forward with his plans and during 1865 he achieved interesting results. He built a small refinery in Porto Recanati, a coastal town only few miles away, which was close to larger towns where the product was sold. The business was doing well, and he invested in new mining concessions in the close areas of Roccamorice and Lettomanopello. Finally, he found a way to reach a few

Fig. 11. 3 March 1866: a section of the rocks crossed during the drilling of wells 1 and 2 by Ribighini using French drilling machinery.

metres down in the ground to the cavern source of the bitumen that was discharged into the creek and, thanks to a system of artificial channels, he managed to increase productivity dramatically. The highest peak of production recorded in Ribighini's concession land was some 800 tons in November 1865. The amount of oil sent to Porto Recanati was exceeding the capacity of the plant, and it had an even greater expectation for the following year. He found a former sugar refinery that had been abandoned since 1859 in the nearby city of Grottammare, in the province of Ascoli Piceno, and purchased it with the plan of setting up a larger refining plant.

In January 1866, Ribighini wrote again to Capellini with the firm intention of convincing him to come in Tocco Casauria to locate finally the real oil deposit, and not just halfway pockets embedded in-between meandering siphons (known also as a Tantalus Cup) and a creek (Gerali 2013).

Capellini accepted the offer and completed the field survey in early 1866 (Capellini 1866), confirming to Ribighini that the main oil deposit could be reached by digging down further to a greater depth, using modern drilling mechanical machinery, right in the proximity of the Colle d'oro and the creek (Fig. 10). He then concluded his task by locating on the map where to drill two exploratory wells and suggested hiring the mine engineer Leopoldo Ferretti to be in charge the operation. Ribighini then ordered French drilling machines, just as his neighbours had the year before. The derrick was set up few months later and the first well of Blumer

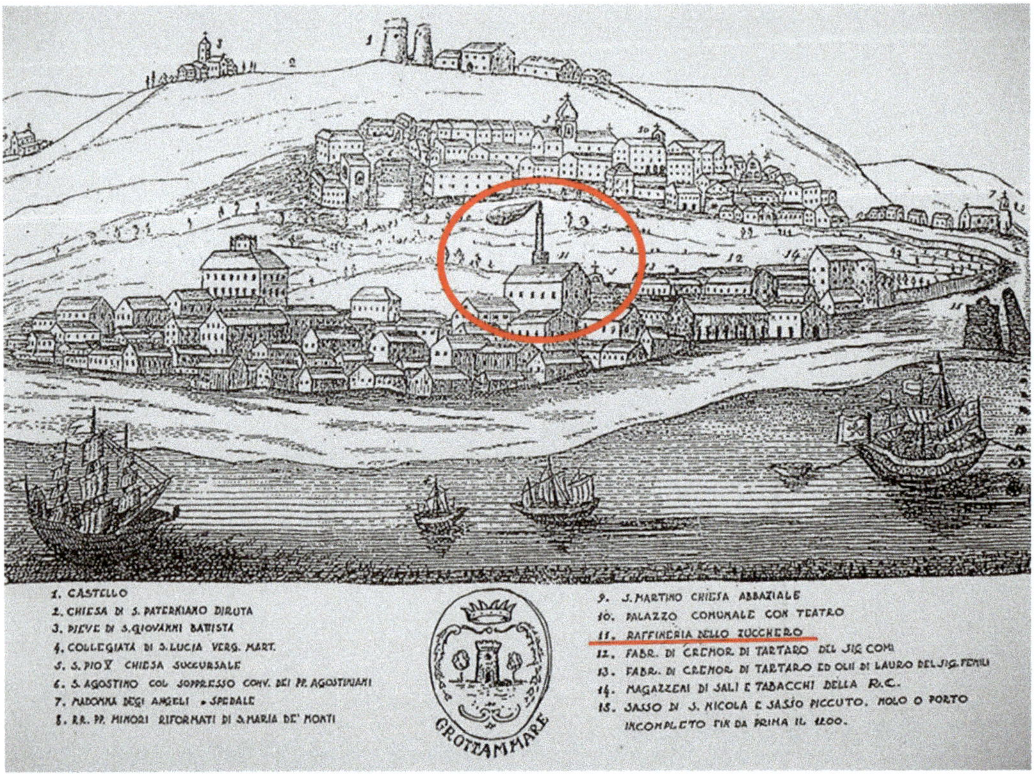

Fig. 12. The refinery at Grottammare, circled in red.

& Jenny reached 60 m, with a yield of up to 500 kg of oil per day. The company ended 1866 with about 300 tons of oil coming from two wells, named the Colle d'oro and De Gasperi (Accademia Lunigianese di Scienze, Personal Archive of Giovanni Capellini, Box XIX, Folder O, papers 4 and 5) (Fig. 11).

The petroleum produced in the Tocco fields was collected in barrels containing 151 l, hauled onto carriages and taken to the refinery in Grottammare (Fig. 12). From 1866, Ribighini decided to utilize the smaller facility in Porto Recanati to refine only the bitumens from the mines in Lettomanopello in the Maiella area, where productivity was quite consistent. The refinery produced a lot of fuel oil and asphalts, and small quantities of lamp oil. The latter was the real target product for Ribighini, who was looking to refine it in larger quantities at the Grottammare plant. In 1867, he successfully introduced a lamp oil branded as 'Toccolina' onto the market; he presented it at the Paris 'Exposition Universelle' – together with the Società Laschi e Trovati – and visited the municipal authorities of numerous cities to propose that his oils be used to fuel the public illumination systems (Biblioteca Archiginnasio, Box CXVIII, Folder 4, papers 1–3).

The same year, Ribighini and his chief engineer Leopoldo Ferretti started to advertise a new mining venture, and to gather capital to establish the Società Anonima Abruzzese dei Minerali della Maiella (SAM) (Anonymous Abruzzese Company for the Minerals of the Maiella). The SAM was the first company in Maiella to be financed by selling shares to the public (Fig. 13).

The SAM was officially constituted by notary on June 1868 in Chieti. Then, in the following July, the government authorities acknowledged and approved the statute. Unfortunately, at the same time, the project of Ribighini and Ferretti was jeopardized by a sudden depletion of production in Tocco, and a halt to the operation in the refinery at Grottammare. The smoke, the sulphur emissions and the acres of fumes generated by the refining process troubled the neighbourhood. The continuous protests by the residents forced the city hall to suspend all activities. The SAM ended 1868 by producing just 30 tons of crude, which was only 10% of the mineral produced just 3 years earlier; it ceased its activities in 1869, and Ribighini continued to run only the asphalt business between Lettomanopello and Porto Recanati.

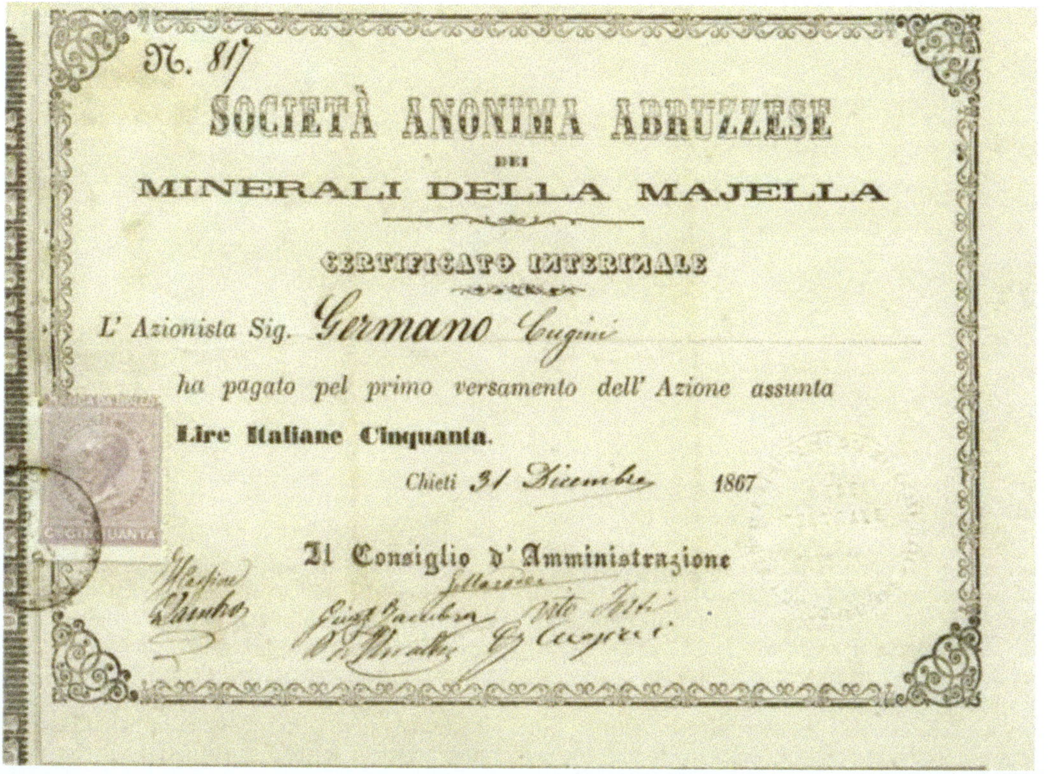

Fig. 13. Stock of the Società Anonima Abruzzese dei Minerali della Maiella, 1867.

The operations of Laschi, Trovati and Calabi (1863–70), and Blumer & Jenny (1864–67) – the latter merged shortly after into the Società Anonima Abruzzere dei Minerali della Maiella (1867–69) – had a central role in the development of oil exploration, production and refining in the area. They are acknowledged to have led the production of oil in Italy at their time, and to have turned a spotlight on Maiella in the European oil panorama. Other operators had commercial interests in the region, like the French firm Asphaltene, founded by Augusto D'Aygesvives, and the Compagnia del Gas della Cittá di Chieti (the Gas Company of the City of Chieti), the public illumination company of the nearby town (Massimi *et al.* 2002; Novelli & Sella 2009). Their primary interest was not to search for oil, but mostly to mine for asphalts and bitumens, and to run a local business. These companies all together gave the momentum to the newborn sector and contributed dramatically to shaping what in the following years was defined as the 'Maiella oil district'.

Fig. 14. Compagnie Italo-Française du Pétrole d'Italie.

The fall and the rising of the Maiella oil district, 1870–1900

In 1870, the oil sector in Maiella could be associated with the metaphor of the phoenix. From the ashes of the businesses of the national small investors, it rose as an international oil district made up of foreign

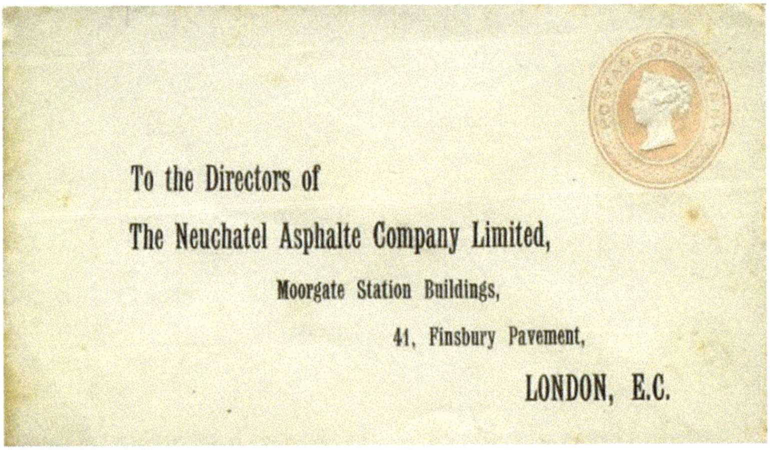

Fig. 15. The Neuchâtel Asphalte Company Ltd.

companies. These new players brought capital and know-how; revitalized old concessions and discovered new mines; invested in mechanization, and the construction of roads and facilities; and made Maiella a point of reference in the geography of the heavy oils in Europe.

In 1873, the early pioneer Silvestro Petrini, no longer a victim of political purges, co-founded with his new British partners the Anglo-Italian Mineral Oils and Bitumen Company, established in London. He opened mines in Scafa, San Valentino and Roccamorice (both close to Lettomanopello) in the Maiella area to extract and refine oil from asphaltic rocks. The Scafa plant was made up of several buildings and halls: two mills, a crushing plant, a foundry, a refinery, a press department, workshops, warehouses and offices. The choice of Scafa was mainly due to the nearby presence of the Pescara–Sulmona railway, then extended towards Rome and Isernia, and the Pescara and Lavino rivers (Petrini 1911; Massimi *et al.* 2002). In 1879, the French Asphaltene company is reported to have still been in business at Lettomanopello. In 1880, a new Italian outsider company Crocco e Giampietro is recorded, but little is known about the extent of its business. The same year, in the local railway hub of Piano d'Orta, close to Bolognano, a refinery with a capacity of ten of tons per day was built by Vittorio Croizat, an entrepreneur from the NW region of Piedmont who was already active in the public illumination business

Fig. 16. Photographs of the *in situ* exploitation carried out at surface in the Maiella oil district.

Fig. 17. (**a**) The 'furnace' used to extract bitumen by heating, from the Acquafredda mine and (**b**) the Fonticelli mine, along the NW flank of the Maiella Mountain. (**c**) Loading of mules for transportation out of the 'S. Spirito mine tunnel' (REH 1911); and (**d**) asphalt hauling using mules (REH 1911).

Fig. 18. (**a**) Cableway system across the valley; (**b**) miners at the Santo Spirito mine entrance; (**c**) railway entry point in the 'Pilone' mine system (1903); and (**d**) 48 m three-arch railbridge built across the Lavinio River.

Fig. 19. The REH plant in the San Valentino, Scafa locality.

(Ministero di Agricoltura, Industria e Commercio 1895). In 1881, Giuseppe Laschi still had some properties in the area. He was co-founder in Paris of the Compagnie Francaise pour l'exploitation des sources petroliferes de l'Italie et de leurs produits. The Compagnie purchased all the assets of the former Anglo-Italian Mineral Oils and Bitumen Company from the British firm Claseen. Apparently, the British associates of Petrini, the chief investors, had arranged to cut Petrini out of the business pretty arbitrarily and changed the name of the company. However, the venture of Claseen had a short life. In the same period, the Compagnie Francaise–Italienne du petrole de Italie also entered the district (Novelli & Sella 2009; Zanoni 2014) (Fig. 14).

The turnover of companies was high, and the quantity of mineral produced varied every year. In 1883, the production of crude oil recorded was 125 tons. Between 1889 and 1893 (in this period the production of crude had fallen to just 12 tons, while the mining of asphalts and bitumen was sky rocketing), two companies arrived in the district that in the years to come would give new momentum to the development of the Maiella oil industry. The first to join was the German company REH & Ci. Società di Asfalti San Valentino, founded in Berlin

Fig. 20. Photographs of first wells drilled in the area in the late 1800s.

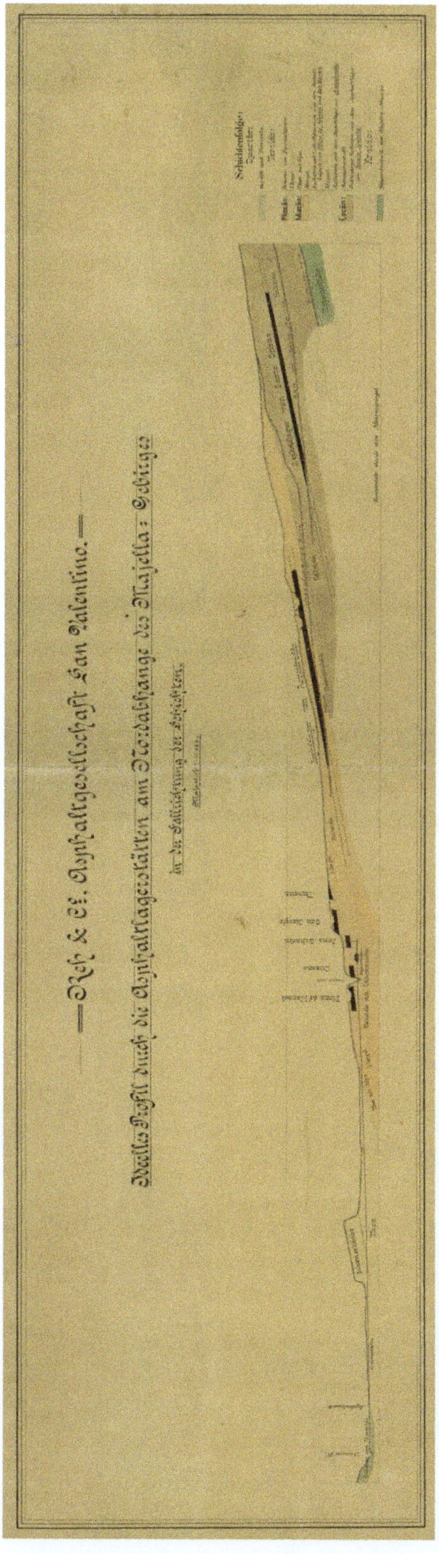

Fig. 21. NW–SE geo-mining original section (ink and watercolour) by the REH company, showing some of the key bitumen-rich intervals exploited at the beginning of the 1900s along the NW flank of the Maiella Mountain.

in 1888 by Adolf Reh; the second was the Neuchâtel Asphalte Company (Giavarini 2011), a British mining company working in the field of asphalt in Europe, established in 1873 (Fig. 15). Both companies established their main facilities in the town of San Valentino, which was close to the railway yard of Scafa. The increasing mining activities required numerous labourers from nearby regions, and Scafa soon became a 'workers village' with 113 inhabitants in 1894 (in 1948 the village would be appointed as an autonomous municipality) (Felice 1990*a*, *b*; De Luca *et al.* 1998). In 1893–94, the Neuchâtel Asphalte Company purchased mining concessions for bitumen mining from firms in difficulty, and started up five new mines in the territories of the municipalities of Abbateggio, Lettomanoppello, Manoppello and Roccamorice (Benegiamo 2012) (Fig. 15). A few years later, the bitumen and asphalt exploitation scenario of Maiella was completed by the French company Società Lyonnaise in 1896, and the German company Valle Romana Asphalte Minendi Lipsia in 1908.

The turn of the new century and the process of industrial transformation across the 'Maiella oil district'

Initially, bitumen and asphalt production in the Maiella district was obtained through *in situ* exploitation activities carried out at surface, from the hydrocarbon-rich reservoirs cropping out along the flank of the mountain (Fig. 16). The bitumen was mainly used for illumination purposes and impermeabilization, while asphalts were used mainly to produce bricks. The workers (including women and children) used to mine and fragment the bitumen-rich rock so that it could be heated in order to extract the bitumen (Fig. 17a, b); transportation was frequently carried out using women and mules (Fig. 17c, d).

With time, a more complex and articulated mining system was created (Colecchia & Agostini 2014), which delivered employment to hundreds of workers in the Maiella area, between the villages of Lettomanoppello, Roccamorice and San Valentino: the exploitation system was extended further, with tunnels, pits and several mines being excavated. Also, as a consequence of all these efforts, by the early 1900s the Maiella area had become the second largest oil district in Italy and an important oil province for future investment in exploration activities. It was defined as the 'Polo minerario della Maiella' (the Maiella oil district), with several production sites over an area of more than 20 km, reported to be active for asphalt and bitumen production; this will eventually deliver an annual production of

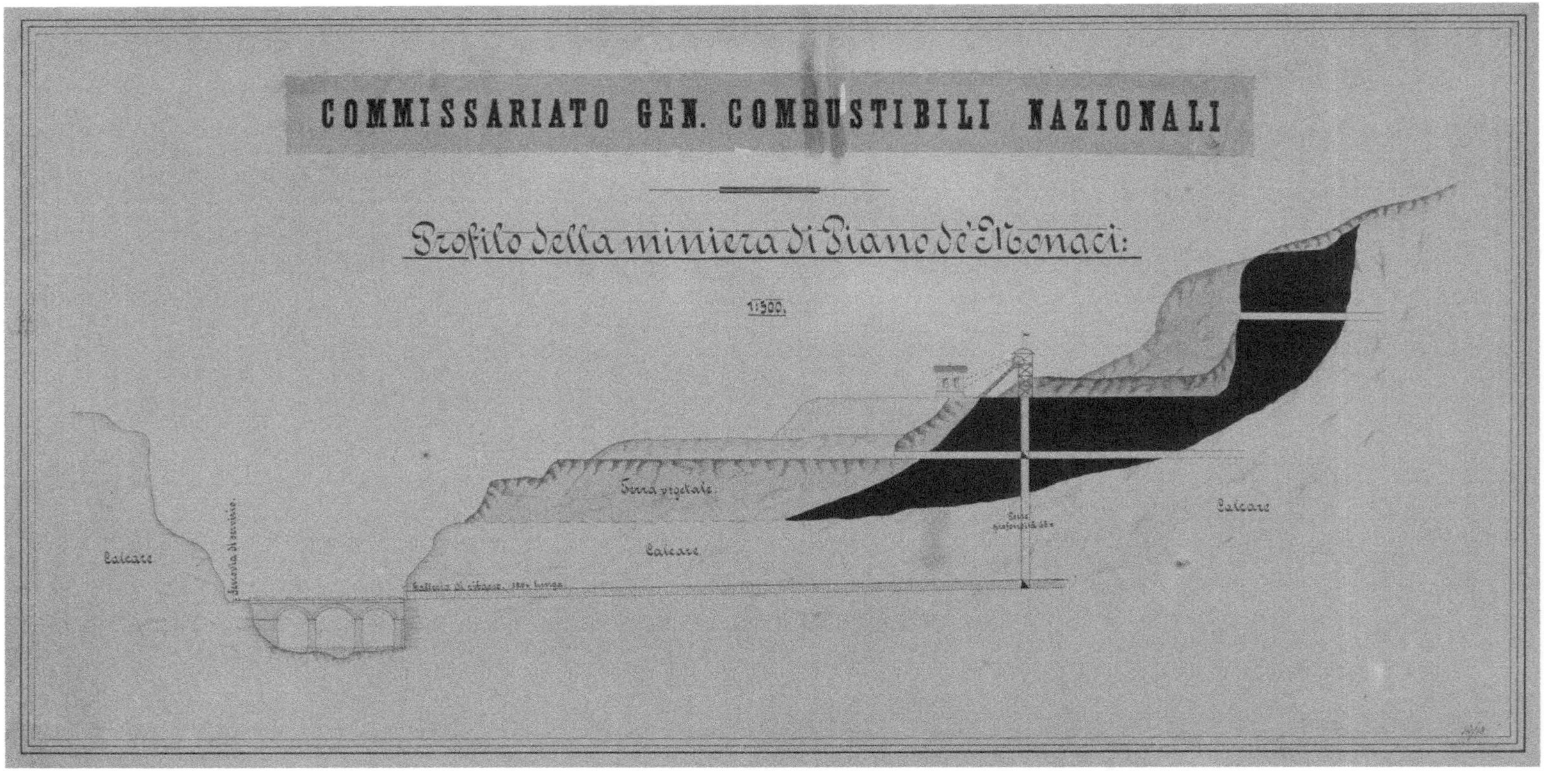

Fig. 22. North–south geo-mining section (ink and watercolour) originally by the REH company showing the gradual 'Italianization' of the mine labels. The section shows the mining system active at the 'Piano de Monaci' mine. The bitumen-rich rocks were displaced along the pit and taken out through carts, across the 'Lavinio' River to the left of the drawing. The same bridge is shown in Figure 18d (courtesy of S.A.M.A. – Italcementi).

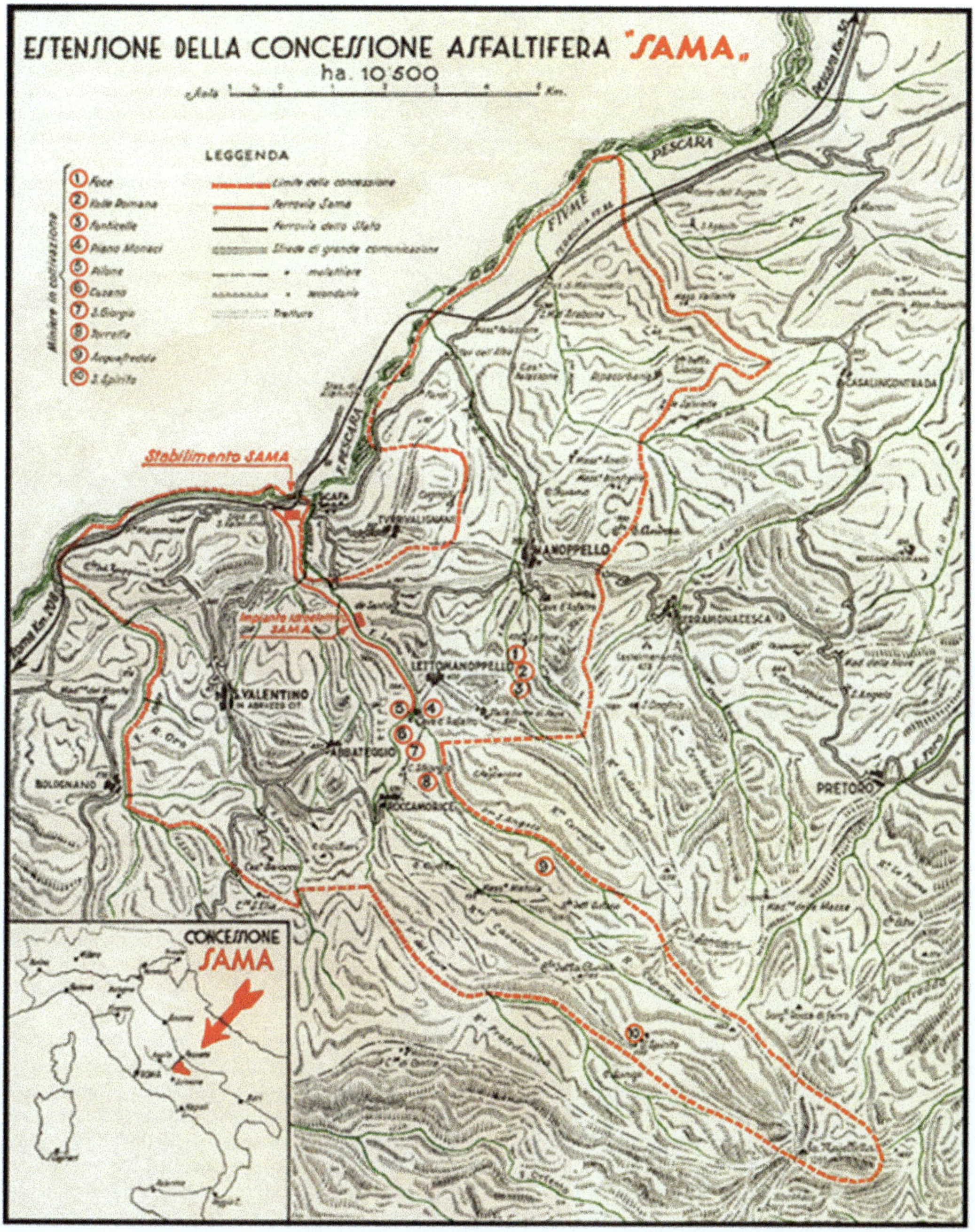

Fig. 23. Map of the S.A.M.A. concession for asphalt and bitumen exploitation (1919–29) (courtesy of the Eni Archives).

more than 20 000 tons of bitumen that would be exported all around the world, covering about 40% of the Italian national production (Ministero di Agricoltura, Industria, e Commercio 1908).

The overall effort was aimed at industrially produced bitumen (directly *in situ*) and asphalt-rich rocks, which, once treated in the factory, could produce asphalt powders, asphalt 'bricks' and refined

Fig. 24. Well Abbateggio-1 drilling derrick (1927).

bitumen (REH 1911), used as products for roads, pavements and the local cement industry.

These products were then exported either by railway or by ship, from the Ancona and Ortona harbours, mainly to Germany, but also to Egypt, Austria, Russia, South Africa, Holland, Denmark and the Americas (REH 1911). The German REH & Ci. and British Neuchâtel Asphalte companies were the key actors in the industrial transformation of the exploitation activity over these years. In 1903, a 6.5 km-long railway was built by REH to collect extracted material from the widespread system of cableways and mine carts that had been constructed (Fig. 18a, b). The material was then directed by rail (Fig. 18c, d) to the Scafa plant facility (Fig. 19), which was expanded by REH as the main logistical centre of collection and transformation of the extracted material, being well positioned close to the main roads, the railway national system and the Pescara River.

As seen, since early times the history of petroleum-related activities in the Maiella oil district was correlated with the ongoing parallel exploration activity in the Tocco da Casauria area, which was continuing with additional efforts to discover hydrocarbons in the subsurface. Unfortunately, only rare and fragmented documentation exists about the exploration and drilling efforts of that time, with only some wells being documented, such as the Arolla, Marano, Casa del Sergente, San Valentino, Piano d'Orta, Torre dei Passeri, Tre Croci, Caracciolo, Cavone and De Gasperis wells (Coppetti 1954): of these, a rare photographic record exists of the Marano, Caracciolo and Cavone wells (Fig. 20), thanks to E. Paparella, a very active businessman in these areas (Paparella 1928; Coppetti 1954; De Luca *et al.* 1998).

The economic interest over the bitumen-rich reservoirs also required the interest and attention of the scientific community, in particular of geologists who contributed to the understanding of the phenomena, evaluating the type, extension and quantity of hydrocarbon resources associated with the reservoirs (e.g. Giattini 1907). As a good example of the geological view over the extension of these bitumen-rich reservoirs, a geo-section from an original ink/watercolour drawing by REH is shown in Figure 21.

This interest, from both an economic and scientific point of view, is also demonstrated by the fact that, in 1913, a field seminar was organized by the Italian Geological Society over the area, associated with its annual congress held in the city of L'Aquila in September of that year (Verri 1913).

Fig. 25. Example of fascist regime propaganda (D'Ascanio 2013).

1917–18: the nationalization of the Abruzzo mining activities

On 21 May in 1918, following the declaration of war with Germany, the properties of German companies such as the REH were confiscated by the Italian government. This is reported in the mining activity official report of 1918 (Ministero dell'Agricoltura 1920), and well documented from the geo-mining section in Figure 22: in this drawing, the same REH style as seen in Figure 21 can be easily recognized, but with the names now all in Italian. This substitution and renaming action is particularly evident from the main label 'Commissariato Gen. Combustibili Nazionali', which was the authority that was entitled to the confiscated activities. As a consequence, in 1918, after many years in which the bitumen-extraction activity was dominated by foreign companies in this area, a new Italian company called S.A.M.A. (Società Abruzzese Miniere Asfaltiche) was created, to which most of the confiscated concession was then assigned (Giavarini 2011).

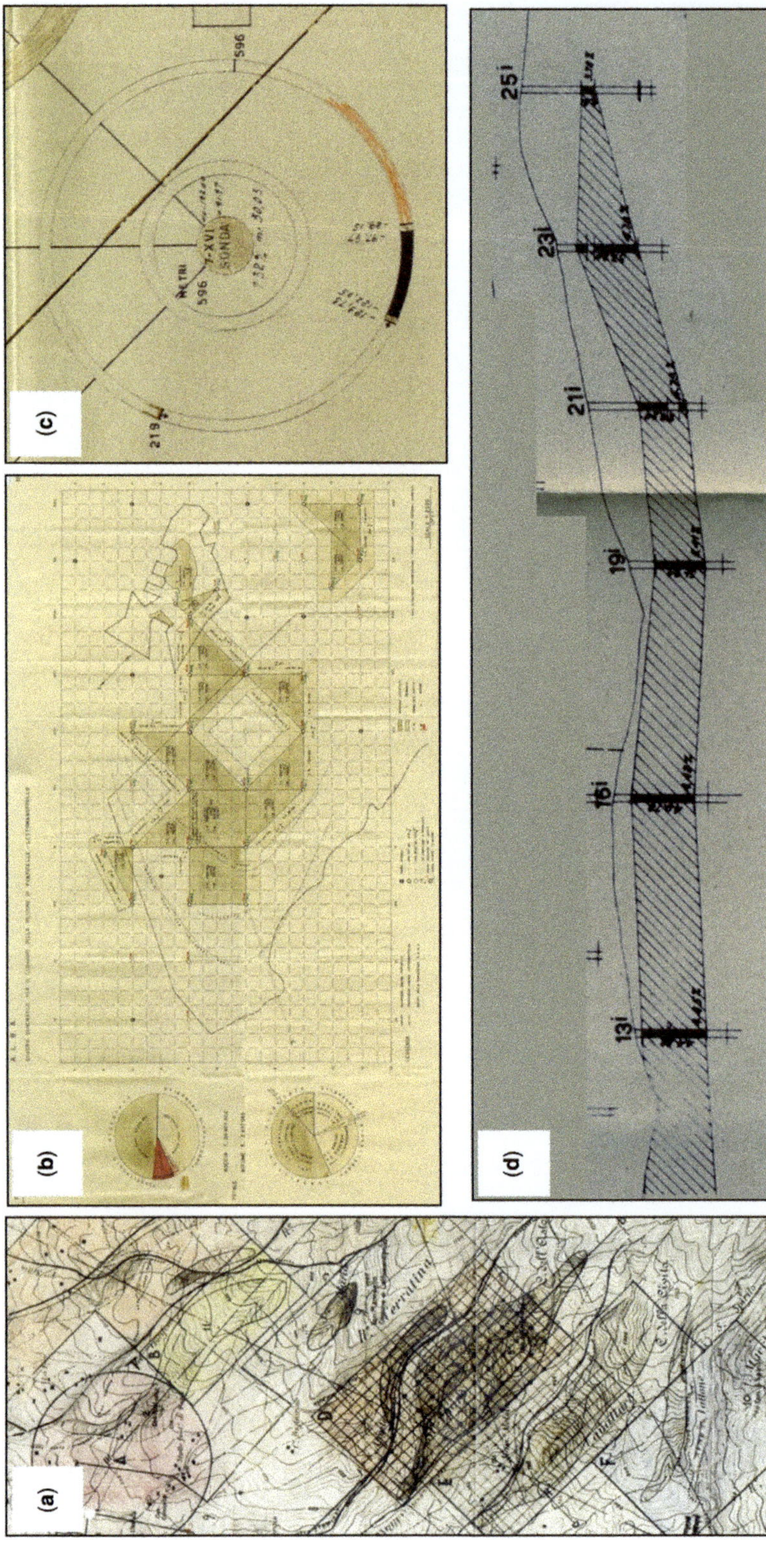

Fig. 26. Example of the A.L.B.A. documentation recovered from the Eni Historical Archive. (**a**) General map of the planned drilling activity over the some key areas along the NW flank of the Maiella Mountain. (**b**) Example of a summary map of the results of the drilling campaign aimed to evaluate bitumen presence and degree. (**c**) Example of a circular diagram used to summarize well information (the hydrocarbon-rich section drilled by this specific well is reported in black). (**d**) Example of a well correlation section showing the hydrocarbon-rich main interval across the Acqufredda area.

The S.A.M.A., which is still operating today, was particularly active in between the two world wars, when it was managing a good part of the existing complex mining and exploitation system (Fig. 23). By 1929, S.A.M.A. was reported to be employing about 950 workers, with the headquarters always in Scafa, at the main plant.

During the fascist period, Italy became in many regards a modern country, also from an oil and gas perspective, and in 1926 the Agip (Azienda Generale Italiana Petroli) company (today ENI S.p.A.) was founded. As a result, a new modern period of exploration and drilling was initiated, in particular with the drilling by Agip and other operators of the Abbateggio-1 (in 1927, reaching a total depth (TD) of 1168 m: ENI S.p.A. Associazione Pionieri e Veterani: http://www.pionierieni.it; (Fig. 24), Lavino-1 (in 1934, TD = 803 m) and Alanno-1 (in 1937, TD = 1185 m) wells, and the drilling of 25 wells in the Tocco Casauria field area between 1933 and 1943. Of these, 14 resulted in producing wells, while 11 others were considered to be uneconomic (Coppetti 1954).

1941–43: the need for national bitumen resources

Approaching 1941, to meet the needs from the industry close to and during the World War II period, a new important effort was dedicated to the exploitation of the bitumen and asphalt resources in the outcrops of the Maiella, and production was increased up to 1000 tons/year.

In 1941, in particular, a new company called A.L.B.A (Azienda Lavorazione Asfalti Bitumi) was created by the fascist regime (Lionel & Bertrand 1941; D'Ascanio 2013) to assess and evaluate bitumen/asphalt resources from the existing oil-rich outcrops known along the Italian Peninsula, mainly from the Sicily, Basilicata, Emilia Romagna and Abruzzi regions, this last one over the 'Bacino minerario asfaltifero abruzzese' ('Abruzzo asphalt-rich petroleum province').

The creation of A.L.B.A. (which also means 'sunrise' in Italian) has to be framed against the national war effort of the country to be as independent as possible from oil imports, and at the same time affirm self-sufficiency (autarky), which was also a top political value of the fascist regime (Fig. 25).

A.L.B.A., which was controlled by the Agip Italian oil company (nowadays ENI) and I.R.I., was assigned to assess the total volume of bitumen/asphalts that could be extracted from the NW flank of the Maiella: this resulted in an extensive geological and drilling campaign carried out between July 1941 and December 1942, with the drilling of about 200 'shallow' vertical wells (drilled using a rotary drilling approach), regularly spaced over grids of 100 × 100 m/200 × 200 m, and with total depths in the range of 80–250 m below surface (A.L.B.A. 1942, 1943). In our research, we were able to find an extensive database of reports, memos, letters and technical drawing from these activities, care of the Eni S.p.A. Historical Archive, some of which are shown in Figure 26.

The whole effort abruptly ended in early 1943, with the arrival of the war front and the retreating defence line from the Nazi army, and the whole site was abandoned for many years thereafter. Even if this attempt was unable to succeed, it did, indeed, represent the first effort to create a national and efficient approach to the development of a key resource, such as bitumen and asphalts (D'Ascanio 2013).

Conclusions

In the 1860s, petroleum in Abruzzo was synonymous with modernity. Many Romanian and Austro-Hungarian fields were a more advanced model of productivity, and this is undisputable. However, in the national contest, the early entrepreneurial protagonists of the 1865–68 boom were forerunners. Blumer & Jenny (later the SAM) can be considered to be the first Italian company operating with an 'integrated model', which included exploration, drilling, production, transportation, refining and retailing. This was a very special and uncommon situation to find, and not many companies at that time in Europe were able to set up and sustain such a complex structure. On the one hand, it was a very effective model: the liquid assets circulated into the company from one stage to another, without substantial external expenditure to third parties. On the other hand, it was a frail equilibrium: every problem in one of the 'ring' components of the productive chain could jeopardize the entire system, and generate losses even in the working sectors. Ribighini first experienced problems with the refinery and then with the yield of the well: the lack of availability of capital worsened the situation, and within about 1 year the company imploded.

A conclusive reflection goes to a historiographical issue that is still not clear in the history of the oil industry in Italy. Was the well dug in December 1865 by the Società Laschi e Trovati the first modern productive well drilled in Italy by means of mechanical equipment and powered by a steam engine? We can state with certainty that it was the first modern well drilled in the Maiella district and, most probably, it was the first – or one of the earliest – drilled by mean of a percussion system

Fig. 27. Pietra Prima – Opere da ammirare: 'Maiella madre di cultura' (Craftwork to admire: 'Maiella, mother of culture') by the sculptor Ettore Altieri.

in Italy (Stoppani 1866*b*; Ribighini & Ferretti 1867; Zanoni 2014).

Until in the late 1860s, the southern oil regions in Sicily and lower Lazio were producing primarily heavy oils, tar and asphalts, and the first tracks of modernization in those areas are documented in later decades. The land on the western slope of the Northern Apennines, included in the provinces of Parma, Piacenza and Pavia, had been known for their oil for centuries, but the documents found so far describe an antiquated industry poor in technology and scarce in production until the 1870s. However, Maiella in the 1860s represented the vanguard of the Italian oil industry in terms of organization, technologies and entrepreneurial organization (Jervis 1873; Magini 1977; Novelli & Sella 2009). This set of conditions was reflected during the period 1865–68, when national oil production was driven by the oil coming from Maiella. These conditions, we think, support our interpretation of the protagonists and the events that occurred in December 1865 close to the Piccolo Arollo Creek.

In Italy, several histories have been written, with which we do not necessarily agree, about the 'first modern oil well of Italy'. Since the 1950s, there have been quite a number of reports on numerous 'first discoveries' accomplished by different people in various places and over several years; absolute statements about 'game-changing' groundbreaking achievements; and improbable numbers of barrels produced. The history of the oil industry is a relatively new field of research. The early oil industry in Italy (1860–70) is still not well documented, and there is much work to be done by historians in the many small, local and badly preserved municipal archives. We are confident that other historians in the future will find further evidence and confirm our thesis (Squarzina 1958).

Maiella has been the most long-running modern oil district in Italy, and the only one that produced considerable quantities of crude, bitumen and asphalts together (Fig. 27). For almost 100 years, oil shaped the classes of skilled workers; created some of the wealth that distinguished eastern Abruzzo from other less fortunate regions of Italy; and contributed to the development of the industrial and civil infrastructure in the Abruzzo territory. The evidence still visible at surface – abandoned mines, pits, quarries and facility ruins currently studied by industrial archaeologists – are the witnesses to the cultural richness and historical relevance of the oil industry in the area, from both a geological and technological point of view. Finally, it would be desirable in the future to cultivate renewed interest and encourage a new

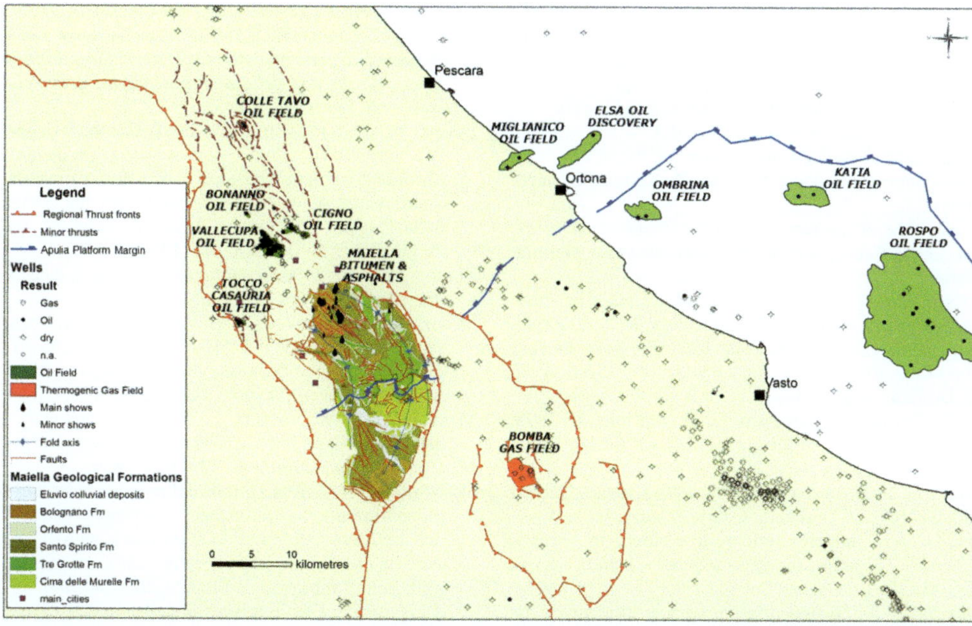

Fig. 28. Regional hydrocarbon occurrence map onshore Abruzzo and offshore the Adriatic Sea (central Italy). The surface geology and main tectonic elements are reported. The whole Maiella structure is represented (in colour), as well as the main Apennine-related compressive thrusts (in red) and the main oil fields discovered in the area (both onshore and offshore).

campaign of exploration in this region, where many opportunities remain to be discovered in the subsurface and others can be better investigated (Fig. 28).

Funding This research was partially funded by Fondi di Dottorato 2016–18, University La Sapienza of Rome.

References

A.L.B.A. 1942. *Campagna geologica A.L.B.A. – Scala 1:25 000, versante Nord-Ovest della Maiella – Quadro indice*. Azienda Lavorazione Asfalti Bitumi (A.L.B.A.), 10 August 1942. Eni S.p.A. Historical Archives, Pomezia, Rome.

A.L.B.A. 1943. *Relazione sulla Campagna dei Sondaggi e Cubaggio accertato e probabile del giacimento asfaltico-bituminoso del versante settentrionale della Maiella*. Campagna Geologica Azienda Lavorazione Asfalti Bitumi (A.L.B.A.), July 1943. Eni S.p.A. Historical Archives, Pomezia, Rome.

Alberti, F.L. 1551. *Descrittione di tutta l'Italia di F. Leandro Alberti bolognese: nella quale si contiene il sito di essa, l'origine, et le Signorie delle Città et più gli huomini famosi che l'hanno illustrata*. Pietro de i Nicolini da Sabbio, Venezia, Italy.

Anon 866–1182. *Liber instrumentorum seu chronicorum Monasterii casauriensis. 'Chronicon Casauriense' or 'Liber instrumentorum seu chronicorum monasterii Casauriensis'*. Compiled by the monk Johannes Berardi, Abbazia di San Clemente a Casauria, Abruzzo, Italy.

Anon 1845. *Lavori delle Società economiche delle province continentali*. Annali civili del Regno delle Due Sicilie, Napoli, 37–39.

Bardini, C. 1998. *Senza carbone nell'età del vapore. Gli inizi dell'industrializzazione italiana*. Mondadori, Milano, Italy.

Benegiamo, M. 2012. L'oro della Maiella: industria mineraria e quadro ambientale nell'Appennino abruzzese dall'Unità all'età giolittiana. *Proposte e Ricerche*, **69**, 187–204.

Biondo, F. 1482. *Blondi Flauii Forliuiensis in Italiam illustratam praefatio*. Impressum in Inclyta ciuitate Veronae, Bonino Bonini.

Brandano, M., Scrocca, D. *et al.* 2013. Physical stratigraphy and tectonic settings of Bolognano Fm (Maiella): a potential carbonate reservoir. *Journal of Mediterranean Earth Sciences*, **5**, (Special Issue), 151–176.

Capellini, G. 1866. *Petrolio di Tocco e Bitumi di Letto Manopello*. Regia Accademia delle Scienze di Torino, **3**, Series 1.

Cazzini, F., Dal Zotto, O., Fantoni, R., Ghielmi, M., Ronchi, P. & Scotti, P. 2015. Oil and gas in the Adriatic Foreland, Italy. *Journal of Petroleum Geology*, **38**, 255–279.

Colecchia, A. & Agostini, S. 2014. Economie marginali e paesaggi storici nella Maiella settentrionale (Abruzzo, Italia). *European Journal of Post Classical Archaeologies*, **4**, 219–258.

Coppetti, P. 1954. *I Pozzi dell Abruzzo Settentrionale*. Azienda Generale Italiana Petroli (Agip), Rome.

Crescenti, U., Crostella, A., Donzelli, G. & Raffi, G. 1969. Stratigrafia della serie calcarea dal Lias al Miocene nella regione marchigiano abruzzese: Parte II. Litostratigrafia, biostratigrafia, paleogeografia. *Memorie della Società Geologica Italiana*, **9**, 343–420.

D'Ascanio, A. 2013. *Storia dell'A.L.B.A. – Un tentativo autarchico di politica petrolifera nell'Italia dei primi anni Quaranta*. Collana 'Faretra', **28**. Solfanelli, Chieti, Italy.

Degousée, J.M. & Laurent, C.A. 1861. *Guide du sondeur; ou, Traité théorique et pratique des sondages*. Garnier, Paris.

Degousée, J.M. & Laurent, C.A. 1868. Detailles des travaux recents de sondages. *In*: Chevalier, M. (ed.) *Exploration Universelle de Paris de 1867, Rapports du Jury International, Groupe VI, Classe 47, Section II*. P. Doupont, Paris, 13–31.

De Luca, G., Di Giandomenico, G. & Martino, N. 1998. *SCAFA 50 ANNI – Storia e Tradizioni*. CEIO, Scafa, Italy.

Donzelli, G. 1968. *Studio geologico della Maiella*. Università Gabriele D'Annunzio, Chieti-Pescara, Italy.

Durini, G. 1837. Industria sullo zolfo e bitume in Abruzzo Citeriore. *Giornale abruzzese di Scienze Lettere ed Arti*, **IV**, 13–21.

Errera, A. 1870. *Documenti alla storia e statistica delle industrie venete e accenni al loro avvenire*. Stabilimento Privilegiato Giuseppe Antonelli, Venezia, Italy.

Felice, C. 1990*a*. Dal sonno di Aligi al grande capitale. La prima industrializzazione della Val Pescara. *Italia contemporanea*, **180**, 505–526.

Felice, C. 1990*b*. Protoindustria o pluriattività rurale? La deindustrializzazione nel secondo Ottocento in Abruzzo e Molise. *Trimestre*, **23**, 273–309.

Ferraris, M.R. 1942. *Lettera dell'Ing. Mario Rosazza Ferraris al Direttore Generale Dott. Ing. Capparucci, sulla Campagna Geologica in Maiella*, 5 October 1942. Eni S.p.A. Historical Archives, Pomezia, Rome.

Gerali, F. 2013. Imprenditoria e scienza nell'industria petrolifera abruzzese del XIX secolo. *Energia*, **2**, 50–55.

Gerali, F. & Gregory, J. 2017. Understanding and finding oil over the centuries: the case of the Wallachian Petroleum Company in Romania. *Earth Science History*, **36**, 41–62.

Ghisetti, F. & Vezzani, L. 2002. Normal faulting, extension and uplift in the outer thrust belt of the central Apennines (Italy): Role of the Caramanico fault. *Basin Research*, **14**, 225–236.

Giattini, G.B. 1907. *Osservazioni geologiche sopra i terreni terziari di San Valentino (Chieti) e sopra i loro giacimenti di bitume*. Giornale di Geologia Pratica (Perugia), **5**, 5–6.

Giavarini, C. 2011. *Civiltà d'asfalto*. Mondatori and 'Sapienza Università di Roma', Rome.

Huguenet, I. & Frojo, G. 1856. *Asfalto. Considerazioni generali sull'origine e formazione degli asfalti e del loro uso come cemento naturale applicati ai lavori di utilità pubblica e privata arricchita di note ed accresciute di un appendice sugli asfalti del Regno di Napoli*. Priggiobba Vincenzo, Napoli, Italy.

Jervis, G. 1873. *I tesori sotterranei dell' Italia: Descrizione topografica e geologica di tutte le località nel Regno d'Italia in cui rinvengonsi minerali, oridnata secondo i bacini idrografici del paese, arricchita di analisi di minerali impiegati nelle arti e nelle industrie, di considerazioni economiche, studi geologici e numerose note su tutti gli argomenti collaterali: Repertorio d'informazioni utili ad uso delle amministrazioni provinciali e comunali*. Loescher, Roma.

Lionel, M. & Bertrand, D.M. 1941. Gauthier Legge 29 Dicembre 1941: 'Norme per favorire la produzione di oli inerali greggi nazionali'. Schema Convenzione Stato A.L.B.A. (S.A.B.A.)', approvato dal Senato e dalla Camera dei Fasci e delle Corporazioni. Legge 29 Dicembre 1941 – XX, No. 1479.

Magini, M. 1977. *L'Italia e il petrolio tra storia e cronologia*. Mondatori, Vicenza, Italy.

Massimi, G., Mazzara, P. & Massimi, T. 2002. *L'Abruzzo, Tocco da Casauria e il Bel Paese: Una Lettura Geografica di Testo e Contesto dall'Esplorazione Naturalistica alla Globalizzazione*. Editrice Itinerari, Lanciano.

Mattavelli, L. & Novelli, L. 1990. Geochemistry and habitat of oils in Italy. *American Association of Petroleum Geologists Bulletin*, **74**, 1623–1639.

Mazzella, S. 1586. Descrittione del regno di Napoli di Scipione Mazzella ... Nella quale s'ha piena contezza così del sito d'esso, de'nomi delle prouincie antichi e moderni, de'costumi de'popoli, delle qualità de'paesi, e de gli huomini famosi, che l'hanno illustrato. *In*: *Napoli: Ad istanza di Gio.* Battista Cappelli.

Ministero di Agricoltura, Industria e Commercio 1895. *Notizie sulle condizioni industriali delle Province di Aquila, Chieti e Teramo*. Tipografia Nazionale Bertero, Rome.

Ministero di Agricoltura, Industria, e Commercio, Direzione generale dell'agricoltura 1908. *Rivista del Servizio minerario*. Pubblicazioni del Corpo reale delle miniere.

Ministero dell'Agricoltura, Ispettorato Centrale Tecnico delle Miniere 1920. *Rivista del servizio minerario nel 1918*. Ministero dell'Agricoltura, Rome.

Neuburger, H., Macintosh, J.G. & Henri, N. 1901. *Technology of Petroleum: the Oil Fields of the world. Their History, Geography and Geology Annual Production Prospection and Development; Oil-Well Drilling, Explosives and the use of the 'Torpedo', Transport of Petroleum by Sea and Land, the Storage of Petroleum Technical Data-Tables, Formulae, Problems, etc.* Scott, Greenwood & Co., London.

Novelli, L. & Sella, M. 2009. *Il petrolio: una storia antica*. Silvana Editoriale, Cinisello Balsamo.

Paparella, E. 1928. *Memoriale per i petroli. Alcuni gravi ingiustizie ai danni della Regione Abruzzese*. Rome.

Petrini, N. 1911. *Memorie politiche di un vecchio patriotta – con prefazione del prof. Giuseppe Mezzanotte*. Tipografia Moderna, Vincenzo Marcantonio, Manoppello, Italy.

Pilla, L. 1841. Lettere geologiche sul M. Maiella negli Abruzzi dirette a L. Pilla da A. C. De Meis. *Lucifero*, **28**, 21–34.

REH & Ci, Società di Asfalti San Valentino 1911. *Miniere Impianti Prodotti della Ditta REH & Ci, Società di Asfalti San Valentino*. REH Company report.

Ribighini, C. & Ferretti, L. 1867. *Memoria diretta al comitato promotore di una societa' anonima abruzzese per l'esercizio delle miniere della Maiella*. Company report.

Scanello, C. 1568. *Cronica o ver compendio dell'antiqua regione et provintia del Abruzzo, nella quale si contiene la origine de detti abruzzesi con la edificatione de tutte le citta e castelle che in essa si ritrovano. Con i nomi antichi, et moderni. Raccolta per Christoforo cieco da Forli.* Per Agostino Colaldi, Viterbo, Italy.

Scrocca, D., Brandano, M., Petracchini, L. & Lipparini, L. 2013. Analysis of an exhumed oil accumulation: the Oligo-Miocene carbonate ramp deposits of the Maiella Mountain (Central Italy). *AAPG-ER Newsletter*, **8**, 2–4.

Squarzina, F. 1958. *Le ricerche del petrolio in Italia, Cenni storici dal 1860 e cronache dell'ultimo decennio.* Jandi Sapi, Rome.

Stoppani, A. 1864. Saggio di una storia naturale dei petrolii. Il Politecnico. *Repertorio mensile di studj applicati alla prosperità e alla coltura sociale*, **23**, 5–94.

Stoppani, A. 1866*a*. I petroli in Italia – First part (I). *Il Politecnico, repertorio di studj letterari, scientifici e tecnici parte letterario-scientifica*, Series 4, 1 (2), 77–93.

Stoppani, A. 1866*b*. I petroli in Italia – First part (II). *Il Politecnico, repertorio di studj letterari, scientifici e tecnici parte letterario-scientifica*, Series 4, 1 (2), 219–230.

Stoppani, A. 1876. *Il bel Paese. Conversazioni sulle bellezze naturali, la geologia e la geografia fisica d'Italia.* Agnelli, Milano, Italy.

Tenore, M. 1828. *Sopra Quattro sostanze fossili della Majella. Lettera di M. Tenore al direttore del Giornale Abruzzese.* Catalogus plantarum Horti regii neapolitani ad annum.

Toppi, E. 2017. *Lettomanoppello nelle fonti storiche.* Lulu, Morrisille, NC, USA.

Trapani, F. 2003. *L'Oleum Petronicum, un antico rimedio galenico delle falde settentrionali della Maiella. Annuario del Club Alpino Italiano (CAI), Sezione di Pescara.* Club Alpino Italiano (CAI), Pescara, Italy.

Verlengia, F. 1956. Bitumi e Petrolio nella Valle della Pescara attraverso i secoli. *Attraverso l'Abruzzo*, **2**, 1.

Verri, A. 1913. Programma e adunanze del Congresso della Congresso della Società Geologica Italiana, Aquila, Settembre 1913. *Bollettino della Società Geologica*, **32**, XIL–LXXXII.

Vezzani, L. & Ghisetti, F. 1998. *Carta geologica dell'Abruzzo 1:100 000.* S.EL.CA., Firenze, Italy.

VIDEPI Project 2012. *Visibility of Petroleum Exploration Data in Italy.* Ministero dello sviluppo economico DGRME–Società Geologica Italiana–Assomineraria, http://unmig.sviluppoeconomico.gov.it/videpi/videpi.asp

Zambra, G. 1869. *Un antico pezzo di asfalto.* Tipografia Marchi-Vella, Chieti.

Zanoni, E. 2014. *Scienza, patria, religione: Antonio Stoppani e la cultura italiana dell'Ottocento.* FrancoAngeli, Milano, Italy.

Zappaterra, E. 1994. Source-rock distribution model of the Periadriatic region. *American Association of Petroleum Geologists Bulletin*, **78**, 333–354.

Historical study of geosciences and engineering in the oilfields of the Emilia-Romagna region in the socio-economic context of post-Unitarian Italy (1861–1914)

FRANCESCO GERALI[1,2], PAOLO MACINI[3]* & EZIO MESINI[3]

[1]*The University of Oklahoma School of Law, Oil & Gas, Natural Resources, and Energy Center, 300 Timberdell Road, Norman, OK 73019-5081, USA*

[2]*The University of Western Australia, FABLE, 35 Stirling Highway, Crawley, Perth, WA 6009, Australia*

[3]*School of Engineering and Architecture, University of Bologna, Viale del Risorgimento 2, 40136 Bologna, Italy*

P.M., 0000-0002-6851-3262

**Correspondence: paolo.macini@unibo.it*

Abstract: The purpose of this paper is to frame selected episodes in the establishment and development of the modern oil and gas industry in the Northern Apennines (Emilia-Romagna region), Italy. The research spans between the political unification of the Italian peninsula into the Kingdom of Italy (1861), and the outbreak of World War I (1914). In the attempt to delve into the socio-economic scenario of those times, we look to contextualize and describe the work of the geologists and engineers who contributed to developing the early scientific knowledge of practical utility on the oil-bearing formations of the Northern Apennines.

The history of the Emilia-Romagna oil industry was influenced and shaped by several episodes, and this paper discusses in particular two events that occurred during 1911. The first is an important and political act: the promulgation from the Italian authorities of the first regulation aimed at fostering the national oil industry, including the generous and criticized subsidies to support oil well drilling. The second is a technical–scientific landmark: the publication by Enrico Camerana and Bartolomeo Galdi (both engineers in the Royal Corps of Mines in Bologna) of the treatise *I giacimenti petroleiferi dell'Emilia* [*The Oilfields of the Emilia Region*], thanks to which Italian oil expertise made an important step forward. This seminal study on the state of the art of geological exploration of the oilfields of northern Italy represents the point of fracture between the empirical knowledge concerning oil and gas exploration and production and the development of the first national original research in the field of geosciences, to be fully developed in the decades to come.

Oil in Emilia-Romagna before Italian unification

From antiquity to the modern age, the geographical distribution and utilization of petroleum, bitumen, pitch, asphalt and natural gas in Emilia-Romagna is documented in numerous primary sources that can be valid references for the interested reader (Ariosto 1690; Fougeroux de Bondaroy 1770; Volta 1777, 1784; Calindri 1781–83; Spallanzani 1792). The aim of this section is to give an account of the 'craft' of oil exploration and production in the first part of the nineteenth century, with special reference to the 'province parmensi' (the present-day province of Parma), one of the most representative micro-case histories of the Emilia-Romagna region.

In the first half of the nineteenth century, petroleum was not produced by an organized mining industry. It was a rudimentary handcraft activity mostly run by the owner of the land in which crude oil was skimmed from the surface of creeks, harvested from natural springs or collected with buckets from shallow hand-dug wells (10–50 m). In the Northern Apennine area facing the Po River, the valuable light–ultralight crude oils (typically white to yellow or orange to red in colour) were collected and sold with profit as fuel for lamps and as medical remedies (Zolesi 1891). The first chemical analyses of several crude oils of the province parmensi were performed by the Italian chemist Mojon (1806) and the Swiss chemist Theodore de Saussure (already one of the major pioneers in the study of photosynthesis at this time) in 1802, and later (1816–17) on samples coming from the hand-dug wells of Miano, province of Parma (Saussure 1816, 1817). Among the few other relevant works on petroleum in the early nineteenth century, we must mention the 1840's treatise by Giuseppe Bianconi (geologist from the

From: Craig, J., Gerali, F., MacAulay, F. & Sorkhabi, R. (eds) 2018. *History of the European Oil and Gas Industry*. Geological Society, London, Special Publications, **465**, 305–332.
First published online May 30, 2018, https://doi.org/10.1144/SP465.21

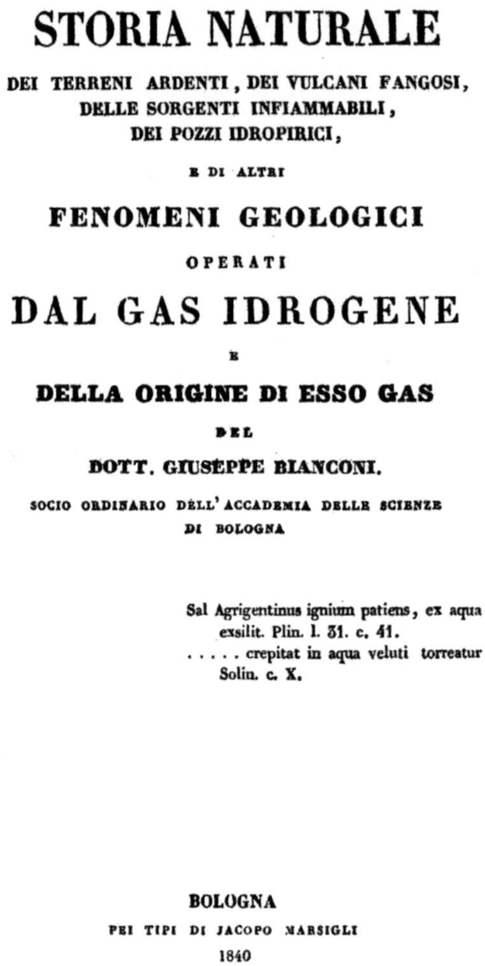

STORIA NATURALE

DEI TERRENI ARDENTI, DEI VULCANI FANGOSI,
DELLE SORGENTI INFIAMMABILI,
DEI POZZI IDROPIRICI,

E DI ALTRI

FENOMENI GEOLOGICI

OPERATI

DAL GAS IDROGENE

E

DELLA ORIGINE DI ESSO GAS

DEL

DOTT. GIUSEPPE BIANCONI.

SOCIO ORDINARIO DELL' ACCADEMIA DELLE SCIENZE
DI BOLOGNA

Sal Agrigentinus ignium patiens, ex aqua exsilit. Plin. l. 31. c. 41.
..... crepitat in aqua veluti torreatur Solin. c. X.

BOLOGNA
PEI TIPI DI JACOPO MARSIGLI
1840

Fig. 1. Front cover of Giuseppe Bianconi's treatise on the origin of natural gas, 1840.

University of Bologna). His study may be considered a precursor to the studies on petroleum geology in the Emilia-Romagna region and in Italy (Fig. 1).

However, lamp fuel and medication apparently were not the only markets for local oils. In the State Archive of Parma are recorded the activities of entrepreneurs looking to undertake other businesses. On 14 August 1846, Pietro Bavagnoli and Giulio Lecompte du Camp filed an application to obtain a 'patente di privilegio' ('licence of exclusivity') to produce for 10 years a new bituminous mastic in the Duchy of Parma and Piacenza, which could be used for several purposes, including the production of asphalts (Fig. 2). Two years later, on 4 February 1848, the engineer Silas Medail, from the Kingdom of Sardinia, placed an application to obtain a 15-year licence to produce in the Duchy the 'lighting gas from the tar obtained by the dehydration of coal' using a new procedure that he had invented (coal is not present in the province of Parma; it is possible that Medail was interested in using the natural bitumen seeping from several sites of the province, or small local deposits of lignite). The document reports that the applicant had already paid half the tariff required to be granted the licence (75 lire, the currency of the Duchy), but it does not report where Medail was planning to obtain the raw material (Fig. 3) (Historical Archive of the City of Parma (HAP), Box Trade: Volpicelli 1848). In any case, this document reveals that even in the oil-rich provinces of Parma there was a market for other means of illumination (gas light for public illumination was the most utilized until late 1890). Probably, the several episodes of oil-illuminated cities in the area often reported by many authors should be considered more isolated cases rather than the ordinary practice.

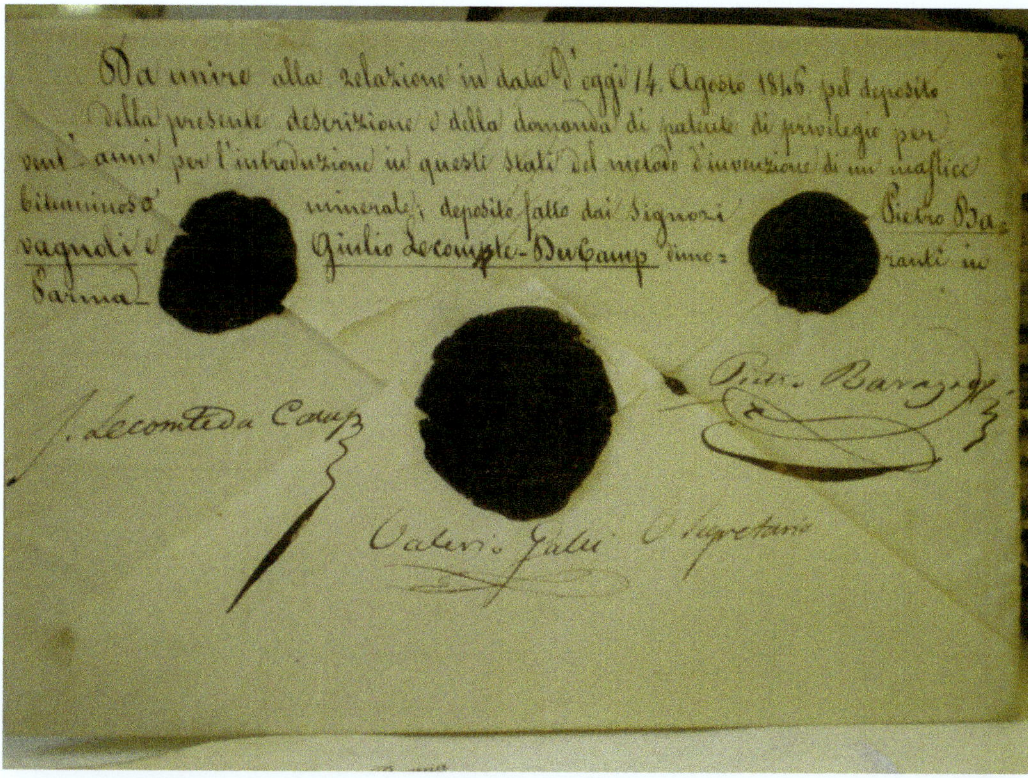
Da unire alla relazione in data d'oggi 14 Agosto 1846 pel deposito della presente descrizione e della domanda di patente di privilegio per venti anni per l'introduzione in questi stati del metodo d'invenzione di un mastice bituminoso minerale, deposito fatto dai Signori Pietro Bavagnoli e Giulio Lecompte-DuCamp dimoranti in Parma

Fig. 2. Patent application of Bavagnoli and Lecompte du Camp, 1846.

In those years, knowledge about the nature of petroleum was quite limited, as much as its commerce and utilization. The development of the early petroleum industry in Emilia-Romagna was hampered by the unpredictability of the availability of the resources. In a letter dated 6 September 1851, from the Governor of the province of Parma to the Podestà (Mayor) of the nearby town of Fornovo, he writes about the new instructions concerning the storage of dangerous fuels. The mayor is informed that, 'due the considerable quantity of oil that is gathered latterly in the province, we are assuming that the local warehouses are storing that dangerous liquid' (Historical Archive of the City Hall of Fornovo (HAF), Box 114, Agriculture, Industry and Commerce; Gerali 2011).

The letter continues informing that municipalities are called to respect: 'article 9 of the decree issue on October 4, 1850 [...] those goods must be transported in safe places and far from the inhabited centres'. The letter concludes with a request for information on the trade and the storage of oil in the area of Fornovo. The Podestà answers on 20 October 1851:

> ... in the area of Neviano de' Rossi there are about three wells from three different owners ... the yield is very limited ... the oil is collected in can of tin and carried in Parma ... .

This exchange of correspondence is very significant and confirms two historical facts. First, the existence of a state decree issued by the Secretary of the Interior (Duchy of Parma and Piacenza) which shows how oil production and trade was rooted in the socio-economic fabric of the area. Second, both letters give an idea of the unpredictability of oil finds, which was still manually harvested (Historical Archive of the City Hall of Fornovo (HAF), Box 114, Agriculture, Industry and Commerce; Gerali 2011).

Unfortunately, the available sources do not allow quantification of what was considered miserable and, conversely, what was considered large in the parameters of that time. There is a lack of reliable data, as is evident when the few existing statistics are compared with the surviving primary sources. The Italian mining engineer Celso Capacci (1905) reported at the Second International Petroleum Congress, held in Liege, that in 1860 the provinces of Parma, Piacenza and Pavia (the latter in Lombardy) together produced 5 tons of oil. On the other hand, the manuscript chart of the production for the mineral springs in the area of Fornovo reports a figure of 5208 tons of crude.

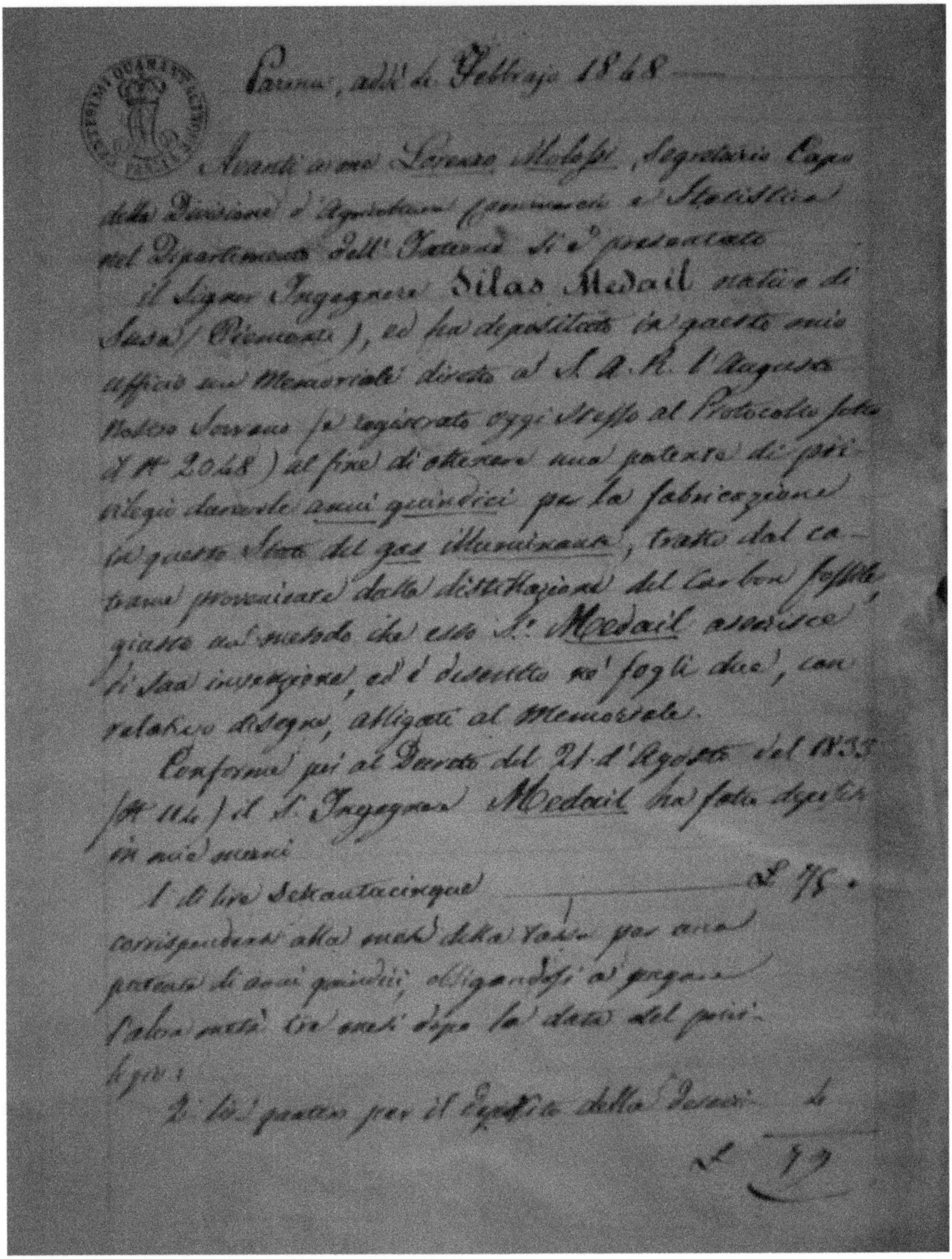
Parma, addì ... di Febbrajo 1848 ——
Avanti a me Lorenzo Molossi, Segretario Capo
della Divisione d'Agricoltura, Commercio e Statistica
nel Dipartimento dell'Interno si è presentato
il Signor Ingegnere Silas Medail nativo di
Susa (Piemonte), ed ha depositato in questo mio
ufficio un Memoriale diretto a S. A. R. l'Augusto
Nostro Sovrano (registrato oggi stesso al Protocollo sotto
il N. 2048) al fine di ottenere una patente di privi-
legio durante anni quindici per la fabbricazione
in questo Stato del gas illuminante, tratto dal ca-
trame proveniente dalla distillazione del carbon fossile,
giusta un metodo che esso S.r Medail asserisce
di sua invenzione, ed è descritto ne' fogli due, con
relativo disegno, allegati al Memoriale.
Conforme poi al Decreto del 21 d'Agosto del 1835
(N. 114) il S. Ingegnere Medail ha fatto deposito
in mie mani
1° di lire settantacinque —— L. 75.
corrispondenti alla metà della tassa per una
patente di anni quindici, obbligandosi a pagare
l'altra metà tre anni dopo la data del privi-
legio;
2° di lire [illegible] per il deposito della Domanda
L. [illegible]

Fig. 3. Patent application of Silas Medail, 1848.

The annual production in Fornovo, as witnessed by a primary manuscript source, overtakes the data reported by Capacci 45 years later with reference to the three provinces. This finding, as for some others in the course of verification, generates uncertainty on the reliability of the pre-unification statistics on oil (Historical Archive of the City Hall of Fornovo (HAF), Box 114, Agriculture, Industry and Commerce; Capacci 1905; Squarzina 1958; Gerali 2011).

While Italy was producing oil in the traditional-centennial way, in 1859 the global oil industry made a huge change in direction. This was not at all an unexpected event, quite probably the opposite. Oil was already produced, refined and traded in significant quantities in Russia (Azerbaijan), Poland, Germany, Romania and Canada. All of these countries were experiencing a technologically transitory phase, taking steps away from the manual collection of crude oil to a more systematic production technology. The real unexpected condition was the actual location where oil found momentum: the remote NE corner of Pennsylvania.

American primacy emerged when the USA started to exploit the oilfields on a large scale for the production of lighting fuels, receiving important lines of credit from entrepreneurs who decided to invest in exploration and production in the rich seepages of the area of Titusville (Pennsylvania). The literature on this topic is vast, and the reader can find detailed accounts in Crum (1911), Bacon & Hamor (1916), Forbes (1958, 1959), Brice (2009) and Gerali & Gregory (2017).

Oil in the early political and economic contest of Italy

On 17 March 1861, Italy became a united country after centuries of divisions and particularisms. The new kingdom that unified the seven states into which the peninsula was divided was born. However, much had to be done to accomplish the real unity of the new nation – that of the social, economic, industrial and mining points of view.

After unification, Italy began the difficult reorganization of the mining sector to combine the pre-existing mining codes into a consolidated body (Fenoaltea 2001). The Kingdom of Sardinia (the state that led the political unification of the peninsula) established the Royal Corps of Mines on 18 October 1822. The authority of the Corps was then extended to the whole country, which was divided into regional areas and then formally established in 'Mining Districts' (1863–71), each one headed and managed by a 'Chief Engineer'. The Corps was at the service of the newly-established Ministry of Agriculture, Industry and Commerce, who entrusted it to study, organize, discipline and survey the mining industry. Oil, bitumen, asphalt and every related mining activity was included under the authority of the Corps of Mines. As for petroleum activities – since the 1880s, it was considered strategic to relieve the Italian chronic energy shortage – the Corps never acted just as a 'bureaucratic leviathan', rather as a producer of knowledge at the service of the operators (Zoppetti 1883; Macini & Mesini 2011). In that sense, in terms of state science at the service of national wealth, Italy experienced the same as Pennsylvania had years earlier, thanks to the Geological Survey directed by John F. Carll (Carll 1880; Brice 2009) and, the same as in Western Australia later, thanks to the joint activity of the Geological Survey and the Bureau of Mines of Andrew Gibb Maitland and Alexander Montgomery (Gerali & Gregory 2016).

Soon after unification of the country, a project involving surveying and producing an official geological map of Italy was begun; an activity that was strongly pursued by Quintino Sella, an experienced mining engineer, mineralogist and influential political player leading the newborn Italian state. The project for the Italian Geological Survey began in 1861 (Corsi 2007; Pantaloni 2014). The purpose was to provide Italy with a tool for accessing knowledge of the territory and its natural resources, which was essential for economic development (Camerana & Galdi 1911; Camerana *et al.* 1926).

A run-down industry: a difficult moment for petroleum in Emilia-Romagna, 1860s–70s

Since the early 1860s, the major activities of oil exploration and production in Emilia had taken place in the same areas that had been known for decades by the presence of several natural seepages. The most relevant oil areas included in the province of Piacenza were Montechiaro, Montechino and Velleja. The oil areas of the province of Parma were Salsomaggiore, Miano di Medesano, Neviano degli Arduini, and the localities of Ozzano and Vallezza in the area of Neviano de' Rossi. In the province of Reggio Emilia, the only oil area worth mentioning was Rivaltella. Lastly, in the province of Modena, many wells and trenches were dug in the areas of Prignano, Montegibbio, Sassuolo, Fiorano and Montefestino (Fig. 4). Simple and scarcely scientific (i.e. mainly based on the localization of seepages), early oil exploration in the Emilia-Romagna region returned negligible results to the Italian economy. Production was scarce, and the low profit derived from crude oil sales did not promote investments to improve and increase production. This condition gave rise to a technological immobilization in the whole of the Emilia-Romagna region (Stoppani 1865, 1866, 1870*a*; Fairman 1868; Gerali 2011).

The production of oil was measured in kilograms and buckets, while the number of wells in each production area ranged from one to a dozen. The depth of the wells constructed during the 1860s and the 1870s averaged between 30 and 50 m: the oil industry in Emilia-Romagna was still in a pre-industrial state (Stoppani 1870*a*, *b*; Scicli 1972; Novelli & Sella 2009).

Fig. 4. Map of the Emilia-Romagna oil areas (Fairman 1868).

An account of the state of the art in the Emilia region is given by the geologist Antonio Stoppani. In 1865, he was commissioned by the entrepreneurs William Mayo and Prof. Vincenzo Botta from New York to study the territory of Parma and Piacenza in order to assess the opportunity of starting a mining company in those areas (Stoppani 1865). They were attracted by the supposed richness of the subsoil, which over the years was even described as the 'Pennsylvania of Europe' – a hazardous comparison with the richest oil area in the world at the time, which was later revealed to be a mermaid-chant for many. Since 1863 Stoppani had been interested in the search for mineral deposits, especially petroleum. He tackled the oil question because he considered it to be a subject of great interest both for science and for industry, and in 1864 he proposed the necessity of applying sound geological principles and scientific interpretation in order to turn the then-primitive and low-yield oilfield technology into a proper industrial activity (Stoppani 1865, 1866, 1876; Zanoni 2014).

From Stoppani's survey of the oil areas of the Taro River near the town of Parma, the flaws of the current systems in use, together with the inherent difficulties in petroleum exploration, emerged. He pointed out the need to develop new and more efficient field practices, which would be the only way that the success of the operations could be guaranteed. Such methods required substantial amounts of capital expenditure and, as many Italian companies had limited technical and economic dimensions, it was suggested that government intervention would be advisable. Stoppani believed that the government should provide the subsidies that would allow operators to compensate for the large expenses incurred with such risky mining enterprises. He underlined the shortcomings and the inadequacies of the existing Italian mining laws, which were outdated and without any specific reference to petroleum. Stoppani concluded the report commissioned by Mayo and Botta by stating that the current operators (small and isolated entrepreneurs who only carried out manual excavations near the natural seeps) were hampered more by the inefficiency of their technical methods rather than by the scarce availability or quality of the mineral. In other words, the negative outcome in most of the cases was dependent on the lack of professionalism, practical knowledge and the scarcity of resources to cope with the 'unforeseen odds' of the new industry (Stoppani 1864, 1865, 1866, 1876; Zanoni 2014).

When Rome became the capital of Italy in 1870, the country had a population of about 27 million people. The gross domestic product (GDP) was mainly based on a very backward agriculture, which absorbed over 40% of the workforce, while industry occupied only a little over 10%. The metallurgical and mechanical industry was taking its first steps, but it could not meet the internal demand driven by the rapidly expanding railway network; and both machinery and technical management staff had to be imported from abroad. The problem was critical, especially for heavy industry (in particular, for steel production), where the incidence of the cost of fuels for the production of energy was very high (Bardini 1998; Fenoaltea 2001).

This situation was also reflected in the oil sector, as was witnessed in 1869 by the mechanical engineer Alessandro Bonariva in his interesting treatise *Memorie per la coltivazione dei petroli nelle province parmensi* [*Memoirs on the Petroleum Production in the Provinces of Parma*]. He was the owner of a service company based in Bologna specializing in water well drilling and was interested in joining the oil business. His company, the Ing. A. Bonariva e C. – Mechanical Workshops, was pioneering the practice of modifying and adapting borehole drilling machinery used for water wells to turn them into oil-well drilling rigs. Bonariva, on the verge of 1870, reported the same discouraging situation described by Stoppani between 1864 and 1865: the lack of know-how and the prevalence of old-fashioned hand-dug wells, although he also mentioned the company 'Cornegliano', based in Genova, which was producing drilling tools that were used mainly in the oil areas of Parma. Bonariva also noted that the majority of the operators were involving foreign capital. He concluded his treatise expressing a bitter remark on the professionalism of the current engineering practices. He wrote that the lack of technology was just a component of the oil problem in the oil areas of the Emilia-Romagna region, and complained about the lack of good education and training for engineers in Italy. Engineering practice, Bonariva concluded, is both a science and an art, which needs special scholarship and field practice (Bonariva 1869, 1882).

In 1871, Italy produced about 38 tons of oil, and only 30 t came from the provinces of Parma, Piacenza and Pavia. That year was the beginning of what the authors named the 'Decennio Nero' (the 'Black decade') of the Emilia-Romagna oil industry, during which less than 50 tons of oil were produced (Capacci 1905). During this period, many companies suffered a shortage of crude production and were forced to go out of business, like the Miano: Societa' Anonima Italiana per l'Estrazione del Petrolio in Miano di Medesano Parmense [the Anonymous Italian Company for the Extraction of Oil in Miano di Medesano Parmense] which with letter No. 319, dated 26 November 1872, announced to their shareholders that the Miano Company was being wound up (Miano 1872; Fig. 5). In spite of the heavy downturn, the industry continued, as the diffusion and the consumption of crude oil in Italy was fed by imports, the only relevant oil business in Italy at that time.

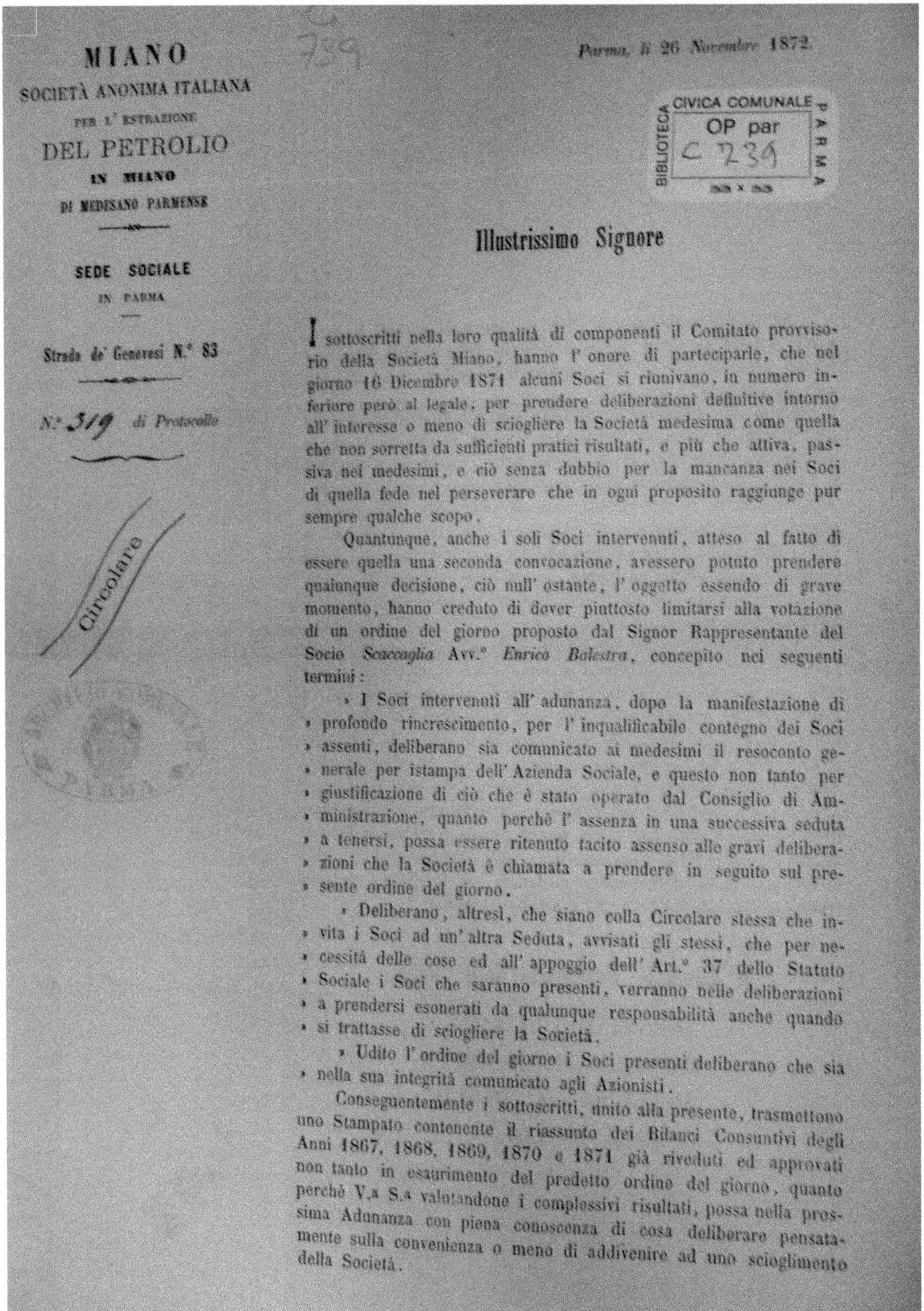

MIANO
SOCIETÀ ANONIMA ITALIANA
PER L' ESTRAZIONE
DEL PETROLIO
IN MIANO
DI MEDESANO PARMENSE

SEDE SOCIALE
IN PARMA

Strada de' Genovesi N.° 83

N.° 319 di Protocollo

Circolare

Parma, li 26 Novembre 1872.

Illustrissimo Signore

I sottoscritti nella loro qualità di componenti il Comitato provvisorio della Società Miano, hanno l' onore di parteciparle, che nel giorno 16 Dicembre 1871 alcuni Soci si riunivano, in numero inferiore però al legale, per prendere deliberazioni definitive intorno all' interesse o meno di sciogliere la Società medesima come quella che non sorretta da sufficienti pratici risultati, e più che attiva, passiva nei medesimi, e ciò senza dubbio per la mancanza nei Soci di quella fede nel perseverare che in ogni proposito raggiunge pur sempre qualche scopo.

Quantunque, anche i soli Soci intervenuti, atteso al fatto di essere quella una seconda convocazione, avessero potuto prendere qualunque decisione, ciò null' ostante, l' oggetto essendo di grave momento, hanno creduto di dover piuttosto limitarsi alla votazione di un ordine del giorno proposto dal Signor Rappresentante del Socio *Scaccaglia* Avv.° *Enrico Balestra*, concepito nei seguenti termini:

» I Soci intervenuti all' adunanza, dopo la manifestazione di » profondo rincrescimento, per l' inqualificabile contegno dei Soci » assenti, deliberano sia comunicato ai medesimi il resoconto ge- » nerale per istampa dell' Azienda Sociale, e questo non tanto per » giustificazione di ciò che è stato operato dal Consiglio di Am- » ministrazione, quanto perchè l' assenza in una successiva seduta » a tenersi, possa essere ritenuto tacito assenso alle gravi delibera- » zioni che la Società è chiamata a prendere in seguito sul pre- » sente ordine del giorno.

» Deliberano, altresì, che siano colla Circolare stessa che in- » vita i Soci ad un' altra Seduta, avvisati gli stessi, che per ne- » cessità delle cose ed all' appoggio dell' Art.° 37 dello Statuto » Sociale i Soci che saranno presenti, verranno nelle deliberazioni » a prendersi esonerati da qualunque responsabilità anche quando » si trattasse di sciogliere la Società.

» Udito l' ordine del giorno i Soci presenti deliberano che sia » nella sua integrità comunicato agli Azionisti.

Conseguentemente i sottoscritti, unito alla presente, trasmettono uno Stampato contenente il riassunto dei Bilanci Consuntivi degli Anni 1867, 1868, 1869, 1870 e 1871 già riveduti ed approvati non tanto in esaurimento del predetto ordine del giorno, quanto perchè V.a S.a valutandone i complessivi risultati, possa nella prossima Adunanza con piena conoscenza di cosa deliberare pensatamente sulla convenienza o meno di addivenire ad uno scioglimento della Società.

Fig. 5. The winding-up act of the Miano Company, 1872.

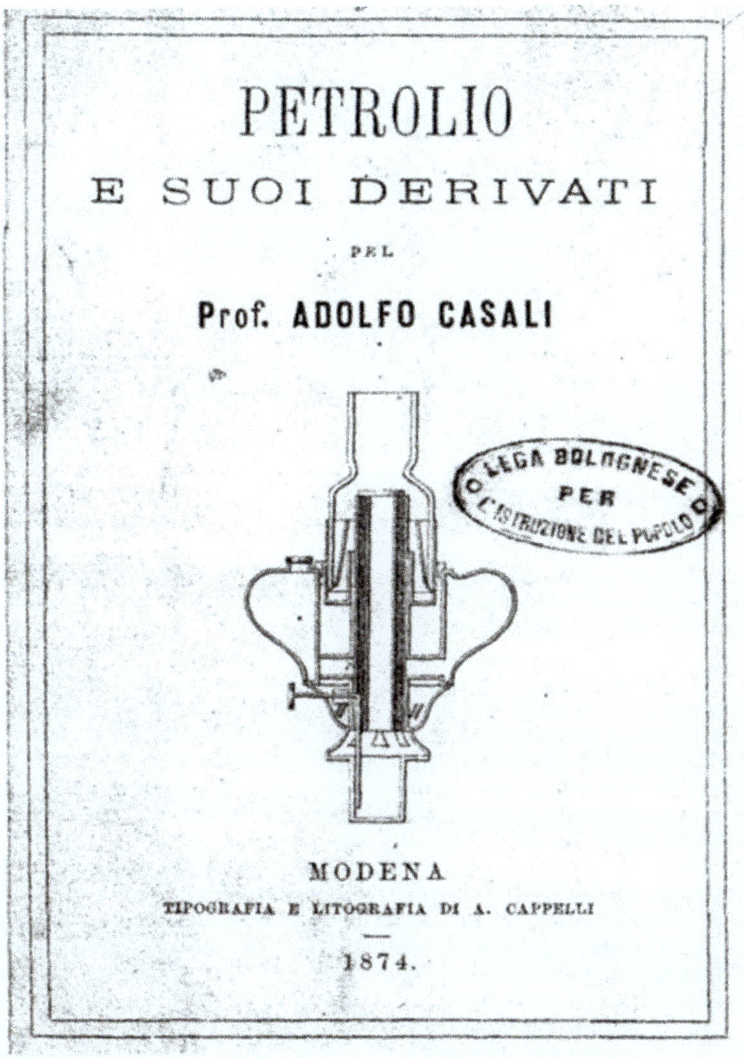

PETROLIO
E SUOI DERIVATI
PEL
Prof. ADOLFO CASALI

MODENA
TIPOGRAFIA E LITOGRAFIA DI A. CAPPELLI
1874.

Fig. 6. Front cover of Adolfo Casali's treatise on petroleum and its derivatives, 1874.

Italians were discovering the potential and value of oil in industrial and civil society, as was explained by Adolfo Casali in his 1874 *Petrolio e i suoi derivati* [*Petroleum and its Derivatives*]. However, the real issue was to find oil and pump it out from the subsoil (Casali 1874; Zoppetti 1883) (Fig. 6).

During the 1860s, the region of Abruzzo, in central Italy, provided the largest share of oil production in Italy, reaching its peak during the years 1865–67. In the 1870s, more than 90% of the oil produced in the peninsula came from the area of Ripi (Fig. 7) and San Giovanni Incarico, in the southern region of Lazio (not far from Rome), and the most productive year was 1878. Finally, in the 1880s, there was a change in the management of the oilfields in the Emilia-Romagna region, which succeeded to boost production up to unprecedented levels (Zoppetti 1883; Capacci 1905; Novelli & Sella 2009).

1880s–90s, technology, production and development in Emilia-Romagna

Twenty years of trials, errors, sacrifices and frustration did not get lost, not by the oilmen, not by the government authorities. During 1880–83 the production in the oilfields of the western part of the Emilia-Romagna region became consistent and competitive with the rest of Italy. Since 1884,

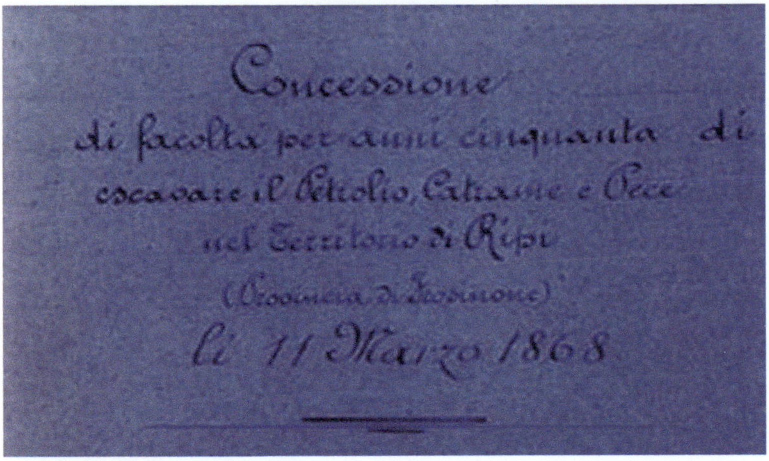
Concessione
di facoltà per anni cinquanta di
escavare il Petrolio, Catrame e Pece
nel Territorio di Ripi
(Provincia di Frosinone)
li 11 Marzo 1868

Fig. 7. Headers of the concession of Ripi, southern Lazio oil district (Italy).

Emilia-Romagna had achieved the position of first and most important oil and gas district in the country, a primacy that lasted for over a century (Novelli & Sella 2009).

Petroleum in Italy at some point was seen as the new liquid coal, which could be burnt in a boiler and produce steam power. It was perceived as an opportunity to start an ambitious process of energy substitution. In Italy, this opportunity did not end well, due to the fact that consumption grew faster than oil production, which remained and will remain minimal (Bardini 1998).

In the 1890s, Italy was still lacking its own 'Aristocrazia del Petrolio' ('oil aristocracy'). In the USA, the cornerstone of the oil industry was the Rockefeller family, which since the late 1860s had been establishing its trade agencies in almost every country of the old and new world. In Sweden, the Nobel brothers were financing and managing the operations in the oilfield of the Caucasian region. The Rothschild family from Frankfurt was also imposing its interest in Russia. In France, the Le Bel family was ruling the production in domestic oilfields, in parallel with their lucrative business in oil shales. In the UK and in The Netherlands, a small circle of investors granted an independent oil supply from the West Indies and were developing a national refining system, which just a few years later laid the basis for the constitution of the Royal Dutch Shell Oil Company (Crum 1911; Scicli 1972; Sédillot 1974).

In Italy, petroleum entrepreneurship was represented by small companies with limited resources, and very often the few successful companies were joint ventures formed with more experienced French and German operators. Finally, from 1890, thanks to technical advances in the oilfields of the province of Piacenza and the beginning of new operations in the province of Bologna, the Emilia-Romagna region achieved the title of being the leading producing district of Italy. In 1890, the region overtook the symbolic figure of 1000 tons; the peak was reached in 1895, with 3532 tons; and the decade concluded with a stable threshold of just a little below 2000 tons. During the 1890s, the domestic production of oil came almost entirely from the Emilia-Romagna fields (Capacci 1905). In spite of these results, internal production was still far from covering the increasing demand – Italy was poor in coal, and heavy oils were a good alternative to fuel the furnaces and boilers of industry and manufacturers. Oil imported from abroad was subject to stiff high tariffs. However, this policy did not have a tangible effect on the equilibrium of the market, which was in any case subject to the prices and rules imposed by foreign companies (Zolesi 1891; Pantanelli 1910).

After years of serious economic crisis, by 1896 industrial production had already recovered, taking an unlikely step forward in the period of high prices that characterized the world economy at the turn of the century. This recovery was to last, with the only interruption occurring in 1907, until the start of World War I. At the same time, large-scale use of the 'new' hydroelectric power generation intervened to integrate and ease the chronic lack of Italian energy sources, encouraging and supporting the development of the national industry. This period is known in Italy as the age of the Giolitti Industrial Revolution, named after the Italian Prime Minister Giovanni Giolitti, which benefited from new relationships established in the previous 20 years between banks and industry, especially with contributions from German investors. The steel and shipbuilding industry developed under the umbrella of the protectionist tariff of 1878, the first specifically addressed the procurement and construction of war material. The mechanical industry developed more slowly, disadvantaged by customs duties on raw materials and low domestic consumption: nevertheless, it received a strong stimulus from the significant demand for manufactured products by the new national railway system. At the beginning of the new century, large mechanical industries were built, paired by the expansion of the new automotive industry (by 1910, the number of cars circulating in Italy was already around 20 000). In this social, political and economic environment, Italy took its first steps on the way to the petroleum age (Sacco & Stella 1926; Bardini 1998; Fenoaltea 2001; Macini & Mesini 2011).

The new century and the old problems

The first decade of the twentieth century saw a rapid decline in the prices of imported petroleum products. In fact, in 1906 Standard Oil of New Jersey won the challenge against the German Trust Europäische Petroleum Union (EPU). In 1908, Standard Oil tried to repeat the round against another EPU company in the UK market, but here it clashed with the rising power of Royal Dutch Shell headed by Henry Deterding. In 1910, the two oil companies, destined to become world powers in petroleum exploration and production, began a tough fight to also seize control of the Chinese market. To undermine the stance of Standard Oil in China, Royal Dutch Shell resorted to fighting price, which ended victoriously for Royal Dutch Shell in 1912, with an agreement that was to assign half of the Chinese market and a substantial stake in the procurement of the British one. The Italian market suffered the effects of these events, although to a lesser extent than the English and French markets. In July 1911, just in the fervour of the price struggle, taking into account the exchange rate, a quintal of duty-free oil in London cost 8 lire and 9 lire in Paris, while in Rome, where no-one seriously threatened the monopoly of Standard Oil, it cost nearly 15 lire. The decrease in prices recorded in 1911 was enough to crush the fragile Italian oil industry, paving the way for government intervention in the market (Squarzina 1958; Sédillot 1974; Magini 1976; Macini & Mesini 2011).

Oil companies operating in Italy obviously supported this intervention. In fact, to deal with the war against Turkey, in 1911 Italy imported about 217 000 tons of petroleum products (64% from the USA, 14% from Russia and 12% from Romania), against a domestic production that had just passed 7000 tons in 1910. Since then, the Italian market, previously controlled by Americans and Germans, became the exclusive domain of the two largest international companies, Standard Oil and Royal Dutch Shell, which owned almost all tankers, coastal reservoirs and distribution networks. This duopoly would implement the most ruthless policy: by 1913, the price of kerosene in Italy increased by 26% compared to 1911 (Macini & Mesini 2011).

1911– The first legislative act on petroleum issued in the Italian kingdom

In 1910, the Italian population reached 36 million. World oil consumption was calculated at 23 million tons per year. Italy was consuming almost 168 000 tons, but producing just 7000 tons: the remaining share was imported mainly from the USA, and a minor part from Russia. These already discouraging numbers were destined to worsen during and after World War I, and are the laconic picture of the lethargy and idleness of the Italian oil sector in the 1910s (Fig. 8). The Italian oil industry started to have a more systemic form, if not from a mere technical standpoint, at least from a strategic and legislative one, in the 1910s.

In 1911, Law 250, of 19 March, was promulgated which granted an economic reward of 30 lire per each metre drilled deeper than 300 m 'in boreholes aimed to search for crude oil'. It was a law of a protectionist nature, which aimed, in the spirit of the time, to promote oil exploration in Italy, where the strategic importance of hydrocarbons for its own industrial development had now been realized. In the same year, in Bologna, the first modern essay on domestic oil exploration was published, written by Camerana & Galdi (1911), chief and field engineer, respectively, at the local mining district of the Corps of Mines.

As noted in the previous paragraph, the intervention by the Italian government to stimulate and encourage petroleum exploration in the country by passing Law 250, of 19 March 1911, was a milestone for the petroleum industry, granting economic rewards to the drilling industry. This law was bitterly criticized by Luigi Einaudi, who was elected President of the Italian Republic in 1948 and was notoriously averse to public and private monopolies. This law, when proposed in 1910, found opposition from both the Socialist Party and the Liberal Party. Among the latter, Luigi Einaudi (Fig. 9) wrote a famous essay, 'The state drillers', published in the Italian journal *Riforma Sociale* in January 1911, where he clearly expressed the reasons for its opposition. Since then, the term 'state drillers' is synonymous with devastators of public money. Einaudi was against all 'attacks on public money', that would take the form of duties (on wheat, textiles, steel products, sugar, chemical fertilizers, etc.) or subsidies to illusory industries, not able to stand in the market without money

Production and consumption of crude oil in Italy 1900-1910-1920

Year	Mt
1900	23
1905	31
1910	43
1915	60
1918	70
1919	110
1920	110

World consumption inceased five-fold from 1900 to 1920

Year	Mt
1910	0.17
1914	0.26
1918	0.45
1919	0.37
1920	0.41

Italian consumption inceased three-fold from 1910 to 1920

Year	Italian prod. (kt)	Import (kt)
1910	7.0	156
1914	5.5	256
1918	4.9	442
1919	4.8	363
1920	7.4	413

Italian production and import of oil, 1910 - 1920

Fig. 8. Charts of production, import and consumption of crude oil in Italy, 1900–20. kt, thousand tons; Mt, million tons.

from taxpayers and consumers. With a good sense of humour, he recalled a newspaper report in which, long ago, to remedy a wine crisis, a Royal Commission of Inquiry had proposed the establishment of 'State Tasters'. Due to their sophisticated palate, they would be able to distinguish adulterated wines from the genuine ones, and so Einaudi wrote:

> Oggi ... sono invece i petrolieri d'Italia ... i quali si lagnano di dover fare delle buche troppo fonde per non trovare petrolio, e chieggono protezione allo Stato contro i produttori a buon mercato di petrolio genuino straniero. Naturalmente il governo interviene e crea una nuova professione, forse ancor più meravigliosa, quella dei trivellatori di Stato.

> Today ... the oil companies operating in Italy ... complain of drilling deep wells and not finding any oil, and so they look for the State protection against the producers of genuine cheap foreign crude oil. Of course, the Government steps in and creates a new profession, perhaps even more wonderful, the one of State Drillers.

Fig. 9. Photograph of Luigi Einaudi (1874–1961), Italian economist and statesman.

Today, we know that Einaudi was right, but mostly for reasons that he did not mention: the managers and engineers of the oil industry over many years of field activity had demonstrated incompetence and improvisation. As a matter of fact, prospectors did not yet have access to any form of geophysical instrumentation, nor to advanced drilling technologies and not even to capital that only later would allow identification and production of the well-hidden Italian oil and gas reserves, and, what is more, located in areas with no geological evidence or surface oil and gas seeps. The economic rewards were therefore inevitably destined to encourage research in the wrong areas and at depths at which oil, given the complex geological structure of the Italian territory, could not exist, or could only exist in very small deposits. Nonetheless, the economic rewards were granted. Since then, in the jargon of the Liberals, the term 'drillers' indicates those who tapped indiscriminately into government coffers, and became synonymous with 'stormtroopers' of public money (Squarzina 1958).

The oil industry in the Emilia region between the nineteenth and early twentieth centuries

> Il petrolio dell'Emilia è conosciuto da parecchi secoli, come ne fanno fede varie notizie storiche nonché le denominazioni di Rivo e Rio dell'Olio che designano vari corsi d'acqua nei cui letti compaiono gemicazioni di petrolio, e il nome di parecchi abitati quali: Ponte dell'Olio, Sassuolo, Sassuno. L'estrazione di petrolio si faceva un tempo per piccole quantità a mezzo di pozzi ordinari a grande sezione, nei quali il petrolio si raccoglieva alla superficie dell'acqua: la resa di tali pozzi raggiungeva talvolta 1000 kg di prodotto al giorno: la profondità loro variava da 40 a 60 e anche a 100 metri. Siffatti centri di produzione erano assai numerosi a Monte Gibbio e Monte Bonello nel Modenese; a Marzolara, Neviano de' Rossi, San Michele di Cavanna, Miano e Salsomaggiore nel Parmense; a Velleja, Montechino, Montechiaro nel Piacentino. Tale coltivazione riusciva però costosissima a causa delle difficoltà che presentava lo scavo dei pozzi per i pericoli di asfissia cui si esponevano i lavoranti non appena lo scavo giungeva in vicinanza dello strato petrolifero.
>
> The crude oil of Emilia is known for several centuries, as is proved by various historical information, as well as the names of Rio dell'Olio (Oil Creek), designating small streams with floating oil seeps, and the name of several villages as Ponte dell'Olio (Oil Bridge), Sassuolo, Sassuno (oil of rock). Oil production was then accomplished by means of large diameter wells in which the oil was collected at the water surface. The production of these wells, hand-dug to depths from 40 to 60 and even 100 m, sometimes reached even 1000 kg per day. Production fields were numerous at Monte Gibbio and Monte Bonello in the province of Modena; at Marzolara, Neviano de' Rossi, San Michele di Cavanna, Miano and Salsomaggiore in the province of Parma; at Velleja, Montechino, Montechiaro in the province of Piacenza. However, this production was very expensive for the difficulties to dig wells and for the risks of suffocation to which the workers were exposed as soon as the excavation reached in the vicinity of the oil layer
>
> (Camerana & Galdi 1911, pp. 111–112).

These are the words used by the aforementioned Enrico Camerana and Bartolomeo Galdi to describe the state of the art of petroleum exploration and production in Emilia in 1911. Apart from the small amount of oil production that came from traditional hand-dug wells (which, in fact, continued in production until the first decades of the twentieth century), the petroleum industry in Italy soon started to deploy the new methods of mechanical drilling. In 1888, a percussion rig drilled a well near Salsomaggiore, in the province of Parma, that reached a total depth of 683 m (Squarzina 1958). These technological achievements also triggered the development of a prolific industrial district, based in the Emilia region, for the construction of drilling rigs and drilling tools (Paoloni 1976).

According to the statistical data of the Royal Corps of Mines, domestic crude oil production increased from the negligible 4 tons in 1861 (estimated data, no official statistics are available), to 315 tons in 1865, 602 tons in 1878, 3594 tons in 1895, 8327 tons in 1907, up to 10 390 tons in 1911. In these years, domestic production came almost exclusively from the aforementioned oilfields of Emilia (Bonariva 1869; Zoppetti 1883; Scotti 1901; Vecchia 1916; Squarzina 1958). These oilfields continued to be the largest Italian producers up to about 1934, when Agip, the national oil company founded in 1926, discovered and/or revamped oil production in southern Italy (Tocco da Casauria in the Abruzzo region, Ripi and San Giovanni Incarico in the province of Frosinone, south of Rome, and Tramutola in the Basilicata region).

The first data on natural gas production date back to 1894: 12 000 m^3 (Paoloni 1976). Then, production increases rapidly, surpassing 1 MMm^3 (million cubic metres) in 1900, 2 MMm^3 in 1904, 3 MMm^3 in 1905, 5 MMm^3 in 1906, 6 MMm^3 in 1908, 8 MMm^3 in 1909 and 9 MMm^3 in 1911. The statistics also report the number of employees involved in the production industry: 50 in 1867, 267 in 1892 and 592 people in 1907. By the number of metres drilled, the first available data are those of 1896, with a total of 4073 m drilled; the maximum was reached in 1910, with 12 600 m drilled; over 15 years (1896–1910) drilling totalled 104 304 m, with a significant increase in the last 5 years. In the same years, there were never more than 10 active oil leases simultaneously. We know that at the turn of the century, there were 142 producing wells in the Montechino lease and 261 producing wells in the Velleja lease, both located in the Emilia-Romagna region (Squarzina 1958; Magini 1976).

Many factors contributed to the numerous fiascos in petroleum exploration in the Emilia-Romagna oil industry in the 1860s and 1870s: the weak financial position of the small oil companies; the low level of investment; the geological inexperience of most researchers; the outdated drilling technology; the lack of technological research; and the absence of an established petroleum geoscience. In Italy at that time, as elsewhere in the world, exploratory wells were bored only near oil seepages. Many oil and gas fields would be discovered years later at greater depth and in geological settings that were impossible to recognize and explore at that time (Novelli & Sella 2009; Gerali 2011; Macini & Mesini 2011).

Prevailing over the Italian oil companies were some small French, German and English independent ones, operating with their engineers and managers, similar to what occurred in almost all industrial sectors in the first 50 years following Italian unification. The industry also attracted the interest of foreign studies, such as the one by Edward John Fairman (1868). It is also indicative that just a few oil companies were capable of maintaining positive financial results and distributing a dividend to shareholders: much more numerous were those, who far from a real return on investments, after a hectic

R. UFFICIO GEOLOGICO

MEMORIE DESCRITTIVE

DELLA

CARTA GEOLOGICA D'ITALIA

VOLUME XIV

I GIACIMENTI PETROLEIFERI DELL'EMILIA

STUDIO

DEGLI

Ing.[i] E. CAMERANA E B. GALDI

DEL R. CORPO DELLE MINIERE

(con un Atlante e 127 incisioni intercalate nel testo)

BOLOGNA

STABILIMENTO POLIGRAFICO EMILIANO

Piazza Calderini, 6 - Palazzo Loup

1911

Fig. 10. Front cover of *I Giacimenti Petroleiferi dell'Emilia* (Camerana & Galdi 1911).

Fig. 11. Photograph of Enrico Camerana (1860–1941).

Fig. 12. Photograph of Bartolomeo Galdi (1873–1962).

handover of exploration permits and production concessions withdrew or ended in bankruptcy.

Unfortunately, none of the large international oil companies joined the oil and gas exploration in Italy, with the late exception of Standard Oil of New Jersey which in 1927 took over the Società Petrolifera Italiana (SPI) (the Italian Society for Petroleum), which was founded in 1905, and its Vallezza oilfield facilities (in the province of Parma). Standard Oil introduced into Italy the oilfield technology used in the USA and boosted oilfield production, attaining an output of 11 000 tons per year in 1932, more than 80% of the total domestic production.

Another aspect is the lack of production capacity and technological development of the refining industry in Italy. While in the USA it was growing at a very fast pace, adopting cutting-edge processes, in Italy the underdevelopment of this industrial sector could be due both to a shortage of domestic crude oil production and to the difficulty of competing with the imported petroleum products. Even Italy might have to import crude oil, as did Germany, but the huge amounts of capital and the expertise necessary to develop the refining industry did not yet exist. Moreover, these possibilities were playing against the large economic interest of foreign companies that had invested heavily in Italy in infrastructure for the import of kerosene and gasoline. These companies, since 1861, had begun to commercialize Pennsylvanian lamp oil in Italy, and to popularize and spread the use of kerosene for home, industry and public illumination. In the following years, they set up modern networks of distribution in Italy, such as Standard Oil of New Jersey who, in 1891, founded the Società Italo-Americana per il Petrolio (SIAP) (the Italian–American Company for Petroleum).

Imported American crude oil in Italy cost 35 lire per quintal in 1866, 30 lire in 1870 and 20 lire from 1877 to 1884; then its price suffered further decline, ranging between 22 lire in 1900 and 15 lire in 1911. National crude oil cost, on average, two to four times more than imported crude. The government, unwilling to introduce customs duties in favour of the refining industry, in fact, imposed taxes more heavily on crude oil compared to the ones applied to imported petroleum products, in defence of Italian oil production (Paoloni 1976), which was still tax-free.

In the Emilia region, the refinery of Borgo San Donnino (the present-day town of Fidenza, in the province of Parma) was operational in 1885 and, a few years later, another plant was built in the nearby town of Fiorenzuola d'Arda. This factory took advantage of an oil pipeline (probably the first one in Italy) constructed for transporting the crude

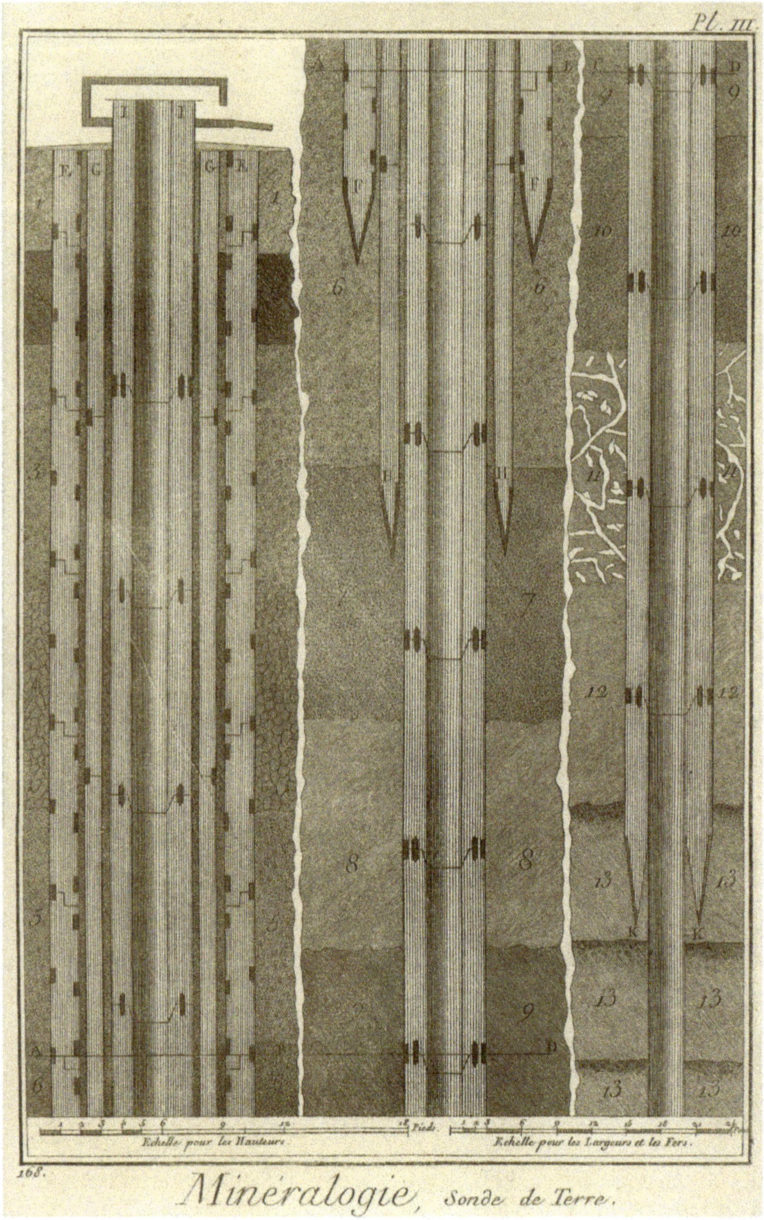

Fig. 13. Cross-sections of a cased well, from the *Encyclopédie* by Diderot & d'Alembert (1772).

produced in the aforementioned Montechino oilfield; the plant treated up to 25 m^3 of crude oil per day (Squarzina 1958; Kovacs 1964).

Throughout these years of 'rising through the ranks' the Italian petroleum industry generated an expert class of technicians and engineers, for geologists worked side by side with mining and petroleum engineers; small companies were leaving space for shareholding companies, all led by a managerial class that had acquired vast experience working in the Italian oilfields and, very often, abroad. The first great outcome of this long process of human, economic and technological growth in a macro-area (the provinces of Parma and Piacenza), by then on the cutting edge of the petroleum industry in Italy, was the founding of the aforementioned SPI (Società Petrolifera Italiana). This was the first integrated oil company that would be in a position to operate at a national level, and it is still currently active with the name of Gas Plus. The circumstances that led to the later creation of Agip are of a different political and social nature. Following the example of other European countries, Agip was created as the national Italian oil company by Mussolini's government in 1926,

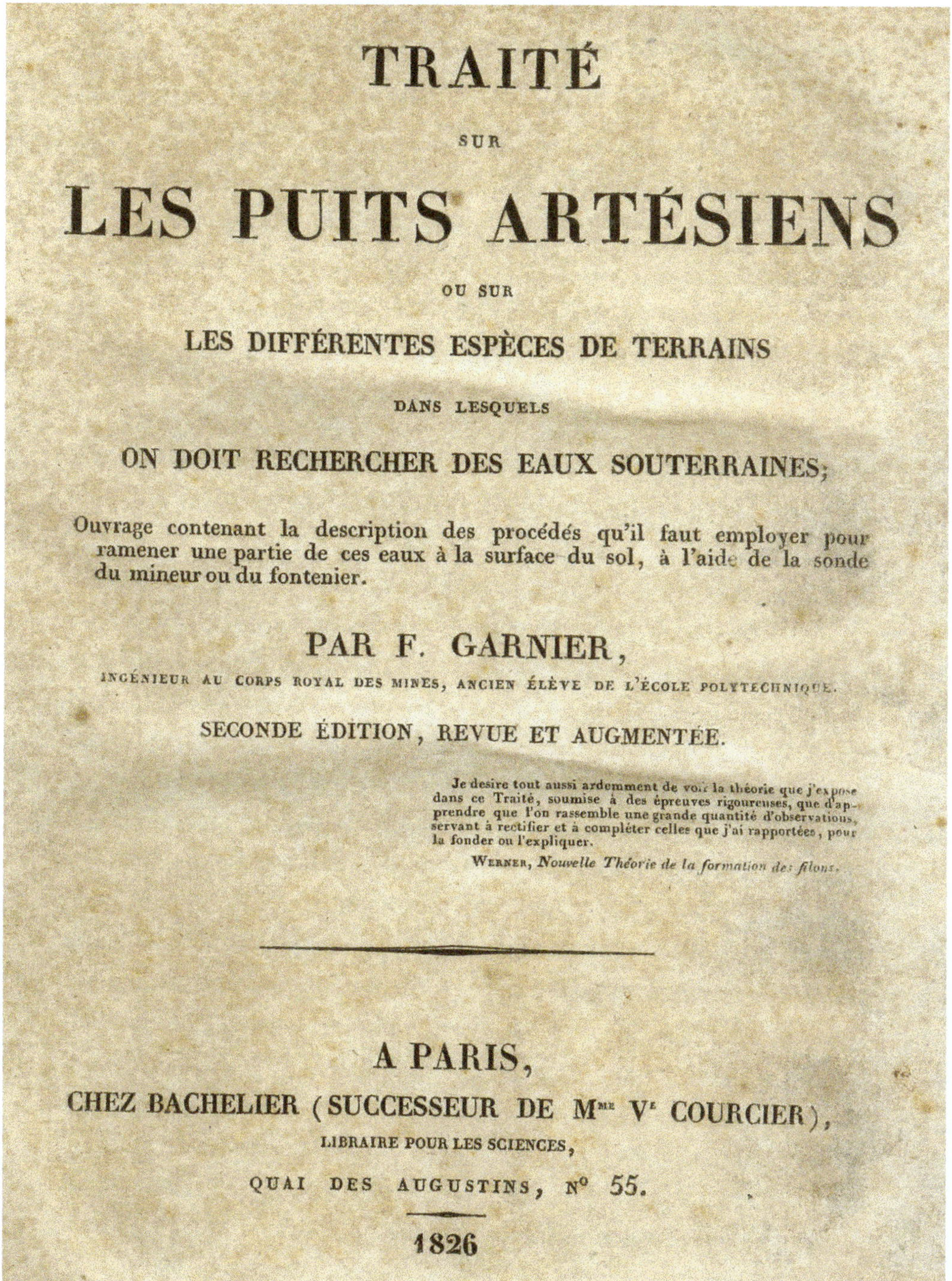
TRAITÉ

SUR

LES PUITS ARTÉSIENS

OU SUR

LES DIFFÉRENTES ESPÈCES DE TERRAINS

DANS LESQUELS

ON DOIT RECHERCHER DES EAUX SOUTERRAINES;

Ouvrage contenant la description des procédés qu'il faut employer pour ramener une partie de ces eaux à la surface du sol, à l'aide de la sonde du mineur ou du fontenier.

PAR F. GARNIER,

INGÉNIEUR AU CORPS ROYAL DES MINES, ANCIEN ÉLÈVE DE L'ÉCOLE POLYTECHNIQUE.

SECONDE ÉDITION, REVUE ET AUGMENTÉE.

Je desire tout aussi ardemment de voir la théorie que j'expose dans ce Traité, soumise à des épreuves rigoureuses, que d'apprendre que l'on rassemble une grande quantité d'observations, servant à rectifier et à compléter celles que j'ai rapportées, pour la fonder ou l'expliquer.

WERNER, *Nouvelle Théorie de la formation des filons.*

A PARIS,

CHEZ BACHELIER (SUCCESSEUR DE MME V^E^ COURCIER),

LIBRAIRE POUR LES SCIENCES,

QUAI DES AUGUSTINS, N° 55.

1826

Fig. 14. Front cover of Jacques Garnier's treatise on artesian wells, 1826.

with the aim of fostering the exploration and production of hydrocarbons within the Italian territories (not only in Italy), and to supply the country with oil and gas (Kovacs 1964).

Camerana & Galdi's *I giacimenti petroleiferi dell'Emilia* (1911)

In 1911, the seminal study of Enrico Camerana and Bartolomeo Galdi, *I giacimenti petroleiferi dell'Emilia* [*The Oilfields of the Emilia Region*], was published in Bologna (Camerana & Galdi 1911). It constitutes an irreplaceable testimony of the state of the art of geological exploration applied to the oil industry in northern Italy (Fig. 10).

Enrico Camerana (1860–1941) (Fig. 11) graduated from the Application School of Turin and specialized at École des Arts et Manufactures et des Mines in Liege from 1881 to 1883. He became Engineering Geologist, officer at the Geological Map Department and later Director of the Mining District

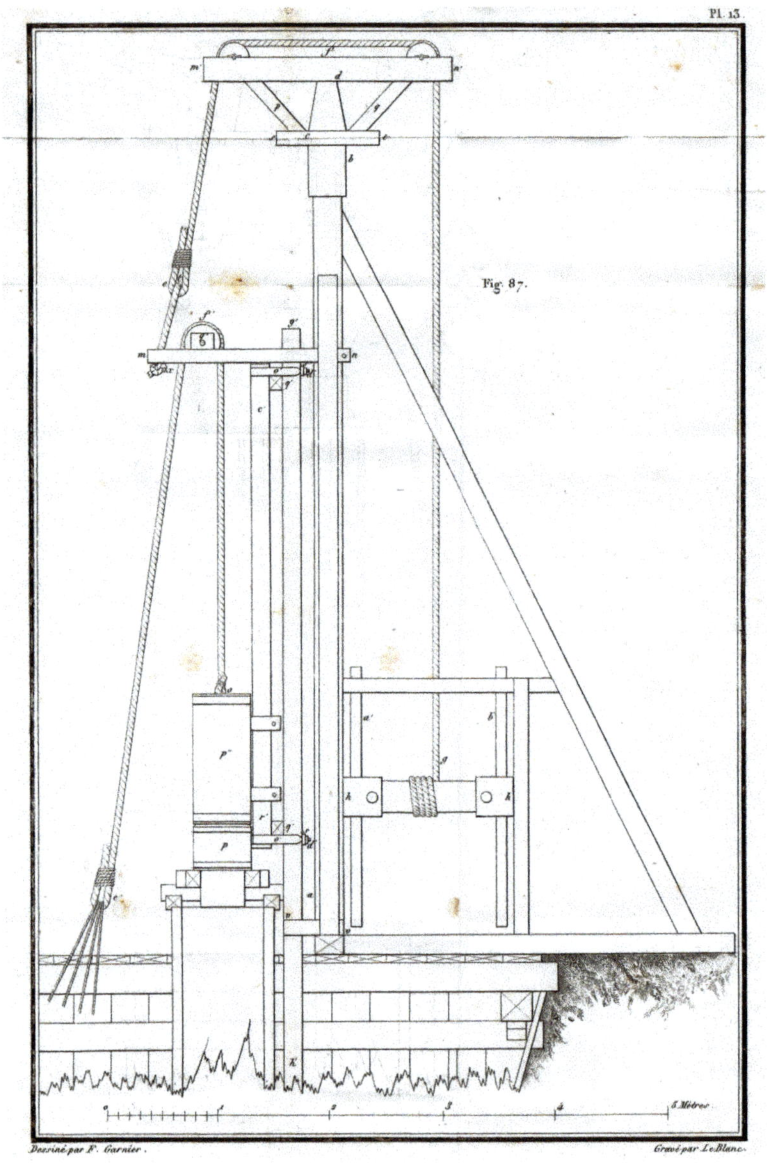

Fig. 15. Jacques Garnier's drilling machine, 1826.

of Bologna in 1902. In 1907, he attended the International Petroleum Congress held in Bucharest. On that occasion, he published a monograph on the oil industry in Italy, pointing also to the probability that:

> [T]he already exploited oil reservoirs of the Emilia Region seems to be of secondary origin, i.e. derived from petroleum concentrations located at the base of the 'argille scagliose' formation, at a depth not yet reached by drilling. [Argille scagliose is a geological term first used in 1840 to describe rocks in the northern Apennines of Italy, also known as Ligurian melange, Chaotic Complex, etc.] [...] Field experience confirms the existence of a correlation between the oil deposits of secondary origin and regional folding. So, it will be of the foremost importance to perform a complete geologic and tectonic survey in that part of the Apennines of the Emilia Region where petroleum seeps turned into an industrial activity
>
> (Camerana 1907, p. 123).

Bartolomeo Galdi (1873–1962) (Fig. 12) collaborated with Enrico Camerana for the entire period of research and writing of the aforementioned study on the Emilia oilfields. His highly detailed work had been appreciated during his former geological

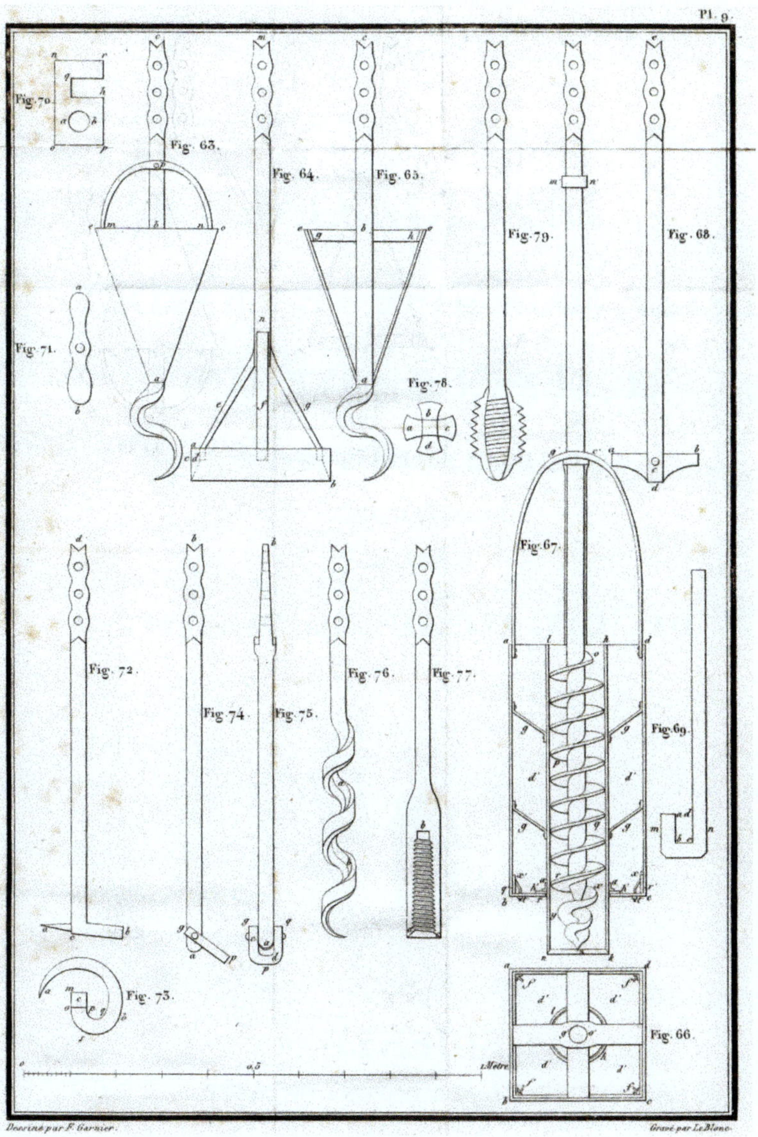

Fig. 16. Jacques Garnier's set of fishing and drilling tools, 1826.

surveys performed in the coal mines of Sardinia, and for this he was assigned as assistant to Camerana. He published several works on Italian hydrocarbons, and his recommendations on the existence of gaseous hydrocarbons in the subsurface of the Po river valley are outstanding. By 1911, he had already noted that very little was known by Italian geologists. Indeed, in those years, the natural gas seepages found in a few springs in the Po river valley area were attributed to modest local accumulations (it was only after World War II that they were fully explored and contributed a great deal to the modern industrialization of post-war Italy).

In May 1910, a preliminary study was presented to the Royal Corps of Mines. However, the Royal Geological Survey had not yet completed its geological mapping of the Emilia region, and new geological and 'tectonic' surveys were necessary, which were perfected by Galdi and allowed the outlining of the final cartographic and geological mapping to be utilized in their final study. The attention of Camerana and Galdi was addressed, in particular, to the precise and complete cartographical location of almost all hydrocarbon seepages. It must be recalled that in those years geophysical prospecting was not yet available to identify potential hydrocarbon reservoirs. The attention of petroleum exploration was then aimed at recognizing any possible surface indication, such as folding, faulting, seepages, stratigraphic reconstruction, mapping, etc.

The 1911 study of Camerana & Galdi was much appreciated in the scientific context of the time. It was completed under the direction of Bernardino Lotti, Director of the Royal Geological Survey,

DESCRIZIONE

DELLE MACCHINE PE' TRAFORI

MODENESI O ARTESIANI

E DEI POZZI FORATI IN TOSCANA

DAL 1829. AL 1833.

PUBBLICATA

DAL CAV. ALESSANDRO MANETTI

FIRENZE

TIPOGRAFIA ALL' INSEGNA DI DANTE

M. DCCC. XXXIII.

Fig. 17. Front cover of Alessandro Manetti's treatise on artesian well drilling rigs, 1833.

who was subsequently elected President of the Italian Geological Society. Camerana and Galdi certainly also used as a source of geological information the work published by Federico Sacco (Sacco 1892) and the conclusions on anticlines elaborated by Thomas Sterry Hunt (Hunt 1862). These latter two are also recalled indirectly in the preface, written by Lucio Mazzuoli and later President of the Italian Geological Society in 1906. The experiences previously gathered by Giovanni Giuseppe Bianconi (Bianconi 1840), Antonio Stoppani (1864, 1865, 1866) and Dante Pantanelli (1910), matured during extensive fieldwork, were the starting point on which were based the numerous surveys carried out by Bartolomeo Galdi.

Concluding notes on the development of mechanical drilling in Europe

In the early eighteenth century, in Europe, percussion drilling started to be developed and to become the preferred technique for constructing small-diameter (less than 50 cm) and 'deep' wells (more than 50 m). Percussion drilling methods changed from the primitive spring pole to develop gradually into the modern drilling rig models (mast and derricks) with a mechanical walking beam and a hoisting system.

The development of modern drilling tools and drilling rigs originated probably in France. In 1753, a well 80 m deep was drilled with a percussion rig at Drancy Pres Les Bourget, France: unfortunately, no technical details are available (de Thury 1829). In 1818, the French government provided financial support to foster the drilling of deep wells, over 600 m in depth, to supply Paris with high-quality freshwater withdrawn from artesian aquifers. The plates of the *Encyclopédie* by Diderot & d'Alembert are well known (Fig. 13), as are the treatise on the drilling of artesian wells by Abdon-Jacques-Frambourg Garnier (Garnier 1822, 1826). First a student at the École Polytéchnique and later a mining engineer, Garnier translated this discipline into the language of modern engineering, studying the construction of drilling rigs and tools, and defining a terminology still used today (sonde, sondage, sondeur, trepan, tige, curette (drill, drilling, driller drill-bit, drill-rod, scraper), etc.) (Figs 14–16).

The technology of Garnier was soon also exported to Italy. In 1833, the Italian hydraulic engineer Alessandro Manetti described the rigs used in the Grand Duchy of Tuscany for the drilling of water wells (mainly near the city of Florence) (Manetti 1833). He reported that the machinery (still operated manually or by means of horse power) was built

Fig. 18. The drilling of the Grenelle well in Paris, 1841.

by the Flachat Company (established in Paris) and designed by the above-mentioned Garnier (Fig. 17).

In 1841, under the impetus of François Arago (who was at the time the Mayor of Paris), after seven years of work, the drilling of the first famous well, the Grenelle well, 548 m deep, was completed (Fig. 18).

In 1845, again in France, Pierre-Pascal Fauvelle took only 22 days to drill a 170 m well in the town

GUIDE

DU SONDEUR

OU

TRAITÉ THÉORIQUE ET PRATIQUE

DES SONDAGES

PAR

MM. DEGOUSÉE ET CH. LAURENT

Ingénieurs civils, fabricants d'équipages de sonde, entrepreneurs de sondages pour les puits artésiens, les desséchements, la recherche des mines, l'étude des chemins de fer et canaux.

OUVRAGE ACCOMPAGNÉ D'UN GRAND NOMBRE DE FIGURES DANS LE TEXTE

Et d'un Atlas.

DEUXIÈME ÉDITION

REVUE, CORRIGÉE ET CONSIDÉRABLEMENT AUGMENTÉE.

TOME PREMIER

PARIS

GARNIER FRÈRES, LIBRAIRES-ÉDITEURS

6, RUE DES SAINTS-PÈRES, ET PALAIS-ROYAL, 215

1861

Fig. 19. Front cover of Degousée and Laurent's drilling handbook, 1861.

of Perpignan using pipes in which water circulated to remove the drill cuttings (Fauvelle 1845; Arago 1846). The invention of the principle of rotary drilling with the circulation of fluid is often attributed to Fauvelle, even though it is not certain whether he actually used a proper rotating drill string. His method

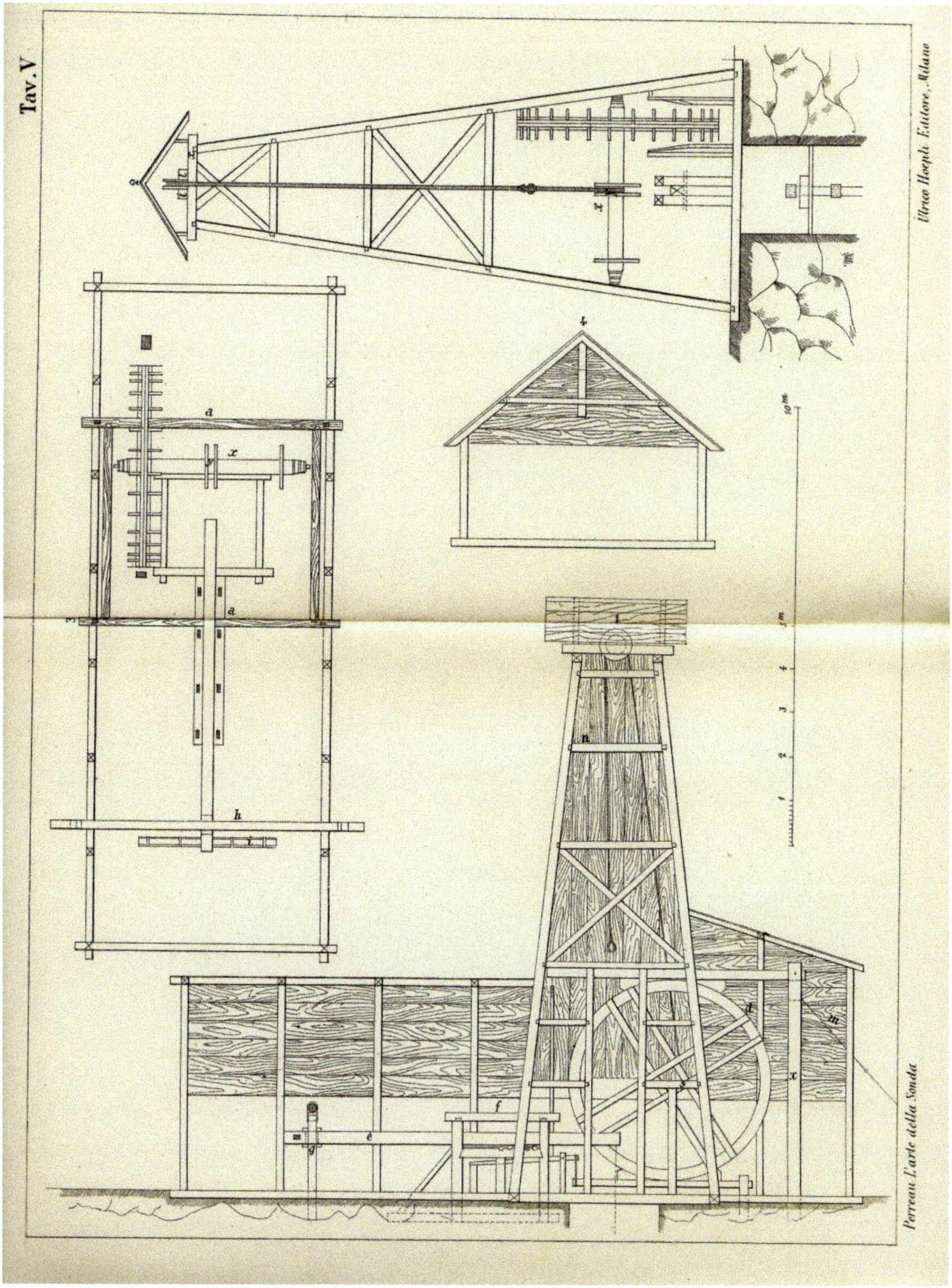

Fig. 20. Percussion drilling rig (Perreau 1885).

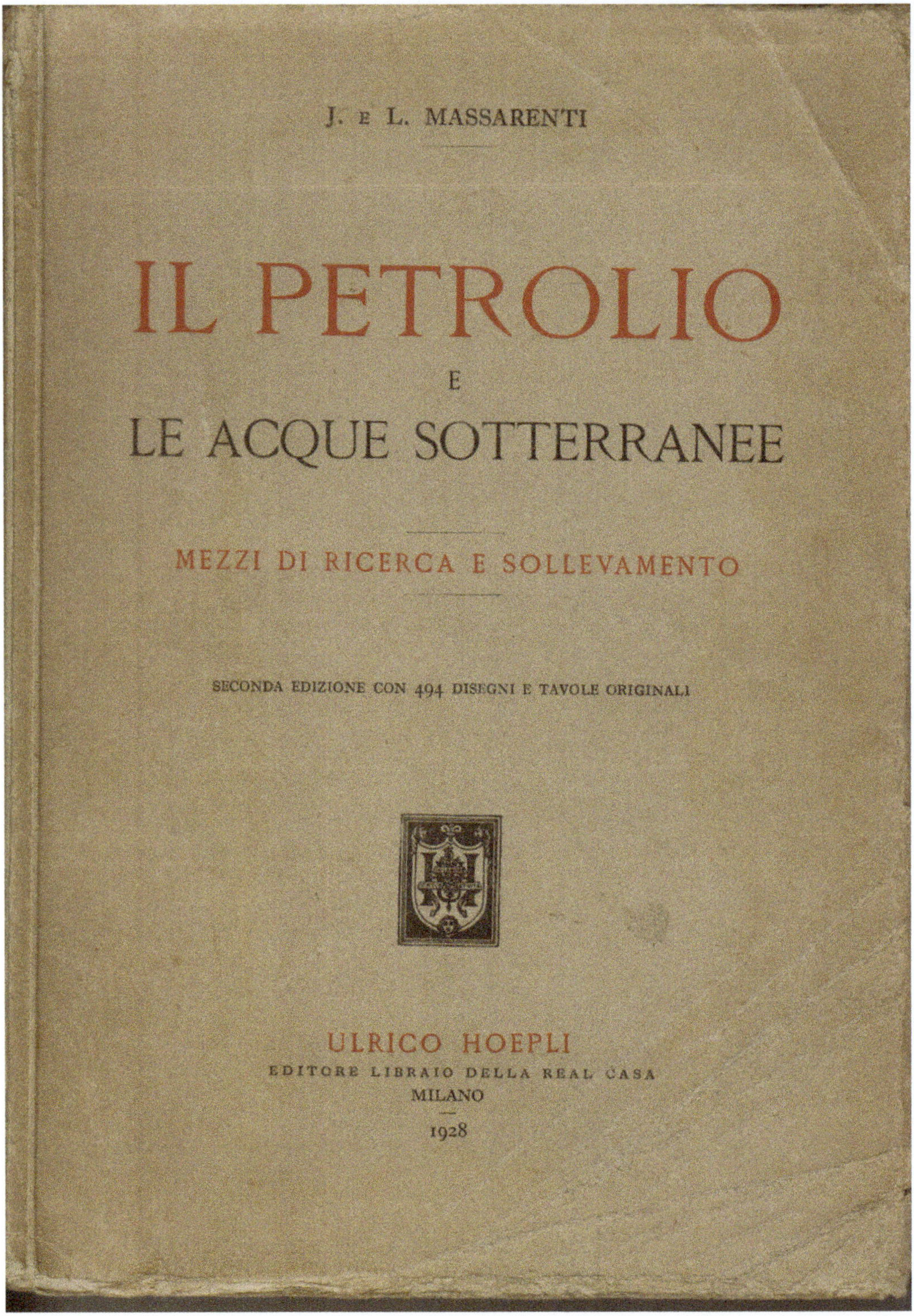

Fig. 21. Front cover of James and Luigi Massarenti's treatise on petroleum and water well drilling, 2nd edition, 1928.

was probably a percussion rig with water circulation but which, however, was a great step forward in well technology. The drill-string rotation was forced by hand during the upstroke. In any case, it is ascertained that Fauvelle conceived the idea of removing the cuttings by circulating a fluid and using drill pipes.

In the same period, we also recall the work of the Paris-based company Degousée–Laurent, founded and owned by Joseph Marie Anne Degousée and Charles Auguste Laurent (Fig. 19). They described themselves as 'civil engineers, drilling tool makers, water well and drainage well drillers, entrepreneurs in the field of mining, railways and canals' (Degousée & Laurent 1861).

In 1844, Robert Beart patented a rotary drilling method in England, while in the USA the earliest-known patent of this type was assigned to Bowles in 1857. Amédée Burat, in his treatise on the art of mining, set out a construction drawing of a steam-driven percussion rig, already modern in its basic architecture and in its operating principles (Burat 1846, 1855). The technique of drilling using a diamond coring bit, devised by the French engineer Rudolph Leschot a few years later, also stems from these early industrial applications (Brighenti & Macini 2008).

It is worth recalling that in Europe throughout the nineteenth century, rotary drilling was considered less reliable than percussion drilling, which appeared more efficient for the analysis of well stratigraphy. Until the second decade of the twentieth century in Europe, a large number of drilling rigs were still of the percussion type, and the predominant models were those of Fauck, patented in 1889, and of Raky; both were Hungarian engineers who used as the pattern of reference the Fauvelle system with water circulation.

In the USA, European technologies made slow progress, and probably evolved independently. The need to drill deeper wells occurred soon in the early part of the nineteenth century. The spring-pole technique spread rapidly in the eastern USA, and then further and further west. This developed into the so-called Canadian method, which was known as the Galician method in Europe, which used a drill-string composed of solid rods. In 1841, William Morris patented an articulated joint for

Fig. 22. The walking beam of a Massarenti drilling rig, 1920.

percussion drilling (drilling jar), and in 1854 a well was drilled at St Louis (Alabama) down to about 650 m to supply a sugar plant with freshwater. In 1858, a similar well was drilled at Louisville (Kentucky) in just 16 months. The technology also spread to Canada where, in 1858, at Petrolia (Ontario), a water well encountered oil and was completed as such. The Pennsylvanian method was a different and more agile cable-tool drilling, which was to become widespread within a few decades, and which later became the commonest method of drilling all over the USA and Europe. Nevertheless, the real development of percussion drilling took place only after the drilling of Drake's well in 1859 (Brighenti & Macini 2008).

From the completion of Drake's well until the end of the 1880s, the technology of percussion drilling made great advances, both in mechanical equipment and technical ability. In 1862, the first drilling service Company, John Eaton Co., was founded based in Oil City (Pennsylvania), a few miles from Titusville, which provided machinery and equipment. Later, machinery and drilling rigs increased in type, shape and principles of operation, which led to the development of large percussion systems used until the first half of the twentieth century. In Italy, we note the publication of the interesting drilling manuals written by Luigi Perreau in 1885, *L'arte della Sonda* [*The Art of Drilling*], an early example of Italian drilling technology (Fig. 20).

Very little is known about rotary tool drilling prior to 1844, although it is likely that the technique was already known and used in various parts of the world. In Europe, machinery and systems for rotary hydraulic drilling with diamond tools had already been patented by Leschot in 1863, and were used in mining and tunnelling. A noteworthy technological improvement was made in 1866, again in the USA, with the patent of P. Sweeney for the construction of a small-scale rotary drilling rig. Sweeney's machine operated with a toothed disc tool, the precursor of the modern rotary bits, which enabled the drilling of hard rock; it was hand operated and drilled with water circulation. From 1880 onwards, numerous patents in the USA and Europe were issued that aimed at improving a technology that, in a more or less rudimentary form, included all the essential elements of the present technology. Just to name but a few: in 1883, the Baker brothers obtained a patent for a new system of rotation and circulation (the Baker company, still in operation today, was founded in 1907 by Reuben C. Baker); in 1889, M.T. Chapman obtained a patent for an injection head and a twin-drum winch; in 1891 A.J. Ross obtained a patent for underreaming equipment, similar to the present-day underreamers; and S.W. Douglas obtained a patent for a downhole sampler (Brighenti & Macini 2008).

The first rotary oil rig was used in 1894 at Corsicana in Texas by John Galey, and then again in Texas, in 1901, to drill the famous giant Spindletop well (the first huge oil discovery made for many years, which awakened the interest in new oil ventures all over the world). The success of these first attempts, with the discovery of new oilfields, caused new rotary systems to be more and more appreciated and developed, even in the European oilfields. Although percussion rigs continued to drill the majority of the wells in the ensuing decades, rotary rigs started gradually to replace percussion ones in areas characterized by soft, unconsolidated formations (Brighenti & Macini 2008). Useful insights into the drilling technology and practices applied in the European oilfields are presented in Neuburger & Noalhat (1900), Beeby-Thompson (1910) and Brantly (1971).

Finally, we would like to remember the 1920s treatise *Il petrolio e le acque sotterranee – mezzi di ricerca e sollevamento* [*Oil and Groundwater – Means of Research and Lifting*] by the Italian brothers James and Luigi Massarenti (Massarenti & Massarenti 1920), the forerunners of all Italian drilling manuals to come for about 40 years (Figs 21 & 22).

Acknowledgements We thank Fabiana Console and the Library of ISPRA for assistance in finding Figure 4, and Marco Pantaloni, ISPRA, for the careful review of the paper.

Funding This research received no specific grant from any funding agency in the public, commercial or not-for-profit sectors.

References

Arago, M. 1846. Sur un nouveau système de forage par M. Fauvelle, de Perpignan. *Comptes rendus des séances de l'Académie des Sciences*, **23**, 438–440.

Ariosto, F. 1690. *Francisci Ariosti De oleo Montis Zibinii, seu petroleo agri mutinensis Libellus.* Bockenhoffer, Copenhagen.

Bacon, R.F. & Hamor, W.A. 1916. *The American Petroleum Industry, Volume 1.* McGraw-Hill, New York.

Bardini, C. 1998. *Senza carbone nell'età del vapore: gli inizi dell'industrializzazione italiana.* Mondadori, Milano, Italy.

Beeby-Thompson, A. 1910. *Petroleum Mining and Oil-Field Development: A Guide to Exploration of Petroleum Lands, and a Study of the Engineering Problems Connected with the Winning of Petroleum.* Crosby Lockwood and Son, London.

Bianconi, G. 1840. *Storia Naturale dei Terreni Ardenti, dei Vulcani Fangosi, delle Sorgenti Infiammabili, dei Pozzi Idropici, e degli Altri Fenomeni Geologici Operati dal Gas Idrogene e dell'Origine del Detto Gas del Dott.* Giuseppe Bianconi. Marsigli, Bologna, Italy.

Bonariva, A. 1869. *Memorie per la coltivazione dei petrolj nelle provincie parmensi*. Tipografi a G. Ferrari e figli Parma, Italy.

Bonariva, A. 1882. *Le perforazioni del suolo per la ricerca d'acque salienti e per esplorazioni minerarie*. Premiato Stabilimento Tipografico Successori Monti, Bologna, Italy.

Brantly, J.E. 1971. *History of Oil Well Drilling*. Gulf Publishing, Houston, TX, USA.

Brice, W. 2009. *Myth, Legend, Reality; Edwin Laurentine Drake and the Early Oil Industry*. Oil Region Alliance, Oil City, PA, USA.

Brighenti, G. & Macini, P. 2008. History of the upstream industry. Encyclopedia of hydrocarbons. *Istituto della Enciclopedia Italiana Treccani*, **V**, 525–548.

Burat, A. 1846. *Géologie appliquée, ou traité de la recherce et de l'exploitation des minéraux utiles*. Langlois et Leclercq, Paris.

Burat, A. 1855. *Traité du gisement et de l'exploitation des minéraux utiles*. Lacroix-Comon, Paris.

Calindri, S. 1781–83. *Dizionario corografico, georgico, orittologico, storico ec. ec. ec. della Italia. Montagna e collina del territorio bolognese*. Stamperia di San Tommaso d'Aquino, Bologna, Italy.

Camerana, E. 1907. *L'industrie des hydrocarbures en Italie: communication présentée au 3eme Congrès international du pétrole à Bucarest*. De J. Bertero, Rome.

Camerana, E. & Galdi, B. 1911. *I giacimenti petroleiferi dell'Emilia*. Stabilimento Poligrafico Emiliano, Bologna, Italy.

Camerana, E., Cotese, E. *et al*. 1926. *Studi geologici per la ricerca del petrolio in Italia*. Memorie Descrittive della Carta Geologica d'Italia, **20**.

Capacci, C. 1905. Apercu sur les gisements petroliferes de la province de Bologne (Italie). Communication présentée au 2me Congres International du Petrole, Liege, Belgium.

Carll, J.F. 1880. *The Geology of The Oil Regions of Warren, Venango, Clarion, and Butler counties: Including Surveys of the Garland and Panama Conglomerates in Warren and Crawford, and in Chautauqua Co., N.Y., Descriptions of Oil Well Rig and Tools, and a Discussion of the Preglacial and Postglacial Drainage of the Lake Erie Country*. Oil Region Alliance, Oil City, PA, USA.

Casali, A. 1874. *Il petrolio e i suoi derivati*. Tipografia e litografia di A. Cappelli, Modena, Italy.

Corsi, P. 2007. Much ado about nothing: the Italian Geological Survey, 1861–2006. *Earth Sciences History*, **26**, 97–125.

Crum, A.R. 1911. *Romance of American Petroleum and Gas*. Romance of American Petroleum and Gas Co., New York.

Degousée, J.M. & Laurent, C. 1861. *Guide du sondeur, ou traité théorique et practique des sondages*. Garnier Frères, Paris.

de Thury, H. 1829. *Considerations geologiques et physiques sur la cause du jaillissement des eaux des puit fores*. F. Didot, Paris

Diderot, D. & d'Alembert, J. Le R. (eds) 1751–72. *Encyclopédie, ou Dictionnaire Raisonné des Sciences, des Arts et des Métiers*. André le Breton, Michel-Antoine David, Laurent Durand and Antoine-Claude Briasson, Paris.

Fairman, E.J. 1868. *Treatise on the Petroleum Zones of Italy*. E.G.F.M. Spon, London.

Fauvelle, A. 1845. *Un nouveau procédé pour le forage de puits artesiens*. Perpignan, France.

Fenoaltea, S. 2001. *La crescita industriale delle regioni d'Italia dall'Unità alla Grande Guerra: una prima stima per gli anni censuari*. Banca d'Italia, Rome.

Forbes, R.J. 1958. *Studies in Early Petroleum Industry*. Brill, Leiden.

Forbes, R.J. 1959. *More Studies in Early Petroleum Industry*. Brill, Leiden.

Fougeroux de Bondaroy, A.D. 1770. Second Mémoire sur le Pétrole et sur des vapeurs inflammables, communes das quelques parties de l'Italie. *In*: *Mémoire de Mathématique et Physique de l'Académie Royale des Sciences*. Imprimerie Royal, Paris.

Garnier, A.J.F. 1822. *L'art du fontainier sondeur et des puits artésiens*. Huzard, Paris.

Garnier, A.J.F. 1826. *Traité sur les puits artésiens*. Bachelier, Paris.

Gerali, F. 2011. Il petrolio in Italia: l'Emilia. *Energia*, **32-4**, 62–67.

Gerali, F. & Gregory, J. 2016. Harsh oil: finding petroleum in early twentieth century Western Australia. *In*: Mayer, W., Clary, R.M., Azuela, L.F., Mota, T.S. & Wołkowicz, S. (eds) *History of Geoscience: Celebrating 50 Years of INHIGEO*. Geological Society, London, Special Publications, **442**, 425–436, https://doi.org/10.1144/SP442.33

Gerali, F. & Gregory, J. 2017. Understanding and Finding Oil over the Centuries: the Case of the Wallachian Petroleum Company in Romania. *Earth Science History*, **36-1**, 41–62.

Hunt, T.S. 1862. Notes on the history of petroleum or rock oil. *In*: *Annual Report of the Board of Regents of the Smithsonian Institution, Showing the Operations, Expenditures, and Condition of the Institution for the Year 1861*. House of Representatives Miscellaneous Document 77, 37th Congress, 2nd Session, Government Printing Office, Washington, 319–329.

Kovacs, G. 1964. *Storia delle raffinerie di petrolio in Italia*. Colombo, Rome.

Macini, P. & Mesini, E. 2011. *L'industria petrolifera in Emilia tra Otto e Novecento*. CLUEB, Bologna, Italy.

Magini, M. 1976. *L'Italia e il petrolio tra storia e cronologia*. Mondadori, Milano, Italy.

Manetti, A. 1833. *Descrizione delle macchine pe' trafori modenesi o artesiani e dei pozzi forati in Toscana dal 1829 al 1833*. Tipografia all'insegna di Dante, Firenze, Italy.

Massarenti, J. & Massarenti, L. 1920. *Il petrolio e le acque sotterranee – mezzi di ricerca e sollevamento*. Hoepli, Milano, Italy.

Miano, La Società. 1872. *Miano: Società Anonima Italiana per l'Estrazione del Petrolio in Miano di Medesano Parmense. Circolare 319, 26 Novembre 1872*. Corporate communication, Parma.

Mojon, G. 1806. *Corso Analitico di Chimica*. Dalla Stamperia di Giovanni Grossi, Genova.

Novelli, L. & Sella, M. 2009. *Il petrolio: una storia antica*. Silvano, Milano, Italy.

Neuburger, H. & Noalhat, H. 1900. *Technologie du Pétrole*. Fritsch, Paris.

PANTALONI, M. 2014. 15 giugno 1873, nasce il R. Ufficio Geologico: 140 anni di geologia in Italia. *Geologia Tecnica & Ambientale*, **1**, 37–44.

PANTANELLI, D. 1910. *Ricerche sul petrolio emiliano.* Regia Accademia di Scienze, Lettere e Arti di Modena, Modena, Italy.

PAOLONI, C. 1976. *Storia del Metano.* Tipografia SAPIL, Milano, Italy.

PERREAU, L. 1885. *L'arte della sonda. Manuale teorico-pratico per gli apparecchi e le opere di trivellazioni del suolo. Corredato di XXXIII tavole figurative. Opera originale italiana.* Hoepli, Milano, Italy.

SACCO, F. 1892. *L'Appennino Settentrionale (parte centrale). Studio geologico.* Regia Accademia dei Lincei, Rome.

SACCO, F. & STELLA, A. 1926. *Studi geologici per la ricerca del petrolio in Italia.* Memorie Descrittive della Carta Geologica d'Italia, Roma, **20**.

SAUSSURE, DE T. 1816. Recherches sur la composition et les propriétés du naphte d'Amiano dans les États de Parme. *Annales de Chimie et de Physique*, **3–4**, 314–321.

SAUSSURE, DE T. 1817. Experiment on the composition and properties of the naphtha of Amiano. Read to the Society of Natural Philosophy and Natural History of Geneva. *Annals of Philosophy*, **10**, 118–127.

SCICLI, A. 1972. *L'attività estrattiva e le risorse minerarie della regione Emilia-Romagna.* Modena.

SCOTTI, L. 1901. *Ricerche di petrolio nel piacentino.* Bosi, Piacenza.

SÉDILLOT, R. 1974. *Histoire du pétrole.* Fayard, Paris.

SPALLANZANI, L. 1792. *Viaggi alle due Sicilie ed in alcune parti dell'Appennino.* Stamperia di B. Comini, Pavia.

SQUARZINA, F. 1958. *Le ricerche del petrolio in Italia. Cenni storici dal 1860 e cronache dell'ultimo decennio.* Jandi-Sapi, Roma.

STOPPANI, A. 1864. Saggio di una storia naturale dei petrolii. *Politecnico. Repertorio mensile di studj applicati alla prosperità e alla coltura sociale XXIII*, Milano.

STOPPANI, A. 1865. *Relazione di un'esplorazione geologica di alcuni distretti petroliferi nelle Provincie di Parma e Piacenza.* Civelli, Milano, Italy.

STOPPANI, A. 1866. I petroli d'Italia. *Il Politecnico, repertorio di studj letterari, scientifici e tecnici parte letterario-scientifica*, Series 4, 1 (2), 77–93, 574–586.

STOPPANI, A. 1870*a*. Le sorgenti di petrolio. *Prime Letture I*, **1870**, 132–136.

STOPPANI, A. 1870*b*. I pozzi di petrolio. *Prime Letture I*, **1870**, 323–326.

STOPPANI, A. 1876. *Il bel paese, conversazioni sulle bellezze naturali la geologia e la geografia fisica d'Italia.* Milano, Italy.

VECCHIA, D. 1916. *Il petrolio in Italia e la Zona Petrolifera di Rivanazzano.* Redaelli, Milano, Italy.

VOLPICELLI, G. 1848. Rapporto sopra un processo del sig Silas Medail di Susa, per ottenere il gas dal goudron, residuo della distilazione del carbon fossile. *Raccolta di lettere ed altri scritti intorno alla fisica ed alle matematiche. Raccolta scientifica di fisica e matematiche*, **4** (1), 139–140.

VOLTA, A. 1777. *Lettere del signor Don Alessandro Volta … sull'aria infiammabile nativa delle paludi.* Marelli, Milano.

VOLTA, A. 1784. *Sopra i fuochi de' terreni e delle fontane ardenti …, Opuscoli scelti sulle scienze e sulle arti*, tom.7-par.5, 321. Marelli, Milano.

ZANONI, E. 2014. *Scienza, patria, religione: Antonio Stoppani e la cultura italiana dell'Ottocento.* Franco Angeli, Milano, Italy.

ZOLESI, G. 1891. *Il petrolio in Italia e specialmente in depositi sotterranei nel sottosuolo di Rivanazzano soprastante il fiume Staffora e la Madonna del Monte.* Stabilimento Tipografico di G. Schenone, Genova, Italy.

ZOPPETTI, V. 1883. *Sull'industria dei petroli nelle provincie di Parma, Piacenza e Pavia.* Relazione del Servizio Minerario nel 1881, Rome.

Ayoluengo's history – Spain's only onshore oil field

JORGE NAVARRO COMET

Department of Geosciences, CEPSA E.P., S.A., Paseo Castellana (Torre Cepsa) 259, Madrid 28046, Spain

0000-0002-9864-8280

jorge.navarro@cepsa.com

Abstract: The Ayoluengo Field, commonly cited as the only onshore oil field in Spain, was discovered in June 1964 by Amospain, a joint venture of Chevron, Texaco and CAMPSA, by then the state-owned company. Ayoluengo is located about 300 km north of Madrid in the southern part of the Basque–Cantabrian Basin, a geological region where natural oil seeps, tar sands and asphalt have been recognized in outcrops since the early twentieth century. Now, 50 years after the first oil in 1967, the field has a cumulated production of 17 million barrels of oil. The 50-year production concession expired at the end of January 2017 and the field is now closed, awaiting a bidding process for a new concession to be awarded.

The Ayoluengo Field consists of a NE–SW-orientated salt-cored anticline, related to Triassic salt movements. The field is divided into two large structural blocks by a normal fault. Oil and gas production comes from a series of thin lenticular fluvio-lacustrine sandstone packages of Late Jurassic–Early Cretaceous age. More than 50 separate oil and gas sandstone beds have been identified by drilling. This multilayer reservoir, together with the structural component, means that Ayoluengo is considered to be a grouping of hundreds of small oil and gas fields.

After years of intense exploration activity, the Ayoluengo Field still, surprisingly, remains a unique oil discovery and is the only onshore commercial oil field in Spain and also the only one in the entire Iberian Peninsula. This geological singularity has resulted in recurrent discussions between petroleum geologists because it is difficult to explain why a petroleum system is working uniquely at this particular spot and nowhere else within such a vast territory.

The early years

The Ayoluengo Field was the first commercial oil discovery in Spain and today, more than 50 years later, remains as the only onshore oil field in the Iberian Peninsula.

The field was discovered in 1964 and first oil was produced in 1967. It is located about 300 km north of Madrid in the southern part of the Basque–Cantabrian Basin (Fig. 1), a geological region where natural oil seeps, tar sands and asphalt have been recognized in outcrops since the early twentieth century (Dupuy de Lôme 1941; Sanz 1967; Marín *et al.* 2014). This region was considered to be highly promising, and most of the hydrocarbon exploration efforts in Spain during the 1940s and 1950s were focused here (Sanz 1967; Navarro 2014, 2016*a*, *b*).

In 1946, the 'Compañía Arrendataria del Monopolio de Petróleos Sociedad Anónima' (CAMPSA), the Spanish government petroleum monopoly created in 1927, was granted the hydrocarbon exploration rights for an area of 2800 km^2 north of Burgos that was declared a state national reserve (Sanz 1948; Almela *et al.* 1952). With light rigs, CAMPSA drilled some shallow stratigraphic wells (Almela *et al.* 1946), all based on surface geological surveys as no reflection seismic data were available then, resulting in many of the outcropping anticlines being pierced.

Later exploration by CAMPSA was focused in the Zamanzas Valley, along the eroded axis of a large surface anticline with outcropping Cretaceous tar sands in its core (Ríos 1955; Puche *et al.* 2011). The shallow wells typically found heavy black oil while drilling the Cretaceous section; traces of gas and very small amounts of lighter oil (26–28° API) along with salt water were occasionally recovered from the Jurassic carbonates, but no commercial flow was established (Fig. 2). Additionally, during the 1940s, rudimentary and experimental underground mining in the Zamanzas Valley was carried out by CAMPSA to exploit the tar sands, a process that was quite primitive. The tar sands were crumbled and dumped into large water tanks that were heated by wood fire (Ayala 2006). The heat facilitated the separation of the bituminous material which floated on the surface, and then was manually collected with dippers and poured into barrels. Oil production was marginal, some 1–3 barrels of oil per day. Eventually this exploitation was abandoned because of the uneven distribution of the tar impregnations and the poor economic results.

After the Spanish Civil War (1936–39), the country faced economic sanctions from the international

From: Craig, J., Gerali, F., MacAulay, F. & Sorkhabi, R. (eds) 2018. *History of the European Oil and Gas Industry*. Geological Society, London, Special Publications, **465**, 333–343.
First published online May 8, 2018, https://doi.org/10.1144/SP465.9

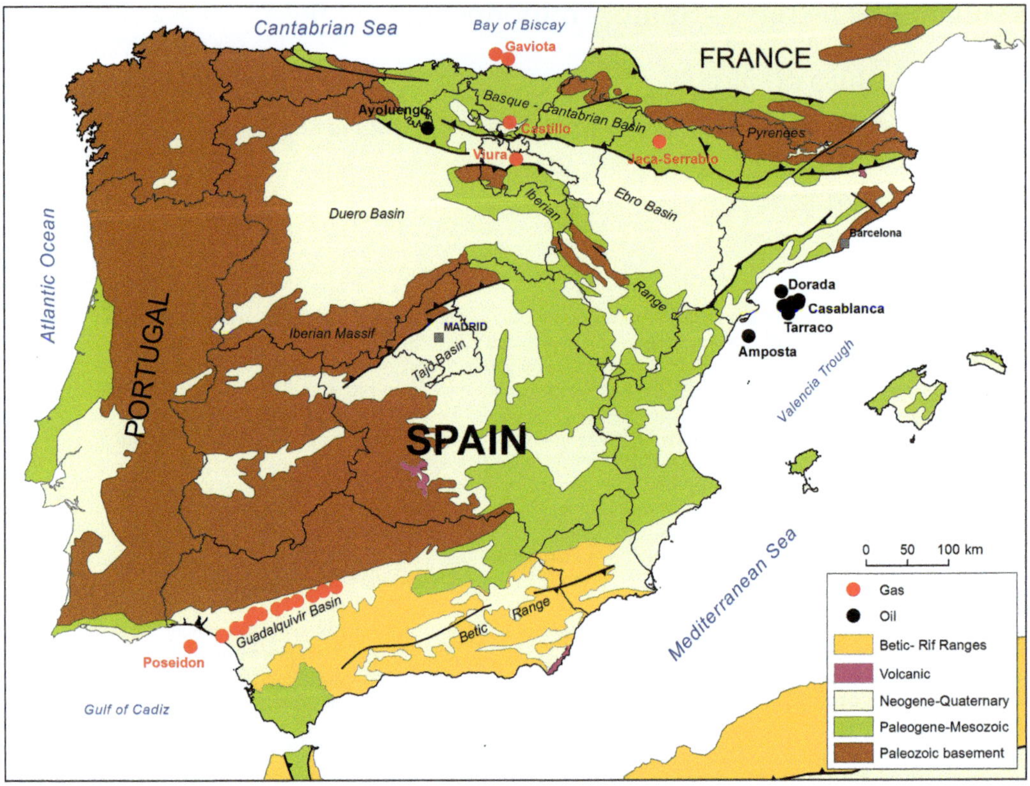

Fig. 1. Simplified geological map of the Iberian Peninsula showing the location of the commercial oil and gas discoveries made to date in Spain. Ayoluengo is the only onshore oil field, located within the Basque–Cantabrian Basin, about 300 km north of the city of Madrid.

community. Economic isolation combined with the nationalist rhetoric committed the Franco dictatorship regime to pursue self-sufficiency (San Román & Sudrià 2003; Tortella *et al.* 2003). By the late 1950s, the failure in self-sufficiency planning seemed clear, and the regime began modifying economic

Fig. 2. A drilling rig operated by CAMPSA in 1946 in the Zamanzas anticline (Burgos). This area was located within the state national reserve north of Burgos (photograph from Almela *et al.* 1946).

policy and lowering restrictions to foreign investment. In December 1958, a new hydrocarbon law was enacted for petroleum exploration and production that sought to encourage foreign companies to explore. As a result, most of the promising Spanish basins were covered with exploration licences (Fig. 3). It allowed companies with as much as 100% foreign capital to work in Spain, while favouring the association of national and foreign companies for the exploration of state reserves, with the state keeping the majority of the shares.

In 1959, CAMPSA signed an agreement with American Overseas Petroleum Ltd, whereby CAMPSA assigned a 50% interest in their exploration permits: 25% went to Standard Oil Company of California (which years later became Chevron) and 25% to Texaco (Ballestero 1989). In 1960, the company named Amospain (CAMPSA 50%, California Oil Company of Spain 25% and Texaco Spain Inc. 25%) was awarded the operatorship of the Ubierna Exploration Permit (Fig. 3), which was located within the former state national reserve north of Burgos, previously held by CAMPSA.

Under this new operatorship, exploration effort was focused on the Ubierna permit, where photogrammetric studies, stratigraphic measurements, detailed surface geological mapping and modern reflection seismic data were acquired during the early 1960s. These studies resulted in the identification of a faulted anticline in the subsurface, below an Upper Cretaceous carbonate flat plateau, an agricultural terrain mostly dedicated to growing potatoes and locally known as 'Loras'. Although this faulted anticline was suspected from surface mapping, it was the first properly matured seismic prospect to be tested in the north of the Burgos area (Fig. 4).

The Ayoluengo discovery

The location of the first exploration well was carefully chosen jointly by Amospain and CAMPSA's engineers and geologists. It was named Ayoluengo-1, as per the small village nearby, and located some 15 km SW of the Zamanzas Valley, where previous shallow exploration drilling by CAMPSA had been concentrated. The well was designed as a 3500 m-deep test of the Cretaceous sandstones, the Lower Jurassic marine carbonates, which had commonly recorded oil shows in the old CAMPSA wells, and the Triassic section (Sanz 1967).

The well was spudded on 5 May 1964 into limestones of Upper Cretaceous age, and was terminated at 2397 m in salt and anhydrite of Triassic age (Amospain 1964). From 990 to 1346 m depth, numerous poor shows of oil were observed, none worthy of further interest. On 2 June, at 1346 m, the well penetrated a 5 m-thick sandstone bed of Late Jurassic–Early Cretaceous age with significant oil and gas shows. The geological personnel assigned permanently to the well at that moment were William Thornton (Bill) Stoeckinger, an American geologist working for Standard Oil Company of California, and Cristóbal Racero, a Spanish mining engineer working for CAMPSA. A first drill stem test (DST) was run, but the testing tool failed. In order to improve the borehole conditions for testing, the well was deepened a by few metres and a conventional core was cut over a very hard shaly section, just below the oil-bearing sandstones (Amospain 1964). Then, a second DST (1348–1361 m) was run on Saturday 6 June. The tool was opened at 08:55 a.m. with an immediate good blow that rapidly increased to strong. The oil rushed up the hole and flowed up to a height of some 30 m in the air, spraying oil across the drilling site, vehicles and nearby cropland. The well was rapidly brought under control and flowed naturally for 1 h. Flow rate was estimated at 85 bbl (barrels) of 36° API oil per day (Amospain 1964; Sanz 1967). Euphoria spread at the rig site, especially among the Spanish technicians from CAMPSA who had witnessed for the first time such an oil flow in their country. The church bells in nearby villages began to ring, and inhabitants came to the well location to observe the oil flow and celebrate the discovery by toasting with champagne, shooting fireworks and collecting some oil samples as souvenirs. Meanwhile, CAMPSA employees were carefully planning how to relay this exciting news to Ruperto Sanz, the CAMPSA exploration manager, a Spanish mining engineer who had been one of the petroleum exploration pioneers in Spain, and a strong supporter of foreign investment, and was fully persuaded of the commercial oil possibilities in this region.

Spain's 'Oklahoma'

Ayoluengo was the first commercial oil discovered in Spain after drilling more than 100 holes. It brought great expectations to the region, which was presumed to become a prolific 'black gold' area. The oil discovery gained national attention, with large coverage in the media and attracting many visitors to the well site, as shown in the photograph taken in June 1964 (Fig. 5). Rumours and speculation abounded in the national press as discussions of the importance of the discovery, the size of the reserves and the quality of the oil became headline news. An 'Oklahoma Oil Boom in Spain' was headlined by some national newspapers (Martínez & Moreno 2014). To avoid speculation, trading of CAMPSA shares on the Madrid Stock Exchange was suspended for several days. The radio, television and the popular NO-DO, a state-controlled series of

(a) (b)

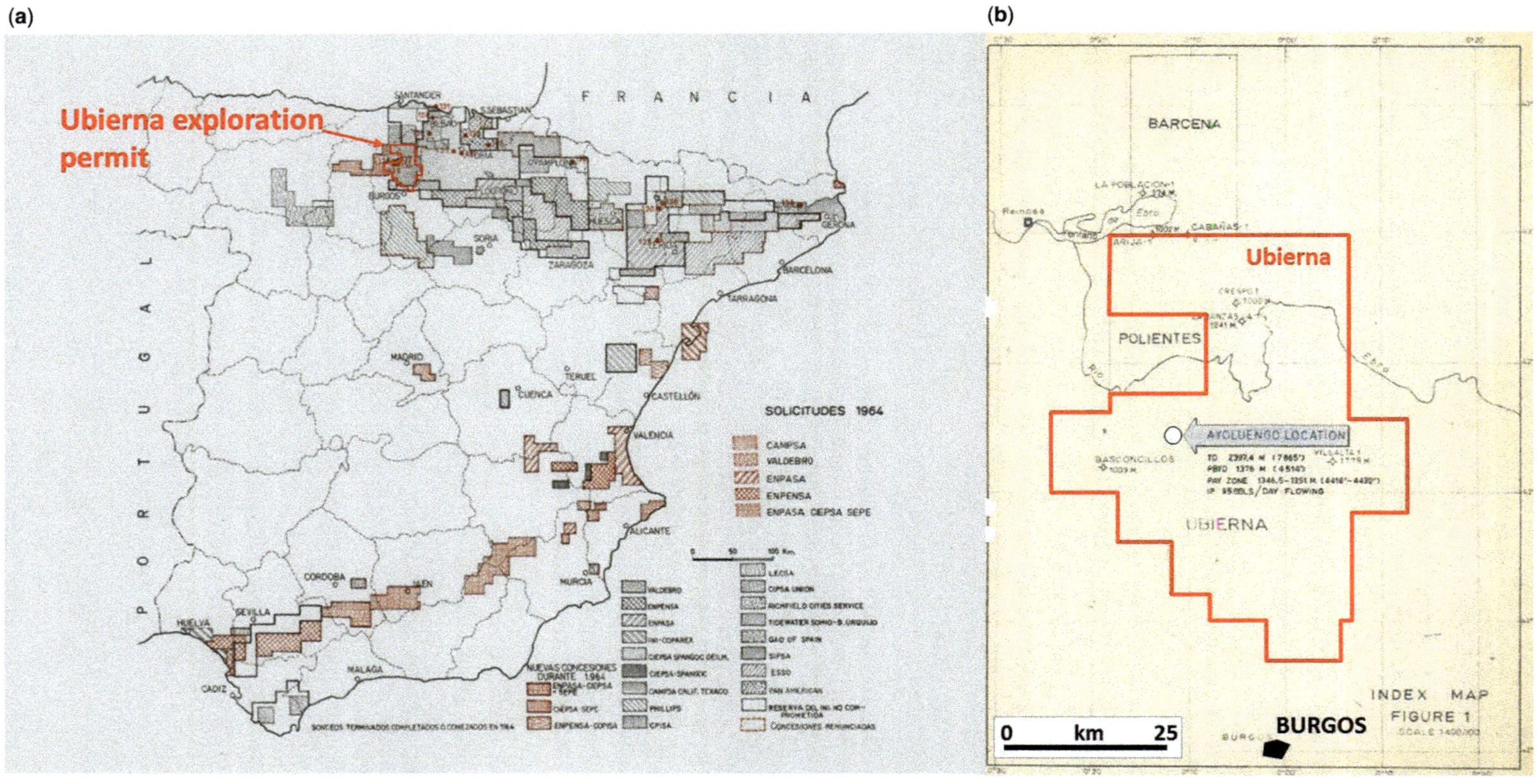

Fig. 3. (**a**) Map of Spain showing the active exploration licences in 1964 (Ríos 1965). The main exploration activity in 1964 was concentrated in the southern Pyrenees, in the Basque–Cantabrian, Ebro and Guadalquivir basins. (**b**) The Ubierna exploration permit showing the location of the Ayoluengo-1 discovery well (Amospain 1964).

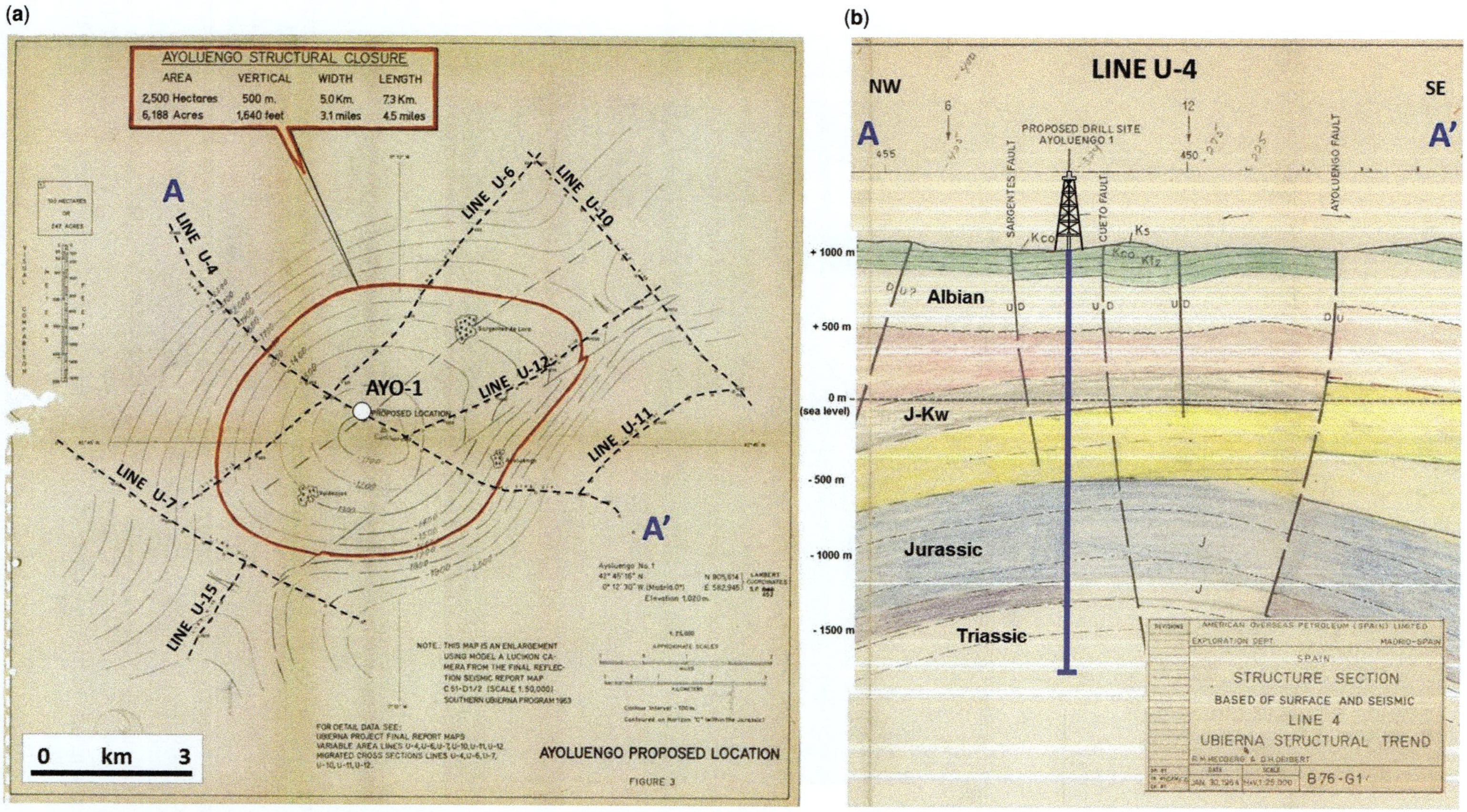

Fig. 4. (**a**) Structural map of an intra-Jurassic horizon showing the location of the Ayoluengo-1 well proposal. The well was located on seismic line U-4 at the highest structural point. (**b**) Structural cross-section based on surface geology and seismic line U-4 (Amospain 1964).

Fig. 5. The Ayoluengo-1 drilling rig. The photograph was taken by Federico Vélez on June 1964, a few days after first oil was tested (Martínez & Moreno 2014).

cinema newsreels that contained servile reporting in favour of the Franco regime, widely broadcast the good news. With news of the discovery of the first oil in Spain being so widespread, well-site geologist Bill Stoeckinger became a sort of celebrity in Madrid. According to Ina Stoeckinger, Bill's wife: 'Bill was often recognized and congratulated on the street by total strangers who had seen him on the television news covering the story'. CAMPSA's Ruperto Sanz was also continuously requested by the media to release data related to the Ayoluengo discovery, even directly contacted by some government members to provide them with 'original, unfiltered information'. The two Spanish mining engineers, Ruperto Sanz and Cristóbal Racero, were decorated in May 1965 by the Spanish government for their outstanding participation in the Ayoluengo discovery.

After well testing, Ayoluengo-1 drilling continued to the top of the Triassic, reaching a total depth of 2397 m on 18 July 1964 (Fig. 6). An oil zone (2150–2164 m) in the Lower Jurassic carbonates was tested, flowing a small amount of 41° API oil that revealed a very tight reservoir which was considered as non-commercial. A total of eight DSTs were run, the well was plugged back and completed in the upper sandstone bed.

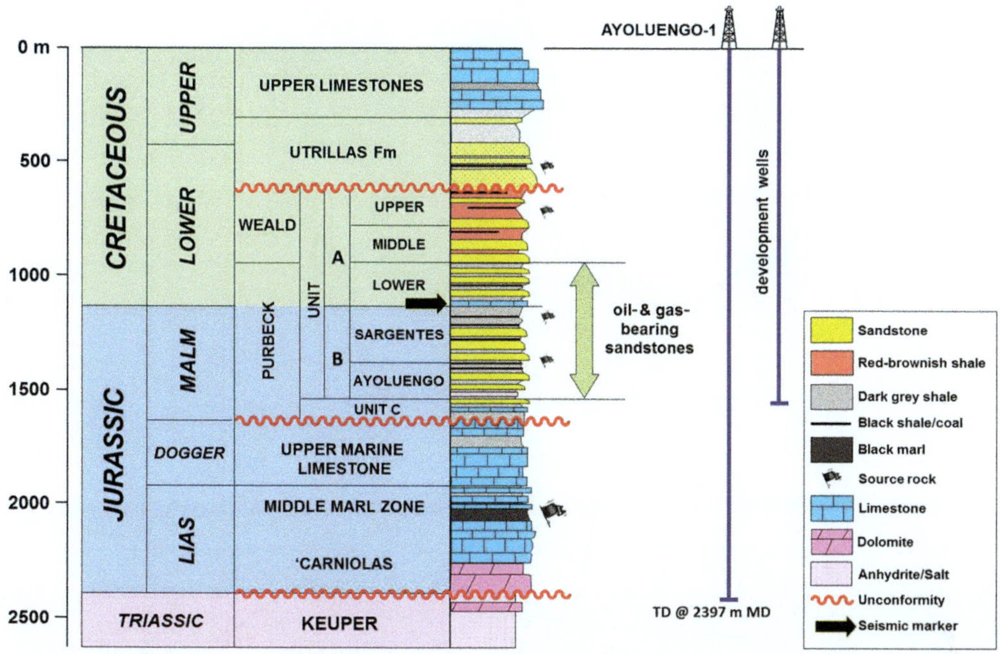

Fig. 6. Stratigraphy of the Ayoluengo oil field, showing the productive interval, and the section penetrated by Ayoluengo-1 and the field development wells.

The Ayoluengo-1 discovery well appeared to have been drilled on the NW flank of the anticline and not on the crest. Further appraisal and delineation drilling at higher structural positions resulted in finding additional and thicker oil-bearing sandstones, but also evidenced a large degree of vertical and horizontal compartmentalization, showing that the structure was a complex anticline divided by several normal faults (Fig. 7).

The Ayoluengo petroleum system

The Ayoluengo Field, as we know today, consists of a NE–SW-elongated salt-cored anticline related to Triassic salt movements (Klimowitz *et al.* 1999). The Ayoluengo trend is viewed as an arch separating the Polientes Trough to the NW and the Sedano Trough to the SE (Beroiz & Permanyer 2011). The Ayoluengo Field covers an area of 10 km^2 and the structure has a vertical closure of some 200 m (Fig. 7). The structure is divided into two structural blocks by the Ayoluengo normal fault, with 250 m of vertical throw that affects the Jurassic carbonates up to the surface (Abeger *et al.* 2006).

The field produces from a series of thin beds of fluvio-lacustrine sandstone packages of Late Jurassic–Early Cretaceous age, with no single reservoir present and is productive over the entire area. Most of the oil production comes from the locally known Unit B or Purbeck, which is further subdivided into two zones: Sargentes and Ayoluengo (Fig. 6). The most prolific reservoir is in the Ayoluengo sandstones on the upthrown side of the NE–SW main fault (Ríos 1965). More than 50 separate oil and gas sandstone beds have been identified (Alvarez & García 1996; García & Alvarez 1996). Some beds are as thick as 10 m, but the average is only 2–3 m. The areal extent of these lenticular sandstone bodies varies widely. Some are quite restricted (less than 2 km^2), while others are laterally continuous. Although many of the sands are silty to shaly, porosity and permeability is typically fair to good, with effective porosities commonly in the 12–18% range and permeabilities up to 1 D. The sandstone beds are friable and composed of quartz. In general, sand development and net sand in the section improves in a northeasterly direction, and tends to shale out to the south and west (Sanz 1967). Because of the number of fault blocks and the rapid lateral changes in sand percentage, the thicknesses of the oil columns are also extremely variable. An exceptional and continuous limestone bed, located at the Cretaceous–Jurassic boundary and varying in thickness from 15 to 30 m, is used as a stratigraphic and seismic marker (Fig. 6), where minor quantities of oil have been also tested if naturally fractured.

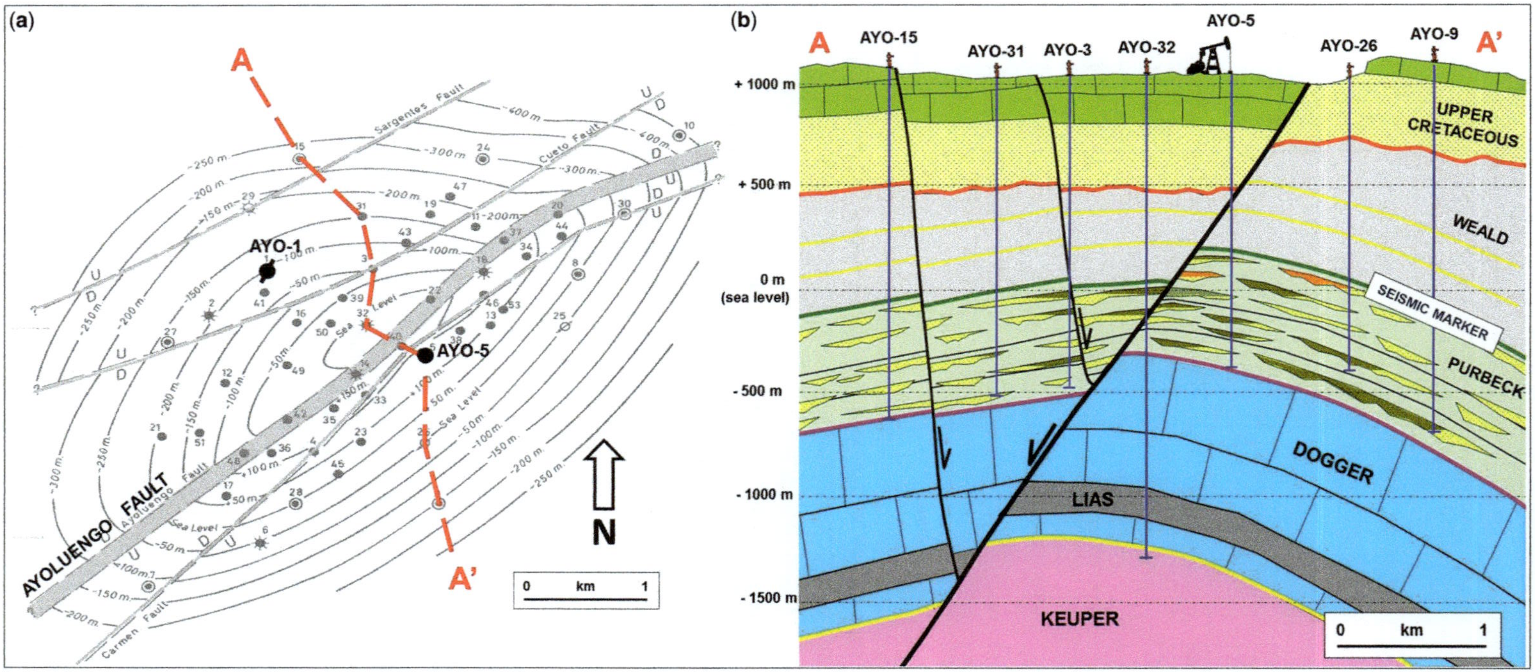

Fig. 7. (**a**) Ayoluengo depth map of the limestone seismic marker (see Fig. 6). Depth is subsea in metres. Note the Ayoluengo-1 well located downdip and the Ayoluengo-5 well located at the structure culmination. (**b**) NW–SE structural cross-section along the Ayoluengo faulted anticline. The location of the cross-section is marked on the map.

Most of the individual reservoir layers are isolated by shales and compartmentalized by faults, making the Ayoluengo not a single field but a grouping of more than 100 independent small accumulations (Alvarez & García 1996; García & Alvarez 1996). The reservoir drive mechanism is primarily gas expansion and gravity drainage. Oils present different physical properties from well to well and from sand to sand, with gravities ranging from 20° to 39° API, which demonstrates the complexity of the petroleum system. The organic-rich marls and black shales of Liassic age have largely been considered as the only source of the Ayoluengo oils (Quesada *et al.* 1997, 2006; Beroiz & Permanyer 2011), but this is still far from clear. The analysis of outcrops of the Liassic organic-rich shales reveal a total organic carbon (TOC) content of around 4%, high Rock-Eval S_2 values ($S_2 < 21$ mg HC/g of rock) and hydrogen indices (HI) greater than 400 mg HC/g of organic carbon (Marín *et al.* 2014), being by far the most organic-rich section in the region. Oil-to-source correlations appear to indicate the Liassic as the only source interval (Quesada *et al.* 1997; Beroiz & Permanyer 2011). However, there are several geological constraints that lead us to question this interval as the only source rock for the Ayoluengo oils. The wide variety of Ayoluengo oils, the puzzling vertical alternation of water, gas and oil, and the complex migration pathways (lateral and vertical) required to charge the lenticular sandstone beds from the Liassic source section lead us to challenge the widely accepted Ayoluengo model. It would first be required to migrate laterally 5–10 km through the tight Jurassic carbonates from the adjacent Sedano or Polientes troughs (Beroiz & Permanyer 2011) and then move updip through the Ayoluengo Fault or associated fractures into the reservoir beds of a particular stratigraphic section, with no further migration updip, even into much better quality and shallower reservoirs. Furthermore, all the oil-to-source correlations are based on mixed oil samples, either coming from a bulk Ayoluengo oil (Quesada *et al.* 1997) or commingled (i.e. from a single well but with several reservoir beds open to production) (Beroiz & Permanyer 2011). Probably the thin beds of coals and carbonaceous pyritic shales, which commonly appear interbedded within the producing sandstones, are responsible for most of the oil and gas in Ayoluengo. However, it is clear that more detailed geochemical analyses are still required to clearly demonstrate the origin of the Ayoluengo oils.

Ayoluengo production

In January 1967, an exploitation concession named Lora derived partially from the Ubierna exploration permit was awarded for 50 years. Some discussions were held concerning the laying of a 150 km oil pipeline to the port city of Bilbao or even building a refinery in Burgos, as was requested by the local authorities. Finally, at the end of 1966, a 11 km pipeline was completed from the field's central facilities to a cargo terminal on the Burgos–Santander highway, 300 m lower in altitude. The first Ayoluengo oil production started in February 1967, at a rate of 3500 bbl of oil/day from 15 wells, reaching the peak production of 5200 bbl of oil/day in 1969. Since then, production has gradually decreased.

The Ayoluengo oil is paraffinic with a relatively high arsenic and vanadium content that damages catalysts and makes it inadequate for refining. An alternative market was found and production was sold as fuel oil to local users in northern Spain, and it continues today. Oil is produced by rod pumps, locally and popularly known in Spanish as 'caballitos'. The small amount of produced natural gas is used to power the rod-pump motors and to generate the electricity used in the field. A gas power plant to generate electricity (2.5 MWh) for sale was installed in 1986, but the gas production decline forced it to be dismantled in 1997. Produced water (around 50 000 ppm NaCl) is injected back into the reservoir in one disposal well.

All of the CAMPSA holdings, including the Lora concession, were assigned in 1981 to the state company Eniepsa, which in 1985 merged with Hispanoil and two years later became Repsol. In 1985, the Chevron Oil Company of Spain (formerly the California Oil Company of Spain) replaced Amospain as the Ayoluengo operator, which in 1990 transferred the Ayoluengo operatorship to Repsol. In 2002, Repsol sold its interest to the British company Northern Petroleum, which in turn sold it to Ascent Resources plc who, finally, in 2007 farmed it out to the latest field operator, the 'Compañía Petrolífera de Sedano S.L. (CPS)', a subsidiary of the British company Leni Gas & Oil plc (now Columbus Energy Resources plc).

A total number of 52 wells have been drilled in the field, the last one in 1990 when the field was operated by Repsol. During the last years of production, only 10 wells were active. Many of the infill wells encountered undepleted oil-bearing sandstone beds, indicating the field's complexity. A 3D vibro seismic of 70 km^2 was acquired in 1988 aimed at identifying undrained reservoir beds and better estimating the remaining reserves but, unfortunately, poor results were obtained. The 11 km pipeline was dismantled in 1993 and from then all the production was transported by tanker trucks.

The 50-year production concession expired at the end of January 2017 and the field is now closed, awaiting a bidding process for a new concession to be awarded. During the last years, average production varied between 100 and 150 bbl of oil/day, and the accumulated oil production is nearly 17 million barrels of oil. The field has been commercially a very profitable asset when compared with other international onshore oil fields, considering that the oil concession terms in Spain are very favourable, with no royalties payable on production, and only corporate income tax was applied until 1 January 2016 when a new tax became applicable for hydrocarbon production. However, Ayoluengo has not met the promising expectations generated immediately after the discovery in 1964. The 50 years of oil production in Ayoluengo has resulted in a very low impact on Spain's economy, a country that currently consumes more than 1 million barrels of oil/day and which almost fully depends on imports for its oil needs. The Ayoluengo economic impact in the local communities was significant during the early years, especially with recruitment of personnel, but as production steadily decreased, it has been reduced considerably.

The Oil Museum in Sargentes de La Lora

The region provides an excellent opportunity for introducing the oil exploration and production industry to the public, now enhanced by the opening (in March 2015) of the Oil Museum in Sargentes de la Lora (Burgos), the first one of this type in Spain, which will introduce visitors to the world of petroleum science and technology, highlighting the local and regional geological complexity together with the history of the Ayoluengo Field.

Deep river erosion in nearby areas allows the spectacular geological exposures of most of the elements of the Ayoluengo petroleum system to be observed: tar-impregnated sandstones, the claimed Liassic source rock and textbook faulted anticlines (Fig. 8). Furthermore, the possibility of seeing rod pumps in the field and of visiting the surface production facilities, together with the large amount of well data and seismic coverage available, has long made the Ayoluengo Field an ideal training ground, providing students and non-technical visitors with an excellent demonstration of a working petroleum system (Quesada *et al.* 2006), which is also attracting tourists to the region.

The museum is sponsored by the municipality of the village of Sargentes de la Lora and the Fundación Repsol in collaboration – among others – with the University of Burgos and the Association of Spanish Petroleum Geologists and Geophysicists (AGGEP). The museum's aim is to enlighten visitors about the oil industry and its products, but mainly is focused on the upstream, introducing the petroleum system concept, and the wide variety of geological, geophysical and engineering techniques used in exploration and production. An important part of the exhibition is dedicated to Ayoluengo Field's geology and history, captured in an excellent collection of photographs provided by the villagers and local newspapers, together with press clippings, documentaries of the mid-1960s, educative panels, geological 3D models, and drilling and production material. The museum also features an outdoor authentic working rod pump, courtesy of the latest field operator.

The Ayoluengo and nearby areas present a rich and varied geological heritage, which allows visitors to see sediments ranging from the Triassic to the Cretaceous in age, exceptional landscapes, spectacular folds, deep canyons, abandoned coal mines, cascades, caves, travertine terraces and a wide variety of impressive karstic landforms. The region also contains interesting palaeontological and archaeological sites, medieval hermitages carved into the rocks, and Romanesque churches.

The Ayoluengo enigma

The Ayoluengo discovery revitalized seismic and prompted drilling activity in the region, but of the

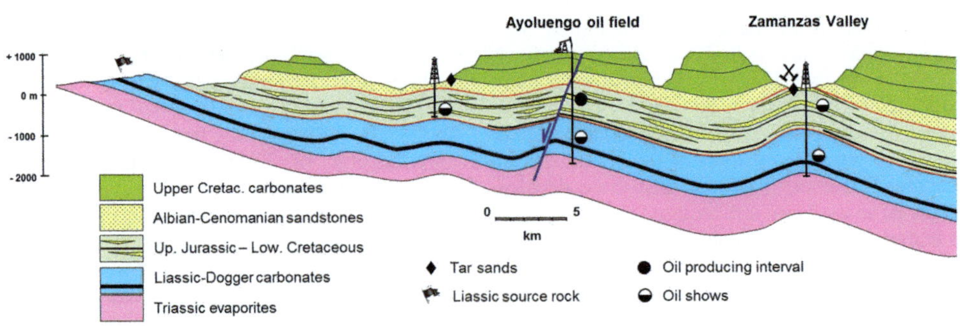

Fig. 8. Schematic regional cross-section through the Ayoluengo oil field. The deep erosion by rivers in the nearby areas allows most of the elements of the Ayoluengo petroleum system to be seen on outcrops: tar-impregnated sands in the core of a spectacular surface anticline can be observed. Also, outcrops of the black shales of Liassic age, which have been largely considered as the only source rock for the Ayoluengo oil, can be visited.

subsequent exploration wells drilled during the 1960s in this geological province only a few tested non-commercial oil flow rates: Tozo-1, Huidobro-2 (12 bbl of oil) and Hontomín-2 (42 barrels of 32° API oil). In these wells, the initial flow decreased rapidly during production testing, indicating oil accumulations that were too small to be economically viable (Beroiz & Permanyer 2011).

After years of intense exploration activity, the Ayoluengo Field, surprisingly, still remains a unique oil discovery as the only onshore commercial oil field in Spain and also the only one in the entire Iberian Peninsula. This geological singularity has brought about recurrent discussions between petroleum geologists because it is difficult to explain why a petroleum system is uniquely working at this particular spot and nowhere else within such a vast territory.

Despite a long history of hydrocarbon exploration, the Basque–Cantabrian Basin is still considered one of the most prospective sedimentary basins in Spain and, at the present time, most of the exploration activity for conventional and non-conventional plays has been mainly concentrated in this region.

Funding This research received no specific grant from any funding agency in the public, commercial, or not-for-profit sectors.

References

ABEGER, G., SERRANO, A. & QUESADA, S. 2006. Petroleum geology of the Onshore Cantabrian Basin. *In*: MARTÍNEZ DEL OLMO, W. (ed.) *Libro XXV Aniversario AGGEP (Asociación de Geólogos y Geofísicos Españoles del Petróleo)*. Madrid, 155–162.

ALMELA, A., RIOS, J.M. & GARRIDO, J. 1946. *Report on the Work and Drillings CAMPSA is Doing at the Present Time in the Valley of Zamanzas (Burgos)*. Geological Report 54. CIEPSA, Madrid (unpublished).

ALMELA, A., LIZAUR, J. & MUÑOZ, C. 1952. Estudio de la reserva petrolífera de Burgos. *Boletín del Instituto Geológico y Minero de España*, **64**, 205–255.

ALVAREZ, E. & GARCÍA, A. 1996. Cálculo de reservas remanentes de hidrocarburos en zonas estructuralmente complejas: aplicación al campo de 'Ayoluengo' (1° parte). *Geogaceta*, **20**, 161–164.

AMOSPAIN 1964. *Ayoluengo No.1. Final Geological and Drilling Reports* (unpublished).

AYALA, F.J. 2006. *El petróleo de la Lora. La esperanza que surgió del páramo*. Editorial Dossoles, Burgos, Spain.

BALLESTERO, A. 1989. *Buscando petróleo*. Espasa Calpe, Madrid.

BEROIZ, C. & PERMANYER, A. 2011. Hydrocarbon habitat of the Sedano Trough, Basque-Cantabrian Basin, Spain. *Journal of Petroleum Geology*, **34**, 387–410.

DUPUY DE LÔME, E. 1941. Las investigaciones de petróleo en España. *Boletín del Instituto Geológico y Minero de España*, **54**, 335–386.

GARCÍA, A. & ALVAREZ, E. 1996. Calculo de reservas remanentes de hidrocarburos en zonas estructuralmente complejas: aplicación al campo de 'Ayoluengo' (2ª parte). *Geogaceta*, **20**, 165–168.

KLIMOWITZ, J., MALAGON, J., QUESADA, S. & SERRANO, A. 1999. Desarrollo y evolución de estructuras salinas Mesozoicas en la parte suroccidental de la Cuenca Vasco-Cantábrica (Norte de España): Implicaciones exploratorias. *In*: MARTÍNEZ DEL OLMO, W. (ed.) *Libro AGGEP (Asociación de Geólogos y Geofísicos Españoles del Petróleo), Homenaje a José Ramírez del Pozo*. Madrid, 159–166.

MARÍN, P., MÁRQUEZ, G., GALLEGO, J.R. & PERMANYER, A. 2014. Characterization of asphaltic oil occurrences from the southeastern margin of the Basque-Cantabrian Basin, Spain. *Geologica Acta*, **12**, 327–342.

MARTÍNEZ, M. & MORENO, J.M. 2014. *Petróleo de la Lora: 50 años de historia*. Ayuntamiento de Sargentes de la Lora, Burgos, Spain.

NAVARRO, J. 2014. Spanish field turns 50. Celebrating Ayoluengo. *AAPG Explorer*, June 2014, 46.

NAVARRO, J. 2016*a*. Ayoluengo – 50th Anniversary of Spain's only onshore oil field. Abstract presented at the European Oil & Gas Industry History Conference, 3–4 March 2016, London.

NAVARRO, J. 2016*b*. Spain's oldest and only onshore oilfield. *AAPG Explorer*, February 2016, 32–33.

PUCHE, O., MAZADIEGO, L.F. & ORTIZ, J.E. 2011. Petroleum in the Spanish Iberian Peninsula. *In*: ORTIZ, J.E., PUCHE, O., RABANO, I. & MAZADIEGO, L.F. (eds) *History of Research in Mineral Resources*. Cuadernos del Museo Geominero, IGME, **13**, 227–238.

QUESADA, S., DORRONSORO, C., ROBLES, S., CHALER, R. & GRIMALT, J. 1997. Geochemical correlation of oil from the Ayoluengo field to Liassic black shale units in the southwestern Basque–Cantabrian Basin (northern Spain). *Organic Geochemistry*, **27**, 25–40.

QUESADA, S., HERNANDEZ, E., ROSALES, I., KLIMOWITZ, J. & SERRANO, A. 2006. A field trip guide book: Two days around Ayoluengo oil field area. *In*: MARTÍNEZ DEL OLMO, W. (ed.) *Libro XXV Aniversario AGGEP (Asociación de Geólogos y Geofísicos Españoles del Petróleo)*. Madrid, 209–222.

RÍOS, J.M. 1955. Prospecciones petrolíferas realizadas por la CAMPSA. *Notas y Comunicaciones del Instituto Geológico y Minero de España*, **38**, 3–13.

RÍOS, J.M. 1965. Relación de las principales actividades para investigación de hidrocarburos llevadas a cabo en España durante 1964. *Notas y Comunicaciones del Instituto Geológico y Minero de España*, **79**, 89–148.

SAN ROMÁN, E. & SUDRIÀ, C. 2003. Synthetic fuels in Spain, 1942–66: the failure of Franco's autarkic dream. *Business History*, **45**, 73–88.

SANZ, R. 1948. *El petróleo*. Colección Memorias del IGME (re-edited in 1979).

SANZ, R. 1967. Ayoluengo field, Southwest Cantabrian Basin, Northcentral Spain. *7th World Petroleum Congress Proceedings, Mexico*, **2**, 251–258.

TORTELLA, G., BALLESTERO, A. & DÍAZ FERNANDEZ, J.L. 2003. *Del Monopolio al libre Mercado. La historia de la industria Petrolera Española*. LID Editorial empresarial, Madrid.

A century of hydrocarbon exploration and production in Spain (1860–1960)

JORGE NAVARRO COMET[1]* & OCTAVIO PUCHE RIART[2]

[1]*Department of Geosciences, CEPSA E.P., S.A., Paseo Castellana (Torre Cepsa) 259, 28046 Madrid, Spain*

[2]*Escuela Técnica Superior de Ingenieros de Minas y Energía, Calle de Ríos Rosas, 21, 28003 Madrid, Spain*

J.N.C., 0000-0002-9864-8280; O.P.R., 0000-0002-6570-7811

**Correspondence: jorge.navarro@cepsa.com*

Abstract: The onset of hydrocarbon exploration and production activity in Spain dates back to 1860, mainly consisting of rudimentary mining of tar sands, asphalts and bituminous shales. Petroleum exploration drilling started properly in 1900 by simply locating wells in the vicinity of petroleum impregnations at the surface, often with little or no geological input. Until 1940, most of these early exploration wells were shallow tests (less than 500 m deep), commonly operated by small private companies with foreign investment, but no oil and gas accumulations were found.

Systematic hydrocarbon exploration in Spain began in 1940 after the Spanish Civil War (1936–39). Two oil companies led these activities: CAMPSA, a state-run petroleum monopoly established in 1927; and CEPSA, a fully private Spanish company founded in 1929, which created the exploration subsidiary CIEPSA in 1940. However, the poor economic situation in Spain, the currency shortage and the international embargo after the Civil War made it very difficult to import any rig capable of deep drilling until the late 1940s. During the early 1950s, the domestic and international position of Spain started to improve. Slowly, foreign hydrocarbon exploration companies (especially German and American) began to form partnerships with national and Spanish private companies. They conducted the acquisition of the first seismic reflection surveys and were responsible for a considerable increase in deep drilling and pioneering borehole logging. Noteworthy amongst these were the intense activities of the VALDEBRO, a consortium formed in 1952 by American companies and the state.

In 1960, 100 years after the initiation of petroleum production in Spain, the first commercial success was obtained with the discovery of dry gas in the Castillo-1 well, drilled by CIEPSA in the onshore Cantabrian–Basque Basin. This find was later complemented by the oil discovery at Ayoluengo in 1964 by a joint venture led by CAMPSA and which still remains the only commercial oil field discovered onshore Spain.

The presence of surface oil seeps, both liquid and solid, has been known in Spain since historical times (Puche *et al.* 2011). In the majority of cases, these take the form of slightly impregnated beds or very minor exudations of oil and asphalt in fractures or cavities, rather than what would be called live oil seepages (Anderson 1936). These surface indications have initially served to determine the geological regions where oil could be most likely present, such as the region of Andalucía in the south and the Basque–Cantabrian Basin in northern Spain (Fig. 1), the latter being the geological region in Spain with the most abundant hydrocarbon indications at the surface (Fig. 2).

During the early years (1860–1900), petroleum activity was focused on the production of oil from bituminous shales, asphalt impregnations or tar sands by subsurface mining and distilling, commonly with poor economic success. The first exploration drilling started early in the twentieth century, by simply locating wells in the vicinity of petroleum impregnations at the surface. Locations were often selected with little or no geological input and limited to sparse surface mapping or rare gravimetry surveys; no seismic data were available at this time. Well penetrations were shallow and, whilst occasionally oil and/or gas shows were reported while drilling, no commercial hydrocarbon accumulations were found.

True systematic hydrocarbon exploration in Spain commenced in 1940 with two Spanish companies: the fully private CEPSA and the state-owned CAMPSA (Ríos 1949). This concluded in 1960 with the discovery of the small Castillo gas field by CIEPSA (the CEPSA subsidiary for exploration), which later culminated in 1964 with the Ayoluengo oil field discovered by CAMPSA, in a joint venture with Chevron and Texaco.

Exploration and production activity in Spain during 1860–1960 can be divided into three

From: Craig, J., Gerali, F., MacAulay, F. & Sorkhabi, R. (eds) 2018. *History of the European Oil and Gas Industry*. Geological Society, London, Special Publications, **465**, 345–360.
First published online May 8, 2018, https://doi.org/10.1144/SP465.10

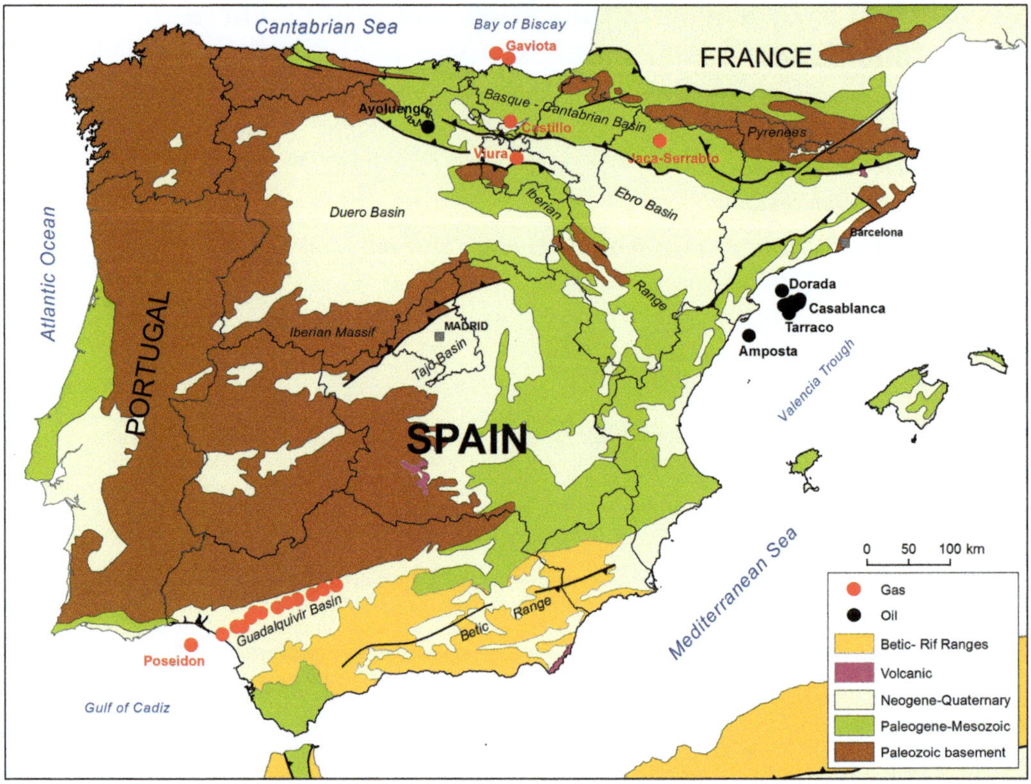

Fig. 1. Simplified geological map of the Iberian Peninsula showing the location of the commercial oil and gas discoveries made to date. The first hydrocarbon discovery in Spain was the Castillo gas field in 1960 by CEPSA.

periods: (i) petroleum mining works, from 1860 to 1900; (ii) early exploration drilling, from 1900 to the end of the Spanish Civil War in 1939; and (iii) systematic exploration, from 1940 to 1960, when eventually the first commercial hydrocarbon discovery was made.

Petroleum mining works (1860–1900)

The onset of hydrocarbon production activity in Spain dates back to the middle of the nineteenth century, mainly consisting of the rudimentary mining of tar sands, asphalts and bituminous shales (Fig. 2). The opening of these mines was based on the presence of hydrocarbons or organic-rich shales at the surface and basic field geology. The application for the mining rights of these hydrocarbon deposits was governed by the Mining Law Act of 1869 that was in force until 19 July 1944.

Some of these surface seeps had been recognized since historical times (Puche *et al.* 2011). Amongst the earliest accounts of hydrocarbons in Spain are those recorded by the Arabic geographer Al-Udri (1003–85), who reported a 'strange mineral' (asphalt) that was extracted from tar sands of Triassic age near the village of Siguenza (Guadalajara). Later news about this deposit is recorded in the second half of the nineteenth century, when several concessions for the exploitation of these tar sands were awarded from 1872 to 1923.

The mining of the asphaltic Cretaceous sands at Fuentetoba, close to the city of Soria, started during the middle of the nineteenth century (Huguenet 1852; Dupuy de Lôme 1937), mainly by the company 'Asfaltos del Volcán', later renamed the 'Sociedad del Verdadero Asfalto Español' (which translates as the 'True Spanish Asphalt Society'). Some facilities were built there to treat and distil the asphaltic sands, mainly aimed at producing asphalt for paving the streets of downtown Madrid. A rudimentary hot-water process enabled the bitumen to be separated from the sand by placing the asphaltic sands into iron boilers. Once the water was boiling, the bitumen was released from the sand, floating to the top of the separation vessel, where the bitumen was manually skimmed off. Production was sporadic and reached a maximum in the year 1861, when more

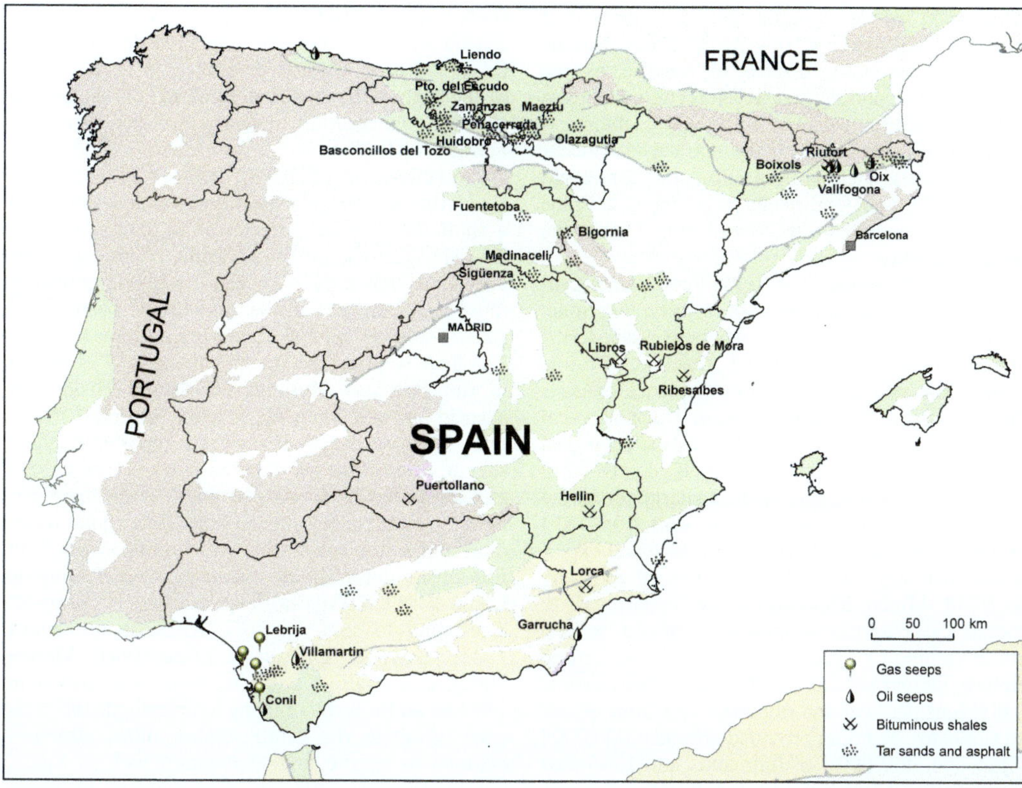

Fig. 2. Map of the Iberian Peninsula showing the location of outcrops in Spain of bituminous shales, tar sands, asphalts, and oil and gas seeps.

than 1000 tons of sand was extracted; thereafter, it decreased annually until the mines were abandoned by the end of the nineteenth century (O'Shea 1918). The Fuentetoba tar-impregnated sands form part of the Utrillas Formation, a very characteristic stratigraphic sequence of Albian–Cenomanian age, widely present over vast areas of northern Spain. It is composed of multicoloured sandstones and shales, interbedded with carbonaceous shales and lignite, deposited in a fluvio-deltaic environment. Similar tar-impregnated sands within the Utrillas Formation have also been found in other areas of northern Spain, such as Bigornia (Zaragoza), Basconcillos del Tozo and Huidobro (both in Burgos).

In Maeztu (Alava), there are numerous asphalt impregnations within Upper Cretaceous carbonates and sandstones, especially in the southern flanks of a semi-circular salt dome of Triassic evaporites. These asphalt impregnations are most likely to be the largest in volume in all of Spain. During the second half of the nineteenth century, several concessions were awarded that had been continuously exploited by underground and opencast mining until recent times. The 'Compañía de Asfaltos de Maeztu' was created in 1892 and has been, until very recently, the main operator of most of these asphalt concessions. The main commercial output was asphalt paving blocks, manufactured by crushing, milling, heating and rolling the impregnated limestone.

The Eocene oil shales in Riutort (Barcelona) are located in the foothills of the Pyrenees and represent the most important occurrence of primary oil in Spain, which had been occasionally mined since 1898. In 1905, interest in these shales was revived. The 'Compañía Minera de Riutort', owned by French entrepreneurs Jules Leon Clavier and Philippe Petit Brice, started mining the steeply dipping oil shales and then treating them in a nearby processing plant by 1906 (Serra 2008). The oil shales were transported in small wagons dragged by animals along a railway from the pithead to the distillation plant. The exploitation was not commercially successful and was soon abandoned. Today, it is possible to visit the oil mine and observe minor oil seeps from these beds in the galleries that were originally dug. The Riutort oil mines produced up to 100–160 l of crude oil per ton of gross rock from a 40–60 m interval, revealing an excellent hydrocarbon source potential, probably the best of any geological horizon in Spain. These Eocene shales are regionally known in outcrop as the Armancies Formation and it is recognized as a world-class source rock. In this region, other less important oil indications at the surface also occur in Eocene beds, such as in Vallfogona (Barcelona) and Oix (Gerona); rudimentary attempts to mine these were abandoned after World War I due to the poor profitability.

The Carboniferous oil shales in Puertollano (Ciudad Real) are the most important of its type in Spain (Tortella *et al.* 2003). The Puertollano Basin forms an east–west-orientated, closed syncline of some 40 km^2. The infill Carboniferous sequence consists of a maximum of 500 m thickness of near-horizontal shales and siltstones, interbedded with several coal seams and bituminous shale horizons, deposited in a lacustrine environment with a marginal-marine influence. Coal subsurface mining started in 1874 but it was not until the 1910s that a basic and experimental distillery was built by the French company 'Sociedad Minero-Metalurgica de Peñarroya' to process the bituminous shales to obtain several petroleum products, such as oil, benzol, ammonium sulphate, tar and natural gas. The plant treated some 60 000 tons per year and obtained 8700 tons of oils and sulphates. Refining provided around 4 000 000 l of gasoline, 400 000 l of lubricants and 1300 tons of asphalts (San Román & Sudrià 2003).

The presence of bituminous shales in the Miocene lacustrine basins of Rubielos de Mora (Teruel) and Ribesalbes (Castellón) have also been well known since the nineteenth century (Fernández 1914; Gavala 1921; Peñalver *et al.* 2016). Both basins are also acknowledged as presenting one of the most exceptional and well-preserved Lower Miocene examples of fossil fauna and flora in Europe (Peñalver *et al.* 2016), similar to other world famous palaeontological sites such as Ediacara (Australia), Burgess Shale (Canada) or Solnhofen (Germany). The bituminous shales in Rubielos cover an area of 5 km^2, with a net average thickness of 10 m of oil shale. Significant variations in oil yields are observed, ranging from 15 to 35 l of oil per ton of rock. These shales were industrially exploited during World War I by the Spanish refining company 'Sabadell y Henry', but were soon abandoned because of the poor economic circumstances prevailing at the end of the war (Marín 1978; Tortella *et al.* 2003). The large ovens used for oil distillation during this period are still standing. The Miocene oil shales in Ribesalbes (Castellón) occur in finely laminated beds up to 2 m thick in an area extending for 3 km^2, yielding a maximum of 60–100 l of oil per ton of rock. These shales were first mined and distilled in 1865 and years later exploited by the British company 'Castellon Oil Company Ltd' until 1914 (Peñalver *et al.* 2016). Mining activity was later reactivated by the 'Compañía Española de Aceites de Esquisto, S.A.', but was abandoned in 1922 because of the high carbonate content of the oil shales, the absence of exporting infrastructure and poor economic results (Hernández-Sampelayo & Cincúnegui 1926; Peñalver *et al.* 2016).

During the early 1880s, oil was found seeping in a shaft dug for the purpose of sulphur mining in Triassic-age rocks in Conil (Cádiz). This mine was very well known by many European natural history museums because of the perfection, beauty and size of sulphur crystals found there. Some of these excellent sulphur specimens can be now observed in the National Museum of Natural History in Madrid, as well as in many other European museums. The petroleum seepage was reported to yield several litres of oil per day during the course of several days after being opened up. A few years later, nothing was to be seen there, other than a natural sulphur and saline spring that was bringing up a faint film of oil and a minute quantity of gas (Dupuy de Lôme 1941). Today, not a drop of oil is observed. According to some authors (Alvarado & Cantos 1941; Grove & Dupuy de Lôme 1941; Merten 2006), one shallow well (<200 m) was drilled in 1899 by an English company to investigate the presence of oil in the Conil sulphur mine, which is claimed to be the first exploration well in Spain, but no official records have been found to document it (Dupuy de Lôme 1937).

Early exploration drilling (1900–40)

The Huidobro well, drilled in the year 1900, is widely recognized as the first relatively deep borehole specifically designed to find hydrocarbons in Spain (Gavala 1916*b*; Dupuy de Lôme 1941; Marín 1978; Ayala 2006). It was located in the Burgos province, in a large natural oval depression (4 km in length by 2 km wide) that was the product of erosion of the crest of a conspicuous surface anticline, with sandstones of Albian age outcropping in the anticline core and bordered by Cenomanian limestones (Fig. 3). The presence of petroleum in the Huidobro anticline had been discovered in the 1860s after detecting some 'mineral oil' and water seeping from a natural source in the sandstones, especially after heavy rains (Fernández 1891; Gavala 1916*b*). These sandstones were mined for copper (malachite and azurite), and occasionally oil and water were also found seeping in the mine galleries (Fernández 1891). A concession named 'El Progreso' was requested in 1866, being one of the first oil exploitation licences officially awarded in Spain. Some work to exploit these sandstones was carried out to obtain oil during the late years of the

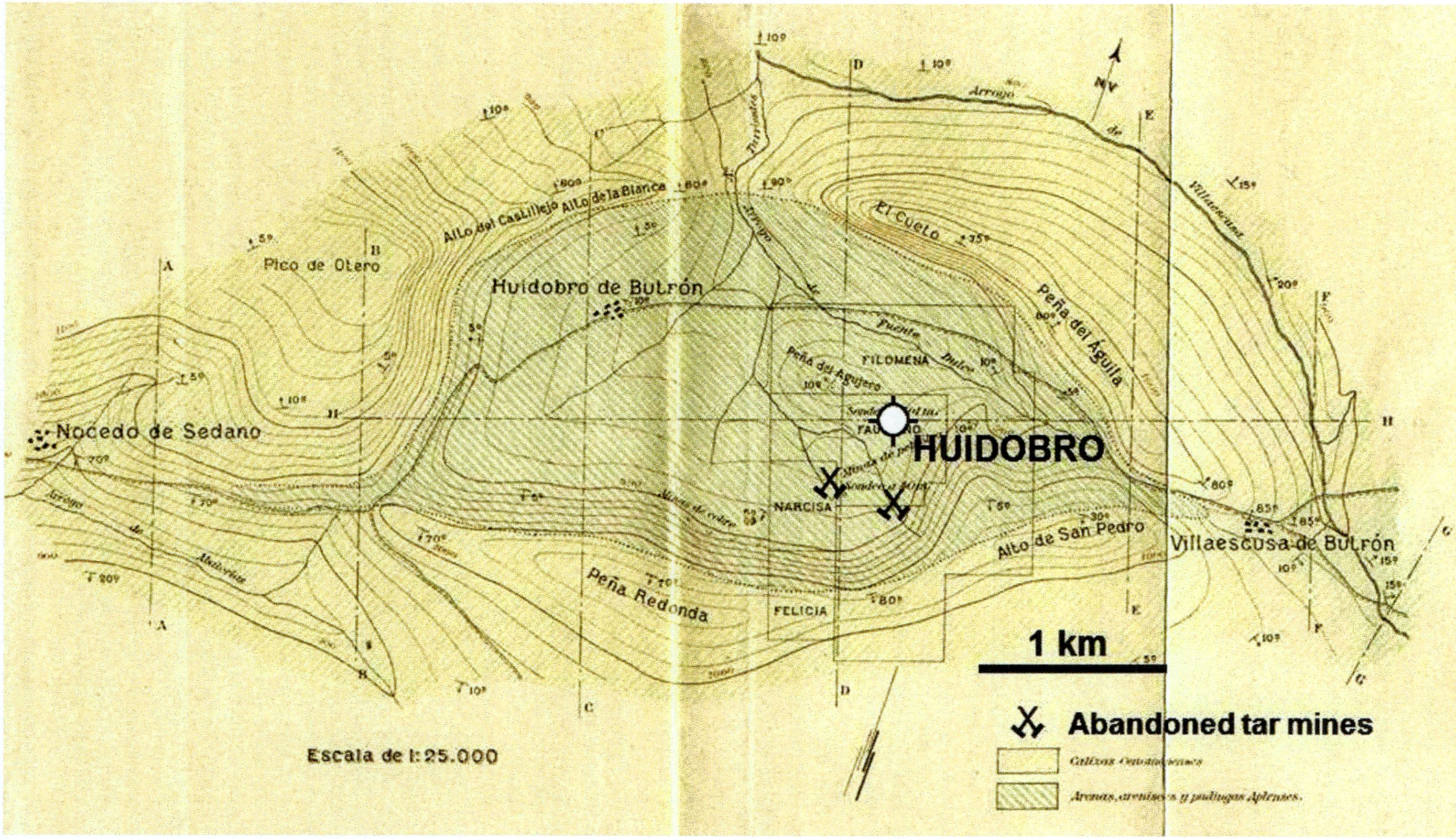

Fig. 3. Geological map and topography by Gavala (1916*b*) showing the location of the Huidobro well (Burgos), drilled in 1900 and claimed as the first hydrocarbon exploration well in Spain. It was located close to the mined tar sands from the Utrillas Formation of Cretaceous age. The Huidobro well reached a total depth of 523 m, recording bitumen and weak oil shows in three sandstone beds (Comité Nacional de Sondeos 1929).

nineteenth century (Fernández 1891; Dupuy de Lôme 1941). Shallow wells (20 m deep) were excavated that found oil, and a mine gallery 50 m long was dug, but the work was soon abandoned because of a lack of capital and poor knowledge of how to get the oil out of the ground (Fernández 1891). Rudimentary oil exploitation and refining was sporadically carried out by the local people who sold it to neighbouring villages for lighting (Fernández 1891).

The Huidobro well was spud in 1900, operated by 'Sociedad de Sondeos de Huidobro' (Gavala 1916*b*; Ayala 2006; Puche *et al.* 2011) and drilled by the French drilling contractor Hulster Frères. The well was properly situated at the bottom of the depression along the east–west-trending anticline axis (Fig. 3). It reached a total depth of 523 m after penetrating a sequence of sandstones and shales, only recording bitumen and weak oil shows in three sandstone beds (Comité Nacional de Sondeos 1929). Years later, during the 1960s, the Huidobro anticline was further investigated by two deeper exploration wells (to 900+ m) located 550 and 1150 m west of this first well location. Both wells were able to recover small amounts of oil with a high water cut by swabbing from the Lower Cretaceous sandstones below 550 m depth: that is, only 27 m deeper than the total depth of the 1900 Huidobro well.

Spain's neutrality during World War I facilitated a short economic boom, with increases in industrial production and export. It also reactivated mining activity to exploit the bituminous shales, which was strongly encouraged by the Spanish government. However, this re-energized activity was soon revealed to be non-commercial and, consequently, abandoned, such as the above-mentioned mines at Riutort, Ribesalbes and Rubielos de Mora.

The largest surface oil seep known in Spain was reported in Villamartín (Cádiz). This seep was discovered in 1907 during excavation works to enlarge a canal for a mill near the Guadalete River (Mallada 1909; Gavala 1916*a*; Gómez 2009). Oil (reported as 'ozokerite') was found within large vugs of a limestone block of presumably Triassic age. A local company, 'Sociedad Petrolífera de Villamartín', was founded, obtained a concession in the area and started reconnaissance works by digging a pit close to the point where oil was first noted. The pit was deepened to some 14 m, finding water containing oil films and bubbles of H_2S, and thus making further digging impossible. The company acquired a small hand rig and drilled several shallow wells to test this occurrence. The wells were spud and finished in Triassic rocks, which locally contained considerable oil saturation close to the surface. Later, the company acquired a small rotary rig that was used to deepen existing holes and drill new wells. These wells appear to have been located arbitrarily with little technical or geological support. At least one of them yielded a few litres of oil at a depth of only 5–10 m. In a number of others, a mixture of brine, petroleum and associated gas (probably H_2S) was recovered below 100 m depth, with a maximum flow rate of barely 1 bbl (barrel) of oil per day. In one case, the gas ejected a column of water and oil to a height of 15 m above ground level, and small eruptions of oil and salt water were frequent. In another well, the tools were blown out of the hole. Some wells went to depths as great as 700 m, but without any commercial success. In 1913, all petroleum concessions in this region were relinquished with no further exploration activity. The oil presence, commonly associated with carbonates, shales and sulphur-rich beds of Triassic age, suggested that it was generated by an intra-Triassic source rock. As a consequence, the Triassic was later considered as an important exploration target, not only in the southern region but also in northern Spain (Gavala 1916*a*).

Gas seeps have been known in the marshes of the Guadalquivir Basin from historical times. They were sufficient to regularly kill the swamp vegetation for several metres around the seep site, giving rise to circular areas called 'ojos' ('eyes') by the local peasants (Grove & Dupuy de Lôme 1941). In 1909, the company 'Petrolífera de Pambanco' was created by Spanish entrepreneurs to investigate gas shows that had been reported a year earlier in a natural water source in Lebrija (Sevilla), close to the swamps of the Guadalquivir River (Tortella *et al.* 2003). According to Professor Vivian B. Lewes, a British expert on petroleum who had visited the area, the gas presence was a strong indication of an oil accumulation in the subsurface; all that was required, was a drilling rig to untap it, and to determine accumulation depth and volume. However, after drilling more than 40 shallow wells (the deepest was 75 m), only 'marsh gas' (biogenic gas) was encountered (Fig. 4). A project to build a small gas power plant had been initially considered, but the gas supply was insufficient and too irregular for the commercial success of the project and it was abandoned in 1913.

The first systematic geological studies focused on the petroleum potential in Spain date back to the 1910s and were run by the 'Instituto Geológico de España' (IGE: the Spanish Geological Survey). In 1911, the IGE appointed three young Spanish mining engineers with the remit to specialize in petroleum geology: Juan Gavala, Pedro Novo and Enrique Dupuy de Lôme. In 1920, Novo and Dupuy de Lôme were sent to America to become familiar with the petroleum industry, and visited the great oil boom regions of Virginia, Pennsylvania, Oklahoma, California and Mexico (Dupuy de Lôme & Novo 1925). Once back, they initiated a reconnaissance study by comparing the 'petroliferous terrains' of America with Spanish geological formations, establishing a regional ranking between areas

Fig. 4. Photograph taken in 1912 of one of the multiple shallow wells, drilled in Lebrija (Sevilla) by the company 'Petrolífera de Pambanco' to explore for gas in the Guadalquivir Basin.

with oil impregnations at the surface, areas with no oil indications that deserved further investigation and areas with no interest at all (Novo 1925). They also started the first catalogue of oil surface indications in Spain, with a crude analysis and identification of the petroleum-bearing formations (Dupuy de Lôme 1941). This catalogue later served to locate and drill some of the wells operated by the IGE and private petroleum companies. In 1931, Dupuy de Lôme also visited the most important oil fields in Romania, by then one of the major world oil producers and the main oil supplier to Spain (Dupuy de Lôme 1933). The international oil industry expertise gained during his trips abroad and the good knowledge he had of Spanish geology made Dupuy de Lôme an expert, pioneering the petroleum exploration effort in Spain.

The first oil well drilled by the IGE was Caldones, aimed to test the gas associated with the Carboniferous coal beds, widely mined in the Asturias region in northern Spain (Ruiz 1916). Methane, locally named 'grisú' by the miners, was occasionally detected during coal-mining work, and is today exploited as Coal Bed Methane (CBM) in many countries. The Caldones well was drilled by contractor Hulster Frères with a rig capable of drilling by rotary or percussion and enough power to reach 800 m in depth. The well was spud in April 1914 and penetrated a 330 m-thick sequence of Triassic red beds lying unconformably on a Carboniferous section of mixed black shale and limestone, but with no coal beds. In January 1915, the well reached a depth of 563 m and took a strong gas kick, resulting in a blow out of gas mixed with water that prevented the drillers from evacuating the area quickly. Immediately after the blow out, a big explosion took place and the derrick caught fire but, fortunately, nobody was injured. It took five months to suppress the fire, before the well was finally plugged and abandoned.

The increase in oil consumption during the 1920s, due to automobile proliferation and the conversion of ships from coal to oil, triggered the worldwide fever to find more oil. Spain was no exception. The IGE and various local private companies with foreign investment set about exploring and drilling for oil (Ríos 1949). Most of the private companies were formed for a single project and rarely drilled more than one or two wells; some went bankrupt even before any drilling had started (Alvarado & Cantos 1941). Noteworthy among the private companies are 'Compañía Franco-Española de Petróleos', 'Compañía Hispano Británica de Petróleos', 'Sociedad Española de Petróleos', 'Sociedad Petrolera Ibero-Americana', 'Petrolera del Jaizquibel, S.A.', 'Sociedad Minera Arganmel', 'Sociedad Explotadora de Petróleo Begoña' and 'Sociedad de Petróleos del Ebro' (Álvarez 1923). These companies commonly drilled shallow tests (less than 500 m deep), with light rigs and scarce resources. Occasionally, oil and gas shows were reported while drilling, but no commercial accumulations were found. With rare exceptions, the exploration drilling strategy was simply to locate wells in the vicinity of surface petroleum seeps, generally with little or no other geological input and, astonishingly, some wells located using a dowser (Ariza 1924).

The American company 'Sociedad Petrolera Ibero-Americana', founded in 1921, drilled a well in the Gastiáin anticline (Navarra) in 1923 (Dixon 1921; Baquedano 2007). It appears to have been the only deep well drilled during this period that can be considered a reasonably satisfactory test of a subsurface structure, based on its location and depth. The Gastiáin well was located by the American petroleum geologist Abner Faison Dixon in the Lana Valley, in the eroded core of a prominent and well-exposed surface anticline, east–west-trending and extending 12 km in length by 4 km wide (Fig. 5). This anticline is in a zone containing nearby asphalt seeps, and it was assumed that the structure as seen on the surface corresponded to that at depths (Dixon 1921). The Gastiáin well was drilled by a cable tool rig owned by the 'Sociedad Petrolera Ibero-Americana'. The well started in Upper Cretaceous strata and is reported to have been drilled entirely through Upper Cretaceous marls until reaching a total depth of 1610 m in a record time of eight months, by then the deepest well in Spain. The stratigraphic prognosis completely failed, the well did not reach its objective and no oil was found. However, the well encountered gas shows (98% methane) which reached their maximum through the interval 400–500 m depth, sufficient to be utilized for the boilers of the drilling engine over the course of several months (Ariza 1924).

The 'Sociedad Petrolera Ibero-Americana' rig was subcontracted by the IGE (in January 1927, the IGE was re-named the IGME, 'Instituto Geológico y Minero de España') for a drilling campaign in northern Spain (Fig. 5). In 1928, the IGME finished the Robredo-Ahedo well (Burgos), reaching a total depth of 1061 m (Dupuy de Lôme & Novo 1925). The well location was based on surface oil seeps but in an unlikely structural closure. Three sandstone beds with oil shows were encountered and several tens of litres of oil were retrieved from

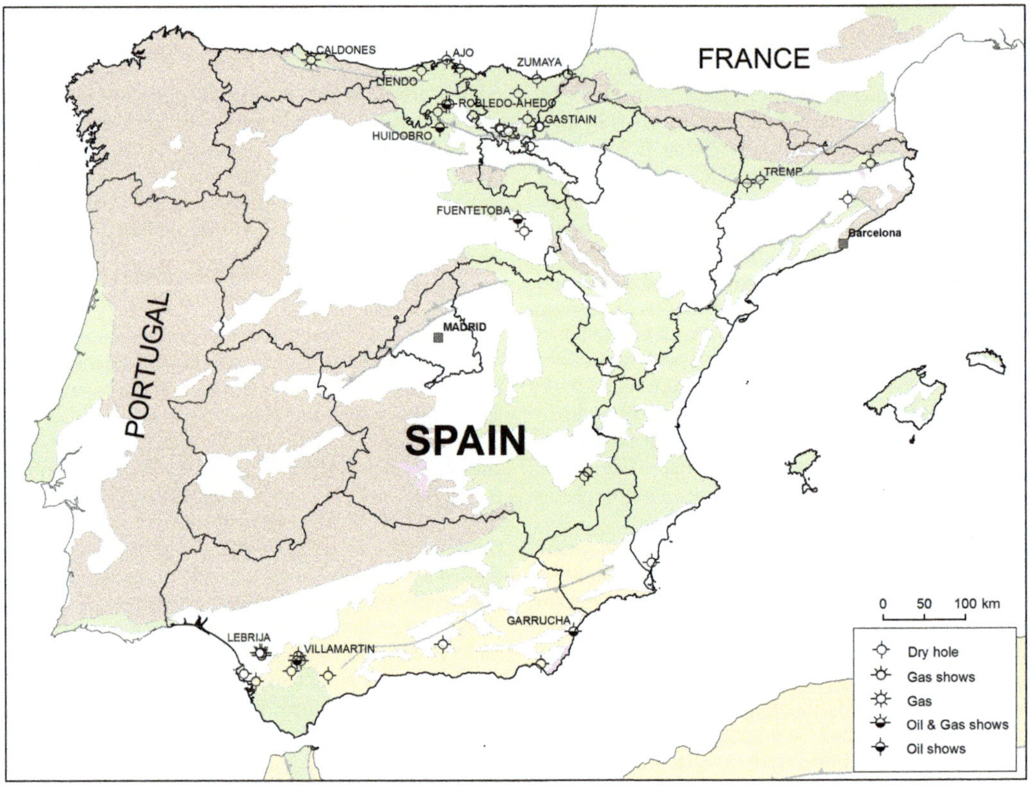

Fig. 5. Map of the Iberian Peninsula showing the location of hydrocarbon exploration wells drilled in Spain during the period 1900–39.

the bailer, but there was no 'gas pressure' (Dupuy de Lôme 1941).

The IGME ran the first geophysical work aimed at hydrocarbon exploration during the late 1920s, which mainly consisted of gravity and refraction seismic surveys over surface anticlines (Anduaga 2009). A well was drilled in 1928 by the IGME close to the Fuentetoba (Soria) tar sands of the Utrillas Formation (Fig. 5). The well objective was to test an asphaltic sand occurrence in the subsurface (Alvarado & Cantos 1941). The Fuentetoba well penetrated a sequence of shales, sands and shaly sandstones, reaching a depth of 398 m, and encountered a dozen or more beds of asphaltic sandstones, but failed to find any producible horizon. Increasing bitumen fluidity with depth was reported but it was, nevertheless, much too viscous to be produced by techniques other than mining. The bituminous material could have been generated in the Utrillas organic-rich beds and, consequently, was too limited in quantity to have given rise to commercial quantities of oil (Grove 1940). The presence of oil in vugs within foul-smelling carbonate beds of Triassic age at a beach east of Laredo (Cantabria) attracted the IGME to this area. As a result of these oil shows, a deep well was drilled with a cable tool rig by the IGME in 1928 at a location in Ajo (Cantabria). The well reached a depth of 1200 m, but encountered some heavy oil in Aptian carbonates and tar sands with oil shows in the Neocomian section (Alvarado & Cantos 1941). The well had been located in an apparent surface closure. A second well drilled by the IGME, named Liendo (Cantabria), was spud in 1928 in a small surface anticline with the same rig but, after reaching a depth of only 120 m into the Lower Cretaceous, it had to be abandoned due to a lack of funds. This was the last well drilled by the IGME aimed at hydrocarbon exploration.

Moving to the Mediterranean coast (Fig. 5), there is a remarkable account of the sudden appearance in 1929 of oil globules and gas bubbles in four shallow-water wells in the village of Garrucha (Almeria). The oil and gas occurrence followed a severe earthquake shock and within a few months it had completely disappeared (Dupuy de Lôme 1941). The oil source was not clear, but today a sample of the Garrucha oil is exhibited in the cabinet of the Geological Museum inside the Spanish Geological Survey (IGME) building in Madrid.

During the Primo de Rivera dictatorship (1923–30) (who had been appointed Prime Minister by King Alfonso XIII), a petroleum monopoly was established in Spain by the Royal Decree of 28 June 1927, which led to the creation of two major petroleum Spanish companies. The 'Compañía Arrendataria del Monopolio de Petróleos, Sociedad Anónima' (CAMPSA) was a state-owned company, dependent on the Ministry of Finance, and held the monopoly for hydrocarbon import, refining, storage, transport, handling, exploration and production for 20 years on the Spanish mainland and the Balearic Islands, but excluding the Canary Islands (Tortella *et al.* 2003). Two years later, on 26 September 1929, the 'Compañía Española de Petróleos, Sociedad Anónima' (CEPSA) was formed by a group of private Spanish investors. CEPSA was founded as an integrated company by acquiring, in 1929, production and exploration acreage in Lake Maracaibo and Monagas state in Venezuela, and in 1930 inaugurated a refinery in Tenerife, one of the Canary Islands whose territory had not been included in the 'monopoly Royal Decree' (Marín 1978).

The creation of CAMPSA and CEPSA resulted in the beginning of a more rigorous and technical approach to petroleum exploration. Both companies initiated a systematic field reconnaissance of the most hydrocarbon prospective areas across the whole of Spain by sending out geology crews to the field to locate and map key structures, and to work out the stratigraphy from outcrops (Ballestero 1989).

Curiously, CEPSA commenced exploration in Venezuela long before doing so in Spain. In 1931, CEPSA began geophysical work in Venezuela and drilled various shallow wells from 1933 to 1937. In 1935, CEPSA decided that it would be necessary to work with an American company to explore for oil in Spain and started a geological field reconnaissance aided by Robert Anderson, an American geologist from the company Socony Vacuum. In the spring of 1936, Anderson commenced geological fieldwork through Spain accompanied by Dupuy de Lôme, visiting the regions where oil indications were known: that is, the Basque–Cantabrian, Pyrenees and the Andalucía region (Anderson 1936). However, the political situation by early June 1936 was so unstable that Anderson had to rush to leave Spain. The Spanish Civil War started 1 month later, on 18 July 1936.

During the years of the Spanish Civil War (1936–39), only minor drilling activity took place in the nationalist-held territory. A well in Tremp (Lerida) that was spud during the Spanish Civil War was not completed until 1941, reaching a total depth of 1700 m (Fig. 5). Another well was spud in November 1939 in Zumaya (Guipúzcoa), close to the sea. The well was drilled by the Belgian firm Foracky and penetrated a sequence in the Upper Cretaceous flysch, recording only minor gas shows before reaching 425 m depth, where the drill pipe became stuck and the well was abandoned.

Whilst it is difficult to be precise about the number of exploration wells and meterage drilled during the period 1900–39, it is estimated that some 30 wells and over 11 000 m in total were drilled (Fig. 5). From all wells drilled in Spain during this period, only four of them (Gastiáin, Robredo-Ahedo,

Ajo and Tremp) penetrated below 1000 m, a very poor record for such a long period of time and for such a vast territory, especially when compared with the activity carried in other European countries. Perhaps the biggest benefit of all these exploration wells was the improved knowledge of subsurface stratigraphy (Rogers 1958).

The potential for commercial petroleum accumulations in Spain was perceived as very low due to the complex geology and relatively poor oil indications at the surface. Exploration failure was further justified by the lack of expertise in petroleum geology, the poor technical rigour used to site well locations, the lack of modern technology, the reluctance to use the latest geophysical techniques and the absence in-country of rigs capable of deep drilling (Anduaga 2009). The investment made in oil exploration was practically negligible compared with other more profitable and less risky mining activities (coal, iron, copper and zinc) in Spain during the same period of time (Marín 1978; Anduaga 2009).

Systematic exploration (1940–60)

After three years of devastating Civil War, Spain had been economically ruined, its productive industrial capacity crippled and most of the infrastructure across the whole country needed to be rebuilt. As if things could not get worse, World War II began in September 1939, just five months after the end of the Spanish Civil War. Although the Franco dictatorship was officially non-belligerent during World War II, the regime supported Germany and Italy, in recognition of the heavy assistance it had received during the Spanish Civil War. As a consequence, when World War II ended in 1945, Spain was internationally isolated and excluded from the plans for the economic recovery of Europe and a strict control on Spanish imports of petroleum products was imposed (San Román & Sudrià 2003). This international isolation allowed the Franco regime to justify a strategy of economic self-sufficiency that lasted for at least a decade, and petroleum exploration was to be part of it.

In spite of the poor economic situation, oil exploration was resumed with greater vigour and using more technical expertise than in any previous period. Activities were renewed in 1940, immediately after the end of the Spanish Civil War (Ríos 1949). The two companies operating in Spain, CAMPSA and CEPSA, intensified their efforts and reinforced their exploration departments in strong collaboration with the IGME. Geological field crews commenced detailed and systematic geological mapping, especially focusing on the largest surface anticlines with a view to making applications for exploration licences. However, the poor economic situation in Spain, the currency shortage and the international embargo after the Civil War made it very difficult to import any rig capable of deep drilling.

Petroleum investigation and exploitation was still regulated by the Mining Law of 1869. Rights were for an unlimited period of time, only expiring by non-payment of the surface fee. A Royal Decree was issued in September 1939, later modified in May 1942. Both decrees were aimed at regulating and encouraging hydrocarbon exploration by limiting the effective time for exploration to five years and, in case of unsuccessful exploration, it allowed the operator to recover from the state, all fees and guarantees incurred (Merry del Val 1951).

In 1940, CAMPSA approved a preliminary plan aimed at identifying potential exploration acreage in the peninsula (Anduaga 2009). CAMPSA focused investigations in the Zamanzas Valley (Burgos), in the core of a large eroded anticline, located in the Cantabrian–Basque Basin. This is a region characterized by the richest surface petroleum indications in Spain and where great expectations had emerged as early as the 1920s (Fig. 2). Zamanzas is a wide valley, crossed by the Ebro River and flanked on all sides by outcrops of Late Cretaceous carbonates. In the bottom of the valley, close to the river, sandstone beds of Early Cretaceous age appear impregnated with tar. During the early 1940s, rudimentary and experimental underground mining in Zamanzas was carried out by CAMPSA to exploit the tar sands using a primitive process. The mining work consisted of excavating galleries no more than 500 m in length. The tar sands were then crumbled and dumped into large water tanks that were heated using wood fire. The heat facilitated the separation of the bituminous material that floated to the surface, and where it was manually collected with dippers and poured into barrels (Ayala 2006). Oil production was marginal at some 1–3 bbl/day. Eventually, this exploitation was abandoned in 1945 because of the uneven distribution of the tar impregnation and the poor economics.

During the period 1941–45, CAMPSA drilled four shallow exploration wells in the Zamanzas Valley using two light rigs (Ríos 1955; IGME 1987). Drilling was slow, with very poor performance and frequent break downs. The wells recorded oil and gas shows in Early Cretaceous–Late Jurassic-age sandstones, and a small amount of oil was recovered from Jurassic carbonates. The deepest well reached only 860 m. In early 1944, CAMPSA was requested by the government to accelerate the exploration of deeper targets and, for this purpose, three drilling rigs were contracted in Germany (Anduaga 2009). However, following the Allied invasion of Normandy in June 1944, CAMPSA cancelled the contract. In 1946, CAMPSA was granted the hydrocarbon exploration rights for a vast acreage north of Burgos, which was declared a state national reserve

(Sanz 1948; Almela *et al.* 1952). The reserve extended over 2800 km^2 and included the areas with the best surface seeps. During 1949–53, CAMPSA drilled three deep wells here with a more powerful rig that was finally imported from the USA in 1948 (Ríos 1949). One of the wells reached a maximum depth of 2177 m in Triassic evaporites. Oil and gas shows were commonly recorded through the Lower Jurassic carbonate section but, once again, no commercial flow rates were established.

The 'Compañía de Investigaciones y Explotaciones Petrolíferas, Sociedad Anónima' (CIEPSA), a subsidiary of CEPSA, was created in 1940 in partnership (50:50) with the Socony Vacuum Oil Company (later ExxonMobil) with the scope of exploring and producing hydrocarbons in Spain. The initial work started with geological mapping in the field of large portions of Spanish territory, which was carried out by American geologists and Spanish mining engineers, such as Fath, Anderson, Grove, Jones, Bray, Trzesniowski, Dupuy, Ríos, Almela and Garrido (Dupuy de Lôme 1955). However, due to World War II, all of CIEPSA's American geologists abandoned Spain in 1943 until 1945; during their absence, the fieldwork was conducted by the Spanish staff who performed their work in spite of all kinds of supply shortages and difficulties in travelling around Spain (Dupuy de Lôme 1955). The early exploratory work was aimed at mapping surface anticlines, and identifying potential reservoir and source rocks from stratigraphic field observations. Notwithstanding, drilling had to be delayed until a suitable drilling rig was available. In 1947, after World War II and when the international situation was no longer blocked, Socony managed to import a used Emsco rotary rig from the USA that was capable of drilling to 2500 m. CIEPSA drilled two dry holes with this rig, determining the locations based solely on surface anticline mapping (Ríos 1949). This first well tested a large anticline in Oliana (Lérida), where a refraction seismic survey had been carried out during the summer of 1942 by García Siñeriz, an expert on geophysical methods from the IGME (Anduaga 2009). Despite the seismic survey being inadequate to resolve the subsurface structure, which had resulted in a complete technical failure (Dupuy de Lôme 1955), the Oliana-1 well was spud in 1947 (Fig. 6). The well drilled a

Fig. 6. Map of the Iberian Peninsula showing the location of hydrocarbon exploration wells drilled in Spain during the period 1940–60.

continuous Eocene marl sequence throughout, to a total depth of 2323 m (Ríos 1958; IGME 1987). This was the drilling depth record in Spain at the time. No shows were registered and the well was abandoned because of rig limitations. The rig moved to Burgo de Osma (Soria) in the southern part of the Duero Basin, where another prominent surface anticline was identified (Fig. 6). The Burgo de Osma-1 well was aimed at testing the Utrillas Formation and the Triassic carbonates, both with tar impregnations in nearby outcrops. The well was spud in December 1948 and completed in January 1950, reaching a total depth of 2211 m. The stratigraphy was as expected, the well started in carbonates of Late Cretaceous age, and continued through a normal sequence of Early Cretaceous, Jurassic and Triassic age sediments (Ríos 1958; IGME 1987). Large mud losses were encountered while drilling but no oil or gas shows were found, and the well was plugged and abandoned. The Burgo de Osma-1 was the first borehole to be electrically logged in Spain (by Schlumberger). In 1950, Socony decided to break its connection with CEPSA and pulled out of Spain against the opinion of their geologists, who considered that two wells were not enough to evaluate Spain's hydrocarbon potential. The decision was solely based on financial reasons. Socony had used blocked Spanish currency (pesetas at that time), which had been generated by marketing oil products. Once this money was spent, Socony decided not to bring dollars into Spain, especially whilst there were prevailing rumours of an imminent devaluation of the Spanish currency.

The 'Instituto Nacional de Industria' (INI) was created in September 1941 as part of the economic and political self-sufficiency strategy of the Franco regime (San Román & Sudrià 2003; Tortella *et al.* 2003; Anduaga 2009). The INI was a single entity dependent on the Ministry of Industry, encompassing the management, co-ordination and funding of all industrial sectors in Spain, including mineral and hydrocarbon resources. The INI had great political and economic empowerment, and soon entered into competition with the Ministry of Finance and its subsidiary CAMPSA. The company Adaro, controlled by the INI, was created in 1942 to manage, operate and investigate all mineral resources in Spain, including hydrocarbons (Anduaga 2009). Adaro focused their exploration in southern Spain until the mid-1960s, where extinct mud volcanoes were identified and sporadic gas shows had been recorded in shallow-water wells drilled in the Miocene section of the Guadalquivir Basin (Gavala & Ríos 1955). As part of this effort, several seismic surveys were acquired in collaboration with the Italian company AGIP and a total of 25 shallow wells were drilled (only a few exceeding 1000 m). Gas shows were common whilst drilling and some gas was occasionally flared, but there was no commercial success. Years later, in the early 1980s, some small but commercial discoveries of biogenic gas were made in stratigraphic traps of turbiditic sandstones in the Tertiary section of the Guadalquivir Basin.

In 1943, construction of an industrial distillation plant was started with German technological support in Puertollano (Ciudad Real). The 'Empresa Nacional Calvo Sotelo' (ENCASO), a company fully controlled by the INI, aimed to process the bituminous shales from the Puertollano Basin at a larger scale (Fig. 2). However, the defeat of Nazi Germany in World War II left ENCASO without suppliers and technology (San Román & Sudrià 2003) and forced the company to overhaul the entire project. After a series of technical and economic difficulties the oil shale distillation plant became operative in 1952, being exploited by the Franco regime in the propaganda campaign for the country's self-sufficiency policy, becoming the most emblematic project regardless of economic considerations (San Román & Sudrià 2003). The net oil shale thickness was 7 m, with oil yields varying from 70 to 150 l of oil per ton of rock. A total of 9 Mt (million tons) of oil shale was processed by the distillation plant. Production continued for 14 years under the protection of the government, which passed on the high production cost to the consumer. Eventually, the situation became untenable and, by 1966, the bituminous shales project was abandoned due to poor economic returns, and the plant was replaced by a conventional crude oil refinery. Currently, the Puertollano refinery is operated by Repsol and it is the only one located in inland Spain, and treats imported crude piped across the country from the port of Málaga in southern Spain.

A new mining law was issued on 19 July 1944 which largely overlooked the needs of the petroleum exploration and exploitation sector, and superseded the Royal Decree of 1942. Foreign participation was constrained to 25%, dissuading companies from investing and clearly meaning a move backwards (Merry del Val 1951).

A peculiar character participating in the exploration scene during this period was Domènec Palet i Barba, a geologist, lawyer and a very active republican politician in Catalonia during the country's turbulent early years of the twentieth century (Puy 2015). He developed a device in 1927 that – according to him – was able to locate water and minerals in the subsurface by measuring electromagnetic waves named 'rabdicas' (from the Greek 'rhabdos'). Immediately after the Spanish Civil War, given his republican background, he set aside his political career in an effort to go unnoticed. He became a 'prospector geologist' focused in petroleum exploration using his own device (Puy 2015). He was fully convinced that the device could locate not only oil, but also the oil column thickness and depth at which the oil accumulation was to be found. He made several field reconnaissance surveys all over Spain, using his device on moving trains, airplanes and even in hotel rooms in the middle of a city. He claimed to have identified oil in 17 provinces of Spain (Puy 2015). He was also hired by disreputable local oil companies to make reconnaissance surveys in its exploration acreage. In 1947, Palet founded his own oil company named 'Petróleos Españoles, S.A. (PESA)', which was awarded exploration rights in Catalonia (Puy 2015). An exploration well was finally drilled by PESA in La Bisbal (Gerona) in 1952 (Fig. 6), at a location selected using his cryptic device (Ríos 1958). The La Bisbal-1 well reached a total depth of 640 m in Paleozoic rocks, after penetrating an Eocene marl section (IGME 1987), and unsurprisingly resulted in a dry hole.

By the early 1950s, after a decade of trade embargo and severe depression in the Spanish economy, the international attitude towards Spain started to change. Spain was gradually admitted into international forums but, most importantly, Franco signed a series of accords with the USA that provided Spain with a much needed economic and diplomatic boost. A Royal Decree issued on 12 December 1952 declared that all hydrocarbon exploration activities were of national interest and commissioned the INI to govern these. The INI was now allowed to be associated with other entities or private societies. Finally, the government was beginning to realize that any prospect of exploration success would require substantial investment and high technical expertise from outside. Nonetheless, there was still distrust of foreign intervention and so all national territory was established as a national reserve in favour of the state, except for any permits and concessions in force.

In 1950, the American entrepreneur George Cranmer visited Spain as a tourist, but he was also interested in looking for business opportunities in oil exploration. He was aware that no oil was produced and that exploration activities were relatively low in Spain (Rogers 1958). He visited the Navarra region in the Ebro Basin, where a large surface anticline had been identified (Rogers 1958; Ballestero 1989). He took photographs of the outcrops and gathered all available geological information, and, in the process, decided to come to Spain to invest in oil exploration. Cranmer made contacts with CAMPSA and the INI before returning to the USA where he secured the interest of three American companies: General American Oil Company of Texas (GAO), Delta Drilling Company and the seismic contractor Geophysical Services Inc. (GSI). Finally, in 1952 and after long negotiations, the company 'Comisión de Investigaciones Petrolíferas VALDEBRO' was created as a consortium formed by the INI and the three American companies. One of the difficulties during the negotiations was the poor familiarity of the Spanish with the American oil industry legal terms: in particular, the concept of overriding royalties (Gavala & Ríos 1955; Rogers 1958). The agreement established the INI as 100% holder of the exploration permits, but the exploration investments were to be paid 50% by INI and 50% by the three American companies, who then had a 30% overriding royalty in the case of production and Cranmer would receive a percentage of the benefits for his promotional work. The consulting firm DeGolyer and MacNaughton was awarded the geological fieldwork at cost price. The large surface anticline in the Ebro Basin, 66 km long and 12 km wide, that closely resembled to the surface expression of the Kettleman North Dome Field in California was selected as first drilling test (Gavala & Ríos 1955). After acquiring some dynamite-sourced seismic lines in 1952, the location for the first VALDEBRO exploration well was constructed close to the Marcilla village in Navarra. A fully equipped 1500 HP rig owned by Delta Drilling Company, along with 1000 sacks of bentonite and barite, hundreds of metres of casing, and heavy-trucks were shipped from the USA to Bilbao, and then transported by train and trucks to the drilling site. This intense traffic flow aroused curiosity amongst the local community, who had never seen such a display of resources; it convinced them that a significant oil accumulation was surely to be found. The locals watched with surprise how the Americans wore gloves to work, drunk only mineral water and ate their cold food in only 15 min in the same workplace; they reasoned that, perhaps, it was not so strange that this hard life was rewarded with very high salaries. In March 1953, the Marcilla-1 well (Fig. 6) was spud in the presence of Spanish authorities and George Cranmer, who at the spud ceremony declared to the press that he was fully convinced of the presence of oil. Surprisingly, the well penetrated a monotonous stratigraphic sequence of nearly 3000 m of Oligocene anhydrite and salt, reaching a total depth of 3416 m (Gavala & Ríos 1955; IGME 1987). It was the deepest well drilled in Europe at that time, but not a single oil show was recorded while drilling and the well was abandoned (Rogers 1958). In spite of this first failure, the VALDEBRO company moved on and became the leading explorer in Spain, performing a large number of seismic surveys, and drilling 42 exploration wells between 1953 and 1970, representing more than 100 000 m drilled. Unfortunately, not even one was a commercial success. Today, we acknowledge with admiration and surprise the persistence of this company, which pioneered the reflection seismic acquisition in Spain and, despite failure after failure, did not give up on exploration until 1970 when the company was finally dissolved.

The 'Compañía Petrolífera Ibérica, S.A.' (COPISA), a Spanish company in partnership with the German company Deutsche Schachtbau und Tiefbohr GmbH drilled several shallow stratigraphic wells during the mid-1950s in the flanks of the Orduña diapir (Vizcaya) south of Bilbao. The fact that nearly all known oil fields in NW Germany by that time were associated with salt domes made this company keen on pursuing – well after well – this play concept in the Basque–Cantabrian Basin, where diapirs of Triassic evaporites are very common. Several stratigraphic wells named Elvira (IGME 1987) were drilled back to back (Fig. 6), aimed at identifying the Orduña salt dome limits and a possible flank accumulation (Ríos 1960). One well had a strong kick of H_2S at a depth of 170 m, but the exploration drilling campaign was ultimately unsuccessful (Baquedano 2007).

In 1953, CIEPSA entered into an agreement with the German company Deilmann Bergbau GmbH (Dupuy de Lôme 1955). Germany had made a big exploration effort after World War II, using modern geophysical methods and drilling a significant number of wells. CIEPSA was keen to benefit from this expertise, especially acknowledging that Germany and Spain shared some similar geology. The discoveries of the oil fields in Lacq and Saint Marcet in southern France encouraged CIEPSA to drill the well Boltaña-1 (Huesca), situated in a large surface anticline in the southern Pyrenees (Dupuy de Lôme 1955). The well was situated with no seismic control (Fig. 6). The well penetrated a sequence of very tight carbonates and sandstones of Late Cretaceous–Eocene age (IGME 1987). A major reverse fault was penetrated and the total depth (TD) was reached at 2164 m in June 1954 with no hydrocarbon shows, and the well was plugged and abandoned.

As well as extensive surface geological mapping, detailed stratigraphic and micropalaeontological studies, CIEPSA carried out its first seismic acquisition in Spain in 1953 in the Ebro Basin (Dupuy de Lôme 1955). During the mid–late 1950s, intensive drilling was carried out by CIEPSA in the Basque–Cantabrian Basin based on abundant petroliferous outcrops and because of the existence of many of what we now call play elements: 'excellent for source, reservoir and cover rocks' (Grove 1941). CIEPSA came to have three rigs in nearly continuous operation. Most of the CIEPSA wells recorded oil and gas shows throughout the Upper Cretaceous section, which steadily encouraged further drilling. Among the geologists working for CIEPSA during these years could be found the Spaniard Oriol Riba (Dupuy de Lôme 1955), who years later made a major contribution to geological sciences by introducing the concept of progressive or syntectonic unconformity in 1973, establishing the interactions between tectonics and sedimentation.

The Gastiáin anticline in the Navarra region that had been drilled in 1923 was tested again by CIEPSA in 1954 with the Zúñiga-1 well (IGME 1987). The new well was located in the axis of the Gastiáin surface anticline and was drilled with a new brand rig imported from Germany by Deilmann Bergbau (Fig. 6). The well penetrated a monotonous 2000 m sequence of Upper Cretaceous marls with scattered sandstone layers and ended in Albian-age sandstones at 3127 m after a fish was stuck in the borehole. The well recorded gas shows throughout nearly the entire stratigraphic section and was intensively tested, establishing a maximum gas flow of 40 000 cfg/day (cubic feet of gas per day) (91% CH_4) from thin sandstone beds of Cenomanian age located within the interval 900–1000 m. A second well named Alda-1 was drilled by CIEPSA in 1956 (IGME 1987). The well was located in the northern flank of the surface Gastiáin anticline with the objective of testing deeper reservoirs. It reached 5024 m, at that time a record depth in Spain and one of the deepest wells in Europe. After several production tests in the Albian sandstones, only minor gas was recovered. Finally, in 1961, a third well by CIEPSA named Gastiáin-1 was drilled only 200 m away from the location of the original 1923 Gastiain hole aimed at testing deeper objectives. The well reached a total depth of 3445 m within Aptian carbonates, recording only minor gas shows in sandstone beds of Albian–Cenomanian age (IGME 1987). Several drill stem tests (DSTs) were run, but only formation water with minor gas was recovered. All boreholes drilled in the same Gastiáin anticline revealed not only poor reservoir properties, but also a complex uneven and unconventional gas accumulation.

A new law to regulate the investigation and exploitation of hydrocarbons in Spain was issued on 26 December 1958. It was the first time that hydrocarbon exploration was regulated by a law and independently of other mineral resources. This new law established fixed terms, distinct administrative processes, established a minimum amount of work and monetary investment, and included area relinquishments during the exploration phase. The state reserved for itself significant exploration acreage, awarding it to the INI, but, in turn, the law clearly opened up the possibility of foreign investment. The opening up of the Franco regime to the international community was now clearly favouring the interest and participation of foreign companies in the petroleum activities in Spain, even with examples of 100% participation of foreign capital (Tortella *et al.* 2003).

In 1959, CAMPSA, the Spanish oil-marketing monopoly, signed an agreement with American

Overseas Petroleum Limited, whereby CAMPSA was assigned a 50% interest in their exploration permits: 25% went to Standard Oil Company of California (which years later became Chevron) and 25% to Texaco. The joint venture was named Amospain, which made the Ayoluengo (Burgos) discovery in 1964, which still remains the only commercial oil field discovered onshore Spain.

In 1960, the exploration effort by the joint venture of CIEPSA (52%), Deilmann Bergbau GmbH (24%) and Gulf Oil Corp. (24%), which had joined the association in 1958, resulted in the first commercial discovery of hydrocarbons in Spain, after having drilled 10 wells in the Cantabrian–Basque region (Fig. 1). Commercial flow rates of natural gas (2 MMcfg/day (million cubic feet of gas per day)) were established in the Castillo-1 well (Fig. 6), located in a small village on the outskirts of the city of Vitoria (Alava). The gas was produced in different reservoirs from a 1000 m section of fractured dark shales, carbonates and tight sandy shales of Albian–Turonian age (IGME 1987). Gas flow rates were initially high, but declined rapidly (García-Portero & Esteban 2014). All subsequent wells tested gas, independently of the structural location, indicating that there was no evident trapping mechanism. Years later, the Castillo Field (Fig. 1) was recognized as the first unconventional gas produced in Europe (Selley 2013). Although the gas accumulation was ultimately relatively small, the production was connected through a 4 km gas pipeline to a factory for sheet iron and bicycle manufacture in the city of Vitoria, and produced a total of 1.2 Bcfg (billion cubic feet of gas) until 1981 when the field was finally shut-in.

Epilogue

If we only consider the period of systematic exploration beginning in 1940, it took more than 20 years to find a commercial hydrocarbon discovery in Spain, and a 100 years, if we consider the onset of the petroleum mining activity in 1860. The Castillo gas discovery was later complemented in 1964 by the oil discovery at Ayoluengo (Burgos), which today, more than 50 years later and after intense drilling activity during the 1960s, still remains the only commercial oil field discovered onshore Spain (Fig. 1). The only other discoveries onshore Spain are of gas: biogenic gas in the Guadalquivir Basin, and thermogenic in the foothills of the Pyrenees (Jaca/Serrablo) and the Ebro Basin (Viura). During the 1970s and 1980s, the exploration was mainly focused offshore: giving place to significant oil discoveries in the Valencia Trough, and gas in the Bay of Biscay and in the Gulf of Cadiz (Fig. 1); but this is another story that would deserve another paper.

Acknowledgments Most of this material is based on well-documented internal reports and published papers by geologists and mining engineers working for the Spanish Geological Survey (IGME), CEPSA, CAMPSA and VALDEBRO during the early years of hydrocarbon exploration in Spain. In particular, we would like to highlight the work and excellent reports by Enrique Dupuy de Lôme Vidiella (1885–1965), Director of the Spanish Geological Survey during the Spanish Civil War (1936–39), CEPSA exploration manager and a pioneer of hydrocarbon exploration in Spain. We thank Chris Carr for his comments and in-depth revision of the manuscript, and Fernando Berlanga who greatly assisted in the drawing of all the maps of Spain.

Funding This research received no specific grant from any funding agency in the public, commercial, or not-for-profit sectors.

References

Almela, A., Lizaur, J. & Muñoz, C. 1952. Estudio de la reserva petrolífera de Burgos. *Boletín del Instituto Geológico y Minero de España*, **64**, 205–272.

Alvarado, A. & Cantos, J. 1941. Notas referentes a investigaciones petrolíferas en España y algunos otros países. *Notas y Comunicaciones del Instituto Geológico y Minero de España*, **8**, 163–169.

Álvarez, E. 1923. Las investigaciones de petróleo. Empresas que actúan en España. *Ingeniería y Construcción*, **I-6**, 281–282.

Anderson, R. 1936. *Oil Possibilities in Spain*. Internal Report. CEPSA, Madrid (unpublished).

Anduaga, A. 2009. *Geofísica, Economía y Sociedad en la España Contemporánea*. Consejo Superior de Investigaciones Científicas, Madrid.

Ariza, R. 1924. Las concesiones de petróleo en España. Provincia de Guipúzcoa. *Revista Minera, Metalúrgica y de Ingeniería*, **74**, 677–678.

Ayala, F.J. 2006. *El petróleo de la Lora. La esperanza que surgió del páramo*. Editorial Dossoles, Burgos, Spain.

Ballestero, A. 1989. *Buscando petróleo*. Espasa Calpe, Madrid.

Baquedano, K. 2007. *La búsqueda del petróleo en Álava*. Eusko Ikaskuntza, Donostia, Spain.

Comité Nacional de Sondeos 1929. Catalogación provisional de los sondeos efectuados en España. *Publicaciones del Comité Nacional de Sondeos*, **I-2**, 56–58.

Dixon, A.F. 1921. *Geological Report on the Oil Possibilities of North Western Spain*, Brokaw, Dixon, Donnelly, Garner & McKee (unpublished).

Dupuy de Lôme, E. 1933. Nota acerca de la Industria Petrolífera Rumana y aplicación de sus datos geológicos a la investigación de algunos yacimientos petrolíferos españoles. *Boletín del Instituto Geológico y Minero de España*, **54**, 138–217.

Dupuy de Lôme, E. 1937. Las investigaciones de petróleo en España. *Boletín del Instituto Geológico y Minero de España*, **54**, 335–386.

Dupuy de Lôme, E. 1941. *Oil indications and oil wells in Spain*. Geological Report 26. CIEPSA, Madrid (unpublished).

Dupuy de Lôme, E. 1955. Labor de la Compañía de Investigación y Explotaciones Petrolíferas, S.A. *Notas y Comunicaciones del Instituto Geológico y Minero de España*, **38**, 15–95.

Dupuy de Lôme, E. & Novo, P. 1925. Datos para investigación de petróleo en España. *Boletín Oficial de Minas y Metalurgia*, **92**, 23–47.

Fernández, L. 1914. La cuenca petrolífera de Rubielos de Mora. *Revista Ciencias Exactas, Academia de Ciencias*, **13**, 237–255.

Fernández, P. 1891. *Memoria que acerca de la cuenca minera de Huidobro, Ayuntamiento de Villaescusa del Butrón, provincia de Burgos, y en razón de los productos de petróleo y cobre.* Imprenta y Estereotipia de Polo, Burgos, Spain.

García-Portero, J. & Esteban, J.M. 2014. Exploración para hidrocarburos en la Cuenca Vasco-Cantábrica: pasado, presente y expectativas de futuro. *In*: Bodego, A., Mendia, M., Aranburu, A. & Apraiz, A. (eds) *Geología de la Cuenca Vasco-Cantábrica.* Universidad del País Vasco Servicio Editorial, Bilbao, Spain, 217–236.

Gavala, J. 1916*a*. Regiones petrolíferas de Andalucía. *Boletín del Instituto Geológico de España*, **37**, 27–213.

Gavala, J. 1916*b*. Yacimientos de petróleo en Huidobro (Burgos). *Boletín del Instituto Geológico de España*, **37**, 265–278.

Gavala, J. 1921. Nota acerca de los yacimientos de lignitos y pizarras bituminosas de Rubielos de Mora (Teruel). *Boletín del Instituto Geológico de España*, **42**, 263–302.

Gavala, J. & Ríos, J.M. 1955. Actividades de la Empresa Nacional 'Adaro' y la Comisión de Investigaciones Petrolíferas 'Valdebro' en la investigación de hidrocarburos. *Notas y Comunicaciones del Instituto Geológico y Minero de España*, **38**, 121–175.

Gómez, J.J. 2009. *La Otra Pennsylvania. Las minas de petróleo de Villamartín entre 1906 y 1914.* Ayuntamiento de Villamartín, Cádiz, Spain.

Grove, B.H. 1940. *The Golmayo Area, Province Soria.* Geological Report 10. CIEPSA, Madrid (unpublished).

Grove, B.H. 1941. *Geological Programme for July–December, 1941.* Geological Report 29. CIEPSA, Madrid (unpublished).

Grove, B.H. & Dupuy de Lôme, E. 1941. *Reconnaissance in Southwestern Spain.* Geological Report 22. CIEPSA, Madrid (unpublished).

Hernández-Sampelayo, P. & Cincúnegui, M. 1926. Cuenca de esquistos bituminosos de Ribesalbes (Castellón). *Boletín del Instituto Geológico y Minero de España*, **46**, 1–88.

Huguenet, I. 1852. Asfaltos. Consideraciones generales sobre el origen y la formación de los asfaltos y de su empleo como elemento natural aplicado a las obras de utilidad pública y privada. *Establecimiento Tipográfico Militar de los Sres.* Mateo y Torrubia, Madrid.

IGME 1987. *Contribución de la exploración petrolifera al conocimiento de la geología de España.* Instituto Geológico y Minero de España (IGME), Madrid.

Mallada, L. 1909. Nota sobre los yacimientos de petróleo y de azufre de la provincia de Cádiz. *Boletín de la Comisión del Mapa Geológico de España*, **10**, 53–65.

Marín, J.M. 1978. *Politica petrolera Española.* Confederación Española de Cajas de Ahorros, Madrid.

Merry del Val, F. 1951. Necesidad de promulgar una ley de hidrocarburos en España. *In*: *II Congreso Nacional de Ingeniería celebrado en Madrid durante los días 28 de mayo a 3 de junio de 1950. Tomo II: Combustibles.* Instituto de Ingenieros Civiles de España, Madrid, 167–173.

Merten, R. 2006. Petroleum exploration and production in Spain. *Zeitschrift der Deutschen Gesellschaft für Geowissenschaften*, **157**, 717–732.

Novo, P. 1925. La investigación de petróleo en la península Ibérica. *Revista Minera, Metalúrgica y de Ingeniería*, **76**, 402–411.

O'Shea, G. 1918. Informe del Instituto Geológico sobre las pizarras bituminosas de la provincia de Soria. *Boletín Oficial de Minas y Metalurgia*, **2**, 29–40.

Peñalver, E., Barron, E., Postigo, J.M., Garcia, J.A. & Saura, M. 2016. *El paleolago de Ribesalbes. Un ecosistema de hace 19 millones de años.* Diputación de Castellón, Instituto Geológico y Minero de España.

Puche, O., Mazadiego, L.F. & Ortiz, J.E. 2011. Petroleum in the Spanish Iberian Peninsula. *In*: *History of Research in Mineral Resources.* Cuadernos del Museo Geominero, IGME, **13**, 227–238.

Puy, J. 2015. *Domènec Palet i Barba.* Prohom de la Terrasa republicana. Fundació Josep Irla, Barcelona, Spain.

Ríos, J.M. 1949. Las investigaciones de petróleo en España. *Notas y Comunicaciones del Instituto Geológico y Minero de España*, **19**, 3–20.

Ríos, J.M. 1955. Prospecciones petrolíferas realizadas por la CAMPSA. *Notas y Comunicaciones del Instituto Geológico y Minero de España*, **38**, 3–13.

Ríos, J.M. 1958. Relación de los principales sondeos para investigación de petróleos llevados a cabo en España desde 1939. *Notas y Comunicaciones del Instituto Geológico y Minero de España*, **50**, 47–73.

Ríos, J.M. 1960. Relación de los principales sondeos para investigación de petróleos llevados a cabo desde 1939 hasta Mayo de 1960. *Notas y Comunicaciones del Instituto Geológico y Minero de España*, **59**, 133–166.

Rogers, J.W. 1958. *Sobre el petróleo en España.* La historia de Valdebro, Madrid.

Ruiz, M. 1916. El sondeo de Caldones en Asturias. *Boletín del Instituto Geológico de España*, **37**, 3–26.

San Román, E. & Sudrià, C. 2003. Synthetic fuels in Spain, 1942–66: The failure of Franco's autarkic dream. *Business History*, **45**, 73–88.

Sanz, R. 1948. *El petróleo.* Colección Memorias del IGME (re-edited in 1979).

Selley, R. 2013. *Perspectives on Future Shale Gas Developments in NW Europe*, https://www.geolsoc.org.uk/~/media/shared/documents/Events/100%20Years%20Prentations/20%20Selley_Centenary%20Mtg_Sept13.pdf?la=en

Serra, R. 2008. *La mina de petroli de Riutort (Guardiola de Berguedà).* Zenobita Edicions, Barcelona, Spain.

Tortella, G., Ballestero, A. & Díaz Fernandez, J.L. 2003. *Del Monopolio al libre Mercado. La historia de la industria Petrolera Española.* LID Editorial empresarial, Madrid.

Entangled between worlds: Swiss petroleum geologists, *c.* 1900–50

MONIKA GISLER

Unternehmen Geschichte, Zurich, Switzerland

0000-0002-7507-3592

info@unternehmengeschichte.ch

Abstract: This chapter presents a collection of first-generation Swiss geologists, who, around 1900, went abroad in order to work for oil organizations, both public and private. With the rise of oil exploration after 1900, the demand for experienced oil geologists grew rapidly. Oil companies started hiring trained geologists from different parts of the world, which led to an increasing number of Swiss geologists finding employment abroad. One of the first to start his career abroad was Carl Schmidt (1862–1923) from Basel, highly esteemed and later renowned for his achievements as a teacher of younger oil geologists. Another was Josef Erb (1874–1934), who accomplished an unparalleled career at Royal Dutch Shell. Other examples include Hans Hirschi (1876–1964), who worked for the Union des Pétroles d'Oklahoma for a few years, and Arnold Heim (1882–1965), working nearly 50 years for major oil companies. Many more followed over the decades to come, so that virtual successions took place between them: specific territories and specific companies were 'handed down' between geologists of the same nationality. Where once the Swiss were employed, others took over; the pioneers acted as door-openers for future generations.

In 1919, Arnold Heim published his study on the question of whether there was any oil to be found in Switzerland. As a side-remark, he mentioned that there were as many as 50 geologists working in the oil sector outside Switzerland at the time:

> It is little known that no other country – the U.S.A. exempt – has trained as many oil geologists as our small country [Switzerland]. There are about 50 [geologists], their sphere of action covering the entire world
>
> (Heim & Hartmann 1919).

In the original German this is:

> Eine Tatsache ist noch wenig bekannt, dass mit Ausnahme der Vereinigten Staaten wohl kein Land der Erde so viele tüchtige Petrolgeologen ausgebildet hat, wie unsere kleine Schweiz. Es sind ihrer bis heute etwa 50; ihr Arbeitsfeld reicht über die ganze Erde.

Many of these geologists we know by name, others we do not. Some of them spent just a few months abroad, others their entire lives. This was true not only around 1900, as in Venezuela alone, we know of 20 Swiss geologists who were working for Royal Dutch Shell after World War II. This is because the Swiss were respected as experts. They were extremely well educated; thanks to their studies of the Alps, they could grapple with complex geological environments, and they were flexible. It was relatively easy for them to find jobs in international oil companies, both in the public and private sectors. Thus, the Swiss played as significant a role in worldwide oil exploration as they had previously contributed to the colonization endeavours of the British and the Dutch.

Prior to 1900, there were only a handful of oil specialists worldwide – the first ever job in the oil industry is said to have come about in 1882 (Owen 1975). At this time, the demand was not so much for geologists, but rather for practitioners who had to deal with near-surface exploration. With the rise of oil exploration, particularly after 1900, the demand for experienced oil geologists grew rapidly. Oil companies hired trained geologists from different parts of the world, which led to an increasing number of Swiss geologists finding employment abroad.

Among the first to be recruited was the above-mentioned Arnold Heim (1882–1965) from Zurich. With extensive expertise gained from nearly 50 years of working for major oil companies, he was a pioneer and one of the leading oil geologists of his time. Even prior to Heim, however, geologists left Switzerland in search of the black gold in promising regions of the world. Successions have therefore taken place within the countries of origin. Specific territories and specific companies were 'handed down' to geologists of the same nationality. Where once the Swiss were employed, others took over. One of the first to start his career abroad and thus clear the way for many others to succeed was Carl Schmidt (1862–1923) from Basel (Fig. 1), highly esteemed and later renowned for his achievements as a teacher at the University of Basel. He was among the first to work for Royal Dutch Shell (then

From: Craig, J., Gerali, F., MacAulay, F. & Sorkhabi, R. (eds) 2018. *History of the European Oil and Gas Industry*. Geological Society, London, Special Publications, **465**, 361–374.
First published online May 14, 2018, updated May 18, 2018, https://doi.org/10.1144/SP465.17

Fig. 1. Carl Schmidt (1862–1923). Naturhistorisches Museum, Basel, © Monika Gisler.

Royal Dutch Petroleum Company). Following on from Schmidt, the company hired dozens of Swiss geologists, initially to prospect for oil in the Dutch East Indies (now Indonesia).

Another Swiss geologist was Josef Erb (1874–1934), educated in Aarau and Zurich. He had an unparalleled career at Royal Dutch Shell, clearing the way for generations of Swiss oil geologists to follow. After 1960, it became a tradition that a Swiss national was appointed Chief Geologist at Royal Dutch Shell.

Another example is Hans Hirschi (1876–1964), who worked for the Union des Pétroles d'Oklahoma, where he recruited many fellow countrymen. In a similar manner, Arnold Heim enlisted fellow countrymen to join the Anglo-Persian Oil Company in the early 1950s. Many more followed over the decades to come, not only geologists but also engineers, chemists and physicists. Such successions were not specific to Switzerland, however, but were also true of other countries, such as The Netherlands, the UK and the USA.

Some of these Swiss pioneers will be introduced below. The decision of who to portray depends largely on the somewhat random availability of written sources. Whilst comprehensive literary remains can be found in archives and libraries for some protagonists (archival records and literary remains for

Daniel Trümpy, Arnold Heim, Hans Hirschi and Augusto Gansser can be found at the Hochschularchiv der ETH Zurich), for others very little still exists – an obituary, the mention of a name in an article. Apart from biographical details, which are invariably fragmented, the local and international context is also of interest. The pioneers portrayed here were not lone warriors, but rather were part of a dense network of people operating in the oil fields of very different regions of this world.

This paper draws on my book published in German: Gisler (2014); see also Gisler & Ruetz (2014) and Gisler & Trümpy (2016).

Sought-after Swiss geologists

Swiss geologists were exceedingly sought-after from the very beginning of oil exploration activities, by both private companies and national governments. In the early twentieth century, for example, the Dutch government recruited Swiss geologists via newspaper advertisements, since The Netherlands itself had not yet begun to train oil geologists (von Wyss-Giacosa & Isler 2013).

The Swiss, henceforth, were spoilt for choice; did they want to work for some government or for a private company? The decision was usually made for practical reasons. Wolfgang Leupold (1895–1986) from Berne, for example, decided in favour of the Dutch government because he was allowed to bring his family with him from the very beginning of his employment, something that would not have been allowed by Royal Dutch Shell, which usually included a clause in their contracts that forbade expatriates bringing their wives in the first year. Salaries in the early years of employment were low, however, and it was not unusual for families to face financial difficulties in their first years of employment.

Why, then, were the Swiss so highly sought-after, and why did they go abroad as young geologists at all? There were many motivations.

After 1900, the outward-looking focus of Swiss geologists came about partly as a professional and economic necessity, and partly because they were an attractive proposition to international and governmental companies. Coming from a neutral country was considered an asset in post-colonial times in many nations of the world. Furthermore, the Swiss commanded a rare understanding of the spatial arrangement of subterranean geological structures, skills that were as essential in the early years of oil exploration as they are today.

As mentioned previously, Swiss geologists mastered excellent expertise, both theoretical and field-based. After all, Switzerland provided the perfect conditions for training – unlike The Netherlands, for instance. The Alps and the Jura offered outstanding illustrative material for gaining geological know-how; the Swiss thus acquired their skills from scratch in the outdoor laboratory (Trümpy 2014). This was essential, since the techniques of seismic exploration were not yet known. The key to the successful discovery of oil deposits lay in the geometry of the subsurface, analysed by means of a detailed mapping of the surface. Knowledge and expertise that went beyond mere theoretical studies were of profound importance. Creativity was required, as was the willingness to embrace the unknown, uncharted territories, harsh climate conditions and rudimentary living conditions.

How little the education at home had to do with the actual fieldwork, though, is shown in a citation by the young Eduard Blösch (1884–1980), who between 1911 and 1915 had joined Hans Hirschi on explorations in Oklahoma (USA):

> My work in Oklahoma, which began at the Kansas line, proved to be very different from what I had done before in Switzerland. On account of poor roads most areas could be worked only by camping out; I had to learn to handle horses. I had to add to my vocabulary words I had not encountered in the English classics, and I also had to familiarise myself with the methods of drilling and producing. There was very little known about the general geology of the area, and this lack of information made it necessary to trace beds over long distances, measure sections etc.
>
> (Eduard Blösch, cited in Owen 1975).

In Switzerland, on the other hand, there were virtually no job opportunities or earning potential for geologists. Studying geology around 1900 was known to be unprofitable; going abroad was therefore nearly the only prospect for a trained geologist. There were a handful of jobs for schoolteachers, a few opportunities for an academic career and some found work in mining or tunnel construction. These jobs were limited, however, a fact of which students were well aware (von Wyss-Giacosa & Isler 2013).

The prospect of finding a decent job as a geologist in Switzerland did not change until the 1960s, when more positions were created at universities (Trümpy 2014). The career profile for field geologists changed eventually too, as jobs in resource exploration were increasingly replaced by jobs in the environmental field. After 1970, private geology companies, such as geo-environmental consultants, started to emerge that employed ever-increasing numbers of geologists, now both male and female.

Fundamentals

Returning to the early twentieth century, the academic environment was fundamental for an excellent education. Swiss universities and their teachers played a crucial role in the training of future oil geologists; hence, new faculties on the subject matter were founded in Zurich by Albert Heim (1849–1937) and Ulrich Grubenmann (1850–1924), and later Rudolf Staub (1890–1961) and Paul Niggli (1888–1953); in Basel by Carl Schmidt, August Buxtorf (1877–1969) and Heinrich Preiswerk (1876–1940); and in Berne somewhat later by Paul Arbenz (1880–1943), a student of Albert Heim (Fuchs 2001). In fact, all future oil geologists of the time passed through one or more of these schools. In Zurich, the Swiss Federal Institute of Technology (ETH Zurich) and the university were among the first educational establishments in Europe to create separate Institutes of Geology and Crystallography–Petrography, led by one full professor each. This was essential for the development of these disciplines because it facilitated the adaptation to evolving demands and the introduction of new fields of research. Whereas early on the focus lay on descriptive mineralogy, under Ulrich Grubenmann (Fig. 2) it shifted to petrography with a particular focus on metamorphism. With the establishment of laboratories to carry out geometric rock and mineral analysis, Grubermann established petrochemical research, which was to be of tremendous benefit to future oil geologists. No less important were the work and teaching of the great Albert Heim (Fig. 3), renowned geologist of his time and a pioneer in the field of ductile deformation (plasticity). With his theoretical and practical knowledge, Heim added significantly to the education of future oil geologists. Neither Grubermann nor Heim, however, were ever actively involved in oil research and exploration themselves.

Fig. 2. Ulrich Grubenmann (1850–1924), Teufen. (Verhandlungen der Schweizerischen Naturforschenden Gesellschaft, **105**, 1924).

In Basel, the situation was somewhat different. Carl Schmidt, who was pivotal in the education of geologists there, had gained his own experience in the oil fields. He thus played a central role not only as a teacher, but also as a mentor for young geologists. Schmidt, who had studied geology, mineralogy and petrography in Geneva and Strasbourg, was appointed Professor of Mineralogy and Geology in Basel from 1891 to 1923, where he founded the Mineralogisch–Geologische Anstalt (Mineralogical–Geological Institution) (Buxtorf 1923; Mühlberg 1923). Schmidt had worked on the geological cartography of the Alps, namely the Simplon massif. Even prior to 1900, he went abroad to study potential ore, coal and oil reserves, and he was a sought-after expert for tunnel projects.

Schmidt was therefore not only an excellent teacher, but also equipped with practical knowledge. He had worked for Royal Dutch Shell for many years as a consultant. In fact, he was one of the first Swiss geologists to go abroad to work for an oil company. He gained experience in 1899 with consultancy trips to South Sumatra, Mid-Java and West Borneo (Brunei), and then again in 1903 in East Borneo. This experience was key to the establishment of petroleum geology as an independent discipline, opening the doors for future geologists.

In 1898, prior to Schmidt's consultancy work, the Royal Dutch Company turned to him when experiencing problems with failed explorations and declining production in North Sumatra. In 1890, the company had founded one of the 'world's greatest petroleum enterprises' (Owen 1975) in the Dutch East Indies, namely in North and South Sumatra, Java and Borneo. The shallow wells first exploited were soon to run dry, so the company was forced to change their approach. This is when Carl Schmidt, together with his Italian colleague Cesare Porro, was hired to conduct a geological examination of the region (Forbes & O'Beirne 1957; Owen 1975). Both had acquired prior experience in oil geology in Alsace and Galicia, and they happened to be the first explorers with scientific training to join the staff of Royal Dutch Shell.

Porro and Schmidt delivered their preliminary findings a year later, in 1899, describing in detail the geology of the areas they had examined. According to Forbes & O'Beirne (1957), the report was of

Fig. 3. Albert Heim (1849–1937). (Hochschularchiv der ETH Zurich (Bildarchiv) Hs_0494b-0115-054-AL.)

exceptional quality, considering the period in which it was written. Amongst other things, the report advocated an entirely new fieldwork procedure for oil geologists in the Dutch East Indies, a technique that was adhered to, essentially unaltered, for the following 20 years. This new procedure consisted of first traversing the stream courses, where most of the natural exposures occurred. In areas where oil seeps appeared or favourable structures were suspected, tracks were hacked out through the jungle, and pits were dug along them at intervals. The pits usually penetrated through 4–8 m of weathered soil down to bed rock, and were wide enough for the geologists to enter to determine the dip and strike of the strata (Fig. 4). A rough survey was carried out to locate the pits, and a sketch map was drawn showing the topography and structural observations. This kind of work required a vast amount of cheap labour, and the company employed young Dutchmen who had been raised in the Indies and knew the native language to supervise the work crews (Owen 1975).

Fig. 4. Test pits. In areas where oil seeps occurred or favourable structures were suspected, tracks were hacked and pits were dug along them at intervals. The pits were wide enough for geologists to enter (Forbes & O'Beirne 1957, p. 71).

In the years to come, the company expanded rapidly and acquired new territories, thereby attracting more geologists, many of whom were Swiss. One of these was August Tobler (1872–1929), who followed Schmidt to South Sumatra in 1903–04, where he established the tertiary stratigraphy and found several anticlines on which small fields were later developed (Gerretson 1953–57). Another of these Swiss geologists was Josef Erb, who, on the advice of his teacher Albert Heim, travelled to Sumatra in mid-1900, where he engaged in petro-geological exploration first in Palembang (South Sumatra) and later in North Sumatra. He went on to play a leading role in the Royal Dutch Shell Company.

Geology and management: not such an unusual combination

Josef Theodor Erb (Fig. 5), born in 1874 in Volkach (Germany), grew up in Aarau and studied sciences in his hometown of Zurich with Albert Heim at ETH Zurich (Hirschi 1935; Van Waterschoot van der Gracht 1935; Forbes & O'Beirne 1957; Owen 1975; Wannier 2018). He later worked as an assistant to Ulrich Grubenmann at the University of Zurich, where he obtained a doctorate in 1899 with a dissertation on 'Die vulkanischen Auswurfsmassen des Höhgaus (Hegaus)' ('The Volcanic Cone Mountains of Hegau').

Fig. 5. Josef Theodor Erb (1874–1934) (Gerretson 1953–57, vol. 2, p. 86).

It was the above-mentioned recruitment by the Royal Dutch Petroleum Company that brought Erb to the Dutch East Indies. From then on he worked for the company continuously until his early death in 1934.

Erb is said to have left Switzerland 'with enthusiasm for his first journey to distant lands' (Hirschi 1935). It not only released him of the burden of financial constraints, but also provided him with the opportunity to become a renowned geologist and a successful manager in the services of Royal Dutch Shell. Thus, Erb was instrumental in establishing the new field of petroleum geology, paving the way for subsequent Swiss geologists to carve a career at the heart of the company to this very day.

The colonies of Dutch East India were the forerunners of present-day Indonesia. The oil fields were situated on East Java, North and South Sumatra, and on East Seram (Moluccas). Borneo, the third largest island worldwide, was divided between: Kalimantan, today belonging to Indonesia; Sarawak in today's Malaysia; and Brunei, situated on the northern coast of the island (Fig. 6). Over roughly 350 years, The Netherlands colonized the Indonesian part of this enormous island empire between the Asian mainland and Australia.

From the second half of the nineteenth century onwards, geologists searched for coal in the area and in 1885 they started drilling for oil, primarily in North Sumatra and later in East Java. A few years later, oil fields were found in South Sumatra (Palembang), and in 1890 the Royal Dutch Petroleum Company entered the region. In 1900, Josef Erb travelled to Palembang for the first time, and then on to North Sumatra and Java, where he remained for the coming years, interrupted only by short trips to Europe.

In 1907, the Royal Dutch Petroleum Company merged with the British Shell Transport and Trading Company Ltd to become Royal Dutch Shell (Beaton 1957; Forbes & O'Beirne 1957; Van Zanden *et al.* 2007), and appointed Erb as head of the Geological Service. Erb was one of only a few geologists at the time who were both experienced technicians and strategic thinkers. Highly respected, he was entrusted with the task of exploring completely unknown territories, and negotiating new purchase leases and concessions for Royal Dutch Shell.

The appointment of Erb as head of the central offices, and thus the company's first chief geologist, heralded the arrival of the Swiss in the inner circles of Royal Dutch Shell Management. This position also marked the first permanent position in Erb's career. The new department began by exploring new fields and developing its geological understanding of how to maximize oil production in existing fields. Alongside everyday tasks of geological exploration, Erb became increasingly involved in negotiating new concessions and targeting new exploration sites. As chief geologist, he was also responsible for the growing number of geologists in the field. As a scientist, though, Erb left little legacy behind. In his managerial position, he was not allowed to publish any research findings.

His new duties forced Erb to travel widely, to North America (Oklahoma), Egypt, Galicia (today's Poland and Ukraine), Russia, Mexico and back to Sarawak, where he achieved his most important geological accomplishments. The Sarawak region, located on the NW coast of Borneo (part of present-day Malaysia), provided excellent conditions for geological success – a prime example of Swiss participation in the exploitation of oil in Borneo.

From 1910 onwards, the Royal Dutch Shell group was active in Miri, a small fishing village in northern Sarawak, where, as in other oil regions, oil seeps provided initial indications of possible deposits. For centuries, the inhabitants had extracted oil by hand from dug wells, using it for lighting lamps, waterproofing boats and medicinal purposes (Sorkhabi 2012). The first to concentrate on commercial oil exploitation in the region was the self-taught British naturalist Charles Hose, who started mapping oil seeps in and around Miri. Eventually, the management of Royal Dutch

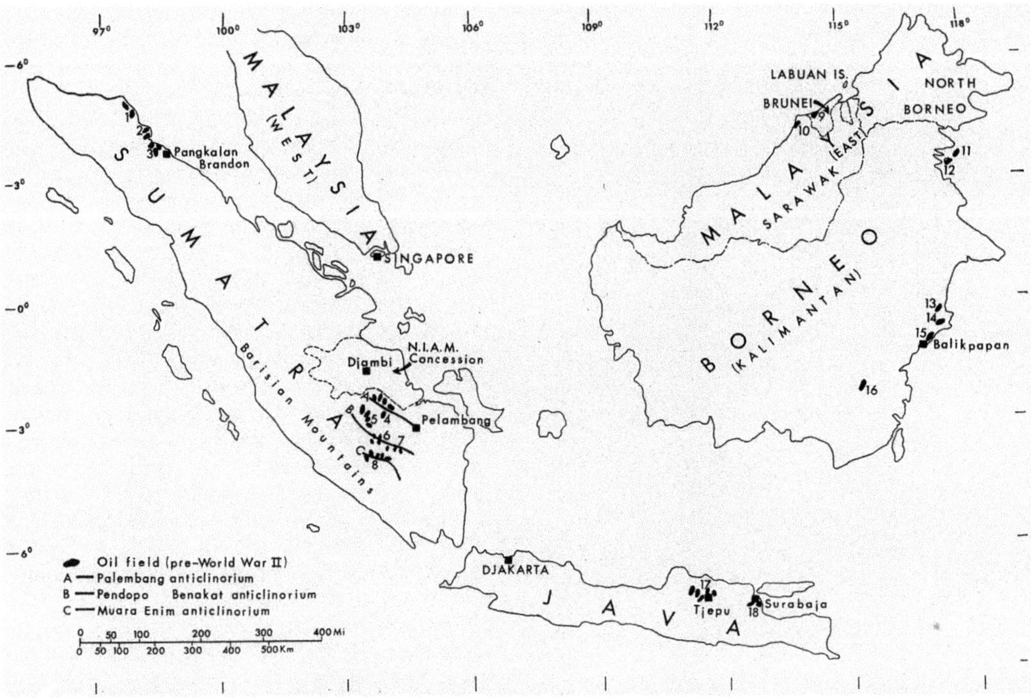

Fig. 6. Map of Java, Sumatra and Borneo (Owen 1975, p. 421).

Shell endorsed the first 'Mining Lease', and put Erb in charge of it.

Arriving in Miri in 1910, Erb immediately began an independent examination of the area, mapping several anticlines where oil seeps were to be expected. Eventually, he located his first exploratory well. This was unique, as until then drilling was always carried out at a visible oil seep. Prior to 1920, practically all major oil fields, with the exception of Mexico, were situated on prominent anticlines, most of which were associated with major Tertiary or Quaternary orogenies. The geologists of the time rarely recommended new prospects without evidence of anticlinal structures, even when located in the vicinity of surface seepages (Owen 1975).

The drilling campaigns conducted by Erb were carried out with a wooden derrick, using a technology developed and used in China for centuries to prospect for brine (Sorkhabi 2012). The drilling operation was so successful that the small village of Miri quickly turned into a booming city. The Miri oil field remained open until 1972, but Erb's work in Miri ended much earlier. After having finished the first drillings, he moved on to the USA in 1911.

When in 1912 the new Central Geological Bureau was established as a consultancy office to the strategic management of Royal Dutch Shell, a new opportunity opened up for Erb. Now one of the most experienced of Royal Dutch's geologists, with 12 years of uninterrupted service to the company, he was appointed director of the new department (Forbes & O'Beirne 1957). The Central Geological Bureau was basically responsible for co-ordinating the geological activities of the various co-operating companies, unifying methods of reporting, exchanging knowledge gained from practical experience and considering the possible introduction into other companies of a particular mode of organization that had been developed and proved satisfactory in one company. In short, the Central Bureau, in addition to its scientific functions, was given an equally important organizational and administrative responsibility.

Ultimately, in April of 1921, Erb was appointed Managing Director of Royal Dutch Shell, one of the most important positions in the oil industry at the time (Forbes & O'Beirne 1957). Erb was now part of the senior management of what was then the largest oil company in the world. At the same time, he and his family were awarded Dutch citizenship (Hirschi 1935).

Even as a director, Erb was in the field quite often. He visited exploration sites in Mexico, Venezuela, East and NW Borneo, and Romania. He was well known and popular. One of the assistants he worked with in Mexico in 1912, Arthur H. Noble – the first

British geologist with Royal Dutch Shell – paid tribute to his mentor, describing Erb as:

> [O]f medium height, stocky, tough and undaunted by difficulty, […] gentle and considerate, with a pleasant sense of humour – a great asset at the end of a tiring or disheartening day
>
> (Noble, cited in Owen 1975).

In 1929, at the peak of his career, yet in poor health and mentally exhausted, Erb resigned. He remained a member of the board of Royal Dutch Shell, and from 1931 until his death he also served as chair of the Geologisch–Mijnbouwkundig Genootschap voor Nederland en Koloniën (Geological Mining Company of the Netherlands and the Colonies). His plans to return to Switzerland and to resume his scientific work remained unfulfilled, however, owing to his early death in 1934.

'Les Hommes du Pétrole'

No other person, however, was more important to early oil exploration – at least from a Swiss point of view – than Arnold Heim. Heim, a citizen of St Gallen, was introduced to geology by two important people, his father Albert Heim (Gansser 1962; Egli 1966; Franks 2011) and Hans Hirschi, who served as both mentor and friend.

Hans Hirschi (Fig. 7), a citizen of Zurich, had studied chemistry, physics and notably geology at the University of Zurich with Albert Heim, as well as mineralogy and petrography with Ulrich Grubenmann (de Quervain 1965, 1972). As early as 1901, he followed his fellow student and friend Josef Erb into the field. He spent his professional life in Borneo, Sumatra, Mexico, the USA, New Guinea, Siam (present-day Thailand) and Iran, either as a senior geologist in a private company or as an independent expert.

Hirschi gained recognition largely for his work in the USA. In 1911, he was appointed chief geologist by the newly founded Union des Pétroles d'Oklahoma. Apparently, Hirschi and his assistant were the first geologists ever to work in this region (Forbes & O'Beirne 1957). Hirschi already had previous experience working and travelling in the USA when he took on the role, which clearly stood him in good stead.

Fig. 7. Hans Hirschi (1876–1964). (Hochschularchiv der ETH Zurich (Bildarchiv) Dia_017-043.)

As an experienced geologist, Hirschi was soon promoted to director of the technological–geological service of the Union. From then on, he spent roughly 4 months a year in the USA and the rest of the time at home. His main task then was to review reports from the field. From the point of view of his colleague and co-worker Eduard Blösch, this was a somewhat complex and exhausting process:

> The organization of the geological department was complicated – my reports went to Dr. Hirschi in Switzerland (he usually made two trips to the States a year). He forwarded them with his comments and recommendations to Holland, from where orders or suggestions went to the local management. The latter did not believe in geology and was slow to act.
>
> (Blösch, cited in Owen 1975).

As chief geologist, Hirschi was now in a position to appoint his own employees, and thus to continue the practice of succession for Swiss geologists. Hirschi initially tried to recruit Heim for the region, but Heim at first declined, being engaged in India and not yet ready to leave. Instead, fellow countrymen Blösch and Emil Kluth arrived in the USA, followed by yet another compatriot, Otto Fischer (1876–43). When Hirschi left the company after a few years, Eduard Blösch succeeded him as chief geologist (Fig. 8).

By the time Arnold Heim arrived in Oklahoma in late 1912 to join Hirschi, he was still an unsteady young geologist, undecided about the shape of his future career. In 1907, he had completed his post-doctoral habilitation (a second degree after a dissertation) in Zurich. To the disappointment of his father, he rejected the option of an academic career and chose a professional occupation in oil (Philipp 1981). As an independent geologist, Heim henceforth undertook many expeditions, starting in Sumatra, then Oklahoma, and from there heading to California in 1913 to supervise the exploitation of an oil field. In early 1915, Heim was back in California, leading a Swiss expedition to evaluate potential territories (Philipp 1983). Between 1917 and 1919, expeditions on his own initiative and for various companies followed, taking him to Galicia, Poland and France. Furthermore, he was commissioned by a Swiss engineering company to explore potential oil reserves in Switzerland, prompted by precarious supply conditions during World War I. Heim also found the time to accompany the Swiss flight pioneer Walter Mittelholzer on his first crossing of the African continent in a seaplane, crossed the Sahara by car

Fig. 8. Eduard Blösch (left) and Arnold Heim (right) working on reports (1912). (Hochschularchiv der ETH Zurich (Bildarchiv) Dia_004-074.)

and even joined an expedition in the mountains on the Chinese–Tibetan border (Philipp 1983).

In-between excursions, Heim was obliged to return to his hometown of Zurich, which remained his primary place of residence throughout his entire life. This was due to the precarious situation affecting the oil industry in those years. Oil from North America flooded the world market, which in the 1920s was still emerging, creating a completely new situation. Exploration activities were being constrained, and at the same time more and more oil geologists were entering the job market (Philipp 1981). In 1923, Heim asked Daniel Trümpy (1893–1971), at the time working in Patagonia, to act as a broker for Royal Dutch Shell, initially without success (Trümpy 1971). Earlier that decade, Heim had complained in his journal that Swiss geologists, once greatly appreciated, now faced stiff competition from colleagues from all over the world. Years later, he had to deal with the situation again when he lost to British and American colleagues when applying for a second expedition tour to the Arabian Peninsula (Philipp 1983). In early 1924, however, he led a research expedition to the Persian Gulf for the Eastern and General Syndicate Ltd, London. The rich oil deposits in the region had not yet been explored because efforts so far had focused on the Dutch East Indies, the Antilles, and parts of North and South America.

Heim, therefore, led a 60-strong expedition into the geologically little-explored areas of Kuwait, al-Hasa (in present-day Saudi Arabia) and Bahrain in NE Arabia in search of oil and water. After a gruelling 2 months of exploration, during which he suffered tremendously from the hot climate of the Persian–Arabian Gulf, he was less convinced of a favourable outcome for future oil drillings than of the success of future water drillings in some places in those regions. The findings of his voluntary water explorations in Bahrain were exploited with great practical benefits not long after Heim's departure. Ironically, Heim had just missed the anticline that 14 years later successfully produced Saudi Arabia's first oil. Thus, it was the New Zealander Frank Holmes who entered the annals of exploration history as the 'father of the Arab oil' (Philipp 1981, 1983).

Heim returned to the Far East several times on exploratory expeditions, one of them an oil exploration for the government of Siam together with Hans Hirschi. He also returned to Sumatra frequently and travelled to different countries in South America, as well as to Russia. When not working on behalf of a private company or a government, he travelled at his own expense. In 1936, he undertook the first Swiss scientific expedition to the Himalayas, together with Augusto Gansser (1910–2012); and in 1947, Heim met up with Daniel Trümpy in South America to explore the oil field in Pirin (Heim 1948). Between 1929 and 1932, he held a professorship at Sun-Yat-Sen-University in Kanton (China).

In the autumn of 1949, Heim, now almost 70 years old, was appointed as the chief geologist of the public Iran Oil Co. – his first ever permanent position. His job was to explore for oil in Iran, excluding the concessionary areas in the southern parts of the country belonging to the British-owned Anglo Iranian Oil Co. (prior to 1935 the Anglo-Persian Oil Company, and BP since 1954). In the first half of 1950, Heim gathered a team of seven oil geologists to be assigned to the service of the Iran Oil Company (Heim 1952; Oberli 2013). The team consisted entirely of Swiss nationals: Jovan Stöcklin (1921–2008) was the first to travel to Iran, followed by Karl Goldschmid (1896–1982), Heinrich Huber (1917–?), Ernst Frei (1894–1974), Max Furrer (1920–2013) and Pierre Soder (1922–2008). The last to arrive was Augusto Gansser, who until then had been employed as Chief Geologist by Royal Dutch Shell in Trinidad. The geologists then focused their main activities around the city of Qum, where major oil deposits were expected (Gansser 1957) (Fig. 9).

It was not entirely by chance that the Iranian government had appointed a group of Swiss geologists to work in the area. With the Cold War evolving and the region suffering from the intense friction between East and West, Switzerland's neutrality was a distinct advantage. The contract that Heim received from the Iranian Planning Administration explicitly requested strict neutrality, and forbade all political activities and any involvement with the internal affairs of the host country.

The first drill results in the region were very promising, hinting at favourable prospects, if not yet an exploitable production. There was, however, a problem: huge salt deposits proved very difficult to bypass, preventing access to the oil beneath for many years.

Heim left Iran in 1952 for health reasons. Thus, he did not witness the huge Qom wildcat blow-out (Alborz-5 well: Gisler & Trümpy 2016) that occurred on 26 August 1956. Augusto Gansser, student, mentee and expedition companion of Heim (Markus & Eichenberger 2008), was now in charge, having succeeded Heim as the chief geologist of the Iran Oil Company. Heim, however, had the privilege of publishing a paper on the tectonics of the region, its major potential for oil and gas, and of describing the event:

> There followed one of the largest eruptions of gas and oil on earth. It was still spewing a plume of oil over 100 m high even after 30 hours. Roads and railways were immediately closed by troops and a 6-km long stream of oil was channelled into a dammed desert basin. A. Gansser estimates production at 10 000 tonnes

Fig. 9. Swiss geologists in Iran: Pierre Soder, Augusto Gansser, Heinrich Huber, Bagher Mostofi (Managing Director of the Iran Oil Co.), Arnold Heim, Jovan Stöcklin, Karl Goldschmid, Ernst Frei and Max Furrer. (Hochschularchiv der ETH Zurich (Bildarchiv) Hs 494: 82.4.)

> a day. A reflective, rippling lake of light oil emerged in the distance, without a drop of water in it. The event that took place at 3 am on 26 August will have incalculable scientific and political consequences as an entire oil field develops around Qum

In the original German this is:

> Es folgte eine der grössten Gas- und Erdöleruptionen der Erde. Noch nach 30 Stunden erreichte der Springbrunnen eine Höhe von über 100 m. Sofort wurden Strasse und Eisenbahn durch Truppen abgesperrt und das Öl in Form eines 6 km langen Baches in ein abgedämmtes Wüstenbecken geleitet. Die Produktion wird von A. Gansser auf täglich 10 000 Tonnen geschätzt. Es entstand ein ferner, spiegelnder, welliger See von Leichtöl ohne einen Tropfen Wasser darin. Das Ereignis vom 26. August morgens 3 Uhr eröffnet noch unabsehbare wirtschaftliche und politische Konsequenzen, denn es lässt sich bei Qum ein ganzes Bohrfeld entwickeln
>
> (Heim 1956).

Only one year later, Gansser also left the country to resume a professorship at both ETH and the University of Zurich. Many of his colleagues, among them Jovan Stöcklin, were to remain in the area until the 1960s (Stöcklin 2006).

For almost his entire career, Heim remained an independent scholar and researcher, repeatedly working for different oil and mining companies over the years on mandate contracts. His principal geographical regions were Mexico, Peru, the Andes in Argentina, West Africa, Iran, the Himalayas, and the south and SW of China (Trümpy 1969). Many extensive trips to all parts of the world not only contributed to his geological research, but also served his interests in environmental and ethnographical studies. Heim developed a scepticism for the Western lifestyle, and became an advocate for decolonialization and nature conservation. He found little interest in purely technical discussions about the new methods of geophysics; preferring instead the thrill of scientific exploration.

Many of Heim's research findings initially remained unpublished as, like many others, he had to subscribe to the policy of non-publication when working for private companies. Travel journals, which combine geological as well as ornithological and anthropological observations, offered a means of circumventing this ban on publication. Heim's overarching work on Central Himalaya (Heim

1939) became a classic, as did his book *Weltbild eines Naturforschers* (*Worldview of a Naturalist*) (Heim 1942), in which he advocated a wholesome lifestyle, and respect for humans and the environment.

Epilogue

'The geologists returning from the tropical forests were trailblazers brimming with professional pride' (Lehner 2004; original German: 'Die aus den tropischen Urwäldern heimgekehrten Geologen waren von Berufsstolz strotzende Pioniergestalten, die auf ihre Welterfahrung pochten'). This is how a geologist of a younger generation described the pioneers of the oil exploration. Shell too, in an illustrated book from 1956, *Les Hommes du Pétrole*, depicted the work of the oil geologists as romantic and adventurous (Melland 1956). The thirst for adventure and discovery may, indeed, have been a rationale for Swiss geologists after 1900 to leave their home country and spend months, years or even an entire lifetime 'in oil'. Discomfort and deprivation for the entire family must have been permanent companions along the way, successes rare and disappointments the norm, rather than the exception. For most of these first-generation geologists, however, it was their only opportunity to make a proper living in their own profession, and they were enthusiastic about the option to do what they knew best, having comprehensive fundamentals from an outstanding education at hand. They were highly esteemed abroad and sought-after experts at home. It is only today that we realize that they were in at the beginning of an oil boom that has not yet come to its end.

Acknowledgements I wish to thank two anonymous reviewers for helpful comments on the first version of the paper.

Funding This research received no specific grant from any funding agency in the public, commercial or not-for-profit sectors.

Correction notice: The original version was incorrect. The caption to Figure 1 had omitted the attribution details.

References

Beaton, K. 1957. *Enterprise in Oil. A History of Shell in the United States.* Appleton-Century-Crofts, New York.

Buxtorf, A. 1923. Professor Dr. Carl Schmidt 1862–1923. *Verhandlungen der Schweizerischen Naturforschenden Gesellschaft = Actes de la Société Helvétique des Sciences Naturelles = Atti della Società Elvetica di Scienze Naturali*, **104**, 43–54.

de Quervain, F. 1965. Dr. Hans Hirschi – 1876–1964. *Schweizerische mineralogische und petrographische Mitteilungen*, **45**, 1073–1075.

de Quervain, F. 1972. Hirschi, Hans. *Neue Deutsche Biographie*, **9**, 227–228, http://www.deutsche-biographie.de/pnd116914599.html

Egli, E. 1966. Arnold Heim. *Mitteilungen der Naturforschenden Gesellschaft des Kantons Solothurn*, **22**, 262–268.

Forbes, R.J. & O'Beirne, D.R. 1957. *The Technical Development of the Royal Dutch/Shell 1890–1940.* E. J. Brill, Leiden, The Netherlands.

Franks, S. 2011. Heim, Arnold. Historisches Lexikon der Schweiz, http://www.hls.ch

Fuchs, T. 2001. Arbenz, Paul. Historisches Lexikon der Schweiz, http://www.hls.ch

Gansser, A. 1957. Die geologische Erforschung der Qum Gegend, Iran. *Bulletin der Vereinigung schweiz. Petroleum-Geologen und -Ingenieure*, **23**, 1–16.

Gansser, A. 1962. Arnold Heim als Geologe. *In*: *Arnold Heim zum achtzigsten Geburtstag am 20.* Schweizerische Stiftung für Alpine Forschungen (ed.), Zurich, **3–5**, 16–18.

Gerretson, F.C. 1953–57. *History of the Royal Dutch.* (**4** vols). Brill, Leiden, The Netherlands.

Gisler, M. 2014. *Die 'Swiss Gang'. Schweizer Pioniere der Erdölforschung.* Verein für Wirtschaftshistorische Studien, Zurich, Switzerland.

Gisler, M. & Ruetz, B. 2014. Aus der Schweiz ins Öl. *Neue Zürcher Zeitung*, 15 January 2014.

Gisler, M. & Trümpy, D. 2016. The rise of the Swiss Mafia. *AAPG Explorer*, April 2016, http://www.aapg.org/publications/news/explorer/column/articleid/28474/the-rise-of-the-swiss-mafia

Heim, A. 1939. *Central Himalaya: Geological Observations of the Swiss Expedition 1936.* Denkschriften der Schweizerischen Naturforschenden Gesellschaft, Zurich, **73**.

Heim, A. 1942. *Weltbild eines Naturforschers: Mein Bekenntnis.* Huber, Bern.

Heim, A. 1948. Die Petrolfelder von Pirin und GAnso Azul in Peru. *Bulletin der Vereinigung schweiz. Petroleum-Geologen und -Ingenieure*, **15**, 10–13.

Heim, A. 1952. Die Schweizer Geologen in Iran. *Bulletin der Vereinigung schweiz. Petroleum-Geologen und -Ingenieure*, **19**, 65–66.

Heim, A. 1956. Tektonische Bemerkungen über den neuen Erdölfund in Zentral-Iran. *Eclogae Geologicae Helvetiae*, **49**, 522.

Heim, A. & Hartmann, A. 1919. *Untersuchungen über die petrolführende Molasse der Schweiz.* Beiträge zur Geologie der Schweiz, Bern.

Hirschi, H. 1935. Josef Theodor Erb 1874–1934. *Schweizerische Mineralogisch-Petrologische Mitteilungen*, **15**, 1–3.

Lehner, P. 2004. Zum 70. Geburtstag der VSP/ASP. *Bulletin für Angewandte Geologie*, **9**, 3–9.

Markus, U. & Eichenberger, U. 2008. *Augusto Gansser. Aus dem Leben eines Welt-Erkunders.* AS, Zurich.

Melland, A.M. 1956. Les Hommes du Petrol. *In*: *50 Jahre Shell in der Schweiz/50 ans Shell en Suisse 1906–1956.* Shell Switzerland (ed.).

Mühlberg, M. 1923. Professor Dr. Carl Schmidt. 23. Juni 1862 –21. Juni 1923. *Mitteilungen der aargauischen Naturforschenden Gesellschaft*, **16**, 149–154.

Oberli, L. 2013. *Auf der Suche nach dem schwarzen Gold – Schweizer Petrolgeologen und die Verstaatlichun des Erdöls im Iran*. ETH Bibliothek/ETHeritage, Zurich, Switzerland, https://blogs.ethz.ch/digital-collections/2013/01/18/auf-der-suche-nach-dem-schwarzen-gold-schweizer-petrolgeologen-und-die-verstaatlichung-des-erdols-im-iran/

Owen, E.W. 1975. *Trek of the Oil Finders: A History of Exploration for Petroleum*. American Association of Petroleum Geologists, Tulsa, OK, USA.

Philipp, H.J. 1981. Arnold Heims erste Arabienreise. *Neue Zürcher Zeitung*, **73**, 81–83, 28–29 March.

Philipp, H.J. 1983. Arnold Heims erfolglose Erdölsuche und erfolgreiche Wassersuche 1924 im nordöstlichen Arabien. *Vierteljahrsschrift der Naturforschenden Gesellschaft in Zurich*, **128**, 43–73.

Sorkhabi, R. 2012. Miri 1910. Geo Expro 7/2 [Online: http://www.geoexpro.com/article/Miri_1910/4c4a4f69.aspx]

Stöcklin, J. 2006. *Persien. Erinnerungen eines Geologen*. Eigenverlag, Seuzach, Switzerland.

Trümpy, R. 1969. Heim, Albert Arnold. *Neue Deutsche Biographie*, **8**, 265, http://www.deutsche-biographie.de/pnd116655844.html

Trümpy, R. 1971. *Memorial to Daniel Trümpy 1893–1971*. Geological Society of America, reprint, pp. 1–4.

Trümpy, R. 2014. *Geologie*. Historisches Lexikon der Schweiz, http://www.hls.ch

Van Waterschoot van der Gracht, W.A.J.M. 1935. Memorial Josef Theodor Erb. *AAPG Bulletin*, **19**, 309–310.

Van Zanden, J.L., Jonker, J., Howarth, S. & Sluyterman, K. (eds) 2007. *A History of Royal Dutch Shell. Volume 1: From Challenger to Joint Industry Leader, 1890–1939*. Oxford University Press, New York.

von Wyss-Giacosa, P. & Isler, A. 2013. *Memori dari Kalimantan/Memories from Borneo 1921–1927. Dokumentasi foto oleh ahli geologi Swiss Wolfgang Leupold/Photographs by the Swiss geologist Wolfgang Leupold*. Ethnographic Museum of the University of Zurich, Zurich.

Wannier, M.M.A. 2018. 'Uniformity in Geological Reports' (1917) by Josef Theodor Erb, petroleum geologist and manager (1874–1934). *In*: Craig, J., Gerali, F., Macaulay, F. & Sorkhabi, R. (eds) *History of the European Oil and Gas Industry*. Geological Society, London, Special Publications, **465**, https://doi.org/10.1144/SP465.7

Myron Kinley and the 'Torch of Moreni'

JEFF SPENCER[1]* & MARIUS FURCUTA[2]

[1]*Petroleum History Institute, Oil City, Pennsylvania, USA*

[2]*Petroleum engineer, Moreni, Romania*

**Correspondence: spencerj320@gmail.com*

Abstract: Myron M. Kinley (1898–1978) has been called the first, the dean, the grandfather and the original pioneer oil-well firefighter. He and his father Karl are often credited with being the first to use explosives to extinguish an oil-well fire in 1913 in a California, USA oil field. Myron formed the M.M. Kinley Company in 1923, specializing in controlling well blowouts and extinguishing well fires. During Kinley's long career, he fought hundreds of oil-well fires throughout the world.

On 28 May 1929, after drilling to a depth of 1453 m, the No. 160 Romano-Americana well in Moreni, Romania had a blowout and caught fire. The derrick was destroyed and the 100 m-high flames could be seen in the city of Ploiesti over 50 km away. Attempts to extinguish the well were unsuccessful and the well burned for over 2 years. More than 100 workers were injured and 14 died in the process of trying to extinguish the fire. By the spring of 1931, the 'Torch of Moreni' had created a crater 76 m wide and 20 m deep. Newspapers covered the attempts to extinguish the fire and many of the photographs of the fire were used to create postcards.

Kinley first visited the No. 160 well in 1930, but was unable to convince the operators to allow him to try his firefighting methods. It took a second visit, a year later, before he was hired by the Romano-Americana Company, an affiliate of Standard Oil. A three-man crew consisting of Americans Kinley and Grady Chupp, and Romanian Costeca Luper battled the fire for 3 months. The fire was finally extinguished and Kinley was paid his $50 000 fee and he returned to the USA. American newspapers recounted his exploits and praised his heroism. Eighty-five years later, oil production facilities and producing wells surround the crater, and patches of carbonized ground attest to the intensity of the 'Torch of Moreni'.

Myron M. Kinley (1898–1978)

Myron Macy Kinley was born is Santa Barbara, California, USA in 1898. As a youngster, Myron longed to end his formal schooling and go to work in the California oilfields where his father Karl worked as a well torpedo shooter. 'Shooting' a well is a method of stimulation to increase the flow of oil. In the early days of oil well completion, this was often accomplished by lowering a shell filled with nitroglycerine into a wellbore and then detonating the explosive material using a variety of methods. Dropping out of school after the eighth grade, 13 year-old Myron began working in the oil fields of Taft, California. One of Myron's jobs was to drive the nitroglycerin truck. He and his father are often credited with being the first to use explosives to extinguish an oil well fire in 1913.

Myron enlisted in the army in 1917 and saw action as an artilleryman in the battlefields of France. At the end of World War I, Kinley returned to the USA, arriving in New York in January 1919. In 1920, J.C. Percival, R.S. Baird and M. M. Kinley incorporated the Western Torpedo Company of Tulsa, Oklahoma with a capital of $25 000 to 'manufacture explosives, metal products, etc.' (*Iron Age* 1920). Within a short time, Myron was back fighting oil-field fires:

> M.M. Kinley, of the Western Torpedo Co., who makes a specialty of fighting oil and gas field fires, conquered the Tom Payne gas well blaze at Bald Hill (Oklahoma) in thirty minutes after he got on the job. The well, with a flow of 20 000 000 cubic feet of gas, was literally blown out by a shot of explosive dragged to the mouth of the well and touched off by electricity
>
> (*Petroleum Age* 1922).

In 1923, he and his younger brother Floyd (1904–38) founded the M.M. Kinley Company and moved to Houston, Texas (Kinley 1996). A rig accident in Goliad, Texas claimed the life of Floyd in 1938. Myron first met Paul Neal 'Red' Adair (1915–2004) on a well in Alice, Texas in the spring of 1941. Red would work for Myron on and off for a few years before becoming a full-time employee in 1946 (Singerman 1990, pp. 242–245). In June of 1959, Red left to form his own company. Myron retired in 1959 and moved to Oklahoma where he drilled and operated oil and gas wells until his death in 1978.

From: Craig, J., Gerali, F., MacAulay, F. & Sorkhabi, R. (eds) 2018. *History of the European Oil and Gas Industry*. Geological Society, London, Special Publications, **465**, 375–380.
First published online May 8, 2018, https://doi.org/10.1144/SP465.8

Moreni, Romania and the Romano-Americana Company

The Romano-Americana Company was founded in 1904 by the Standard Oil Company of New Jersey with a capitalization of 1 250 000 000 Romanian lei. Before Standard Oil's entry into Romania, the German, French and English provided the foreign capital for the country's petroleum industry. The most prolific oil fields were located near the town of Campina in Prahova County and were operated predominantly by Steua Romana, a German company.

Moreni is a town in Dambovita County, Romania, located approximately 110 km (70 miles) NW of Bucharest and 12 km (15 miles) SW of Campina (Fig. 1). The Romano-Americana Company had great success in this area, establishing a large camp and developing extensive oil reserves at Tuscani-Moreni. Initially, shallow oil sands from approximately 365–425 m depths were developed. With the introduction of rotary drilling in the field in 1913, oil sands as deep as 900 m were developed (*Petroleum* 1919). The company's oil refinery was built near the town of Ploiesti, 30 km (19 miles) to the east. From 1903 to 1914, Steua Romana, Romano-Americana and Royal Dutch-Shell's Astra Romana controlled 94.5% of the oil industry in Romania. Romano-Americana was the largest producer of Romanian oil in 1915 at 2 631 000 barrels (bbl), followed closely by Steau Romana Company at 2 572 000 bbl (*Wall Street Journal* 1916).

The German army entered Romania in November 1916. In early December, Romanian soldiers, working with the French and English began destroying the Romanian petroleum industry:

> For probably the first time in the history of the oil industry an effort was made to wipe out completely the producing, transporting, refining, and storage facilities of a large and progressive oil country, so that the enemy could not profit by acquiring it
>
> (*Petroleum* 1919, p. 184).

In late 1916, Romano-Americana owned 46 producing wells with an average daily production of 7800 bbl, 30 000 acres leased and a 10 000 bbl/day refinery. The value of all of these assets was estimated to be approximately 7 million US dollars (*Wall Street Journal* 1920).

During the spring of 1917, the Germans rebuilt the German-owned refineries, scavenging parts and equipment from the other foreign-owned facilities. The Germans also cleaned out and returned to production many of the oil wells that were plugged by the Allies. After the war, Standard Oil claimed

Fig. 1. Map of Romania showing the location of the No. 160 well at Moreni.

damages to its Romano-Americana subsidiary's properties of 10 million US dollars and filed for reimbursement from Germany under the Dawes Plan of 1924. The Dawes Plan was an American-sponsored programme for German reparation payments.

By 1930, the British, French and Belgians each had multiple petroleum investments in Romania, but the only American petroleum company in the country was Standard Oil of New Jersey. In April 1948, the petroleum industry of Romania was nationalized. In July of that year, Romano-Americana's oil fields and refineries, along with the assets of 18 other companies, were placed into a 'petroleum central' named Muntenia (Jordan 1955).

New York Times' early coverage of the No. 160 well

The earliest mention of the No. 160 well fire in American newspapers was a short article in the 29 May edition of the *New York Times* (1929*a*), headlined 'Fire at Rumanian oil well'. The article stated that the damages had already reached several thousand US dollars at the 'Rumano-American Oil Company' well. The article also stated that 10 nearby 'Astra-Rumano Company' wells had caught fire, but that 'heroic work by the fire guards succeeded in localizing the blaze'. By mid-July the well fire was still burning and two men had been burned to death, and damage estimates were one million US dollars (*New York Times* 1929*b*).

'Oil well transformed into a flaming torch. Surrounding Rumanian property threatened by blaze which has burned over 100 days' was the headline of a 22 September *New York Times* (1929*c*) article that stated:

> For more than a hundred days 'the torch of Moreni', a burning geyser of oil, has illuminated the countryside, even to the capital, sixty miles away from 'Well 160, Moreni-Standard'. All efforts to extinguish the flames have been in vain so far, and thousands of dollars' worth of oil and gas are being consumed every day, many lives have been lost and there remains the danger of the fire spreading to hundreds of near-by wells.

The article describes the work being done when the well ignited and the early attempts to extinguish the well:

> The digging had been completed, pumps and pipes had been laid and everything was prepared for the first gush from a depth of 1453 meters. Everything went well, albeit the initial burst was so terrific that the villagers and engineers, although used to happenings of this kind, were anxious. The pressure was two hundred atmospheres, and the first shock carried light machinery and structural scaffolds hundreds of feet into the air to be scattered as much debris. Five minutes later workmen and engineers ran in all directions. The pillar of gas and oil had become one of fire and smoke, so forbidding that it was futile to attempt to get anywhere near it. The steel and wooden towers of six nearby wells ignited almost instantly and were consumed or transformed into writhing metal skeletons.
>
> While the red, black and yellow torch of Moreni is seen day and night within a radius of many miles, 40 000 000 cubic feet of gas have been destroyed since the well caught fire. All methods of extinguishing the blaze have so far proved useless. First an attempt was made to put out the fire by means of a huge iron bell weighing many tons, but this was blown into the air and shattered and torn as if it had been a paper hat. The next step was to tunnel in order that the gas and oil path might be diverted to a new course. The first tunnel was blown in, or blown out, and four workmen were killed. A second tunnel was likewise blown in and three more workers were killed.
>
> For the third time engineers are directing the work of digging a tunnel 250 meters away from the burning well and some eighty meters deep. They hope to divert the underground gases before they can be ignited, and thus not only save the well itself, but practically the entire district.

Kinley at the well

Kinley first visited the No. 160 well in 1930, but was unable to convince the operators to allow him to try his firefighting methods (Kinley 1996). Perhaps the most complete American newspaper account of the well fire was in the *Ames Daily Tribune* (May 5, 1932). The article covers much of the same material as the *New York Times* (1929*d*) and picks up where the *New York Times* article ended. There are also extensive quotes from Kinley himself.

The tunnelling is described differently. The first tunnel is described as 200 feet long and intercepting the well at a depth of 70 feet below the surface. The death of four workmen in that first tunnel explosion is confirmed. A second tunneling attempt was unsuccessful. The third tunnel was dug to a length of 800 feet and reached the well 270 feet below ground, and an explosion in this tunnel took the lives of another four men. This tunnel was then cleaned out and as workmen attempted to drill into the well's casing, another explosion took the lives of a ninth worker and injured several others.

This article described the crater and the flames that greeted Kinley on his second visit to the well on 1 August 1931. The crater dimensions were 65 feet deep, 260 feet in diameter with a 30 foot-high flame above the crater, 60 feet wide and burning 250 million cubic feet of gas a day (Fig. 2). Using explosives, Kinley extinguished the fire, but only for minutes, as the hot material in the crater and associated ground fires reignited the gas. In the article, Kinley described the second attempt:

Fig. 2. Real photo postcard of the 'Torch of Moreni'. Jeff Spencer collection.

I decided that we must get the gas away from the bottom of the crater, as with the terrific heat it was difficult to drag out the hot metal and put out the ground fires. So we installed a 30-foot sheet metal stack with a funnel base over the top of the well and this removed the main flame from the bottom of the crater, allowing us to cool off the metal and put out the ground fires. This done, on August 24 and 25 we set off two shots of 'gelatin' but both missed and the stack fell when guy wires were blown loose. The flame was again at the bottom of the crater.

I knew the flame would require a heavy charge of 'gelatin' so we next directed our attention to the first tunnel in which the 10-inch pipe was still in place. This was cleaned out and a suction fan placed at the end which drew some of the gas from the main flow, reducing the flame at the crater bottom about 50 per cent. But the metal at the bottom of the crater was again glowing with heat, so in order to cool this and the ground, streams of water were turned into the pit. Finally the softened walls slid over the mouth and plugged the opening. This was on September 18. Most of the gas was diverted to the 10-inch pipe, but some broke through the soft ground around the crater.

The fire was out (temporarily as it proved), it is true, but this was but half of our work, for the well had to be put under control. It was feared that the gas would settle and accumulated in the valley surrounding the property and present a very dangerous fire hazard.

Thus mud and water in the crater were removed by a steam injector. This mud boiled like lava as the gas bubbled up through it. It was still hot as the ground had not had time to cool. The mud removed, the gas again caught fire 10 days later; was shot out but caught again. Things looked discouraging.

I charged the banks of the crater, and the slide following the explosion smothered the fire on November 4. This was the last fire in the crater. But we had not yet put the well under control, and we had no assurance that the gas would not again catch fire. Sparks from the iron debris, caused by dragging the metal from the pit, often set the gas again.

We would clean out the crater and snow cavings would refill it. Finally, however, we succeeded and placed a second stack over the mouth of the well. Cement 10 feet deep was poured around the stack and earth to a depth of 25 feet packed on top of that, clear to the depth of the first tunnel. Then at last, we were able to divert all the gas out of the first tunnel. Thus the well had been conquered as the flow of the gas was connected with a gasoline plant.

Other American newspaper headlines and coverage

The 28 August edition of the *El Paso Herald Post* (1931) reported that famed oil-well fighter H.L. Patton (1889–1989) would be joining Kinley in Romania to 'play a leading part' in extinguishing the well fire. There is no reference to suggest that Patton travelled to Romania as part of the team. The first

American newspaper to prematurely cover the end of the fire was the *Corsicana Daily Sun* (1931), a day after the fire was first extinguished. Its headline read, 'Great Rumanian burning gas well was blasted out … M.M. Kinley of Okla. accomplished what others failed to do'.

After Kinley's return to the USA, many newspapers picked up the story of the extinguished Romanian well fire. Over 6 months later, headlines included, 'Fireman back from 7000 mile run' (*Valley Morning Star* 1932) and 'Oklahoma oil expert halts stubborn fire … Myron M. Kinley accomplishes "impossible" in Rumania after 14 had died' (*San Bernardino Daily Sun* 1932).

The Bucharest *Adevaru* and Romanian newspaper coverage

In the 19 September 1931 edition of Bucharest's *Adevarul* (translation: *Truth*), it was reported (translation):

> Moreni, 18 – This morning, at 4, oil well No. 160 Romano-Americana from Tuicani oil field who had burned without stopping two and half years, quenched by herself. A portion of crater shore collapsed on the fire column entirely suppressing. The real story is that not only the game of chance did it, but also the human hand. We know that two American engineers, Kinley and Chupp tried for a few times to blast the well with dynamite but without result
>
> (*Adevarul* 1931*a*).

Ten days later, the newspaper's headline was 'Moreni well reignited' and a portion of the article read (translation):

> Moreni, 28 – Well No. 160 Romano-Americana from Tuicani field who was extinguished 10 days ago caught fire again last night at 1:30. This is what we know about what caused the fire to start again: ten days ago the commission established the following: to clean out the crater from the mud and water; secondly to build over the well mouth a solid concrete funnel and third, to raise a derrick in order to put mud into the hole in Meotian strata. The works started under supervision of American technicians Kinley and Chupp, the same who extinguished the flame 10 days ago. They mounted a pump to get rid of crater mud and water. By doing that they decrease the pressure and the gases increase to

Fig. 3. The present-day 'Torch of Moreni' crater. Photograph by Jeff Spencer (29 October 2015).

came out. Meanwhile a steam choke was kept over the crater. Nevertheless the measure was in vain. The well reignited last night at 1:30. Immediately came Messrs. Eng. Furduescu, Chief of Moreni Mines and Eng. Budurovici –Director of Romano-Americana. At the well is a lot of water and mud bubbling and the fire is small. No fire column was formed. A meeting will take place to discuss the situation. Meanwhile, a lot of water is pumped into the crater

(*Adevarul* 1931*b*).

Kinley and his workers were paid 50 000 US dollars. The contract with Kinley specified that only with the extinguishing of the fire would the crew be paid (Badica 2007).

The crater: today

The crater of the No. 160 well is still visible today after more than 85 years (Fig. 3). Pieces of carbonized ground can be found within the crater. One of the metal bells used in the attempts to extinguish the fire sits approximately 100 m from the crater, near a small OMV-Petrom company field office. Producing wells are still pumping throughout the field area, encircling the rim of the No. 160 well crater.

Acknowledgements JS would like to thank co-author MF, who provided the Romanian to English translations. Marius also was an informative and enthusiastic host during our October 2015 visit to the No. 160 well crater. Marius is the author of an excellent blog (http://furcuta.blogspot.com); with much more information on Romanian oil and gas history.

Funding This research received no specific grant from any funding agency in the public, commercial, or not-for-profit sectors.

References

Adevarul 1931*a*. Oil well from Moreni quenched after two years. 19 September 1931.

Adevarul 1931*b*. Moreni well reignited. 29 September 1931.

Ames Daily Tribune 1932. Greatest of the fire fighters. 5 May 1932, 5.

Badica, P. 2007. *Focul de 850 de zile*, http://www.romanialibera.ro/special/documentare/focul-de-850-de-zile-98639 [last accessed 9 December 2016].

Corsicana Daily Sun 1931. Great Rumanian burning gas well was blasted out. 18 September 1931.

El Paso Herald-Post 1931. Texan will fight fire in Rumania. 28 August 1931.

Iron Age 1920. The Central South. *Iron Age*, **106**, 180, 15 July 1920.

Jordan, C.N. 1955. *The Romanian Oil Industry*. New York University Press, New York.

Kinley, J.D. 1996. *Call Kinley*. Cock-A-Hoop Publishing, Tulsa, OK.

New York Times 1929*a*. Fire at Rumanian oil well. 29 May 1929.

New York Times 1929*b*. Oil well fire burning 42 days. 18 July 1929.

New York Times 1929*c*. Oil well transformed into a flaming torch. Surrounding Rumanian property threatened by blaze which has burned over 100 days. *New York Times*, 22 September 1929.

New York Times 1929*d*. Oil well transformed into blazing torch. 29 September 1929.

Petroleum 1919. Standard plans opening Rumanian works. *Petroleum*, **7**, 40–41, 184–185.

Petroleum Age 1922. Fire at 20,000,000 ft gasser is 'blown out'. *Petroleum Age*, **9**(9) 17.

San Bernardino Daily Sun 1932. Oklahoma oil expert halts stubborn fire. 4 April 1932.

Singerman, P. 1990. *An American Hero – The Red Adair Story*. Little Brown & Co., Boston, MA, USA.

Valley Morning Star 1932. 'Fireman' back from 7,000 mile run. 31 March 1932.

Wall Street Journal 1916. Falls on Jersey Company. 30 December 1916.

Wall Street Journal 1920. Standard Oil of N.J. war claim strengthened. 26 April 1920.

'Uniformity in Geological Reports' (1917) by Josef Theodor Erb, petroleum geologist and manager (1874–1934)

MARIO M. A. WANNIER

Petronas Carigali Sdn Bhd, Petronas Twin Towers, 50088 Kuala Lumpur, Malaysia

Present address: Chemin des Vignes 5, 1806 St-Légier, Switzerland

0000-0002-4184-8839

mwannier@yahoo.com

Abstract: One hundred years ago, standards in geological report writing were formalized within the Royal Dutch Shell Group. Chief Geologist Josef Theodor Erb defined these standards in an internal report aimed at geologists working for Shell subsidiaries. His purpose was to raise the level of scientific reporting and to provide a measure of uniformity in geological reports sent to head office from all parts of the world. Ultimately, 'Uniformity in Geological Reports' was to allow the chief geologist to readily identify and follow-up on the better business opportunities. With time, Erb's report developed into the Shell 'Standard Legend'. Erb's career is symptomatic of a developing company culture, which he embraced as his life's goal. Realizing the shortcomings of this narrow personal philosophy of life may have been at the root of the mental problems that led to his death.

An account of the career of Josef Theodor Erb (Fig. 1) has been given in the form of an Obituary and a Memorial by, respectively, H. Hirschi (1935) and W. van Waterschoot van der Gracht (1935). Here, the life of this Swiss geologist, turned member of the board of directors of the Royal Dutch Shell Group, is considered in more detail (see also Jonker & van Zanden 2007 and Gisler 2014).

Born on 23 February 1874, Erb entered the Swiss Polytechnic School in Zürich at the age of 19 and pursued studies in natural history sciences. In December 1899, he completed a doctorate in geology at the University of Zürich, where he studied under Professors A. Heim and U. Grubenmann; his dissertation was about volcanic rocks ('Die vulkanischen Auswurfsmassen des Höhgaus'). It was during his formative years that the oil industry experienced its first boom and fuelled the nascent car industry.

His university studies completed, Erb entered into short-term employment with the Koninklijke Nederlandsche Petroleum Maatschappij as one of its first recruits and was sent to Sumatra on a mission in May 1900. After three years of fieldwork in the jungle, mapping geological formations and learning about the search and production of oil, Erb fell ill with malaria and was sent back to Europe on medical grounds. While convalescent, he wrote an account of his observations on the geology and coastal morphology of SW Sumatra (Erb 1905) and followed a course in economic geology. In 1905, fully recovered from the bout of malaria, Erb was sent to Java on a six-month mission.

Over the period 1907–11, working now on behalf of the newly formed oil and production company Bataafsche Petroleum Maatschappij (BPM), Erb was as much a world-travelling geologist as one can be today! Employed as a freelancer and hired from one assignment to the next, he visited the main oil-producing regions of the globe to assess prospective sweet spots, scout for concessions and companies, buy properties, and start operations. In a few years, Erb had travelled through Europe (Romania, Poland and Ukraine), North Africa (Egypt), North America (Oklahoma, Mexico), Russia and the Far East again. At the time, the demand for oil was increasing exponentially due, in large part, to the booming automobile industry; by 1910, nearly half a million cars were already on the streets in the USA, up from a mere 8000 in 1910 (US Department of Transportation 1997).

To give today's reader a flavour of what travelling was like for Erb, let us consider his 1909 journey from London to Miri in Borneo, as he accompanied Charles Hose, Resident (Chief Officer) of the Fourth Division of Sarawak. Departing from London, it took 16 days of uninterrupted rail travel, crossing Siberia, to reach Shanghai. From there, a further two weeks of various boat journeys were needed to eventually reach Miri. Inevitably, unexpected things happened: as Erb and Hose travelled on the Trans-Siberian Railway, they shared a compartment with

From: Craig, J., Gerali, F., MacAulay, F. & Sorkhabi, R. (eds) 2018. *History of the European Oil and Gas Industry*. Geological Society, London, Special Publications, **465**, 381–390.
First published online May 8, 2018, https://doi.org/10.1144/SP465.7

Fig. 1. Portrait of Josef Theodor Erb, generally referred to as Dr Erb.

two sleeping berths. Having stopped in Irkutsk for two days, Hose decided to purchase three live seals from Lake Baikal, had them packed in osier cages and then put them on the luggage racks in their cabin. Hose reported (Hose 1927) that Erb 'for one moment … seemed perturbed, but he was a polite man and a good friend of mine'. Later, apparently, 'Erb now showed himself a whole-hearted supporter of the scheme and positively welcomed our strange fellow-passengers'. The three seals died shortly after: Hose skinned them on the go and threw their flesh out of the train window.

In 1910, the Miri oil field (Sarawak, Borneo) was discovered based on Erb's geological mapping and selection of a drilling location (Sorkhabi 2010; Wannier *et al.* 2011). This significant discovery led to the establishment of a new company, the Sarawak Oilfield Ltd. The Miri Field was eventually described as 'the most prolific oil-field in the whole of the British Empire' (Hose 1929), a pompous and self-gratifying comment by the person who alerted the Shell Group to the oil potential in Miri.

In January 1911, recognizing his outstanding services and noting that he was seriously overworked, Henri Deterding, Chairman of the Royal Dutch Shell Group, granted Erb a formal position in the company ('definite arrangement') and at the same time asked him to take four months' leave.

With its business rapidly expanding and more geologists being employed, BPM established a Central Geological Department in 1912, which was headed by Erb. Both the creation of a geological department and that of a chief geologist position at its head were ground-breaking new institutions in the oil industry. In his new role, a pressing issue for Erb was to ensure business continuity in the central office when chief experts were engaged on surveys and visits abroad. During the summer of 1913, between two visits to the Middle East, Erb set up a small department in The Hague in the Netherlands to address this bottleneck. This new department initially concentrated on exploration activities, but with time, as the know-how was expanding, it devoted also much of its activities to development work in producing fields (Jonker & van Zanden 2007).

In 1915, Erb acted as chief negotiator for the Shell Group to acquire private holdings in the Oklahoma Healdton Field. His frequent travels abroad made it difficult to supervise the new staff required by the expanding central geological department. One area of special concern were the shortcomings in the preparedness of geology students and young engineers freshly recruited from universities, which required the company to give them practical training for at least another year. Throughout his tenure as chief geologist and while in central office, Erb personally guided young recruits and company geologists in the practical aspects of petroleum geology.

At the time, a fundamental issue for Erb was the task of evaluating in a coherent manner the largely unstructured and often incomplete geological reports arriving at the central geological department from all parts of the globe. The lack of systematic criteria in compiling geological and petroleum data hampered comparisons between opportunities; it made it difficult for informed decisions to be made and for the better investment proposals to be followed through. It is Erb's greatest technical legacy to have single-handedly defined reporting standards in geological reports: initially published in 1917, his company report 'Uniformity in Geological Reports' was sent to all Shell subsidiaries, eventually leading to streamlined communication with the central office.

By the end of World War I, the number of staff of BPM amounted to nearly 460 and Erb's geological department was the only fully functional unit within the entire Royal Dutch Shell Group (Jonker & van Zanden 2007), proof of his organizational skills. This largely enabled BPM to achieve the extraordinary growth that the company experienced in the 1920s.

In April 1921, in recognition for his outstanding services over 20 years, Erb was elected as one of four new managing directors of the Royal Dutch Shell Group and he took up Dutch nationality on that occasion. In his new capacity, he pursued the application of emerging new technologies to support the search for oil. Following a trip to Germany in 1921, Erb introduced the torsion balance into the

Shell Group as a new geophysical tool to identify subsurface structures. The company organized training in the use of the torsion balance, and survey teams were deployed in all main areas of operation. The use of a torsion balance to detect buried structures with no surface expression was particularly successful in Brunei, where the initial result 'seems to indicate at least one anticline below the swamp deposits' (Harper 1975). This led to the discovery of the giant Seria oil field in 1929. Soon after the introduction of torsion balances in the company, other exploration techniques were introduced through the Central Geological Department in Shell: micropaleontology became an established tool in the early 1920s, formalized in a sub-department in 1924 (Jonker & van Zanden 2007); the refraction seismograph was tested as early as 1923, but was replaced by the reflection seismograph in 1929; the magnetometer and the gravimeter were used in the early 1930s, as was the electrical well-logging method (Beaton 1957).

In 1929, Erb resigned from his position in order to join the supervisory board of the Royal Dutch Shell Group. This required him to travel frequently, particularly to the USA, Mexico and Venezuela, but also to the East Indies and Sarawak. Developing mental problems, leading progressively to a state of depression, marked his final years. The severe economic downturn and financial crisis of the 1930s may have added to his woes. Erb passed away on 24 October 1934, aged 60, while on service in The Hague.

'Uniformity in Geological Reports' (1917)

In his Memorial to Erb, W. van Waterschoot van der Gracht (1935, p. 309) wrote:

> It is greatly to be regretted that a practical scientist of such ability and exceptional experience as Dr. Erb, had practically no opportunity to leave publications accessible to the public.

Indeed, mainly because of his position, Erb was bound by confidentiality and, having only authored internal company reports, his scientific legacy remains unknown. Besides 'Uniformity in Geological Reports', Erb's only other report available in the Shell technical archives is a geological and commercial evaluation of the Wyoming oil fields (Erb 1918). Both of these documents have been 'unearthed' as part of a personal project to unravel Erb's career. Other reports authored by Erb (e.g. on the Miri oil field) are no longer in the company's archives.

Here, his report 'Uniformity in Geological Reports' (1917) is brought to light for the first time; in some way, it is a precursor to the Shell 'Standard Legend', but the emphasis is less on the use of symbols than it is on the organized structure of technical reports and their content. Following the 1917 publication by Erb, Shell issued the 'Standard Legend for Field Sections' (1932), the 'Standard Legend, Exploration and Production Departments' (1954), and further publications of the same with addendums and updates in 1956, 1958, 1976, 1995, 2002, 2012 and 2014.

Erb's report is an internal company document, machine-typed, 37 pages long with 13 folded plates, following American spelling (e.g. 'color', 'organization', etc.). Surprisingly, he finished writing it in Nyon, a town near Geneva, on 17 April 1917 while World War I was still being fought.

The content includes an introduction, a chapter about 'the written report', a 29 page-long chapter about maps, sections and other graphic representations (topographical base maps; representation of geological features on the map; symbols for marking local features on geological maps; local and generalized stratigraphic sections, structure and well sections; diagrams and photographs), and conclusions. As an example of Erb's writing, p. 19 of the report is reproduced in Figure 2.

The 13 tables cover the colour scheme, patterns and abbreviations for time- and rock-stratigraphic units (plates I–IV), symbols for fossils, hydrocarbon (HC) indications, structural geology and wells (plates V–VII: see Figs 3 & 4), examples of a generalized stratigraphic chart, a well penetration log (plates VII–X), examples of production curves of wells, total output, and wells per field and production diagrams (plates XI–XIII: see Fig. 5).

In this publication, Erb defines the editing and geological standards which BPM technical staff should follow while preparing internal company reports. At that time, BPM was rapidly expanding its businesses and the staff at headquarters were trying to keep abreast of technical reports coming from 'all countries where oil is known to exist' (Erb 1917). BPM had five separate operations in the Dutch East Indies, three in Russia, one in Egypt, one in Romania, and five on the North and South American continents.

Erb's report also offers insight into operational procedures in BPM, on the mind-set and the business drivers of this emerging international oil corporation, and its efforts to set up a global geological control-and-command centre in the head offices. We are also reminded that at the time, 'petroleum geologists or oil geologists' were labelled 'economic geologists or practical geologists'!

In the following sections, excerpts from Erb's 'Uniformity in Geological Reports' are quoted that summarize his main recommendations on writing company reports.

Case for standards

> Although confined to one particular branch of the applied sciences, the geological service of our

If any one using the lists should feel inclined to substitute the proposed symbol by one of his own designs, he should remember that it is less important to use the most characteristic symbol than to adhere to standards, which will obviate confusion in the minds of readers, who necessarily have to deal with reports from many sources. Suggestions for better adapted markings might be forwarded to headquarters and we shall be pleased to consider them in a later edition of the Standards.

1) Plate III. Formations and other units of rocks distinguished on the map. It has become general practice to indicate formations and groups of igneous rocks by capital letters. Sub-groups are then characterized by an index to this letter, which index may be either again a letter or a combination of such or a figure. In cases where a formation has become subdivided into well established units and general consent has assigned to the individual groups special letters--like Greek characters for the units of the main Jurassic divisions--such letter should be employed. Where on the other hand the stratigraphical or petrographic unit bears a local name, the letters attached to the capital letter for the formation should be an abbreviation of that local name. "Isl" will thus designate the San Louis group of the Jurassic system in California. "Ilo" the lowermost stage of the Lias on the European Continent. The indexes to the letters for formations on Plate III are of course only examples which shall indicate the principle and they must be varied according to local needs. To fix standards for the thousands of stratigraphical units would, of course, be an impossible task.

2) Plate IV. Sediments. An exploration party in a new district will seldom be in a position to assign at once the strata encountered to

-19-

Fig. 2. An example of Erb's writing (p. 19 of Erb's report).

Company has doubtless become the most extensive or universal organization in the world and it is almost impossible for a few individuals to properly digest the huge amount of interesting and valuable information gathered all over the world by the field staff. Everyone will realize the necessity that certain rules and regulations must be laid down as to the most efficient way to present observations and views of the field geologists to the directors and their advisors. It would otherwise be impossible at headquarters to keep track of the exploration work of the staff, to the detriment not only of the Company, but of the reporting geologist himself.

Only in case the forthcoming information is properly presented the geological department at The Hague will be able to fulfill its most important task, namely to act as a clearing house for general rules governing oil deposits.

The same need for a certain degree of uniformity has been felt long ago by other large geological organizations, particularly that huge Governmental institution, the United States Geological Survey. A good deal of the suggestions below has been adapted from the seasoned principles and methods of that service and from other American sources.

There is however a great difference between the work of an official in a Government geological institution or of another scientific body and the task of an economic or practical geologist, employed by a commercial concern.

The purely scientific worker tends towards a well-rounded complete discussion of the district or problem under review, regardless of the time when his report will be completed.

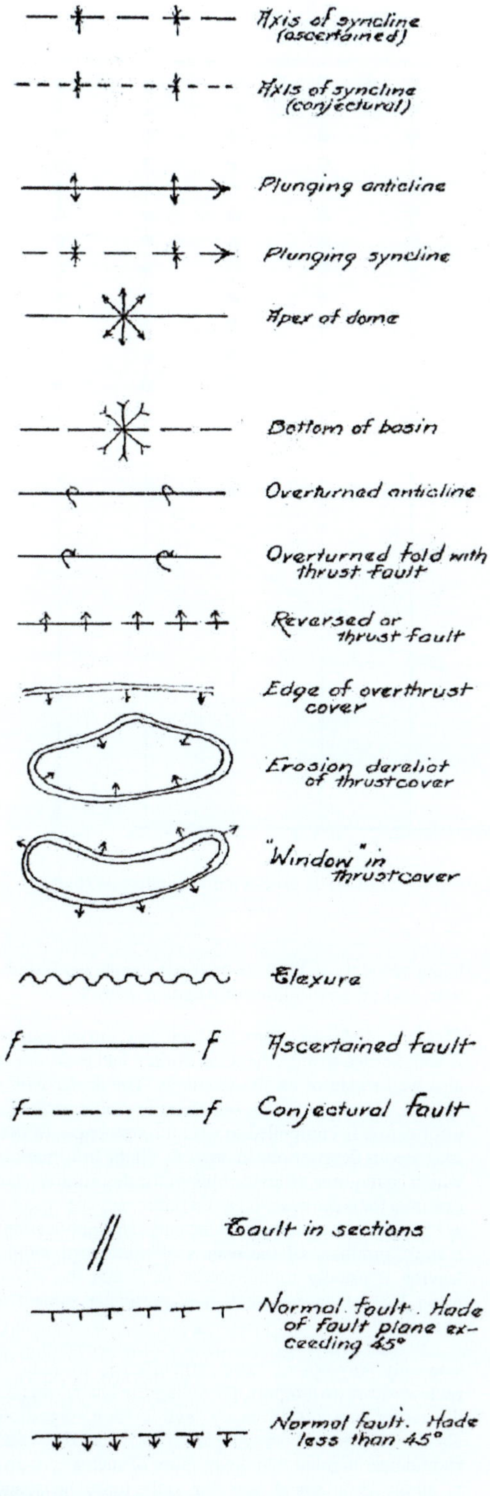

Fig. 3. Structural geology symbols (plate VI of Erb's report).

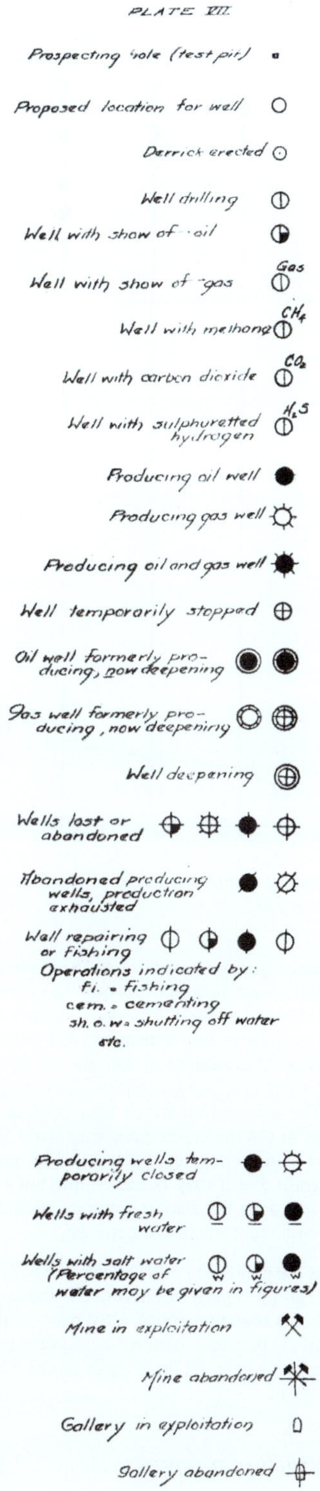

Fig. 4. Well symbols (plate VII of Erb's report).

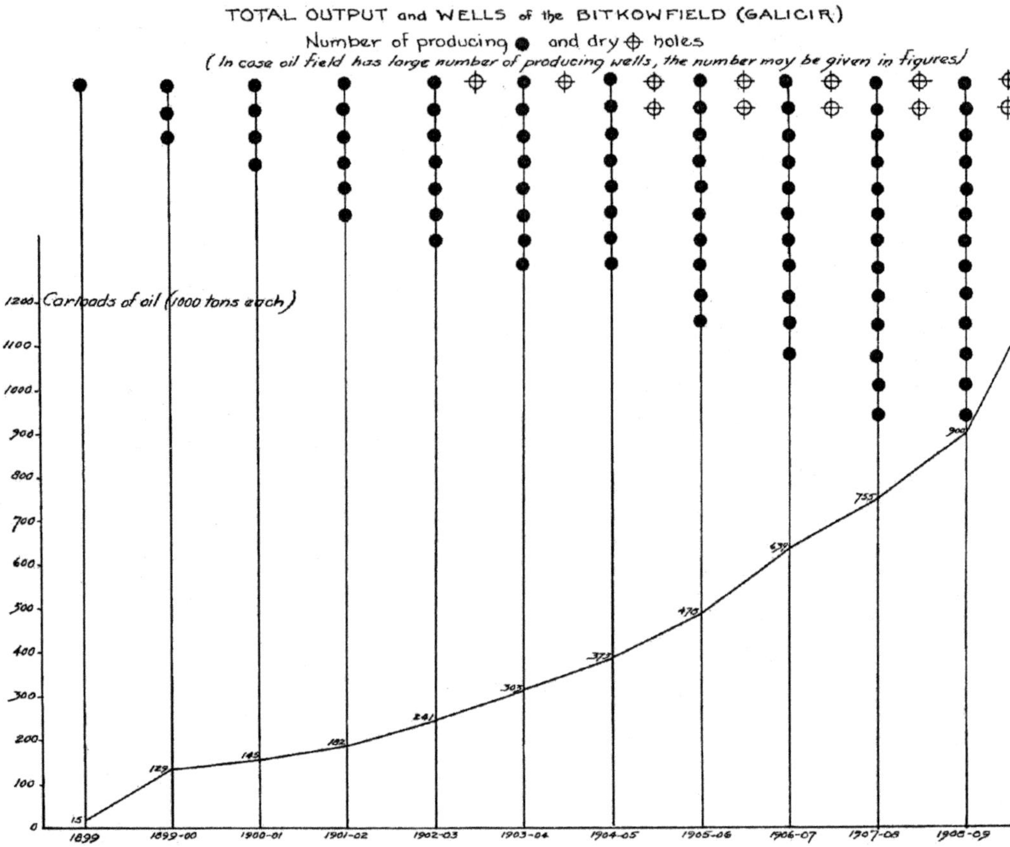

Fig. 5. Graphical display of a field production profile, showing the number of wells producing through time (plate XII of Erb's report).

The practical geologist working for a commercial enterprise must always bear in mind that the company is not entitled to spend money for the advancement of science, but that geology is only a means to secure better financial results. The geological reports are of ephemeral interest rather than being compiled for lasting scientific value. The information which they contain should be available to the managing people as early as possible after the field work is completed and be presented in such a form that it may be accessible not only to the trained geologist but also to commercial and technical people with but a smatter in geology.

Whereas the purely scientific account may only be read by small group of persons interested in the same lines of research, the economic report must necessarily be studied by all the people to whom the task falls to direct the technical work or to act as advisors on the funds appropriated for it.

How to structure a report

The majority of the reports presented to an Oil Company are of a descriptive and geographical nature, being accounts of observations on the geological conditions and oil possibilities of a certain district.

The long established form for such descriptive treatises is well known to any scientific worker, and everyone is also well aware of its shortcomings. The reader with a specific object in mind or searching for some particular information is compelled to read through pages of heterogeneous descriptions in order to sift the information, which he requires, from the bulk of the descriptions, not essential for is purpose. It has therefore become customary that the scientist places at the end of his publications a short summary of the results of his investigations, leaving it usually to the reader to search the whole paper for the pages, where that particular subject is treated.

The only way out of these difficulties is an author's review of his own report, preceding the latter, wherein the writer presents in easily legible form, digestible also for laymen, a short synopsis of his work. The writer should bear in mind that the purpose of such a synopsis is purely economical and that only those facts and observations should find a place therein, which are essential to grasp the practical conclusions, equally presented. Such a review will thus be more comprehensive than the usual summary at the end of a report; and a good plan would be to annotate it with the number of those pages, of the main account, where a more complete treatment of the individual subject may be found ... In adopting this scheme it will become unnecessary for the average reader to plough through the whole report and the latter may then be a complete descriptive account of all the observations made and the views entertained by its writer.

Where however the investigations have brought together a large amount of facts, they should be classed under different headings, such as topography, stratigraphy, structure, oil deposits, etc. and merely the essential observations given in extenso.

The various subjects should equally be balanced and be in harmony with the economic purpose of the account ... Long reports should always be accompanied by a 'Table of Contents', so as to facilitate their consulting.

Standard illustrations

Another point of difference (distinguishing the economic report from the scientific publications) refers to its illustrative parts: maps, sections, etc. The economic report is but seldom printed, the number of copies being too limited. It can therefore be more profusely illustrated than a scientific paper, provided that the graphic representations are of such a nature that they may be manifolded by simple methods. Illustrations have moreover the great advantage over the written words, that they are intelligible even to the layman, whose knowledge of technical terms may be limited.

Geological maps

The size of maps should not exceed 1 × 1.5 meters. In case maps are larger they should be presented in sections.

The groups of rocks distinguished in a territory should be recognizable without difficulty on the plan, the map should be easily legible and the method employed should render quick and accurate reproductions possible.

A third requirement of a geological map or rather of any graphic representation is good taste in its execution and it is in the geologist's own interest that the results of his labors be presented in a pleasing and attractive manner.

A most important rule, which no compiler of a map should forget, requires each unit of strata or group of rocks distinguished on the map to be indicated not only by a color or a separate pattern, but also by a letter or combination of letters to be marked in the legend as well as distributed over the area constituted by it.

Wherever time permits the geologist should endeavor to increase the legibility of the map and section, and avail himself of colors; 'Red' to indicate anticlines (an almost universal association) and for marking strikes and dip of strata, 'Green' – instead of blue, so as not to confound it with water courses – for the axis of a syncline.

In order to show the internal structure of an area the geological plans and maps must be accompanied by sections through the earth's crust.

Columnar stratigraphic sections

The columnar section is a generalization of the various local vertical cuts and intends to show the general sequence and thickness of stratigraphical (...) units of a district. The composition of the rocks is shown by suitable patterns, their petrographic names are marked, the established or proposed designations of the individual groups and the names of the formation, whereof they constitute subdivisions, must also be indicated. Being a generalized picture of the composition of the district also other stratigraphic features like breaks and unconformities may not be omitted.

For the reader with mainly economic interest such columnar sections are most valuable, because they show at a glance in graphic form what otherwise had to be gleaned from tedious reading through pages of descriptions.

In mountainous countries or in desert regions strata weather out from their enclosing rocks in typical form and this feature may become useful in their recognition. The resistance against weathering may then be rendered in the columnar section by drawing it in a step-like arrangement, the steepness of the steps being proportionate to their resistance against atmospheric influences.

Where great changes in the thickness and composition of strata occur in a district, more than one columnar section is needed in order to show also the lateral variation in the stratigraphical conditions.

Structural geology sections

The structure-section, being a generalization from small local exposures, is already more hypothetical than the geological picture of the earth's surface on the map. Its theoretical part lies mainly in the reconstruction of the eroded portion of the surface and in the prolongation of the strata down to depths beyond the range of actual observations. To do both successfully requires a thorough knowledge of the geology of that district as well as general experience in tectonics. It is evident that the oil geologists are practically interested in the continuation of the bedding planes of the strata down to the depth where oil is found.

Well penetration sections

The value of trustworthy well-sections is seldom appreciated. The driller to whom falls the task to collect and preserve the rock samples from the hole considers this work generally a useless and troublesome addition to his daily duties, and the geologist seeing how little reliance can be put on the material, reluctantly collected by untrained people, is often discouraged in their compilation and use. And yet reliable well-sections are the only means to complete the observations at the surface and to carry them down to those depths, where oil-bearing deposits occur. It requires some good will on both sides – the drillers' as well as the geologists' – in order to develop the feeling among the technical staff that a thorough knowledge of underground conditions is necessary for a sound exploitation and that the geologist has the training and wide experience necessary to promote this knowledge. To take successfully rock-samples from a boring is, however, a difficult matter. The instruction usually given state that either samples should be collected as often as the rock drilled through changes or that bits of the formation should be preserved at regular intervals (in meters, sagenes or feet drilled). In the first instance it is left to the driller, with little or no training in the recognition and determination of rocks, to distinguish between various formations, and thus sandstone and limestone of the same color or shales and slates are become often mixed up. In the second case not only important boundaries of strata might be overlooked, but thin and often most typical beds may be passed without recognition. The first named advice is, therefore, the more preferable one, provided that the attention of the technical staff is directed and trained in the distinction of a least the local formation by supervision on the part of the geologist.

An attentive and trained driller will moreover often be able to complete the data from the samples by carefully watching his progress while drilling or the behavior of his tools and the geologist will be glad to avail himself of this knowledge.

Diagrams

Besides maps and sections, there are however other economic features, which must be set forth by the practical geologist and in many cases the written word can advantageously be supplemented or replaced by graphic methods. To them belong large tables or long lists showing the variations in the yield of a well, a field or a district during certain periods, statements regarding drilling progress, lists of figures demonstrating the chemical composition of either the crude or water found in connection with oil. Such series of figures make seldom interesting reading and it usually causes more difficulties to draw from them general conclusions than from diagrams.

If, however, a graphic representation shall be effective it must fulfill the following requirements: (1) It may not be intricate, as it otherwise misses its essential point to be easily legible; (2) all the factors of which the diagram is the result must be embodied therein and their influence be evident from the graphic representation. In case – for example – the yield of a plot or a field is represented by a production curve, the diagram would be of little value if not combined with some kind of graphic display of the variations in the number of wells, which are yielding the oil.

Photographs

In natural science, photography has to a large degree replaced the art of drawing and sketching and an easily portable camera is now-a-day an essential part in the outfit of the traveling geologist. For geographical purposes – the representation of topographical features – the camera is indeed unequalled, furnishing an absolutely true picture of the earth's surface.

In geology, however, it has its shortcomings, because the photographic lense cannot discriminate between essential geological facts and accidental topographical detail or the covering vegetation, both of which may come out more prominent on the pictures than they appeared to the observer in nature.

Artistic qualities are not required from a geologist's photographs. All that is needed is a judicious selection of the point from which the picture is taken, and a great distinctness in its execution.

The photograph itself should bear not only the title of the picture, but the viewer's attention should be directed in a foot-note to its essential points and further mention may therein be made of the point where taken and other photographic details.

Should the picture of an outcrop be less conspicuous than desired, the transparent water colors or indelible inks might advantageously be used to enhance the distinctions. To color the photograph itself is better than a 'retouché' of the negative, because the latter may not be recognizable in the print.'

Conclusions

In conclusion the writer trusts that every reader and every member of our staff will realize the necessity that a considerable degree of uniformity in the geological reports has become necessary through the enormous flight, which our enterprise has taken. To comply in this respect with the proposals contained in the preceding pages will serve to further advance the interests of the Bataafsche and its affiliated concerns. To carry out the wishes and directions of the Central Geological Department will of course no more interfere with the scientific independence of the individual geologist than the submission of a literary worker to the rules of spelling diminishes the value of his production.

Epilogue

'Uniformity in Geological Reports' is in some ways a precursor to the Shell 'Standard Legend' (Shell 1995) but the purpose of these two documents is quite different. Erb's report sets standards about the framing of geological reports, such that data pertaining to the stratigraphy, the tectonic history and the oil manifestations are treated exhaustively in separate chapters, backed up by standard illustrations. Uniformity in the structure of reports sent to head office made it easier for reviewers to focus on the salient strong and weak points of proposals. This purpose is entirely different to that of the Shell 'Standard Legend', which is a collection of symbols to ensure the standardization of data in map form.

It is unclear for how long the reporting standards established by Erb were enforced after he left his position as Head of the Geological Department; while the 1917 version of his 'Uniformity in Geological Reports' had initially been sent to all Shell subsidiaries, we have no information on how this reference publication was disseminated throughout the Shell Group in the 1920s. When issued, Erb's report was part of a process driven by necessity, as surface geology and oil seeps were the only defining criteria to guide exploration efforts. The relevance of these factors faded through time as the oil industry experienced a profound transformation with the advent of seismic, a new technique that provided an image of the subsurface, based on which targets could be defined and measured with accuracy ahead of the drill. Decentralization of the company structure and the creation of largely autonomous operating companies relaxed further the need for uniformity in reporting as defined by Erb.

As stated by W. van Waterschoot van der Gracht (1935, p. 309), Erb:

> grew up with an entirely new science, petroleum geology, and was largely instrumental in establishing it, by the very personal guidance he gave to all the geologists of his company, which was a pioneer in scientific exploration.

In a larger sense, Erb's legacy is more about ensuring focus of economic value while evaluating exploration opportunities, based on the application of sound geological principles backed by technology. His focus on rigorous economics is also reflected in the way he evaluated the purchase of producing fields for Shell and in his recommendations to create new asset-based companies. In so doing, Erb successively guided and directed the company's extraordinary business expansion in the wake of World War I.

In analysing his life and career, Erb appears as the archetype of the stalwart company man, unfaltering to duty, committed to any tasks, ready for all personal sacrifices and giving due deference to his boss, Henry Deterding. Although he had, by all accounts, an extremely successful career in Shell, recognition for his dedication was not fairly translated into monies. As Shell experienced an inordinate phase of growth after World War I, Deterding turned more and more into a despot: board profit-share splitting in 1923 was nearly 97% in favour of Deterding. In 1929, Deterding received 2.3 million guilders in a Royal Dutch bonus, leaving a pitiful 3700 guilders for Erb (Jonker & van Zanden 2007). Fortunately, Erb 'knew how to maintain his attitude under sometimes difficult circumstances, which required great tact and diplomacy, as insiders realize' (van Waterschoot van der Gracht 1935, p. 310).

So little is known about Erb's private life that he remains an obscure character, difficult to decipher. He was described as friendly but had few close friends, and these were mainly other Shell employees; he never married. He liked to be called 'Doctor Erb' and signed his papers as Dr J. Erb. Besides German and Dutch, he could converse in English, French, Malay, Romanian, Spanish and Russian. His hobby was to collect crystals and he assembled a large collection of precious minerals. He made generous donations to geological societies, notably in Switzerland.

Although endowed with immense intellectual powers, an analytical mind and the ability to think strategically, his personal drivers in life remain mysterious. He was so fully embedded into a company culture that the world outside this bubble seems to have had little presence in his daily life. Perhaps, it was the progressive realization of the futility of his all-encompassing dedication to one company that led, ultimately, to a state of deep depression and probable dementia.

Acknowledgements Royal Dutch Shell agreed to Erb's report being made public and supplied the portrait in Figure 1.

Charlie Lee (Shell) and Patrice Wannier helped with data mining. Arthur van Vliet, Marc de Urreiztieta and Phillip G. Cassidy (all at Petronas Carigali Sdn Bhd) and Rasoul Sorkhabi (University of Utah) commented on the original draft. The manuscript greatly benefited from the thorough comments and suggestions made by Tony Doré (Statoil, London), Hans Krause, Cliff Lovelock and Daniel Truempy.

Funding This research received no specific grant from any funding agency in the public, commercial, or not-for-profit sectors.

References

Beaton, K. 1957. *Enterprise in Oil. A History of Shell in the United States*. Appleton-Century-Crofts, New York.

Erb, J. 1905. Beiträge zur Geologie und Morphologie der südlichen Westküste von Sumatra [Contributions to the Geology and Morphology of the Southern West Coast of Sumatra]. *Zeitschrift der Gesellschaft für Erdkunde zu Berlin*, **4**, 251–284.

Erb, J. 1917. *Uniformity in Geological Reports*. Shell Internal Report (written in Nyon, Switzerland).

Erb, J. 1918. *Oil in Wyoming*. Appendix: List of some of the main structures and fields in Wyoming. Shell Internal Report, Shell International, The Hague, The Netherlands

Gisler, M. 2014. *'Swiss Gang' – Pioniere der Erdölexploration*. Schweizer Pioniere der Wirtschaft and Technik, **97**. Verein für wirtschaftshistorische Studien, Zürich, Switzerland.

Harper, G.C. 1975. *The Discovery and Development of the Seria Oilfield*. The Star Press, Bandar Seri Begawan, Brunei.

Hirschi, H. 1935. Orbituary: Josef Theodor Erb: 1874–1934. *Schweizerische mineralogische und petrographische Mitteilungen*, **15**, 1–3.

Hose, C. 1927. *Fifty Years of Romance and Research.* Hutchinson & Co., London.

Hose, C. 1929. *The Field-Book of a Jungle-Wallah. Shore, River and Forest Life in Sarawak.* H.F. & G. Witherby, London.

Jonker, J. & van Zanden, J.L. 2007. *From Challenger to Joint Industry Leader, 1890–1939. A History of Royal Dutch Shell.* Oxford University Press, Oxford.

Shell 1995. *Shell Standard Legend.* Shell International, The Hague, The Netherlands, http://w3.energistics.org/Shell_Standard_Legend/STANDLEG.pdf

Sorkhabi, R. 2010. History of Oil: Miri 1910: The Centenary of Oil Discovery in Sarawak. *GeoExpro*, **7**, 44–49.

US Department of Transportation 1997. *State Motor Vehicle Registrations by Years, 1900–1995*, http://www.fhwa.dot.gov/ohim/summary95/mv200.pdf

van Waterschoot van der Gracht, W.A.J.M. 1935. Memorial: Josef Theodor Erb (1874–1934). *AAPG Bulletin*, **19**, 309–310.

Wannier, M., Lesslar, P., Lee, C., Raven, H., Sorkhabi, R. & Ibrahim, A. 2011. *Geological Excursions around Miri, Sarawak. Celebrating the 100th Anniversary of the Discovery of the Miri Oil Field.* Ecomedia Software, Miri, Sarawak.

German petroleum geologists and World War II

MARTINA KÖLBL-EBERT

Jura-Museum Eichstätt, Willibaldsburg, 85072 Eichstätt, Germany
Koelbl-Ebert@jura-museum.de

Abstract: In the 1930s, Alfred Bentz, August Moos and Karl Krejci-Graf were among the most noteworthy petroleum geologists in Germany. Being scientific modernists, they systematized the search for oil, introducing modern exploration methods. All three, at some stage, worked for the German state on providing the petroleum needed by the German military during World War II. The three colleagues seem to have had an amicable relationship. They were, however, very different. Bentz was not a member of the National Socialist party but obviously lent his expertise to the regime; as did Krejci-Graf, who, although also was not a party member, was a member of the SS, whereas Moos, due to his Jewish background, was murdered in January 1945 in the concentration camp of Buchenwald. This paper endeavours to sketch the lives of the three colleagues, highlighting their relationship and the interconnectedness of contemporary moral issues with professional and scientific demands.

Beginnings

Up to the 1920s, the German oil industry consisted of a diverse conglomerate of companies. Oil was produced by a variety of methods, by quarrying and mining in regions of oil seeps, by drilling for oil in promising localities, and chemically from coal and oil shale. Most companies did not employ a geologist and petroleum geology was not taught at most universities. Geological expertise, if needed, was contributed by the geological surveys, most prominently by Jakob Stoller (1873–1930) from the Preußische Geologische Landesanstalt (Prussian Geological Survey), the mining academies in Clausthal (since 1926) and Freiberg (since 1927), and by Wilhelm Salomon-Calvi (1868–1941), Director of the Institute of Geology and Palaeontology at the University of Heidelberg (Kasper 1973, p. 192; Kockel 2005, chap. 1; Fahrion 2009), who incidentally was dismissed by the Nazi regime in 1934 due to his Jewish descent. He emigrated to Turkey, where he became a professor at the Agricultural High School in Ankara (Guntau 2002, p. 130).

Thus, it had become traditional for the private oil companies to turn to the state for geological expertise. After World War I, demand began to increase with the beginning of private motorization and the interest of the military (Kockel 2005, p. 130). This led to the establishment of a separate department of petroleum geology at the Prussian Geological Survey and the employment of a young geologist, Alfred Bentz (Fig. 1), as assistant to Stoller. Alfred Bentz (1897–1964) had left school in 1916 and became a soldier in World War I. From 1918 to 1922 he studied geology in Munich and Tübingen, obtaining his doctoral degree in 1922 with a stratigraphical thesis (Engelhardt & Hölder 1977, p. 263; Seibold & Seibold 2002):

> On 1 April 1923 I joined the Prussian Geological Survey in Berlin and was sent for geological surface mapping to Western Prussia and the Rhenish Mts. In 1925 already I got a special mapping area along the Prussian–Dutch frontier for my own in which I was actively engaged until the end of 1931.
>
> By the expert in oil-problems with the Geological Survey in Berlin of that time, Professor Dr. Stoller, who was very ill, I was introduced to the dealing with problems of oil-geology as his assistant. Besides this my mapping work was continued. When Professor Stoller died on November 15th 1930 I was called in to take charge of the oil geology with the Geological Survey. In 1932 I started mapping of the sheet Hohenhameln near Hildesheim to keep close contact with the drilling areas. The activities for the oil industry, however, increased to such an extend [sic] that in 1933 I resigned from all mapping work and spent my full time in dealing with problems of oil geology exclusively
>
> (Alfred Bentz: Supplement to my questionnaire, 11 February 1946; National Archives, London, UK: FO 1039/496; see also Owen 1975, p. 1539).

Gifted with outstanding social skills, Bentz established close personal ties with many people interested in petroleum geology. He organized meetings, workshops and round-table discussions, leading to a fruitful exchange between various smaller and larger oil companies and the state geologists. Since the late 1920s, the oil industry increasingly employed their own geologists, who then also profited from these meetings (Schott 1965; Fahrion 2009):

> At regular meetings held under its aegis, the Survey became a forum for inter-company discussion and for the exchange of scientific data and opinion, a most important contribution to subsequent developments not only in Germany but in Europe as well
>
> (Gunther 1947*a*, p. 1096).

From: Craig, J., Gerali, F., MacAulay, F. & Sorkhabi, R. (eds) 2018. *History of the European Oil and Gas Industry*. Geological Society, London, Special Publications, **465**, 391–407.
First published online April 11, 2018, https://doi.org/10.1144/SP465.1

Fig. 1. Alfred Bentz (photo from *Geologisches Jahrbuch* 83 (1965): XV; courtesy of http://www.schweizerbart.de/series/geoljb).

Fig. 2. August Moos (photo from Moos: *Journey of Hope and Despair*, 2010, vol. II, p. 242).

Around 1930–31, there was intense discussion about oil genesis; whether it was a purely stratigraphical phenomenon or whether it migrated – and if so, then how far and why – from an as yet unknown source to where it was found at present. The migration theory was promoted mainly by two geologists, August Moos (Fig. 2) and Karl Krejci-Graf (Fig. 3), who managed to convince Bentz, who then supported his colleagues (Kockel 2005, pp. 46–58).

August Moos (1893–1945) finished school in 1912. He then studied geology and mining for two years in Berlin and Munich. In 1914, he volunteered for World War I and was seriously injured one year later (Bentz 1948):

> He [. . .] endured very harsh circumstances as a prisoner of war in France. He had tried to escape three times, apparently almost made it the third time. But when he was caught again, he was treated like a hardened criminal and severely punished.
>
> For five long years, August had not seen a bed and had to work at hard labor. The worst part of it was that, after he was caught on his third attempt to escape, he was given an extra jail sentence that lasted until well after

Fig. 3. Karl Krejci-Graf (photograph from http://www.dgmk.de/Engler_Medaille/Krejci-Graf.pdf).

> the armistice was signed and all the prisoners of war were freed
>
> (Moos 2010, vol. 2, p. 242).

After being released, Moos continued to study geology and obtained a doctoral degree early in 1922. However, by November 1921, he was already employed by Anton Raky AG, a mining company in Salzgitter, where he supervised the first drillings for oil near Karlsruhe. In 1926, his company sent him to Yugoslavia for oil exploration, which was followed by a drilling campaign for water in Turkey and some coal prospecting in Austria, leading to the first discovery of crude oil in Austria, after which he was sent to northern Germany for research in the oil fields around Hannover. In 1931, Moos was employed by the Preussag Company, whose only geologist he was at that time (Bentz 1948).

His younger colleague, Karl Krejci-Graf (1898–1986), was an Austrian citizen. Due to World War I, he finished school early, in 1916, and then volunteered to become a soldier. From 1919 onwards, he studied geology and mining at the Mining Academy in Leoben (Austria), the University of Vienna, Åbo Akademi (Finland) and the University of Berlin. In 1921, he spent some time ore prospecting in northern Sweden. In 1923, he obtained his PhD from the University of Vienna with a palaeontological thesis. From 1922 to 1930, he was employed as petroleum geologist in Romania; during 1930–36, Krejci-Graf became Professor of Palaeontology and Mining at the Chinese Sun-Yatsen University (Carlé 1988, p. 237, CV by Karl Krejci-Graf, 1 February 1946, AGV #20864; Petrascheck 1989) (AGV is the Archiv der Geologischen Vereinigung, Universitätsbibliothek Freiburg im Breisgau, Germany).

The restricted data from known German oil wells were, as yet, insufficient to solve the scientific controversy over the migration theory. Testing such a theory, however, was costly as it required intense scientific drilling – impossible without financial support from industry or politics.

In May 1932, Alfred Bentz organized a conference of the Deutsche Geologische Gesellschaft (the German Geological Society) on petroleum geology. Social events were added, and representatives of the oil industry, civil servants and politicians were invited in addition to the geologists. The programme was a broad review of the state of the art concerning drilling techniques, promising new geophysical methods, stratigraphy and the potential of the migration theory for prospecting (Kockel 2005, pp. 232–244).

1933: changes

In January 1933, the National socialist party under Adolf Hitler (1889–1945) came into power and immediately began to restructure society to their political agenda: extreme nationalism, anti-Semitism, imperialistic expansion towards Eastern Europe, and revenge for the defeat in World War I and the Treaty of Versailles. The scientific landscape in Germany was affected by a law that allowed the 'removal' of undesired personnel at the German universities and other public employers for racist or political reasons. From 1934 onwards, to gain the qualification for a professorship, a political assessment of the applicant and training in a camp for military sports, which included political indoctrination, became mandatory (Daniels & Michl 2010, p. 55; for a brief history of German geology under the Nazi Regime see Kölbl-Ebert 2017).

As Rudolf Moos, a distant cousin of the petroleum geologist August Moos, noted in his diary in 1938:

> The most difficult circumstances we Jews had to endure, and our worst disappointment, was that our German friends entirely abandoned us. There was a complete and terrifying lack of support from our former colleagues. Again and again we asked ourselves why people did not stand up to counter the worst complaints about Jews. Someone should have stood up and supported us and cried out that what was being said about the Jews was wrong, that the Jews had a more positive side, and that no group of people loved the German Fatherland more and was more concerned with the common welfare
>
> (quoted after Moos 2010, vol. 1, p. 13).

Somewhat later, he noted:

> Everybody is looking for evidence to prove that someone was a socialist or communist before 1933, or that the person had said derogatory things about the Nazis.
>
> This type of denunciation is an outcome of the new system that has awakened the worst instincts of the citizens and used them against the Jews. Nobody trusts anybody else. People on buses or trains sit silently with a grim facial expression. Nobody dares to begin a conversation with anyone else. One does not know whether one's fellow passenger is an informer or a Jew. One is not allowed to have a relationship with a Jew; otherwise one could lose one's job so it is best to be very careful
>
> (quoted after Moos 2010, vol. 1, pp. 14–15).

1934: science

Right from the beginning, the expansionist and revanchist policy of the Nazi regime headed for a new war, and this meant more motorization in the military than ever before: motorcycles, transporters, tanks, ships, submarines and airplanes. Thus, oil as fuel, as well as lubricant, became a crucial resource, and so there was suddenly money for extensive drilling. In January 1934, a German-wide drilling programme – 'Reichsbohrprogramm' – was established under supervision of Alfred Bentz and his team (Owen 1975, p. 1539):

> On 1 April 1934 the Reich decided on the granting of subsidies to exploration drilling, the Geological Survey had to supervise the geological part of this drilling programme. For this purpose the 'Institute for Oil Geology' was founded within the frame of the geological Survey. I was appointed head of this office and three geologists were attached to me to carry out the duties pertaining to the subject. In 1936 when the first successes of the Reich drilling subsidy turned up the then Oberberghauptmann SCHLATTMANN recommended my appointment [i.e. promotion] out of the turn as a 'Landesgeologe und Professor'. On 12 October 1936 this was approved by the Minister of Economics SCHACHT. The party offices did not agree and the 'Reichskanzlei' (Herr HESS) raised objections for the following reason. It was generally known that I was 'far from having adopted the national socialist way of thinking.' SCHACHT, however, turned down these objections. At that time he still was in the position of appointing officials within his department without interference from outside
>
> (Alfred Bentz: Supplement to my questionnaire, 11 February 1946; National Archives, London, UK: FO 1039/496).

Problems with Bentz's promotion were confirmed by Julius Versé, Director of the Prussian Geological Survey from 1936 to 1939. Walter Schriel (1892–1959), who had been the party's official monitor on potential candidates' political reliability and who, consequently, had the right to approve or disapprove advancements and appointments, had reported unfavourably about Bentz (letter from Julius Versé to Alfred Bentz, 13 August 1946, AGV #20490).

Schriel studied geosciences in Leipzig and Göttingen – interrupted by World War I – and obtained his doctoral degree in 1920. Since 1923, he had worked as a geologist at the Prussian Geological Survey in Berlin. In 1935, he became a professor at the University of Berlin, changing to Göttingen in 1940. In 1945, he lost his position due to denazification. He then worked at a brick factory and later as freelance geologist. From 1955 to 1959, he lectured at the Faculty for Agriculture in Göttingen (Szabó 2000, p. 306; Reich *et al.* 2014, pp. 20, 22, letter from Walter Schriel to Prof. Bederke, AGV #20341, 15 February 1946).

The Reichsbohrprogramm was flanked by the 'Geophysikalische Reichsaufnahme' (established in 1935), a nationwide geophysical survey systematizing the search for oil and introducing modern exploration methods, such as seismic investigations with Hermann Reich (1891–1976) as the leading seismologist.

Reich studied geology, mineralogy, chemistry, zoology and botany. In 1914, he obtained a doctoral degree with a geological topic. During World War I, he was a soldier and military geologist specializing in hydrogeology. After the war, he continued to study in Göttingen with a focus on geology and seismology. He then became Assistant at the seismological observatory in Königsberg. During this time, he also took a teacher's exam. In 1921, he was employed as a geologist at the Prussian Geological Survey. In 1925, he qualified for a professorship in applied geophysics at the Technical University in Berlin, where he lectured from 1938 to 1945, in addition to his tasks at the Survey. After World War II, he worked for the Geological Survey of Niedersachsen (Göttingen), where he also lectured. In 1948, he became Chair for Applied Geophysics in Munich and Director of the Geophysical Observatory in Fürstenfeldbruck (Closs 1977).

In the early phase of the programme, a gravimetric survey was the foremost method used, which later, up to about 1940, was increasingly superseded by seismic investigations. The belated use of seismic methods was due to protected patents and, therefore, expensive equipment for reflection seismics had to be purchased in the USA; which, however, due to the economic crisis was frowned upon. Consequently, only a few devices were available. Refraction seismics were easier to get, as the patent was held by a German company. The method was preferred financially, but could not be developed further without consent of the company, which held the patent and, thus, the monopoly. All this slowed the application of seismic investigations (Kasper 1973, pp. 208–217).

Bentz and Reich closely co-operated, and in 1937 the first maps of the underground structures of Schleswig-Holstein were published. Similar reports for other areas followed. Between 1933 and 1939, the number of known salt domes doubled (Kasper 1973, p. 130; Closs 1977, p. 549).

The Reichsbohrprogramm consisted of loans made by the state to the German oil industry for extensive exploration drilling, provided that the drill sites were selected by Alfred Bentz and his team at the Prussian Geological Survey after a thorough geophysical assessment, and provided that the state geologists had full access to all data obtained by the drilling. The loans were only to be paid back if a new array of boreholes struck oil and thus brought actual financial rewards (Kasper 1973, p. 195):

> Advised by the Prussian Geological Survey as governmental central authority, drilling localities were prescribed including programmes for drilling depth, stratigraphy and coring. The Institute for Petroleum Geology [...] provided constant geological supervision of the drillings. The companies informed on the progress of the *Reich* drillings in weekly and monthly reports. After completion, the stratigraphic report including samples was sent to the archive in Berlin.

> Up to 1939, the initiative for drillings and contracts with the state rested with the companies. After that, the *Reichsamt für Bodenforschung* and the office of the *Vierjahresplan* decided all drillings, even when financed by the industry
>
> (Kasper 1973, pp. 295–296; translated from German).

There were monthly meetings of state geologists and representatives of the various participating companies, among them August Moos representing Preussag, for the exchange of data and experience, as well as discussions of results (Kasper 1973, p. 195; Fahrion 2009).

The whole programme functioned from a win–win situation, with three interested partners: the German petroleum industry, the Nazi state and the geologists. For the industry, exploration drilling guided by scientific expertise and outside the established oil fields became possible, because the risk of dry boreholes was taken over by the state. The Nazi state, on the other hand, was interested in a significant increase in domestic oil production for economic reasons but also in preparation of what was to become World War II, while the financial risk was quite calculable as long as the drilling was guided by scientific expertise. For the geologists, the Reichsbohrprogramm was the perfect tool to learn about petroleum geology and to test the various theories on petroleum genesis.

In 1934, the Institute for Micropalaeontology was founded within the Prussian Geological Survey, whose task was also to support petroleum research. Up to 1939, more than 60 000 drill samples were investigated (Kasper 1973, p. 195).

Thus, the Reichsbohrprogramm, guided by geophysical investigation, not only led to a significant increase in oil production but also widened the knowledge about structural controls of crude oil reservoirs and confirmed its migration from a deeper source rock (Kasper 1973, pp. 129–130, 201).

The technological side of drilling profited as well. Up to the early 1930s, percussion drilling was still mostly used. The Reichsbohrprogramm now significantly increased the use of rotary drills and improved on other technical aspects as well (Kasper 1973, pp. 217, 227).

1936: preparations

In 1936, the so-called 'Vierjahresplan' (Four Years' Plan) came in action, which endeavoured to prepare the country, its military and economy for a war beginning at the end of the allocated timespan. Hermann Göring (1893–1946) became the 'Beauftragter für den Vierjahresplan', the head of the organization. Wilhelm Keppler (1882–1960) was commissioned to have the German territory prospected for mineral resources. Because of the strategic importance, crude oil prospecting, as well as production, was regulated by the Vierjahresplan. The drilling programme was further intensified and plans were made for the war scenario (Kasper 1973, pp. 52–53, 130, 195).

Keppler had studied mechanical engineering in Karlsruhe and Danzig (Gdańsk). From 1921 to 1932, he was a co-director of a chemical factory. By 1927, he was already a member of the NSDAP, the Nazi Party, and in 1931 became Hitler's personal advisor for economic policy. Consequently, he was appointed Reich Commissioner for Economic Affairs in July 1933. In 1934, he was commissioned by Hitler to seek possibilities for replacing foreign imports with German resources. In October 1936, his former responsibilities were taken over by the Vierjahresplan, and Göring became his superior. Simultaneously, he ceased to be Hitler's personal advisor. In 1938, Keppler was sent to Austria to prepare the annexation. From autumn 1938, he was a secretary of state at the Foreign Office, and became President of the 'Reichsamt für Bodenforschung' (i.e. the National Geological Survey, the successor organization to the Prussian Geological Survey) in March 1939 until 1945. In 1949, at the Nuremberg Trials, Keppler was sentenced to ten years' imprisonment. He was, however, released in February 1951 (Kasper 1973, p. 195; Riedel 1977; Klee 2013):

> [...] Keppler fused the economics of the Four Years' Plan with the science of the Landesanstalt [i.e. the Survey]. Thus for several years before the war Keppler had assisted in giving German geology a very considerable stimulus, so that when war was declared there were geologists sufficient to supervise and to undertake the widespread commitments which Germany's occupation of Europe entailed.
>
> [...] That he was interested in science is doubtful: but he was very interested in what science could do or what he thought it should do. He undoubtedly saw it as his duty to use his authority with the Party to see that science got its proper place, for Greater Germany depended ultimately upon its application
>
> (Gunther 1947*b*, p. 1149).

In 1938, the chemist Carl Krauch (1887–1968) became Commissioner General of the Vierjahresplan for the chemical industry, which also involved synthetic fuels and refineries for crude oil, and Alfred Bentz was named Commissioner of the Vierjahresplan for petroleum exploitation, although he still worked within the administrative structure of the Prussian Geological Survey (Gunther 1947*b*, p. 1151; Kasper 1973, p. 54; Seibold & Seibold 2002, p. 1090):

> In view of the growing importance of the Reich drilling programme the Four years plan office directed by GÖRING wished to be kept informed directly on the

> results. He had hitherto only been informed via the Ministry of Economics which up to 1939 had in hand the administration of the drilling programme. On July 28th 1938 I was commissioned by Göring to act as his commissioner for the crude oil exploitation in order 'to carry out the search for new oil fields and prepare their exploitation'. A task I actually carried out already for years. The point was stressed that this was a personal mission to be carried out within the frame of the Reichsamt für Bodenforschung [...].
>
> My activities were restricted to scientific-technical problems in connection with the Reich drilling programme. [...] In fall 1942 I engaged Herr Artur Groote assisted by a small office staff. Herr Groote, in carrying out my instructions could not get along with the head of the office of the Ministry Speer, Herr Schieber, being called a 'Defaitist' by the latter. I therefore replaced him in fall 1943 [...]. This office never comprised more that 5 to 6 employees only dealing with questions pertaining to the drilling- and exploitation programme
>
> (Alfred Bentz: Supplement to my questionnaire, 11 February 1946; National Archives, London, UK: FO 1039/496).

Although Bentz still did what he had 'actually carried out already for years', he and his staff no longer were merely scientists supporting the industry but obviously lent their expertise to the Nazi regime helping to 'lubricate' the German war machine.

Regarding oil, preparation for war meant that basically four strategies where followed simultaneously (Kockel 2005; Eichholtz 2006; Toprani 2015):

- Import of surplus petroleum from Iraq and Mexico, and storage for the time when the UK or the USA might enter the war and block supplies.
- Gaining allies who would be able to supply oil even under the conditions of war: Austria, Hungary and especially Romania.
- Chemically synthesizing liquid fuels (particularly kerosene) from coal and oil shale – an important part of the Four Years' Plan, but which was, however, the domain of chemistry and chemical engineering, and not of geology.
- Prospecting and drilling for oil within Germany to increase the production there (gas was not yet important at this time).

Alfred Bentz was involved in the first idea. He travelled to both countries, assessed the situation, helped in negotiations, and also learned a lot about how things were done in British and American companies. He was also mainly responsible for the fourth strategy via the Reichsbohrprogramm, which now had a definitive political edge. Production from new finds within Germany tended to be withheld as secret reserves, leading to a significant production increase with the beginning of World War II (Kasper 1973; Kockel 2005; Eichholtz 2006; Toprani 2015). As Director of the Institute for Oil Geology within the Survey, Bentz became responsible for the crude oil supply of the Third Reich, at times reporting directly to Hermann Göring (Seibold & Seibold 2002).

Nevertheless, Bentz never became a member of the NSDAP. Later, he loved to tell the following anecdote: Göring reproached him: 'Bentz, it is high time that you become a party member'. – 'High Commander, do you want me or do you want petroleum?' – 'Oh, well, then stay out, you dammed Swabian democratic strong head: I need oil' (Carlé 1988, p. 132; translated from German).

Even though the truth of such a story cannot be verified, we may learn perhaps that a certain amount of coquettish dissent was quite in order with the very powerful, as long as the person in question was needed and doing his job. Göring needed oil and, thus, Bentz, because he had high-flying and quite unrealistic ideas: in October 1938, he announced plans to increase the number of aircraft in the German air force five-fold, which would have required 85% of the world's total oil production of the time to get them actually off the ground. In the end, Göring settled for 4000 instead of 20 000 airplanes (Evans 2010*a*, p. 443).

Karl Krejci-Graf was to become one of the petroleum geologists responsible for the second strategy. In 1936, he had finished his qualification for a professorship, at the Technical University in Berlin-Charlottenburg. In spring 1937, he returned from China to Germany and became Chief Geologist for the Preussag Company, as a successor to August Moos. With the support of Bentz and others, Krejci-Graf was appointed Professor for the Geology of Combustibles at the Mining Academy in Freiberg in August 1937, although he was not a member of the Nazi Party and insisted on retaining his Austrian citizenship (Petrascheck 1989, CV by Karl Krejci-Graf, 1 February 1946, AGV #20864).

1938: despair

Meanwhile, August Moos faced increasing difficulties due to his Jewish descent. Since he was employed by the Preussag Company and not by the state, he managed to keep working with the support of his employers, even though pressure increased significantly and from 1936 his employment had to be disguised by the company. In 1938, however, employment of a Jew was no longer possible and Moos was reduced to a freelance consultant (Bentz 1948).

Moos tried desperately to emigrate to Australia with the help of the geologist Curt Teichert (1905–96), who had left Germany in 1933 to accompany his Jewish wife. The ensuing odyssey led the couple via Denmark and Australia to the USA (Seibold & Seibold 2008). Moos contacted Teichert in

September 1938 for the first time (letter from August Moos to Curt Teichert, 15 September 1938, GA15/1066) (GA is the Geologenarchiv, Universitätsbibliothek Freiburg im Breisgau, Germany). Teichert, however, had not much hope:

> Unfortunately, the need of Australia for geologists is rather small and the demand seems generally not greater than the natural issue from within the country. That I was called here must not mislead, because they wanted a palaeontologically educated man, of which there is a lack here. On the other hand, practical geology, which you represent, here is very much in focus of the training.
>
> Next year, I am going to drive to eastern Australia. So far, I have little overview over the conditions there. I shall, however, keep your case in mind and let you know, if any possibility should seem to present itself
>
> (letter from Curt Teichert to August Moos, 15 October 1938, GA15/1067; translated from German).

Shortly after the pogrom of the 'Reichskristallnacht', 9–10 November 1938, Moos and his wife reconfirmed their urgent interest in leaving the country: 'Since here I now have no longer any possibility for job or existence [...]' (letter from August and Beata Moos to Curt Teichert, 16 November 1938, GA15/1068). Roughly a month later, they tried again:

> Since my last letter to you, I have now, on 18 December, submitted an 'Application for Permit to enter Australia' to the Australian government in Canberra. I have allowed myself to name you, very honoured Doctor, as my acquaintance in Australia.
>
> Since an emigration from here in the near future becomes ever more urgent, I beg you, if possible, to second my immigration or even to hasten it, should this be the proper thing to do.
>
> In mid-December, I have also written to Mr Walter George Woolnough, Commonwealth geological Adviser, Canberra, with the request to support my immigration.
>
> I'd be exceedingly thankful, if you could name me other offices or people, to whom I might practically turn because of my immigration or possibility of a job. You yourself had the friendliness to assure me that you would investigate in this direction on your planned journey to eastern Australia.
>
> At the same time, with separate post, I am sending the doctoral dissertation of my wife, which as palaeontological work on the Dutch-East-Indies [Indonesia], might be of interest to you
>
> (letter from August Moos to Curt Teichert, 28 December 1938, GA15/1069; translated from German).

However, neither August Moos nor his wife, the palaeontologist Dr Beata Moos (1902–84), were able to obtain a job in Australia, and thus they did not receive an immigration permit (Moos 2010, vol. 2, p. 363). As August's distant cousin wrote in his diary:

> But how should one get out and where could one go? The Nazis confiscated our passports to prevent us from traveling to other countries. And almost all other countries were closed to Jewish immigration. Since 1933, the most important topic of conversation among Jews focused on emigration. Whichever way we turned, there were problems, and one of the biggest problems was how to finance emigration
>
> (Moos 2010, vol. 1, p. 17).

Time was rapidly running short for the Moos family because, in September 1939, World War II started with the German invasion into Poland and emigration became impossible. In this desperate situation, help came from Alfred Bentz and his connections to the oil industry:

> Dr. August Moos, a jewish [sic] oil geologist, was employed on my intervention and intercession by the Preussag and was supported by this company as long as possible. When his employment was no longer possible in Germany, he was transferred to Gewerkschaft Elwerath, also on my personal intercession and protection. He then was set to work at Agram [Zagreb], Yugoslavia, where he could work without being disturbed up to summer 1944, though the SS again and again asked for his removal. I opposed against it stating that Moos had an experience which was irrecoverable
>
> (Alfred Bentz: Supplement to my questionnaire, 11 February 1946; National Archives, London, UK: FO 1039/496).

The petroleum company Gewerkschaft Elwerath, indeed, hired Moos as an oil expert for their business expansion into Yugoslavia, and so August Moos and his family managed to leave Germany in December 1939 and settle in Croatia. This was, however, but a temporary relief, as the country's independence existed only on paper. It also had a fascist regime and, beginning in August 1942, Croatian Jews were deported to German extermination camps. For the Moos family, there was nowhere to go and officials of the Elwerath Company tried to camouflage Moos' existence. Officially, he worked for a Croatian bank but still delivered reports as to petroleum until summer 1944 (Bentz 1948, letter from Hans Brockhaus to Alfred Bentz, 12 February 1947, AGV #20520).

1939: war

The war essentially stopped the supply of oil from the Near East and Mexico, putting additional pressure on Alfred Bentz and his growing department of petroleum geologists to supply enough oil from within Germany, from the allies on the SE flank and from newly invaded territory.

Domestic oil production increased from *c.* 238 000 t (tons) per year in 1933 (i.e. little more than 10% of the total German petroleum consumption in 1933) to 737 000 t in 1939 (now including Austria) to 2 002 000 t in 1944, in which year an additional 863 000 t were produced in the occupied eastern territories and 1 915 000 t were imported mainly from Romania until August 1944, when the German army had to withdraw. On top of that, the chemical industry produced 3 975 000 t of synthetic fuels in 1944 (Kasper 1973, pp. 117, 128; Eichholtz 2003, p. 493):

> From the outbreak of the war in September 1939, emphasis was upon production, and upon intensive exploitation to get it. Regardless of technical advice, the Reichswirtschaftsministerium […] ordered first the cessation of exploration drilling, and second the concentration of drilling in productive areas. The experts argued that reserves could not be maintained without exploration, and as a compromise a few of the more important wildcats were allowed to continue […] In 1940 the German Supreme Commander ordered the concentrated exploitation of the Reitbrook field and all the rigs that could be freed for drilling were concentrated there, seven being brought from iron-ore drilling to the Brunswick area. In April 1940, forty-two drilling rigs were running at the same time. The protests of the technical men, if they amounted to protests, were overruled. Though the reserves in this field were known to be large (judged by German standards) […] the result of the policy of all-out production soon showed up. But it was necessary to wait until gas and water were being blown from the wells at the expense of oil that the Party could be brought to realise that their methods of exploitation were likely to lead to a serious loss both of immediate and of ultimate production
>
> (Gunther 1947*a*, p. 1097).

In mid-October 1939, a German administration, the so-called 'Generalgouvernement', was established in the occupied Poland, which also included the management of petroleum reserves and refineries. In order to avoid internal conflicts, representatives of the German Oil industry – among others Dr Erich Will (who died in 1942), a member of the executive board of the Gewerkschaft Elwerath – were directly commissioned with the task. Consequently, in November 1939, a consortium of several interested companies was formed, the Beskiden-Erdölgewinnungs-GmbH. By the end of January 1940, all petroleum facilities within the Generalgouvernement had been confiscated and Erich Will was appointed Special Commissioner for the petroleum economy (Kasper 1973, pp. 81, 105).

The major part of the Polish oil fields was, however, not yet in German hands due to the Molotov–Ribbentrop Pact, a non-aggression pact between Nazi Germany and the Soviet Union (signed 23 August 1939). The treaty included a secret protocol dividing various territories including Poland into German and Soviet spheres of interest. Thus, when Germany attacked Poland, Stalin ordered a Soviet invasion of Poland on 17 September 1939 seizing the eastern part of Poland (Evans 2010*b*, p. 28).

With the beginning of the war and having worked as a petroleum geologist in Romania before, Krejci-Graf was suspended from his university position and, instead, commissioned by the representative of the Vierjahresplan and President of the German Geological Survey, Wilhelm Keppler, to conduct oil exploration in Romania and organize the supply for Germany (Carlé 1988, p. 237, affidavit by Karl Krejci-Graf, 11 February 1946, AGV #20866; Petrascheck 1989):

> My employment as Professor in Freiberg required me to work for other official bodies of the Reich as well; among others for the *Reichsamt für Bodenforschung*, whose President was *SS-Obergruppenführer* Keppler
>
> (affidavit by Karl Krejci-Graf, 11 February 1946, AGV #20866; translated from German).

> He [Keppler] was very much interested in the question of fuels and supported me in my research via recommendations, allocation of labs etc; [he] promised considerable [financial] means also for the future. – Requests to become a member of the NSDAP I have always evaded
>
> (affidavit by Karl Krejci-Graf, 9 February 1946, AGV #20867; translated from German).

As a consequence, Krejci-Graf became a technical advisor at Kontinentale Oel AG, a consortium of the German state, several petroleum companies and banks. As such, Krejci-Graf was responsible for 25% of the Romanian crude oil production and 30% of the refinery capacity (Gunther 1948*a*, p. 66; Carlé 1988, p. 237, CVs by Karl Krejci-Graf, 1 February 1946, AGV #20864 and 8 February 1946, AGV #20865, affidavit by Karl Krejci-Graf, 11 February 1946, AGV #20866; Petrascheck 1989):

> The upper structure of Kontinental was vast and impressive, but it was no oil company. […] the bias of its aim was towards money. […] In Bucarest [sic] the director was a banker and his German colleague a political diplomat. Neither knew anything about oil. Consequently there was constant friction between the directors and their technical advisers in Roumania backed by the Reichsamt für Bodenforschung in Berlin, whose job it was to get the crude oil out of the ground. The Berlin directors of Konti wished to see the colour of their RM. 80 000 000 without the long-term investment crude oil requires
>
> (Gunther 1947*c*, pp. 1212–1213).

Romania was, and remained, the most important foreign source of petroleum for the Nazi regime and attempts were made to force other economic contenders out of the oil industry. However, the aim of increasing production significantly was not achieved as exploration drilling was delayed by a lack of equipment and local resistance against ruthless exploitation. It took until the turn of the year 1942–43 before a greater number of outfits for rotary drilling were delivered – after they were no longer needed in the Caucasus oil fields (see below) due to the German retreat. By then, it was too late to make any difference (Alfred Bentz: Supplement to my questionnaire, 11 February 1946, National Archives, London, UK: FO 1039/496; Eichholtz 2006, pp. 22, 31, 32, 37, 39).

Since Krejci-Graf had to deal with all sorts of bureaucrats and the German military, not being a member of the Nazi Party sometimes proved detrimental to his authority, and so, as the story is told, in May 1939 Keppler surprised Krejci-Graf with a high 'honour'. He promoted him 'SS-Hauptsturmführer'. The SS, short for 'Schutzstaffel', was a major paramilitary organization of the Nazi Party and responsible for many of the crimes against humanity during Nazi time. Thus, Krejci-Graf was saddled with an officer's rank in the SS, comparable to a captain, although he never applied for it:

> The fact that I was the only university teacher [in Freiberg], who was not a member of the NSDAP, began to cause me trouble, especially also for getting permission to work in secret [petroleum] facilities such as Wietze, Schwarzheide etc. The appointment to *Hauptsturmführer* was meant as a surprise and honour for me, which should remove all obstacles from my way. A refusal would not only have hurt the feelings of a man to whom I owed thanks, but would not have been justifiable without direct attack on the ruling regime, after my instinctive objection that I had no time for it was dismissed with the mention that no service would be expected of me [...]
>
> (affidavit by Karl Krejci-Graf, 9 February 1946, AGV #20867; translated from German).

> I did not report to any SS-office, never swore an oath, never wore uniform and never did service
>
> (affidavit by Karl Krejci-Graf, 11 February 1946, AGV #20866; translated from German).

In March 1939, the geological surveys of the various German regions lost their independence. They were centralized and fused with Keppler's office to become the 'Reichsstelle für Bodenforschung' in December 1941, promoted to 'Reichsamt für Bodenforschung': that is, the National Geological Survey with Wilhelm Keppler as its president. The new administrative body was subordinate to the Reich Ministry of Economics. This organization's task was also the takeover of geological surveys and organizations in conquered territories in order to exploit the local mineral resources for the benefit of Germany. Branch offices were opened in Vienna, Prague, Oslo, Krakow and elsewhere. The Reichsamt für Bodenforschung also closely collaborated with the offices of military geologists, even providing staff (Kasper 1973, p. 195; Seibold & Seibold 2002, p. 1083).

With the occupied territories, Bentz's sphere of office expanded as well. He had to draw up drilling plans and determine where to send the drilling equipment (Kasper 1973, p. 143). He also had to report on the potential of the various oil fields within German reach and the availability of petroleum resources for the war effort while simultaneously assessing logistical difficulties (Eichholtz 2006, pp. 46–47). Thus, Alfred Bentz definitely supported the Nazi regime with his work as a geologist:

> The crude oil industry, for one, was able to produce ample evidence to show how reluctant a supporter of Party it had been [...].
>
> But the preservation of the oilfield records [...] worked in another way too. The advent of Military Government brought out of industrialists, company directors and of technocrats, all the fragments of evidence available in their ample and well-ordered files of 'opposition' to a régime to which, for 12 years, they had given undeviating support
>
> (Gunther 1948*d*, p. 436).

Nevertheless, it can be shown that Bentz managed to save a number of geologists, engineers and skilled workers prosecuted for political or racist reasons in Eastern Europe from concentration camps by claiming them for his petroleum drilling campaigns (cf. Martini 1965; Seibold & Seibold 2002). The Polish mining engineer Fingerchut, who after the war continued his geological career in the Polish oil industry and administration, might serve as a well-documented example: Maximilian [Maksymilian] Fingerchut (1891–1955) studied at the Mining Academy in Leoben (Austria). He began his professional career as a mining engineer. When Poland was occupied by German troops in World War II, he was employed by the Beskiden-Erdölgewinnungs-GmbH, later Karpathen Öl AG (Carpathians Oil Consortium) as a section director. In May 1940, he was arrested along with over 700 other Polish intellectuals and imprisoned in the Auschwitz Concentration Camp (https://pl.wikipedia.org/wiki/Maksymilian_Fingerchut, 15 March 2016):

> A Polish mine-engineer and head of a mine, FINGERCHUT, had been put in a concentration camp by the SS. He asked for my assistance by smuggled letters, knowing my personal way of thinking, although I had met him only once for a short interview. After energetical efforts I succeeded in getting him free. After some time when his case was forgotten he could return to Poland
>
> (Alfred Bentz: Supplement to my questionnaire, 11 February 1946; National Archives, London, UK: FO 1039/496).

Fingerchut was released from Auschwitz possibly in November 1942. His health had deteriorated so much that he had to ask for, and was granted, an extended leave of absence from his duties at Karpathen Öl AG:

> Unfortunately, I have to prolong my vacation; I am so weak and also tired that I still have to stay at home for a while.
>
> I send you, most honoured professor, my heartiest greetings for Christmas and the New Year 1943. I wish you much good and much luck, much joy and much sunshine; much warmth for your good, noble heart, which has so much understanding for the misery and hardship of others.
>
> It is so exceedingly sad here, but the presence of my family makes me forget all this and one savours the joy of home and native soil.
> One lives and feels how beautiful live might be; how beautiful it is in reality.
>
> With pleasure, I remember that I owe you my life […]
>
> (letter from Maximilian Fingerchut to Alfred Bentz, 18 December 1942, AGV #20188; translated from German).

Fingerchut then apparently volunteered to work as a petroleum geologist within Germany and applied to be officially recognized as 'Volksdeutscher' (Nazi terminology meaning German in terms of race, regardless of citizenship). This would have meant considerably higher provision rations (more calories) for him and his family had he succeeded. It also seems that it was a prerequisite for a job position with more independence and responsibilities. His application, however, was rejected in March 1943 despite attempts of his employer, the petroleum company DEA, to sway the decision:

> Today, DEA in Wietze has received this communication:
>
> '[…] For given reason it is brought to attention that the above mentioned cannot be considered for the proceedings of re-admittance to German citizenship, because he does not conform to the conditions of selection. Therefore, F. must still be employed as Pole'.
>
> Since my work here is impossible with a P., I have for certain to come home and seek [work] there. I see only one chance, most honoured Professor, to continue my work, and that is your help.
>
> On Monday, our Mr Zipfel will be in Berlin, maybe he will have the possibility to speak with you. I have heard that he also wants to talk with Secretary of State Keppler; maybe it can still be achieved that I need not wear the P. […] I am completely shattered, don't know what to do and ask you from my heart, most honoured Professor, for your help, should this be possible.
>
> Please excuse that I bother you with my matters, but I have nobody here, who could possibly help me. I have come voluntarily to Germany and wanted to work here honestly until the end of the war.
>
> But this is impossible
>
> (letter from Maximilian Fingerchut to Alfred Bentz, 18 March 1943, AGV #20189; translated from German).

Bentz helped him to return to Karpathen Öl AG in Poland (letter from Alfred Bentz to Maximilian Fingerchut, 23 March 1943, AGV #20190). After the war, Fingerchut sought contact again with Alfred Bentz thanking him for his help:

> I have very often thought of you, as one never forgets people who have saved our lives; and only yesterday I learned that you are alive and healthy. […] I have often wondered, when I shall see you to once again express my thanks; thanks to the man, who has rescued me for my Fatherland and my family from the 'death camp' Oswięcim (Auschwitz). I work, as has been my custom, in my beloved petroleum industry and I – as well as my family – shall never forget that I owe this to you
>
> (letter from Maximilian Fingerchut to Alfred Bentz, 6 July 1946, AGV #20533; translated from German).

Apparently, Bentz also used his influence to 'free' another Polish petroleum and mining geologist, Józef Zwierzycki (1888–1961):

> Professor Dr. ZWIERZICKY [sic], the director of the Polish Geological Survey at Warsaw, was sent to a concentration camp for reasons I do not know. I tried by all means to get him free and finally succeeded by stating that ZWIERZICKY could give valuable information on the oil fields of the Dutch-East-Indies due to his former activities over there. ZWIERZICKY was released, I kept him busy in my department in Berlin for years without asking for a report
>
> (Alfred Bentz: Supplement to my questionnaire, 11 February 1946; National Archives, London, UK: FO 1039/496).

Zwierzycki studied mining and geology at the Mining Academy and the University in Berlin, obtaining his PhD in 1914. He then worked for the Geological Survey in the Dutch-East-Indies and eventually became its director. From 1933 to 1938, he was also a geology professor at the Technical University in Bandung. In 1938, he returned to Poland as an expert in the Polish mining administration. In April 1941, he was arrested and transferred to the Auschwitz Concentration Camp, where he remained until July 1942. He was then taken to Berlin to work as an interpreter for the German Geological Survey. In July, he was transferred to Warsaw. He took the opportunity to flee to Krakow, where he went into hiding. After the war, he lectured at the University of Wrocław, becoming a full professor in 1948 (https://pl.wikipedia.org/wiki/J%C3%B3zef_Zwierzycki, 15 March 2016).

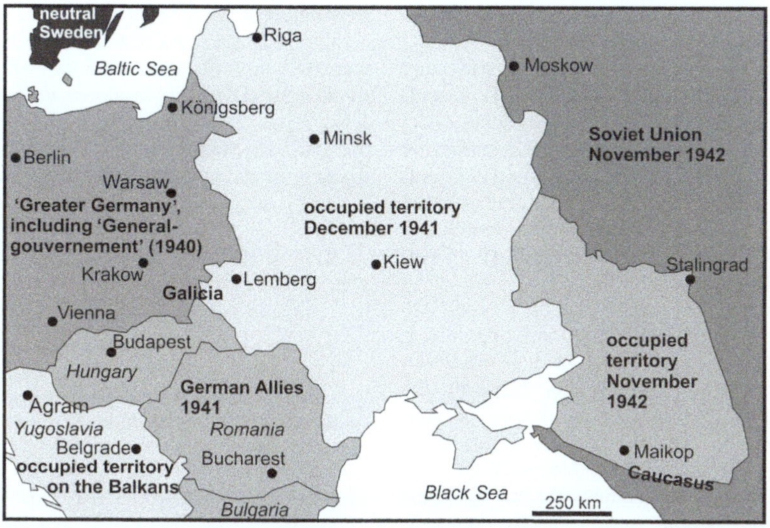

Fig. 4. Eastern Europe in World War II, 1941–November 1942 (redrawn after Putzger 1965, p. 128).

1941: Holocaust

On 22 June 1941, the German invasion into the Soviet Union in direct violation of the Molotov–Ribbentrop Pact began. In the newly invaded territory, the petroleum geologists followed the advancing troops as quickly as possible in order to gain access to the petroleum fields of eastern Galicia (Fig. 4), formerly a Polish province, which had been occupied by Soviet troops since mid-September 1939. There were basically three organizations involved in the operations at this stage (Kasper 1973; Kockel 2005; Eichholtz 2006; Toprani 2015): the military itself, the petroleum specialists under Alfred Bentz and a consortium of major German petroleum companies.

First, the so-called petroleum command under 'Rittmeister' Erich Will with 10 petroleum and refinery experts, and additional pioneers, technicians and military geologists supported and protected by fighting troops under General Thomas and assisted by Ukrainian collaborators followed the 17th German Army in July 1941 to secure the petroleum operations around Borysław and Drohobycz (Kasper 1973, pp. 19, 71, 105; Eichholtz 2006, pp. 19, 85):

> If there were oilfields or refineries to be secured it was the aim to bring these into immediate operation. For this purpose Crude Oil Commissions were formed which followed behind the front. The responsibility for these Commissions lay jointly with the OKW (Oberkommando der Wehrmacht) and with the Wehrwirtschafts-rüstungsamt (General Thomas). They were recruited by the Reichswirtschaftsministerium, usually from the Reichsamt and from the industry jointly. The first of these Commissions was the Erdöl Kommission Galizien, formed in August 1939, for the technical occupation of the West Galician (Polish) oilfields. Commissions were formed subsequently for France, Iraq, East Galicia, Ukraine, and finally the Caucasus
>
> (Gunther 1947*d*, p. 1274).

Their task was to secure fuels and lubricants for the fighting troops, to take over the oil fields and, if possible, continue production, and to assess local storage and reserves. This time, they were lucky. The oil fields were close to the border and the Russian forces were taken by surprise. Thus, the facilities of the local petroleum industry were taken nearly undamaged (Kasper 1973, pp. 19, 81, 83).

From the first day onwards, the large Jewish minority of the area was terrorized, driven into ghettos and systematically mass-murdered – first shot locally and later transported to extermination camps. The Polish population, especially the intellectuals among them, Soviet prisoners of war and others suspected of 'communism' suffered as well (Sandkühler 1996, p. 88).

Straight after the military came the team of Alfred Bentz, whose task it was to assess the situation, and repair what damage the retreating enemy might have done to the derricks and oil wells. They also had to find ways to increase the output by increasing the number of wells, even though it might jeopardize long-term total productivity of the field, and prospect for potential new sites. Then followed the oil companies, in the form of the Beskiden-Erdölgewinnungs-GmbH, who were responsible for the actual production, refining and transport logistics (Eichholtz 2006, p. 18).

In March 1941, a new company, Kontinentale Oel AG had been founded (see above). Several daughter companies with a specific local focus operated under this heading, among them Karpathen Öl AG, which was established on 28 August 1942 and replaced the Beskiden-Erdölgewinnungs-GmbH (Kasper 1973, pp. 72–74, 84). Alfred Bentz was a member of its executive board (Gunther 1947*c*, pp. 1211–1212; Eichholtz 2006, p. 50).

German geologists visiting the area obviously witnessed the mistreatment of the local population, although they did not report on it in their preserved letters: early in May 1942, Walter Kauenhowen (1900–42), Chief Geologist at Karpathen Öl AG, was sent to Galicia to set up a production timetable, to estimate the expected output and to assess the need for additional drilling capacities. He set up plans which, however, would soon become obsolete again, because industrial capacities were insufficient to produce the drilling equipment and casings necessary to meet the demand (Sandkühler 1996, pp. 297–300). Two months later, on 3 July 1942, Kauenhowen died (*Zeitschrift der Deutschen Geologischen Gesellschaft*, **94**, p. 488, 1942). He had been infected by epidemic typhus, which ran its course among the Jewish slave labourers under the horrible unhygienic conditions of the local concentration and work camps (letter from Mrs L. Kauenhoven to Alfred Bentz, 22 July 1942, AGV #20256).

The atrocious decimation of the Galician Jews, many of whom had worked in the Polish oil industry, was felt to be a logistical handicap by the petroleum business, because the number of available Polish and Ukrainian workers was insufficient to keep the oil production going. Consequently, Karpathen Öl AG tried to keep as many Jewish workers as possible – Bentz intervening directly with Göring, who then bargained with Heinrich Himmler (1900–45). Thus, Karpathen Öl AG was the only company allowed to keep a larger number of their Jewish workforce (Sandkühler 1996, p. 381). Locally responsible was the administrator Berthold Beitz (1913–2013), who, together with his wife, Else Beitz (1920–2014), tried to save not only the petroleum specialists but also their families and unskilled workers as well, and even hid some of them in their own house. Both were later honoured by Yad Vashem as Righteous Among the Nations (Berthold in 1973, Else in 2006). Beitz supported his workers even against constant reprimands from his superiors in Berlin; according to Sandkühler (1996, pp. 386, 396), citing sources from an archive in Lwów (present-day Ukraine) including Bentz's office, who every now and then demanded that the workforce should be 'combed': that is, the number of Jews be diminished and consequently given up to extermination. In spring 1944, the oil industry's concentration camp for a few hundred Jewish workers was the only place in Galicia, where Jews still lived legally. In Galicia alone, more than 520 000 Jews had been murdered (Sandkühler 1996, pp. 386, 396; Schmalhausen 2006, https://de.wikipedia.org/wiki/Liste_der_Gerechten_unter_den_V%C3%B6lkern_aus_Deutschland, accessed 14 June 2016).

In an affidavit for the British authorities after the war, Alfred Bentz claimed an active role in the protection of the Jewish workers of Karpathen Öl AG:

> The 'Karphathenöl Aktiengesellschaft' employed 1,000 to 2,000 jews [sic] which should be picked up several times by the SS. On account of my repeated interventions I succeeded under personal risks that the larger part of them was spared. This also holds true of the librarian of the Karpathenöl A.G. a Jewess, Frau Hager, saved in the same way.
>
> The Polish workmen of the Karpathenöl A.G. received very bad food. I asked the Four Years Plan office to give out better food explaining that the fulfillment of the production plan could only be achieved by the good will of all workmen. This was accepted and the Polish workmen in the oil industry received better food
>
> (Alfred Bentz: Supplement to my questionnaire, 11 February 1946; National Archives, London, UK: FO 1039/496).

Bentz's version of the story, however, needs further verification, especially in light of the incriminating material that Sandkühler (1996, pp. 386, 396) cites from Lwów (see above). An inquiry by a Ukrainian colleague in the Archive in Lwów failed to produce the cited letters (Oleksandr Petrynko pers. comm. 5 December 2016).

Initially, Karpathen Öl AG had planned to withdraw their remaining Jewish workforce in mid-May 1944 – 433 men and 89 women enduring under horrible living conditions – together with machines and other equipment to the refinery in Jasło. These plans, however, were made obsolete by Himmler's orders, who needed workers for a manganese project, and a secret weapons and ammunition factory. When these plans failed, the workers were deported to Płaszów on 21 June 1944 to slave in a salt mine and an aircraft factory. When the front approached, the Jews of Płaszów were transported via Hungary to the extermination camp Auschwitz-Birkenau, where most were immediately killed by gas; some were transferred to other camps or sent on a death march in August 1944. It is estimated that in the end no more than 100 of the Jewish petroleum workforce from Borysław survived the Holocaust (Sandkühler 1996, pp. 403–405).

Elsewhere, the situation was no better. On 26 July 1944, August Moos and his family were arrested by the SS and, over the course of 3 weeks, were transported through nine prisons and the SS jail in Prague to the concentration camp Bergen-Belsen, near Celle (Germany) (Bentz 1948). In December 1945, Alfred Bentz reported:

> The fate of the Moos family is quite horrible. They all (Dr Moos, Mrs Moos, the mother of Mr Moos and both children) were arrested in July 1944 in Agram and then at first transported to the concentration camp Belsen. There, in autumn 1944 the old Mrs Moos died of starvation and in December 1944, Dr Moos was separated from his family and brought to the Buchenwald camp, where he died early in 1945 (probably shot or gassed). Mrs Moos with the children has suffered horrible things in Belsen. They all starved and had severe typhoid fever and when they were rescued, they could only crawl. Early in July, they were on the mend so far that they could be released to her mother in Hamburg, but then some 14 days ago, the 14 year old boy still died from complications of his illnesses. Unfortunately, Mrs. Moos [...] as well, has such a weak constitution that I fear the worst
>
> (letter from Alfred Bentz to Professor Dr Haack, 5 December 1945, AGV #20210; translated from German).

Beata Moos and her daughter, however, survived, and for many years Beata Moos worked as librarian for Alfred Bentz (letter from Alfred Bentz to Walter Kranz, 11 January 1947, GA 19340).

1942: Caucasus

The German invasion plans on the SE were dictated by the presence of known petroleum resources. After Galicia followed the Soviet oil fields north of the Caucasus in 1942 (cf. Fig. 4):

> On 10 July 1942 I was ordered to attend a conference in Göring's Headquarters in East-Prussia. [...] I had to report again on the development of the [petroleum] production in Southeast-European countries and then was ordered to prepare the drilling campaign in the Caucasus from the geological point of view. Herr Schlicht was to take upon him the technical part
>
> (Alfred Bentz: Supplement to my questionnaire, 11 February 1946; National Archives, London, UK: FO 1039/496).

The idea was to eventually reach Baku, and also to cross the Caucasus and spill into Iran and Iraq from the north. The German vanguard included a 'Technical Petroleum Brigade' with some 6000 men commanded by General Erich Homberg, supported by Major Erich Will as the petroleum expert, to secure the Caucasus oil fields near Maikop. Will, however, died in an air crash on 8 August, the day before the German troops reached Maikop (Gunther 1948*b*, p. 116; Kasper 1973, pp. 105, 107; Eichholtz 2006, pp. 89, 104, 108, 117):

> It is an interesting reflection of the attitude of the Wehrmacht to a non-military unit like the Technical Battalion that the civilian was given only half the military ration, later increased under protest to threequarters
>
> (Gunther 1948*b*, p. 116).

The technical petroleum brigade was immediately followed by some 500–600 civilian experts and workers headed by Alfred Bentz, who was to develop the 'drilling programme with emphasis upon immediate and large-scale exploitation' (Gunther 1948*b*, p. 114) and by Günter Schlicht, a mining engineer and director of the petroleum company DEA, as his local representative (Kasper 1973, p. 105; Seibold & Seibold 2002; Eichholtz 2006, pp. 89, 116–117). He 'was put in charge of technical operation in the oilfields, to supervise drilling and production' (Gunther 1948*b*, p. 114):

> When the question of transport was mentioned Goering was empathic that the military victory would be won; that the Black Sea would be open, and that barges, being made by a Bulgarian subsidiary of Kontinental Oel, would be waiting in the western ports for the transport of the oilfield equipment. [...] by the spring of 1943 Russian oil would be flowing from Constanza up the Danube to Vienna, would be piped to the Elbe and brought into the heart of Germany
>
> (Gunther 1948*b*, p. 114).

For this unrealistic gamble, steel and human resources were diverted from crude oil operations already up and running:

> The proposed brigade was no longer the loan by the industry of a few technicians for a temporary project. It meant the drain of a high proportion of staff and of materials from an essential war service. And seen from the angle of the private company, it was one thing for the crude oil industry to accept (as the result of the Party's political success), a present of the Vienna Basin, of exploration rights in the Balkans, or of unlimited drilling subsidies; but quite another to expect it to contribute to the untechnical dreams of the Party leaders
>
> (Gunther 1948*b*, p. 114).

Nevertheless, when the Caucasus campaign was planned, there was only minimal resistance:

> When, after the outbreak of war, the leaders of the industry put forward objections on more technical grounds, it was too late for their influence to have effect. Their advice was ignored; and there was no further opposition. The Party had played its cards well. By the outbreak of war it had led German industry so far up the garden path that there was no turning back – unless by leave of the Gestapo
>
> (Gunther 1948*d*, p. 424).

Disappointment was great when it was realized that there was no oil in Maikop directly; instead, there were at least another 20 km to go with no proper roads. It took until 17 August to reach the oil field, constantly threatened by Soviet military and partisans (Gunther 1948*b*, p. 119; Eichholtz 2006, pp. 124–125):

> Dear Sir, Director!
>
> With train and airplane, I am now 'pretty' close. Schlicht is expected back from the front tonight and will bring back a first report of the situation. Mr Maiweg has been wounded before reaching point, shot to the shoulder, but already up and about and quite the old fellow. Of the batt[alion] (Captain Schmidt) 18 have fallen.
>
> With hearty greetings and *Heil Hitler*
>
> Yours sincerely Mayer-Gürr
>
> (postcard from Alfred Mayer-Gürr to Alfred Bentz, 18 August 1942, AGV #20304; translated from German).

Alfred Mayer-Gürr (1912–2008) studied geology, mineralogy and chemistry in Kiel and Tübingen. He obtained his doctoral degree in 1934, and then worked as geologist in Tanzania and Iraq. In 1938, he joined the petroleum department of the Prussian Geological Survey under Alfred Bentz. During World War II, he was sent to oil fields in Poland and Romania before he was transferred to the Petroleum Brigade of the German military for the attempted conquest of the Caucasus region. After this campaign failed, he was sent to oil fields in Austria. In 1944, he wrote a book on petroleum geology. After the war, Mayer-Gürr followed Bentz to Celle, then worked shortly for the British Oilfield Investigation, before he found employment with the gas producer Gewerkschaft Brigitta until his retirement in 1975 (http://www.dgmk.de/allgemeines/Engler_Medaille/Mayer_Guerr.pdf; accessed 17 November 2015).

This time, most of the oil-field installations had been destroyed by retreating Soviet troops, and it took weeks of toiling against rain, mud and enemy fire to get railway lines, bridges and roads into working order, which were necessary to transport the desperately needed equipment to the oil fields. Mostly, it was necessary to put down new drill holes. The drill rigs had to be surrounded by barbed wire fences so that they could not be overrun by partisans. Soviet air raids were also a constant threat (Kasper 1973, p. 107; Seibold & Seibold 2002, p. 1085; Eichholtz 2006, p. 125). In October, the Technical Petroleum Brigade reached a second, more productive oil field but, again, the widespread damage prevented immediate production (Eichholtz 2006, pp. 125, 130).

On 20–21 November, Bentz and Schlicht reported to Göring that under these circumstances effective drilling was not possible before spring 1943 and recommended increasing productivity in Austria rather than setting hopes on Maikop (Gunther 1948*b*, p. 120; Seibold & Seibold 2002, p. 1085):

> In November 1942 I carried out an 8-days inspection together with Herr SCHLICHT of the Maikop fields (Caucasus) as far as they were in German hands. We then were ordered to GÖRING to Berchtesgaden on 20 November 1942. I informed GÖRING that the exploitation even of a reduced production in the Caucasus would be badly hampered and delayed by the huge transport difficulties. He was very furious on this information and the conference was broken up. Next day he accommodated to the hard facts and acknowledged the very low figures of estimated production. Obviously he had been informed in the mean time [sic] that it was impossible to remove the transport difficulties
>
> (Alfred Bentz: Supplement to my questionnaire, 11 February 1946; National Archives, London, UK: FO 1039/496).

They never got the field back into production; even though they continued working often under constant enemy fire, they finally had to abandon much of their gear. The long retreat began in January 1943, which ended in May 1945 with the total defeat of Germany (Gunther 1948*b*, p. 120; Seibold & Seibold 2002; Eichholtz 2006, pp. 108, 121, 131–132).

1944: last straws

Parallel to the world of geologists reporting to the ministries of research, economics or the military, there was a second world reporting to Heinrich Himmler, chief of the SS, which operated in accordance with different rules, because Himmler was greatly interested in esoteric practices and thus promoting 'alternative' methodologies in institutions within his sphere of influence. When the petroleum supply of the German military and, especially, its air force became increasingly precarious, Bentz and his colleagues had to deal with growing interference from rival projects:

> At the end of 1944 HIMMLER started a drilling programme based on phantastic [sic] statements of a dowser on rare ores in the Harz Mts. and oil in Schleswig-Holstein. Four outfits were provided for this job but drilling did not start on account of the military situation. I tried in vain to stop this nonsense.
>
> [...] Certain high Party offices openly sympathized with the divining rod and with other unproved pseudogeophysical outfits trying again and again to induce them to the industry. HESS etablished [sic] a particular institute to investigate the divining rod problems, the SS interceded for the unproved 'Geoskop' and GÖRING for the 'geophysical' fantasies of Major SCHRÖDER-STRANTZ. Opposing recklessly against the Pseudoscientists I made myself more and more unpopular with the Party offices
>
> (Alfred Bentz: Supplement to my questionnaire, 11 February 1946; National Archives, London, UK: FO 1039/496).

Meanwhile, equipment that had been intended for the Caucasus mission, such as new rotary drills and a refinery, which had been disassembled in France, was redirected to Galicia, which was still firmly in German hands (Kasper 1973, p. 100; Sandkühler 1996, p. 300).

In August 1944, with the retreat of German troops from Romania, Krejci-Graf was arrested in Bucharest and imprisoned as a prisoner of war (postscript to an affidavit by Karl Krejci-Graf, 4 April 1946, AGV #20866).

By now, the military lived from hand to mouth in terms of fuel. Nevertheless, up to summer 1944, there was enough to keep the German military operational. With the import stop from Romania, and the increasing destruction of refineries, chemical factories, and transport and storage facilities within Germany by British and American air raids, however, the petroleum supply dwindled rapidly (Kasper 1973, pp. 75, 117, 143). All major cities suffered severely from bombardment. Among the victims was Bentz's 16 year-old son, who had died in 1941 in an air raid on Berlin (letter from Alfred Bentz to Hubert Guyod, 21 April 1947, AGV #20513; Seibold & Seibold 2002, pp. 1082, 1084).

In 1943–44, the increasing threat of air raids on Berlin led Bentz to think about the wealth of scientific information that his office had collected over the last decade via the Reichsbohrprogramm and the Geophysikalische Reichsaufnahme. Despite the custom of sharing information at the monthly exchange meetings, they were considered classified and had not been published. Thus, a loss of reports and samples would have destroyed the fruits of years of labour, not counting the considerable financial resources:

> Being always convinced that Germany would loose [sic] this war I was intent on the bringing away of irrecoverable scientific files, assembled by myself. Already in August 1943 I brought this material to Dessow (Mark) a small village 90 km off Berlin. In fall 1944 I intended to remove the material to the Hannover oil district, to have it at hand if the industry would badly want it after the war
>
> (Alfred Bentz: Supplement to my questionnaire, 11 February 1946; National Archives, London, UK: FO 1039/496).

This sensible act, however, was dangerous as it was considered defeatist by ardent Nazis, a 'crime' which was punishable by death (Fischer 2012, p. 5). A denunciation by Dr Norbert Lauffs, editor of the journal *Oel und Kohle* since 1940, against Bentz and two others (Gunther 1948*c*, p. 372) was thwarted by a combination of good connections and luck:

> This plan [of removing material] was frustrated by a denunciation with the Gestapo informed by an employee of the publishers of 'Oel und Kohle' against me, my head clerk Henning Koch von Hernhaussen and his brother [...]. In the beginning of December 1944 the Gestapo issued a warrant of arrest against us on account of 'Deterioration of the *Wehrmacht* and malice'. The Hernhaussen brothers were arrested, I escaped from it being absent incidentally. The evidence given by the Hernhaussen brothers and their employees however was highly incriminating for the informer. They were released, the whole case should be reconsidered by the '*Sicherheitshauptamt*' (SS-headquarters). Incidentally the respective files were lost there and in February 1945 I removed to Celle. In spite of the express orders by Party- and Government offices I did not destroy anything of my files which were considered secret documents. The files were secured and were fully at the disposal of the occupation forces after their marching in. They form an essential base for the carrying on of the exploration activities
>
> (Alfred Bentz: Supplement to my questionnaire, 11 February 1946; National Archives, London, UK: FO 1039/496).

As late as March 1945, miracles were still expected from the petroleum geologists:

> On 5 March 1945 Minister Speer wanted to see me and informed me that returning from Hungary he had got the impression that far more outfits could be kept drilling there. I replied that from a geological point of view the Hungarian fields could not produce larger quantities without endangering their production within a very short limit. But apart from that the present transport situation would prevent the sending of outfits to Hungary, a manoever [sic] which would take months to be carried out
>
> (Alfred Bentz: Supplement to my questionnaire, 11 February 1946; National Archives, London, UK: FO 1039/496).

Aftermath

In April 1945, Bentz managed to abscond himself from the final crash of the Nazi regime by allegedly being off on an official visit to Vienna. Instead, he went into hiding. In June, he returned to his office in Celle, to where his institute had been evacuated. There he successfully engaged himself and his colleagues in fruitful collaboration with the British military authorities. A Geological Survey was established in autumn 1945 with Bentz as its director, a position he held until his retirement in 1962. Especially in the early years after the war, Bentz proved his considerable social skills in actively re-establishing contact with emigrated and foreign colleagues, who seemed to have welcomed him back amiably. In addition, a geophysical map of northern Germany was published in the late 1940s, which proved to be a valuable tool in restoring severed connections with foreign institutions. In 1964, Bentz died suddenly from a heart attack (Seibold & Seibold 2002; Bentz correspondence in AGV).

At the end of 1946, Krejci-Graf was released to return to Vienna, where the authorities in the Soviet

Zone were already waiting for arrival of the well-known petroleum geologist. Thus, he immediately found employment in his field of expertise. Whereas Krejci-Graf worked in the Soviet Sector of Vienna, he lived in the British Sector, and he was arrested there one month later and handed over to the American authorities, who kept him in custody until 1 July 1947; the reason for his imprisonment being his SS membership and – as was suspected – the reluctance of American authorities to lose this petroleum expert to the Russians (postscript to an affidavit by Karl Krejci-Graf, 4 April 1946, AGV #20866; letter from Emilie Krejci to Alfred Bentz, 12 May 1946, AGV #20860).

Alfred Bentz organized a series of petitions to free Krejci-Graf. Among the people who responded was the celebrated geochemist Victor Moritz Goldschmidt (1888–1947), who had left Germany in 1935 owing to his Jewish descent and in the meanwhile had ended up in Britain after a most adventurous escape: he had first moved back to Norway as he was a Norwegian citizen; but there as well, the advancing German troops reached him. In the end of October 1942, he was arrested by the Norwegian police, who meanwhile were collaborating with the Nazis, for being a Jew and imprisoned in the Berg concentration camp. A few days later, however, Goldschmidt was released as a result of the intervention of friends. On 26 November, he was arrested again and taken to a ship bound for Poland and transported to Auschwitz. Upon the appeal of Professor Halvor Solberg (1895–1974: a meteorologist and, at that time, the University Dean) and University Rector Adolf Hoel (1879–1964: a Norwegian geologist and polar researcher, and member of the Norwegian fascist party and a high-ranking spy for Germany), however, Goldschmidt was released again. Three weeks later, he was smuggled over the Swedish border by the Norwegian resistance in a haywagon. Early in March 1943, he emigrated to the UK (http://www.wikipedia.org/wiki/Victor_Moritz_Goldschmidt; accessed 7 April, 2011; Kranish 1987; https://de.wikipedia.org/wiki/Adolf_Hoel, accessed 9 June 2016).

Krejci-Graf had corresponded and worked with Goldschmidt on trace elements right up to the beginning of the war (letter from Karl Krejci-Graf to Alfred Bentz, 10 July 1946, AGV #20623), and so Goldschmidt was happy to speak up for Krejci-Graf:

> When I was Professor at the University of Göttingen (1929–35) I had an extensive cooperation in the scientific sense with Professor Krejci-Graf, especially about the rarer metallic components of crude oil, asphaltites and oil shales; a cooperation which was carried on – after my return to Norway – into the year 1939. I think back to this cooperation with particular pleasure; a cooperation which always has been especially pleasant due to the honesty of Professor Krejci; as likewise I always had and also have today highest regards for his particular scientific competence and exceptional experience. I know quite well that he never was a National Socialist and has always retained an independent attitude. Being at present probably the best authority on European petroleum reservoirs, his work is of the utmost importance for general geology as well. It would be my genuine pleasure, if I might contribute to help that liberty will be restored to this excellent man
>
> (letter from Victor Moritz Goldschmidt to Alfred Bentz, 29 October 1946, AGV #20857; translated from German).

After his release in 1947, Krejci-Graf again worked for the Soviet Petroleum Administration in Vienna and then in the Austrian petroleum industry. From 1951 to 1953, he went to Portugal for a Swedish petroleum company. In 1953, when denazification ended in Germany by a concluding law, he was appointed Chair of Geology and Palaeontology at the University of Frankfurt am Main. He retired officially in 1963, but continued to work at the institute up to his death in 1986 (Carlé 1988, pp. 237–238).

Acknowledgements I am indebted to the University Library of Freiburg im Breisgau and the National Archives in London, as well as their staff members, for granting me generous access to their repositories. Many thanks also to Oleksandr Petrynko (Collegium Orientale, Eichstätt) for negotiating access to Ukrainian archive material. I would also like to thank Uwe Altenberger (Potsdam) and two anonymous reviewers.

Funding This research received no specific grant from any funding agency in the public, commercial or not-for-profit sectors.

References

Bentz, A. 1948. Dr. August Moos zum Gedenken. *Erdöl und Kohle*, **1**, 55–56.

Carlé, W.E.H. 1988. Werner – Beyrich – von Koenen – Stille. Ein geistiger Stammbaum wegweisender Geologen. *Geologisches Jahrbuch, Reihe A*, **108**, 3–499.

Closs, H. 1977. In Memoriam Hermann Reich. *Journal of Geophysics*, **42**, 547–550.

Daniels, M. & Michl, S. 2010. Strukturwandel unter ideologischen Vorzeichen – Wissenschafts- und Personalpolitik an der Universität Tübingen 1933–1945. *Tübinger Beiträge zur Universitäts- und Wissenschaftsgeschichte*, **73**, 13–73.

Eichholtz, D. 2003. Öl, Krieg, Politik: Deutscher Ölimperialismus (1933–1942/43). *Zeitschrift für Geschichtswissenschaft*, **51**, 493–510.

Eichholtz, D. 2006. *Krieg um Öl. Ein Erdölimperium als deutsches Kriegsziel (1938–1943)*. Leipziger Universitätsverlag, Leipzig, Germany.

Engelhardt, W. von & Hölder, H. 1977. *Mineralogie, Geologie und Paläontologie an der Universität Tübingen von den Anfängen bis zur Gegenwart*. J.C.B. Mohr (Paul Siebeck), Tübingen, Germany.

Evans, R.J. 2010*a*. *Das Dritte Reich: Diktatur*. DTV, München, Germany.

Evans, R.J. 2010*b*. *Das Dritte Reich: Krieg*. DTV, München, Germany.

Fahrion, H. 2009. Geschichten und Geschichtchen aus der Frühzeit der deutschen Erdölgeologie – Persönliche Erinnerungen aus den Jahren 1936 bis 1939. *Nachrichtenblatt zur Geschichte der Geowissenschaften*, **19**, 69–84.

Fischer, M. 2012. *Dr. phil. habil. Hans Jüngst 1901–1944. Ein Leben im deutschen Zeitalter der Extreme*. KIT Scientific Publishing, Karlsruhe, Germany.

Guntau, M. 2002. Zu einigen Aspekten der Geologie in der Zeit des Nationalsozialismus in Deutschland (1933–1945). *Geohistorische Blätter*, **5**, 125–150.

Gunther, A.E. 1947*a*. The German war for crude oil in Europe 1934–1945: Chapter I – The north German industry: struggle of party and technologist. *The Petroleum Times*, **29**, 1095–1099.

Gunther, A.E. 1947*b*. The German war for crude oil in Europe 1934–1945: Chapter II – German geological organisation in Europe 1939–1945. *The Petroleum Times*, **29**, 1147–1152.

Gunther, A.E. 1947*c*. The German war for crude oil in Europe 1934–1945: Chapter III – German structure of the European crude oil industry. *The Petroleum Times*, **29**, 1211–1215.

Gunther, A.E. 1947*d*. The German war for crude oil in Europe 1934–1945: Chapter IV – Mobilisation plans. *The Petroleum Times*, **29**, 1272–1274.

Gunther, A.E. 1948*a*. The German war for crude oil in Europe 1934–1945: Chapter VI – Failure in Roumania. *The Petroleum Times*, **30**, 65–84.

Gunther, A.E. 1948*b*. The German war for crude oil in Europe 1934–1945: Chapter VII – The Caucasus expedition. *The Petroleum Times*, **30**, 113–121.

Gunther, A.E. 1948*c*. The German war for crude oil in Europe 1934–1945: Chapter XIII – Petroleum literature in Grossdeutschland, 1933–1945. *The Petroleum Times*, **30**, 371–373.

Gunther, A.E. 1948*d*. The German war for crude oil in Europe 1934–1945: Chapter XV – The crude oil industry, the Party and Grossdeutschland. *The Petroleum Times*, **30**, 422–436.

Kasper, H.-H. 1973. *Die Erdölgewinnung Deutschlands in der Zeit von 1933 bis 1945*. Dr. sc. thesis, University of Dresden, Dresden, Germany.

Klee, E. 2013. *Das Personenlexikon zum Dritten Reich. Wer war was vor und nach 1945*. Kramer, Koblenz, Germany.

Kockel, T. 2005. *Deutsche Ölpolitik 1928–1938*. Jahrbuch für Wirtschaftsgeschichte, Beiheft, **7**.

Kölbl-Ebert, M. 2017. Geology in Germany 1933–1945: people, politics and organization. *Earth Sciences History*, **36**, 63–100, https://doi.org/10.17704/1944-6178-36.1.63

Kranish, A. 1987. *Der Greif. Paul Rosbaud. Der Mann der Hitlers Atompläne scheitern ließ*. Kuidler, Munich.

Martini, H.J. 1965. Ansprache anläßlich der Trauerfeier für Alfred Bentz am 10. 7. 1964 in Hannover. *Geologisches Jahrbuch*, **83**, XXI–XXVIII.

Moos, R.H. (ed.) 2010. *Journey of Hope and Despair. Memoirs of Rudolf Moos*. Xlibris Corporation, Bloomington, IL, USA.

Owen, E.W. 1975. *Trek of the Oil Finders: A History of Exploration for Petroleum*. American Association of Petroleum Geologista, Tulsa, OK, USA.

Petrascheck, W.E. 1989. Karl Krejci-Graf 15. 4. 1898–8. 8. 1986. *Mitteilungen der österreichischen geologischen Gesellschaft*, **81**, 259–260.

Putzger, F.W. 1965. *Historischer Weltatlas*. 87th edn. Velhagen & Klasing, Bielefeld, Germany.

Reich, M., Gehler, A. & Stegemann, T.R. 2014. *Eine Chronik der Göttinger geowissenschaftlichen Institute und Sammlungen*. Geowissenschaftliches Museum Göttingen, Göttingen, Germany.

Riedel, M. 1977. Keppler, Wilhelm. *Neue Deutsche Biographie*, **11**, 509–510.

Sandkühler, T. 1996. *'Endlösung' in Galizien. Der Judenmord in Ostpolen und die Rettungsinitiativen von Berthold Beitz 1941–1944*. Dietz, Bonn, Germany.

Schmalhausen, B. 2006. *A Man of Courage in an Inhuman Time: Berthold Beitz in the Third Reich*. Yad Vashem, Jerusalem.

Schott, W. 1965. Alfred Bentz (1897–1964). *AAPG Bulletin*, **49**, 464–466.

Seibold, E. & Seibold, I. 2002. Alfred Bentz – Erdölgeologe in schwieriger Zeit, 1938–1947. *International Journal of Earth Sciences (Geologische Rundschau)*, **91**, 1081–1093.

Seibold, E. & Seibold, I. 2008. Curt Teichert – Dokumente zu einer Emigration (Dänemark – Australien – USA). *International Journal of Earth Sciences (Geologische Rundschau)*, **97**, 665–673.

Szabó, A. 2000. *Vertreibung, Rückkehr, Wiedergutmachung. Göttinger Hochschullehrer im Schatten des Nationalsozialismus*. Wallstein, Göttingen, Germany.

Toprani, A. 2015. 'The Navy's success speaks for itself'? The German Navy's independent energy security strategy, 1932–1940. *Naval War College Review*, **68**, 91–118.

Simon Papp, a prominent Hungarian petroleum geologist: how to run exploration projects from a prison cell

GÁBOR TARI[1]* & ISTVÁN BÉRCZI[2]

[1]*OMV Exploration & Production GmbH, Trabrennstraße 6–8, 1020 Vienna, Austria*

[2]*MOL Hungarian Oil and Gas plc, Október huszonharmadika u. 18, 1117 Budapest, Hungary*

**Correspondence: Gabor.Tari@omv.com*

Abstract: Simon Papp (1886–1970) is regarded as the 'father' of Hungarian petroleum exploration. He was also considered internationally as an exceptional oil finder, as his exploration career took him from Hungary to Romania, Serbia, Croatia, Turkey, Albania, Canada, Australia and Papua New Guinea.

In 1932, Dr Papp returned to Hungary and became engaged in exploration in western Hungary, working for EUROGASCO (an international consortium headed by Standard Oil of New Jersey). His prospecting work was very successful, resulting in the discovery of four oil fields with reserves totalling about 100 MMbbl (million barrels).

In 1948, after returning from a trip to the USA, Dr Papp was arrested by the communist secret police in Hungary, allegedly because of sabotage and attempting to overthrow the communist government. After being forced to 'confess', he was sentenced to death but, probably due to worldwide protest by scientific organizations, the sentence was changed to life in prison.

As his exceptional knowledge of the petroleum industry was a major asset, the communist regime arranged for him to continue his work from his prison cell. He was finally released from prison in 1955 and lived in retirement in Budapest until his death in 1970.

Simon Papp was not only regarded as a prominent oil explorer in Hungary but was also recognized as such internationally in the petroleum industry (e.g. Humble 1970; Dank 1996). However, in this paper we describe his professional career not only as an oil finder but also his plight in the historical changes behind the 'Iron Curtain'. In this entire area, which became known as 'Eastern Europe', communist regimes were established with Soviet support shortly after the end of World War II. The fact that Dr Papp worked for the 'capitalist Americans' until his retirement in 1947, gave an excuse to the newly established communist government in Hungary to drag politics into hydrocarbon exploration.

To provide a technical background to this primarily biographical paper focusing on Dr Papp, we also briefly summarize the history of Hungarian petroleum exploration and production as a contribution to this volume dedicated to the history of European oil and gas exploration. The summary below is largely based on the second, updated edition of the autobiography of Simon Papp (Papp 2000), compiled and edited for a posthumous publication by Lajos Srágli and János Tóth. The additional biographical highlights are based on the same book and a recent summary by Nyári (2016).

Brief history of Hungarian oil and gas exploration and production

Oil production started in Hungary in the late 1700s, mostly out of tar sands in the vicinity of Oravica (Oravita) and Stájerlak (Steierdorf) in the Southern Carpathian foothills (Fig. 1a) and in the Mura River Valley (Papp 1963, 1964). The cumulative production until 1905 is estimated to have been about 300 Mbbl (thousand barrels). The lack of a proper mining law, specifically dealing with oil production, was recognized as a hindrance. Therefore, after the recommendation of János Böckh, the Director of the Hungarian Geological Institute, the government decided to look into this issue and a Petroleum Code was thus put in place in 1911. Thereafter, oil and gas exploration licences could be obtained from the state by both domestic and foreign companies. A state exploration company was also established at this time which was able to find the first gas field in Transylvania in 1909 at Kissármás (Sărmăşel, now in Romania) and the first oil field at Egbell (Gbely, now in Slovakia) just before the start of World War I (Fig. 1a).

After the war, the peace treaty signed at the Little Trianon Palace in Versailles (near Paris) in 1920

From: Craig, J., Gerali, F., MacAulay, F. & Sorkhabi, R. (eds) 2018. *History of the European Oil and Gas Industry*. Geological Society, London, Special Publications, **465**, 409–422.
First published online May 18, 2018, https://doi.org/10.1144/SP465.19

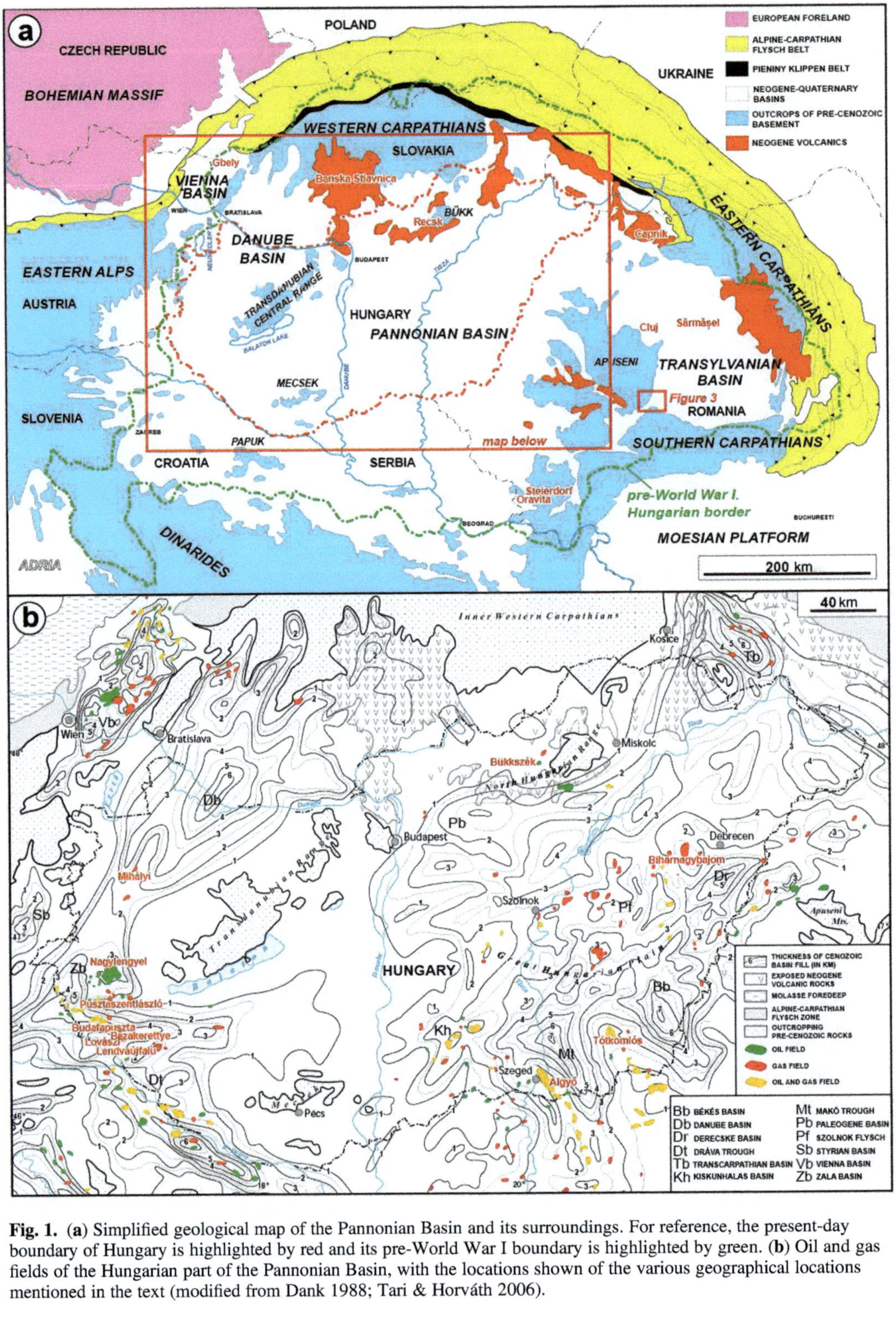

Fig. 1. (**a**) Simplified geological map of the Pannonian Basin and its surroundings. For reference, the present-day boundary of Hungary is highlighted by red and its pre-World War I boundary is highlighted by green. (**b**) Oil and gas fields of the Hungarian part of the Pannonian Basin, with the locations shown of the various geographical locations mentioned in the text (modified from Dank 1988; Tari & Horváth 2006).

redefined the boundaries of the country and, as such, two-thirds of the territory of Hungary (the Hungarian half of the Austro-Hungarian monarchy) was assigned to the adjacent countries of Austria, Czechslovakia, Romania and Yugoslavia (Fig. 1a). Czechslovakia became a new state by merging Bohemia and Slovakia together, where the easternmost part of Slovakia was ceded to the Soviet Union in 1945. After the collapse of the Soviet Union in 1991, this area became the SW corner of Ukraine. The Czech Republic and Slovakia became independent states in 1993. Yugoslavia was split up after 1991 to Serbia (the southern part of which segregated as Kosovo in 2008), Croatia, Slovenia, Montenegro, Bosnia-Herzegovina and Macedonia.

The impact of this dramatic reduction in the size of Hungary in 1920 (Fig. 1a) also had an immediate effect on petroleum exploration and production, as all the hydrocarbon fields and prospective areas, at least as they were perceived at that time, were lost. Regardless, the Hungarian state gave an exploration licence to the Anglo-Persian Oil Company in 1920. Their daughter company, D'Arcy Exploration Co. Ltd, established a local company called the Hungarian Oil Syndicate Ltd. This company explored mostly in western Hungary but, after a few years without success, it withdrew from exploration in 1927.

After a short break, the state itself initiated a new exploration programme in northern Hungary in 1933 following the recommendation of Lajos Lóczy, a

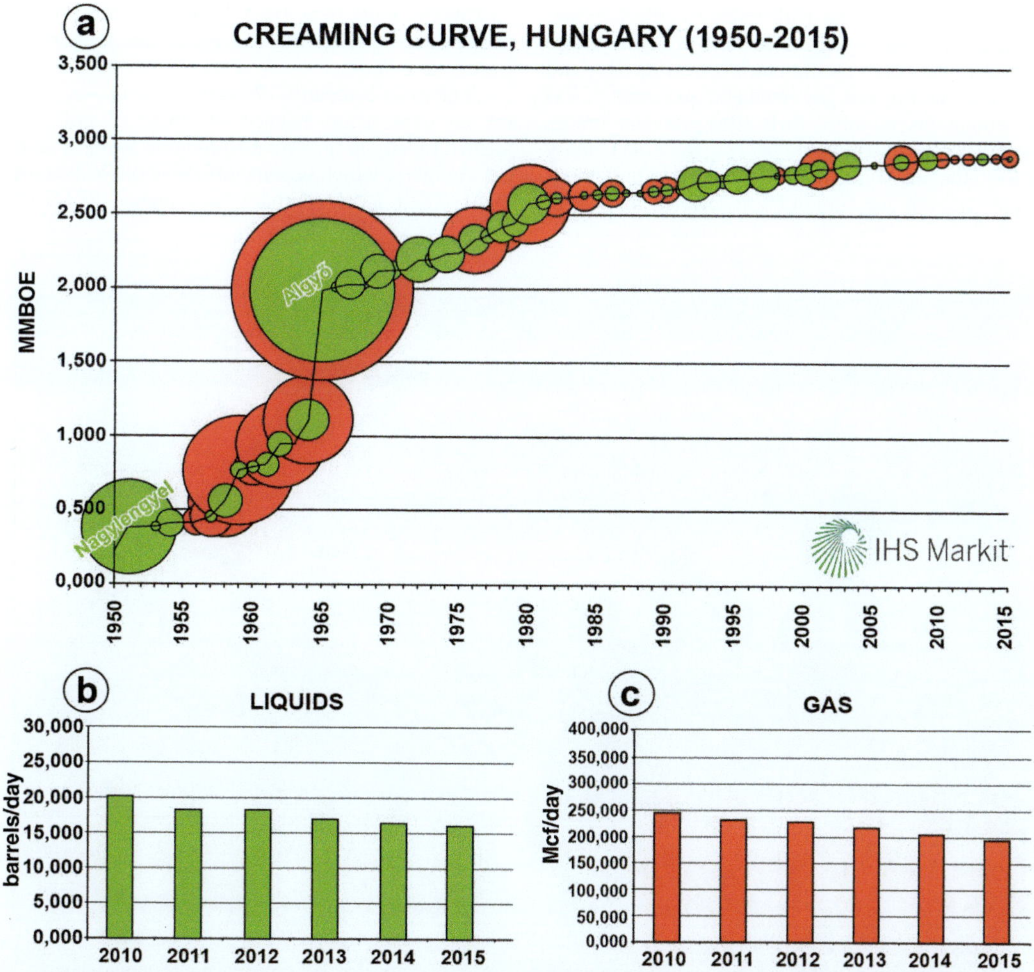

Fig. 2. (**a**) Creaming curve of Hungarian hydrocarbon finds between 1950 and 2015, and daily production rates for (**b**) liquids and (**c**) gas between 2010 and 2015. Note the two largest oil fields: the Nagylengyel Field (western Hungary) found in 1951, and the Algyö Field (SE Hungary) found in 1965 (Fig. 1b). Data courtesy of IHSMarkit. Mcf, thousand cubic ft.

prominent geologist. Along the northern margin of the Pannonian Basin, in the area of Recsk (Fig. 1a), 50 shallow wells were drilled between 1933 and 1938. In one of these wells, Bükkszék-1, multiple oil shows were found in 1937. The appraisal wells established a small oil field here (Fig. 1b), with daily production rates in the range of 80–100 barrels.

An American–English company, EUROGASCO (European Gas and Electric Company, an international consortium headed by Standard Oil of New Jersey), applied for exploration in western Hungary in 1932. Their application for the entire Transdanubia (i.e. the area west of the Danube in Hungary) was approved in July 1933. They started their exploration programme in September 1933, followed by drilling in November 1934.

EUROGASCO utilized state-of-the-art technology (i.e. rotary drilling with mud, electric logging, etc.). They found a large CO_2 field at Mihályi (the largest in the entire Pannonian Basin: Fig. 1b), and oil shows and mixed gas (methane and inert gases) accumulations at other drill sites. As the breakthrough well, Budafapuszta (B-1) in the Zala County of SW Hungary (Fig. 1b), found oil shows and good-quality natural gas, Simon Papp selected the site of the follow-up B-2 well which started to produce oil on 21 November 1937.

Following the first commercial oil discovery at Budafapuszta in post-war Hungary, EUROGASCO was replaced by MAORT (Magyar–Amerikai Olajipari Részvénytársaság (i.e. Hungarian–American Oil Industry Incorporated)) on 15 July 1938. Ninety per cent of the initial capital of this new company was owned by Standard Oil of New Jersey. The oil production by MAORT was able to provide 80% of the domestic consumption in 1939 and 100% by 1940 by applying water injection – one of the first such operations in Europe. Additional oil discoveries followed: Lovászi in amongst the first in 1940 and Lendvaújfalu in 1941 (Fig. 1b). The refining of an increasing amount of oil demanded a solution other than transporting the crude by rail. An oil pipeline, *c.* 250 km long, was built between Bázakerettye and Budapest (Fig. 1b) in 1941, the longest in Europe at that time.

A German company – Wintershall A.G. – applied for an exploration licence of about 30 000 km^2 located mostly in the Great Hungarian Plain in eastern Hungary, which became effective on 26 August 1940. Their local company was named MANÁT

Fig. 3. Geological map of the Gyulafehérvár (today Alba Iulia, Romania) area, Transylvania, hand-coloured by Simon Papp, *c.* 1910. For the location, see Figure 1a. Courtesy of the Hungarian Oil and Gas Industry Museum.

(Magyar–Német Ásványolajmüvek K.F.T. (i.e. Hungarian–German Oilworks Ltd)). Following the exploration style of MAORT, they also acquired geophysical datasets first and then started to drill with modern rotary technology. The 22 wells drilled by MANÁT did not provide any commercial discoveries; however, the blow-out at the Tótkomlós-1 well (Fig. 1b) showed for the first time that there was a petroleum system in the Great Hungarian Plain area of the Pannonian Basin.

As Hungary temporarily regained its pre-World War I territory partially during the course of World War II, exploration efforts expanded into the Transylvanian Basin and into the Drava/Sava sub-basins of the SW Pannonian Basin. One of the new companies engaged in the exploration in northern Transylvania was MOLÁRT (Magyar–Olasz Ásványolajipari Rt. (i.e. Hungarian–Italian Oil Industry Incorporated)), funded 55% by the Hungarian state and 45% by Azienda Generali Italiana Petroli (AGIP). MAORT also became active in northern Transylvania and drilled 31 wells, making numerous commercial gas discoveries between 1940 and 1944.

Due to the termination of diplomatic ties between Hungary and the USA, the operational control of the largely American-owned MAORT was taken over by the state and the American management had to leave the country on 20 December 1941.

The Hungarian oil and gas industry sustained heavy damage as a result of the massive amount of bombing carried out by the Allied air forces between April 1944 and April 1945. Shortly after the war, exploration and production efforts restarted relatively quickly in western Hungary. However, the property of the mostly German-owned MANÁT in eastern Hungary was transferred to the Soviet Union based on the Potsdam Agreement. Therefore a new company was set up, MASZOVOL (Magyar–Szovjet Nyersolaj R.T. (i.e. Hungarian–Soviet Crude Oil Incorporated)) with a 50%–50% Hungarian and Soviet ownership.

MASZOVOL started to explore in the Biharnagybajom area in the Great Hungarian Plain (Fig. 1b) and managed to find the first oil field in this part of the Pannonian Basin on 8 February 1947. Until 1953, 42 wells were drilled over this

Fig. 4. Simon Papp among papuas, in their festive sing-sing dress, New Guinea, 25 December 1928. Courtesy of the Hungarian Oil and Gas Industry Museum.

Fig. 5. Simon Papp among the drilling crew of the Mihályi-1 well in western Hungary, 7 August 1935. Courtesy of the Hungarian Oil and Gas Industry Museum.

Fig. 6. Simon Papp was a real petroleum geologist: that is, he did not mind getting his hands dirty – literally. This photograph was taken on 22 August 1935 at the Mihályi-1 well in western Hungary (Fig. 1b), when he had a close look at the drilling mud. Courtesy of the Hungarian Oil and Gas Industry Museum.

trend, of which 17 became producers with a typical 20–100 BOPD (barrels of oil per day) flow rate.

Also, the fields operated in west Hungary by MAORT were put back into production in record time after the war. For example, in the Lovászi Field, production was restarted on 10 April 1945, just a few days after the withdrawal of the last German troops from the territory of Hungary. However, production rates were forcefully greatly increased in order to supply the Soviet army, and this overproduction in the post-war years had a major toll on most of the fields operated by MAORT. As a result, by 1948 the production rates had started to drop drastically, which provided an excuse for the communist government to claim sabotage and to arrest most of the managers of MAORT, including Simon Papp (see later).

After a number of adjustments in their organizations, MASZOVOL and MAORT were consolidated into a new 50%–50% Hungarian and Soviet company, MASZOLAJ, in October 1952, which covered the exploration and production activities over the whole of Hungary by 1954. The discovery of the Nagylengyel Field in SW Hungary (e.g. Clayton & Koncz 1994; Figs 1b & 2a) in 1952 was a major development as it was the largest oil find at that point in the Pannonian Basin (Fig. 2). This oil field still produces oil at a rate of about 2500 BOPD.

The already maximized production rates at Nagylengyel were expected to be tripled by the communist state as part of their 5 year planning cycle, but the water-cut in the field had already started to increase dramatically in 1954. When it became obvious that the expected production plans were impossible, the Communist Party fell back on the same scape-goating as earlier with the American MAORT and Simon Papp (i.e. claiming sabotage by industry managers and experts). Another wave of investigation started in the summer of 1956, which, fortunately, came to end due to the Hungarian Revolution in October 1956. However, after the anti-communist and anti-Soviet uprising was stopped by the Soviet tanks, the purported court cases against the oil industry spilled over into 1957. Béla Majerszky, a petroleum engineer, and his 12 colleagues in Zalaegerszeg, and Ernő Buda and his colleagues in Szolnok had to face almost identically made-up, entirely politically-driven, charges (Papp 2000), which were compounded with additional extra charges for their pro-revolution roles.

Fig. 7. Simon Papp himself selected the well sites of all the early oil discoveries in Transdanubia, SW Hungary. He also carefully designed the tests in the exploration, appraisal and production wells drilled in these fields. Courtesy of the Hungarian Oil and Gas Industry Museum.

After the political situation was consolidated and the hard-line communism driven by Moscow was replaced by the much more pragmatic and relaxed 'gulyás-communism' of the socialistic János Kádár government, the Hungarian oil and gas industry was once again reorganized. The new organization, OKGT (Országos Kőolaj es Gázipari Tröszt (i.e. National Oil and Gas Trust)) was responsible for all exploration and production in Hungary starting in 1960. The new organization proved to be very efficient as many major discoveries were made in the subsequent few years (Fig. 2a). The largest field in Hungary, the oil and gas Algyő Field in SE Hungary (Fig. 1b), was discovered in 1965 and developed by OKGT (Magyar *et al.* 2006). The production started in 1966, 1 year after the discovery. Present-day production is still about 8500 barrels/day.

After the fall of the Iron Curtain in 1990, OKGT was replaced by MOL plc (Hungarian Oil Company) on 1 October 1991. Since then, MOL has become an international, vertically integrated oil company. The current MOL Group is a leading integrated Central and East European oil and gas corporation headquartered in Budapest. It has operations in

Fig. 8. The photograph illustrates a moment from the grotesque trial where Dr Papp was forced to give a horrifying and absurd testimony about his intentional sabotage of the Hungarian oil industry. Ironically, this industry, to a large degree, was created by him. Dr Kertai, a witness for the defence and also a prominent geologist, is listening to Dr Papp's self-destroying testimony in confusion not understanding what is happening. 'Dramatic Testimony': the front page of a communist weekly newspaper, *The Observer*, 4 December 1948. Courtesy of the Hungarian Oil and Gas Industry Museum.

over 33 countries and employs 26 000 people worldwide.

As to domestic exploration, the creaming curve for the period between 1950 and 2015 shows the area of post-war Hungary becoming a very mature petroleum province (Fig. 2a). The total discovered reserves are close to 3 BBOE (billion barrels of oil equivalent), and the flattening of the curve since about 2000 is consistent with all the known conventional play types in the Pannonian Basin (e.g. Dank 1988; Tari & Horváth 2006) being nearly exhausted. The rates for the last 5 years also show a steady decline of domestic oil and gas production (Fig. 2b). Clearly, new exploration play types and/or development technology improvements are required, and are underway to rejuvenate production in the Hungarian part of the Pannonian Basin.

Unconventional exploration of the potentially large basin-centred, tight gas-sand accumulation in the Makó Trough (Fig. 1) was attempted during the last decade; this project, however, failed for various reasons (Badics *et al.* 2011). Several unconventional shale-gas and shale-oil exploration targets have been considered in Hungary but all of these have considerable risk (Badics & Vető 2012).

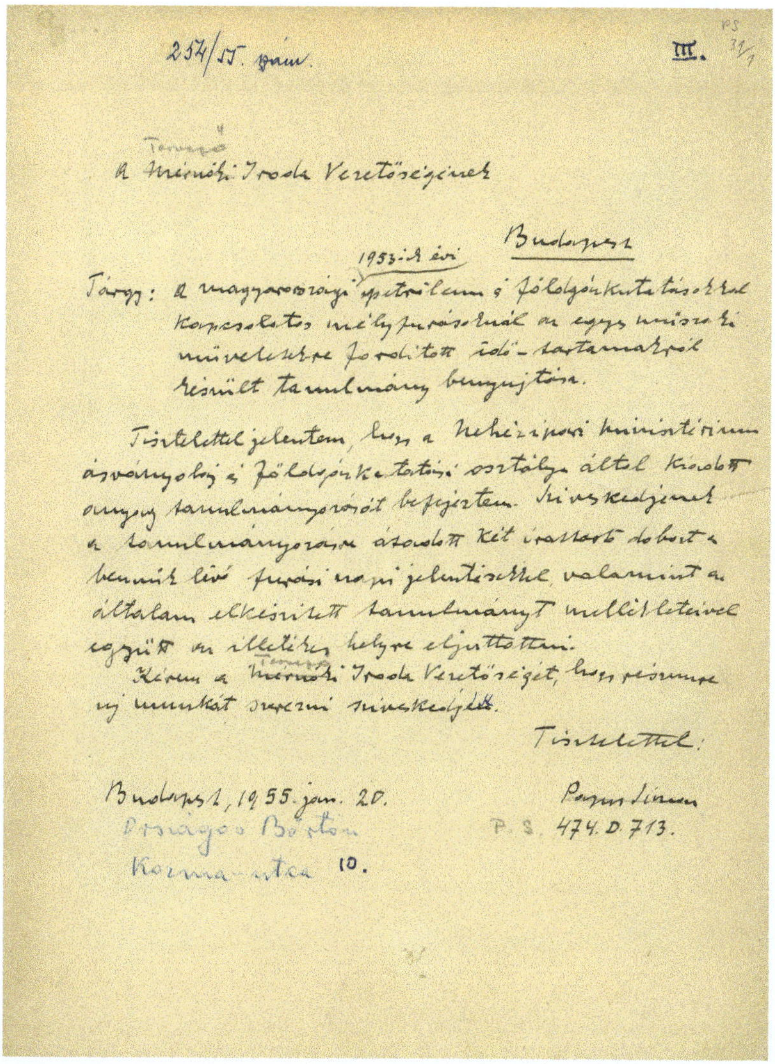

254/55. szám. III. PS 31/1

A Tervező Mérnöki Iroda Vezetőségének

Budapest

Tárgy: A magyarországi 1953-ik évi petróleum és földgázkutatásokkal kapcsolatos mélyfúrásoknál az egyes műszaki műveletekre fordított idő-tartamokról készült tanulmány benyujtása.

Tisztelettel jelentem, hogy a Nehézipari Minisztérium ásványolaj és földgázkutatási osztálya által kiadott anyag tanulmányozását befejeztem. Szíveskedjenek a tanulmányozásra átadott két irattartó dobozt a bennük lévő fúrási napi jelentésekkel, valamint az általam elkészített tanulmányt mellékleteivel együtt az illetékes helyre eljuttatni.

Kérem a Tervező Mérnöki Iroda Vezetőségét, hogy részemre új munkát szerezni szíveskedjék.

Tisztelettel:

Budapest, 1955. jan. 20. Papp Simon
Országos Börtön P. S. 474. D. 713.
Kozma-utca 10.

Fig. 9. Example of a technical study handwritten by Simon Papp in his prison cell. Courtesy of the Hungarian Oil and Gas Industry Museum.

The early years of Simon Papp in Hungary (1886–1920)

Simon Papp was born in Kapnikbánya, Transylvania, Hungary (today Capnic in Romania: Fig. 1a) in 1886. His mother came from an Austrian (Tirolian) mining family, whereas his Hungarian father was the head of the the school in town. He attended the university in Kolozsvár (today Cluj-Napoca, Romania: Fig. 1a), and obtained a doctoral degree in geology, petrology and physical geography (Fig. 3). Dr Papp was also an assistant professor

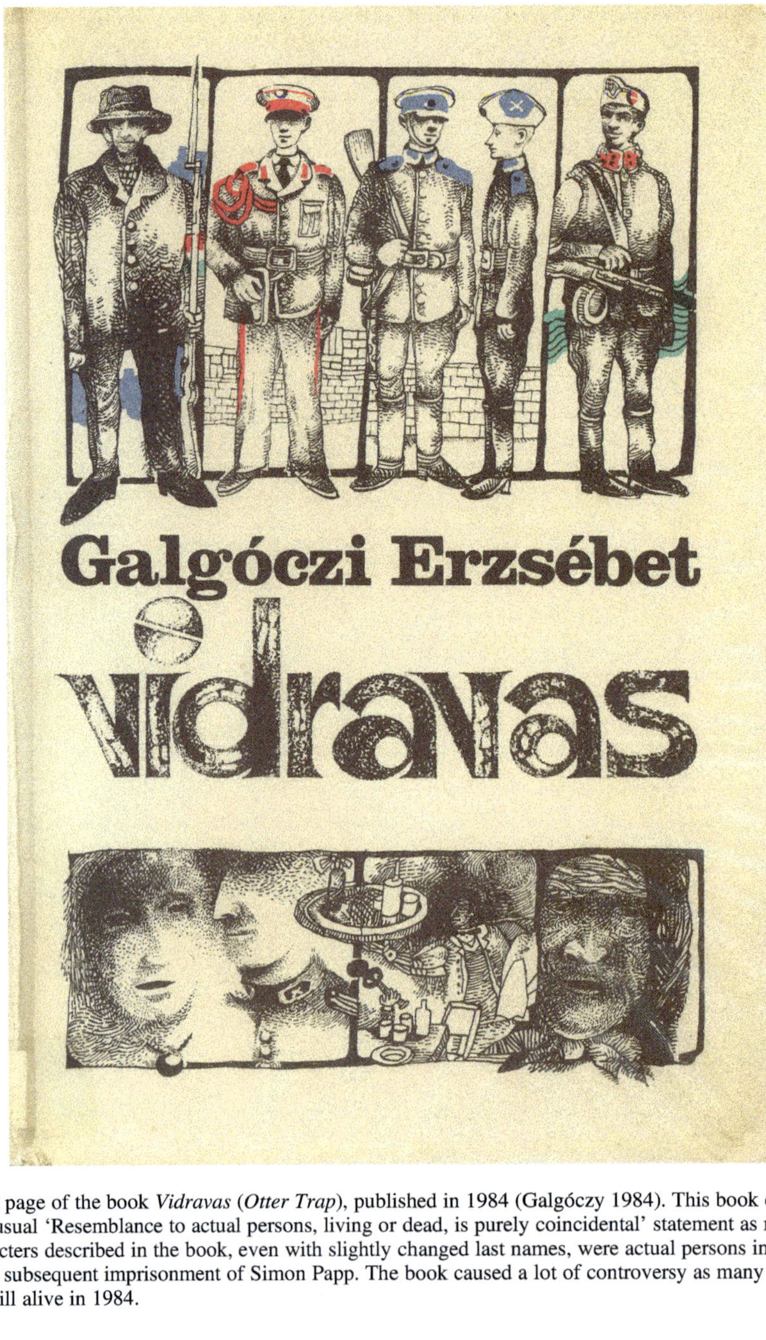

Fig. 10. Front page of the book *Vidravas* (*Otter Trap*), published in 1984 (Galgóczy 1984). This book clearly could not claim the usual 'Resemblance to actual persons, living or dead, is purely coincidental' statement as many of the negative characters described in the book, even with slightly changed last names, were actual persons involved in the court case and subsequent imprisonment of Simon Papp. The book caused a lot of controversy as many of these people were still alive in 1984.

at the University of Kolozsvár from 1909 to 1911. However, he had an inclination towards applied geology and found his job at the university too academic. Therefore, he joined the Academy of Mining and Metallurgical Engineers at Selmecbánya (now Banska Stiavnica, Slovakia: Fig. 1a), where he became a lecturer on economic geology and palaeontology. In 1914, the Egbell oil field in the Vienna Basin (now Gbely in Slovakia: Fig. 1a) was discovered, and Papp divided his time between working at the university and studying this oil field.

During World War I, in 1915, Papp left the Mining Academy and worked as chief state geologist with the Mining Department of the Hungarian Ministry of Finance. From 1915 to 1920, Papp, as an associate of Hugo von Böckh, carried out intensive geological surveys, field mapping and subsurface geological work in connection with oil and gas prospecting for the Hungarian government. His area of interest extended from Transylvania (eastern Hungary), to upper (or northern) Hungary to Galicia (today shared by Poland and Ukraine).

Exploring for petroleum internationally (1920–32)

From 1920 to 1930, Papp was employed by the Anglo-Persian Oil Company and its subsidiaries (D'Arcy Exploration Company and Bitumen) as a

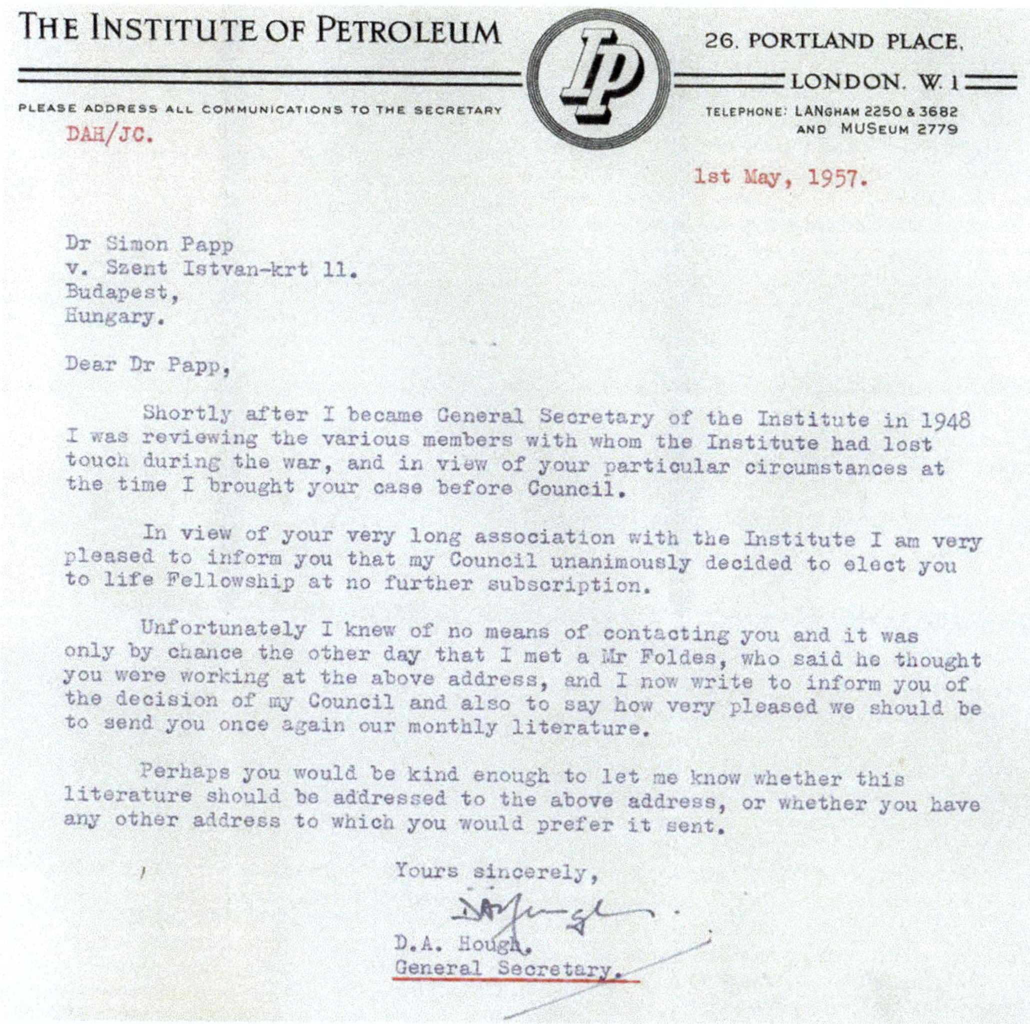

THE INSTITUTE OF PETROLEUM

IP

26. PORTLAND PLACE,
LONDON. W. 1

PLEASE ADDRESS ALL COMMUNICATIONS TO THE SECRETARY

TELEPHONE: LANGHAM 2250 & 3682
AND MUSEUM 2779

DAH/JC.

1st May, 1957.

Dr Simon Papp
v. Szent Istvan-krt 11.
Budapest,
Hungary.

Dear Dr Papp,

Shortly after I became General Secretary of the Institute in 1948 I was reviewing the various members with whom the Institute had lost touch during the war, and in view of your particular circumstances at the time I brought your case before Council.

In view of your very long association with the Institute I am very pleased to inform you that my Council unanimously decided to elect you to life Fellowship at no further subscription.

Unfortunately I knew of no means of contacting you and it was only by chance the other day that I met a Mr Foldes, who said he thought you were working at the above address, and I now write to inform you of the decision of my Council and also to say how very pleased we should be to send you once again our monthly literature.

Perhaps you would be kind enough to let me know whether this literature should be addressed to the above address, or whether you have any other address to which you would prefer it sent.

Yours sincerely,

D.A. Hough.
General Secretary.

Fig. 11. Example of personal support for Simon Papp from beyond the Iron Curtain. Courtesy of the Hungarian Oil and Gas Industry Museum.

senior geologist, advisor and, finally, as chief geologist. He worked for these companies in Yugoslavia, Albania, Turkey, Australia and New Guinea. He was one of the pioneering geologists to explore the tribal and remote land of Papua (Fig. 4). From 1930 to 1931, Papp was chief geologist of the International Prospecting Co., London, and from 1931 to 1932 of the Land and Royalty Co., London. He worked for the latter company in Germany and Canada.

Exploring back in Hungary and 'retirement' (1932–47)

At the end of 1932, Dr Papp returned to Hungary. On 13 July 1933, he became affiliated with EUROGASCO (European Gas & Electric Company) as the chief geologist for Hungary. He organized and directed the exploration of this company's concession in Transdanubia, SW Hungary. This prospecting work was very successful, resulting in the discovery of four oil fields in western Hungary (Budafa, Lovászi, Lendvaújfalu and Pusztaszentlászló: Fig. 1b) with a total reserve of about 100 MMbbl (million barrels) (Figs 5–7). As EUROGASCO was taken over by the Standard Oil Company of New Jersey, Dr Papp became director and later vice-president and manager of the successor company, MAORT. He retired from this assignment on 31 December 1947, but was retained as a consulting geologist.

In 1938, he became Professor of Oil Technology (Exploration and Production) at the Mining and Metallurgical Academy, Sopron, Hungary (Fig. 1). However, he began lecturing only after he retired from the MAORT. Dr Papp has been recognized in Hungary as an authority on oil and gas exploration, and his name is linked inseparably to the Hungarian oil industry.

Prison years (1948–55)

In 1948, upon returning from a study trip to the USA, Dr Papp was arrested by the communist secret police in Budapest, Hungary, allegedly because of sabotage and attempting to overthrow the communist government. There were many politically-driven 'show' trials during this time, to provide ammunition to the strong propaganda machine of the new communist government in order to nationalize well-established private companies (i.e. confiscate and transform them into state-owned entities) to be managed mostly by loyal Party members irrespective of their – frequently – inadequate professional credentials.

After being forced to 'confess' to multiple sabotage acts (Fig. 8), Dr Papp was sentenced to death. However, probably due to worldwide protest by scientific associations, the sentence was changed to life in prison.

As Dr Papp's exceptional knowledge of the petroleum industry was a major asset, the communist secret police arranged for him to continue his work from his prison cell. Not only was he asked to follow and provide advice on the oil and gas exploration and production projects in Hungary but he also had to provide expert opinion on other raw material exploration projects, as the government planned to build heavy industry in communist Hungary.

One of the allegations against Dr Papp during his 1948 trial was that he intentionally picked well locations to produce poor results as an act of sabotage. Ironically, during his imprisonment, he was not only asked by the communist secret police to select new well locations but also to recommend testing intervals and production schemes (Fig. 9).

After the death of Joseph Stalin in 1953, many political prisoners in the communist regimes throughout the Soviet-influenced Eastern Europe were released. Dr Papp's turn came as late as 1955. But he has never been formally pardoned, rehabilitated or compensated for his 7 years in prison, a sentence that was absolutely a politically-driven injustice to this great Hungarian scientist.

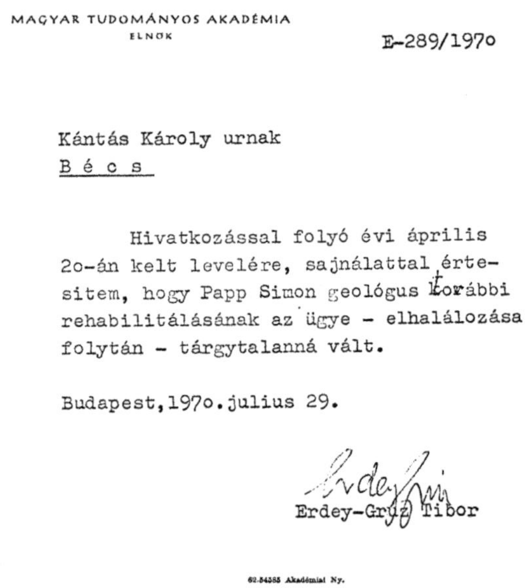
MAGYAR TUDOMÁNYOS AKADÉMIA
ELNÖK

E-289/197o

Kántás Károly urnak
B é c s

Hivatkozással folyó évi április 2o-án kelt levelére, sajnálattal értesitem, hogy Papp Simon geológus korábbi rehabilitálásának az ügye – elhalálozása folytán – tárgytalanná vált.

Budapest,197o.julius 29.

Erdey-Grúz Tibor

62.84388 Akadémiai Ny.

Fig. 12. A letter of response from the President of the Hungarian Academy of Sciences at that time re the reinstatement of Dr Simon Papp to the same organization. Translation from Hungarian: 'Hungarian Academy of Sciences, President. To: Mr. Károly Kántás, Vienna. With reference to your letter dated April 20 of this year, I regretfully inform you that the rehabilitation case of Simon Papp geologist – due to his death – became pointless. Budapest, 1970, July 29. Tibor Erdey-Grúz'.

A famous Hungarian book, *Vidravas* (*Otter Trap*: Fig. 10), written by Erzsébet Galgóczi in 1984, was a semi-historical account of the bizarre story in which the communist political police had arranged for fake personal letters to be written and sent to the prison pretending to be from Dr Papp's wife. In fact, Ms Irmi Papp had passed away in 1953 while her husband was still in prison. Indeed, the communist political police did not want Dr Papp to become distressed about his wife's death and so had faked the letters which were produced on a typewriter without a signature. According to his

Fig. 13. A photograph of Dr Simon Papp. Translation from Hungarian: 'Born 1886. II. 14. Photograph: 1962 November 27'. Courtesy of the Hungarian Oil and Gas Industry Museum.

posthumous autobiography (Papp 2000), it was clear to Papp that these letters were not being sent from his wife. In the book *Vidravas*, the communist police find a young girl who can copy the handwriting of Ms Papp and then for a period of 2 years these faked letters are received by Dr Papp in his prison cell. The book caused a lot of controversy in Hungary as many of the negative characters described in it, and only given slightly different last names, were still alive when it was published in 1984.

Dr Papp was a member of numerous scientific and cultural societies. He was elected as a corresponding and then a regular member of the Hungarian Academy of Sciences in 1946, he was president of the Hungarian Association of Mining and Metallurgical Engineers, and president and life member of the Hungarian Geological Society before his imprisonment in 1948. Most international organizations, like the Institute of Petroleum, tried to reach out and reinstate him after his prison years (Fig. 11). However, he was not rehabilitated by the Communist-Party-controlled Hungarian Academy of Sciences, even after his release from prison in 1955. Although many people attempted his reinstatement to the Academy, this did not happen during his life time (Fig. 12). He was rehabilitated in 1989, once the political changes reached the Hungarian Academy of Sciences. In addition to the posthumous rehabilitation of Simon Papp by the Academy, he also received the prestigious Széchenyi Award in 1990. As he had no surviving relatives, the award and cash prize were received by MOGIM, which is the second largest oil and gas industry museum in Europe after the Polish one in Bóbrka. MOGIM not only has an exceptionally large collection of documents (reports, books, maps and photos) related to the more than 100 years of oil and gas exploration in Hungary, but also has similar historical documents from many other countries of the world.

Conclusions

Dr Simon Papp was the most prominent geologist in the history of Hungarian petroleum exploration and production before and after World War II (Fig. 13). His post-1948 exploration 'career' sheds light on the anatomy of the communist regimes behind the Iron Curtain (1945–90), which remains, unfortunately, largely unknown to, and/or still neglected by, the Western public.

Acknowledgements The authors are most grateful for the various materials, including most of the photographs in this paper, provided by János Tóth and András Szép of the Hungarian Oil and Gas Industry Museum (MOGIM), Zalaegerszeg, Hungary (http://www.olajmuzeum.hu/cid2/). Diana Hay kindly provided documents from the library of the Hungarian Academy of Sciences. Piotr Gawenda of IHSMarkit is thanked for the creaming curve and the current Hungarian production rates shown in Figure 2. Special thanks are due to Jonathan Craig for inviting this contribution to this volume and for his editorial patience. This paper was reviewed by Rasoul Sorkhabi, Balázs Badics and Andrew Zolnai; their suggestions substantially improved the text.

Funding This research received no specific grant from any funding agency in the public, commercial or not-for-profit sectors.

References

Badics, B. & Vető, I. 2012. Source rocks and petroleum systems in the Hungarian part of the Pannonian Basin: the potential for shale gas and shale oil plays. *Marine and Petroleum Geology*, **31**, 53–69.

Badics, B., Uhrin, A., Vető, I., Bartha, A. & Sajgó, C. 2011. Assessment of the basin-centered gas accumulation in the Makó Trough. *Földtani Közlöny*, **141**, 445–468 [in Hungarian with an English abstract].

Clayton, J.L. & Koncz, I. 1994. Petroleum geochemistry of the Zala Basin, Hungary. *AAPG Bulletin*, **78**, 1–22.

Dank, V. 1988. Petroleum geology of the Pannonian basin, Hungary: An overview. *In*: Royden, L. & Horváth, F. (eds) *The Pannonian Basin: A study in Basin Evolution*. AAPG Memoirs, **45**, 319–331.

Dank, V. 1996. Review of the book '*My Life*' by Simon Papp. *Földtani Közlöny*, **126**, 148–149 [in Hungarian].

Galgóczy, E. 1984. *Vidravas [Otter Trap]*. Szépirodalmi Kiadó, Budapest [in Hungarian].

Humble, E.A. 1970. In memoriam of Simon Papp. *AAPG Bulletin*, **54**, 2448–2450.

Magyar, I., Fogarasi, A., Vakarcs, G., Buko, L. & Tari, G. 2006. The largest hydrocarbon field discovered to date in Hungary: Algyö. *In*: Golonka, J. & Picha, F. (eds) *The Carpathians and their Foreland: Geology and Hydrocarbon Resources*. AAPG Memoirs, **84**, 619–632.

Nyári, K. 2016. Portrait of Simon Papp. *BBC History (Hungarian edition)*, **6**, 58–61.

Papp, S. 1963. The Hungarian oil and gas exploration between 1780 and 1945, part I. *Transactions of the Department of Technical Sciences of the Hungarian Academy of Sciences*, **32**, 449–465 [in Hungarian].

Papp, S. 1964. The Hungarian oil and gas exploration between 1780 and 1945, part II. *Transactions of the Department of Technical Sciences of the Hungarian Academy of Sciences*, **33**, 421–437 [in Hungarian].

Papp, S. 2000. My life. *In*: Srágli, L. & Tóth, J. (eds) *My Life*. 2nd edn. Hungarian Oil and Gas Industry Museum, Zalaegerszeg, Hungary, 19–322 [in Hungarian].

Tari, G. & Horváth, F. 2006. Alpine evolution and hydrocarbon geology of the Pannonian Basin: an overview. *In*: Golonka, J. & Picha, F. (eds) *The Carpathians and their Foreland: Geology and Hydrocarbon Resources*. AAPG Memoirs, **84**, 605–618.

Sir Thomas Boverton Redwood (1846–1919): a watershed in the British oil industry

RASOUL SORKHABI

University of Utah, Energy & Geoscience Institute, Salt Lake City, UT 84108, USA

rsorkhabi@egi.utah.edu

Abstract: Sir Thomas Boverton Redwood (1846–1919) stands as a giant in the history of petroleum science and technology. This paper pieces together scattered information about his life and work, and discusses his contributions, directly or indirectly, to petroleum exploration and production in various parts of the world, especially Burma (Myanmar), Persia, Mexico and the West Indies with which British colonial and commercial interests were related. Redwood established one of the first and most successful consulting firms for petroleum companies, and, in this way, trained generations of petroleum geologists and engineers. His masterpiece, *Petroleum: A Treatise*, which went through five editions from 1896 to 1926, summed up the knowledge of petroleum of its age, and still remains a valuable book for historical reference. Redwood's mediatory position between scientific, industrial and political circles in Britain enabled him to play a leading role in the development of petroleum technology, as well as Britain's oil operations around the world during the 1880s–1910s. He served as a technical advisor to many government committees and oil companies, and his total dedication to Britain's oil security during World War I and during the transformation of the Royal Navy's fleet from coal to oil fuel were the climax of his professional life.

Great Britain (which became the United Kingdom in 1800) played a leading role in the Industrial Revolution of Europe, beginning with James Watt's steam engine in 1769 and continuing through the nineteenth century (Allen 2010; Wrigley 2010). One consequence of the Industrial Revolution was the massive use of coal resources, which are indeed abundant in the UK.[1] The British also pioneered the production of coke and coal gas derived from pyrolysis of coal in the nineteenth century. British coal production drastically increased from 12 Mt (million tonnes) in 1800 to its peak of 287 Mt in 1913 (Mitchell 1984). In the 1860s, following the successful Drake oil well in Titusville, Pennsylvania, as well as similar developments in Russia (Azerbaijan) and Romania, the industrial nations realized the value and supremacy of oil as a fuel. Unlike coal, the UK does not possess rich sources of conventional oil in her onshore basins. Despite the success of oil-shale (Allen 1921; McKay 2012) and shale-gas (Williams 1981) industries in the country, the history of the conventional petroleum industry in Great Britain during the late nineteenth and early twentieth centuries was largely associated with the country's colonial influence in various parts of the world. A towering figure of this period was Sir Thomas Boverton Redwood. Despite his many contributions and his pioneering role in the early development of the petroleum industry, a definite volume on the life and works of Redwood is yet to be written.[2] This paper pieces together scattered information about him from a variety of sources. We first review Redwood's life and published works, and then discuss his involvement in oil exploration and production in several countries with which British enterprises were active.

[1]The history of the British coal industry should be considered as background for the history of the oil and gas industry in the country. The work by Nef (1932) in two volumes, as well as a five-volume Oxford University Press series entitled *The History of the British Coal Industry* (Flinn & Stoker 1984, vol. 2; Ashworth 1987, vol. 5; Roy *et al.* 1987, vol. 3; Supple 1988, vol. 4; Hatcher 1993, vol. 1) are comprehensive and authoritative sources of information in this subject.

[2]For oil historians and scholars who want to undertake more extensive research on the life and work of Sir Boverton Redwood, the following archives offer original documents related to Redwood or the early history of the British oil industry: Burmah Castrol plc, Swindon, UK (which was acquired by BP in 2000); University of Warwick Modern Records Centre (http://www2.warwick.ac.uk/services/library/mrc/); BP Archive (www.bparchive.com); Museum of the Scottish Oil Industry (www.scottishshale.co.uk); The Admiralty papers (Royal Navy Department, Ministry of Defence, London); and www.archive.org

From: Craig, J., Gerali, F., MacAulay, F. & Sorkhabi, R. (eds) 2018. *History of the European Oil and Gas Industry*. Geological Society, London, Special Publications, **465**, 423–451.
First published online May 25, 2018, https://doi.org/10.1144/SP465.22

A sketch of his life

This biographical sketch is largely based on Anon (1900*a*, 1920), A.C.K. (1919), Perkin (1920), Black (1926), Jones (1981), Jones & Goodall (1984–86), Kidd & Williamson (1990), Mosley (2003), Corley (2004) and Grace's Guide (2016*a*).

Thomas Boverton Redwood (Figs 1 & 2) was born on 26 April 1846 in London (at 19 Montague Street, Russell Square, to be exact). His father, Theophilus Redwood (1806–92) (Figs 3 & 4), of Boverton in Glamorganshire and of Montague Street in London, was a renowned pharmacist: he was a professor in the School of Pharmacy of the Pharmaceutical Society of Great Britain and editor of the *Pharmaceutical Journal.*[3] His mother, Charlotte Elizabeth (who died in 1868, when Boverton was 22), was the daughter of Thomas Newborn Robert Merson (of Hornsey in Middlesex), who owned a pharmaceutical company in London. Boverton was the eldest in the family of six boys and two girls.[4] He studied at the University College School, London

[3]Theophilus Redwood was born on 2 March 1806 in Boverton and died on 5 March 1892 also in Boverton (pronounced 'Trebefered' in Welsh) – a village located in the Vale of Glamorgan in South Wales. The Redwood family originated from Bristol: in 1613, Robert Redwood donated his lodge to the City of Bristol to build a library. One of his descendants, Isaac Redwood moved from Bristol to Cowbridge, also in the Vale of Glamorgan, in the early part of the eighteenth century. Isaac's grandson, Thomas Redwood, a tanner and schoolmaster, settled at Orchard House, Boverton after his second marriage, and this is where Theophilus Redwood, the third child of the family, was born. His siblings included Charles, Elizabeth and the younger Lewis. At the age of 14, Theophilus became an assistant to Charles Vacehell, an apothecary. (Vacehell was the brother-in-law of Theophilus, having married the eldest daughter of Thomas Redwood by his first marriage.) Theophilus spent 3 years with Vacehell and then went to London as an apprentice to the pharmaceutical firm of John Bell and Co., where he spent 7 years, while also taking courses in chemistry at the Royal Institution and medicine at King's College. In 1830, Theophilus set up his own dispensary in Crawford Street, London. In 1841, Theophilus Redwood, together with the leading pharmacists of the day, founded the Pharmaceutical Society of Great Britain, where Redwood served as its first professor (until his retirement in 1885), librarian (until 1888), curator (until 1861), sub-editor (and, from 1859, editor) of the *Pharmaceutical Journal.* In 1844, Redwood became Director of the Chemical Laboratories at the Society, and from 1846 to 1881 he held the Society's joint Chair of Pharmacy and Chemistry. In 1845, he married Charlotte Elizabeth Morson, daughter of Thomas Newborn Robert Morson, Vice-President of the Pharmaceutical Society. Redwood contributed to many publications: He co-edited (with Dr Mohr of Coblentz) a *Manual of Practical Pharmacy* based on the German text (1840, 1847 and 1857), edited *British Pharmacopoeia* (1867, 1874 and 1885), abridged Pereira's classical work on *Materia Medica* (1872), and regularly wrote for the *Pharmaceutical Journal* and *The Lancet.* He also served as Secretary of the Cavendish Society, Secretary of the Chemical Society and as President of the Society of Public Analysts (now called The Society for Analytical Chemistry) in London. For his life-long contributions to pharmacy, he was awarded an honorary doctorate from the University of Giessen, Germany. In 1885, Redwood retired and was given an Emeritus Professor title by the Pharmaceutical Society. Upon his death at the age of 86, Theophilus Redwood was buried at Llantwit Major churchyard in his native village of Boverton. His grave, covered with an inscribed pink granite, lies to the north of the church. His obituary was published in *Western Mail* (7 March 1892), *Pharmaceutical Journal and Transactions* (12 March 1892, p. 763) and *The Lancet* (26 March 1892, p. 727). For more information refer to Anon (1873, 1938, 1972) and Davies *et al.* (2008). Note that Theophilus Redwood's younger brother Lewis Redwood and his son Thomas Redwood were renowned medical doctors ('Medical Officer of Rhymney Iron Works'). The Redwood Building of Cardiff University (in the Cathays Park area of Cardiff in Wales) (Fig. 4) was named in 1979 after the Redwood family of Orchard House, Boverton, in honour of Theophilus Redwood and his son Boverton Redwood, as well as Lewis Redwood and his son Thomas Redwood. (The Redwood Memorial Hospital, Rymney, Monmouthshire was named in honour of Lewis Redwood and his son, and successor, Thomas Redwood in 1948.)

[4]Boverton Redwood's siblings included the following: Mary Ann Redwood (born 1848); Theophilus H. Redwood (born 1850); George Herbert Redwood (born 1852), Charles L. Redwood (1855–1907), Robert Redwood (born 1857), Charlotte E.M. Redwood (born 1859) and Iltyd Isaac Redwood (1863–1910). Iltyd Redwood was also a chemist and a member of the Society of Chemical Industry (England), the Royal Society of Arts, the Aeronautical Society of Great Britain and the American Society of Mechanical Engineering. From 1879 to 1881, he resided in London and worked as an assistant to his father at the Pharmaceutical Society of Great Britain. In 1882, he became an assistant chemist in the laboratory of Young's Paraffin Light and Mineral Oil Company in Scotland, and, in 1887, he joined the Queen's County works of the Standard Oil Company, where he had several positions. In July 1892, he went to New York and became a US citizen but returned to the UK in August, and the following month married Katherine Horner (third daughter of W. Sims Horner of Fernhill, Walthamstow in Essex). The couple had two daughters: Katherine E. Redwood and Mary Isabella Redwood. Iltyd Redwood made a second trip to New York in early 1897

Fig. 1. A portrait of Sir Boverton Redwood that has appeared in various places including on the front page of his book *Petroleum: A Treatise* (Redwood 1926) and Grace's Guide (2016*a*).

from 1857 to 1862, and then began to work in his father's pharmaceutical laboratory. In 1866, he became a member of the Petroleum Association, and 3 years later he was appointed its Secretary; this new opportunity drastically changed his life career. The modern oil industry was then emerging, and Redwood's expertise as an analytical chemist and his dedication to the oil industry soon made him the leading authority on matters of oil in Great Britain. Redwood married Mary Elizabeth, the eldest daughter of Frederick Letchford, of Walthamstow in Essex. The marriage took place on 17 April 1873.

As early as in 1872, Redwood appeared as an expert witness on the issue of flash points of oils before a Select Committee of the House of the Lords. In 1877, Redwood joined Sir Frederick Augusts Abel (1827–1902), a leading British chemist, to conduct laboratory research in order to standardize the methods of testing of flash points of various oil types. In the same year, Redwood went to the USA to conduct further research on the subject (Truth's Investigator 1911, pp. 192–194). All of this research transformed the Abele's untrustworthy 'open-test' apparatus into a new 'closed-test' apparatus. Redwood, through more than 1000 tests, determined 73°F as the equivalent of the previous open-test value of 100°F; the new value was adopted in amending the Petroleum Act of 1879, and the improved Abel instrument was established as a legal basis for testing the flash points.

In 1881, Redwood investigated the effect of barometric pressure on dissolved gases in oil. He conducted these tests in a pressure chamber in Berlin and at various altitudes in the Alps. To study the influence of climate on these tests, Redwood travelled to India. Based on this research, Redwood, in conjunction with F. Abel and Dr Kellner, provided expert advice to the Government of India in amending the Petroleum Act.

In 1883, Redwood appeared as an expert witness before the Select Committee of the House of Lords. In the same year, he accompanied Colonel Sir Vivian Majendie (1836–98), Chief Inspector of Explosives to Queen Victoria, in a tour of Europe to study the methods of oil storage, and a few years later he made a trip to the USA for the same purpose.

Writing in 1884, Charles Marvin, author of *The Region of the Eternal Fire: An Account of a Journey to the Petroleum Region of the Caspian in 1883*, mentions that: 'Mr. Boverton Redwood, Chemist to the London Petroleum Association, instituted tests a short time ago with some refined petroleum of Nobels' brand, which had found its way from Baku'. He then gives a summary of Redwood's report on the physical properties of the oil sample (Marvin 1884, pp. 247–249).

In 1885, Abel and Redwood made recommendations for the manufacturing of safe oil lamps; these recommendations were submitted to the Metropolitan Board of Works. Four years later, they were requested by Colonel Majendie to report on how the manufacturers had adopted these recommendations. Then, in 1896, Redwood submitted the Committee Report of the Lancet Commission on the

(Redwood 1897). From 1897 until retirement, he served as a technical manager and advisor of the English works of Borax Consolidated, Ltd. He died in Dartford, Kent, England. Iltyd Redwood was author of several works including *Theoretical and Practical Ammonia Refrigeration* (Redwood 1895), *A Practical Treatise on Mineral Oils and Their By-Products* (Redwood 1897), and *Lubricants, Oils and Greases* (Redwood 1898). His obituaries were published in the *Journal of the American Society of Mechanical Engineers Engineers*, **32**, (7), 24–25 (1910) and in the *Journal of the Royal Society of Arts*, **58**, 613 (1910). For more information about Iltyd Isaac Redwood, refer to Dulally's Family Tree, http://gw.geneanet.org/dulally?lang=en&pz=henri&nz=durlac&ocz=0&p=iltyd+isaac&n=redwood).

Fig. 2. Sir Boverton Redwood in a cartoon by *Spy*, 25 March 1908.

accidents arising from the use of oil lamps to a Select Committed of the House of Commons, presided over by Mr Mundella.[5]

During 1885–86, Redwood invented a viscometer to measure the fluidity of various oils, which became the standard instrument in the British oil

[5]This information is taken from *The Lancet*, **133**, (3412), 152–155 (19 January 1889) and **147**, (3794), 1401–1403 (16 May 1896). *The Lancet* is a weekly medical journal published in England; it was founded in 1823 by the English surgeon Thomas Wakley, who named it after a surgical instrument. One of the oldest medical magazines, *The Lancet* was taken over by Elsevier in 1991.

Fig. 3. Theophilus Redwood (1806–92), father of Boverton Redwood (source: *Proceedings of the Society for Analytical Chemistry*).

industry (Fig. 5). As Redwood (1926, vol. 3, p. 823) mentions, his viscometer was a modification of the instrument formerly used at the Battersea works of Price's Patent Candle Company. Years later, Redwood modified his viscometer for use with heavy oil per the specification of the Admiralty. The Admiralty Pattern was sold by Messers Baird & Tatlock, Ltd, London. In 1886, Abel, Redwood and Charles James Hislop Warden prepared the 'Rules for Testing Petroleum', which were incorporated into the Petroleum Act of 1886 (Anon 1900*b*).

In 1889, Redwood and Sir James Dewar patented a process for distilling and condensing oil under high pressures. This process enabled the production of lighter oil from the heavier grades. Due to the scarcity of demand in those days, the patent was not used on a large industrial scale. However, it laid the foundation of the modern cracking process, as acknowledged by Bacon & Hamor (1916) in *The American Petroleum Industry*.

In 1892, Redwood visited Egypt to investigate the possibility of the transport of oil through the Suez Canal. In 1894, Redwood designed an apparatus for testing the pressure of oil vapour in tanks, which was an important issue for the transport of oil on ships. This won him the Telford Medal from the Institution of Civil Engineers. In the same year, Redwood, together with his brother Robert and Mr H. Barringer, invented the Redwood water finder to determine the amount of water collected beneath the oil in tanks. In 1899, Redwood was awarded an honorary doctorate from the National Normal University in Ohio.

Redwood set up a petroleum consulting and service company in London, and employed geologists and engineers who provided expertise for many petroleum companies, or in the words of Arthur Beeby-Thompson (1873–1968), a British petroleum engineer, Redwood: 'immaculately attired, with an orchid in his buttonhole, speaking in a slightly affected but impressive manner with a provincial accent, … monopolised petroleum consulting work in England' (Beeby-Thompson 1961, pp. 75, 80). Redwood's work took him to various parts of the world – continental Europe, Russia, India, Egypt and the USA. He served as a technical advisor for a number of companies, most notably Burmah Oil (from 1893), D'Arcy's Concession in Persia (from 1900, which developed into the Anglo-Persian Oil Company in 1908) and Weetman Pearson's oil exploration in Mexico (1901–09).

Redwood was also an advisor to several government agencies, including the Corporation of the City of London under the Petroleum Acts and the Thames Conservancy (set up in 1857 for the management of the River Thames – a task which was transferred to the Port of London Authority in 1909). In 1899, Redwood was a member of the Home Office Committee on Acetylene Generators and, together with Professor Vivian B. Lewis, examined the working of the acetylene generators exhibited at the Imperial Institute (later Imperial College London).

Redwood served as an honorary advisor to the Home Office, the Admiralty, the India Office and the Colonial Office. In 1903, he served on the Royal Commission on Coal, as well as the Royal Commission on the Supply of Food and Raw Materials in Time of War. In 1908, he was as a member of the committee, appointed by the Home Secretary and chaired by Sir Henry Cunynghame, to prepare a report on the state of oil legislation in the country. He served as chairman of the chemical section of the Royal Commission for the Exhibitions at Brussels (1910) and Turin (1911), and a member of the Committee for the Exhibition at St Louis (1904). In 1911–12, Redwood was appointed a member (technical advisor) of the Royal Commission on Oil-fuel for the Admiralty presided over by Lord Fisher.

Lord Curzon of Kedleston (1859–1925) famously remarked that the Great Britain and the Allies won World War I (1914–18) because they 'floated to victory on a sea of oil' (Black 1926). During the

Fig. 4. Redwood building at Cardiff University, situated in the Cathays Park area (King Edward VII Avenue) of Cardiff, Wales (source: Wikipedia). Named in honour of the Redwood family (see footnote 3 in this chapter), it is home to the School of Pharmacy and Pharmaceutical Sciences, one of the top in the UK. The building was part of the University of Wales Institute of Science and Technology, which in 1988 was incorporated in the University of Wales College Cardiff (renamed Cardiff University in 1999). The relief sculpture (in the photograph) was done by Edward Bainbridge Copnall, and shows an elderly toga-clad person, surrounded by books and a globe, reaching out protectively over a youthful scientist and a nurse.

war Redwood's expertise was much in demand. He served as Director of the Petroleum Research Department (involved in tests on the extraction of liquid oil from bituminous materials) for the Petroleum Supply Branch at the Ministry of Munitions, which, in 1917, was renamed Director of Technical Investigations on the Petroleum Executive. He also served on several committees, including as Assistant Controller in the Trench Warfare Research, and was a member of the Committee on Mining for the Advisory Council (set up in 1915) that provided expert counselling to the British government and parliament in matters of scientific and industrial research (Varcoe 1970; Grace's Guide 2016*a*). In 1917, he also served as the chairman of the Gas Traction Committee and of the Committee on the Use of Alcohol as Fuel. Indeed, exhaustion from the work during the war years was largely responsible for his illness and death in 1919 (Corley 2004).

Redwood was affiliated with a number of professional societies, including the Pharmaceutical Society of Great Britain, the Petroleum Association (London), the Institute of Chemistry (council member and Vice-President), the Chemical Society of London, the Society of Chemical Industry (Vice-President, and President 1907–08), the Royal Society of Edinburgh, the Royal Geographical Society, the Geological Society, the Institute of Mining Engineers, the Imperial Russian Technical Society (honorary member), the American Philosophical Society (honorary member), the Junior Institute of Engineers (President 1913–14) and the Institute of Petroleum Technologists (founding member and President 1914–15).[6]

[6]The Chemical Society of London (formed in 1841), the Society for Analytical Chemistry (formed in 1874 as the Society of Public Analysts), the Royal Institute of Chemistry (formed in 1877 as the Institute of Chemistry of Great Britain) and the Faraday Society (founded in 1903) were all amalgamated in 1980 to form the Royal Society of Chemistry. The Society of Chemical Industry was founded in London in 1881 and is still an active society, publishing several peer-reviewed journals. Redwood was a founding member of Institute of Petroleum Technologists, of which he served as the first president. He was succeeded by Professor Sir John Cadman, Sir Charles Greenway, Sir Frederick W. Black and Professor John Samuel Strafford Brame. In 1939, the organization was renamed The Institute of Petroleum, which

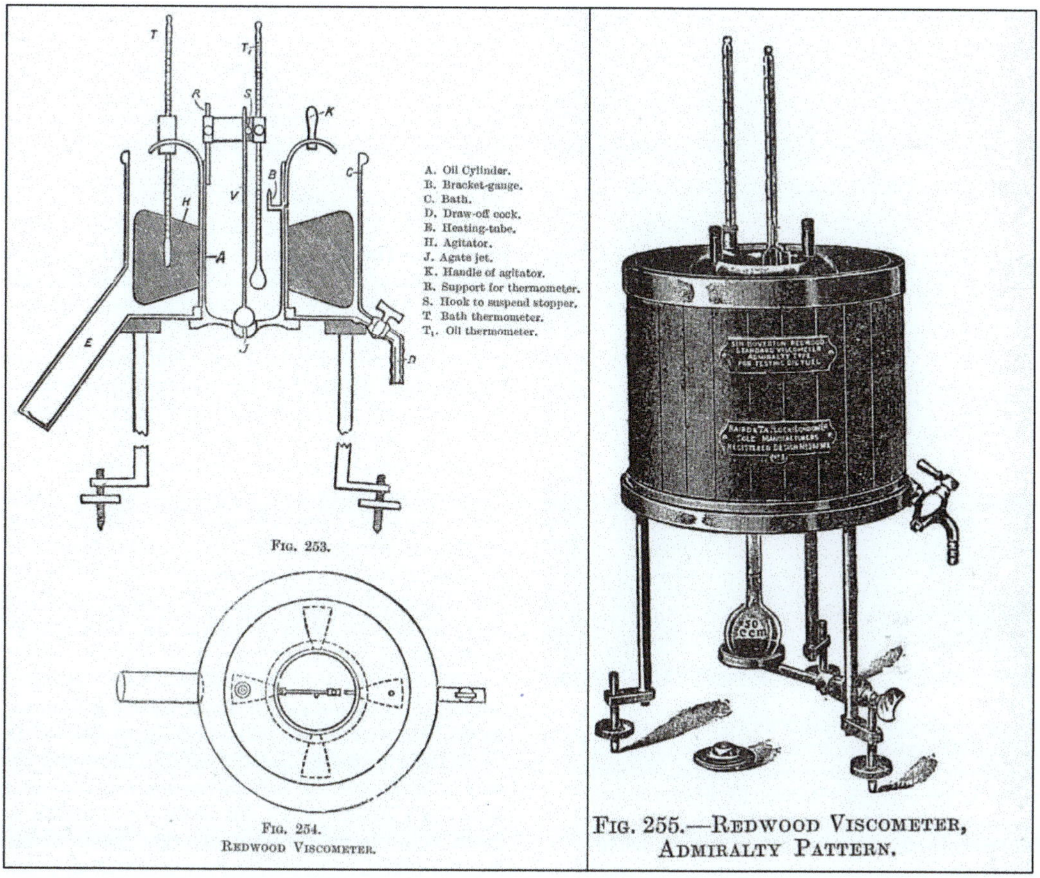

Fig. 253.

Fig. 254.
Redwood Viscometer.

Fig. 255.—Redwood Viscometer, Admiralty Pattern.

Fig. 5. Redwood's viscometers as shown in his book *Petroleum: A Treatise* (Redwood 1926, vol. 3, pp. 824–825).

Not surprisingly, Redwood loved automobiles, along with his hobby of steam-yachting; his first car was a Bolide. In 1897, he bought the first four-cylinder 12 h.p. Daimler manufactured at Coventry according to his own specification, and it was shown to King Edward VII, then Prince of Wales. He also owned a De Dion voiturette, a 7 h.p. Panhard, and a 16 h.p. De Dietrich car (Grace's Guide 2016*a*). Redwood was one of the first members of the Self-propelled Traffic Association (Redwood 1895), founded in 1895 by Sir David Salomons (1851–1925), then Mayor of Tunbridge Wells, shortly after a successful motor-car exhibition in the town. The association was incorporated in 1898 into the Automobile Club of Great Britain and Ireland in London – an organization of which Redwood was a founding member. Redwood played a major role in this merger. The club was renamed the Royal Automobile Club (R.A.C.) in 1907 by King Edward VII (Grace's Guide 2016*a*).

Boverton Redwood was knighted on 30 June 1905; and on 24 July 1911, the Redwood Baronetcy, of the Avenue Road in St Marylebone, was created for Boverton Redwood (a title in the Baronetage of the UK). Redwood was also a Chevalier of the Order of Leopold (a Belgian national honorary order of knighthood established by King Leopold I) by the royal licence of Her Majesty Marie Henriette of Austria (1836–1902). Redwood resembled the British actor Sir Henry Irving (Fig. 6) and it gave him pleasure when he was mistaken for the actor.

In August 1911, the Kelham well drilled near Newark, Nottingham struck oil at a depth of 2540 ft. It yielded several barrels of oil, and the oil

still operates at 61 New Cavendish Street in London. The institute published the *Journal of the Institute of Petroleum Technologists* until 1973 (website: www.petroleum.co.uk).

Fig. 6. A portrait of British actor Sir Henry Irving (1835–1905) (photograph taken by Herbert Rose Barradu in the late 1880s (source: Wikipedia)). Corley (1983, p. 39) writes that Redwood's 'imposing stature and profile often led him to be mistaken for the actor Sir Henry Irving, something he was vain enough to encourage'.

samples were examined by Redwood. The initial results were so encouraging that Charles E. Best of London obtained a government permit and in 1918 established The Oilfields of England. The company continued drilling until 1922 when the well was abandoned as a dry hole (Huxley 1983, p. 30).

World War I motivated the British to explore for oil in their own soil. A pioneer of this venture was Lord Cowdray (formerly Weetman Pearson). The Petroleum Act of 1918, while leaving the question of final ownership of oil in Britain open, gave the government the necessary power to control the oil production procedure by the issuing of licences. Cowdray through his company S. Pearson and Son obtained a £1 million grant from the government to drill for oil at Hardstoft near Chesterfield, Derbyshire. The drilling began on 15 October 1918, and among the attendees in the opening ceremony was Redwood (Jones 1981, p. 203).

Sir Boverton Redwood died on 4 June 1919 at his residence, The Cloisters, Avenue Road, Regent's Park, London. He was a hardworking scientist and engineer. Only a few days before his death, he was examining the results and samples of the first successful oil well in England – the Hardstoft No. 1 in the parish of Tibshelf in the county of Derbyshire, which was struck on 17 May 1919 (see Craig *et al.* 2013 for a detailed account of this well).

Redwood's obituaries were published in the *Journal of the Institution of Petroleum Technology* (July 1919), *Nature* (12 June 1919, pp. 287–288), *Institute of Metals: Proceedings* (1919, p. 426), *Quarterly Journal of the Geological Society* (vol. 55, 1919), *The Times* (5 June 1919), *Oil, Paint and Drug Reporter* (9 June 1919, vol. 95, p. 26), *Proceedings of the Royal Society of Edinburgh* (January 1921, vol. 40, pp. 191–195) and *Journal of the Chemical Society, Transactions of the Royal Society of Chemistry* (F. Mollwo Perkin, 1920, vol. 117, pp. 467–470).

According to the *Calendars of the Grants of Probate and Letters of Administration England and Wales*, 15 August 1919, Sir Redwood's wealth at death was £16 501 314s.5d. (Corley 2004). (Note that the *CGPLA England and Ireland* was an index of probate documents such as wills published from 1858 to 1920.) Sir Redwood's wealth would be about US $448 000 today (using the Historical Currency Conversions at https://futureboy.us/fsp/dollar.fsp).

Sir Boverton and his wife Mary Elizabeth had two daughters and one son, Bernard Boverton Redwood[7] (Fig. 7), who died in 1911 leaving a son, Sir Thomas Boverton Redwood (1906–74, second Baronet). Mary Elizabeth died in 1937, the same year that Peter Boverton Redwood (third Baronet) was born. The third Baronet served as a Colonel in the King's Own Scottish Borders.

[7]Bernard Boverton Redwood, born on 28 November 1874 at Finchley, in Middlesex, was the only son of Sir Thomas Boverton Redwood. He was educated at Bath and Peterhouse, Cambridge, where he took a B.A. degree in Natural Science, and obtained a Half Blue for cycle racing, being captain of the Cambridge team. He travelled in America, Russia, Turkey, Algeria and the West Indies in connection with his father's professional duties. From an early age, Bernard Redwood developed an interest in mechanics. He was the owner of one of the first motor tricycles in England, manufactured by the French company De Dion. He was also a British motorboat racer who competed in the 1908 Summer Olympics, and as crew member of the *Gyrinus* he won two gold medals in the only motor boat competitions at the Olympics. He drove various kinds of cars and drove up to 10 000 miles per year. Yachting and marine motoring were his other hobbies. He died on 28 September 1911 (source: Grace's Guide 2016*b*).

Fig. 7. Bernard Redwood (1874–1911), the only son of Sir Boverton Redwood (source: Grace's Guide to British Industrial History (Grace's Guide 2016*b*)).

His works

Redwood's published works may be categorized into (1) book volumes and (2) articles in journals and conference proceedings. In addition, he wrote forewords to a number of books including Anon (1910), Cunningham-Craig (1912), Campbell (1918) and Greene (1919). In this section, we review his book publications. Table 1 lists his published articles.

- *Cantor Lectures on Petroleum and Its Products* by Boverton Redwood (Redwood 1886*a*) ('F.I.C., F.C.S., member of the Committee of the London Section of the Society of Chemical Industry, member of the American Chemical Society, corresponding member of the Imperial Russian Technical Society/Baku Division'), published by the Society for the Encouragement of Arts, Manufactures and Commerce, London, 1886, 85 pp.; price 'Half-a-Crown'; published in the USA as *Petroleum: Its Production and Uses* (Redwood 1886*b*), Van Nostrand, New York. Redwood (1926, vol. 3, p. 1273, Bibliography) comments that the US print in the same year was 'published without authority from the author'. The book of *Cantor Lectures* (Fig. 8) were based on Redwood's lectures 'delivered before the Society of Arts on 8, 15, 22, 29 March 1886'; they were first published in the *Journal of the Society of Arts*, volume 34 (1886), pp. 817–826 (No. 1752, 18 June 1886), 831–846 (No. 1753, 25 June 1886), 867–878 (No. 1755, 9 July 1886), 899–913 (No. 1757, 23 July 1886), 915–918 (No. 1758, 30 July 1886). The Lectures (chapters) include: I. Historical account of petroleum; II. Production, transportation and storage; III. Manufacture of petroleum products; IV. The uses of petroleum and its products.
- *The Detection and Measurement of Inflammable Gas and Vapour in the Air* by Frank Clowes, D.Sc., with a chapter on the 'Detection and Measurement of Petroleum Vapour', by Boverton Redwood, F.R.S.E., published by Lockwood & Son, London, 1896, xii + 206 pp., 1 plate and 49 diagrams (Clowes & Redwood 1896).
- *Petroleum: A Treatise* by Boverton Redwood ('F.R.S.E., F.I.S., Assoc. Inst. C.E., Vice-President of the Society of Chemical Industry, Fellow of the Chemical, Geological, and Royal Geographical Societies, Member of the American Chemical Society, Honorary corresponding member of the Imperial Russian Technical Society, Baku Division, Consulting advisor to the Corporation of London under the Petroleum Acts, Chemical advisor to the Oil Trade Section of the London Chamber of Commerce, etc.'), assisted by George T. Holloway, in two volumes, published by Charles Griffin and Co., London, and J.B. Lippincott, Philadelphia, 1896, 900 pp.; 2nd edition 1906, xxxi + 1064 pp.; 3rd edition 1913, 3 volumes, xxxii + 1167 pp.; 4th edition 1922, 3 volumes, xxx + 1353 pp.; 1926 fifth edition, 1353 pp., reprint of the 4th edition, hardcover.

Petroleum: A Treatise was probably the first book that dealt with various aspects of the petroleum industry in a systematic manner; its subsequent editions are a testimony to its success in training generations of petroleum engineers at the turn of the twentieth century (Sorkhabi 2015).

The 1896 edition (Fig. 9) is dedicated to: 'the memory of my father Professor Theophilus Redwood, Ph.D. of Boverton, Glamorganshire, in whose laboratory I received my early instruction in practical chemistry'. Volume 1 includes sections 1–6 (xxv + 403 pp.); volume 2 includes sections 7–11 and appendices (pp. 404–900) (Section 1. General historical account of petroleum industry; 2. Geological and geographical distribution of petroleum and natural gas; 3. Chemical and physical properties of petroleum and natural gas; 4. Origin of petroleum and natural gas; 5. Production of petroleum, natural gas and ozokerite; 6. Refining of petroleum; 7. Shale oil and allied industries; 8. Transport, storage and distribution of petroleum; 9. Testing of crude petroleum, petroleum and shale-oil products, ozokerite and asphalt; 10. Uses of petroleum and its products; 11. Statutory, municipal and other regularities

Table 1. *A list of Sir Boverton Redwood's articles and reports*

Year	Author(s)	Title	Publication
1871	Redwood, B.	New readings of the Petroleum Acts	*Pharmaceutical Journal*, series 3, **1**, 1025–26
1871	Redwood, B.	The estimation of colour in liquids	*Pharmaceutical Journal*, series 3, **1**, 344
1872	Redwood, B.	*Evidence before the Select Committee of the House of Lords on the Petroleum Bill 1872 (135) IX*	House of Lords, London, privately printed
1875–76	Redwood, B.	Testing paraffin oil – explosion of lamps	*English Mechanics*, **22**, 355, 376, 458
1876	Redwood, B.	Description of Mr. R. P. Wilson's Chrono-thermometer, for use in testing mineral oils	*English Mechanics*, **22**, 496
1876	Redwood, B.	*The Secretary's Reports of his Mission to the United States and Canada*	London, privately printed
1879	Redwood, B.	Life in the more recently developed oil fields of the United States	*Pharmaceutical Journal and Transactions*, **38**, 753–755.
1883	Redwood, B.	*Report of a Special General Meeting of the Members of the Petroleum Association and those Engaged in the Trade, 8 August 1883 (189) IX*	House of Lords, London, privately printed
1883	Redwood, B.	*The Petroleum Production Company. Nobel Brothers Ltd., St. Petersburg*	London, privately printed
1883	Redwood, B.	*Evidence before the Select Committee of the House of Lords on the Petroleum Bill, 1882*	House of Lords, London, 32–45, privately printed
1883	Redwood, B.	*On the Present Position & Future Prospects of the Russian (Caucasian) Petroleum Industry*	Watelow & Sons, London
1884	Abel, F.A. & Redwood, B.	Notes on the employment of the Abel petroleum using apparatus in tropical climates	*Chemical News*, **49**, 196–197
1884	Redwood, B.	The determination of the percentage of oil in paraffin scale	*Journal of the Society of Chemical Industry*, **3**, 430–432
1885	Redwood, B.	The Russian petroleum industry	*Journal of the Society of Chemical Industry*, **4**, 70–81
1885	Redwood, B.	International Inventions Exhibition. Group XIV. apparatus, processes, and appliances connected with applied chemistry and physics	*Journal of the Society of Arts*, **33**, (1709), 966–969
1885	Redwood, B.	International Inventions Exhibition. Group XIV. Apparatus, processes, and appliances connected with applied chemistry and physics, Part II	*Journal of the Society of Arts*, **33**, (1710), 977–979
1886	Redwood, B.	Petroleum and its products (Cantor Lectures)	*Journal of the Society of Arts*, **34**, 805–814, 817–826 (I), 831–846 (II), 867–878 (III), 883–890, 899–913 (IV), 915–928 (IV)
1886	Redwood, B.	Ueber erdol und seine producte	*Dingler's Polytechnisches Journal*, **262**, 462–468, 531–542
1886	Abel, F.A. & Redwood, B.	On petroleum lamps and their danger	*Daily News*, 22 September 1886; *Polytechnic Journal*, **262**, 416–418

1886	Redwood, B.	On viscosimetry, or viscometry	*Journal of the Society of Chemical Industry*, **5**, 121–133.
1887	Redwood, B.	Notes of a recent visit to some of the petroleum producing territories of the United States and Canada	*Journal of the Society of Chemical Industry*, **6**, 405–412
1888	Redwood, B. & Salamon, A.G.	Institute of Chemistry	*Nature*, **37**, (956), 393
1889	Redwood, B.	*On the Domestic use of Petroleum Products*	The sixth lecture in the series of lecture on *Domestic Hygiene*, specially intended for ladies, delivered at the Parkers Museum, London. Announced in *The British Medical Journal*, **1**(1471), 542 (9 March 1889). Printed by Kenny & Co., London
1889	Abel, F.A. & Redwood, B.	*Suggestions in Reference to the Construction and Use of Mineral Oil Lamps*	London; privately printed leaflet
1890	Abel, F.A. & Redwood, B.	*Accidents with Mineral Oil Lamps*	Report to Home Office by Col. V.D. Majendie, London
1890	Redwood, B.	The oil-fields of India: Notes on the petroleum deposits of Burma, Assam, the Punjab and Baluchistan	*Journal of the Society of Chemical Industry*, **9**, 359–370, 441–443
1891	Redwood, B.	Comment on Ross' paper on the origin of petroleum	*Chemical News*, **64**, 215–216
1891	Redwood, B.	*Report to the Petroleum Defence Committee on the Home Memorandum of 25th March 1891, relative to the Inflammable Liquid Bill, 1891*	London, privately printed
1891	Redwood, B. & Topley, W.	Report on the Riacho-doce and Camarajibe shale deposits on the coast of Brazil, near Mexico	Corporate Publication, London. Summary is given in the *Transactions of the American Institute of Mining, Metallurgical and Petroleum Engineers*, 1921, **65**, 71–72
1892	Redwood, B.	The Galician petroleum and ozokerite industries	*Journal of the Society of Chemical Industry*, **11**, 93–119.
1892	Redwood, B.	Burning oils for lighthouses and lightships	*Journal of the Society of Arts*, **40**, (2049), 334–335
1892	Abel, F.A. & Redwood, B.	*Report on the Transport of Petroleum through the Suez Canal*	British Shipowners, London, privately printed
1893	Redwood, B.	The testing of crude petroleum and of petroleum products	Thorpe, T.E. (ed.) *Dictionary of Applied Chemistry, Volume 2*. Longmans, Green & Co., London, 1912–1913
1893	Redwood, B.	The Russian Petroleum Industry	Thorpe, T.E. (ed.) *Dictionary of Applied Chemistry, Volume 3*. Longmans, Green & Co., London, 134–169
1894	Redwood, B.	*Report on India Rubber Solution, Appended to Instructions to Principal and Other Detaining and Emigration Officers*	Board of Trade Marine Department, No. 1240, May 1894, privately printed
1894	Giles, A., Redwood, B. *et al.*	Discussion: The transport of petroleum in bulk	*Minutes of the Proceedings of the Institution of Civil Engineers*, **116**, (1894), 230–252

(*Continued*)

Table 1. *A list of Sir Boverton Redwood's articles and reports (Continued)*

Year	Author(s)	Title	Publication
1894	Redwood, B.	The transport of petroleum in bulk	*Minutes of the Proceedings of the Institution of Civil Engineers*, **116**, 177–259; Telford Premium by the British Institute of Civil Engineers
1895	Redwood, B.	Section IV. The Petroleum Industry and Lamps	Groves, C.E. & Thorp, W. (eds) *Chemical Technology, Volume 2, Lighting*. J. & A. Churchill, London, 97–242
1895	Redwood, B.	Section V. Lamps	Groves, C.E. & Thorp, W. (eds) *Chemical Technology, Volume 2, Lighting*. J. & A. Churchill, London, 243–334
1895	Redwood, B. & Louis, D.A.	Miners' safety lamps	Groves, C.E. & Thorp, W. (eds) *Chemical Technology, Volume 2, Lighting*. J. & A. Churchill, London, 335–386
1898	Redwood, B.	*Evidence before the Select Committee of the House of Commons on Petroleum, 1894–96, (311) XII*	House of Commons, London, 92–177, privately published
1898	Redwood, B.	The petroleum sources of the British Empire	*Coll. Guard*, **75**, 204, 205, 658
1899	Redwood, B.	Report on petroleum fields of Assam	*Petroleum Review*, **i**, 201, 202, 238, 239
1899	Redwood, B.	Report on the petroleum lands of Galicia	*Petroleum Review*, **i**, pp. 9.03, 204, 237, 238
1899	Burne, W.T., Bramwell, F. *et al.*	Report of Committee on Acetylene Generators	*Journal of the Society of Arts*, **47**, (2414), 289–291 (24 February 1899)
1900	Redwood, B.	Uniformity in methods of testing petroleum and its products (Unification des methodes d'essai du petrole et de ses produits)	Congress international du petrole, 1st International Petroleum Congress, 9–14, 31–33, France
1902	Redwood, B.	The acetylene light in omnibuses	Downes, L., Lewes, V.B. & Redwood, B. *The Lancet*, **159**, (4102), 993
1904	Redwood, B.	*Royal Committee Report on Coal Resources of the United Kingdom 1903–5*	Privately published
1904–05	Redwood, B.	*Evidence on Possible Substitutes for Coal for Production of Heat and Power: Alcohol, Use of Liquid and Solidified Petroleum on Ships and as Domestic Fuel*	Second Report of Royal Commission on Fuel Supply, 190–214; Summary in Final Report, 15–16, London, privately printed
1905	Redwood, B.	School of Pharmacy of the Pharmaceutical Society	*The Lancet*, **166**, (4284), 1022–1023
1905	Redwood, B.	Pharmaceutical Society of Great Britain: general study and specialism	*British Medical Journal*, **2**, (2336), 884

1907	Redwood, B.	Report on a sample of N'hangellite from Inhambane, Portuguese East Africa	*Bulletin of Miscellaneous Information, Royal Botanic Gardens, Kew*, **1907**, (5), 151–153
1908	Redwood, B.	Presidential Address	*Journal of the Society of Chemical Industry*, **27**, 779–799
1909	Redwood, B.	Questions and Answers	*Journal of the Royal Society of Arts*, **57**, (2936), 310
1910	Redwood, B.	Liquid fuel	*Comptes Rendus Congress International Chimie Appliquée* Session 7, Organisation, 61–102
1910	Redwood, B.	Natural gas	*Encyclopaedia Britannica*, 11th edn, vol. 19, 274
1910	Redwood, B.	Asphalt	*Encyclopaedia Britannica*, 11th edn, vol. 2, 768
1910–11	Redwood, B.	Petroleum	*Encyclopaedia Britannica*, 11th edn, vol. 21, 316–322; first published in 1902 in *The Times Supplement* to the *Encyclopaedia Britannica*
1911	Redwood, B.	*Present and Prospective Supplies of Petroleum and Shale Oil available for use as a Source of Power*	*Natural Sources of Energy*, British Science Guild, London, 7–14
1911	Redwood, B.	Turin International Exhibition, 1911: The British Section	*The Lancet*, **177**, (4563), 399–400
1912	Ramsay, W., Strutt-Rayleigh, R.J. *et al.*	*Natural Sources of Energy*	Report of the Committee of the British Science Guild, 1907. Reviewed in *Nature*, **89**, (2214), 117 (4 April 1912); and *Scientific American*: Supplement, **73**, (1898), 311–312 (18 May 1912)
1913	Redwood, B. & Lewes, V.B.	Petrol substitutes for use in motor vehicles	*Petroleum World*, **10**, 477–479
1913	Redwood, B.	Petroleum industry	*The Britannica Year-Book 1913*, Chishols, H. (ed.) *The Encyclopaedia Britannica*, London, 409–411
1913	Redwood, B.	The future of oil	*Journal of the Royal Society of Arts*, **62**, (3188), 103–116
1914	Redwood, B.	The future of oil fuel	*Journal of the Junior Institute of Engineers*, **24**, 190–224
1914	Redwood, B.	Presidential Address	*Journal of the Institution of Petroleum Technologists*, **1**, 9–12
1916	Redwood, B.	Valedictory Address	*Journal of the Institution of Petroleum Technologists*, **2**, 192–195
1918	Redwood, B.	The romance of petroleum	*Journal of the Royal Institute*
1919	Redwood, B.	*Report of the Inter-Departmental Committee on Gas Traction*	His Majesty's Stationery Office (HMSO), London; privately published
1919	Redwood, B.	*Report of the Inter-Departmental Committee on Alcohol Motorfuel (Great Britain)*	His Majesty's Stationery Office (HMSO), London; privately published

Partly based on '*List of Works of Sir Boverton Redwood*', in the *Journal of the Institution of Petroleum Technologists*, **5**, 246–248 (1918–19).

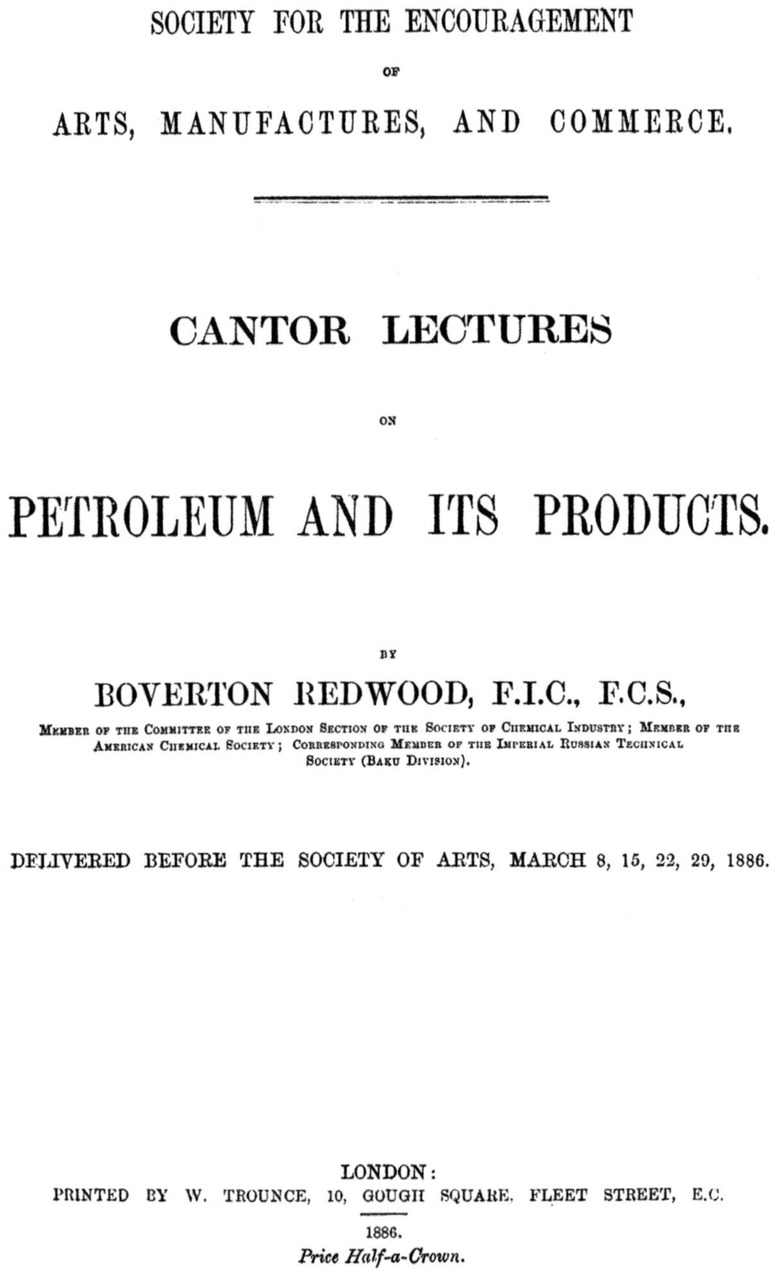

SOCIETY FOR THE ENCOURAGEMENT

OF

ARTS, MANUFACTURES, AND COMMERCE.

CANTOR LECTURES

ON

PETROLEUM AND ITS PRODUCTS.

BY

BOVERTON REDWOOD, F.I.C., F.C.S.,

MEMBER OF THE COMMITTEE OF THE LONDON SECTION OF THE SOCIETY OF CHEMICAL INDUSTRY; MEMBER OF THE AMERICAN CHEMICAL SOCIETY; CORRESPONDING MEMBER OF THE IMPERIAL RUSSIAN TECHNICAL SOCIETY (BAKU DIVISION).

DELIVERED BEFORE THE SOCIETY OF ARTS, MARCH 8, 15, 22, 29, 1886.

LONDON:
PRINTED BY W. TROUNCE, 10, GOUGH SQUARE, FLEET STREET, E.C.

1886.
Price Half-a-Crown.

Fig. 8. Title page of *Cantor Lectures on Petroleum and Its Products* – Boverton Redwood's first book.

relating to the testing, storage, transport and use of petroleum and its products; Appendix A. Statistics; Appendix B. Marine transport of petroleum; Appendix C. Import duties levied on petroleum; Index).

In the second edition (1906), the author is named 'Sir Boverton Redwood, D.Sc., Adviser on petroleum to the Admiralty and the Home Office, etc.'. The book also includes a bibliography complied by W. H. Dalton. The third (1913) and fourth (1922) editions are in three volumes, expanding on the content but basically keeping the same table of contents as in the first edition. Volume I includes sections 1–4, pp. XXX. + 364;

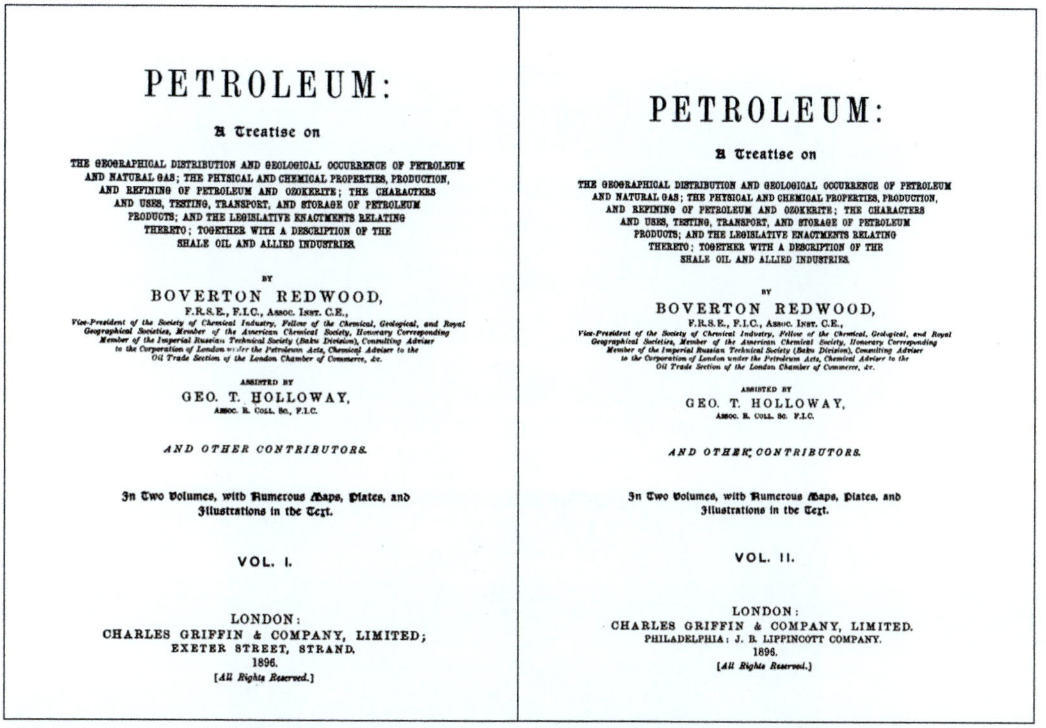

Fig. 9. Cover pages of the first edition of *Petroleum: A Treatise* (in two volumes), Boverton Redwood's most celebrated work which trained generations of petroleum scientists and engineers.

volume II includes sections 5–8, pp. 365–740; and volume III includes sections 9–11 plus appendixes, pp. 741–1353. The fourth edition is priced £5, 5s. It was published posthumously; therefore, the preface is co-signed by Arthur W. Eastlake and Robert Redwood, and a foreword by Sir Frederick W. Black (Past President of the Institution of Petroleum Technologists) includes 'historical notes on the life of the late Boverton Redwood, Bart'. This massive work includes 165 tables, 31 plates and 359 figures. The 5th edition (1926) (Fig. 10), in three volumes, has 1353 pages, and is essentially a reprint of the fourth edition. The book is subtitled, '*A Treatise on the Geographical Distribution and Geological Occurrence of Petroleum and Natural Gas; the Physical and Chemical Properties, Production, and Refining of Petroleum and Ozokerite; the Characters and Uses, Testing, Transport, and Storage of Petroleum Products, and the Legislative Enactments Relating Thereto; Together with a Description of the Shale-oil and Allied Industries; and a Bibliography*'. The Bibliography section in the third edition (volume III) 'compiled by W. H. Dalton, with assistance of L. V. Dalton, S. L. James, and E. S. Ward' is reproduced in the fourth and fifth editions, and includes 8804 references.

- *The Petroleum Lamp: Its Choice and Use* (*A Guide to the Safe Employment of Mineral Oil in What is Commonly Termed the Paraffin Lamp*) by Jocelyn Home Thomson and Boverton Redwood (Thomson & Redwood 1902), published by Charles Griffin & Co., London, 104 pp., 1902; second edition 1906, third edition 1913. A history of the development of oil-burning lamps, their construction, choice, care and use.
- *Handbook on Petroleum for Inspectors under the Petroleum Acts* (*and for Those Engaged in the Storage, Transport, Distribution, and Industrial Use of Petroleum and its Products and Calcium Carbide, With Suggestions on the Construction and Use of Mineral Oil Lamps*) by Jocelyn Home Thomson, HM Chief Inspector of Explosives, and Sir Boverton Redwood, Bart., published by Charles Griffin & Co., London, 1901, xix + 298 pp.; second edition 1906 xix + 324 pp.; third edition 1913 revised by Major Aston Cooper-Key and Sir Boverton Redwood, xix + 340 pp. hardcover (Thomson & Redwood 1913). Contents: Introductory; Sources of supply; Production,

Fig. 10. The fifth edition (1926) of *Petroleum: A Treatise* in three volumes. It is essentially a reprint of the 1922 edition published posthumously.

refining, etc.; Commercial products; Flash point and fire test; Testing; Legislation; Precautions; Oil lamps; Calcium carbide; Appendices; Index.

- *Petroleum Technologist's Pocket-Book*, by Sir Boverton Redwood, Bart., D.Sc., etc., and Arthur Eastlake, M.I.M.E., A.M.I.Mech.E., etc. with a Foreword by Sir Thomas H. Holland, published by Charles Griffin & Co., London, 1915, 454 pp.; second edition 1923, xxiv + 546 pp., hardcover (Redwood & Eastlake 1923). Contents: Part I. General information about petroleum (origin, occurrence, prospecting, acquiring land, raising, storage, refining, etc.). Part II. Geological (identification of rocks, angle of dip, maps, oil-bearing areas, oil per acre, bitumens, etc.). Part III. Physical and chemical (specific gravity, analysis of gas, viscometry, calorific value, flash point, distillation, candle power, etc.). Part IV. Production (drilling, casing, water, raising oil, plugging, cost of drilling, etc.). Part V. Refining, transport, storage and testing (tanks, flow of gas in pipes, pipe lines, pumping, railway cars, barrels, etc.). Part VI. Uses (liquid fuel, air required, oil engines, natural gas, asphalt, etc.). Part VII. Weights and measures (English and foreign). Part VIII. Miscellaneous. Part IX. Statistics (of production, asphalt, oil shale, ozokerite, natural gas).

Burmah Oil Company

The history of the Burmah Oil Company is the subject of a two-volume work by Thomas Anthony Buchanan Corley (1983, 1988). Burmah Oil Company was an important Scottish enterprise founded in 1886 by David Sime Cargill (1826–1904) in Glasgow, while another Glaswegian, Kirkman Finlay (1847–1903), was his right-hand man. Both Cargill and Finlay had been involved in businesses in Burma for years. Cargill had worked for the Glasgow-based Rangoon Oil Company, founded in 1871 by James Galbraith (1818–85). In 1876, Cargill acquired the Rangoon Oil Company. Finlay was co-founder (with Matthew T. Fleming) of Finlay Fleming & Co., which acted as Burmah Oil's agent in Rangoon until 1928. The overthrow of King Thibaw of Burma following the Third Anglo-Burmese War of 1885 paved the way for total British control over Burmah, and the establishment of Burmah Oil a year later. The granting of oilfield concessions to the company was a sign of this colonial takeover. D.S. Cargill remained Burmah Oil's chairman until his death in 1904; he was succeeded by his son Sir John T. Cargill as the chairman from 1904 to 1943. Kirkham Finlay served as Secretary and then Director of Burmah Oil until his death in 1903; his position was then given to his son Sir Campbell Kirkman Finlay, who remained Director until his death in 1937. Burmah Oil operated in Burma until 1963, when the Burmese ruler Ne Win nationalized the country's oil industry, as a result of which the Myanmar Oil and Gas Enterprise was formed. In 1966, the Scottish branch of Burmah Oil acquired Castrol (a lubricant company founded in 1899 by Charles Wakefield) and was renamed Burmah-Castrol, which was eventually purchased by BP in 2000, although the brand name Castrol has been retained.

In this section, we review Redwood's advisory role in the activities of Burmah Oil based on the information given in Corley (1983). Redwood:

> [C]hose his friends carefully; Kirkman Finlay was one of them, and it was he who persuaded Redwood to become part-time consultant to the company, a post he held from 1893 until his death in 1919
>
> (Corley 1983, p. 39).

although Redwood never visited Burma. Corley continues:

> The company was fortunate to acquire a refining patent from Dr. Boverton Redwood and (Sir) James Dewar, which allowed more of the lighter fractions to be extracted and hence markedly increased kerosene yields from 45% in 1889 to 57% in 1897.

This is emphasized a few pages later in Corley's book:

> Kerosene continued to be the most important of the company's products at a time when technical improvements, notably the Redwood–Dewar process, were satisfactorily stepping up yields of the Victoria process
>
> (Corley 1983, p. 49).

Corley (1983, pp. 51–52) gives an overview of the oil market in 1897 through Redwood's eyes. Redwood's forecast was as oil consumption increases every year globally, American oil would increasingly be devoted to domestic demand, and Russian production would be gradually absorbed by domestic development or the neighbouring countries. Since the USA and Russia accounted then for 95% of the world's oil production, Redwood 'assumed that all the prospecting in progress would discover new fields in widely scattered parts of the globe to redress the production imbalance'.

In providing this picture of the world oil to Finlay, Redwood was also persuading him to buy up oil interests in the Indian province of Assam. Back in 1890, Redwood had published a paper on the oilfields of India, for which Finlay had provided data (Redwood 1890). Now, Finlay's response to Redwood revealed 'how far he [Finlay] already dominated company policy'. David Cargill, as chairman, was enthusiastic about going into Assam, only to be overridden by Finlay who believed in concentrating rather than dispersing the company's efforts. As he explained to Redwood, 'we had already quite as much in our hands as would occupy us for the rest of our lifetime' (Corley 1983, pp. 51–52).

Corley (1983, p. 54) mentions the 'persistent rumours of a possible merger between Standard Oil and Shell', in the 1890s, which worried many oilmen in Britain including Finlay:

> Redwood tried to calm his agitation by explaining that to the best of his knowledge relations between the presumed suitors 'only amounted to a sort of mutual offensive-defensive understanding of playing to each other's hands', whatever that meant.

In 1896, the Minbu Oil Company which held a number of blocks in Yenangyat in Burma was founded. The main promoters were also partners in Ogilvy Gillanders & Co., owned by Henry Neville Gladstone (son of the former prime minister of Britain) and his two distant Gladstone relatives. In 1898, having discovered oil in Yenangyat, the Minbu Oil

Company decided to expand their business in Burma; Redwood arranged a meeting between the Gillanders and Burmah Oil, represented by Finlay. Burmah Oil eventually bought their concessions and plant for £30 000. Finlay, 'assigned the major credit to Redwood, the company gladly paying the £3000 fee due to him from the Minbu Oil Company': 'The price', in Finlay's appreciative words, 'due to the prince of oil diplomatists' (Corley 1983, p. 57).

In 1901, Burmah Oil became worried about intense water seepage at their oilfields. One solution was to cement the wells and then drill them when necessary. Redwood suggested the company seek the expertise of Thomas Hayes, who had worked in the Californian fields. Hayes, however, proved to be incapable of doing the job and was described as a fraud by Finlay in Burma: 'Redwood took full responsibility and agreed to give Hayes his *congé*. Yet Redwood at 65 was not the man he had been in his prime' (Corley 1983, p. 175).

In 1902, Standard Oil in New York decided to start oil exploration in Burma. The company tried to secure concessions through its British-registered Anglo-American Oil Company, but the effort was not successful. The company decided to send over:

> [I]ts most powerful troubleshooter, William Libby … When Libby passed through London that June on his way to India, he called on Redwood, who learnt that he intended to visit Burma as well and therefore advised Finlay to offer a letter of introduction addressed to Finlay Fleming & Co. At least it would mean that Libby would be conducted round the fields by company employees rather than sneaking round
>
> (Corley 1983, p. 66).

Libby did not strike any concession in Burma.

In 1902, the Oil Fuel Committee, comprising E.G. Pretyman (the Chairman; Parliamentary Secretary), Gordon Miller (Director of Navy Contracts) and Boverton Redwood, was set up to examine the availability of oil supplies to Great Britain:

> In the early months of 1903, Miller and Redwood gave important evidence before the Royal Commission on Coal Supplies, set up as a result of fears, later found to be much exaggerated, of Britain's coal reserves running out … Whether as a result of Redwood's evidence to the Royal Commission or of his prodding on the Oil Fuel Committee, in May 1903 a high official of the Admiralty privately requested from the government of Burma precise information on oil supplies there
>
> (Corley 1983, pp. 80–81).

(The Oil Fuel Committee was disbanded in 1906 after a change in the government of the UK.)

When John Cargill took over his father's position in 1904:

> [T]he burden on the 36-year-old John Cargill of carrying the whole company and the related trading activities must have been formidable. Almost painfully diffident, he later admitted that at times in that period he had felt overwhelmed by his responsibilities, but drew strength above all from the help and encouragement of Boverton Redwood
>
> (Corley 1983, p. 79).

In 1907, Basil Macrorie, Burmah Oil's resident geologist, wrote a letter to Redwood:

> [C]omplaining that officials in Rangoon, who bore none of the exploration costs, knew little about the needs of the industry and had scant regard for its future. In particular, they laid down no regulations about the space between each well
>
> (Corley 1983, p. 121).

Macrorie suggested well spacing of at least 600 ft, while some drillers dug their wells some 60 ft apart. Redwood showed the letter to John Cargill who seconded its content. As a result, Redwood wrote a memorandum, 'pointing out that the arrival in the fields of new companies, with their disruptive policies for drilling, threatened Burmah Oil's ability to honour the fuel oil contract'. He sent the memorandum to the Admiralty: 'Their Lordships noted Redwood's points and asked the Indian authorities to arrange to regulate drilling for oil on a scientific basis' (Corley 1983, p. 122).

In 1908, Redwood advised the directors of Burmah Oil to use electric power in their field operations in Burma as it had been done successfully in the USA, Russia and Persia. Running separate steam boilers to power the drilling and pumping engines burned up 10% or even more of the produced oil from the field, while electrification provided safer, cleaner and more convenient work (Corley 1983, p. 233).

In 1915, Sir Robert Waley Cohen of the Royal Dutch-Shell Group submitted a proposal to John Cargill and R.I. Watson of Burmah Oil involving a merger between the two companies. His argument was that the merger, 'would secure British control over the whole group, then 60% Dutch-owned and 40% British'. In 1916, Ernest George Pretyman, then Parliamentary Secretary at the Board of Trade and a close friend of Redwood, called a conference of representatives from Parliament, the Admiralty and the India Office, 'with the experienced Redwood in attendance'. The group advised ministers to turn down Cohen's proposal (Corley 1983, p. 248).

Persia

Persia (Iran) was the first country in the Middle East where the modern oil industry began in the early twentieth century. Although bitumen, oil and natural gas were produced from seeps and hand-dug wells for millennia in Iran (Persia) (Sorkhabi 2005), the 1908 discovery of the Masjid Suleiman Field in

SW Iran by a British enterprise led by the businessman Knox D'Arcy and the oil explorer George Bernard Reynolds (Sorkhabi 2010) was a turning point in the history of oil industry in the Middle East. This venture led to the formation of the British-dominated Anglo-Persian Oil Company which not only helped Britain win World War I but eventually led to the formation of British Petroleum (BP) and the National Iranian Oil Company (NIOC) in the 1950s (Bamberg 1994, 2000). In this section, we review Sir Boverton Redwood's influential role in the Persian oil venture in the 1900s, largely drawing on the information in Owen (1975),[8] Jones (1981),[9] Ferrier (1982) and Corley (1983).

On 28 May 1901, William Knox D'Arcy (1849–1917), who had made a great wealth from gold mining in Australia, acquired an oil concession from the government of Iran (then ruled by Mozaffar al-Din Shah Qajar). Right from the beginning, Redwood worked as a consultant for D'Arcy. In 1900, at the Paris Exposition, Antoine Kitabchi Khan, Iran's representative at the Exposition, approached European businessmen for oil exploration in Iran. He talked with Monsieur Edouard Cotte, a European business agent, and Sir Henry Drummond Wolff (1830–1908), who was recovering from a mental breakdown that had ended his career as British Minister in Tehran a few years earlier. Wolff introduced Kitabchi Khan to D'Arcy. Ferrier (1982, p. 32) notes:

> In early January 1901 Cotte visited D'Arcy in London and persuaded him to return to Paris with him and Wolff to meet Kitabgi on 8 January for the first time. Over the next two months ten more meetings were held in Paris and London often with D'Arcy's solicitors, Nicholl Manistry & Company and his geological advisor, Dr. (later Sir) Boverton Redwood.

In March 1901, D'Arcy sent his agent Alfred L. Marriott, together with Cotte and Kitabchi Khan, to Tehran for talks with the Persian government. In a separate move, on the advice of Redwood, a geologist named H.T. Burls was dispatched to Iran to map Chiah Surkh, an area in western Iran where abundant oil seeps had been reported and also to examine Ahwaz in SW Iran. Burls wrote a report in favour of Chiah Surkh (Owen 1975, p. 1254; Ferrier 1982). On 28 May 1901, the D'Arcy concession granted by the Persian government was signed. D'Arcy owed this opportunity to Kitabgi and Wolff 'and he depended on upon Dr. Redwood for technical advice, as the most distinguished oil technologist of his time' (Ferrier 1982, p. 48).

Corley (1983) reports that soon after winning the oil concession in Persia in 1901, D'Arcy planned to do the same in Mesopotamia, which was then part of the Ottoman Empire. But first he wanted to know if oil could be found there. He used the services of Redwood's consulting firm. Redwood arranged for his geologist T.H. Burls to visit Mesopotamia, who came up with a positive report citing the importance of oil seeps in the region: 'That survey encouraged D'Arcy to send H.E. Nichols, a manager in his gold mining company, out to Constantinople in 1903 to negotiate' (Corley 1983, p. 200). However, in contrast to the Persian concession, Nichols found negotiations too hard with the Ottoman sultan.

For executing his oil concession in Persia, D'Arcy hired George Bernard Reynolds, an English engineer who had worked in the Indian railways and in the oilfields of Sumatra (Sorkhabi 2010). Reynolds was suggested by Redwood (Jones 1981, p. 132; Corley 1983, p. 96). Reynolds reached Tehran in September 1901. Kitabchi Khan wanted Reynolds to explore the Shushtar area in SW Iran but Reynolds, on the basis of Redwood's recommendation, visited Chiah Surkh – on the Zuhab river in the Qasr-i Shirin district of western Iran (bordering Iraq). In November 1902, drilling began at Chiah Surkh. Meanwhile, D'Arcy sent another geologist from Redwood's firm, W.H. Dalton, to inspect the area of Shushtar (in the winter of 1902–03), possibly to show Kitabchi Khan that his proposal was not ignored: 'Redwood was not convinced of the need to go elsewhere in Persia' (Ferrier 1982, p. 58). In the summer of 1903, gas and oil shows were encountered at Chiah Surkh and, as the concession agreement demanded, D'Arcy founded the First Exploration Company for the venture in May 1903 (with D'Arcy as chairman, two directors and Redwood as technical adviser: Corley 1983, p. 97). During this time, Reynolds had to go London as his first contract with D'Arcy had expired. In his absence, C.B. Rosenplaenter, an engineer with experience in the oilfields of Assam and Baku and

[8]Note that Owen (1975) includes many valuable notes on Persia written by Peter T. Cox, a native of New Zealand, who worked for the Anglo-Persian Oil Company, in the Masjid Suleiman Field, from 1928 to 1936, and then again from 1946 to 1952, when he moved to London as exploration manager of D'Arcy Exploration Company. Later, he worked as managing director of the British Exploration Company and director of the Kuwait Oil Company until his retirement in 1962. His personal communication with Owen is dated 1965–66 (Owen 1975, pp. 1253–1254).

[9]Jones (1981) reports that many of the original documents, including H.T. Burls' geological report to Boverton Redwood on Persian oilfields (13, 20 and 27 July 1901), report of E.H. Cunningham-Craig (2 January 1908) and a report of Boverton Redwood (26 February 1909) were printed in the First Interim Report for the Admiralty Commission on Petroleum Resources of the Countries Adjoining the Persian Gulf, 28 November 1913.

recommended by Redwood, was in charge (Ferrier 1982, p. 62). In January 1904, a second well at Chiah Surkh hit oil, but the excitement was short lived as the oil flow declined rapidly and brine flow increased. Redwood suggested that D'Arcy had the wells at Chiah Surkh shut down (Corley 1983, p. 100).

D'Arcy, now running out of money, was considering selling his concession even to non-British enterprises. In October 1904, Sir John Fisher became First Lord of the Admiralty and was determined to transform the Royal Navy from coal to oil fuelled. This meant access to oil and keeping the D'Arcy concession in British hands. D'Arcy applied for a loan from the Admiralty – a suggestion made to him by Redwood (Yergin 2008, p. 123). However, Burmah Oil was also approached to join the Persian venture. John Cargill had confided to Redwood in October 1904 that Burmah Oil's exploration in Burma had not been successful for years (Ferrier 1982, p. 69). Redwood recommended that Burmah Oil join hands with D'Arcy (Ferrier 1982, p. 70; Corley 1983, p. 102) (although at the 5th annual meeting of the Burmah Oil General in 1907, Redwood still expressed high hopes of oil discoveries in Burma: Ferrier 1982, p. 83), and also suggested including Lord Strathcona, a retired wealthy Englishman, then 84 years old (Corley 1983, p. 101). A new company, the Concessions Syndicate Ltd was formed in 1905 in Glasgow to take over D'Arcy's concession in Persia with financial support from Burmah Oil and Lord Strathcona. D'Arcy, of course, remained the director of the new company.

Jones (1981, p. 133) elaborates on Redwood's role in the establishment of the Concessions Syndicate:

> After persuading D'Arcy to give the British Admiralty the first chance of saving his Persian oil company, Redwood then put the matter of Persian oil before the Burmah Oil directors. He subsequently brought the three parties together by arranging a meeting between J.T. Cargill, C.M. Wallace, D'Arcy and Pretyman at the Admiralty.

In this meeting, Redwood argued that it was important for Britain to exert influence in Persia as it was located close to India; moreover, the Persian oilfields would prove to be very productive.

The Concessions Syndicate ceased the operation at Chiah Surkh and moved the drilling to the southern province of Khuzestan, where W.H. Dalton in 1903 had particularly selected a prospect at Shardin (Marmatain), about 90 km east of Ahwaz city. Based on Dalton's report, Reynolds was instructed to drill two wells at Shardin. The first well was drilled in 1906 and the second in 1907: both were dry. It seems that Reynolds was not enthusiastic at all about drilling at Shardin; nevertheless, he was under pressure from the Burmah Oil's management in Glasgow as to 'When do you expect to strike oil?'. He had written a letter to Redwood anticipating moral support from him, to which Redwood replied that the Glasgow directors were 'fully entitled to more information than they had hitherto obtained' and added that 'I cannot do for the Company any better than I am doing now and have been doing' (Ferrier 1982, p. 84; Corley 1983, p. 130).

Reynolds decided to cease operations at Shardin and move the drilling to his most favourite place, Masjid Suleiman, located about 90 km NE of Ahwaz. Reynolds had visited this area, rich with oil seeps, in December 1906, and had described it in a letter (with no maps) to London. Redwood 'hailed it as the most important find in Persia to date' (Corley 1983, p. 129). Reynold's decision to move from Shardin to Masjid Suleiman was supported by Edward Hubert Cunningham-Craig, a geologist with Burmah Oil, who visited Reynolds in November 1907 and made a survey of the Masjid Suleiman area. (Owen 1975, p. 1257, remarks that Cunningham-Craig was actually sent by Redwood to Persia.) Redwood and Dalton disagreed with Reynolds, but to no avail. In January 1908, Reynolds began drilling at Masjid Suleiman. After drilling began, and possibly being better informed by Cunningham-Craig, there seems to have been a change of heart on the part of Redwood, for in mid-1908 both Cunningham-Craig and Redwood assured Reynolds that oil would be struck at a moderate depth at Masjid Suleiman (Ferrier 1982, p. 87). On 16 May 1908, the well hit a gusher that changed the Middle East forever. Well No. 2 in June and well No. 3 in September that year were even bigger gushers. For future drills in Persia, Redwood recommended that the wells should be at least 500 ft apart (Corley 1983, p. 181).

Late in 1908, Basil F.N. Macrorie, on the recommendation of Redwood, was dispatched to Persia to conduct a survey of Masjid Suleiman; his report was to be included in the company's prospectus for potential investors (Ferrier 1982, p. 96). The Persian venture was successful at higher costs than expected:

> Redwood had initially forecast an expenditure of £10 000 to drill two wells in Persia, but in four years D'Arcy had spent some £200 000 all told and in three years the Concession Syndicate has spent some £130 000
>
> (Ferrier 1982, p. 107).

Corley (1983, p. 104) mentions that as soon as the Concessions Syndicate formed, Reynolds was summoned to England to discuss with Redwood how much money would be needed for the oil prospecting in Persia. Reynolds estimated £40 000 at most and the company increased it to £70 000 as a safeguard.

In April 1909, a new company, the Anglo-Persian Oil Company (APOC), replaced the Concessions Syndicate Ltd. Charles Greenway (1857–1934) was appointed Director of the APOC. Apparently, the Burmah Oil managers had concerns about the future of the new company as it was becoming apparent to them that the company's operations would focus on providing cheap oil to the Royal Navy. Even back in November 1908, Redwood, on behalf of Burmah Oil, had asked Pretyman of the Admiralty how much the new company would receive from the Admiralty for its fuel oil (Corley 1983, p. 140). Ferrier (1982, p. 106) reports that Redwood acted on the behalf of Burmah Oil and confronted Pretyman stating that Burmah Oil did not consider itself indebted to the Admiralty for putting them into the Persian venture.

In 1912, Redwood was asked by the APOC to identify fuel markets for the Persian oil, and his suggestions included both the Admiralty and the Indian railways (Ferrier 1982, pp. 194, 168). How exactly the Admiralty would buy oil from the APOC was an issue for negotiation. One suggestion was to buy refined oil at Abadan, where a refinery was being built in 1912, rather than transporting the crude. Redwood wished to meet with the First Lord of the Admiralty, Winston Churchill, to seek his support for the APOC case, and especially for the purchase of refined oil at Abadan. Unable to meet Churchill, however, Redwood talked to Frederick Black, Director of Navy Contractors. Black had initially been sceptical of APOC–Admiralty co-operation, but was now a firm supporter; he assured Redwood that Prime Minister Herbert Henry Asquith's cabinet was still considering the proposal. In November 1912, we find Redwood as a member of the Oil Fuel Committee (along with Sir Thomas Holland, Head of the Geological Survey of India, and Frederick Black) discussing with Greenway the advantages of oil arrangements between the APOC and the newly established Royal Commission on Oil Supply, and the dangers of foreign control of the Persian oilfields (Ferrier 1982, p. 171). The Royal Commission on Fuel and Engines was established by Winston Churchill in July 1912; Fisher was its chairman and the members included George Lambert, Civil Lord of the Admiralty, Admiralty naval and civilian officials, as well as Sir Boverton Redwood and Sir Thomas Holland (Corley 1983, pp. 186–187).

The APOC was not certain about a contract from the Admiralty to buy its oil. In January 1913, however, Greenway was informed by Redwood to 'sit tight' for the contract because, 'the Admiralty are beginning to realise that it would be unsafe for them to rely on Mexico as a source of supply' (Ferrier 1982, p. 177). In order to secure an oil supply for the Royal Navy at a low price, Asquith's government decided in April 1914 to buy 51% of the APOC and appointed two government directors (with veto power) on its board.

During World War I, fuel oil from the Persian fields proved to be critical for the Royal Navy; however, the British Government was not paying the full amount: 'By mid-1917, the outstanding calls due from the Government still amounted to £949 000' (Ferrier 1982, p. 220). In September 1917, at a meeting of the Inter-Departmental Committee on Petroleum Products (which had been founded as a war-time measure in February that year), Admiral Sir Edmond Slade, Vice-Chairman of APOC, appealed for financial support from the Admiralty to expand the APOC's operations in Persia. The meeting was presided over by Pretyman and also included Redwood as an advisor (Ferrier 1982, p. 221).

Soon after oil production began in Persia, Burmah Oil realized that the properties of Persian and Burmese oils were somewhat different. In 1909, Redwood analysed the properties of Persian oil samples, but his observations were limited to specific gravity, sulphur content, etc. In 1915, the Admiralty, the APOC and Redwood were confronted with the viscosity problem of fuel oil from Persia. Redwood sought the help of A.E. Dunstan, Head of the Chemical Department at the East Ham Technical College. Greenway hired Dunstan, who, together with his colleague Dr F.B. Thole, set up the APOC's research laboratory in 1917 to study the physical and chemical properties of Persian oil and how to refine it better (Ferrier 1982, p. 277).

The APOC wanted to discover more oilfields in Persia. At the end of 1912, Redwood sent S. Lister James, a geologist trained by Dalton and Cunningham-Craig, to Persia. Lister James selected drill sites at Ahwaz. He was joined in early 1913 by H.T. Mayo, a Burmah geologist and in 1914 by H.G. Busk. Drilling at Ahwaz began in August 1913, but was abandoned in September 1915 at 3317 ft (a deep well for its time) after encountering difficult rocks. Lister James, however, did not give up. In 1913–14, Lister James was accompanying The Admiralty Commission to Persia (headed by Slade), which encouraged Lister James to find other drill sites in Persia. He selected a site at Naft Safid and two wells drilled there (in 1914? and 1918) were dry. He also proposed a site on Qishm Island in the Persian Gulf but a well drilled in 1918 was dry (Owen 1975, p. 1262; Ferrier 1982, p. 282). Lister James was appointed the APOC's chief geologist in Persia in 1919, succeeding Macrorie (who had served as the chief geologist since 1915). Jones (1981, p. 104) observes:

> Burmah pioneered the use of geologists in the search for oil, many of them supplied by Boverton Redwood's consulting firm. In 1907 the company began to

employ full-time geologists, many of whom, like E.H. Cunningham-Craig, Basil Macrorie and Lister James, were later to use the experience they gained in Burma in the service of Anglo-Persian. Yet they could find no oil.

A view from Royal Dutch Shell

The Royal Dutch Shell Group formed in February 1907 through the combination of two formerly rival companies: (1) The Royal Dutch Petroleum Company of The Netherlands, which was founded in 1890 to develop oilfields in Sumatra (Dutch East Indies) and was led by August Kessler, Hugo Loudon and Henri Deterding; and (2) The 'Shell' Transport and Trading Company Limited of the United Kingdom, which was founded in 1897 by Marcus Samuel and his brother Samuel Samuel. (The quotation mark on 'Shell' was part of the company's legal name.) The merger of these two companies was largely due to their need to compete with the American company Standard Oil around the world. Three major works have chronicled the history of the Royal Dutch Shell Group: (1) Gerretson (1958), author of the four-volume *History of the Royal Dutch*; (2) a more recent work, also four volumes, titled *A History of Royal Dutch Shell* written by a team of scholars from Utrecht University (vol. 1 by Jonker & van Zanden 2007; vol. 2 by Howarth & Jonker 2007; vol. 3 by Sluyterman 2007; vol. 4 by van Zanden 2007); and (3) Howarth (1997), author of *A Century in Oil: The 'Shell' Transport and Trading Company*.

Curiously, Howarth (1997) does not refer to Boverton Redwood, and Jonker & van Zanden (2007) refer to him only once in the context of the late nineteenth century:

> [G]eologists were rarely consulted by the oil industry, the technical side of which was dominated by mining engineers such as Stoop, chemical engineers like Dr. Paul Dvorkovitz and Sir Boverton Redwood, and drilling engineers in the field. All regarded geologists, perhaps out of professional jealousy, as a waste of money
>
> (Jonker & van Zanden 2007, p. 51).

Gerretson (1958) mentions Redwood's involvement in the shipment of oil from the Suez Canal (constructed from 1859 to 1869 by the French company of Ferdinand de Lesseps). Standard Oil was looking for a shorter and cheaper route to the Far East than the long way round the Cape of Good Hope in southern Africa. Meanwhile, a Russian company, The Caspian and Black Sea Society, was seeking permission to use the Suez Canal to transport oil. This company was owned by Rothschild and a number of Russian oligarchs, and operated from Batum on the SE coast of the Black Sea. The opening of the Suez Canal to oil tankers alarmed the British, who had a big stake in maintaining the transportation system. For one thing, the fleet of sailing vessels and steamers were nearly all British property. Moreover, the tin plate in which both American and Russian oil was packed was largely manufactured in Wales and from there exported to the USA and Russia. The British preferred to keep the trading system as it was. The British shipping industry approached the Foreign Office and wrote to Prime Minister Lord Salisbury not to let the Suez Canal be used for oil transportation. To support their case, they:

> [W]ere able to obtain a report favourable to themselves from two leading experts in the petroleum world, Abel and Boverton Redwood, who pointed out that visibility was very poor in the Suez Canal during the periodical sandstorms and there was great likelihood that tankers would founder or collide. In the event of the ship's side being holed as a result of such an accident, large quantities of petroleum would spread over the closed basin of the Canal and, at a high temperature, spontaneous combustion might easily occur
>
> (Gerretson 1958, vol. 1, p. 216).

Salisbury, however, aiming for a closer relationship with Russia, refused to support these trade interests and opined that neither the British government nor the Suez Canal Company was qualified to close the canal to any kind of commercial traffic. Meanwhile, the directors of the Suez Canal Company, who were reviewing applications, had received a very attractive offer from M. Samuel & Co. (forerunner of Shell formed in 1897), a London syndicate composed of several major Eastern business companies and British shipbuilders. This import–export company was run by the brothers Marcus and Samuel, who had important trade with the Far East. M. Samuel & Co. proposed to construct a new tankship, the *Murex*, according to special regulations of the Suez Canal Company. In 1892, the *Murex* (a 338 ft long and 3564 gross ton M class oiler built by William Gray & Co.) became the first tank steamer to sail through the Suez Canal *en route* to Thailand.[10]

In a 1911 book entitled *The Great Oil Octopus*, written by the anonymous Truth's Investigator (apparently critical of the oil industry worldwide), we read the following regarding Redwood's involvement in the Suez Canal issue:

[10]During World War I, the *Murex* was used first by the Royal Australian Navy and later by the Royal Admiralty to attack the Germans. In December 1916, the *Murex* was torpedoed by a German submarine in the Mediterranean Sea and sunk offshore Egypt. For more information refer to Hope (1948) and Howarth (1992).

Sir Frederick Abel and Mr. (now Sir) Boverton Redwood prepared a report for those British shipowners who were hostile to the bulk carriage of oil through the Canal ... While this gentleman [Redwood] was still in Egypt Sir Marcus Samuel artfully published in the *Times* an extract from a paper Mr. Redwood read to the Institution of Civil Engineers, in which he said: 'The tank storage of kerosene oil has undoubtedly a great advantage over barrel or case storage in the event of fire'. Mr. Redwood was thus rather neatly cornered, for he had to admit in his report that this statement was still true. So he had to lay the chief stress on the danger of burning oil escaping on to the water – which the experience safe transit of nearly twenty years has proved to be a very trifling risk. The directors of the Suez Canal Company took a very accurate measure of this report when they replied: *Without entering into the question whether the work of Sir F. Abel and Mr. Boverton Redwood is not merely a criticism of our regulations bearing too exclusively the impression of the anxiety of parties interested in the present mode of transporting petroleum to the East, we, etc.* After this the agitation fizzled out, and the transport of oil in bulk still continues. The subject was referred to at the Institution of Civil Engineers in February, 1894, when Mr. (now Sir) Fortescue Flannery invited Mr. Boverton Redwood to state how his prophecies on the carriage of bulk oil through the Canal had been fulfilled. Mr. Redwood replied thus (*Proceedings Inst. C.E.*, vol. cxvi. p. 250): *He could only say that if, as appeared to be the case, the transport of petroleum through the canal had been going on with entire absence of anything approaching to an accident, he was very glad to hear it. He did not know, however, that that was to be taken as absolute proof that no risk existed. Time alone, and a longer time than had as yet elapsed, would demonstrate that.* Nearly twenty years have now elapsed; the Standard Oil Trust itself has tank steamers which convey oil through the Canal, and Mr. Henry in his work shows that between 1892 and 1906, 2 000 000 tons of oil were thus transported

(pp. 159–161).

A view from the Standard Oil Company

The Standard Oil Company was founded by John Rockefeller in Cleveland, Ohio in 1870. It grew to become a powerful trust in 1882, and because of its ethically controversial practices in eliminating its competitors and dominating the oil industry, the United States Supreme Court ruled in 1911 to dismantle Standard Oil. The company was split into 34 separated companies; it is regarded as the mother company of all existing major oil companies in the USA including ExxonMobil and Chevron. The comprehensive history of Standard Oil is given in a four-volume book series (Hidy & Hidy 1955; Gibb & Knowlton 1956; Larson *et al.* 1971; Wall *et al.* 1988). In the first volume, *Pioneering in Big Business 1882–1911*, Hidy & Hidy (1955) refer to Boverton Redwood twice. The first instance is Standard Oil's attempt to offer the best American kerosene to consumers in the UK:

> Standard Oil officials arranged for the manufacture of wicks designed to provide the proper drawing power for American kerosene. On January 13, 1885, representatives of New York Standard signed a contract (which Jersey Standard took over in 1899) with the largest maker of wicks in England, Joseph Morgan & Son of Manchester. The latter agreed to manufacture at least ten thousand pounds per week of the 'ordinary super Petroleum' or 'Paraffin Wick' of all appropriate sizes from the three-eighths of an inch to two inches to fit accurately the large variety of plain and ornamental English lamps. The wicks were to be subject to the rigid testing of Boverton Redwood
>
> (Hidy & Hidy 1955, p. 140).

The second instance concerns Standard Oil's attempts to expand its activities in the Far East:

> Standard Oil executives negotiated with several companies which had properties in the Dutch East Indies and by February, 1989, had reached a tentative agreement with one, the Moeara Enim Petroleum Company. While negotiations with Shell quickly broke down, those with Moeara Enim proceeded smoothly. Although the company's concession for exploration was on land deep in a 'dense swampy jungle,' Fertig [John H. Fertig of Standard Oil] recommended allying with the company; its production was promising, and plans were under way to build a pipeline to a site below Palembang on the Moesi River for which Boverton Redwood, assisted by practical refiners from the United States and Russia, was designing a refinery on the continuous distillation principle
>
> (Hidy & Hidy 1955, pp. 265–266).

Paul Giddens in his book *Standard Oil Company (Indiana): Oil Pioneer of the Middle West* (Giddens 1955) recounts how a group of Standard Oil chemists including Dr Robert Humphreys, a Johns Hopkins PhD in chemistry and chief chemist of Standard Oil's laboratory at Whiting, conducted experiments in order to increase the yield of gasoline from crude. Since their laboratory had a small library, the researchers did not have access to much of the scientific literature published on their experiments before. Therefore, they started from scratch, going with hunches and the trial and error method:

> If an extended search and study on the literature on the subject or the Patent Office records had been made, Humphreys would have undoubtedly come across the work of Sir Boverton Redwood and James Dewar in attempting to crack oil, which was patented in the United States on April 22, 1890. They were the first to patent a pressure distillation process for the purpose of increasing the yield of gasoline from Russian crude oil, but it had never been applied industrially on a large scale (Giddens 1955, p. 142).

Salt Creek Field, Wyoming

Owen (1975, p. 23) writes that Redwood submitted a geological report on Salt Creek in 1899 which apparently formed a basis for the Dutch company – Petroleum Maastschappji Salt Creek – to buy a claim in Wyoming and send an Italian geologist named Cesare Porro in 1906 to examine the company's property in 1906. Porro selected a location for test drilling, and subsequent drilling in 1908 resulted in an oil discovery at 1175 ft as Porro had anticipated. The story is also recounted by Greesley (1970):

> Sir Boverton Redwood, an internationally renowned petroleum geologist and engineer, visited Wyoming on his American tour of 1899. While the terse, location statements in his published account (*A Treatise on Petroleum*, 1913, p. 98) gave little hint as to his later enthusiasm, Wyoming apparently greatly impressed by the good scientist, for he furnished rhapsodic prose to the press (via letters to Rudi Landeur) about tremendous petroleum potentialities of Wyoming. Redwood went so far as to offer to select the drilling sites from his London office! To buttress Sir Boverton's zealousness, the Lobells selected a Wyoming geologist, Dr. Wilbur C. Knight, to sanction the plans.

Mexico

During the rule of President Porfirio Diaz, Mexico's dictator, the British firm of S. Pearson and Son, Ltd (later named El Aguila, Mexican Eagle, Ltd) led by Sir Weetman Dickinson Pearson (1856–1927) entered Mexico as contractors for the construction of the National Railway of Tehuantepec. The company employed:

> Boverton Redwood's consulting firm, of London, to undertake petroleum exploration in the surrounding territory. W.H. Dalton, one of Redwood's senior geologists, made a reconnaissance survey in 1901 and found abundant surface indications which led to commencement of a drilling program in 1902. Commercial production was developed on the San Cristobal salt dome in 1904–05, and the Pearson interests built a small refinery at Minatitlan in 1908 to process the output
>
> (Owen 1975, pp. 251–252).

(see also Jones 1981, p. 68).

Geologists who were dispatched by Redwood to explore for oil for Pearson during 1901–09 included: W.H. Dalton, Leonard V. Dalton, Grimely, George Jeffrys, F. Laurie, George Madgwick, F.W. Moon and P.C.A. Stewart: 'They made the first comprehensive study of the petroleum geology of Mexico under adverse circumstances' (Owen 1975, p. 252). In his *Treatise on Petroleum*, Redwood devotes several pages to the discovery, geology and distribution of oilfields in Mexico (Redwood 1926, pp. 132–137, 245–246).

In 1909, Pearson suddenly fired the English geologists from Redwood's firm and instead hired American geologists, most notably Everette DeGolyer. In 1910, Pearson's company discovered Potrero del Liabo 4, one of the world's most productive wells. In the same year, Pearson was knighted as Lord Cowdray (after the Cowdray Estate in Sussex that he had acquired). Mexican Eagle made other oil discoveries along the 'Golden Lane' (a rim of buried Cretaceous atolls along the western Gulf of Mexico), which propelled Mexico into a world-class oil producer. In 1919, Pearson sold Mexican Eagle Ltd to Shell (Yergin 2008, p. 214; Pfiester 2011).

Knowles (1978, p. 59) writes that after two British companies tried their luck in Mexico and failed, 'Sir Boverton Redwood, one of the world's most eminent geologists, then wrote an obituary for Mexico's oil, concluding that it would never be found in quantities'. She does not give any citations for this obituary.[11] Of course, as Knowles recounts, American explorers like Edward Doheny and Everette DeGloyer had better luck in Mexico.

West Indies

The first island of the West Indies to come to the attention of British oil exploration was Barbados, when a small enterprise, the West India Petroleum Company, began in the late 1890s. In 1903, the company approached the Admiralty and the colonial government for funds in return for fuel. In order to support its case, the company pointed out the dangers of the West Indies' oil falling under the domination of Standard Oil, and also obtained a

[11]It is possible that Knowles was referring to the following description in Redwood's *Treatise on Petroleum* (Redwood 1926, vol. 1, p. 245): 'The State of Durango and Chihuahua are vaguely reported to possess evidence of the presence of petroleum, presumably in insignificant amount ... There is, however, little probability of a yield of commercial value, as the district [Oaxaca] consists of granite-gneiss, granulite, quartzite, marble, and other rocks, of probably Archaean age, covered by sandy and loamy matter, partly derived from disintegration *in situ* of the subjacent rocks, partly rearranged as river- and lagoon-deposits. These appear to have been locally saturated with oily hydrocarbons from the decomposition of organic matter in the lagoons'. But these districts mentioned by Redwood lie away to the west of the Golden Lane oilfields, Mexico's oil-rich province.

favourable report from Redwood on the oil geology of Barbados (Jones 1981, p. 106). The Admiralty, however, declined the proposal.

In 1910, the British and Foreign Oil and Rubber Trust, chaired by Charles Conway (Governor of Trinidad) and comprising the leading British oil companies, including Burmah Oil, Anglo-Persian, Assam Oil, Lobitos Oilfields, California Oilfields and Pacific Oilfields, began to express an interest in the oilfields of Trinidad. In November 1910, the Trust held a meeting in London with Greenway and representatives from the Colonial Office and the Admiralty in order to gain favourable concessions in Trinidad. Redwood attended this meeting and supported the Trust's position of getting, 'a *de facto* monopoly such as which the Burmah Oil Company had secured through their agreement with the Secretary of State for India' (Jones 1981, p. 109). Eventually, a consortium of Burmah Oil (25%), the Shell Group (25%), Greenway's Trinidad Syndicate (25%), and the Paris Rothschilds and Frederick Lane (25%) formed.

Corley (1983, p. 284) recounts Burmah Oil's involvement in Trinidad: in 1909, the company was offered a concession in Trinidad, but Cargill had refused primarily because the island was 'really quite outside our sphere'. However, in 1912, Burmah Oil took a quarter share, together with Shell, Greenway's syndicate and the Paris Rothschild, in the United British West Indies Petroleum Syndicates Ltd. In reality, 'Cargill had been forced into it partly by Redwood and partly by the anti-Shell Greenway'. The venture was not successful and Burmah Oil lost £150 000 before pulling out in 1923.

Timor

The Pualaca oil seeps on the north coast of Timor attracted overseas oil explorers in the late nineteenth century. In 1892, a Manila-based English engineer, W.A. Duff, drilled a shallow well (less than 30 ft deep) at Pualaca, from which he obtained 470 gallons (about 2100 l) of crude in a few hours. Samples of Duff's oil were analysed by Redwood, who reported on 22 December 1892 that:

> Results of analyses on crude oils from Pualaca cannot be regarded otherwise than of a most favourable character … On the whole, I regard these samples of crude petroleum as of a remarkably high quality. They yield at least as much Kerosene as the best descriptions of United States crude petroleum, together with a large proportion of paraffin, and in comparison to standard Baku crude, the comparison is still more striking. In terms of the proportion of paraffin present, the samples resemble the crude petroleum of Java

(Ministerio das Colonias, Portugal, 1915; Charlton 2017).

Sakhalin

The island of Sakhalin, which spans 900 km northwards from Hokkaido to the Siberian coast, was explored by Russian and British geologists in the 1900s, and by the Japanese between 1918 and 1925, who drilled the first commercial oil well in 1921. Geologists for British interests included W.H. Dalton in 1903, F. Anderson and L.V. Dalton in 1909, and F.A. Holiday in 1913 working for Redwood's firm (Owen 1975, p. 428).

The Admiralty: from coal to oil

Gerretson (1958, vol. 4, p. 250) in discussing the Royal Navy's decision to switch from coal to oil fuel refers to Redwood as follows:

> John Fisher was appointed First Sea Lord in the summer of 1903. He immediately began to fight for his ideal in the Royal Commission which was appointed in 1904 to advise on matters of Imperial defence. The result, however, was not a complete victory. Boverton Redwood, a well-known expert of that period, thought it out of question that fuel oil could replace coal to any appreciable extent in the foreseeable future. Nevertheless, Fisher managed to have his way: in 1908 all British torpedo boats were already burning fuel oil; in 1909 the first destroyer followed their example.

Gerretson continues:

> Redwood had good reasons for his opinion. His main objection concerned the quantity available: in 1904 the production of fuel oil did not approach that of coal. Nevertheless, this objection has lost of much of its force. The free world oil industry, which was never far behind when there was money to be made anywhere, had quickly begun not only to supply the existing needs, but also to stimulate the demand and furnish the additional requirements.

This report is in agreement with Redwood's own article, 'The future of oil', published in 1913. He begins:

> In 1903, when giving evidence before the Royal Commission on Coal Supplies, I expressed the view that 'there will be certain selected application of liquid fuel where the advantages of employing such a fuel are especially obvious, but for anything like general employment I cannot see where we are to look for adequate supplies'
>
> (Redwood 1913).

Redwood then continues:

> This view, which was adopted by the Commissioners, and embodied in their final report, was largely based upon a comparison of the world's output of petroleum with that of coal, but it should be added that the conditions in regard to supply have greatly improved since that time
>
> (Redwood 1913).

The rest of his article deals with the geographical distribution of coal and oil production, and in doing so he makes an estimate that heralded today's shale revolution: 'It may also be found possible to utilize the bituminous clay or Kimmeridge shale of Dorsetshire as source or oil fuel' (Redwood 1913).

Ellwell-Sutton (1955) in the book *Persian Oil: A Study in Power Politics* connects three vital elements in the development of the British oil industry just before World War I. Redwood was the connecting line between these three elements: Burmah Oil, the D'Arcy Concession in Persia and the Admiralty. Ellwell-Sutton writes:

> In 1904 Sir John Fisher had become the First Lord of the Admiralty, and on taking up office began to push vigorously the conversion of the Royal Navy from coal to oil fueling, a project in which he had been interested as early as 1882 ... On Fisher's initiative a four-man committee was set up in November, 1904, under the chairmanship of E. G. Pretyman, Secretary to the Admiralty, to investigate the whole problem. The consultant to this committee was Sir Boverton Redwood, who had been associated with D'Arcy in some of his petroleum interests.
>
> (Ellwell-Sutton 1955, p. 16).

Black (1926, p. vii) expresses more direct admiration for Redwood's contribution to the Admiralty:

> There was no service that he performed and no honour that came to him that he valued so much as his advisory work for the Admiralty on the use of fuel oil in British warships. He was either a member or one of the most important advisors of every committee that considered the question of oil fuel for the Navy from about 1901 onwards. His crowning in this connection was his membership in 1913 of the Royal Commission on Oil Fuel under the chairmanship of Admiral Lord Fisher of Kilverstone. To the work of that Commission he gave literally the best that was in him, and that best was something which only those who saw his work could truly appreciate. Redwood had been one of the earliest and most consistent advocates of the advantages of liquid fuel in warships. For other purposes oil might or might not be *fuel de luxe*, time would show, but for the British Navy it was a necessity. Lord Fisher was no less enthusiastic.

Conclusions

The contributions of Sir Thomas Boverton Redwood (1846–1919) to the petroleum industry were manyfold. First, as a chemist he helped advance petroleum science and engineering. In the words of Corley (1983, p. 30), Redwood was: 'a walking encyclopedia of oil knowledge, some of which he put into this authoritative works on petroleum, while some he kept in his head'. Redwood's magnum opus, *A Treatise on Petroleum*, which went through several editions was a benchmark in the knowledge of petroleum; it trained generations of petroleum geologists, chemists and engineers, and can still be useful for its historical information.

Second, from his base in London, Redwood established one of the first and few consulting companies in the world, linking academic science and engineering to the business-orientated and practical industry: 'At a time when few oil companies employed full-time geologists, his consulting firm was used by virtually every British oil company' (Jones 1981, p. 97).

Third, from his position that saddled politics and industry, he played an important role in the advancement of British petroleum interests around the world. He was, 'the *éminence grise* of British oil policy before the First World War' (Jones 1981, p. 98). In 1912, Redwood told an Admiralty Committee: 'Every reasonable encouragement that is possible should be given to the development of source of supply of fuel oil under the British flag' (quoted in Jones 1981, p. 98). For this reason, he believed strongly:

> [T]hat this oil should be developed by independent British oil companies and not by the foreign combines. He developed a strong dislike for both Shell and Standard Oil, an attitude probably related to his long association with smaller British oil companies threatened by these giants
>
> (Jones 1981, p. 98).

Owen (1975, p. 1259) observed:

> Boverton Redwood's consulting connection with the oil-producing companies and his equally intimate association with government agencies must have been an important factor in the adequacy of the British Navy's fuel supply at this critical time.

Several obituaries have also remarked on Redwood's charming, polite and friendly character (e.g. A.C.K. 1919). So, it was not merely his expert knowledge but also his personality that enabled him to win the respect and trust of peer scientists, business leaders and high officials in the government:

> As a man, Redwood had a most charming manner with an old-world courtesy. He was always willing to help the younger generation and would take an infinite amount of trouble to help young men commencing their career. He was a staunch friend, and never spoke unkindly even of those with whom he did not agree
>
> (Perkin 1920).

Redwood played a leading role in the development of the British petroleum industry and policy from the 1870s to the 1910s; he was the watershed or, in the words of Daniel Yergin in *The Prize*, Redwood was 'Britain's premier oil expert' at the turn of the twentieth century (Yergin 2008, p. 123). It is thus surprising to see that Henry Longhurst in his *Adventure in Oil: The Story of British Petroleum*

(Longhurst 1959) does not mention Redwood, even though he was a watershed in the history of it. It is appropriate to end this article with a quote from a person who knew Redwood personally – Sir Frederick Black, Past President of the Institution of Petroleum Technologists:

> That one brain could absorb, assimilate and deal with data so varied and so complex as those which Sir Boverton Redwood handled in his daily work as a consultant, and expressed in the books and pamphlets, nearly fifty in number, which stand in his name, was a subject of wonder even to those who worked with him and knew what his powers of concentration, quick and orderly thinking, and sure methods of work were capable of accomplishing
>
> (Black 1926, p. vi).

Acknowledgements Reviews by William Brice and Raymond Sorenson improved this paper. I also thank Jonathan Craig for editorial handling of the paper.

Funding This research received no specific grant from any funding agency in the public, commercial or not-for-profit sectors.

References

A.C.K. 1919. Sir Boverton Redwood, Bart. *Nature*, **103**, 287–288, 12 June 1919.

Allen, R. 1921. *Oil Shales: Monographs on Mineral Resources with Special Reference to the British Empire*. Murray, London.

Allen, R.C. 2010. *The British Industrial Revolution in Global Perspective*. Cambridge University Press, Cambridge.

Anon 1873. Theophilus Redwood, Ph.D., F.C.S. *Chemist and Druggist*, **15**, 336, 15 October 1873.

Anon 1900*a*. 'The Automotor' Portrait Gallery: Professor Boverton Redwood. *The Automotor Journal*, **IV**, 291–292.

Anon 1900*b*. Charles James Hislop Warden. *The Lancet*, **1**, 632, 25 August 1900.

Anon 1910. *Guide to the Oil Companies, with Introduction by Sir Boverton Redwood*. Pall Mall Gazette, London.

Anon 1920. *Redwood, Sir Boverton. The New International Year Book*. Dodd, Mead & Co., London.

Anon 1938. Theophilus Redwood, Biographical sketch. *Pharmaceutical Journal*, **88**, 31 December 1938, 677.

Anon 1972. Theophilus Redwood. *Proceedings of the Society for Analytical Chemistry*, **9**, 49b–52.

Ashworth, W. 1987. *The History of the British Coal Industry. Volume 5. 1946–87. The Nationalized Industry*. Clarendon Press, Oxford.

Bacon, R.F. & Hamor, W.A. 1916. *The American Petroleum Industry*. McGraw-Hill, New York (2 vols).

Bamberg, H.H. 1994. *The History of the British Petroleum Company, Volume 2. The Anglo-Iranian Years 1928–54*. Cambridge University Press, Cambridge.

Bamberg, H.H. 2000. *British Petroleum and Global Oil, 1950–75. The Challenges of Nationalism*. Cambridge University Press, Cambridge.

Beeby-Thompson, A. 1961. *Oil Pioneer: Selected Experiences and Incidents Associated with Sixty Years of World-wide Petroleum Exploration and Oilfield Development*. Sidgwick and Jackson, London.

Black, F.W. 1926. Foreword with historical notes on the life of the late Sir Boverton Redwood, Bart. *In*: Redwood, B. (ed.) *Petroleum: A Treatise, Volume 1*. Charles Griffin & Co., London, v–viii.

Campbell, A. 1918. *Petroleum Refining, with a Foreword by Sir Boverton Redwood, Bart*. Charles Griffin & Co., London.

Charlton, T. 2017. History of petroleum exploration in Timor-Leste, http://www.timcharlton.co.uk/history/ [last accessed 30 January 2017].

Clowes, F. & Redwood, B. 1896. *The Detection and Measurement of Inflammable Gas and Vapour in the Air*. C. Lockwood & Son, London.

Corley, T.A.B. 1983. *A History of the Burmah Oil Company, 1886–1924*. Heinemann, London.

Corley, T.A.B. 1988. *A History of the Burmah Oil Company, 1924–66*. Heinemann, London.

Corley, T.A.B. 2004. Redwood, Sir Thomas Boverton, first baronet (1846–1919). *In*: Matthew, H.C.G. & Harrison, B. (eds) *Oxford Dictionary of National Biography*. Oxford University Press, https://doi.org/10.1093/ref:odnb/47996

Craig, J., Gluyas, J., Laing, C. & Schofield, P. 2013. Hardstoft – Britain's first oil field. *Oil-Industry History*, **14**, 97–116.

Cunningham-Craig, E.H. 1912. *Oil Finding: An Introduction to the Geological Study of Petroleum: with an Introduction by Sir Boverton Redwood*. Edward Arnold, London (2nd edn, 1920: Longmans, Green & Co., New York).

Davies, J., Jenkins, N., Baines, M. & Lynch, P.L. (eds) 2008. Redwood, Theophilus (1806–92) pharmacist. *In*: *The Welsh Academy Encyclopedia of Wales*. University of Wales Press, Cardiff.

Ellwell-Sutton, L.P. 1955. *Persian Oil: A Study in Power Politics*. Lawrence & Wishart, London (reprinted in 1975 by Greenwood Press, Westport, CT, USA).

Ferrier, R.W. 1982. *The History of the British Petroleum Company, Volume 1. The Developing Years 1901–32*. Cambridge University Press, Cambridge.

Flinn, M.W. & Stoker, D. 1984. *The History of the British Coal Industry, Volume 2. 1700–1830. The Industrial Revolution*. Clarendon Press, Oxford.

Gerretson, F.C. 1958. *History of the Royal Dutch*. E.J. Brill, Leiden, The Netherlands (4 vols).

Gibb, G.S. & Knowlton, E.H. 1956. *History of Standard Oil Company (New Jersey): The Resurgent Years 1911–27*. Harper & Brothers, New York.

Giddens, P.H. 1955. *Standard Oil Company (Indiana): Oil Pioneer of the Middle West*. Appelton-Century-Crofts, New York.

Grace's Guide 2016*a*. *Thomas Boverton Redwood*, http://www.gracesguide.co.uk/Thomas_Boverton_Redwood [last accessed 15 August 2016].

Grace's Guide 2016*b*. *Bernard Boverton Redwood*, http://www.gracesguide.co.uk/Bernard_Boverton_Redwood [last accessed 15 August 2016].

GREENE, J.A. (ed.) 1919. *A Treatise on British Mineral Oil*, foreword by Sir Boverton Redwood, Bart., D.Sc. Charles Griffin & Co., London.

GREESLEY, G.M. 1970. The French, Belgians, and Dutch Come to Salt Creek. *The Business History Review*, **44**, 498–519.

HATCHER, J. 1993. *The History of British Coal Industry. Volume 1. Before 1700. Toward the Age of Coal*. Clarendon Press, Oxford.

HIDY, R.W. & HIDY, M.E. 1955. *History of Standard Oil Company (New Jersey). Pioneering in Big Business 1882–1911*. Harper & Brothers, New York.

HOPE, S. 1948. *Tanker Fleet: The War Story of the Shell Tanker and the Men Who Manned Them*. The Anglo-Saxon Petroleum Company, London.

HOWARTH, S. 1992. *Sea Shell: The Story of Shell's British Tanker Fleet, 1892–1992*. Thomas Reed, London.

HOWARTH, S. 1997. *A Century in Oil: The 'Shell' Transport and Trading Company, 1897–1997*. Weidenfield & Nicolson, London.

HOWARTH, S. & JONKER, J. 2007. *A History of Royal Dutch Shell, Volume 2. Powering the Hydrocarbon Revolution, 1939–73*. Oxford University Press, Oxford.

HUXLEY, J. 1983. *Britain Onshore Oil Industry*. Palgrave Macmillan, Basingstoke, UK.

JONES, G. 1981. *The State and the Emergence of the British Oil Industry*. Macmillan, London.

JONES, G. & GOODALL, F. 1984–86. Redwood, Sir Thomas Boverton. *In*: JEREMY, D.J. (ed.) *Dictionary of Business Biography: A Biographical Dictionary of Business Leaders Active in Britain in the Period 1860–1980*. Butterworths, London, **4**, 862–864.

JONKER, J. & VAN ZANDEN, J.L. 2007. *A History of Royal Dutch Shell, Volume 1. From Challenger to Joint Industry Leader, 1890–1939*. Oxford University Press, Oxford.

KIDD, C. & WILLIAMSON, D. (eds). 1990. *Debrett's Peerage and Baronetage*. St Martin's Press, New York.

KNOWLES, R.S. 1978. *The Greatest Gamblers: The Epic of American Oil Exploration*, 2nd edn. University of Oklahoma Press, Norman, OK, USA.

LARSON, H.M., KNOWLTON, E.H. & POPPLE, C.S. 1971. *History of Standard Oil Company (New Jersey): New Horizons 1927–50*. Harper & Row, New York.

LONGHURST, H. 1959. *Adventure in Oil: The Story of British Petroleum*. Sidgwick & Jackson, London.

MARVIN, C. 1884. *The Region of the Eternal Fire: An Account of a Journey to the Petroleum Region of the Caspian in 1883*. W.H. Allen & Co., London.

MCKAY, J. 2012. *Scotland's First Oil Boom: The Scottish Shale-Oil Industry 1851–1914*. John Donald Short Run Press, London.

MINISTERIO DAS COLONIAS, PORTUGAL. 1915. *Provincia de Timor*. Informações relitivas aos jazigos de petróeo e à agricultura. Sociedade de Geografia de Lisboa.

MITCHELL, B.R. 1984. *Economic Development of the British Coal Industry 1800–1914*. Cambridge University Press, Cambridge.

MOSLEY, C. 2003. *Baker's Peerage, Baronetage and Knightage*. 107th edn. Burke's Peerage, London (3 vols) (1st edn, 1826).

NEF, J.U. 1932. *The Rise of the British Coal Industry*. Routledge and Sons, London (2 vols reprinted in 2016) (previous reprint by Cass, London, 1966).

OWEN, E.W. 1975. *Trek of the Oil Finders: A History of Exploration for Petroleum*. American Association of Petroleum Geologists Memoirs, **6**.

PERKIN, F.M. 1920. Sir Boverton Redwood, Bart. *Journal of the Chemical Society, Transactions of the Royal Society of Chemistry*, **117**, 467–470.

PFIESTER, S.L. 2011. *The Golden Lane*. Chengalera Press, Georgetown, TX, USA.

REDWOOD, B. 1886*a*. *Cantor Lectures on Petroleum and Its Products*. Society for the Encouragement of Arts, Manufactures and Commerce, London.

REDWOOD, B. 1886*b*. *Petroleum: Its Production and Uses*. Van Nostrand, New York.

REDWOOD, B. 1890. The oil-fields of India: notes on the petroleum deposits of Burma, Assam, the Punjab, and Baluchistan. *Journal of Society of Chemical Industry*, **9**, 359–370, 441–443.

REDWOOD, B. 1913. The future of oil. *Journal of the Royal Society of Arts*, **62**, 103–116, 26 December 1913.

REDWOOD, B. 1926. *Petroleum: A Treatise*, 5th edn. Charles Griffin and Co., London (3 vols (reprint of 4th edn)) (1st edn, 1896 (2 vols), assisted by George T. Holloway; 2nd edn, 1906 (2 vols); 3rd edn, 1913 (3 vols); 4th edn, 1922 (3 vols)).

REDWOOD, I. 1895. *Theoretical and Practical Ammonia Refrigeration – A Work of Reference for Engineers and Others Employed in the Management of Ice and Refrigeration Industry*. Spon & Chamberlain, New York; E. & F. N. Spon, London.

REDWOOD, I. 1897. *A Practical Treatise on Mineral Oils and Their By-Products – Including a Short History of the Scotch Shale Oil Industry, the Geological and Geographical Distribution of Scotch Shales, Recovery of Acid and Soda Used in Oil Refining, and a List of Patents Relating to Apparatus and Processes for Obtaining and Refining Mineral Oils*. Spon & Chamberlain, New York; E. & F. N. Spon, London.

REDWOOD, I. 1898. *Lubricants, Oils and Greases – Treated Theoretically and Giving Practical Information Regarding Their Composition, Uses and Manufacture. A Practical Guide for Manufacturers, Engineers and Users in General of Lubricants*. Spon & Chamberlain, New York; E. & F. N. Spon, London.

REDWOOD, B. & EASTLAKE, A. 1923. *Petroleum Technologist's Pocket-Book*. 2nd edn. Charles Griffin & Co., London (1st edn, 1915).

ROY, C., HALL, A. & KANEFSKY, J. 1987. *The History of the British Coal Industry, Volume 3, 1830–1913. Victorian Pre-Eminence*. Clarendon Press, Oxford.

SLUYTERMAN, K. 2007. *A History of Royal Dutch Shell, Volume 3. Keeping Competitive in Turbulent Markets, 1973–2007*. Oxford University Press, Oxford.

SORKHABI, R. 2005. Pre-modern history of bitumen, oil and gas in Persia (Iran). *Oil-Industry History*, **6**, 153–177.

SORKHABI, R. 2010. George Bernard Reynolds: a forgotten pioneer of oil discoveries in Persia and Venezuela. *Oil-Industry History*, **11**, 157–172.

SORKHABI, R. 2015. A history of petroleum geoscience textbooks. *Oil-Industry History*, **16**, 165–208.

SUPPLE, B. 1988. *The History of the British Coal Industry. Volume 4. 1913–46. The Political Economy of Decline*. Clarendon Press, Oxford.

THOMSON, J.H. & REDWOOD, B. 1902. *The Petroleum Lamp, Its Choice and Use. 'A Guide to the safe employment of Mineral Oil in what is commonly termed the Paraffin Lamp'*. Charles Griffin & Co., London (2nd edn, 1906; 3rd edn, 1913).

THOMSON, J.H. & REDWOOD, B. 1913. *Handbook on Petroleum for Inspectors under the Petroleum Acts and for Those Engaged in the Storage, Transport, Distribution, and Industrial Use of Petroleum and Its Products and Calcium Carbide, with Suggestions on the Construction and Use of Mineral oil Lamps*. Charles Griffin & Co., London (3rd edn. revised by Aston Cooper-Key) (1st edn, 1901; 2nd edn, 1906).

TRUTH'S INVESTIGATOR 1911. *The Great Oil Octopus*. T. Fisher Unwin, London.

VAN ZANDEN, J.L. 2007. *A History of Royal Dutch Shell, Volume 4. Appendices. Figures and Explanations, Collective Bibliography, and Index (including 3 DVDs)*. Oxford University Press, Oxford.

VARCOE, I. 1970. Scientists, government and organized research in Great Britain 1914–16: The early history of the DSIR. *Minerva*, **82**, 192–216.

WALL, B.H., CARPENTER, C.G., YEAGER, G.S. & HARBERT, E.N. 1988. *Growth in a Changing Environment: A History of Standard Oil Company (New Jersey) Exxon Corporation 1950–75*. McGraw-Hill, New York.

WILLIAMS, T.I. 1981. *A History of the British Gas Industry*. Oxford University Press, Oxford.

WRIGLEY, E.A. 2010. *Energy and the English Industrial Revolution*. Cambridge University Press, Cambridge.

YERGIN, D. 2008. *The Prize: The Epic Quest for Oil, Money, and Power*. 2nd edn. Simon & Schuster/Free Press, New York.

Bibliography of additional selected key publications on the history of the European oil and gas industry

FRANCESCO GERALI[1,2], JONATHAN CRAIG[3]*, FIONA MacAULAY[4] & RASOUL SORKHABI[5]

[1]*The University of Oklahoma School of Law, Oil & Gas, Natural Resources, Energy Center, 300 Timberdell Road, Norman, OK 73019-5081, USA*

[2]*The University of Western Australia, FABLE, 35 Stirling Highway, Crawley, Perth, WA 6009, Australia*

[3]*Eni Upstream & Technical Services, Via Emilia 1, 20097 San Donato Milanese, Milan, Italy*

[4]*Echo Energy plc, 40 George Street, London W1U 7DW, UK*

[5]*Energy & Geoscience Institute, University of Utah, 423 Wakara Way, Salt Lake City, UT 84108, USA*

**Correspondence: jonathan.craig@eni.it*

Abel, F. 1876. *Report to the Secretary of State for the Home Department on the Subject of the Testing of Petroleum*. HMSO, London.

Abel, F. & Redwood, B. 1884. Petroleum testing in tropical climates with Abel's apparatus. *Chemical News*, **49**, 196.

Abel, F. & Redwood, B. 1892. *Report of Frederick Abel and Boverton Redwood on the 'Proposed transport of petroleum in tank steamships through the Suez Canal'*. Worral & Robey, London.

Abich, H. 1879. Ueber die Producktivitat und die geotektonischen Verhaltnisse der kaspischen Naphtaregion. *Jahrbuch der Kaiserlich Königlichen Geologischen Reichsanstalt*, **29**, 165–189.

Academia Romana (ed.) 1998. *Dicționarul explicativ al limbii române, ediția a II-a. Academia Română, Institutul de Lingvistică 'Iorgu Iordan*. Editura Univers Enciclopedic, Bucharest.

Academy of Sciences of Azerbaigian Institute of Geography 1957. Mineral Resources of Azerbaydzhan [in Russian]; English translation. *The American Geological Institute's International Geology Review*, **1**, 80–85.

Agout, M. 1949. *Bibliographie des livres, thèses et conférences relatifs a l'industrie du pétrole*. Standard Francaise des Petroles, Paris.

Agricola, G., Hoover, H. & Hoover, L.H. 1912. *Georgius Agricola De Re Metallica, Translation from the 1st Latin Edition of 1556, with Biographical Introduction, Annotations and Appendices Upon the Development of Mining Methods, Metallurgical Processes, Geology, Mineralogy & Mining Law, from the Earliest Times to the 16th Century*. The Mining Magazine, London.

Aikin, A. 1811. Observations on the Wrekin, and on the great coal-field of Shropshire. *Geological Society of London Transactions*, **1**, 191–212, https://doi.org/10.1144/transgsla.1.191.

Alberg, M. 1929. *Le pétrole en Pologne*. Agence Générale de Librarie, Paris.

Albrecht, M. 1874. *Das Paraffin und de Mineralole*. E. Schweizerbartsche Verlagshandlung, Stuttgart, Germany.

Albritton, C.C., Jr. (ed.) 1963. *The Fabric of Geology*. Addison-Wesley, Reading.

Allest, J.D. 1887. Le Chauffage des Torpilleurs au moyen des hydrocarbures liquides. *Annales de Génie Civile*, **XI**, 344–348, 362–363, 397–402.

Andes, L. 1920. *Praktisches rezeptbuch fur die gesamte. Fett, ol, seifen und schmiermittel industrie*. Buchhandlung Schulze & Co., Leipzig, Germany.

Andouin, P. 1868. Application des hydrocarbures liquides (petrole, goudron, huile lourde) a l'obtention des hautes temperatures et au chauffage des machines a vapeur. *Annales de Chimie Physique*, Series 4, **XI**, 30–40.

André, R. 1910. *L'industrie et le commerce du pétrole en France*. Rousseau, Paris.

Angeli, L. 1795. *De'Bollitori di Bergullo e suoi fanghi*. Imola.

Angenot, C. 1885. *Leçons sur le pétrole et ses dérivés donnees a l'Institut Superieur de Commerce*. S.N., Paris.

From: Craig, J., Gerali, F., MacAulay, F. & Sorkhabi, R. (eds) 2018. *History of the European Oil and Gas Industry*. Geological Society, London, Special Publications, **465**, 453–462.
https://doi.org/10.1144/SP465.25

Angermann, C. 1889. Die Naphtafelder in Wietrzno. *Jahrbuch der Kaiserlich Königlichen Geologischen Reichsanstalt*, **39**, 281–288.

Anon 1863. *The Oil Industry in Rumania*. Meridiane Publishing House, Bucharest.

Anon 1914. *Zakon o povlasticama koja se daju g. J. Kocare-u, industrija lcu iz Haga, holandskom podatku, za eksploataciju petrolejska majdana 'Smrdež'*. Drž. Štamp, Cetińe.

Archiac, Vicomte d' 1847. *Histoire des progrès de la géologie de 1834 a 1845*. Société Géologique de France, Paris.

Ariosto, F. 1690. *De oleo Montis Zibinii seu petrolio agri mutinensis. Libellus e Manuscriptis membranis editus ab Oligerus Jacobaeo. Hafniae, Literis Reg. Maj.& Univ.* Typogr Joh. Phil. Bockenhoffer, Copenhagen.

Bachmayer, F. 1957. *Erdol in Osterreich*. Natur und Technik, Wien.

Bailey, N. (ed.) 1730. *Dictionaium Britannicum: or, A More Compleat Universal Etymological English Dictionary than any Extant*. T. Cox, London.

Balfour, A.J., Société des Nations & Conférence économique internationale 1927. *Situation mondiale en ce qui concerne le pétrole brut et ses dérivés, 1922–26*. Société des Nations, Genève.

Banca di Piacenza 1999. *Dall'Oglio di Sasso alla Supercortemaggiore. In Relazioni e Bilancio 1998*. Corporate Technical Report, Piacenza.

Barnett, D.W. 1864. *Mr. Barnett's Report to the Directors Wallachian Petroleum Company Limited*. By the Company, London.

Baron, C. 1934. *Au pays de l'or noir. Bakou-Grozny-le pétrole russe*. Liège, Paris.

Barrabe L. & Schneegans D. 1935. Rapport d'ensemble sur le gisement pétrolifère de Gabian. *Annales de l'Office National des combustibles liquides* No. 4 (juillet–août) et No. 5 (septembre–octobre).

Beaujeu-Garnier, J. 1952. La France et le pétrole. *L'information Géographique*, **16**, no. 2, 65–71.

Beeby-Thompson, A. 1910. *Petroleum Mining and Oil-Field Development: A Guide to Exploration of Petroleum Lands, and a Study of the Engineering Problems Connected with the Winning of Petroleum*. Crosby Lockwood and Son, London.

Beeby-Thompson, A. 1916. *Oil-Field Development and Petroleum Mining, A Practical Guide to the Exploration Of Petroleum Lands, and a Study of the Engineering Problems Connected with the Winning Of Petroleum, Including Notes on Petroleum Legislation and Customs and a Discussion of the Origin of Petroleum*. Van Nostrand, New York.

Bentz, A., Herrmann, R. & Kraiss, A. 1931. *Deutsches Erdöl. Schriften aus dem Gebiet der Brennstoff-Geologie; Heft 7*. Ferdinand Enke, Stuttgart, Germany.

Bergman, T. 1784. *Manuel du minéralogiste, ou sciagraphie du règne minéral, distribué d'après l'analyse chimique*. Chez Cuchet, Paris.

Berthelot, M. 1866. Sur l'origine des carbures [hydrocarbonés] et des combustibles mineraux. *Annales de Chimie et des Physique*, 4th séries, **9**, 481–483.

Berthelot, C. 1935. *L'industrie des combustibles solides et gazeux. Fasc 1 des monographies de la Revue de Chimie Industrielle*. Gauthier-Villars, Paris.

Bérenger, H. 1920. *Le petrole et la France*. E. Flammarion, Paris.

Bianconi, G. 1840. *Storia Naturale dei Terreni Ardenti, dei Vulcani Fangosi, delle Sorgenti Infiammabili, dei Pozzi Idropici, e degli Altri Fenomeni Geologici Operati dal Gas Idrogene e dell'Origine del Detto Gas del Dott. Giuseppe Bianconi*. Marsigli, Bologna, Italy.

Bini, E. 1973. *La potente benzina italiana: Guerra Fredda e consumi di massa tra Italia, Stati Uniti e Terzo mondo (1945–73)*. Carocci, Roma.

Bolton, E.R. & Pelly, R.G. 1924. *Oil, Fats, Waxes and Resins*. E. Benn Ltd, London.

Bonand, de R. 1921. *Le petrole: origine, gisements, commerce, succedanes*. Librairie Polytechnique Ch. Beranger, Paris.

Bonarelli, L. 2001. *Guido Bonarelli. La vita e l'opera scientifica*. Associazione Pionieri e Veterani Agip. Eni, Roma.

Bondaroy Fougeroux, de A.D. 1770. Second Mémoire sur le Pétrole et sur des vapeurs inflammables, connues das quelques parties de l'Italie. *Histoire de l'Académie Royale des Sciences Paris*, **91**, 23–39.

Bossi, L. 1817. *Spiegazione di alcuni vocaboli geologici, litologici, mineralogici, per ordine d'alfabeto*. Tipografia Sonzogno e comp., Milano, Italy.

Boussingault, J.B. 1883. Sur la composition des substances minérales combustibles. *Annales de Chimie Physique*, Series 5, **XXIX**, 363–392.

Brantly, J.E. 1971. *History of Oil Well Drilling*. Gulf Publishing Company, Houston, TX, USA.

Bremner, D. 1869. *The Industries of Scotland: Their Rise, Progress, and Present Condition*. A. & C. Black, Edinburgh.

Brusa, A. 1961. *Gli idrocarburi in Italia: saggio geografico di un mercato*. M. Bozzi, Genova, Italy.

Buzatu, Gh. 2011. *A History of Romanian Oil, Volumes I & II*. Mica Valahie Publishing House, Bucharest.

Calmel, A. 1901. Le Transport du petrole. *Journal du Pétrole*, **1**, 3–5, 21–23, 40–41, 55–56, 71–73.

Camerana, E. 1924. *Utilizzazione dei gas naturali*. Stabilimento topografico per l'amministrazione dello Stato, Roma.

Camerana, E. 1924. *L'olio di asfalto*. Stabilimento tipografico per l'amministrazione dello Stato, Roma.

Canning, S., Kelvin, W.T. & Redwood B. 1888. *Priestman's Petroleum Engine and Eteve's Patent Petroleum Engine*. Promotional leaflets. Gatehouse & Co., London.

Capacci, C. 1896. *Perizia extra giudiziale nella Causa tra Il Sig. Giorgio Arimondi e il Sig. Conte Enrico Sergardi*. Tipografia Luigi Niccolai, Firenze, Italy.

Capacci, C. 1905. *Deuxième Congrèsse internationale du pétrole. Aperçu sur les gisements pétrolifères de la province de Bologne (Italie)*. The Congress, Liegi.

Capellini, G. 1864. *Report on the Petroleum Districts in Wallachia, Belonging to the Wallachian Petroleum Company, Limited*. By the Company, London.

Capellini, G. 1866. *Petrolio di Tocco e Bitumi di Letto Manopello*. Regia Accademia delle Scienze di Torino, Série I, **III**.

Capellini, G. 1868. Giacimenti Petroliferi di Valacchia e loro Rapporti coi Terreni Terziari dell'Italia Centrale. *Memorie della Regia Accademia delle Scienze dell'Istituto di Bologna, 3rd Series*, **7**, 156–202.

Caracristi, C.F.Z. 1900. The World's production of petroleum. *Mining Industry*, **8**, 454–466.

Casali, A. 1874. *Il petrolio e suoi derivati*. Tipografia e litografia A. Cappelli, Modena, Italy.

Cayeux, L. 1916. Introduction à l'étude pétrographique des roches sédimentaires. *Mémoires pour servir à l'explication de la carte géologique détaille de la France*.

Chadinoff, C.I. 1936. *Exposé de la situation actuelle du pétrole brut*. J. Mersch, Paris.

Chambers, E. (ed.) 1748–49. *Dizionario Universale delle Arti e delle Scienze*. (9 vols). Presso Giambatista Pasquali, Venezia, Italy.

Chambre syndicale de l'industrie des pétroles (Paris) 1892. *Démonstration de l'écart compensateur nécessaire à l'industrie du pétrole en France*. J. Kugelmann, Paris.

Chambrier, P. de 1919. *Historique de Pechelbronn, 1498–1918*. Attinger Freres, Paris.

Charrin, V. 1921. *Les innovations alsaciennes dans la technique du petrole*. Dunot, Paris.

Chartrou, J.J. 1931. *Petroles naturels et artificiels*. A. Colin, Paris.

Chivot-Naudé, L. 1866. Rapport sur l'huile de petrole. Ses proprietes et ses Applications. *L'Invention*, **20**, 43–52.

Christian, H.E., Jr. 1969. Some observation on the initiation of salt structures of the southern British North Sea. *In*: Hepple, P. (ed.) *The Exploration for Petroleum in Europe and North Africa*. The Institute of Petroleum, London, 231–250.

City of Modena 1865. *Instruzione popolare sulla vendita e sull'uso del petrolio/compilata dal sig. professore Manzini, per incarico del Consiglio Provinciale Sanitario, approvata e mandata a divulgarsi per le stampe dalla Deputazione Provinciale nella sua seduta delli 22 aprile 1865*. Tip. Moneti, Modena, Italy.

Clarke, W.B. 1866. On the occurrence and geological position of oil-bearing deposits in New South Wales. *Geological Society of London Quarterly Journal*, **22**, 439–448, https://doi.org/10.1144/GSL.JGS.1866.022.01-02.31.

Clowes, F. & Redwood, B. 1896. *Detection and Measurement of Inflammable Gas and Vapour in the Air; with a Chapter on the Detection and Measurement of Petroleum Vapour*. Crosby Lockwood, London.

Coles, E. (ed.) 1713. *An English Dictionary, Explaining the Difficult Terms that are used in Divinity, Husbandry, Physic, Philosophy, Law, Navigation, Mathematics, and Other Arts and Sciences*. Printed by F. Collins for R. Bonwick, London.

Comitato Scientifico del Museo del Petrolio di Vallezza 1993. *Museo del petrolio di Vallezza. Studi Propedeutici*. Centro stampa AGIP, San Donato Milanese, Italy.

Cook, E.E. 1965. Geophysical operations in the North Sea. *Geophysics*, **30**, 495–510.

Coquand, H. 1867. Sur les gites de petrole de la Valachie et de la Moldavie et sur l'âge des terrains qui les contiennent. *Société Géologique de France Bulletin*, 2nd séries, **24**, 505–569.

Corporate Author 1916. *Koninklijke metaalwarenfabrieken J.N. Daalderop en Zonen, Tiel, (Holland): [verlichtingsartikelen, gas en petroleum, no. 7]*. Koninklijke Metaalwarenfabrieken J.N. Daalderop en Zonen, Tiel.

Cortesi, G. 1819. *Saggi geologici degli Stati di Parma e Piacenza*. Dai Torchj del Majno, Piacenza.

Courau, R. 1921. *Technique des petroles*. Doin, Paris.

Coustau, H., Gautier, J., Kulbicki, G. & Winnock, E. 1969. Hydrocarbon distribution in the Aquitaine basin of SW France. *In*: Hepple, P. (ed.) *The Exploration for Petroleum in Europe and North Africa*. The Institute of Petroleum, London, 73–85.

Croker, T.H. (ed.) 1764–66. *The Complete Dictionary of Arts and Sciences, in which the Whole Circle of Human Learning is Explained, and the Difficulties Attending the Acquisition of Every Art, Whether Liberal or Mechanical, are Removed*. 3 vol. Printed for the authors, and sold by J. Wilson & J. Fell; J. Fletcher & Co.;

J. Coote; Mess. Fletcher & Hodson, Cambridge; and W. Smith & Co. Dublin, London.

Cunningham-Craig, E.H. 1912. *Oil-Finding; An Introduction to the Geological Study of Petroleum*. E. Arnold, London.

Cunningham-Craig, E.H. 1917. *Oil-Finding: An Introduction to the Geological Study of Petroleum*. E. Arnold, London.

Dabrowski S. 1982. *Jan Jozef Ignacy Lukasiewicz, 1822–82*. Interpress, Buffalo, NY, USA.

Dalton, L.V. 1909. A sketch of the geology of the Baku and European oil-fields. *Economic Geology*, **4**, 89–117.

Dank, V. & Kokai, J. 1969. Oil and gas exploration in Hungary. *In*: Hepple, P. (ed.) *The Exploration for Petroleum in Europe and North Africa*. The Institute of Petroleum, London, 131–145.

Degousèe, J.M.A. & Laurent, C.A. 1861. *Guide du sondeur; ou, Traité théoriques et pratique des sondages*. Garnier Fréres, Paris.

Deterding, S.H. 1934. *An International Oil Man*. Ivor Nicholson & Watson, London.

Draghicenu, M.M. 1890. Erlauterungen zur geologischen Uebersichtskarte des Konigreiches Rumanien. *Jahrbuch der Kaiserlich Königlichen Geologischen Reichsanstalt*, **20**, 399–420.

Dunstan, A.E. 1922. *The Petroleum Industry. A Brief Survey of the Technology of Petroleum, Based upon a Course of Lectures Given by Members of the Institution of Petroleum Technologists on the Occasion of the Petroleum Exhibition, Crystal Palace, 1920*. Institution of Petroleum Technologists, London.

Dunstan, A.E., Brooks, B.T. & Benjamin, T. 1938. *The Science of Petroleum: A Comprehensive Treatise of the Principles and Practice of the Production, Refining, Transport and Distribution of Mineral Oil*. Oxford University Press, London.

Dupaigne, A. 1872. *Le Pétrole, son histoire, sa nature, ses usages et ses dangers, par Albert Dupaigne*. Palmé, Paris.

Dücker, F.F. 1880. *Petroleum and Asphalt, Rock-Salt &c., in Germany*. Escherich, Bückeburg, Germany.

Edeleanu, L. & Ministère de l'industrie et du commerce 1912. *Fabrication du petrole brut: Les propiétés du pétrole brut par L. Edeleanu*. Independenţa, Bucarest.

Eldridge, G. 1892. On the weak points in steamers carrying oil in bulk. *Transactions of the Institute of Naval Architect*, **XX**, 88–103.

Engler, C. 1888. Zur Bildung des Erdols. *Berichte der Deutschen Chemischen Gesellschaft*, **21**, 1816.

Engler, K. 1887. *Die deutschen erdöle*. Druck von L. Simion, Berlin.

Engler, K. & Hofer, H. 1913. *Das Erdol, seine Physik, Chemie, Geologic, Technologie und sein Wirt- schaftsbetrieb*. (4 vols). Hirzel, Leipzig, Germany.

Ente Nazionale Idrocarburi 2002. *Glossario dell'industria petrolifera*. Eni, Roma.

Ferrier, R. 1982. *The History of the British Petroleum Company*. Cambridge University Press, Cambridge.

Filjak, R., Pletikapic, Z., Nikolis, D. & Aksin, V. 1969. Geology of petroleum and natural gas from the Neocene complex and its basement in the southern part of the Pannonian basin, Yugoslavia. *In*: Hepple, P. (ed.) *The Exploration for Petroleum in Europe and North Africa*. The Institute of Petroleum, London, 113–130.

Fitz, J. 1907. Das Rohol als Heizmaterial fur Locomotiven. *Naptha*, **15**, 485–488.

Flannery, J.F. 1902. On liquid fuels for ships. *Transactions of the Institute of Naval Architect*, **44**, 53–75.

Fonvielle, W. & Harlow, J.G., Sr 1888. *Le pétrole (Bibliothèque des merveilles)*. Librairie Hachette, Paris.

Forbes, R.J. 1948. *Short History of the Art of Distillation: From the Beginnings up to the Death of Cellier Blumenthal*. E. J. Brill, Leiden, The Netherlands.

Forbes, R.J. 1959. *More Studies in Early Petroleum History*. E. J. Brill, Leiden, The Netherlands.

Frank, A.F. 2005. *Oil Empire: Visions of Prosperity in Austrian Galicia*. Harvard University Press, London.

Frederick, M. (ed.) 1866. *The Statesman's Year-Book*. Macmillan & Co., London.

Froom, G.B. 1897. Oil burners as used on the River Volga. *Transactions of the Institute of Marine Engineers*, **9**, 50.

Gehrke, F. 1906. *Die neuere Entwicklung des Petroleumhandels in Deutschland*. H. Laupp, Tübingen, Germany.

Gerali, F. 2010. Geology and oil exploration; the studies of Giovanni Capellini in Romania. *Oil-Industry History*, **10**(1), 121–131.

Gesner, A. 1861. *A Practical Treatise on Coal, Petroleum, and Other Distilled Oils*. Bailliere Brothers, New York; H. Bailliere, London.

Golubyatnikov, D. 1906. La Region naphtifere de Berekei. *Russia Geological Committee Bulletin*, **25**, 361–412.

Goodenough, W.J. 1903. Memorandum regarding fuel oil aboard ship. *Marine Engineering*, **8**, 4–6.

Gorgoni, C. 2003. *Le salse di Nirano e le altre salse emiliane – I segreti di un fenomeno tra mito e realtà. Comune di Fiorano Modenese*. Tip. ABC, Sesto Fiorentino, Firenze, Italy.

Gortani, M. 1937. *Problemi aperti alla ricerca di petrolio in Italia: comunicazione presentata al IIo Congresso Mondiale del Petrolio di Parigi per conto dell'Azienda Generale Italiana Petroli*

(A.G.I.P.). Anonima Tipografica Editrice Laziale, Roma.

Goubkin, I.M. 1904. Principaux resultats des travaux geologiques effectues en 1903 dans le Péninsule d'Apscheron. *Russia Geological Committee Bulletin*, **23**, 289–330.

Goubkin, I.M. 1934. *Tectonics of the South-Eastern Caucasus and Its Relation to the Productive Oil Fields*. U.S.S.R. State Scientific-Technical Publishing Office for Mining, Geology, and Oil Industry, Leningrad, Moscow.

Graf, G.E. 1925. *Erdöl, Erdölkapitalismus und Erdölpolitik*. Urania, Jena, Germany.

Grigoras, N. & Constantinescu, N. 1969. Geological activity for oil and gas in the Socialist Republic of Rumania. *In*: Heppel, P. (ed.) *The Exploration for Petroleum in Europe and North Africa*. The Institute of Petroleum, London, 147–159.

Groves, C.E., Thorp, W. *et al.* 1895. *Chemical Technology; or Chemistry in its Applications to Arts and Manufactures*. P. Blakison, Philadelphia, PA, USA.

Grzybowski, J. 1920. The geology of Polish oil fields. *Petroleum World*, **17**, 389–390.

Harper, L. 1872. *A Geological Report on Most Important Petroleum Lands Situated in the Royal Prussian Province of Hannover in North-Germany and on Petroleum in General*. Weissenbruch, Brussels.

Harris, J. (ed.) 1704. *Lexicon Technicum, or, an Universal English Dictionary of Arts and Sciences: Explaining Not Only the Terms of Art, But the Arts Themselves*. (2 vols). Printed for Dan Brown, Tim Goodwin, John Wahler, The Newburugh, John Newbolson, London.

Harris, W.J. 1970. The French oil industry. *Petroleum Review*, **24**, 248–250.

Hassmann, H. 1950. *Erdöl in Deutschland: Geschichte, Gebiete, Probleme*. Industrieverlag von Hernhaussen, Hamburg, Germany.

Heck, P. 1907. *Die deutsche Erdölindustrie*. Aachener Verlags und Druckerei-Gesellschaft m.b.H, Aachen, Germany.

Heise, G., Schmidt, W. & Friedrich, E. 1927. *Erdöl und Benzin: Produktion, Handel und Konsum*. Columbus, Berlin.

Henry, J.D. 1914. *History and Romance of the Petroleum Industry*. Bradbury, Agnew & Company Ltd, London.

Hicks, J.A., Berry, A.G.V. & Redwood, B. 1918. *The Laboratory Book of Mineral Oil Testing*. Charles Griffin & Co., London.

Himus, G.W. 1932. *Fuel Testing: Laboratory Methods in Fuel Technology*. L. Hill, London.

Hinde, M. (ed.) 1770–71. *A New Royal and Universal Dictionary of Arts and Sciences or Complete System of Human Knowledge*. (2 vols). Printed for J. Cooke, London.

Holub, L., Švajgl, O., Nevosad, M., Soukup, A. & Kopal, R. 2005. *The Century of Petrol. The History of the Refining Industry in the Czech Lands*. The Ceská Rafinérská, Prague.

Huguenet, I. & Giuseppe, F. 1856. *Asfalto. Considerazioni Generali sull'Origine e Formazione degli Asfalti e del Loro Uso come Cemento Naturale Applicati ai Lavori di Utilità Pubblica e Privata Arricchita di Note ed Accresciute di un Appendice sugli Asfalti del Regno di Napoli*. Stabilimeto tipografico di Vincenzo Priggiobba, Napoli, Italy.

Hynek, C. E. 1906. *Ueber die Petroleumindustrie und der Petroleumhandel: mit besonderer Berücksichtigung der rumänischen Oelfelder*. Segall & Marcu, Botosani, Romania.

Imperial Institute of HM's Petroleum Department & Cronshaw, H.B. 1921. *Petroleum*. John Murray, London.

International Petroleum Congress 1905. *Compte rendu memoires, 2me Congres International du Petrole: Sous le haut patronage du Gouvernment Belge, Liege, 1905*. Meert & Cie, Bruxelles.

International Petroleum Congress 1910. Compte Rendu. *3rd International Petroleum Congress*, 1907, Bucharest.

Janoschek, R. & Gotzinger, K.G.H. 1969. Exploration for oil and gas in Austria. *In*: Hepple, P. (ed.) *The Exploration for Petroleum in Europe and North Africa*. The Institute of Petroleum, London, 161–180.

Jauch, L. 1921. *Le petrole et son industrie*. Challamel, Paris.

Jefferys, T. (ed.) 1754. *A New and Complete Dictionary of Arts and Sciences: Comprehending All the Branches of Useful Knowledge, With Accurate Descriptions as Well of the Various Machines, Instruments, to Oils, Figures, and Schemes Necessary for Illustrating Them*. Printed for W. Owen, London.

Jervis, G. 1873. *I tesori sotterranei dell'Italia, parte II*. Loescher, Torino, Italy.

Jimenez, M. 1840. *Tratado de Farmacia Esperimental*. Imprenta de Don Narciso Sanchiz, Madrid.

Johnston, F. 1839. On the origin of petroleum, and on the nature of the petroleum from Whitehaven. *British Association for the Advancement of Science Repertoire*, 8th Meeting, **7**, 60.

Jung, J. 1935. *Principes de geologie du petrole*. Librairie Polytechnique Ch. Beranger, Paris.

Jung, J. 1938. Historique des recherches de pétrole en Limagne. *Revue du Science Naturelle d'Auvergne*, **IV**, 129–135.

Jung, J. 1946. *Géologie de l'Auvergne et de ses confins bourbonnais et limousins*. Mémoire de la Carte Géologique de France, **11**. Société Geologique de France, Paris.

Kalichevsky, V. 1938. *The Amazing Petroleum Industry*. Chapman & Hall, London.

Kirwan, R. 1784. *Elements of Mineralogy. By Richard Kirwan, Esq; F.R.S.* Printed for P. Elmsly, in the Strand, London.

Kirwan, R. 1797. Of the composition and proportion of carbon in bitumens and mineral coal. *The Transactions of the Royal Irish Academy*, **6**, 141–167.

Kissling, R. 1923. *Das Erdöl.* Wissenschaftliche Verlagsgesellschaft, Stuttgart, Germany.

Knab, C.R. 1868. Notice sur l'emploi des combustibles de l'huile mineral pour le chauffage des navires a vapeur. *Annales de Genie Civil*, 7, 305–321.

Lees, G.M. & Cox, P.T. 1937. The geological basis of the recent search for oil in Great Britain by the D'Arcy Exploration Co., Ltd. *Geological Society of London Quarterly Journal*, **93**, 156–194, https://doi.org/10.1144/GSL.JGS.1937.093.01-04.09

Lees, G.M. & Taitt, A.H. 1946. The geological results of the search for oilfields in Great Britain. *Geological Society of London Quarterly Journal*, **101**, 255–317, https://doi.org/10.1144/GSL.JGS.1945.101.01-04.12

Leeuw, W.C. 1919. *Petroleum.* Uitgevers-Maatschappij, 'Elsevier', Amsterdam.

Lévi, G. 1924. *Eléments de la technique du pétrole (recherché et exploitation).* Institut du Petrole, Université de Strasbourg, Strasbourg, France.

Libavius, A. 1601. *D. O. M. A. Singularium Andreae Libavii pars Tertia. Continens octo libros Bituminum et Affinium, Historicè, Physicè, Chymicè, Cum Controversiis Difficilimus, Expositorum Indicatorumque (…).* Impressa typis Ioannis Suarii, impensis Petri Kopffij, Francofurti.

Lidgett, A. 1919. *Petroleum.* I. Pitman & sons Ltd, London.

Lumet, M. 1909. *Utilisation du pétrole lampant dans les moteurs destinés aux embarcations et navires pratiquant la pêche maritime.* A. Gout, Orléans, France.

Macini, P. & Mesini, E. 2011. *L'industria petrolifera in Emilia tra Otto e Novecento.* CLUEB, Bologna, Italy.

Macovei, G. 1938. *Les gisements de petrole: Geologie, statistique, economie.* Masson et Cie, Paris.

Macquer, P.J. 1769. *Dictionnaire de chymie: Contenant la theorie & la pratique de cette science, son application a la physique, a l'histoire naturelle, a la medecine & a l'economie animale : Avec l'explication detaillee de la vertu & de la maniere d'agir des medicamens chymiques, et les principes fondamentaux des arts, manufactures & metiers dependans de la chymie.* Chez Lacombe, Paris.

Magini, M. 1977. *L'Italia e il petrolio tra storia e cronologia.* Mondatori, Vicenza, Italy.

Martens, E. 1936. *Manuel du laboratoire de chimie appliquée de l'Université de Louvain. Eaux, combustibles, solides, carburants et huile minérales, analyse de gaz par liquéfaction, réaction d'équilibre.* Uystpruyst, Louvain, Belgium.

Marvin, C. 1884. *The Petroleum of the Future: Baku: The Petrolia of Europe: A Historical Sketch, Showing the Immense and Inexhaustible Character of the Petroleum Deposits of the Caspian Region, From the Earliest Times.* R. Anderson, London.

Marvin, C. 1884. *The Region of the Eternal Fire: An Account of a Journey to the Petroleum Region of the Caspian in 1883.* W.H. Allen, London.

Marvin, C. 1886. *The Petroleum Question: The Coming Deluge of Russian Petroleum and its Bearing on British Trade.* R. Anderson, London.

Martin, G. 1923. *Le petrole, son origine, sa composition, sa recherche.* Alcan, Paris.

Masini, R. 1939. Lazzaro Spallanzani e i suoi viaggi sull'Appennino e sulle Alpi Apuane. *In*: *'Commemorazioni Spallanziane', Vol. II, scritti geologici e bibliografici.* Bianchi, Pavia, Italy, 57–98.

Massimi, G. 2002. *L'Abruzzo, Tocco da Casauria e il Bel Paese.* Itinerari, Lanciano, Italy.

Meyn, R.A. 1882. *Die Oel-Kreide bei Heide in Holstein: deren bergmännische Gewinnung, ihre Nutzanwendung zur Mineralöl-Erzeugung und Asphalt-Fabrication, und ihre Verwendung als Brennmaterial für industrielle Zwecke unter gleichzeitiger Herstellung von gebranntem Kalk.* Druck von Pontt & v. Döhren, Hamburg, Germany.

Migliorini, C. 1937. *Cenno sullo studio e sulla prospezione petrolifera di una zona dell'Italia meridionale: comunicazione presentata al IIo Congresso Mondiale del Petrolio di Parigi per Conto dell'Azienda Generale Italiana Petroli (A.G.I.P.)* [*Short Account of the Geological Conditions and of the Oil Prospection of a Zone in Southern Italy*]. Anonima Tipografica Editrice Laziale, Roma.

Ministere de l'Industrie et du Commerce, Direction Generale des Mines, Royaume de Roumanie 1928. *Les mines at le petrole en Roumanie. Aperqu generale sur le regime minier. Situation de l'industrie du petrole. Industrie miniere metallique.* Socec. & Co., Bucarest.

Ministero Agricoltura, Industria e Commercio 1890. *Repertorio delle miniere. Concessioni e permissioni delle miniere vigenti al primo gennaio 1875 con l'aggiunta di quelle accordate dal primo gennaio 1875 al primo gennaio 1890. Serie 2, Vol. IV.* Tipografia italiana, Roma.

Ministero di Agricoltura, Industria e Commercio, Direzione Generale Dell'agricoltura 1879. *Annali di Agricoltura. Relazione sul servizio minerario nel 1877.* Tipografia Eredi Botta, Roma.

Mitzakis, M. 1922. *The Oil Encyclopaedia.* Chapman & Hall, London.

Mojon, G. 1806. *Corso Analitico di Chimica*. Dalla Stamperia di Giovanni Grossi, Genova.

Molinari, F. 1893. *Combustibili industriali e petrolio: con appendice sui medesimi in rapporto all'igiene*. Fratelli Dumolard Editori, Milano, Italy.

Moniteur du Petrole Roumain 1941. Developments in the principal oil fields of Rumania during 1940. *Translation in the American Institute of the Mining Engineers Transactions*, **142**, 547–553.

Montagnani, P. 1955. *Il petrolio italiano*. Edizioni cultura sociale, Milano, Italy.

Moore, H. 1936. *Petroleum, Charts and Tables*. Moore & George Ltd, London.

Mrazec, L. 1902. Apercu geologique sur les formations saliferes et les gisements de sel en Roumanie. *Moniteur des Interests Petroliferes Roumains*, 43–51.

Mrazec, L. 1910. Uber die Bildung der Rumanischen Petroleumlagerstatten. *3rd International Petroleum Congress (1907)*, Bucharest, Comte Rendu, **2**, 80–134.

Munro, D. 1914. The proposed German petroleum monopoly. *The American Economic Review*, **4** (2), 315.

Nauwelaerts, L. 1936. *Petroleum*. Boekh, Antwerp, Belgium.

Nederlandse Aardolie Maatschappij 1968. *Symposium on the Groningen gas field*. Koninklijk Nederlands Geologisch Mijnbouwkunding Genootschap Verhandelingen, *Geological Series*, **25**, 1–65.

Neuburger, H., Macintosh J.G. & Noalhat, H. 1901. *Technology of Petroleum. The Oil Fields of the World: Their History, Geography and Geology, Annual Production, Prospection and Development, Technical Data-tables, Formulæ and Problems*. Scott, Greenwood and Co., London.

Neville, R.G. & Smeaton W.A. 1981. Macquer's *Dictionnaire de Chymie*: A bibliographical study. *Annals of Sciences*, **38**, 613–662.

Nowak, J. & Tołwiński, K. 1921. *KARPATY. Nafta Karpat polskich w świetle geologji regjonalnej (Le pétrole des Carpathes polonaises, sous le point de vue de la géologie regionale); Dyslokacje poprzeczne oraz kierunki tektoniczne w Karpatach polskich (Dislocations transversales et directions tectoniques des Carpathes polonaises)*. Towarzystwo nauczycieli szkół wyższych, Warszawa, Lwów.

Olteanu, Gh. 1966. The salt-raising mechanism in the Subcarpathian zone of the Ploiesti region [in Rumanian]. *Petrol si Gaze*, **17**, 213–216.

Oszlaczky, S. 1959. Past and present trends of the geophysical exploration for hydrocarbons in Hungary [in Hungarian, English summary]. *Hungarian Geological Society Bulletin*, **89**, 351–363.

Owen, E.W. 1975. *Trek of the Oil Finders: A History of Exploration for Petroleum*. American Association of Petroleum Geologists Memoirs, **6**.

Pallmann, R. 1882. *Petroleum in der Mark Brandenburg: Eine vergleichende geographische studie. Vortrag gehalten in der Fortbildungsschule des Aeltesten-collegiums der kaufmannschaft von Berlin*. J. Klönne & G. Müller, Berlin.

Paoloni, C. 1976. *Storia del Metano*. Tipografia Sapil, Milano, Italy.

Parker, W. 1887. On the progress and development of marine engineering. *Transactions of the Institution of Naval Architect*, **28**, 125–146.

Paul, C.M. 1881. Die Petroleum—und Ozokerit-Vorkommnisse Ostgaliziens. *Jahrbuch der Kaiserlich Königlichen Geologischen Reichsanstalt*, **31**, 131–168.

Paul, C.M. 1887. Zur Wieliczka Frage. *Jahrbuch der Kaiserlich Königlichen Geologischen Reichsanstalt*, **37**, 109–116.

Paul, C.M. & Tietze, E. 1879. Neue Studien in der Sandsteinzone der Karpathen. *Jahrbuch der Kaiserlich Königlichen Geologischen Reichsanstalt*, **29**, 189–304.

Peckston, T. 1823. *The Theory and Practice of Gas Lighting*. Thomas and George Underwood, London.

Perutz, H. 1868. *Die Industrie der Mineralöle: Des petroleums, paraffins und der harze*. Carl Gerold's Sohn, Wien.

Petit, V. 1907. *Guide de sondeur au pétrole: traité théorique et pratique des sondages appliqués à la recherche et à l'exploitation du pétrole*. A. Castaigne, Paris; C. Béranger, Bruxelles.

Pettus, J. & Ercker, L. 1683. *Fleta Minor. The Laws of Art and Nature, in Knowing, Judging, Assaying, Fining, Refining and Inlarging the Bodies of Confin'd Metals*. (2 vols). Printed for the author by T. Dawks, London.

Phillips, E. (ed.) 1706. *The New World of Words: Or, Universal English Dictionary*. Printed for J. Phillips, London.

Pickering, G.F. 1919. *Aids in the Commercial Analysis of Oil, Fats, and Their Commercial Products. A Laboratory Handbook*. C. Griffin & Co, London.

Pieri, M. 1969. Exploration for oil and gas in Italy. *In*: Hepple, P. (ed.) *The Exploration for Petroleum in Europe and North Africa*. The Institute of Petroleum, London, 87–111.

Pizanty, M. & Institut Economic Romanesc 1933. *Le petrole en Roumanie: (édition completée) apercu general sur la situation economique, statistique, juridique ainsi que sur les perfectionnements dans le domaine de exploitation et du raffinage*. 'Moderna', Bucarest.

Poggi, J. 1802. Account of a new spring of petroleum discovered in Italy; in a letter to the editors of the Annales de Chimie. *Philosophical Magazine*, **XVI**, 321–324.

Poggi, J. 1802. Sul petrolio di amiano. *Annales de Chimie*, **XLV**, 171–176.

Popescu, B.M. 1994. *Hydrocarbons of Eastern Central Europe: Habitat, Exploration and Production History*. Springer, Berlin.

Possanner von Ehrenthal, B. 1893. *Chemische Technologie der landwirtschaftlichen Gewerbe nebst einer kurzen Abhandlung über Mineralöle etc.* Druck und Verlag der Kaiserlich-Königlichen Hof und Staatsdruckerei, Wien.

Predescu, C. 1922. *Etude physique sur le petrole Roumain: Brut, fractions et lampants*. Cartea Romaneasca, Bucarest.

Priabinin, A. 1906. Geological operations in 1905 in the Binagadi oil-field, Baku. *Russia Geological Committee Bulletin*, **25**, 145–182.

Prudhomme, E. A. 1927. *Le petrole-roi. Sa fabrication industrielle*. Dunod, Paris.

Querol, R. 1969. Petroleum exploration in Spain. *In*: Hepple, P. (ed.) *The Exploration for Petroleum in Europe and North Africa*. The Institute of Petroleum, London, 49–71.

Rankine, W.J.M. 1868. On the estimation of the evaporative power of fuel. *Proceedings of the Philosophical Society of Glasgow*, **5**, 123–126.

Redwood, B. 1883. *On the Present Position & Future Prospects of the Russian (Caucasian) Petroleum Industry*. Waterlow & Sons Limited, London.

Redwood, B. 1884. The determination of the percentage of oil in paraffin scale. *Journal of the Society of Chemical Industry*, **3** (8), 430–432.

Redwood, B. (ed.) 1886. *Cantor Lectures on Petroleum and its Products*. W. Trounce, London.

Redwood, B. 1887. *Petroleum: Its Production and Use*. Van Nostrand, New York.

Redwood, B. 1887. Notes of a recent visit to some of the petroleum-producing territories of the United States and Canada. *Scientific American*, **24** (610), 9743–9744; (611), 9765.

Redwood, B. 1889. *On 'The Domestic Uses of Petroleum Products'*. Kenny & Co., London.

Redwood, B. & Topley, W. 1891. *Report by Boverton Redwood ... and William Topley ... on the Riacho-doce and Camarajibe Shale Deposits on the Coast of Brazil, near Maceió*. Corporate publication, London.

Redwood, B. 1894. *The Transport of Petroleum in Bulk*. Institution of Civil Engineers, London.

Redwood, B. 1896. *Petroleum; A Treatise on the Geographical Distribution and Geological Occurrence of Petroleum and Natural Gas; The Physical and Chemical Properties, Production, and Refining of Petroleum and Ozokerite; The Characters and Uses, Testing, Transport, and Storage of Petroleum Products; And the Legislative Enactments Relating Thereto; Together With a Description of the Shale Oil and Allied Industries; And a Full Bibliography*. C. Griffin & Co., London (2nd edn, 1906; 3rd edn, 1913; 1922, 4th edn, 1926).

Redwood, B. 1910. *Guide to the Oil Companies. Particulars of the Principal Oil Companies, with Alphabetical List of Directors. With Introduction by Sir Boverton Redwood*. Pall Mall Gazette, London.

Redwood, B. 1913. The future of fuel oil. *Journal of the Royal Society of Arts*, **62** (3188), 103–116.

Redwood, B. (ed.) 1919. *A Treatise on British Mineral Oil*. G. Griffith & Co., London.

Redwood, B. & Inter-Departmental Committee on Alcohol Motorfuel (Great Britain) 1919. *Report of the Inter-Departmental Committee of Various Matters Concerning the Production and Utilization of Alcohol for Power and Traction Purposes*. HMSO, London.

Renner, J. 1953. Die geophysikalischen Forschungen von Eötvös im Dienste der praktischen Erdöl und Gas Schürfung. *In*: Selenyi, P. (ed.) *Roland Eötvös Gesammelte Arbeiten*. Akademiai Kiado, Budapest, 379–384.

Ribighini, C. & Ferretti, L. 1867. *Memoria diretta al comitato promotore di una società anonima abruzzese per l'esercizio delle miniere della Majella*. Company report.

Rittenhouse, J.D. 1954. *The History of Drilling Mud*. Baroid Sales Division, National Lead Company, Houston, TX, USA.

Roll, A. 1969. Recent development in German exploration for oil and gas. *In*: Hepple, P. (ed.) *The Exploration for Petroleum in Europe and North Africa*. The Institute of Petroleum, London, 221–229.

Rosenthal, T. 1910. Die Entwasserung von rohem Erdol. *Petroleum; Zeitschrift für die gesamten Interessen der Erdöl-Industrie und des Mineralöl-Handels*, **5**, 315–316.

Rowland, J. & Baron Cadman, B. 1960. *Ambassador for Oil; the Life of John, First Baron Cadman*. Herbert Jenkins, London.

Sage, B.G. 1772. *Élémens de mineralogie docimastique*. Imprimeur Libraire de l'Académie royale de musique, rue du Foin, chez P. de Lormel, Paris.

Sage, B.G. 1784. *Description Méthodique du Cabinet de l'École Royale des Mines*. Imprimerie Royale, Paris.

Sainte-Claire Deville, H. 1870. Emploi de l'huile de pétrole pour le chauffage des locomotives. *Annales de la Science et de l'Industrie*, **14**, 89–92.

Saul, N. 1991. *Distant Friends: The United States and Russia, 1763–1867*. University Press of Kansas, Wichita, KS, USA.

Schestakow, P. 1928. *Course sommaire sur la fabrication du gas d'eclairage, la technologie des derives du petrole (solvants et huiles), et des corps gras*. Goschimisdat, Leningrad [in Russian].

Schilling, L. 1867. Die Beleuchtung mit Gas aus Petroleum Riickstanden in der Locomotive Fabrik von Krauss & Co. zu Munchen. *Journal fur Gas beleuchtung*, **10**, 152–157.

Schneider, R. 1902. *Der Petroleumhandel*. H. Saupp, Tübingen, Germany.

Schwarz, R. 1931. *Petroleum-Vademecum: international petroleum tables; Tafeln für die Erdölindustrie und den Mineralölhandel; tables pour l'industrie et le commerce du pétrole*. Verlag für Fachliteratur. Berlin.

Scicli, A. 1972. *L'attività estrattiva e le risorse minerarie della regione Emilia Romagna*. Poligrafico Artioli, Modena, Italy.

Scott, J. (ed.) 1765–66. *A General Dictionary of Arts and Sciences, or, A Complete System of Literature*. (2 vols). S. Crowder, London.

Scotti, L. 1901. *Ricerche di Petrolio nel Piacentino*. Bosi, Piacenza, Italy.

Selley, R.C. 1992. Petroleum seepages and impregnations in Great Britain. *Marine and Petroleum Geology*, **9**, 226–244.

Selwyn, J.H. 1868. Further information on the employment of mineral oils as fuel for steamships. *Journal of the Royal United Services Institute for Defence Studies*, **12**, 28–43.

Selwyn, J.H. 1868. On liquid fuel. *Transactions of the Institute of Naval Architect*, **9**, 88–103.

Selwyn, J.H. 1869. On the progress of liquid fuel. *Transactions of the Institute of Naval Architect*, **10**, 32–46.

Selwyn, J.H. 1870. On liquid or concentrated fuel. *Transactions of the Institute of Naval Architect*, **11**, 160–174.

Selwyn, J.H. 1885. Liquid fuel for marine purposes. *Journal of the Royal United Services Institute for Defence Studies*, **29**, 689–701.

Severin, E. 1933. *Petolul. Studio fizic, chimic, geologic, technologic si economic*. Imprimerie Nationale, Bucarest.

Sędziwój, M. 1650. *A New Light of Alchymie: Taken Out of the Fountaine of Nature, and Manuall Experience. To Which is Added a Treatise of Sulphur*. Printed by R. Cotes, for T. Williams, London.

Sielemann Sperat, C.A. 1993. *The impact of the Imperial Russian and Soviet oil industry on the city structure of Baku*. Master's thesis, University of Oklahoma, Norman, AZ, USA.

Sjogren, H. 1888. Ueber das transkaspische Naphtaterrain. *Jahrbuch der Kaiserlich Königlichen Geologischen Reichsanstalt*, **37**, 47–62.

Smith, N.J. 2011. *The Sea of Lost Opportunity, Volume 7. North Sea Oil and Gas, British Industry and the Offshore Supplies Office*. Elsevier, Amsterdam.

Smith, P.G.A. 1920. *The Shell That Hit Germany Hardest*. 'Shell' Marketing Co., London.

Società Petrolifera Italiana 1976. *SPI, settant'anni*. Corporate publication, Parma, Italy.

Soldani, A. 1794. Memoria sopra il Terreno Ardente del Portico di Romagna (Piccolo Castello della Romagna Granducale) e di simili fuochi. *Atti dell'Accademia dei Fisiocratici, Pisa*, **VII**, 200–214.

Sorgenfrei, T. 1969. A review of petroleum development in Scandinavia. *In*: Hepple, P. (ed.) *The Exploration for Petroleum in Europe and North Africa*. The Institute of Petroleum, London, 191–208.

Soulie, E. & Haudouin, H. 1865. *Le petrole; ses gisements son exploitation son traitement industriel, ses produits derives, ses applications a l'eclairage et au chauffuage*. Librarie Scientifique, Industrielle et Agricole, Paris.

Spadoni, P. 1802. *Osservazioni mineralovulcaniche fatte in un viaggio per l'antico Lazio*. Presso Bartolomeo Capitani, Macerata, Italy.

Spallanzani, L. 1792–97. *Viaggio nelle due Sicilie e in alcune parti dell'Apennino*. Nella Stamperia di Baldassare Comini, Pavia.

Squarzina, F. 1958. *Le ricerche del petrolio in Italia, Cenni storici dal 1860 e cronache dell'ultimo decennio*. Jandi Sapi, Roma.

Stewart, J.L. 1886. On the use of petroleum as fuel in steamships and locomotives, based on its employment in that way on the Caspian Sea and in the TransCaspian Region. *Journal of the Royal United Service Institution*, **30**, 927–949.

Stheeman, H.A. 1963. Petroleum development in The Netherlands. *Koninklijk Nederlands Geologisch Mijnbouwkunding Genootschap Verhandelingen, Geological Series.*, **21-1**, 58–96.

Stigand, A. & Muhlberg, M. 1925. *Outlines of the Occurrence and Geology of Petroleum*. Charles Griffin and Company, London.

Stoppani, A. 1864. Saggio di una Storia Naturale dei Petroli per A. Stoppani, Letto al Reale Istituto Lombardo. *Il Politecnico: Repertorio Mensile di Studi Applicati alla Prosperità e Cultura*, Series 2, **23**, 5–95.

Stoppani, A. 1866. I petroli d'Italia. *Il Politecnico, repertorio di studj letterari, scientifici e tecnici parte letterario-scientifica*, Series 4, 1 (2), 77–93, 574–586.

Stoppani, A. 1871–73. *Corso di geologia del Professor Antonio Stoppani*. (3 vols). G. Bernardoni e G. Brigola, Milano, Italy.

Stoppani, A. 1876. *Il Bel Paese. Conversazioni sulle bellezze naturali, la geologia e la geografia fisica d'Italia*. Ditta Giacomo Agnelli, Milano, Italy.

Strippelmann, L. 1878. *Die Petroleum-Industrie Oesterreich Deutschlands dargestellt zur Klarstellung deren Wichtigkeit und Zukunft, und zur Aufklarung des, fur diese Industrie sich interessenden Capital's in geschichtler, geologischbergmannischer, wirtschaftlicher und technischer Beziehung*. Knapp, Leipzig, Germany.

Thiadens, A.A. 1969. A history of the exploration for hydrocarbons within the territorial boundaries of The Netherlands. *In*: Hepple, P. (ed.) *The Exploration for Petroleum in Europe and North Africa*. The Institute of Petroleum, London, 259–269.

Thomas, W.H. & Withers, J.G. 1945. *Deutsche Erdöl A.G. Hamburg – Germany: Crude Oil and Products*. Final report, British Intelligence Objectives Sub-Committee 638, Item 30. BIOS, London.

Thompson, A.B. 1910. *Petroleum Mining and Oil-Field Development*. Crosby Lockwood & Son, London.

Thomson, J.H. & Redwood, B. 1901. *Handbook on Petroleum for Inspectors Under the Petroleum Acts: And for Those Engaged in the Storage, Transport, Distribution and Industrial Use of Petroleum and its Products and Calcium Carbide, With Suggestions on the Construction of Use of Mineral Oil Lamps*. Griffin & Co., London (2nd edn, 1906; 3rd edn, 1913; 4th edn, 1922).

Thomson, J.H. & Redwood, B. 1902. *The Petroleum Lamp, its Choice and Use: A Guide to the Safe Employment of Mineral Oil in What is Commonly Termed the Paraffin Lamp*. Griffin, London.

Thuloup, A. 1904. *Le petrole*. Impr. Adm., Melun, France.

Tolf, R.W. 1976. *The Russian Rockfellers*. Hoover Institution Press, University of Stanford, Stanford, CA, USA.

Torrontegui, de S. 1929. *Tratado de quimica industrial. La gran industria química. Industria de los acidos. Industria de los petroleos. Materiales empleados en la industria quimica.* Feliu y Susanna, Barcelona.

Trapani, F. 2003. L'Oleum Petronicum, un antico rimedio galenico delle falde settentrionali della Majella. *In*: *Annuario del Club Alpino Italiano – Sezione di Pescara*. CAI, Pescara, Italy, 61–76.

Trevithick, F.H. 1886. *A Sketchy Report on the Petroleum Industry at Baku, May 1886*. National Printing Office, Cairo.

Tsenkov, T. 1966. The Bulgarian oil and gas industry. *Institute of Petroleum Review*, **20**, 199–202.

Turyn, I.K. 1940. *Petroleum-Vademecum. Tafen f.d. Erdolindustrie u.d. Mineralolhandel. Bd. I Chemish-technisch-physikalischer Teil. Hilfstabellen f.d. Bohrtechnik, d. Erdolfelder d. Welt, d. Erdolraffinerien d. Welt.* Bohmann, Wien.

Tweddle, H., Jr. 1886. The petroleum industries of Europe. *Engineering*, **51**, 99–100, 149–152, 193–194, 239–243, 289–290, 341–342.

Ugolini, G. 1924. *Il petrolio e noi*. T.U., Roma.

Urquhart, T. 1890. On the compounding of locomotives burning petroleum refuse in Russia. *Proceeding of the Institute of the Mechanical Engineer*, **43–44**, 47–111.

Ussing, A. 1915. *Petroleumspriser Og Petroleumsforbrug: Med særligt Henblik Paa Danmarks Motorfiskeri*. Corporate publication.

Vecchia, D. 1896. *Il petrolio italiano e la zona petroleifera di Rivanazzano*. Tipografia G. Redaelli, Milano.

Virlet d'Aoust, T. 1834*a*. Note sur les sources et les mines d'asphalte. *Société Géologique de France Bulletin*, **4**, 203.

Virlet d'Aoust, T. 1834*b*. Nouvelles notes relatives a l'origine volcanique des bitumes mineraux. *Société Géologique de France Bulletin*, **4**, 372–376.

Volck, J. 1625. *Hanawischen Erdbalsams Petrolei oder Weichen Agsteins Beschreibung*. Jezua Rihel, Strassburg.

Volta, A. 1776. *Lettere del Signor Volta Alessandro, Patrìzio comasco e Decurione, sull'aria infiammabile nativa delle paludi (I a e II a lettiera)*. Giuseppe Marelli, Milano, Italy.

Volta, A. 1777. *Lettere dell'Illustrìssimo Signor Volta Alessandro sull'aria infiammabile nativa dette paludi (7 lettere)*. Giuseppe Marelli, Milano, Italy.

von Goulishambaroff, S. 1883. *Bibliographie der Petroleum-Industrie, Versuch einer allgemeinen (pt. I)*. H. Schmitzdorff, St. Petersbourg.

von Goulishambaroff, S. 1884. *Bibliographie der Petroleum-Industrie, Versuch einer allgemeinen (pt. 2)*. A. Tranchell, St. Petersbourg.

Walker, P.F. 1902. Stability of an oil-tank steamer. *Marine Engineering*, **91** (7), 492–495.

Wall, M. 1789. Some observations on the phenomena which take place between oil and water. *Memoirs of the Manchester Literary and Philosophical Society*, **2**, 435–444.

Walter, B. 1880. Die Chancen einer Erdolgewinnung in der Bukowina. *Jahrbuch der Kaiserlich Königlichen Geologischen Reichsanstalt*, **30**, 115–148.

Walters, R.P. 1940. La presence du petrole dans le Dacien de Gura Ocnitei-Est et la question de l'origine de petrole en Romanie. *Romanian Institute of Geology Comptes Rendus*, **24**, 10–17.

Yates, D.A. 2009. *The French Oil Industry and the Corps des Mines in Africa*. Africa World Press, Trenton NJ, USA.

Yuschkin, E. 1903. Recherches geologiques dans la region naphtifere de Grozny en 1901–1902. *Russia Geological Committee Bulletin*, **22**, 619–644.

Zambrini, A. & Mariani M. 2005. *Alla scoperta dell'Appennino. Sulle tracce degli studiosi del passato*. Club Alpino Italiano, Sezione di Imola, Imola, Italy.

Zolesi, G. 1891. *Il petrolio in Italia*. Stabilimento Tipografico di G. Schenone, Genova, Italy.

Zuber, R. 1881. Detail-Studien in den ostgalizischen Karpathen zwischen Delatyn und Jablon6w. *Jahrbuch der Kaiserlich Königlichen Geologischen Reichsanstalt*, **32**, 351–372.

Gazetteer of oil and gas museums in Europe

FRANCESCO GERALI[1,2], JONATHAN CRAIG[3]*, FIONA MACAULAY[4] & RASOUL SORKHABI[5]

[1]*The University of Oklahoma School of Law, Oil & Gas, Natural Resources, Energy Center, 300 Timberdell Road, Norman, OK 73019-5081, USA*

[2]*The University of Western Australia, FABLE, 35 Stirling Highway, Crawley, Perth, WA 6009, Australia*

[3]*Eni Upstream & Technical Services, Via Emilia 1, 20097 San Donato Milanese, Milan, Italy*

[4]*Echo Energy plc, 40 George Street, London W1U 7DW, UK*

[5]*Energy & Geoscience Institute, University of Utah, 423 Wakara Way, Salt Lake City, UT 84108, USA*

**Correspondence: jonathan.craig@eni.it*

Austria

Technisches Museum Wien – Vienna Technical Museum
(Section 'Mines and Energy')
Mariahilfer Straße 212, 1140 Wien
Phone: +43 1 899980
Email: museumsbox@tmw.at
Website: http://www.technischesmuseum.at

Czech Republic

Muzeum naftového dobývání a geologie v Hodoníně – Museum of Oil Mining and Geology of Hodonin
Kasárenská ul. 1022, 695 01 Hodonín
Phone: +420 604 413 665
Email: mng.hodonin@seznam.cz
Website: http://mng.webz.cz

Denmark

GASmuseet, The Danish GAS Museum
Gasværksvej 2, 9500 Hobro
Phone: +45 5371 5120
Email: gas2gas@gasmuseet.dk
Website: http://www.gasmuseet.dk/en

Estonia

Põlevkivimuuseum – Museum of Oil Shale
Lehe 10a, Kukruse, 30621 Kohtla-Järve
Phone: +372 332 1350
Email: info@pkm.ee
Website: http://www.pkm.ee/en

France

Musée Français du Pétrole – French Museum of Oil
4 rue de l'école, 67250 Merkwiller-Pechelbronn
Phone: +33 03 88809108
Email: musee.petrole@musees-vosges-nord.org
Website: http://www.musee-du-petrole.com

Germany

Deutschen Erdölmuseum Wietze – German Petroleum Museum of Wietze
Schwarzer Weg 7–9, 29323 Wietze
Phone: +49 5146 92340
Fax: +49 514 92342
Email: info@erdoelmuseum.de
Website: http://www.erdoelmuseum.de

Erdöl-Erdgas-Museum Twist – Oil and Gas Museum of Twist
Flensbergstraße 13, 49767 Twist
Phone: +49 5936 93300
Fax: +49 5936 933044
Email: museum@twist-emsland.de
Website: http://www.erdoel-erdgas-museum-twist.de

Greece

Technopolis (Gazi) – City of Technology (Gas)
100 Pireos Street, 118 54 Gazi, Athens
Phone: +30 210 347551-589
Fax: +30 210 3413228
Email: info@athens-technopolis.gr,
gasmuseum@athens-technopolis.gr
Website: http://www.technopolis-athens.com/web/guest/home

From: CRAIG, J., GERALI, F., MACAULAY, F. & SORKHABI, R. (eds) 2018. *History of the European Oil and Gas Industry*. Geological Society, London, Special Publications, **465**, 463–465.
https://doi.org/10.1144/SP465.26

Hungary

Magyar Olajipari Múzeum (MOIM) – Hungarian Oil Industry Museum
H-8900 Zalaegerszeg, Wlassics Gyula u. 13, Zalaegerszeg, Magyarorszag
Phone: +36 (0) 92 313-632
Fax: +36 (0) 92 311-081
Email: moim@olajmuzeum.hu
Website: http://www.olajmuzeum.hu

Italy

Museo dell'Energia di Ripi
Via Merigo Alto, 03027 RIPI (FR)
Phone/Fax: 0775 285160
Email: museoenergia.info@gmail.com
Website: http://www.museoenergiaripi.it

Museo Fisogni – Fisogni Museum
Via Bianchi 25/b, 21049 Tradate, Varese
Phone: +39 3396217531
Email: fisogni@museo-fisogni.org
Website: http://www.museo-fisogni.org/en/

Parco Museo del Petrolio Vallezza – Vallezza Park Museum of Petroleum
Ex Cantiere SPI, Via Maiatico, Vallezza, Fornovo di Taro, Parma
Phone: +39 0525 2599
Email: info@museodelpetrolio.it
Website: http://www.museodelpetrolio.it/

Piacenza Oil & Gas Museum (POGaM)
Via Scalabrini, 107, Piacenza 29121
Phone: +39 0523 450411
Fax: +39 0523 452680
Email: info@pogameschool.org
Website: http://www.pogameschool.org/museo/piacenza-oil-gas-museum.html

Norway

Norsk Oljemuseum – Norwegian Petroleum Museum
Kjeringholmen, Postboks 748, Sentrum 4004, Stavanger
Phone: +47 5193 9300
Fax: +47 5193 9301
Email: post@norskolje.museum.no
Website: http://www.norskolje.museum.no

Poland

Borysławska Rada Miejska – Boryslavska City Council
ul. Szewczenki 42, 82300 Borysław
Phone: +38 32 485 2952
Fax: +38 32 4852552
Email: starosta@powiat.jaslo.pl

Muzeum Gazownictwa w Warszawie – Warsaw Gasworks Museum
Marcina Kasprzaka 25, 01-224 Warszawa
Phone: +48 22 589 42 41
Fax: +48 22 589 48 69
Email: muzeum.warszawa@pgnig.pl
Website: http://muzeum.gazownictwa.pl/

Muzeum Skansen Przemyslu – Museum of the Oil and Gas Industry
38-458 Chorkówka, Bobrka
Phone: +38 (013) 4333 489
Fax: +38 (013) 4333 478
Email: muzeum@bobrka.home.pl
Website: http://www.bobrka.pl/en/oil-museum

Romania

Petrom (Muzeul Național al Petrolului) – National Museum of Petroleum
Strada Doctor Dumitru Bagdazar 8, 100575 Ploieşti (see Fig. 1)
Phone: +40 244 597585
Fax: +40 244 519542
Email: muzeu.petrol@petrom.com
Website: www.petrom.ro
http://www.ploiesti.ro/mpetrolului.php

Sweden

Oljön i Ängelsberg – Oil Factory in Engelsbergs
Ängelsbergsvägen 15, 737 90 Ängelsberg
Phone: +44 223 13100
Email: info@oljeon.se
Website: http://www.oljeon.se

UK

Dukes Wood Oil Museum
Kelham Hall, Main Road, Newark, Nottinghamshire NG23 5QX
Phone: +44 115 9588242
Email: info@dukeswoodoilmuseum.co.uk
Website: http://www.communityarchives.org.uk/content/organisation/dukes-wood-oil-museum

National Gas Museum
195 Aylestone Road, Leicester LE2 7QJ
Phone: +44 07557 612 340
Email: nationalgasmuseum@gmail.com
Website: http://www.nationalgasmuseum.org.uk

Aberdeen Maritime Museum
Shiprow, Aberdeen AB11 5BY
Phone: +44 (0) 1224 337700
Fax: +44 (0) 1224 213066
Email: info@aagm.co.uk
Website: http://www.aagm.co.uk/Visit/AberdeenMaritimeMuseum/amm-overview.aspx

Museum of the Scottish Shale Oil
Almond Valley Heritage Centre,
Millfield, Livingston, West Lothian EH54 7AR
Phone: +44 1506 414957
Email: info@almondvalley.co.uk
Website: www.almondvalley.co.uk
http://www.scottishshale.co.uk

Flame: The Gasworks Museum of Ireland (formerly the Carrickfergus Gasworks Museum)
44 Irish Quarter West, Carrickfergus BT38 8AT, Northern Ireland
Phone: +44 28 9336 9575
Email: info@flamegasworks.co.uk
Website: http://www.flamegasworks.co.uk

Fig. 1. Ploieşti, front view of Romania's *Petrom* – National Museum of Petroleum (established in 1961).

Index

Page numbers in *italics* refer to Figures. Page numbers in **bold** refer to Tables.